Lecture Notes in Computer Science   14134

Founding Editors

Gerhard Goos
Juris Hartmanis

Editorial Board Members

The series Lecture Notes in Computer Science (LNCS), including its subseries Lecture Notes in Artificial Intelligence (LNAI) and Lecture Notes in Bioinformatics (LNBI), has established itself as a medium for the publication of new developments in computer science and information technology research, teaching, and education.

LNCS enjoys close cooperation with the computer science R & D community, the series counts many renowned academics among its volume editors and paper authors, and collaborates with prestigious societies. Its mission is to serve this international community by providing an invaluable service, mainly focused on the publication of conference and workshop proceedings and postproceedings. LNCS commenced publication in 1973.

Ignacio Rojas · Gonzalo Joya · Andreu Catala
Editors

Advances in Computational Intelligence

17th International Work-Conference
on Artificial Neural Networks, IWANN 2023
Ponta Delgada, Portugal, June 19–21, 2023
Proceedings, Part I

Springer

Editors
Ignacio Rojas (iD)
University of Granada
Granada, Spain

Gonzalo Joya
University of Malaga
Málaga, Spain

Andreu Catala
Polytechnic University of Catalonia
Vilanova i la Geltrú, Spain

ISSN 0302-9743   ISSN 1611-3349 (electronic)
Lecture Notes in Computer Science
ISBN 978-3-031-43084-8   ISBN 978-3-031-43085-5 (eBook)
https://doi.org/10.1007/978-3-031-43085-5

This Springer imprint is published by the registered company Springer Nature Switzerland AG
The registered company address is: Gewerbestrasse 11, 6330 Cham, Switzerland

Paper in this product is recyclable.

Preface

We are proud to present the set of final accepted papers for the 17th edition of the IWANN conference - the International Work-Conference on Artificial Neural Networks - held in Ponta Delgada, São Miguel, Azores Islands, (Portugal) during June 19–21, 2023.

IWANN is a biennial conference that seeks to provide a discussion forum for scientists, engineers, educators, and students about the latest ideas and realizations in the foundations, theory, models, and applications of hybrid systems inspired by nature (neural networks, fuzzy logic, and evolutionary systems) as well as in emerging areas related to these topics. As in previous editions of IWANN, it also aims to create a friendly environment that could lead to the establishment of scientific collaborations and exchanges among attendees. The proceedings include the communications presented at the conference.

Since the first edition in Granada (LNCS 540, 1991), the conference has evolved and matured. The list of topics in the successive Call for Papers has also evolved, resulting in the following list for the 2023 edition:

1. *Deep Learning*
2. *Learning and adaptation*
3. *Emulation of cognitive functions*
4. *Bio-inspired systems and neuro-engineering*
5. *Advanced topics in computational intelligence*
6. *Agent-based models*
7. *Time series forecasting*
8. *Robotics and cognitive systems*
9. *Interactive systems and BCI*
10. *Machine Learning for Industry 4.0 solutions*
11. *AI Health*
12. *AI in 5G technology*
13. *Social and Ethical aspects of AI*
14. *General applications of AI*

At the end of the submission process, and after a careful peer review and evaluation process (each submission was reviewed by at least 2, and on the average 2.7, program committee members or additional reviewers), 108 papers were accepted for oral or poster presentation, according to the reviewers' recommendations.

During IWANN 2023 several special sessions were held. Special sessions are a very useful tool for complementing the regular program with new and emerging topics of particular interest for the participating community. Special sessions that emphasize multi-disciplinary and transversal aspects, as well as cutting-edge topics are especially encouraged and welcome, and in this edition of IWANN comprised the following:

- **SS01: Ordinal Classification**
  Organized by: Victor M. Vargas, David Guijo-Rubio, Pedro A. Gutiérrez
- **SS02: Machine Learning in Mental Health**
  Organized by: Pepijn van de Ven
- **SS03: Interaction with Neural Systems in both Health and Disease**
  Organized by: Pablo Martínez Cañada, Jesus Minguillón Campos
- **SS04: Deep Learning applied to Computer Vision and Robotics**
  Organized by: Enrique Dominguez, José García-Rodríguez, Ramon Moreno Jiménez
- **SS05: Applications of Machine Learning in Biomedicine and Healthcare**
  Organized by: Miri Weiss Cohen, Daniele Regazzoni, Catalin Stoean
- **SS06: Neural Networks in Chemistry and Material Characterization**
  Organized by: Ruxandra Stoean, Patricio García Báez, Carmen Paz Suárez Araujo
- **SS07: Real-World Applications of BCI Systems**
  Organized by: Ivan Volosyak
- **SS08: Spiking Neuron Networks: Applications and Algorithms**
  Organized by: Elisa Guerrero Vázquez, Fernando M. Quintana Velázquez
- **SS09: Deep Learning and Time Series Forecasting: Methods and Applications**
  Organized by: Francisco Martínez Álvarez, Verónica Bolón-Canedo, David Camacho
- **SS10: ANN HW-Accelerators**
  Organized by: Mario Porrmann, Ulrich Rückert

In this edition of IWANN, we were honored to have the presence of the following invited speakers:

1. Alberto Bosio, Full Professor at the INL – École Centrale de Lyon, France. *Title of the presentation: Reliable and Efficient Hardware for Trustworthy Deep Neural Networks*
2. Amaury Lendasse, Department Chair Information and Logistics Technology Faculty, University of Houston, USA. *Title of the presentation: Metric Learning with Missing Data*

It is important to note that for the sake of consistency and readability of the book, the presented papers are not organized as they were presented in the IWANN 2023 sessions, but classified under 14 chapters. The organization of the papers is in two volumes arranged basically following the topics list included in the call for papers. The first volume (LNCS 14134), entitled "Advances in Computational Intelligence. IWANN 2023. Part I", is divided into five main parts and includes contributions on:

1. Advanced Topics in Computational Intelligence
2. Advances in Artificial Neural Networks
3. ANN HW-Accelerators
4. Applications of Machine Learning in Biomedicine and Healthcare
5. Applications of Machine Learning in Time Series Analysis

In the second volume (LNCS 14135), entitled "Advances in Computational Intelligence. IWANN 2023. Part II", is divided into nine main parts and includes contributions on:

1. Deep Learning and Applications
2. Deep Learning Applied to Computer Vision and Robotics
3. General Applications of Artificial Intelligence
4. Interaction with Neural Systems in Both Health and Disease
5. Machine Learning for Industry 4.0 Solutions
6. Neural Networks in Chemistry and Material Characterization
7. Ordinal Classification
8. Real-World Applications of BCI Systems
9. Spiking Neural Networks: Applications and Algorithms

The 17th edition of the IWANN conference was organized by the University of Granada, University of Malaga, and Polytechnical University of Catalonia.

We would like to express our gratitude to the members of the different committees for their support, collaboration and good work. We specially thank our Honorary Chairs (Joan Cabestany, Alberto Prieto and Francisco Sandoval), the Technical Program Chairs (Miguel Atencia, Francisco García-Lagos, Luis Javier Herrera and Fernando Rojas), the Program Committee, the Reviewers, Invited Speakers, and Special Session Organizers. Finally, we want to thank Springer LNCS for their continuous support and cooperation.

June 2023   Ignacio Rojas
                                                                   Gonzalo Joya
                                                                  Andreu Catala

Organization

Steering Committee

Davide Anguita	Università degli Studi di Genova, Italia
Andreu Catalá	Universitat Politècnica de Catalunya, Spain
Marie Cottrell	Université Paris 1 Panthéon-Sorbonne, France
Gonzalo Joya	University of Málaga, Spain
Kurosh Madani	Université Paris-Est Créteil, France
Madalina Olteanu	Université Paris Dauphine – PSL, France
Ignacio Rojas	University of Granada, Spain
Ulrich Rückert	Universität Bielefeld, Germany

Program Committee

Kouzou Abdellah	Djelfa University, Algeria
Vanessa Aguiar-Pulido	University of Miami, USA
Arnulfo Alanis	Instituto Tecnológico de Tijuana, Mexico
Ali Alkaya	Marmara University, Turkey
Amparo Alonso-Betanzos	University of A Coruña, Spain
Gabriela Andrejkova	.
Davide Anguita	University of Genoa, Italy
Cecilio Angulo	Universitat Politècnica de Catalunya, Spain
Javier Antich Tobaruela	University of the Balearic Islands, Spain
Alfonso Ariza	University of Málaga, Spain
Corneliu Arsene	SC IPA SA, Romania
Miguel Atencia	University of Málaga, Spain
Jorge Azorín-López	University of Alicante, Spain
Halima Bahi	Badji Mokhtar – Annaba University, Algeria
Juan Pedro Bandera Rubio	University of Málaga, Spain
Oresti Banos	University of Granada, Spain
Bruno Baruque	University of Burgos, Spain
Lluís Belanche	Universitat Politècnica de Catalunya, Spain
Francisco Bonnín	University of the Balearic Isles, Spain
Julio Brito	University of la Laguna, Spain
Pablo C. Cañizares	Universidad Complutense de Madrid, Spain
Joan Cabestany	Universitat Politècnica de Catalunya, Spain
Eldon Glen Caldwell	University of Costa Rica, Costa Rica

Luis Herrera	University of Granada, Spain
Cesar Hervas	.
Wei-Chiang Hong	Asia Eastern University of Science and Technology, Taiwan
Petr Hurtik	University of Ostrava, Czech Republic
M. Dolores Jimenez-Lopez	Rovira i Virgili University, Spain
Juan Luis Jiménez Laredo	Université Le Havre Normandie, France
Gonzalo Joya	University of Málaga, Spain
Vicente Julian	Universitat Politècnica de València, Spain
Otoniel Lopez Granado	Miguel Hernández University de Elche, Spain
Rafael Marcos Luque Baena	University of Málaga, Spain
Ezequiel López-Rubio	University of Málaga, Spain
Kurosh Madani	Lissi/Université Paris-Est Créteil, France
Bonifacio Martin Del Brio	University of Zaragoza, Spain
Luis Martí	Inria Chile Research Centre, Chile
Pablo Martínez Cañada	University of Granada, Spain
Francisco Martínez Estudillo	Universidad Loyola Andalucía, Spain
Francisco Martínez-Álvarez	Universidad Pablo de Olavide, Spain
Montserrat Mateos	Universidad Pontificia de Salamanca, Spain
Jesús Medina	University of Cádiz, Spain
Salem Mohammed	Mustapha Stambouli University, Algeria
Jose M. Molina	Universidad Carlos III de Madrid, Spain
Miguel A. Molina-Cabello	University of Málaga, Spain
Juan Moreno Garcia	University of Castilla-La Mancha, Spain
John Nelson	University of Limerick, Ireland
Alberto Núñez	Universidad Complutense de Madrid, Spain
Madalina Olteanu	SAMM, Université Paris Dauphine - PSL, France
Alfonso Ortega	Universidad Autónoma de Madriod, Spain
Alberto Ortiz	Universitat de les Illes Balears, Spain
Osvaldo Pacheco	University of Aveiro, Portugal
Esteban José Palomo	University of Málaga, Spain
Massimo Panella	University of Rome "La Sapienza", Italy
Miguel Angel Patricio	Universidad Carlos III de Madrid, Spain
Jose Manuel Perez Lorenzo	University of Jaen, Spain
Irina Perfilieva	University of Ostrava, Czech Republic
Vincenzo Piuri	University of Milan, Italy
Hector Pomares	University of Granada, Spain
Mario Porrmann	Osnabrück University, Germany
Alberto Prieto	University of Granada, Spain
Alexandra Psarrou	University of Westminster, UK
Fernando M. Quintana	University of Cádiz, Spain
Pablo Rabanal	Universidad Complutense de Madrid, Spain

Contents – Part I

Advances in Artificial Neural Networks

ANN HW-Accelerators

Applications of Machine Learning in Biomedicine and Healthcare

Applications of Machine Learning in Time Series Analysis

Contents – Part II

Deep Learning Applied to Computer Vision and Robotics

General Applications of Artificial Intelligence

Interaction with Neural Systems in Both Health and Disease

Machine Learning for 4.0 Industry Solutions

Neural Networks in Chemistry and Material Characterization

Advanced Topics in Computational Intelligence

Energy Complexity of Fully-Connected Layers

Jiří Šíma[1]([⊠])[ID] and Jérémie Cabessa[1,2][ID]

[1] Institute of Computer Science of the Czech Academy of Sciences, Prague, Czechia
sima@cs.cas.cz
[2] DAVID Laboratory, UVSQ – University Paris-Saclay, Versailles, France
jeremie.cabessa@uvsq.fr

Abstract. The energy efficiency of processing convolutional neural networks (CNNs) is crucial for their deployment on low-power mobile devices. In our previous work, a simplified theoretical hardware-independent model of energy complexity for CNNs has been introduced. This model has been experimentally shown to asymptotically fit the power consumption estimates of CNN hardware implementations on different platforms. Here, we pursue the study of this model from a theoretically perspective in the context of fully-connected layers. We present two dataflows and compute their associated energy costs to obtain upper bounds on the optimal energy. Using the weak duality theorem, we further prove a matching lower bound when the buffer memory is divided into two fixed parts for inputs and outputs. The optimal energy complexity for fully-connected layers in the case of partitioned buffer ensues. These results are intended to be generalized to the case of convolutional layers.

Keywords: Convolutional neural networks · Energy complexity · Dataflow

1 Energy Complexity Model for CNNs

Deep neural networks (DNNs) represent a cutting-edge machine learning technology, with countless applications in computer vision, natural language processing, robotics, etc. These models are typically composed of hundreds of thousands of neurons and tens of millions of weights, and are thus computationally demanding and highly energy-consuming. With the ever-growing use of mobile devices, like smartphones or smartwatches, comes the issue of the implementation, deployment, and portability of already trained DNNs on low-power hardware. Recently, extensive research has been conducted on techniques that enable energy-efficient DNN processing on a variety of hardware platforms and architectures (e.g., GPUs, FPGAs [4], memory hierarchies) [8]. The proposed techniques reduce the computational cost via hardware design (including massive parallelism) and/or approximation of DNN models. For example, in error-tolerant applications such as image classification, the use of approximate computing methods [3] (e.g. low

I. Rojas et al. (Eds.): IWANN 2023, LNCS 14134, pp. 3–15, 2023.
https://doi.org/10.1007/978-3-031-43085-5_1

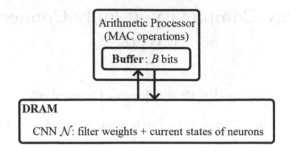

Fig. 1. The energy complexity model.

float precision, approximate multipliers) can save an enormous amount of energy at the cost of only a small loss in accuracy.

For a particular DNN hardware implementation, the power consumption of the inference process can be either practically measured or analytically estimated using physical laws. This power consumption depends on parameters and constants related to the hardware architecture, and hence, its evaluation varies for different hardware implementations. Some computer programs [5,9] can optimize the power consumption of a particular DNN on various hardware platforms [2,6]. It has been empirically observed that the energy cost of DNN processing mainly consists of two components: the computation energy, and the data energy which represents around 70% of the total cost [10]. The *computation energy* is needed for performing arithmetic operations, especially the so-called multiply-and-accumulate (MAC) operations ($S \leftarrow S + wx$ on floats S, w, x), used to compute the weighted sums of inputs of the neurons. The *data energy* is required for moving the data inside the memory hierarchy of the hardware (dataflow), and is related to the number of memory accesses.

In a recent paper [7], we have introduced a simplified hardware-independent model of energy complexity for convolutional neural networks (CNNs). This model abstracts from the hardware implementation details related to different platforms, and preserves the asymptotic energy complexity of the CNN inference. It is composed of only two memory levels called *DRAM* and *Buffer*, illustrated in Fig. 1. The network parameters and states are stored in DRAM, and the arithmetic operations are performed over numerical data stored in Buffer, which is of a limited capacity of B bits. The transfer of data between the two memories is the dataflow. The main idea behind this model is that, for a given CNN stored in DRAM, the three arguments of all the MAC operations (i.e., input x, weight w and accumulated output S of operation $S \leftarrow S + wx$) employed for the evaluation of the network must occur together at the same time in Buffer. This process requires a certain number of data transfers between DRAM and Buffer (i.e., the number of DRAM accesses multiplied by the number of bits in a float number), which corresponds to our measure of the data energy.

For simplicity, we assume that the energy cost is not optimized across multiple CNN layers, as for instance in [1]. Hence, the energy complexity is defined as a simple sum over separate convolutional and fully-connected layers only, while the less energy-intensive max pooling layers are omitted. Formally,

$$E = \sum_{\text{non-pooling layer } \lambda} \left(E^{\lambda}_{\text{comp}} + E^{\lambda}_{\text{data}} \right) \tag{1}$$

where the computation energy $E^{\lambda}_{\text{comp}}$ and the data energy $E^{\lambda}_{\text{data}}$ for evaluating a non-pooling layer λ is proportional to the corresponding numbers of MACs and DRAM accesses, respectively.

The energy complexity model of CNNs has been exploited for calculating the theoretical energy of processing convolutional layers in the context of two common dataflows and under realistic buffer capacity constraints [7]. For the first dataflow, any input to each neuron is read into Buffer only once. For the second one, any accumulated output of each neuron is written to DRAM only once. In both cases, each weight of the CNN is read into Buffer only once. These dataflows provide upper bounds on the energy complexity of CNNs, which have been compared to the real power consumptions estimated for Simba [6] and Eyeriss [2] architectures by using the Timeloop/Accelergy software tool [5, 9]. As it turns out, the theoretical upper bounds fit asymptotically very well the empirical optimal power consumptions, when individual parameters such as the height, width, depth, kernel size, and stride of a convolutional layer are varied [7]. Hence, the introduced energy complexity model appears to be capable of asymptotically capturing all important sources of energy consumption that are common to the diverse CNN hardware implementations.

The model can also be exploited for proving lower bounds on the energy complexity of CNNs, in order to establish asymptotic limits on the energy efficiency of any CNN hardware accelerators. Here, we start this study by investigating the case of *fully-connected layers*, as a specific case of convolutional layers. We first present two types of dataflows in which each weight and each output (or alternatively each input) are read into Buffer only once. In the first dataflow, the Buffer memory is assumed to be partitioned into two fixed parts of given capacities for inputs and outputs, respectively. The second dataflow is parameterized by the maximum number of inputs residing in Buffer at the same time. We determine the data energy complexity of both dataflows, which provides upper bounds for the optimal energy complexity. For the first dataflow, we further prove a matching lower bound by means of the weak duality theorem from linear programming. The optimal energy complexity for fully-connected layers in situations where Buffer is partitioned into two fixed parts ensues. The results are partially generalized to contiguous Buffer and are intended to be extended to convolutional layers in a future research.

The paper is organized as follows. Section 2 formally defines the energy complexity for fully-connected layers, and derives a general lower bound on the energy. Section 3 present two dataflows with their associated upper bounds on the energy. In Sect. 4, a matching and thus optimal lower bound is derived for the case of partitioned Buffer, and a partial generalization to contiguous Buffer is provided. Section 5 summarizes the results and discusses open problems.

2 Energy Complexity of Fully-Connected Layer

For simplicity, we consider a fully-connected CNN layer λ, which is composed of m neurons (units), each of which receiving connections labeled with real weights from all the n neurons in the previous layer $\lambda - 1$. This can be viewed as a complete weighted bipartite graph $G = (X, Y, E)$ where $X = \{x_1, \ldots, x_n\}$ and $Y = \{y_1, \ldots, y_m\}$ are disjoint sets of $n = |X|$ *inputs* and $m = |Y|$ *outputs*, respectively, and $E = X \times Y$ is a set of directed edges (x_i, y_j) leading from input x_i to output y_j, each labeled with a real *weight* w_{ji}, for every $j = 1, \ldots, m$ and $i = 1, \ldots, n$. The fully-connected CNN layer is evaluated as follows:

$$y_j = \text{ReLU}\left(w_{j0} + \sum_{i=1}^{n} w_{ji}x_i\right) \quad \text{for every } j = 1, \ldots, m, \quad (2)$$

where $\text{ReLU}(x) = \max(0, x)$ is the rectified linear unit activation function and w_{j0} is a *bias* of output neuron y_j, for every $j = 1, \ldots, m$.

To evaluate the computation energy $E_{\text{comp}}^{\lambda}$ of fully-connected layer λ in (1), note that the total number of MAC operations needed for computing (2) is mn: each output y_j is initialized with bias w_{j0} and requires n MAC updates to be computed. The computation energy is thus given by

$$E_{\text{comp}}^{\lambda} = C_b \, mn \quad (3)$$

where C_b is a non-uniform parameter depending on the number of bits b in floating-point MAC operations, since the design of a MAC circuit inside a microprocessor differs for each b.

We now focus on the data energy $E_{\text{data}}^{\lambda}$ of fully-connected layer λ in (1). This energy cost can be split into three components that count the DRAM accesses separately for the outputs, inputs, and weights:

$$E_{\text{data}}^{\lambda} = E_{\text{outputs}}^{\lambda} + E_{\text{inputs}}^{\lambda} + E_{\text{weights}}^{\lambda}. \quad (4)$$

In order to evaluate the sums in (2), all the mn couples of inputs and (accumulated) outputs (i.e. partially evaluated sums) need to occur in Buffer at least once. Each such pair (x_i, y_j) is associated with the unique weight w_{ji} that can be read from DRAM when the pair meet in Buffer for the first time. This means that each of the mn weights is read only once. Let ν and μ be the numbers of DRAM accesses to read inputs and outputs (or biases when initialized), respectively, and b be the number of bits in the floating point representation of outputs, inputs, and weights. The data energy (4) can thus be rewritten as

$$E_{\text{data}}^{\lambda} = b\,(2\mu + \nu + mn) \quad (5)$$

since each output that is read into Buffer is later written back to DRAM, which corresponds to two DRAM accesses, whereas each input and weight are only read into Buffer. In order to optimize the data energy (4), it is thus sufficient to minimize $2\mu + \nu$.

We will now derive a simple general lower bound on the data energy (4) for fully-connected layers. Assume that Buffer has a size of $B = b(\beta + 1)$ bits, where $\beta > 1$ floats are reserved for storing inputs and outputs, and the remaining capacity of one float is dedicated to the weights. For notational simplicity, suppose that $\beta - 1$ divides m. In addition, for any dataflow, let r be the minimum number of phases during which either only inputs or only outputs are read into Buffer consecutively. Note that by reading a single input or output into Buffer, one can get at most $\beta - 1$ new input-output pairs in Buffer. Since all mn pairs need to meet in Buffer, we obtain the following trivial lower bound on the number of DRAM read accesses:

$$\mu + \nu \geq \frac{mn}{\beta - 1}. \tag{6}$$

Moreover, in order to keep generating new pairs in Buffer, at most n inputs or m outputs can be read during each phase. This ensures that $r(\beta - 1)\max(m, n) \geq mn$ which implies

$$r \geq \frac{\min(m, n)}{\beta - 1}. \tag{7}$$

Observe that, when a next phase begins, the reading of an input immediately after an output has been read (or vice versa) provides at most β new pairs in Buffer through these two DRAM read accesses (cf. the trivial upper bound $2(\beta - 1)$ of new pairs counted in (6) for two reads). Indeed, if there are k inputs ($1 \leq k \leq \beta - 1$) and $\beta - k$ outputs in Buffer, the reading of an input yields at most $\beta - k$ new pairs, while the subsequent reading of an output generates at most k new pairs, which sums up to at most β new pairs in total. Let s be the number of readings that do not occur at the beginning of a new phase. The following lower bound on the number of DRAM read accesses ensues:

$$\mu + \nu \geq 2r + s + 1 \tag{8}$$

with

$$\beta r + (\beta - 1)s \geq mn \tag{9}$$

because all the mn pairs have to occur in Buffer, the two readings at the beginning of each of the r phases generate at most β new pairs, and each of the remaining s readings produces at most $\beta - 1$ new pairs, except for the very first DRAM read access providing no pair.

Inequality (9) can be rewritten as

$$(\beta - 1)(2r + s) \geq mn + (\beta - 2)r \tag{10}$$

which implies

$$\mu + \nu \geq \frac{mn}{\beta - 1} + \frac{\beta - 2}{\beta - 1}r + 1 \geq \frac{mn}{\beta - 1} + \frac{\beta - 2}{(\beta - 1)^2}\min(m, n) + 1 \tag{11}$$

according to (8) and (7). Since the biases of all m outputs must first be read into Buffer, we have $\mu \geq m$, and thus

$$2\mu + \nu \geq \frac{mn}{\beta - 1} + m + \frac{\beta - 2}{(\beta - 1)^2}\min(m, n) + 1. \tag{12}$$

This provides a general lower bound on the data energy of fully-connected layer λ:

$$E_{\text{data}}^{\lambda} \geq b \left(mn + \frac{m(n-1)}{\beta - 1} + \frac{\beta}{\beta - 1}m + \frac{\beta - 2}{(\beta - 1)^2} \min(m, n) + 1 \right) \qquad (13)$$

according to (5).

3   Upper Bounds on Energy Complexity

Any correct dataflow for processing a fully-connected layer can be described by a sequence of p sets $B_0, B_1, \ldots, B_p \subseteq X \cup Y$, each of which being composed of vertices in G, that represent the successive contents of Buffer (excluding weights) after each DRAM access to read an input or output, in the course of evaluating the sums in (2). The sequence satisfies the following conditions:

1. $B_0 = \emptyset$
2. $|B_i| \leq \beta$ for every $i = 1, \ldots, p$
3. $|B_i \setminus B_{i-1}| = 1$ and $|B_{i-1} \setminus B_i| \leq 1$ for every $i = 1, \ldots, p$
4. $Y \subseteq \bigcup_{x \in B_i} B_i$ for every $x \in X$,

and its length p is the total number of DRAM read accesses,

$$p = \mu + \nu. \qquad (14)$$

Condition 1 assumes empty Buffer at the beginning, and Condition 2 guarantees that its size is not exceeded. Condition 3 ensures that, by reading a single input or output into Buffer, at most one input or output is overwritten. Condition 4 ensures that all of the outputs meet every input in Buffer.

In the two following subsections, we present two dataflows for fixed and bounded number of inputs in Buffer, respectively, such that each output is read into Buffer only once (i.e., when initialized by a corresponding bias), which means that

$$\mu = m. \qquad (15)$$

Clearly, the role of inputs and outputs can be reversed in these dataflows.

3.1   Fixed Number of Inputs in Buffer

For the first dataflow, we assume that Buffer is partitioned into two fixed parts for inputs and outputs, respectively, and contains one more float for reading the weights. One part is reserved for storing d inputs and the second one to store $\beta - d$ outputs, where d is a fixed parameter such that $1 \leq d \leq \beta - 1$. The dataflow can be described by the following sequence of sets $B_0, B_1, \ldots, B_p$ that meet Conditions 1–4, $|B_i \cap X| \leq d$, and $|B_i \cap Y| \leq \beta - d$ for every $i = 1, \ldots, p$:

$$\emptyset, \{x_1\}, \{x_1, x_2\}, \ldots, \{x_1, \ldots, x_d\}, \qquad (16)$$

$$\{x_1, \ldots, x_d, y_1\}, \{x_1, \ldots, x_d, y_1, y_2\}, \ldots, \{x_1, \ldots, x_d, y_1, \ldots, y_{\beta - d}\}, \qquad (17)$$

$$\{x_{d+1}, x_2 \ldots, x_d, y_1, \ldots, y_{\beta-d}\}, \{x_{d+1}, x_{d+2}, x_3, \ldots, x_d, y_1, \ldots, y_{\beta-d}\}, \ldots,$$

$$\{x_{n-d+1}, \ldots, x_n, y_1, \ldots, y_{\beta-d}\}, \quad (18)$$

$$\{x_{n-d+1}, \ldots, x_n, y_{\beta-d+1}, y_2, \ldots, y_{\beta-d}\},$$

$$\{x_{n-d+1}, \ldots, x_n, y_{\beta-d+1}, y_{\beta-d+2}, y_3, \ldots, y_{\beta-d}\}, \ldots,$$

$$\{x_{n-d+1}, \ldots, x_n, y_{\beta-d+1}, \ldots, y_{2(\beta-d)}\}, \quad (19)$$

$$\{x_{n-d+1}, \ldots, x_{n-1}, x_{n-d}, y_{\beta-d+1}, \ldots, y_{2(\beta-d)}\},$$

$$\{x_{n-d+1}, \ldots, x_{n-2}, x_{n-d-1}, x_{n-d}, y_{\beta-d+1}, \ldots, y_{2(\beta-d)}\}, \ldots,$$

$$\{x_1, \ldots, x_d, y_{\beta-d+1}, \ldots, y_{2(\beta-d)}\}, \quad \ldots \quad (20)$$

After an initialization where the first d inputs are read into Buffer (16), the dataflow alternates between two phases of reading $\beta - d$ outputs (17) (or (19) etc.) and reading $n - d$ inputs (18) (or (20) etc.), respectively, while overwriting the outputs in Buffer by new outputs (cf. (19)) and the inputs in Buffer by new inputs (cf. (20)). Apart from d reads at initialization, $\frac{m}{\beta-d}$ changes from the first phase to the second one are performed before each of the m outputs has been read into Buffer once, which implies

$$p = d + \frac{m}{\beta - d} \left((\beta - d) + (n - d) \right) = \frac{m(n - d)}{\beta - d} + m + d. \quad (21)$$

Hence, this dataflow provides an upper bound on the data energy of fully-connected layer λ:

$$E_{\text{data}}^{\lambda} \leq b \left(mn + \frac{m(n - d)}{\beta - d} + 2m + d \right) \quad (22)$$

according to (5), (14), and (15). This upper bound takes the smallest value for $d = 1$, provided that $n \geq \beta$, since $n \geq \beta$ is equivalent to

$$\frac{m(n - 1)}{\beta - 1} \leq \frac{m(n - d)}{\beta - d}.$$

Furthermore, an alternative upper bound to (22) is obtained when the roles of the inputs and outputs are reversed in the dataflow (16)–(20):

$$E_{\text{data}}^{\lambda} \leq b \left(mn + \frac{2n(m - (\beta - d))}{d} + n + 2(\beta - d) \right). \quad (23)$$

This upper bound has the smallest value for $d = \beta - 1$, provided that $m \geq \beta$, since $m \geq \beta$ is equivalent to

$$\frac{2n(m - 1)}{\beta - 1} \leq \frac{2n(m - (\beta - d))}{d}.$$

Finally, assuming $n \geq \beta$ and $m \geq \beta$, we can compare (22) and (23) for their smallest values, namely $d = 1$ and $d = \beta - 1$, respectively:

$$b \left(mn + \frac{m(n - 1)}{\beta - 1} + 2m + 1 \right) \overset{?}{\leq} b \left(mn + \frac{2n(m - 1)}{\beta - 1} + n + 2 \right) \quad (24)$$

which can be rewritten as

$$0 \overset{?}{\leq} m(n - 2\beta + 3) + n(\beta - 3) + \beta - 1. \tag{25}$$

This inequality holds for $n > 2\beta - 3$ implying $n \geq \beta$ due to $\beta \geq 2$. Therefore, we can conclude that for sufficiently large $n > 2\beta - 3$ and $m \geq \beta$, the minimal energy for fully-connected layers achieved by the dataflow (16)–(20) is obtained when $d = 1$, i.e., when Buffer is partitioned to $\beta - 1$ outputs, one input, and one weight. This situation leads to the following upper bound:

$$E_{\text{data}}^{\lambda} \leq b \left(mn + \frac{m(n-1)}{\beta - 1} + 2m + 1 \right). \tag{26}$$

3.2   Bounded Number of Inputs in Buffer

The second dataflow is parameterized by the maximum number k of inputs that can simultaneously occur in Buffer, where $1 \leq k \leq \beta - 1$. The dataflow is described by the following sequence of sets $B_0, B_1, \ldots, B_p$ satisfying Conditions 1–4 and $|B_i \cap X| \leq k$ for every $i = 1, \ldots, p$:

$$\emptyset, \{x_1\}, \{x_1, x_2\}, \ldots, \{x_1, \ldots, x_k\}, \tag{27}$$

$$\{x_1, \ldots, x_k, y_1\}, \{x_1, \ldots, x_k, y_1, y_2\}, \ldots, \{x_1, \ldots, x_k, y_1, \ldots, y_{\beta-k}\}, \tag{28}$$

$$\{x_1, \ldots, x_{k-1}, y_1, \ldots, y_{\beta-k+1}\}, \{x_1, \ldots, x_{k-2}, y_1, \ldots, y_{\beta-k+2}\}, \ldots,$$
$$\{x_1, y_1, \ldots, y_{\beta-1}\}, \tag{29}$$

$$\{x_n, y_1, \ldots, y_{\beta-1}\}, \{x_{n-1}, y_1, \ldots, y_{\beta-1}\}, \ldots, \{x_{k+1}, y_1, \ldots, y_{\beta-1}\}, \tag{30}$$

$$\{x_k, x_{k+1}, y_2, \ldots, y_{\beta-1}\}, \{x_{k-1}, x_k, x_{k+1}, y_3, \ldots, y_{\beta-1}\}, \ldots,$$
$$\{x_2, \ldots, x_{k+1}, y_k, \ldots, y_{\beta-1}\}, \tag{31}$$

$$\{x_2, \ldots, x_{k+1}, y_{k+1}, \ldots, y_\beta\}, \{x_2, \ldots, x_{k+1}, y_{k+2}, \ldots, y_{\beta+1}\}, \ldots,$$
$$\{x_2, \ldots, x_{k+1}, y_\beta, \ldots, y_{2\beta-k-1}\}, \tag{32}$$

$$\{x_2, \ldots, x_k, y_\beta, \ldots, y_{2\beta-k}\}, \{x_2, \ldots, x_{k-1}, y_\beta, \ldots, y_{2\beta-k+1}\}, \ldots,$$
$$\{x_2, y_\beta, \ldots, y_{2\beta-2}\}, \tag{33}$$

$$\{x_1, y_\beta, \ldots, y_{2\beta-2}\}, \{x_n, y_\beta, \ldots, y_{2\beta-2}\}, \{x_{n-1}, y_\beta, \ldots, y_{2\beta-2}\}, \ldots,$$
$$\{x_{k+2}, y_\beta, \ldots, y_{2\beta-2}\}, \tag{34}$$

$$\{x_{k+1}, x_{k+2}, y_{\beta+1}, \ldots, y_{2\beta-2}\}, \{x_k, x_{k+1}, x_{k+2}, y_{\beta+2}, \ldots, y_{2\beta-2}\}, \ldots,$$
$$\{x_3, \ldots, x_{k+2}, y_{\beta+k-1}, \ldots, y_{2\beta-2}\}, \ldots \tag{35}$$

After an initialization when the first k inputs are read into Buffer (27), the dataflow alternates between two phases of reading $\beta - 1$ outputs (28)–(29) (or (32)–(33) etc.) and reading $n-1$ inputs (30)–(31) (or (34)–(35) etc.), respectively. In the general first phase (32)–(33) (when outputs are read into Buffer), $\beta - k$ outputs currently stored in Buffer are first replaced by new ones (32), and only then the $k - 1$ inputs residing in Buffer are overwritten by outputs (33) until one input remains in Buffer. During the second phase (34)–(35) (when inputs are read into Buffer), the remaining input is being replaced one by one with $n - k$ inputs (34), and then the last $k - 1$ read inputs overwrites the outputs stored in Buffer, so that k inputs and $\beta - k$ outputs are left in Buffer at the end of the second phase. This phase can again be followed by the first phase, etc. Apart from k reads at initialization, the first phase changes to the second one $\frac{m}{\beta-1}$ times before each of the m outputs is read into Buffer once, which implies

$$p = k + \frac{m}{\beta - 1}\left((\beta - k) + (k - 1) + (n - k) + (k - 1)\right) = \frac{m(n - 1)}{\beta - 1} + m + k. \quad (36)$$

Hence, this dataflow provides an upper bound on the data energy of fully-connected layer λ:

$$E_{\text{data}}^{\lambda} \leq b\left(mn + \frac{m(n - 1)}{\beta - 1} + 2m + k\right) \quad (37)$$

according to (5), (14), and (15). Note that the first dataflow (16)–(20) for $d = 1$ coincides with the second dataflow (27)–(35) for $k = 1$, producing the same upper bound (26).

4 Lower Bounds on Energy Complexity

4.1 Partitioned Buffer

We now study the case where Buffer is divided into two fixed parts dedicated to the reading of d inputs and $\beta - d$ outputs, respectively, plus one float for weights, where d is a fixed parameter such that $1 \leq d \leq \beta - 1$. In this context, we improve the general lower bound (13) on the data energy $E_{\text{data}}^{\lambda}$ of fully-connected layer λ so that it matches the upper bounds (22) and (23), up to an additive constant. We distinguish two cases according to whether d is at most or at least $\frac{2}{3}\beta$.

Case $1 \leq d \leq \frac{2}{3}\beta$. Assume first that

$$1 \leq d \leq \tfrac{2}{3}\beta. \quad (38)$$

We formulate a linear program of finding μ and ν that

$$\text{minimize} \quad 2\mu + \nu \quad (39)$$
$$\text{subject to} \quad d\mu + (\beta - d)\nu \geq mn \quad (40)$$
$$\mu \geq m \quad (41)$$
$$\nu \geq 0, \quad \mu \geq 0. \quad (42)$$

Constraint (40) expresses the fact that all mn input-output couples have to occur in Buffer, since by reading one output or input, at most d or $\beta - d$ new pairs meet in Buffer, respectively. Constraint (41) ensures that at least m outputs are read into Buffer. We convert the linear program (39)–(42) to the corresponding dual linear program of finding ϕ and ψ that

$$\text{maximize} \quad mn\,\phi + m\,\psi \quad (43)$$
$$\text{subject to} \quad d\phi + \psi \leq 2 \quad (44)$$

$$(\beta - d)\phi \leq 1 \qquad (45)$$
$$\phi \geq 0, \quad \psi \geq 0. \qquad (46)$$

Observe that $\phi_0 = \frac{1}{\beta-d}$ and $\psi_0 = 2 - \frac{d}{\beta-d}$ is a feasible solution for the dual program, satisfying (44)–(46) due to (38). By the weak duality theorem, the objective function value of the primal (39) at any feasible solution is lower bounded by the objective function value of the dual (43) at any feasible solution, that is,

$$2\mu + \nu \geq mn\,\phi_0 + m\,\psi_0 = \frac{m(n-d)}{\beta-d} + 2m. \qquad (47)$$

According to (5), inequality (47) provides the following lower bound on the data complexity of fully-connected layer λ:

$$E_{\text{data}}^{\lambda} \geq b\left(mn + \frac{m(n-d)}{\beta-d} + 2m\right) \qquad (48)$$

when Buffer is divided into two parts for d inputs and $\beta - d$ outputs, and the fixed parameter d meets (38). This lower bound matches the corresponding upper bound (22) achieved by the dataflow (16)–(20), up to the additive constant d.

Case $\frac{2}{3}\beta \leq d \leq \beta - 1$. Similarly, for

$$\frac{2}{3}\beta \leq d \leq \beta - 1, \qquad (49)$$

we have a linear program of finding μ and ν that minimize $2\mu + \nu$ subject to $d\mu + (\beta - d)\nu \geq mn$, $\nu \geq n$, $\nu \geq 0$, and $\mu \geq 0$. This is converted to the corresponding dual linear program of finding ϕ and ψ that maximize $mn\,\phi + n\,\psi$ subject to $d\phi \leq 2$, $(\beta - d)\phi + \psi \leq 1$, $\phi \geq 0$, and $\psi \geq 0$, which has a feasible solution $\phi_1 = \frac{2}{d}$ and $\psi_1 = 1 - \frac{2(\beta-d)}{d}$ due to (49). By the weak duality theorem we have

$$2\mu + \nu \geq mn\,\phi_1 + n\,\psi_1 = \frac{2n(m - (\beta - d))}{d} + n \qquad (50)$$

which provides the following lower bound on the data complexity of fully-connected layer λ:

$$E_{\text{data}}^{\lambda} \geq b\left(mn + \frac{2n(m - (\beta - d))}{d} + n\right) \qquad (51)$$

when Buffer is divided into two parts for d inputs and $\beta - d$ outputs, and the fixed parameter d meets (49). This lower bound matches the corresponding upper bound (23) achieved by the dataflow (16)–(20) with the reversed role of inputs and outputs, up to the additive constant $2(\beta - d)$.

We can conclude that the data energy for fully-connected layers achieved by the dataflow (16)–(20) when Buffer is partitioned to d inputs, $\beta - d$ outputs, and one weight, is optimal for any fixed d, and the minimum of data energy (26) is achieved for $d = 1$.

4.2   Partial Generalization

In general case when Buffer is not divided into separate parts, the lower bound (13) on the data energy complexity still differs from the upper bound (26) by linear additive term $\frac{\beta-2}{\beta-1}\left(m - \frac{\min(m,n)}{\beta-1}\right)$, which can further be improved in some special cases. In particular, denote by μ_k and ν_k for $1 \leq k \leq \beta - 1$ the number of accesses to DRAM for reading outputs and inputs at the points when exactly k inputs reside in Buffer. The linear program (39)–(42) can be generalized to the following program of finding μ_k and ν_k for $1 \leq k \leq \beta - 1$ that

$$\text{minimize} \quad 2\mu + \nu = 2\sum_{k=1}^{\beta-1} \mu_k + \sum_{k=1}^{\beta-1} \nu_k \tag{52}$$

$$\text{subject to} \quad \sum_{k=1}^{\beta-1} k\mu_k + \sum_{k=1}^{\beta-1}(\beta - k)\nu_k \geq mn \tag{53}$$

$$\sum_{k=1}^{\beta-1} \mu_k \geq m \tag{54}$$

$$\mu_k \geq 0, \quad \nu_k \geq 0 \quad \text{for } k = 1, \ldots, \beta - 1. \tag{55}$$

By applying the weak duality theorem to this program, one can achieve only the trivial lower bound (6). Nevertheless, this lower bound can be improved when the following, yet somewhat artificial, condition is added to (53)–(55):

$$\nu_k - \mu_k \geq 0 \quad \text{for } k = 1, \ldots, \beta - 1, \tag{56}$$

that is, $\nu_k \geq \mu_k$ for $1 \leq k \leq \beta - 1$. This condition states that input readings into Buffer is preferred over more expensive output readings, since outputs need to be written back to DRAM. Note that this condition is satisfied by the dataflows presented in Sect. 3.

Thus, we convert the linear program (52)–(56) to the corresponding dual linear program of finding ϕ, ψ, and χ_k for $1 \leq k \leq \beta - 1$, that maximize $mn\,\phi + m\,\psi$ subject to $k\phi + \psi - \chi_k \leq 2$, $(\beta - k)\phi + \chi_k \leq 1$, $\phi \geq 0$, $\psi \geq 0$, and $\chi_k \geq 0$ for every $k = 1, \ldots, \beta - 1$. Observe that $\phi_0 = \frac{1}{\beta-1}$, $\psi_0 = 2 - \frac{1}{\beta-1}$, and $\chi_{k0} = \frac{k-1}{\beta-1}$ for $1 \leq k \leq \beta - 1$, is a feasible solution for this dual. By the weak duality theorem, we have

$$2\mu + \nu = 2\sum_{k=1}^{\beta-1} \mu_k + \sum_{k=1}^{\beta-1} \nu_k \geq mn\,\phi_0 + m\,\psi_0 = \frac{m(n-1)}{\beta-1} + 2m \tag{57}$$

which proves the optimality of the data energy (26) (up to 1) also for contiguous Buffer, provided that condition (56) holds.

5   Conclusion

In this paper, we have theoretically analyzed the energy complexity model for CNNs introduced in our previous work [7], which is asymptotically consistent with estimates of power consumption for different CNN hardware implementations. We have confined ourselves to fully-connected layers as a starting point for the future analysis of convolutional layers. We have shown a simple general lower bound on energy complexity of fully-connected layers. We have presented two dataflows for fixed and bounded numbers of inputs residing in Buffer, respectively, and computed their energy costs to obtain upper bounds on the energy. We have then proven a matching lower bound on the energy for the first dataflow, which in turn provides the optimal energy complexity for fully-connected layers when Buffer is partitioned into two fixed parts for inputs and outputs.

In future research, the lower bound is intended to be generalized to contiguous Buffer, namely Buffer without partition. The partial generalization presented here shows that a linear program formulation seems to be not strong enough to achieve this goal, meaning that a detailed analysis of DRAM accesses would be needed. This analysis could then be used to prove the optimal energy complexity for convolutional layers, which represents the main challenge of this research.

Acknowledgements. The research was partially supported by the institutional support RVO: 67985807 and the Czech Science Foundation grant GA22-02067S. We thank Petr Savický for inspiring discussions in the early stages of this research.

References

1. Alwani, M., Chen, H., Ferdman, M., Milder, P.A.: Fused-layer CNN accelerators. In: Proceedings of IEEE/ACM MICRO 2016, pp. 22:1–22:12 (2016). https://doi.org/10.1109/MICRO.2016.7783725
2. Chen, Y., Emer, J.S., Sze, V.: Eyeriss: A spatial architecture for energy-efficient dataflow for convolutional neural networks. In: Proceedings of ACM/IEEE ISCA 2016, pp. 367–379 (2016). https://doi.org/10.1109/ISCA.2016.40
3. Mittal, S.: A survey of techniques for approximate computing. ACM Comput. Surv. **48**(4), 62:1–62:33 (2016). https://doi.org/10.1145/2893356
4. Mittal, S.: A survey of FPGA-based accelerators for convolutional neural networks. Neural Comput. Appl. **32**(4), 1109–1139 (2020). https://doi.org/10.1007/s00521-018-3761-1
5. Parashar, A., et al.: Timeloop: A systematic approach to DNN accelerator evaluation. In: Proceedings of IEEE ISPASS 2019, pp. 304–315 (2019). https://doi.org/10.1109/ISPASS.2019.00042
6. Shao, Y.S., et al.: Simba: Scaling deep-learning inference with multi-chip-module-based architecture. In: Proceedings of IEEE/ACM MICRO 2019, pp. 14–27 (2019). https://doi.org/10.1145/3352460.3358302
7. Šíma, J., Vidnerová, P., Mrázek, V.: Energy complexity model for convolutional neural networks. In: Proceedings of ICANN 2023. LNCS, Springer (2023)
8. Sze, V., Chen, Y., Yang, T., Emer, J.S.: Efficient Processing of Deep Neural Networks. Synthesis Lectures on Computer Architecture, Morgan & Claypool Publishers (2020). https://doi.org/10.2200/S01004ED1V01Y202004CAC050

9. Wu, Y.N., Emer, J.S., Sze, V.: Accelergy: An architecture-level energy estimation methodology for accelerator designs. In: Proceedings of IEEE/ACM ICCAD 2019 (2019). https://doi.org/10.1109/ICCAD45719.2019.8942149
10. Yang, T., Chen, Y., Emer, J.S., Sze, V.: A method to estimate the energy consumption of deep neural networks. In: Proceedings of IEEE ACSSC 2017, pp. 1916–1920 (2017). https://doi.org/10.1109/ACSSC.2017.8335698

Low-Dimensional Space Modeling-Based Differential Evolution: A Scalability Perspective on *bbob-largescale* suite

Thiago Henrique Lemos Fonseca[1]($\boxtimes$), Silvia Modesto Nassar[1],
Alexandre César Muniz de Oliveira[2,3], and Bruno Agard[4]

[1] Universidade Federal de Santa Catarina (UFSC), Florianópolis, SC, Brazil
thiagolemos_@outlook.com
[2] Universidade Federal do Maranhão (UFMA), São Luís, MA, Brazil
[3] University College Cork, Cork, CO, Ireland
[4] Polytechnique Montréal, Montreal, QC, Canada

Abstract. Scalability is a challenge for Large Scale Optimization Problems (LSGO). Improving the scalability of efficient Differential Evolution algorithms (DE) has been a research focus due to their successful application to high-dimensional problems. Recently, a DE-based algorithm called LSMDE (Low-dimensional Space Modeling-based Differential Evolution) has shown promising results in solving LSGO problems on the CEC'2013 large-scale global optimization suite. LSMDE uses dimensionality reduction to generate an alternative search space and Gaussian mixture models to deal with the information loss caused by uncertainty from space transformation. This paper aims to extend the initial research through the scalability analysis of the LSMDE's performance compared with its main competitors, SHADE-ILS and GL-SHADE, on *bbob-largescale* suite functions. The results show that although all competing algorithms perform worse as dimensionality increases, LSMDE outperforms the competition and is robust to dimensionality expansion in search spaces with diverse characteristics, achieving a target hit rate between 40% and 80%.

Keywords: Gaussian Mixture Model · Dimensionality Reduction · Differential Evolution · Singular Value Decomposition

1 Introduction

Large Scale Global Optimization (LSGO) addresses optimization modeling involving many decision variables strongly related to real problems in engineering and scientific computing problems. It has been studied mainly for designing robust, scalable, and parallelizable algorithms, also promoting research lines related to large-scale computational projects, such as Distributed Systems and Parallel Algorithms [9,25]. The applicability of LSGO to subjects of intensive studies, such as Deep Learning and Big Data, also attracts interest in developing ever-more-effective algorithms [24,28].

I. Rojas et al. (Eds.): IWANN 2023, LNCS 14134, pp. 16–28, 2023.
https://doi.org/10.1007/978-3-031-43085-5_2

Nonetheless, several factors make LSGO problem-solving extremely difficult. The curse of dimensionality significantly contributes to algorithm performance degradation with increasing dimensionality. The curse of dimensionality refers to the fact that high-dimensional space grows exponentially with the number of dimensions, dramatically increasing the number of samples required to model the search space accurately, which can make the optimization problem intractable. In addition, high-dimensional problems often have several local optima, making it difficult for algorithms to converge to the global optimum and escape getting trapped in local optima [16]. Another factor contributing to algorithm performance degradation with increasing dimensionality is the exponential increase in computation time, making it difficult for algorithms to keep up with the growing complexity of the problem, leading to a decrease in performance [16].

Scalability is crucial for designing and evaluating LSGO algorithms. Scalability analysis determines an algorithm's efficiency as the problem size grows, which is essential for real-world applications and improvement areas [6, 32]. This analysis identifies bottlenecks and informs algorithm performance enhancement efforts.

Metaheuristic algorithms have been introduced to address LSGO issues in the past decade [3, 18]. Differential Evolution (DE) has been chosen for LSGO due to its optimization benefits [21]. However, DE performance suffers from the curse of dimensionality in LSGO [19]. Recent papers [1, 2] demonstrated that optimizing in lower-dimensional search spaces can mitigate high dimensionality's adverse effects, potentially hybridizing with metaheuristics like DE. Building on this, [12] developed Low-dimensional Space Modeling-based Differential Evolution (LSMDE), a hybrid algorithm combining DE benefits and dimensionality reduction advances for LSGO problem-solving.

LSMDE uses a Singular Value Decomposition (SVD) method to create a low-dimensional search space based on DE-generated candidate solutions. Gaussian Mixture Model (GMM) is employed alongside DE to enhance the search for optimal solutions in low-dimensional space [12].

The results presented in [12] show that LSMDE performs better or is similar to the main algorithms for LSGO in different search space characteristics, especially for partially separable functions. However, a comparative scalability analysis still needs to be performed to understand the effects of the change in dimensionality on LSMDE and its main competitors.

This work extends [12] with a more in-depth comparative scalability analysis of LSMDE and competitors on the *bbob-largescale* suite.

The paper is organized as follows: Sect. 2 reviews related works; Sect. 3 presents LSMDE; Sect. 4 outlines the experimental setup and scalability analysis results; and Sect. 5 discusses findings and concludes the paper.

2   Related Work

Differential Evolution (DE) has proven to be a promising approach for LSGO problems, with several significant improvements made over the last few years

[27]. Many features contribute to the use of this class of metaheuristics; for example, difference vectors can be correlated with the search space; it uses only $O(Np)$ processes (where Np is the population size); it does not need a predefined probability distribution for generating offspring; the objective functions do not need to be differentiable; it can provide multiple solutions from a single run of the algorithm; it is straightforward to implement, and is a parallel optimization procedure like many other population-based schemes [13]. However, the effectiveness and scalability of Differential Evolution algorithms are still challenges that can be seen with the rapid performance deterioration by increasing the dimension and complexity of the problems [17, 21]. Given these limitations and the growing trend of complex LSGO problems, designing and developing effective and scalable DE algorithms is an emerging issue.

An essential contribution to this effectivity and scalability can be observed in one of the most successful DE for high dimensionality problems, the Success-History Based Parameter Adaptation for Differential Evolution (SHADE) [29]. SHADE is a DE variant with a self-adaptive mechanism applied to the crossover rate (CR) and scaling Factor (F) parameters. Additionally, SHADE incorporates an external archive (A) built from the defeated parents throughout generations since such individuals are used during the application of the mutation operator to promote greater diversity [29].

Several SHADE-based algorithms have been ranked among the best for LSGO in recent years; for example, [22] proposed the SHADE-ILS. Winner of the *IEEE CEC'2018 Special Session and Competition on Large-Scale Global Optimization*, SHADE-ILS combines an exploration stage with the modern SHADE and an intensification stage with a local search method dynamically chosen from two different methods, MTS-LS1 [30], and L-BFGS-B [23]. Experimental results have shown that SHADE-ILS is especially good for high-dimensional space function with overlapping and non-separable components.

A recent hybrid algorithm called Global and Local search using Success-History Based Parameter Adaptation for Differential Evolution (GL-SHADE) was proposed by [26]. GL-SHADE consists of three stages: (1) initialization, (2) global search, and (3) local search. During the initialization stage, a gradient-free non-population-based local search method is applied to one of the best individuals generated to make an early enhancement. Afterward, the global and local search stages are repeated one after another. Moreover, GL-SHADE consists of two populations that collaborate. The first population presents a search scheme specialized in exploration (thus carrying out the global search stage). The second offers a search engine specialized in exploitation (carrying out the local search stage). The first population evolves according to SHADE's algorithm, and the second one is according to a newly developed SHADE variant named eSHADE. GL-SHADE outperformed the SHADE-ILS in many IEEE LSGO test problems [26].

Well-known studies show that dimensionality reduction techniques and space projection are alternatives to deal with the scalability issue of many essential algorithms [14]. It is possible to preserve important features of the original space (such as Euclidean distances or dot products) in the reduced space within a reasonable tolerance [7]. It has also been shown that the distribution of the sample points becomes more Gaussian in the reduced space [10]. These features make it possible to capture the variable correlation of the high-dimensional space using a lower-dimensional subspace. From this premise, many non-based DE algorithms have successfully mapped the optimization process to a more straightforward alternative space [1,2]. LSMDE differs from previous approaches by not requiring prior knowledge about the search space, making it more flexible and applicable to high-dimensional problems. The results on the CEC'2013 large-scale global optimization suite, one of the main benchmarks for LSGO, show that LSMDE presented promising results in solving complex continuous problems with 1000 dimensions and with the most varied characteristics of search spaces. The experiments also showed that LSMDE is especially good for partially separable functions, the category to which most real high-dimensional problems belong.

Despite the technical rigor in conducting the latest experiments with LSMDE on the CEC'2013 large-scale global optimization suite, CEC suites are set up with a fixed number of dimensions (although the problems are, in principle, scalable). The performance assessment is prescribed for a few given budgets. This setup does not allow for reliably measuring scaling behavior with dimension, one of the most important characteristics a benchmarking experiment for large-scale algorithms should investigate [31]. Based on that, this work analyzes LSMDE and its main competitors, SHADE-ILS and GL-SHADE, from the scalability perspective on the *bbob-largescale* suite.

3   LSMDE's Review

LSMDE is a non-decomposition-based method proposed by [12] that uses Gaussian mixture models to find the best solution in the reduced space after optimizing the entire search space using an adaptive dimensionality reduction methodology. LSMDE consists of three primary steps: population initialization, dimension reduction, and search space exploration [12]. LSMDE's population initialization procedure is based on a variant of partial opposition learning inspired by [20] with three different opposition operators for each problem dimension, as defined below.

Definition 1 (LSMDE Opposition Strategy). *Let $\vec{x} = (x_1, ..., x_d)$ be a point in d-dimensional space and $x_j \in [a_j, b_j], j = 1, 2, ..., d$. The opposite of $\vec{x}$ is defined by $\vec{z} = (z_1, ..., z_d)$ where each z_j is selected randomly among three opposition operators, as follows:*

$$z_j = \begin{cases} a_j + b_j - x_j \\ x_j + a_j - ((a_j + b_j)/2) \mod (b_j - a_j) \\ random(a_j + b_j - x_j, a_j + b_j/2) \end{cases} \tag{1}$$

The objective is to produce a diverse starting population so that subsequent optimization processes can more effectively explore the d-dimensional search space [12].

Given a minimization problem, a population of individuals $\vec{x}_i \in P$ generated randomly and its opposite population $\vec{z}_i \in Z$ generated according to Eq. 1, each new individual $\vec{x}_i$ of the new initial population P is formed by:

$$\vec{x}_i = \min(f(\vec{x}_i), f(\vec{z}_i)) - \forall \vec{x}_i, \vec{z}_i \in (P \cup Z) \text{ and } i = 1, 2, ..., n \qquad (2)$$

in which f is a fitness function and n is the number of individuals in the population. In other words, the initial population is formed by the best individuals between the random population P and the opposite population Z.

In the dimension reduction step, LSMDE employs an adaptive singular value decomposition strategy to map a subset of the search space represented by the population of candidate solutions $P \in \mathbb{R}^d$ into a low-dimensional subset $\Psi \in \mathbb{R}^k$, where d is the dimension of the problem and $k < d$.

According to Singular Value Decomposition (SVD) [15], any matrix $P^{n \times d}$ can be decomposed into three matrices U, S, V, where V^T is the transpose of matrix V as follows:

$$P = U \cdot S \cdot V^T \text{ or } P = \Sigma_{i=1}^{\min(n,d)} s_i \cdot \vec{u}_i \cdot \vec{v}_i^T \qquad (3)$$

in which s_i is the ith singular value and $\vec{u}_i, \vec{v}_i^T$ are the corresponding left singular vectors and right transpose singular vectors. In other words, the SVD expresses P as a non-negative linear combination of $\min(n, d)$ rank-1 matrices, with the singular values providing the multipliers and the outer products of the left and right singular vectors providing the rank-1 matrices. As the SVD of P represents the subspace P as a multiplication of matrices ordered by importance, LSMDE keeps only the most important k singular vectors to obtain a projection $U \cdot S \in \mathbb{R}^k$ from subspace $P \in \mathbb{R}^d$ where $k < d$. The general idea to reduce the dimensionality of the matrix P is to keep only the first top k terms on Eq. 4 in order to generate a low-rank approximation $\widetilde{P}$ similar to the original P [15] as defined below:

$$\widetilde{P} = U_k \cdot S_k \cdot V_k^T \text{ or } \widetilde{P} = \Sigma_{i=1}^{k} s_i \cdot \vec{u}_i \cdot \vec{v}_i^T \qquad (4)$$

Matrix $\Psi = U_k S_k$ gives low dimensional representations of the search space P from its k principal components scores (PC), and matrix V_k^T can be interpreted as a reconstruction matrix that projects these low dimensional points back into the approximate high-dimensional space $\widetilde{P}$ (Fig. 1).

LSMDE uses the Johnson-Lindenstrauss lemma (JLL) [8] to infer k for dimensionality reduction by Singular Value Decomposition to preserve the distances of individuals by a factor of $(1 \pm \epsilon)$. By making no assumptions about the topology of the search space, the method is more adaptable and applicable to a variety of search spaces. Based on JLL, it is possible to develop an adaptive ϕ function that maps each high-dimensional element $\vec{x}_i \in P$ into its corresponding element $\vec{\psi}_i \in \Psi$ and an inverse function ϕ' that projects these low dimensional individuals back into the high-dimensional space.

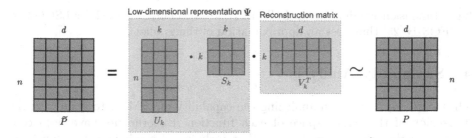

Fig. 1. Illustration of a low-dimensional representation from a d-dimensional random P matrix to a k-dimensional Ψ matrix [12].

In the final step, LSMDE employs GM-SHADE, a SHADE algorithm hybridized with a Gaussian Mixture Model and a variational Bayesian estimation mechanism, in the low dimensionality space to identify regions of optimality that can be converted back to the original space. As part of the variational estimation procedure, GM-SHADE estimates the model parameters and automatically detects the number of clusters using a parameter based on a finite mixture model with the Dirichlet process [4].

GM-SHADE searches for promising regions in the low-dimensional space Ψ by partitioning it into clusters represented by a function composed of multiple Gaussians, each identified by $\lambda \in 1,...,\Lambda$, where Λ is the current number of clusters. The Gaussian λ determines the cluster coverage, composed of a mean individual $\vec{\mu}$, a covariance Υ, and a mixing probability π. The goal is to ensure each Gaussian fits all candidate solutions $\vec{\psi}_i$ belonging to each cluster. Using the variational expectation-maximization algorithm (Variational EM) [4], GM-SHADE finds the best parameters $\theta = [\pi_\lambda, \vec{\mu}_\lambda, \Upsilon_\lambda]$ for each Gaussian λ. The mixing coefficient π_λ represents the overall probability of observing an individual $\vec{\psi}_i$ that comes from Gaussian λ.

The promising Gaussian defined by $\theta = [\pi_p, \vec{\mu}_p, \Upsilon_p]$ represents the region of search space where there are higher chances of finding meaningful solutions to the problem and $\vec{\mu}_p$ is the average individual that represents the promising Gaussian.

When a stopping condition is reached, GM-SHADE returns the $\theta = [\pi_p, \vec{\mu}_p, \Upsilon_p]$ of the promising Gaussians from the current population Ψ. In a multimodal problem, more than one Gaussian may be promising. The average individual $\vec{\mu}_p$ is returned to the original search space using the inverse of the adaptive SVD function ϕ' as follows:

$$\phi'(\vec{\mu}_p) = \vec{\mu}_p \cdot V_k^T \tag{5}$$

Promising Gaussian $\theta = [\pi_p, \vec{\mu}_p, \Upsilon_p]$ reconverted in a promising high-dimensional neighborhood can be explored by a specialized local search algorithm aiming to find the optimum solution for the LSGO problem in a limited region of the approximated original search space. As [22], GM-SHADE uses a combination of two local search algorithms to explore the promising high-dimension neighborhood: MTS LS1 [30], specially designed for LSGO and L-BFGS-B [23] that uses an approximation of the gradient.

4   Scalability Analysis

This section is focused on analyzing the capability of LSMDE to scale with the dimension of the search space of each function. Experiments have been conducted on the *bbob-largescale* suite that includes 24 single-objective functions in the continuous domain, with different features: low or moderate conditioning, separability and non-separability, multimodality, weak global structure, and others. An overview of these functions can be found in the *bbob-largescale* test suite. The control parameters adopted by the competing algorithms are described in their respective papers, and the control parameters adopted by LSMDE are the same as described in [12] and shown in Table 1. Tolerance ϵ specifies the allowed distortion when constructing the low-dimensional space Ψ. Higher ϵ values indicate a greater tolerance for distortion and lower k values. However, as information loss increases, the optimization procedure becomes more difficult. Population size was determined as in competing shade-based algorithms, and ϵ was determined experimentally.

Table 1. Parameters used in LSMDE

Parameter	Value	Description
n	100	Size of the Population
d	1000	Dimension of the problem
ϵ	0.03	Johnson-Lindenstraus tolerance error

The performance of LSMDE in terms of scalability was compared to the two best competitors detailed in [12], SHADE-ILS and GL-SHADE. The results are presented by visualizing bootstrapped Empirical Cumulative Distribution Functions (ECDF) provided by the COCO framework as default for new test suites. We keep the standard 51 targets uniformly chosen on a log scale between 100 and 10^{-8}. All experiments have been run on 24 functions. ECDF show the percentages of target values reached given a budget of function and constraint evaluations (shown on the x-axis normalized by the search space dimension and in logarithmic scale). The targets are defined as certain distances from the optimal value.

In Fig. 2, we can observe the performance of LSMDE, SHADE-ILS, and GL-SHADE over the aggregate of all functions as the dimensionality increases. We follow that LSMDE can have a success rate between 40% and 80% while SHADE-ILS has a success rate between 38% and 60% and GL-SHADE has a success rate between 20% and 50%. This result shows that LSMDE is more effective in solving problems when there is an increase in dimensionality.

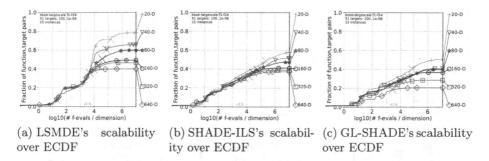

(a) LSMDE's scalability over ECDF   (b) SHADE-ILS's scalability over ECDF   (c) GL-SHADE's scalability over ECDF

Fig. 2. ECDF graph showing the performance of LSMDE, SHADE-ILS and GL-SHADE on the *bbob-largescale* suite for different dimensions across all 24 problems.

Figure 3a to Fig. 3f show the ECDF plot for some selected functions with different characteristics. We can see the LSMDE's good performance on the separable Rastrigin function (Fig. 3a), even for high dimensionalities. However, the algorithm converges towards worse solutions for the non-separable version of the Rastrigin function (Fig. 3c). In this case, the number of local optima is approximately 10^d (d is the problem's dimensionality) [11]. We can conclude that non-separable Rastrigin is too complex for LSMDE to solve without further hyperparameter tuning.

Regarding the Rosenbrock functions (Fig. 3d and 3e), we observe good results, even with increasing dimensionality and complexity additions (such as rotations and low and moderate conditioning). An exception occurs when the dimensionality increases from 320 to 640 on Rosenbrock rotated and Attractive sector functions (Fig. 3e and Fig. 3b), where we can observe a significant performance degradation. There is likely a threshold in the algorithm's ability to maintain the quality of candidate solutions in the alternative low-dimensional space, which is influenced by the intrinsic characteristics of the search space. In Fig. 3f, we can highlight the results obtained in the Different-Powers function in low and high dimensions. An interesting feature of this function is the need for more function evaluations (*f-evals*) for solving the problem in higher dimensions. ECDF for the three selected algorithms aggregated over all 24 *bbob-largescale* functions can be found in Fig. 4 for dimensions 20-D, 40-D, 80-D, 160-D, 320-D, and 640-D. Based on Fig. 4 we can see that:

- Over all *bbob-largescale* problems, the algorithms perform worse with increasing dimension;
- GL-SHADE is most affected by the increase in the dimension of such a manner that only approximately 20% of targets were reached compared to approximately 40% reached by LSMDE for 640-D;
- LSMDE outperforms all competing algorithms with the increase in dimensionality for different categories of functions; however, as the dimensionality increases, the performance of both LSMDE and SHADE-ILS become similar with a difference of performance of approximately 5% for 640-D.

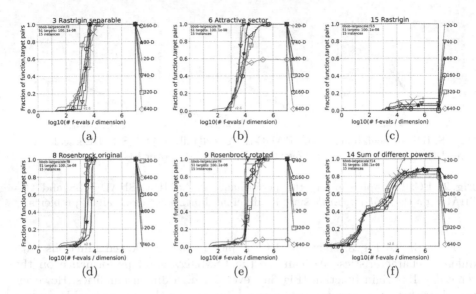

Fig. 3. LSMDE ECDF plots for functions $f_3, f_6, f_8, f_9, f_{14}$ and f_{15} in the *bbob-largescale* suite, aggregated across all instances of each function.

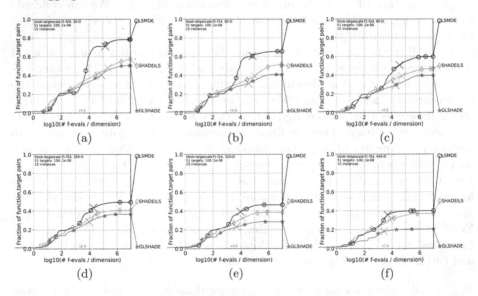

Fig. 4. ECDF for the three selected algorithms, aggregated over all 24 *bbob-largescale* functions.

An important aspect to highlight is the low performance of LSMDE on the weak global structure functions for 160-D, 320-D, and 640-D (maximum of 20% success), as shown in Fig. 5. A function has a weak global structure if the qualities (the f-values) of the local optima are only weakly related to their locations in the search space. For example, when neighboring optima do not generally have similar quality values [5]. This behavior makes searching for an optimal solution more difficult, especially in high dimensionality. Since LSMDE uses a continuous mapping of the high-dimensional search space to a low-dimensional space, weak global structure functions could not produce an alternative search space that guaranteed the position of candidate solutions, which caused a deterioration in the quality of the generated space exploration. However, none of the examined algorithms performed satisfactorily in weak global structure functions.

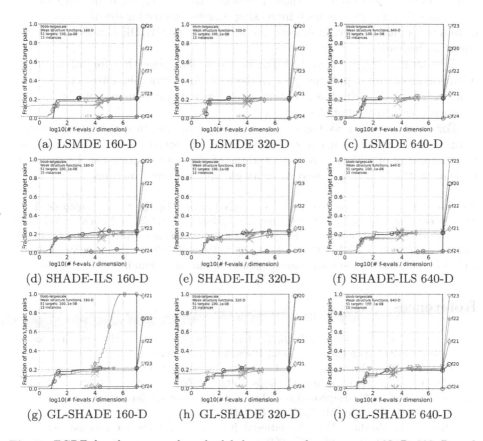

(a) LSMDE 160-D   (b) LSMDE 320-D   (c) LSMDE 640-D

(d) SHADE-ILS 160-D   (e) SHADE-ILS 320-D   (f) SHADE-ILS 640-D

(g) GL-SHADE 160-D   (h) GL-SHADE 320-D   (i) GL-SHADE 640-D

Fig. 5. *ECDF for the group of weak global structure functions in 160-D, 320-D and 640-D.*

5   Conclusions

In this paper, we extended the analysis of the Low-dimensional Space Modeling-based Differential Evolution algorithm (LSMDE) through new tests with scalable functions between dimensions 20 and 640. The experiments were conducted on the *bbob-largescale* suite, which includes 24 single-objective functions in the continuous domain, with different features such as low or moderate conditioning, separability, multimodality, and weak global structure.

LSMDE shows promising results in the scalability analysis compared to the best competitors, SHADE-ILS and GL-SHADE. It was observed that LSMDE has a higher success rate than SHADE-ILS and GL-SHADE when the problem's dimensionality increases. However, its performance on non-separable and weak global structure functions was limited, indicating that further hyperparameter tuning may be necessary for these problems. In general, LSMDE outperforms the competing algorithms with increasing dimensionality for different categories of functions, but its performance becomes similar to SHADE-ILS as the dimensionality increases.

The results suggest that LSMDE is a practical algorithm for solving optimization problems, significantly when the problem's dimensionality is increasing. However, further research is necessary to improve its performance on challenging functions with weak global structures.

For future work, LSMDE can be improved by fine-tuning its performance parameters to solve even non-separable functions better. Additionally, alternative mappings or different representations of the search space can be explored to address the challenge posed by weak global structure functions.

Acknowledgment. This research was financed in part by the *Coordenação de Aperfeiçoamento de Pessoal de Nível Superior - Brasil (CAPES)* - Finance Code 001 and by a research grant from Science Foundation Ireland (SFI) under grant no. SFI/16/RC/3918 (CONFIRM) and Marie Sklodowska-Curie grant agreement no. 847.577 co-funded by the European Regional Development Fund.

References

1. Bagattini, F., Schoen, F., Tigli, L.: Clustering methods for the optimization of atomic cluster structure. J. Chem. Phys. **148**(14), 144102 (2018)
2. Bagattini, F., Schoen, F., Tigli, L.: Clustering methods for large scale geometrical global optimization. Optim. Methods Softw. **34**(5), 1099–1122 (2019)
3. Baş, E., Ülker, E.: Improved social spider algorithm for large scale optimization. Artif. Intell. Rev. **54**(5), 3539–3574 (2021)
4. Blei, D.M., Jordan, M.I.: Variational inference for Dirichlet process mixtures. Bayesian Anal. **1**(1), 121–143 (2006)
5. Brockhoff, D., Auger, A., Hansen, N., Tušar, T.: Using well-understood single-objective functions in multiobjective black-box optimization test suites. Evol. Comput. **30**(2), 165–193 (2022)
6. Chen, M., Du, W., Song, W., Liang, C., Tang, Y.: An improved weighted optimization approach for large-scale global optimization. Complex Intell. Syst. **8**(2), 1259–1280 (2022)
7. Dasgupta, S.: Learning mixtures of gaussians. In: 40th Annual Symposium on Foundations of Computer Science (Cat. No. 99CB37039), pp. 634–644. IEEE (1999)
8. Dasgupta, S., Gupta, A.: An elementary proof of the johnson-lindenstrauss lemma. International Computer Science Institute, Technical Report 22(1), pp. 1–5 (1999)

9. De Falco, I., Della Cioppa, A., Trunfio, G.A.: Investigating surrogate-assisted cooperative coevolution for large-scale global optimization. Inf. Sci. **482**, 1–26 (2019)
10. Diaconis, P., Freedman, D.: Asymptotics of graphical projection pursuit. The annals of statistics, pp. 793–815 (1984)
11. Finck, S., Hansen, N., Ros, R., Auger, A.: Real-parameter black-box optimization benchmarking 2009: presentation of the noiseless functions. Technical report, Citeseer (2010)
12. Fonseca, T.H.L., Nassar, S.M., de Oliveira, A.C.M., Agard, B.: Low-dimensional space modeling-based differential evolution for large scale global optimization problems. IEEE Trans. Evol. Comput. (2022)
13. Iorio, A.W., Li, X.: Improving the performance and scalability of differential evolution. In: Li, X., Kirley, M., Zhang, M., Green, D., Ciesielski, V., Abbass, H., Michalewicz, Z., Hendtlass, T., Deb, K., Tan, K.C., Branke, J., Shi, Y. (eds.) SEAL 2008. LNCS, vol. 5361, pp. 131–140. Springer, Heidelberg (2008). https://doi.org/10.1007/978-3-540-89694-4_14
14. Kabán, A., Bootkrajang, J., Durrant, R.J.: Towards large scale continuous EDA: a random matrix theory perspective. In: Proceedings of the 15th Annual Conference on Genetic and Evolutionary Computation, pp. 383–390 (2013)
15. Kishore Kumar, N., Schneider, J.: Literature survey on low rank approximation of matrices. Linear Multilinear Algebra **65**(11), 2212–2244 (2017)
16. Li, L., Fang, W., Mei, Y., Wang, Q.: Cooperative coevolution for large-scale global optimization based on fuzzy decomposition. Soft. Comput. **25**(5), 3593–3608 (2021)
17. Li, X., Tang, K., Omidvar, M.N., Yang, Z., Qin, K., China, H.: Benchmark functions for the CEC 2013 special session and competition on large-scale global optimization. Gene **7**(33), 8 (2013)
18. Long, W., Wu, T., Liang, X., Xu, S.: Solving high-dimensional global optimization problems using an improved sine cosine algorithm. Expert Syst. Appl. **123**, 108–126 (2019)
19. Ma, Y., Bai, Y.: A multi-population differential evolution with best-random mutation strategy for large-scale global optimization. Appl. Intell. **50**(5), 1510–1526 (2020)
20. Mahdavi, S., Rahnamayan, S., Deb, K.: Partial opposition-based learning using current best candidate solution. In: IEEE Symposium Series on Computational Intelligence, pp. 1–7 (2016)
21. Maučec, M.S., Brest, J.: A review of the recent use of differential evolution for large-scale global optimization: an analysis of selected algorithms on the cec 2013 lsgo benchmark suite. Swarm Evol. Comput. **50**, 100428 (2019)
22. Molina, D., LaTorre, A., Herrera, F.: Shade with iterative local search for large-scale global optimization. In: IEEE Congress on Evolutionary Computation, pp. 1–8 (2018)
23. Morales, J.L., Nocedal, J.: Remark on "algorithm 778: L-bfgs-b: Fortran subroutines for large-scale bound constrained optimization". ACM Trans. Math. Softw. (TOMS) **38**(1), 1–4 (2011)
24. Omidvar, M.N., Li, X.: Evolutionary large-scale global optimization: an introduction. In: Proceedings of the Genetic and Evolutionary Computation Conference Companion, pp. 807–827 (2017)
25. Omidvar, M.N., Li, X., Yao, X.: A review of population-based metaheuristics for large-scale black-box global optimization: Part b. IEEE Trans. Evol. Comput., 1 (2021). https://doi.org/10.1109/TEVC.2021.3130835

26. Pacheco-Del-Moral, O., Coello, C.A.C.: A shade-based algorithm for large scale global optimization. In: International Conference on Parallel Problem Solving from Nature, pp. 650–663. Springer (2020)

27. Segredo, E., Paechter, B., Segura, C., González-Vila, C.I.: On the comparison of initialisation strategies in differential evolution for large scale optimisation. Optim. Lett. **12**(1), 221–234 (2018)

28. Tabernik, D., Skočaj, D.: Deep learning for large-scale traffic-sign detection and recognition. IEEE Trans. Intell. Transp. Syst. **21**(4), 1427–1440 (2019)

29. Tanabe, R., Fukunaga, A.: Success-history based parameter adaptation for differential evolution. In: 2013 IEEE Congress on Evolutionary Computation, pp. 71–78. IEEE (2013)

30. Tseng, L.Y., Chen, C.: Multiple trajectory search for large scale global optimization. In: IEEE Congress on Evolutionary Computation (IEEE World Congress on Computational Intelligence), pp. 3052–3059 (2008)

31. Varelas, K., et al.: Benchmarking large-scale continuous optimizers: the BBOB-largescale testbed, a coco software guide and beyond. Appl. Soft Comput. **97**, 106737 (2020)

32. Zamani, H., Nadimi-Shahraki, M.H., Gandomi, A.H.: Qana: quantum-based avian navigation optimizer algorithm. Eng. Appl. Artif. Intell. **104**, 104314 (2021)

Fair Empirical Risk Minimization Revised

Danilo Franco[✉], Luca Oneto, and Davide Anguita

University of Genoa, Via Opera Pia 13, 16145 Genoa, Italy
danilo.franco@edu.unige.it, {luca.oneto,davide.anguita}@unige.it

Abstract. Artificial Intelligence is nowadays ubiquitous, thanks to a continuous process of commodification, revolutionizing but also impacting society at large. In this paper, we address the problem of algorithmic fairness in Machine Learning: ensuring that sensitive information does not unfairly influence the outcome of a classifier. We extend the Fair Empirical Risk Minimization framework [10] where the fair risk minimizer is estimated via constrained empirical risk minimization. In particular, we first propose a new, more general, notion of fairness which translates into a fairness constraint. Then, we propose a new convex relaxation with stronger consistency properties deriving both risk and fairness bounds. By extending our approach to kernel methods, we will also show that the proposal empirically over-performs the state-of-the-art Fair Empirical Risk Minimization approach on several real-world datasets.

Keywords: Machine Learning · Algorithmic Fairness · In-processing Fairness · Consistency Results · Convex Constrained Optimization · Kernel Methods

1 Introduction

Artificial Intelligence is nowadays ubiquitous thanks to massive investments in its development and in the development of products able to make it a commodity. In some applications, e.g., games [39], healthcare [8], and text generation [33], these tools have been shown to compare to human capabilities. These achievements are accompanied by increasing concerns about their impact on society [14,24]. In this paper, we deal with the problem of ensuring that Machine Learning (ML) models do not discriminate subgroups in the population based on, e.g., gender, race, or political and sexual orientation, namely to develop Fair ML models [29]. In fact, real-world datasets often reflect historical biases in society and, when these data are fed to ML algorithms, they often result in models which actually exacerbate these biases [27].

The first step toward fairer ML is to formally define a notion of fairness [14]. For this purpose, scholars have defined a series of notions based on the applications and the type of biases [41]. These notions are often incompatible with each other [16] but, many works, try to generalize them under a common framework [2,10,31].

© The Author(s), under exclusive license to Springer Nature Switzerland AG 2023
I. Rojas et al. (Eds.): IWANN 2023, LNCS 14134, pp. 29–42, 2023.
https://doi.org/10.1007/978-3-031-43085-5_3

Then, once clear how fairness can be measured, scholars have identified three families of fairness mitigation strategies [27,30,34,41]. The first one, pre-processing, aims to manipulate the data to train models on datasets where the discrimination has been reduced as much as possible. Earlier examples of this family often required manipulating the labels of some instances [26] or reweighing them [21], while more recent works aim at tweaking the feature space to learn a fairer data representation [4,13,25,36,45]. The second group, in-processing, enforces bias mitigation requirements during the training phase to impose fairness guarantees within the models' inner structures. These works usually propose to either add fairness optimization constraints [10,11,18,23,43] or penalize discrimination through specific regularization terms [1,2,15,22]. This family of methods are often paired with consistency results that bind the potential discrimination which is allowed in the trained model. The last family, post-processing, finally aims at mimicking the behaviours of high-performing trained models and directly addressing the discrimination present in their predictions. These methodologies often resort to outcomes randomization [7,19,28] or model decoupling [12] for satisfying precise fairness definitions while theoretical and consistency results are provided to some extent.

In this work, we built on the Fair Empirical Risk Minimization (FERM) [10] where different notions of fairness are generalized under the common definition of ϵ-Fairness. This definition is translated into a fairness constraint, which is exploited for defining a fair constrained empirical risk minimization. Building upon this work, we first further extend ϵ-Fairness into a more general notion able to encompass the most used notions of statistical (also known as group) fairness [41]. Then we translate this definition into a new constraint whose convex relaxation possesses stronger consistency properties in terms of both risk and fairness bounds. Our new framework, the Revised Fair Empirical Risk Minimization (R-FERM), apart from having stronger consistency guarantees, also shows to empirically over-perform the original FERM frameworks when declined into kernel methods and tested on a series of real-world fairness-related datasets.

The rest of the paper is organized as follows. Section 2 provides the theoretical frameworks of R-FERM. Section 3 instantiates the R-FERM to the case of kernel methods. Then, in Sect. 4, we present an empirical comparison between FERM and R-FERM in the case of linear and kernel methods. Finally, Sect. 5 concludes the paper.

2   Fair Empirical Risk Minimization Revised

In this section, we present our approach to learning with fairness. We begin by introducing our notation. We let $\mathcal{D} = \{(X_1, s_1, y_1), \ldots, (X_n, s_n, y_n)\}$ be a sequence of n samples drawn independently from an unknown probability distribution μ over $\mathcal{X} \times \mathcal{S} \times \mathcal{Y}$, where $\mathcal{Y} = \{-1, +1\}$ is the set of binary output labels[1],

---

[1] The extension to multiclass classification and regression can be obtained with some technical steps that are not reported to simplify the presentation.

$\mathcal{S} = \{a, b\}$ represents group membership among two groups[2] (e.g. 'female' or 'male'), and $\mathcal{X}$ is the input space. We note that the input $X \in \mathcal{X}$ may further contain or not the sensitive feature $s \in \mathcal{S}$ in it[3]. Let us consider a function (or model) $f : \mathcal{X} \to \mathbb{R}$ chosen from a set $\mathcal{F}$ of possible models. The error (risk) of f in approximating μ is measured by a prescribed loss function $\ell : \mathbb{R} \times \mathcal{Y} \to \mathbb{R}$ which is defined as $L(f) = \mathbb{E}\left[\ell(f(X), y)\right]$. When necessary, we will indicate with a subscript the particular loss function used, i.e. $L_p(f) = \mathbb{E}\left[\ell_p(f(X), y)\right]$.

The purpose of a learning procedure is to find a model that minimizes the risk. Since the probability measure μ is usually unknown, the true risk cannot be computed. However, we can compute the empirical risk $\hat{L}(f) = \hat{\mathbb{E}}[\ell(f(X), y)]$, where $\hat{\mathbb{E}}$ denotes the empirical expectation. Then, a natural learning strategy, called Empirical Risk Minimization (ERM), is to minimize the empirical risk within a prescribed set of functions.

2.1 ϵ-Fairness Revised

In the literature, there are different definitions of fairness of a model or learning algorithm [5,6,17,27,30] and, depending on the application and the discrimination, one may be better suited than another.

In the past, some attempts have tried to build up a notion of fairness that can encompass multiple notions. For example, the ϵ-Fairness [10] can generalize Equal Opportunity [12,19] and uncorrelation between model outputs and sensitive feature.

In this paper, we introduce a slightly more general notion of fairness, the ϵ-Fairness Revised, which naturally encompasses the ϵ-Fairnes (and consequently all the notions above) plus the Equal Odds [19] and the Demographic Parity [3].

Definition 1. *We say that a function f is ϵ-fair if $F(f) \leq \epsilon$ where*

$$F(f) = \frac{1}{g} \sum_{\breve{y} \in \{\breve{y}_1, \breve{y}_2, \cdots, \breve{y}_g\}} \left| \mathbb{E}_{(X,s,y)|s=a, y \in \breve{y}} \{\ell(f(X), \cdot)\} - \mathbb{E}_{(X,s,y)|s=b, y \in \breve{y}} \{\ell(f(X), \cdot)\} \right|,$$

with $\epsilon \in [0, 1]$, $\breve{y}, \breve{y}_1, \breve{y}_2, \cdots, \breve{y}_g \subseteq \mathcal{Y}$, and $\cdot \subset \{y, \pm 1\}$.

It is trivial to note that by setting $\{\breve{y}_1, \breve{y}_2, \cdots, \breve{y}_g\} = \breve{y}_1 = \{+1\}$ and $\cdot = y$ in Definition 1 we get the ϵ-Fairness [10]

$$F(f) = \left| \mathbb{E}_{(X,s,y)|s=a, y=+1} \{\ell(f(X), y)\} - \mathbb{E}_{(X,s,y)|s=b, y=+1} \{\ell(f(X), y)\} \right|. \quad (1)$$

Instead by setting $\{\breve{y}_1, \breve{y}_2, \cdots, \breve{y}_g\} = \breve{y}_1 = \mathcal{Y}$, $\cdot = 1$, and $\ell(f(X), y) = [y \neq f(X)]$ in Definition 1 we obtain the Demographic Parity [3]

$$F(f) = \left| \mathbb{P}_{(X,s,y)} \{f(X) = 1 | s = a\} - \mathbb{P}_{(X,s,y)} \{f(X) = 1 | s = b\} \right|. \quad (2)$$

---

[2] The extension to multiple groups can be obtained with some technical steps that are not reported to simplify the presentation.

[3] The sensitive feature may not be available in the testing phase or it might not be possible to use it as a predictor in the model due to legal requirements [12].

Finally, is we set $\{\breve{\mathcal{Y}}_1, \breve{\mathcal{Y}}_2, \cdots, \breve{\mathcal{Y}}_g\} = \{\{-1\}, \{+1\}\}$, $\cdot = y$, and $\ell(f(X), y) = [y \neq f(X)]$ in Definition 1 we obtain the Equal Odds [19]

$$F(f) = \frac{1}{2} \sum_{\breve{y} \in \{\pm 1\}} \left| \mathbb{P}_{(X,s,y)}\{f(X) = \breve{y} | s = a, y = \breve{y}\} - \mathbb{P}_{(X,s,y)}\{f(X) = \breve{y} | s = b, y = \breve{y}\} \right|. \quad (3)$$

2.2 Fair Empirical Risk Minimization Revised (R-FERM)

In this paper, we aim to minimize the risk subject to a fairness constraint. Specifically, we consider the problem

$$\min_{f \in \mathcal{F}} L(f), \quad \text{s.t. } F(f) \leq \epsilon, \quad (4)$$

where $\epsilon \in [0, 1]$ is the amount of unfairness that we are willing to bear. Since the measure μ is unknown we replace the deterministic quantities with their empirical counterparts. That is, we replace Problem 4 with

$$\min_{f \in \mathcal{F}} \hat{L}(f), \quad \text{s.t. } \hat{F}(f) \leq \hat{\epsilon}, \quad (5)$$

where $\hat{F}(f)$ is the empirical counterpart of $F(f)$, namely we have to replace $\mathbb{E}$ with $\hat{\mathbb{E}}$ in Definition 1, and $\hat{\epsilon} \in [0, 1]$. We will refer to Problem 5 as Fair Empirical Risk Minimization Revised (R-FERM) since it is the same optimization proposed in FERM [10] where we replaced the old definition of the ϵ-Fairness with the one of Definition 1.

2.3 Consistency Results

Let us denote with f^* a solution of Problem (4) and by $\hat{f}$ a solution of Problem (5). In this section, we will show that these solutions are linked one to another. In particular, if the parameter $\hat{\epsilon}$ is chosen appropriately, we will show that, in a certain sense, the estimator $\hat{f}$ is consistent. In order to present our observations, we require that it holds with probability at least $1 - \delta$ that

$$\sup_{f \in \mathcal{F}} |L(f) - \hat{L}(f)| \leq B(\delta, n, \mathcal{F}) \quad (6)$$

where the bound $B(\delta, n, \mathcal{F})$ goes to zero as n grows to infinity if the class $\mathcal{F}$ is learnable with respect to the loss [37].

We are ready to state the first result of this section which follows easily from the results of FERM [10].

Theorem 1. *Let $\mathcal{F}$ be a learnable set of functions with respect to the loss function $\ell : \mathbb{R} \times \mathcal{Y} \to \mathbb{R}$, let f^* be a solution of Problem (4) and let $\hat{f}$ be a solution of Problem (5) with*

$$\hat{\epsilon} = \epsilon + \frac{1}{g} \sum_{\breve{y} \in \{\breve{\mathcal{Y}}_1, \breve{\mathcal{Y}}_2, \cdots, \breve{\mathcal{Y}}_g\}} \sum_{\breve{s} \in \{a,b\}} B\left(\delta, n(\breve{s}, \breve{\mathcal{Y}}), \mathcal{F}\right).$$

where $n(\check{s}, \check{y}) = |(X, s, y) \in \mathcal{D}_n| s = \check{s}, y \in \check{y}|$. *With probability at least* $1 - 6g\delta$ *it holds simultaneously that*

$$
\begin{cases}
L(\hat{f}) - L(f^*) \leq 2B(\delta, n, \mathcal{F}), \\
F(f) \leq \epsilon + \frac{2}{g} \sum_{\check{y} \in \{\check{y}_1, \check{y}_2, \cdots, \check{y}_g\}} \sum_{\check{s} \in \{a, b\}} B(\delta, n(\check{s}, \check{y}), \mathcal{F}).
\end{cases}
$$

Note that Theorem 1 degenerates in Theorem 1 of FERM [10] if we replace Definition 1 with ϵ-Fairness [10].

A consequence of the first statement of Theorem 1 is that as n tends to infinity $L(\hat{f})$ tends to a value which is not larger than $L(f^*)$, that is, Problem (5) is consistent with respect to the risk of the selected model. The second statement of Theorem 1, instead, implies that as n tends to infinity we have that $\hat{f}$ tends to be ϵ-fair. In other words, Problem (5) is consistent with respect to the fairness of the selected model.

Thanks to Theorem 1 we can state that f^* is close to $\hat{f}$ both in terms of its risk and its fairness.

Problem 4 cannot be solved since it involves unknown quantities but thanks to Theorem 1 we know that Problem 5, which involves only empirical quantities. The final goal is to find an f_h^* which solves the following R-FERM optimization problem

$$
\min_{f \in \mathcal{F}} \hat{\mathbb{E}}_{(X,s,y)}\{[f(X) \neq y]\}, \quad \text{s.t. } \hat{F}(f) \leq \epsilon, \tag{7}
$$

where, as $\hat{F}(h)$, we can use, for example, Demographic Parity or Equal Opportunity. For example, for Demographic Parity we have to set

$$
\hat{F}(f) = \left| \hat{\mathbb{E}}_{(X,s,y)|s=a}\{[f(X) \neq 1]\} - \hat{\mathbb{E}}_{(X,s,y)|s=b}\{[f(X) \neq 1]\} \right|, \tag{8}
$$

calling this quantity Difference of Demographic Parity (DDP), and for the Equal Opportunity (on the class $\diamond \in \{\pm 1\}$) we have to set

$$
\hat{F}(f) = \left| \hat{\mathbb{E}}_{(X,s,y)|s=a,y=\diamond}\{[f(X) \neq \diamond]\} - \hat{\mathbb{E}}_{(X,s,y)|s=b,y=\diamond}\{[f(X) \neq \diamond]\} \right|, \tag{9}
$$

calling this quantity Difference of Equal Opportunity (DEO$^\diamond$). Namely, Problem 7 tries to minimize the classification error with DDP or DEO$^\diamond$ less than ϵ, namely to create accurate and fair models[4].

---

[4] Fron now on we will focus on DDP or DEO since to simplify the presentation but the results holds also for the more general Definition 1.

34   D. Franco et al.

Problem 7 is a difficult non-convex non-smooth problem, then NP-hard, which is extremely hard to solve. For this reason, it is more convenient to solve its convex relaxation. Addressing Problem 7 with a convex relaxation opens another issue, namely how close the function which solves Problem 7 is to the function which solves its convex relaxation.

In fact, the approach of FERM [10] lacks in this aspect providing guaranties under an assumption which needs to be empirically verified and, in general, not always true.

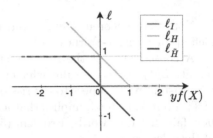

In this work, we propose a new approach. In particular, let us define the following losses which are also represented in Fig. 1: $\ell_I(f(X), y) = [yf(X) \leq 0]$, $\ell_H(f(X), y) = \max[0, 1 - yf(X)]$, and $\ell_{\tilde{H}}(f(X), y) = \min[1, -yf(X)]$. Note that the ℓ_I is the Hard loss function [37] (or

Fig. 1. Convex upper bound and concave lower bound of the Hard loss function.

Indicator loss), which counts the number of errors, the ℓ_H is the Hinge loss function [40], namely the simplex convex upper bound, while, analogously to the ℓ_H, the $\ell_{\tilde{H}}$ is the simplex concave lower bound of the Hard loss function. Note also that

$$\ell_{\tilde{H}}(f(X), y) \leq \ell_I(f(X), y) \leq \ell_H(f(X), y). \tag{10}$$

Thanks to this definition we can reformulate Problem (7) as follows

$$\min_{f \in \mathcal{F}} \hat{L}_I(f), \quad \text{s.t. } \hat{F}_I(f) \leq \epsilon, \tag{11}$$

where $\hat{L}_I(f) = \mathbb{E}_{(X,s,y)}\{\ell_I(f(X), y)\}$ and when we use the Demographic Parity

$$\hat{F}_I(f) = \left|\hat{F}_I^{\pm}(f)\right| = \left|\hat{\mathbb{E}}_{(X,s,y)|s=a}\{\ell_I(f(X), 1)\} - \hat{\mathbb{E}}_{(X,s,y)|s=b}\{\ell_I(f(X), 1)\}\right|, \tag{12}$$

while for the Equal Opportunity (on the class $\diamond \in \{\pm 1\}$)

$$\hat{F}_I(f) = \left|\hat{F}_I^{\pm}(f)\right| = \left|\hat{\mathbb{E}}_{(X,s,y)|s=a,y=\diamond}\{\ell_I(f(X), \diamond)\} - \hat{\mathbb{E}}_{(X,s,y)|s=b,y=\diamond}\{\ell_I(f(X), \diamond)\}\right|. \tag{13}$$

Then can reformulate Problem (11) in its convex relaxation as follows

$$\min_{f \in \mathcal{F}} \hat{L}_H(f), \quad \text{s.t. } \hat{F}_H(f) \leq \epsilon, \tag{14}$$

where $\hat{L}_H(f) = \mathbb{E}_{(X,s,y)}\{\ell_H(f(X), y)\}$ and when we use the Demographic Parity (see Eq. (12))

$$\hat{F}_H(f) = \begin{cases} \hat{\mathbb{E}}_{(X,s,y)|s=a}\{\ell_H(f(X), 1)\} - \hat{\mathbb{E}}_{(X,s,y)|s=b}\{\ell_{\tilde{H}}(f(X), 1)\}, & \text{if } \hat{F}_I^{\pm}(f) \geq 0 \\ \hat{\mathbb{E}}_{(X,s,y)|s=b}\{\ell_H(f(X), 1)\} - \hat{\mathbb{E}}_{(X,s,y)|s=a}\{\ell_{\tilde{H}}(f(X), 1)\}, & \text{if } \hat{F}_I^{\pm}(f) < 0 \end{cases} \tag{15}$$

while, analogously, for the Equal Opportunity (on the class $\diamond \in \{\pm 1\}$, see Eq. (13))

$$\hat{F}_H(f) = \begin{cases} \hat{\mathbb{E}}_{(X,s,y)|s=a,y=\diamond}\{\ell_H(f(X),\diamond)\} - \hat{\mathbb{E}}_{(X,s,y)|s=b,y=\diamond}\{\ell_{\tilde{H}}(f(X),\diamond)\}, & \text{if } \hat{F}_I^{\pm}(f) \geq 0 \\ \hat{\mathbb{E}}_{(X,s,y)|s=b,y=\diamond}\{\ell_H(f(X),\diamond)\} - \hat{\mathbb{E}}_{(X,s,y)|s=a,y=\diamond}\{\ell_{\tilde{H}}(f(X),\diamond)\}, & \text{if } \hat{F}_I^{\pm}(f) < 0 \end{cases} \quad (16)$$

Note that Problem (11) needs to solve two convex problems (one for each of the constraints of Eqs. (15) or (16)).

Let us denote with $\hat{f}_I$ a solution of Problem (11), and by $\hat{f}_H$ a solution of Problem (14). The questions that arise here are whether, $\hat{f}_I$ is to $\hat{f}_H$, how much, and under which assumptions. The following theorem sheds some light on these issues.

Lemma 1. *By definition, we can state that*

$$\hat{L}_I(f) \leq \hat{L}_H(f), \quad \hat{F}_I(f) \leq \hat{F}_H(f). \quad (17)$$

The statement of Lemma 1 tells us that properly exploiting ℓ_H and $\ell_{\tilde{H}}$ instead of ℓ_I in $\hat{L}(f)$ and $\hat{F}(f)$ is a good approximation when $\hat{L}_H(f)$ and $\hat{F}_H(f)$ is small which is often the case. In fact, our purpose is to find functions in Problem (14) with small $\hat{L}_H(f)$ and $\hat{F}_H(f)$ namely accurate and fair models.

Note that other approaches can be used to approximate $\hat{F}_I(f)$ like the one of [10] which matches its first-moment or [42] which matches its second-moment or [35] that matches infinitely many moments. Unfortunately, the first work does not provide so strong results of consistency while the others result in non-convex approaches. Moreover, Lemma 1 may be tightened by using different nonlinear approximations (e.g., [4]), however, the approximation proposed in this work gives a convex problem like the one of [10], but it has better theoretical properties with respect to [10] which already outperform of matches the performance of state of the art approaches [20, 44].

3   From Theory to Practice

In this section, we will first specify the R-FERM framework to the case that the underlying space of models is a Reproducing Kernel Hilbert Space [38]. We let $\kappa : \mathcal{X} \times \mathcal{X} \to \mathbb{R}$ be a positive definite kernel and let $\phi : \mathcal{X} \to \mathbb{H}$ be an induced feature mapping such that $\kappa(X, X') = \langle \phi(X), \phi(X') \rangle$, for all $X, X' \in \mathcal{X}$, where $\mathbb{H}$ is the Hilbert space of square summable sequences. Functions in the RKHS can be parametrized as $f(\boldsymbol{x}) = \langle W, \phi(X) \rangle + b$ with $X \in \mathcal{X}$ for some vector of parameters $W \in \mathbb{H}$ and the parameter $b \in \mathbb{R}$.

We solve Problem (14) with $\mathcal{F}$ a ball, i.e., $\|W\|_2 \leq A \in [0, \infty)$, in the Reproducing Kernel Hilbert Space substituting in is the previous definitions

$$\min_{W,b} \sum_{i=1}^{n} \xi_i, \quad \text{s.t.} \quad y_i(\langle W, \phi(X_i) \rangle + b) \geq 1 - \xi_i, \; \xi_i \geq 0, \; i \in \{1, \cdots, n\},$$

$$\|W\|_2^2 \leq A, \quad \hat{F}_H(W, b) \leq \epsilon, \quad (18)$$

where when we use the Demographic Parity (see Eq. (15))

$$\langle W, \phi(X_i)\rangle + b \geq 1 - \eta_i, \quad \langle W, \phi(X_i)\rangle + b \leq -\tilde{\eta}_i, \tag{19}$$
$$\eta_i \geq 0, \quad \tilde{\eta}_i \leq 1, \quad i \in \{1, \cdots, n\},$$

$$\hat{F}_H(W, b) = \begin{cases} \frac{1}{|\mathcal{I}_a|}\sum_{i\in\mathcal{I}_a}\eta_i - \frac{1}{|\mathcal{I}_b|}\sum_{i\in\mathcal{I}_b}\tilde{\eta}_i, & \text{if } \hat{F}_I^{\pm}(f) \geq 0 \\ \frac{1}{|\mathcal{I}_b|}\sum_{i\in\mathcal{I}_b}\eta_i - \frac{1}{|\mathcal{I}_a|}\sum_{i\in\mathcal{I}_a}\tilde{\eta}_i, & \text{if } \hat{F}_I^{\pm}(f) < 0 \end{cases},$$

where $\mathcal{I}_\circ = \{i : i \in \{1, \cdots, n\}, s_i = \circ\}$ where $\circ \in \{a, b\}$. Analogously, for the Equal Opportunity (on the class $\diamond \in \{\pm 1\}$, see Eq. (16))

$$y_i(\langle W, \phi(X_i)\rangle + b) \geq 1 - \eta_i, \quad y_i(\langle W, \phi(X_i)\rangle + b) \leq -\tilde{\eta}_i,$$
$$\eta_i \geq 0, \quad \tilde{\eta}_i \leq 1, \quad i \in \{1, \cdots, n\}, \tag{20}$$

$$\hat{F}_H(W, b) = \begin{cases} \frac{1}{|\mathcal{I}_{a,\diamond}|}\sum_{i\in\mathcal{I}_{a,\diamond}}\eta_i - \frac{1}{|\mathcal{I}_{b,\diamond}|}\sum_{i\in\mathcal{I}_{b,\diamond}}\tilde{\eta}_i, & \text{if } \hat{F}_I^{\pm}(f) \geq 0 \\ \frac{1}{|\mathcal{I}_{b,\diamond}|}\sum_{i\in\mathcal{I}_{b,\diamond}}\eta_i - \frac{1}{|\mathcal{I}_{a,\diamond}|}\sum_{i\in\mathcal{I}_{a,\diamond}}\tilde{\eta}_i, & \text{if } \hat{F}_I^{\pm}(f) < 0 \end{cases},$$

where $\mathcal{I}_{\circ,\diamond} = \{i : i \in \{1, \cdots, n\}, s_i = \circ, y = \diamond\}$ where $\circ \in \{a, b\}$.

Note that Problem (18) is convex and needs to actually solve two convex problems (one for each of the constraints of Eqs. (19) or (20)) and then selecting the best fitting solution.

Note also that Problem (18), which is an Ivanov-type regularized problem, can be solved more easily in its Tikhonov version [32] for some $C \in [0, \infty)$

$$\min_{W, b} \quad \frac{1}{2}\|W\|_2^2 + C\sum_{i=1}^{n}\xi_i, \tag{21}$$

$$\text{s.t.} \quad y_i(\langle W, \phi(X)\rangle + b) \geq 1 - \xi_i, \ \xi_i \geq 0, \ i \in \{1, \cdots, n\},$$
$$\hat{F}_H(W, b) \leq \epsilon,$$

which is our proposal for a Fair, in the sense of Demographic Parity or Equal Opportunity, Support Vector Machine [38].

Finally, note that exploiting the representer theorem [9] it is possible to get the kernelized version of Problem (18) by minimizing over $\alpha \in \mathbb{R}^n$ simply making the following now-classical substitutions: $f(X) = \sum_{i=1}^{n}\alpha_i\kappa(X_i, X) + b$, $\langle W, \phi(X_i)\rangle = \sum_{j=1}^{n}\alpha_j\kappa(X_j, X_i)$, and $\|W\|_2^2 = \sum_{i=1}^{n}\sum_{j=1}^{n}\alpha_i\alpha_j\kappa(X_i, X_j)$.

4   Experimental Results

We conducted experiments on 7 fairness-related datasets[5] (i.e., Adult, Arrhythmia, COMPAS, Credit, Drug, German, Taiwan) to evaluate the performances of R-FERM in terms of utility and fairness. The datasets concerned with criminal justice, financial credit, or healthcare applications where the task is to predict a

---

[5] The datasets are available at the UCI ML Repository https://archive.ics.uci.edu/ml/index.php.

binary target variable while avoiding generating discriminatory outcomes against protected populations based on binary sensitive attributes such as sex (i.e., males and females for Adult, Arrhythmia, German and Taiwan), age (i.e., youngs and olds for Credit) and ethnicity (i.e., dark and light skinned people for COMPAS).

To measure the fairness of our algorithm, we evaluated it on the test set following three different fairness criteria: two statistical fairness metrics, i.e., DDP and $DEO^\diamond$, and one individual fairness metric, i.e., the Counterfactual Fairness (CF) [23]. For $DEO^\diamond$ we select $\diamond \in \{\pm 1\}$ which gives the largest discrimination. The CF measures what happens if just the sensitive attribute is changed. For this reason, we will compute the following quantity $\hat{\mathbb{E}}_{(X,s)}\{[f(X_s)f(X_{\neg s}) \leq 0]\}$, where X_s is X with no changes in s while $X_{\neg s}$ is X with a switch in the sensitive feature s.

To measure the quality of the model we will use, instead the Balanced Accuracy (BA), namely the average accuracy of the two classes.

In the results we will compare: Classical Linear and RBF Support Vector Machine (SVM) [40], with the FERM based SVM [10], with the R-FERM based SVM which is our proposal in Eq. (21).

We selected 1000 random samples from each dataset (if available) and we ran complete cross-validation with 30 Monte Carlo rounds of random splits between train (70%) validation (15%) and test (15%). For each combination of the hyper-parameters[6], we average each metric for the validation and test splits. In order to select the best configuration, we selected the hyperparameters whose average on the validation split showed the best fairness (according to DDP, $DEO^\diamond$, and CF) while simultaneously exhibiting a BA which is at least 90% of its best value across the whole hyperparameters space. Note that this is the same hyperparameters selection strategy proposed in [10]. In Table 1 we reported the metrics computed on the test set for these best-found hyperparameters configurations.

All our experiments have been carried out using the Python programming language and, in particular, exploiting the Gurobi[7] python library for numerical optimization. For time constraints, Table 1b reports the results for kernel methods only on three datasets, namely Arrhythmia, COMPAS, and German, which consist of a number of samples small enough to perform the experiments in a reasonable time (less than 3 days). To improve the readability of the results, we also report the normalized results of Table 1a in Fig. 2 for the Linear SVM, FERM SVM, and R-FERM SVM, plotting BA against DP, CF, and $EO^\diamond$. The same plot is reported in [10] across the different approaches.

From Table 1 and Fig. 2, it is possible to observe that R-FERM is able to successfully deliver fairer models when compared to the no-fairness baseline both for linear and non-linear methods while preserving utility. Moreover, R-FERM systematically outperforms FERM when measuring fairness through the notion of CF. This behaviour is mostly due to the local constraints of Eqs. (19) and (20) which are absent in the original FERM formulation. Interestingly enough, these

---

[6] For SVM, FERM SVM, and R-FERM SVM we searched for C and γ over 15 values equally distributed in a logarithmic space between 10^{-4} and 10^3. For FERM we set $\epsilon = 0$, and for R-FERM we searched $\epsilon \in \{.25, .5, .75, \cdots, 2\}$.

[7] www.gurobi.com.

Table 1. BA, DDP, DEO°, and CF for Linear and RBF SVM, FERM SVM, and R-FERM SVM on 7 fairness-related datasets. Bold indicates the best value among the competing methods. The green background indicates an improvement in fairness concerning the no-fairness method, while the orange background indicates a degradation in fairness but an improvement in utility.

DDP Constraint

Method	Adult BA	Adult DDP / CF	Arrhythmia BA	Arrhythmia DDP / CF	COMPAS BA	COMPAS DDP / CF	Credit BA	Credit DDP / CF	Drug BA	Drug DDP / CF	German BA	German DDP / CF	Taiwan BA	Taiwan DDP / CF
Lin. SVM	.67±.05	.63±.18 / **.05±.43**	.68±.05	.61±.11 / .17±.11	**.65±.05**	.79±.24 / .17±.16	.63±.04	.14±.1 / .0±.0	**.65±.03**	1.09±.14 / .68±.13	**.65±.06**	.38±.21 / .25±.11	.55±.04	.1±.08 / .01±.01
Lin. SVM (FERM)	.67±.05	.16±.11 / .7±.08	**.7±.05**	.14±.1 / .41±.05	.6±.05	.1±.07 / .36±.04	.63±.04	.13±1 / .1±.04	.61±.04	.07±.04 / .25±.03	.64±.06	.21±.15 / .09±.06	.5±.0	.0±.0 / .0±.0
Lin. SVM (R-FERM)	**.70±.05**	.28±.2 / .06±.05	.69±.05	.16±.11 / **.05±.04**	.6±.05	.1±.07 / **.03±.02**	.63±.04	.14±.1 / .0±.0	.61±.04	.07±.04 / .02±.01	**.65±.06**	.23±.15 / .07±.04	**.57±.05**	.13±1 / .02±.02

EO° Constraint

Method	Adult BA	Adult DEO⁻ / CF⁻	Arrhythmia BA	Arrhythmia DEO⁺ / CF⁺	COMPAS BA	COMPAS DEO⁺ / CF⁺	Credit BA	Credit DEO⁻ / CF⁻	Drug BA	Drug DEO⁻ / CF⁻	German BA	German DEO⁻ / CF⁻	Taiwan BA	Taiwan DEO⁻ / CF⁻
Lin. SVM	.67±.05	.48±.17 / **.05±.04**	.68±.05	.64±.23 / .17±.11	**.65±.05**	.8±.26 / .17±.16	**.65±.03**	.31±.2 / .0±.0	**.65±.03**	1.04±.2 / .68±.13	**.65±.06**	.65±.04 / .25±.11	.55±.04	.14±.13 / .01±.01
Lin. SVM (FERM)	**.69±.05**	.16±.11 / .5±.07	**.71±.05**	.27±.26 / .3±.07	.61±.05	.13±.08 / .33±.05	.62±.04	.35±.23 / .07±.06	.62±.04	.11±.07 / .17±.03	**.65±.06**	.42±.35 / .11±.08	.5±.0	.0±.0 / .0±.0
Lin. SVM (R-FERM)	.68±.06	.15±.11 / .06±.06	.68±.05	.3±.2 / **.06±.04**	.61±.05	.12±.08 / **.03±.02**	.62±.04	.31±.2 / .0±.0	.62±.04	.11±.06 / .03±.02	.62±.06	.32±.31 / .11±.08	**.56±.05**	.16±.15 / .01±.01

(a) Linear methods.

DDP Constraint

Method	Arrhythmia BA	Arrhythmia DDP / CF	COMPAS BA	COMPAS DDP / CF	German BA	German DDP / CF
RBF SVM	.72±.04	.44±.21 / .11±.04	.65±.04	.56±.16 / .17±.05	**.66±.07**	.22±.14 / .17±.05
RBF SVM (FERM)	.72±.04	.44±.21 / .1±.04	.65±.04	.56±.16 / .17±.05	.63±.06	.18±.11 / .12±.04
RBF SVM (R-FERM)	**.76±.05**	.16±.11 / .03±.01	**.69±.01**	.44±.01 / .09±.01	.63±.01	.53±.01 / .06±.04

EO° Constraint

Method	Arrhythmia BA	Arrhythmia DEO⁺ / CF⁺	COMPAS BA	COMPAS DEO⁺ / CF⁺	German BA	German DEO⁻ / CF⁻
RBF SVM	**.74±.01**	1.08±.01 / .11±.04	.65±.04	.56±.2 / .2±.07	.65±.06	.22±.16 / .16±.05
RBF SVM (FERM)	**.74±.01**	.73±.01 / .1±.04	.64±.04	.59±.19 / .22±.09	.64±.07	.22±.15 / .14±.01
RBF SVM (R-FERM)	.72±.01	.25±.01 / .06±.02	**.66±.01**	.34±.01 / .11±.04	**.66±.04**	.21±.13 / .09±.05

(b) RBF kernel methods.

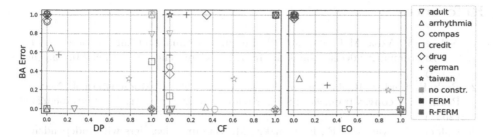

Fig. 2. BA, DDP, DEO°, and CF results from Table 1a normalized in [0,1]. Different shapes represent different datasets, while the colours encode the competing methods. Values closer to the origin are better (low fairness and utility errors). R-FERM (green) achieves better fairness than the no-fairness baseline (red) for most datasets while preserving utility. (Color figure online)

results might suggest that R-FERM can be the preferable choice for those applications where individual fairness (CF) is of greater importance than group statistical notions (DDP and DEO°).

Finally, we made freely available all the code to reproduce the experiments of this work[8].

5   Conclusions

Artificial Intelligence commodification allowed it to become ubiquitous and revolutionize our society. From ludic activities to scientific research, Artificial Intelligence shows great potential for improving our lives and empowering human intelligence. Nonetheless, these achievements are accompanied by increasing concerns about their possible downsides. For this reason, in this paper we addressed the problem of algorithmic fairness in Machine Learning: ensuring that sensitive information does not unfairly influence the outcome of a classifier. We extend the Fair Empirical Risk Minimization framework [10] where the fair risk minimizer is estimated via constrained empirical risk minimization. In particular, we first propose a new, more general, notion of fairness which translates into a fairness constraint. Then, we propose a new convex relaxation with stronger consistency properties deriving both risk and fairness bounds. By extending our approach to kernel methods, we will also show that the proposal empirically over-performs the state-of-the-art Fair Empirical Risk Minimization approach on seven fairness-related datasets.

Acknowledgments. This work is supported in part by ELSA - European Lighthouse on Secure and Safe AI funded by the European Union under grant agreement No. 101070617. This work is also partially supported by project SERICS (PE00000014) under the MUR National Recovery and Resilience Plan funded by the European Union - NextGenerationEU.

---

[8] https://github.com/danif93/Fair-Empirical-Risk-Minimization-Revised.

References

1. Aghaei, S., Azizi, M.J., Vayanos, P.: Learning optimal and fair decision trees for non-discriminative decision-making. In: AAAI Conference on Artificial Intelligence (2019)
2. Berk, R., Heidari, H., Jabbari, S., Joseph, M., Kearns, M., Morgenstern, J., Neel, S., Roth, A.: A convex framework for fair regression. arXiv preprint arXiv:1706.02409 (2017)
3. Calders, T., Kamiran, F., Pechenizkiy, M.: Building classifiers with independency constraints. In: IEEE International Conference on Data Mining (2009)
4. Calmon, F., Wei, D., Vinzamuri, B., Ramamurthy, K.N., Varshney, K.R.: Optimized pre-processing for discrimination prevention. In: Neural Information Processing Systems (2017)
5. Chiappa, S., Isaac, W.S.: A causal bayesian networks viewpoint on fairness. In: Privacy and Identity Management. Fairness, Accountability, and Transparency in the Age of Big Data (2018)
6. Corbett-Davies, S., Goel, S.: The measure and mismeasure of fairness: a critical review of fair machine learning. arXiv preprint arXiv:1808.00023 (2018)
7. Corbett-Davies, S., Pierson, E., Feller, A., Goel, S., Huq, A.: Algorithmic decision making and the cost of fairness. In: ACM SIGKDD international conference on knowledge discovery and data mining (2017)
8. De Fauw, J., Ledsam, J.R., Romera-Paredes, B., Nikolov, S.: Others: clinically applicable deep learning for diagnosis and referral in retinal disease. Nat. Med. **24**, 1342–1350 (2018)
9. Dinuzzo, F., Schölkopf, B.: The representer theorem for hilbert spaces: a necessary and sufficient condition. In: Neural Information Processing Systems (2012)
10. Donini, M., Oneto, L., Ben-David, S., Shawe-Taylor, J., Pontil, M.: Empirical risk minimization under fairness constraints. In: Neural Information Processing Systems (2018)
11. Dwork, C., Hardt, M., Pitassi, T., Reingold, O., Zemel, R.: Fairness through awareness. In: Innovations in Theoretical Computer Science Conference (2012)
12. Dwork, C., Immorlica, N., Kalai, A.T., Leiserson, M.D.M.: Decoupled classifiers for group-fair and efficient machine learning. In: Conference on Fairness, Accountability and Transparency (2018)
13. Feldman, M., Friedler, S.A., Moeller, J., Scheidegger, C., Venkatasubramanian, S.: Certifying and removing disparate impact. In: ACM SIGKDD International Conference on Knowledge Discovery and Data Mining (2015)
14. Floridi, L.: Establishing the rules for building trustworthy AI. Nature Mach. Intell. **1**, 261–262 (2019)
15. Franco, D., Navarin, N., Donini, M., Anguita, D., Oneto, L.: Deep fair models for complex data: graphs labeling and explainable face recognition. Neurocomputing **470**, 318–334 (2022)
16. Friedler, S.A., Scheidegger, C., Venkatasubramanian, S.: On the (im) possibility of fairness. arXiv preprint arXiv:1609.07236 (2016)
17. Friedler, S.A., Scheidegger, C., Venkatasubramanian, S.: The (im)possibility of fairness: different value systems require different mechanisms for fair decision making. Commun. ACM **64**(4), 136–143 (2021)
18. Garg, S., Perot, V., Limtiaco, N., Taly, A., Chi, E.H., Beutel, A.: Counterfactual fairness in text classification through robustness. In: AAAI/ACM Conference on AI, Ethics, and Society (2019)

19. Hardt, M., Price, E., Srebro, N.: Equality of opportunity in supervised learning. In: Neural Information Processing Systems (2016)
20. Hort, M., Chen, Z., Zhang, J.M., Sarro, F., Harman, M.: Bias mitigation for machine learning classifiers: a comprehensive survey. arXiv preprint arXiv:2207.07068 (2022)
21. Kamiran, F., Calders, T.: Data preprocessing techniques for classification without discrimination. Knowl. Inf. Syst. **33**(1), 1–33 (2012)
22. Kamishima, T., Akaho, S., Asoh, H., Sakuma, J.: Fairness-aware classifier with prejudice remover regularizer. In: Machine Learning and Knowledge Discovery in Databases: European Conference (2012)
23. Kusner, M.J., Loftus, J., Russell, C., Silva, R.: Counterfactual fairness. In: Neural Information Processing Systems (2017)
24. Liang, W., Tadesse, G.A., Ho, D., Fei-Fei, L., Zaharia, M., Zhang, C., Zou, J.: Advances, challenges and opportunities in creating data for trustworthy AI. Nature Mach. Intell. **4**, 669–677 (2022)
25. Louizos, C., Swersky, K., Li, Y., Welling, M., Zemel, R.: The variational fair autoencoder. In: International Conference on Learning Representations (2016)
26. Luong, B.T., R., S., Turini, F.: K-NN as an implementation of situation testing for discrimination discovery and prevention. In: ACM SIGKDD International Conference on Knowledge Discovery and Data mining (2011)
27. Mehrabi, N., Morstatter, F., Saxena, N., Lerman, K., Galstyan, A.: A survey on bias and fairness in machine learning. ACM Comput. Surv. **54**(6), 1–35 (2021)
28. Menon, A.K., Williamson, R.C.: The cost of fairness in binary classification. In: Conference on Fairness, Accountability and Transparency (2018)
29. Mitchell, S., Potash, E., Barocas, S., D'Amour, A., Lum, K.: Algorithmic fairness: Choices, assumptions, and definitions. Ann. Rev. Stat. Appl. **8**, 141–163 (2021)
30. Oneto, L., Chiappa, S.: Fairness in machine learning. In: Recent Trends in Learning From Data (2020)
31. Oneto, L., Donini, M., Pontil, M.: General fair empirical risk minimization. In: International Joint Conference on Neural Networks (2020)
32. Oneto, L., Ridella, S., Anguita, D.: Tikhonov, Ivanov and Morozov regularization for support vector machine learning. Mach. Learn. **103**(1), 103–136 (2016)
33. OpenAI: GPT-4 technical report. arXiv preprint arXiv:2303.08774 (2023)
34. Pessach, D., Shmueli, E.: A review on fairness in machine learning. ACM Comput. Surv. **55**(3), 1–44 (2022)
35. Quadrianto, N., Sharmanska, V.: Recycling privileged learning and distribution matching for fairness. In: Neural Information Processing Systems (2017)
36. Samadi, S., Tantipongpipat, U., Morgenstern, J.H., Singh, M., Vempala, S.: The price of fair pca: one extra dimension. In: Neural Information Processing Systems (2018)
37. Shalev-Shwartz, S., Ben-David, S.: Understanding Machine Learning: From Theory to Algorithms. Cambridge University Press (2014)
38. Shawe-Taylor, J., Cristianini, N.: Kernel Methods for Pattern Analysis. Cambridge University Press (2004)
39. Silver, D., Huang, A., Maddison, C.J., Guez, A., et al.: Mastering the game of go with deep neural networks and tree search. Nature **529**, 484–489 (2016)
40. Vapnik, V.N.: The nature of statistical learning theory. Springer science & business media (1998)
41. Verma, S., Rubin, J.: Fairness definitions explained. In: International Workshop on Software Fairness (2018)

42. Woodworth, B., Gunasekar, S., Ohannessian, M.I., Srebro, N.: Learning non-discriminatory predictors. In: Computational Learning Theory (2017)
43. Zafar, M.B., Valera, I., Gomez Rodriguez, M., Gummadi, K.P.: Fairness beyond disparate treatment & disparate impact: Learning classification without disparate mistreatment. In: International Conference on World Wide Web (2017)
44. Zafar, M.B., Valera, I., Gomez-Rodriguez, M., Gummadi, K.P.: Fairness constraints: a flexible approach for fair classification. J. Mach. Learn. Res. **20**, 2737–2778 (2019)
45. Zemel, R., Wu, Y., Swersky, K., Pitassi, T., Dwork, C.: Learning fair representations. In: International Conference on Machine Learning (2013)

Scalable Convolutional Neural Networks for Decoding of Terminated Convolutional Codes

Werner G. Teich$^{(\boxtimes)}$ and Weikun Pan

Ulm University, 89069 Ulm, Germany
werner.teich@uni-ulm.de

Abstract. We present a convolutional neural network (CNN) for the decoding of a terminated convolutional code (CC). For this use cases, an unlimited amount of labeled training data can be generated. However, the number of code words, i.e., pattern, to be learned by the CNN increases exponentially with the dimension of the code. Therefore, scalability of the neural network is of critical importance. The CNN is trained with a CC with small code word length, i.e., a limited number of code words. Therefore the training complexity is feasable. As the CNN is properly matched to dimension and structural properties of the CC, it can actually learn the structure of the CC. This allows to upscale the CNN decoder to match to CCs with larger code dimension. A retraining of weights is not required. The upscaled CNN successfully decodes code words it has never seen during the training process (generalization).

Keywords: convolutional neural network · scalability · terminated convolutional codes

1 Introduction

Deep neural networks (DNNs) are known to be able to approximate highly complex functions. Structure and size of a DNN are typically found by trial and error. The weights of the DNN are obtained by a supervised or unsupervised training process, requiring a sufficiently large amount of (labeled) training data. In the last decade, DNNs and deep learning (DL) have shown impressive results in various fields such as image classification, speech recognition, or playing the abstract strategy board game Go. Recently, an increased interest in the application of DNNs and DL to physical layer problems in digital communications can be observed [1–4]. Gruber et al. [3,5] revisit the problem of using neural networks for decoding of forward error-correcting codes. For these applications the amount of labeled training data is pratically unlimited. However, in contrast to other applications, the number of possible patterns (code words) increases exponentially with the dimension of the code. Gruber et al. [3] have shown that DNN do not only learn individual code words, but, to some extend, are able

I. Rojas et al. (Eds.): IWANN 2023, LNCS 14134, pp. 43–54, 2023.
https://doi.org/10.1007/978-3-031-43085-5_4

to learn the structure of the code. That is they are able to successfully decode noisy code words the DNN has never seen during the training process. However, this generalization property of the DNN is only very limited. The performance of the DNN decoder degrades substantially, if only a small fraction of all code words is used during the training [3,4].

We revisit the problem of decoding convolutional codes (CCs) with neural networks. Three decades ago, Caid and Means [6] and Marcone et al. [7] have proposed a multilayer perceptron with a single hidden layer, i.e., a shallow neural network, for the decoding of convolutional codes. Teich et al. [8,9] have improved this concept by introducing a feedback component to the network. Later, also high-order recurrent neural networks (HORNNs) have been successfully applied to decode special classes of CCs [10–12]. Based on HORNN, Teich et al. [13] proposed an extremly energy efficient analog decoder implementation for the case of convolutional self-orthogonal codes (CSOCs). In [4] Teich et al. used unfolded HORNNs for the decoding of CCs. Here, the structure of the neural network is derived directly from the underlying coding problem. Only few parameters are left, which are optimized by supervised training. The unfolded HORNNs show a good scaling property for the special case of CSOCs. Unfortunately, the decoding performance and the upscaling property degrade severely for general CCs, especially for CCs with nonsystematic encoding.

Here, we propose to use fully convolutional neural networks (CNNs) to decode terminated CCs. Due to their special structure and the limited number of weights to be optimized, CNNs have also been applied successfully to many applications in, e.g., pattern or speech recognition. We show, that a properly designed CNN can actually learn the special structure of a convolutional code. The weights of the CNN are found by a supervised training process for a low code dimension. Similar to other approaches like a fully connected DNN or an (unfolded) HORNN [4], the proposed CNN decoder shows a close to maximum likelihood performance. Furthermore, as the constraint length and the connectivity of the code are fixed, the CNN decoder can be upscaled to any arbitrary code dimension. A retraining of the scaled CNN is not required. Specifically, the CNN is trained for a small code dimension with a code word length of, e.g., $N = 16$ or a total of 256 code words. Upscaling the network to larger code dimensions, e.g., a code word length of $N = 128$ or about $18 \cdot 10^{18}$ code words, the scaled CNN is able to successfully decode all code words of the longer code, none of them it has seen in the training process. That is the CNN shows an excellent generalization property. In contrast to an unfolded HORNN [4], upscaling of our proposed CNN decoder does not only work for self-orthogonal CCs, but also for general CCs, including CCs with nonsystematic encoding.

Zhang et al. [14] have also proposed to use CNN to decode CCs. They do not terminate the CCs but insted used a sliding window approach. They have also shown, that the performance of the CNN decoder is equivalent to a Viterbi soft decoding algorithm.

The remaining of the paper is organized as follows. In Sect. 2 we introduce the system model and in Sect. 3 we describe the scalable CNN decoder. Simulation results are given in Sect. 4 and Sect. 5 concludes the paper.

2   System Model

Figure 1 shows the discrete-time transmission model. A sequence of binary source (or information) symbols $q[k] \in \mathbb{F}_2$ is encoded by a convolutional encoder (COD). The resulting sequence of binary code symbols $c[k] \in \mathbb{F}_2$ is mapped (MAP) to the sequence of transmit symbols $x[k] \in \{\pm 1\} \in \mathbb{Z}$ (binary phase shift keying (BPSK) modulation)

$c[k] \in \mathbb{F}_2$	$x[k] \in \mathbb{Z}$
0	$+1$
1	-1 .

The transmit symbols are sent over a discrete-time additive white Gaussian noise (AWGN) channel, i.e., $y[k] = x[k]+n[k]$. Here, $n[k] \in \mathbb{R}$ is an i.i.d. Gaussian noise sequence with variance σ^2. The sequence of received symbols $y[k] \in \mathbb{R}$ is the input to the channel decoder (DEC), which returns the sequence of detected source symbols $\hat{q}[k] \in \mathbb{F}_2$. We use here the discrete-time AWGN channel, a model commonly applied in coding theory [15]. However, the results can easily be extended to other channels, such as the Rayleigh fading channel.

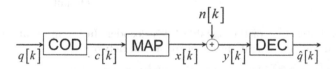

Fig. 1. Discrete-time transmission model.

We consider terminated CCs with a finite block length. To prevent a loss of data rate, we employ tail-biting [16], which also avoids a different protection level of source symbols at the edges of the block. This results in a cyclic code $\mathcal{C}(N, K, d)$, where N is the length of a code word c, K the dimension of the code, respectively, the length of the information word q, and d the minimum distance of the code. The code rate is given by $r_c = \frac{K}{N}$ and the generator as well as the parity check matrix have the structure of a Toeplitz matrix. As the code word is closed to a ring, there is no distinguished position in a code word. All code symbols see the same neighborhood. The main difference between this terminated CC and a "classical" block code is the internal structure of the code. The relation between information symbols and code symbols is fixed and can graphically be represented by the shift-register representation of the encoder [15]. An important parameter of any CC is the number of register elements of the shift-register or the memory of the code. The memory is related to the constraint length of the CC. The constraint length characterizes the number of information symbols which influence the code symbols at a specific point in time. On the other hand, the constraint length is also related to the number of code symbols which are influenced by a given information symbol.

3   Scalable Convolutional Neural Network Based Decoder

For the DL based channel decoding, we follow the general approach as given by Gruber et al. [3]. However, insted of using a fully connected DNN as a decoder, we use a CNN which picks up the structural property of the tail-biting CC. This allows an easy upscaling of the CNN decoder. The proposed CNN decoder consists of an input and an output layer and one to three hidden layers, depending on the used CC (see Fig. 2).

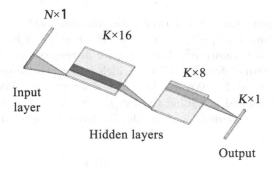

Fig. 2. General structure of the scalable CNN decoder. In the application phase, the signal flow is from left to right (feedforward neural network).

To allow an easy upscaling, all layers of the neural network decoder are convolutional layers, i.e., in contrast to [14], the NN contains no fully connected layer. Furthermore, a circular padding is applied to all convolutional layers. This accounts for the tail-biting property of the CC (circular code). Note, that for terminated CCs with a given fixed initial and final state of the shift register, e.g., the all zero state, the padding has to be adjusted accordingly. In this case a fixed padding should be used. Padding and kernel size have to be adopted to the constraint length of the CC. These structural properties of the CNN decoder allow to train the CNN for a terminated CC with a small code word length N and find the best weights for a specific code with a given connection set. As these weights are code specific and represent the connectivity between code and information symbols, the CNN decoder can easily be upscaled to larger code word length N. This is simply done by adjusting the size of the convolutional layers to the modified code word length N. Padding size, as well as kernel size and weights remain the same. A retraining of the network is neither necessary nor desirable. Due to the large number of possible code words a suitable training of the weights would not be feasable anymore.

The input layer has one input channel, corresponding to one received word y or N noisy code symbols, and 16 output channel. The convolutional kernel length is ten and the padding length eight. The stride of the input layer is two. This performs the reduction from N code symbols at the input of the input layer to $K = \frac{N}{2}$ detected information symbols at the output of the output layer. Note, that we have assumed here a CC with a coding rate of $r_c = \frac{1}{2}$. For the hidden layers the stride is one, the kernel size is four and the padding length is three. All hidden layers have 16 input channels and, except for the last hidden layer, all have 16 output channels. The last hidden layer has only eight output channels. The varying number of hidden layers accounts for the unequal decoding complexity of the different CCs. The output layer therefore has eight input channels and one output channel, corresponding to the decoded information word q. Kernel size and stride of the output layer are both one. Circular padding is applied in all layers. Also the tanh is used as the activation function in all layers. In our simulations the tanh provided a comparable or better performance than the rectified linear unit and did not suffer from a gradient dispersion. The main parameters of the different layers are summarized in Tab. 1.

Table 1. Parameters for the different layers of the CNN decoder.

Layer	Kernel size	Padding size	Stride	Input channel	Output channel
input	10	8	2	1	16
hidden (last)	4	3	1	16	16 (8)
output	1	0	1	8	1

Weights and bias are obtained by a suitable training process. To achieve that, we introduce a loss function, also called objective function, which measures the difference between the true value and the prediction by the CNN. During the training with labeled data, weights and bias of the CNN are adjusted in such a way as to minimize the loss function. Gruber et al. [3] have shown, that the choice of the loss function is not critical for DNN-based decoding of polar codes. Also, simulations have shown, that using the binary cross entropy as loss function does not improve the training process of DNN-based decoders for convolutional codes [4]. Similar as in [4] we therefore employ in this paper the mean squared error (MSE) [17] as loss function. Furthermore, we use the backpropagation algorithm with gradient descent [17], i.e., weights and bias are iteratively corrected in the direction of the negative gradient of the loss function. The gradient with respect to every node in every layer is obtained by using the chain rule. In order to avoid to get stuck in local minima, minibatch stochastic gradient descent is used, i.e., the gradient is averaged over a subset of the training data called the batch size. To provide a fast convergence of the gradient descent and to avoid an overshooting, a proper learning rate has to be chosen. We could obtain good results with a learning rate $\epsilon = 0.002$.

The training data set is comprised of noisy code words, i.e., the received vectors y together with the corresponding information vector q (labeling). Generally, one training epoch contains all possible code words. Thus it consists of 2^K vector pairs (y, q). Note, that each vector pair (y, q) is applied only once to the network. In the next epoch, a different noise vector is added to the code word, leading to a different vector pair. Similar to [4] we used a small code word length of $N = 16$ for the training of the CNN. The signal-to-noise ratio (SNR) during the training is set to 3 dB and all possible $2^K = 256$ code words are used for the training. In total we used 2^{15} training epochs.

The test set contains noisy received vectors y which are not part of the training set and thus have never been seen by the CNN. It is used to evaluate the performance of the CNN. In the application phase, the input of the CNN decoder gets N noisy code symbols and puts out $K = \frac{N}{2}$ decoded information symbols.

The CNN decoder can be upscaled to decode any terminated CC of dimension N by simply adjusting the respective number of neurons in each layer.

4   Simulation Results

We investigated the performance and scaling property of a CNN decoder for four different CC. All four CC are terminated with tail-biting. We consider three CCs with systematic encoding, two of them being CSOCs. Finally, we consider also a CC with non-systematic encoding. Table 2 summarizes the code parameters. Figure 3 shows the shift-register representation of the four convolutional encoders.

Table 2. Code parameters.

Code	Memory	Constraint length	Self-orthogonal	Minimum distance	Encoding
$\mathcal{C}_1$	1	2	yes	3	systematic
$\mathcal{C}_2$	3	4	yes	4	systematic
$\mathcal{C}_3$	2	3	no	4	systematic
$\mathcal{C}_4$	2	3	no	5	non-systematic

To keep the training complexity low and to be able to compare performance and scaling property of our proposed CNN with the results obtained for DNN and (unfolded) HORNN decoder [4], we trained all CNN with a code word length of $N = 16$. All possible 256 code words have been used for the training.

Figure 4 shows the bit error ratio (BER) and the word error ratio (WER) for the most simple CC with memory one ($\mathcal{C}_1$), which is also a CSOC. As the decoding complexity is comparably low, the CNN decoder has only one hidden layer. The BER performance of the CNN decoder is close to the performance of a maximum likelihood (ML) decoder. This is in line with the results for decoders

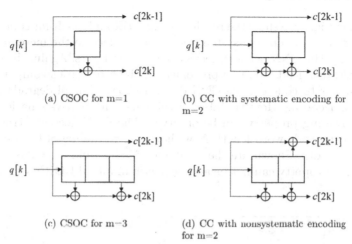

(a) CSOC for m=1

(b) CC with systematic encoding for m=2

(c) CSOC for m=3

(d) CC with nonsystematic encoding for m=2

Fig. 3. Shift-register representation of the encoders for the four codes $\mathcal{C}_1$ (a), $\mathcal{C}_2$ (c), $\mathcal{C}_3$ (b), $\mathcal{C}_4$ (d).

based on fully connected DNNs [4]. The simulation results further show that the CNN decoder can be adjusted to code word length of $N = 32$ and $N = 128$. These upscaled CNN decoders have no performance loss. That is the CNN decoder successfully decodes code words it has never seen during the training process. This clearly demonstrates the strong generalization property of the proposed CNN decoder and is in line with the results obtained for a decoder based on an unfolded HORNN [4]. The WER gets worse as the code word length is increased. This is expected, as with increasing code word length N the probability that at least one of the code symbols in a code word is wrong also increases.

Similar, Fig. 5 shows BER and WER for the code $\mathcal{C}_2$ with memory $m = 3$. $\mathcal{C}_2$ is also a CSOC. As the minimum distance of this code is increased to $d_{min} = 4$, the asymptotic decoding performance compared to $\mathcal{C}_1$ has slightly improved. The CNN decoder also shows a decoding performance close to ML and a perfect upscaling property. For CSOC (with systematic encoding) a comparable error rate performance for $N = 16$ can be obtained with DNN or HORNN. However, in general, upscaling the neural network is only possible for an unfolded HORNN. Similar as for code $\mathcal{C}_1$, the WER gets worse as the code word length is increased.

Next we consider CCs with memory $m = 2$. With systematic encoding (code $\mathcal{C}_3$), the resulting minimum distance is also $d_{min} = 4$. Figure 6 shows the BER and WER for this code $\mathcal{C}_3$. To cope with the increased decoding complexity, two hidden layers are required for the CNN decoder. The decoding performance shows a small degradation compared to ML decoding. However, it still shows a perfect upscaling property, i.e., the BER performance is independent of the code length N. Shown are the results for $N = 32$ and $N = 128$. A comparable error rate performance for a code length of $N = 16$ can be obtained with properly designed DNN or an unfolded HORNN [4]. However, upscaling the neural network is only possible for an unfolded HORNN [4]. Similar as for code $\mathcal{C}_1$, the WER gets worse as the code word length is increased.

Finally, in Fig. 7 we show the results for the BER and WER for code $\mathcal{C}_4$, a CC with memory $m = 2$ and non-systematic encoding. Due to the increased decoding complexity, the CNN decoder needs three hidden layers. Again, the increased minimum distance leads to an improvement of the error ratios compared to the code $\mathcal{C}_1$. Similar to code $\mathcal{C}_3$, the CNN decoder shows a small degradation compared to ML decoding. However, also for this code with nonsystematic encoding a perfect upscaling property can be observed. The CNN decoder structure can be scaled to any code word length N without any retraining of the weights and bias of the network. Shown are the results for $N = 32$ and $N = 128$. Note, this generalization property can not be achieved with unfolded HORNN for this type of CCs.

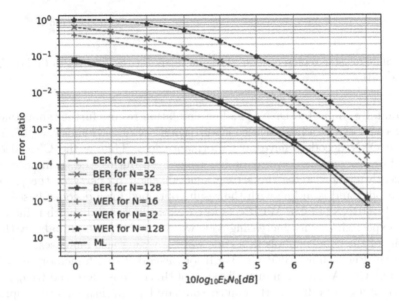

Fig. 4. BER (solid line) and WER (dashed line) as a function of the SNR of the CNN decoder for code $\mathcal{C}_1$. Code parameter: $m = 1$, $d_{\min} = 3$, systematic encoding. Training performed for code length $N = 16$.

Note, that for code $\mathcal{C}_4$, the minimum distance of the terminated CC has not yet reached its maximum value for a block length of $N = 16$. Therefore we show in Fig. 8 the BER and WER for code $\mathcal{C}_4$ when the CNN decoder has been trained with a code word length $N = 20$ (insted of $N = 16$ as in the previous four examples). This corresponds to a total of 1024 code words and the maximum value of the minimum distance for this CC, $d_{\min} = 5$, has been reached. The increased number of code words leads to an increase in the training complexity. As Fig. 8 shows, training the CNN decoder with longer code words and an increased minimum distance leads to an improvement of the error rate performance. Similar as before, the CNN decoder can be upscaled to any code word length N without any need for retraining the network. Shown are the results for $N = 32$ and $N = 128$.

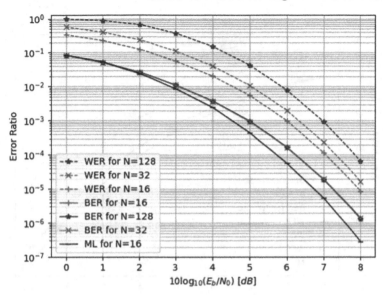

Fig. 5. BER (solid line) and WER (dashed line) as a function of the SNR of the CNN decoder for code C_2. Code parameter: $m = 3$, $d_{\min} = 4$, systematic encoding. Training performed for code length $N = 16$.

5   Summary

We have revisited the problem of decoding CCs with feedforward neural networks. We proposed a scalable CNN decoder for decoding of CCs. The CNN decoder is taylored to the specific structural properties of CCs. It consists only of convolutional layers and a proper circular padding is applied. The complexity of the CNN decoder is adjusted to the varying decoding complexity of different codes by changing the number of hidden convolutional layers. The CNN decoder is trained for a small code dimension $N = 16$, i.e., a code with only 256 code words. This keeps the training complexity low. We have shown by simulation, that the CNN decoder can be upscaled to basically any code length N without the need for a retraining of the weights and bias of the network. That is the CNN decoder has properly learned the structure of the code. It is able to decode code words it has never seen during the training process. A similar generalization property can be achieved by unfolded HORNN. However, unfolded HORNN can only be applied to CCs with systematic encoding. In contrast to that, our proposed scalable CNN decoder works also for general CCs with nonsystematic encoding.

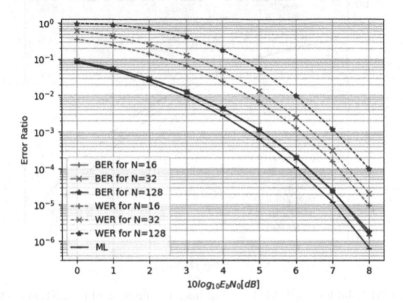

Fig. 6. BER (solid line) and WER (dashed line) as a function of the SNR of the CNN decoder for code $\mathcal{C}_3$. Code parameter: $m = 2$, $d_{\min} = 4$, systematic encoding. Training performed for code length $N = 16$.

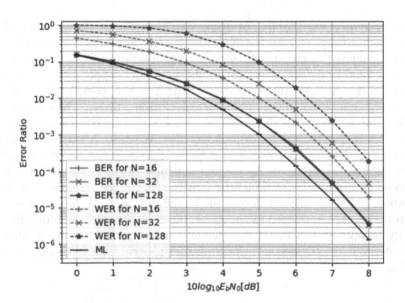

Fig. 7. BER (solid line) and WER (dashed line) as a function of the SNR of the CNN decoder for code $\mathcal{C}_4$. Code parameter: $m = 2$, $d_{\min} = 5$, nonsystematic encoding. Training performed for code length $N = 16$.

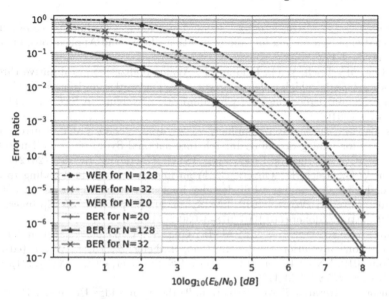

Fig. 8. BER (solid line) and WER (dashed line) as a function of the SNR of the CNN decoder for code $\mathcal{C}_4$. Code parameter: $m = 2$, $d_{\min} = 5$, nonsystematic encoding. Training performed for code length $N = 20$.

References

1. O'Shea, T., Hoydis, J.: An introduction to deep learning for the physical layer. IEEE Trans. Cognitive Commun. Networking **3**(4) (2017)
2. Farsad, N., Goldsmith, A.: Detection Algorithms for Communication Systems using Deep Learning, arXiv preprint arXiv:1705.08044 (2017)
3. Gruber, T., Cammerer, S., Hoydis, J., ten Brink, S.: On deep learning-based channel decoding. In: 51st Annual Conference on Information Sciences and Systems (CISS), Baltimore (2017)
4. Teich, W.G., Liu, R., Belagiannis, V.: Deep learning versus high-order recurrent neural network based decoding for convolutional codes. In: GLOBECOM 2020–2020 IEEE Global Communications Conference, pp. 1–7 (2020). https://doi.org/10.1109/GLOBECOM42002.2020.9348117
5. Cammerer, S., Gruber, T., Hoydis, J., ten Brink, S.: Scaling deep learning-based decoding of polar codes via partitioning. In: GLOBECOM 2017–2017 IEEE Global Communications Conference, Singapore (2017)
6. Caid, W.R., Means, R.W.: Neural network error correcting decoders for block and convolutional codes. In: GLOBECOM 1990, San Diego, CA 2, pp. 1028–1031 (1990)
7. Marcone, G., Zincolini, E., Orlandi, G.: An efficient neural decoder for convolutional codes. Eur. Trans. Telecommun. **6**, 439–445 (1995)
8. Teich, W.G., Boettrich, M.: Optimal Decoding of Convolutional Codes using a Multilayer Perzeptron, ITG-Fachtagung Codierung für Quelle, Kanal und Übertragung, München, ITG-Fachbericht 130, pp. 73–80. VDE Verlag GmbH, Berlin (1994)
9. Teich, W.G., Lindner, J.: A novel decoder structure for convolutional codes based on a multilayer perceptron. In: IEEE ICNN'95, Perth, Western Australia 1, pp. 449–454 (1995)

10. Teich, W.G.: Iterative decoding of one-dimensional convolutional codes. In: 4th International Symposium on Communication Theory and Applications, pp. 52–53. Ambleside, UK (1997)
11. Mostafa, M., Teich, W.G., Lindner, J.: Analog realizatiuon of iterative threshold decoding based on high-order recurrent neural networks. In: International Conference on Signal Processing and Communication Systems (ICSPCS), Gold Coast, Australia (2010)
12. Mostafa, M., Teich, W.G., Lindner, J.: Analysis of high order recurrent neural networks for analog decoding. In: 7th International Symposium on Turbo Codes and Iterative Information Processsing. ISTC), Goeteborg, Sweden (2012)
13. Teich, W.G., Teich, H., Oliveri, G.: From iterative threshold decoding to a low-power high-speed analog VLSI decoder implementation. In: Rojas, I., Joya, G., Catala, A. (eds.) IWANN 2019. LNCS, vol. 11507, pp. 615–628. Springer, Cham (2019). https://doi.org/10.1007/978-3-030-20518-8_51
14. Zhang, Z., Yao, D., Xiong, L., Ai, B., Guo, S.: A convolutional neural network decoder for convolutional codes. In: Gao, H., Feng, Z., Yu, J., Wu, J. (eds.) ChinaCom 2019. LNICST, vol. 313, pp. 113–125. Springer, Cham (2020). https://doi.org/10.1007/978-3-030-41117-6_10
15. Johnson, S.: Iterative Error Correction: Turbo. Cambridge University Press, Low-Density Parity-Check and Repeat-Accumulate Codes (2010)
16. Hagenauer, J., Offer, E., Méasson, C., Mörz, M.: Decoding and equalization with analog non-linear networks. Eur. Trans. Telecommun. **10**, 659–680 (1999)
17. Goodfellow, I., Bengio, Y., Courville, A.: Deep Learning. MIT Press (2016)

Optimizing an IDS (Intrusion Detection System) by Means of Advanced Metaheuristics

A.M. Mora[1]([🖂])(iD), M.G. Arenas[2](iD), A. Romero-Horno[1], J. Camacho-Páez[1](iD), and P.A. Castillo[2](iD)

[1] Department of Signal Theory, Telematics and Communications, ETSIIT and CITIC-UGR, University of Granada, 18071, Granada, Spain
{amorag,josecamacho,mgarenas,pacv}@ugr.es, andresromero@correo.ugr.es
[2] Department of Computer Engineering, Automation, and Robotics. ETSIIT and CITIC-UGR, University of Granada, 18071, Granada, Spain

Abstract. Intrusion Detection Systems (IDSs) are a primary research area in Cybersecurity nowadays. These are programs or methods designed to monitor and analyze network traffic aiming to identify suspicious patterns/attacks. MSNM (Multivariate Statistical Network Monitoring) is a state-of-the-art algorithm capable of detecting various security threats in real network traffic data with high performance. However, semi-supervised MSNM heavily relies on a set of weights, whose values are usually determined using a relatively simple optimization algorithm. This work proposes the application of various Evolutionary Algorithm approaches to optimize this set of variables and improve the performance of MSNM against four types of attacks using the UGR'16 dataset (includes real network traffic flows). Furthermore, we analyzed the performance of a Particle Swarm Optimization approach and a Simulated Annealing algorithm, as a baseline. The results obtained are very promising and show that EAs are a great tool for enhancing the performance of this IDS.

1 Introduction

Anderson [1] defined long time ago an *intrusion* as an unauthorized attempt to access, manipulate, or disrupt a system to render it useless. In the same line, a computer attack or *cyberattack* is known as any event or circumstance that can have a negative impact on an organization's operations, assets, or users, through unauthorized access, destruction, disclosure, or modification of information, and/or a denial of services [2]. These network attacks usually gain privileged access to a host by exploiting known vulnerabilities. Therefore, it is crucial to prevent or detect cyberattacks as soon as possible to avoid significant losses for affected companies. A considerable amount of effort and investment is put into research in this domain, with the design of effective Intrusion Detection Systems (IDSs) being one of the main topics.

I. Rojas et al. (Eds.): IWANN 2023, LNCS 14134, pp. 55–67, 2023.
https://doi.org/10.1007/978-3-031-43085-5_5

An *Intrusion Detection System (IDS)* [3] is a defense mechanism that detects malicious activity on a network. Its main goal is to detect and prevent activities that could compromise the security of the system, including unauthorized recognition or data collection phases, such as port scans or ongoing hacking attempts. IDSs are equipped with a crucial feature to provide insight into unusual activity and produce warnings to notify administrators and block suspicious connections. IDSs can be HIDSs, which monitor critical hosts in the network, such as accesses to their resources, modification of internal data, or running programs; or NIDSs, which focus on monitoring data transported by the network, such as streams, datagrams, or packets. NIDSs are essentially algorithms that can identify malicious or suspicious patterns in real-time network traffic analysis. These systems are usually pre-trained and tuned using datasets of gathered network traffic flows and belong to one of two main types: sign-based detectors (which consider a database of known attacks) or anomaly-based detectors (which consider a model of normal traffic and try to detect unusual variations in traffic).

MSNM (Multivariate Statistical Network Monitoring) [4] is a cutting-edge anomaly-based IDS that employs a variation of Principal Component Analysis [5], known as Multivariate Statistical Process Control, to detect anomalies in a massive dataset of network traffic flows called UGR'16 [6]. This approach analyzes a set of data related to normal traffic and establishes a set of thresholds so that new network traffic data can be classified as anomalous if the derived variables exceed these thresholds. The method is effective even when there are a large number of input variables. However, it has a limitation in that it considers all input variables equally, making it difficult to detect attacks that alter only some of these variables. To address this issue, the authors of the MSNM algorithm developed a semi-supervised version [7], which assigns different weights to each variable to determine its importance in detection. This generates the problem of optimizing the weights for each type of attack. In this regard, José Camacho developed a specific optimization algorithm named run to run PLS [8], which achieved excellent results in three of the four types of attacks tested in the dataset (Denial of Service and two kinds of Port Scan) but performed considerably worse in the case of the fourth attack (BotNet).

The present study proposes the application of different metaheuristics to optimize the set of weights on which MSNM depend for the detection of anomalous patterns in the considered dataset. These anomalies are identified as (potential) attacks and categorized later as one of the four aforementioned types. Namely, we will apply several approaches of Evolutionary Algorithms [9] (using different configuration parameters), as well as a Particle Swarm Optimization method and a simple Simulated Annealing implementation, taken as baseline. The obtained results will be analysed in order to select the best approach overall, considering their detection performance on the different types of attacks.

2   Considered Optimization Metaheuristics

2.1   Evolutionary Algorithms

Evolutionary Algorithms (EAs) [9] are a type of metaheuristic that belongs to the field of Evolutionary Computation. These algorithms are stochastic optimization techniques that mimic the process of natural selection. EAs work with a population of individuals, which represent potential solutions to a given problem. Each individual is evaluated using an objective function, and the fittest ones are more likely to reproduce and generate offspring with a similar structure. Over time, the selection pressure leads to the emergence of better solutions. In each generation, parents are chosen to recombine through crossover, and mutations can also be applied to modify individuals. The worst-performing individuals are eliminated at the end of each generation. This process continues until a stopping criterion is met, such as a fixed number of generations. Figure 1 illustrates the general EA process.

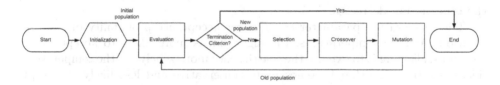

Fig. 1. EA General Process

EAs can find optimal or near-optimal solutions to complex problems that are difficult to solve by human experts, in a reasonable amount of time. Genetic Algorithms (GAs) are the most famous type of EAs, in which individuals are encoded as a vector of genes to which crossover and mutation operators are applied. Other EAs evolve different structures, such as Genetic Programming, where trees representing programming code are evolved, and Evolution Strategy, which evolves genes with the probability of mutating each of them.

2.2   Particle Swarm Optimization

Particle Swarm Optimization (PSO) [10] is a population-based optimization metaheuristic that mimics the social behavior of animal movement in nature, like bird flight or fish schools. PSO has been widely used in different scientific fields and there have been countless different approaches proposed. The algorithm initializes with a random population of candidate solutions, formed by particles that move through the search environment with a velocity and direction that depend on several parameters [11]. The cooperation and competition of the particles in the population are important, as their movement guides the search for better positions. PSO calculates the speed of the population based on individual and group quality parameters, ensuring diversity and adaptiveness.

Each particle's search references its best-found position and the best global positions found by the rest of the particles. PSO tries to converge to good solutions while avoiding local optima, but this is not guaranteed (since it is a metaheuristic) as it depends on several factors.

2.3 Simulated Annealing

Simulated Annealing (SA) [12] is a stochastic optimization technique used to solve complex problems. It is inspired by the annealing process used in metallurgy, where metals are heated and then slowly cooled to obtain desirable properties. Simulated Annealing is based on a probabilistic approach and is useful when the search space is very large, and finding the global optimum is difficult.

The algorithm starts with an initial solution and then perturbs it randomly to generate a new solution. The quality of the new solution is evaluated by calculating its cost or fitness function. If the new solution is better than the current one, it is accepted as the new current solution. However, if the new solution is worse, it may still be accepted with a probability that decreases as the temperature decreases.

The temperature parameter controls the acceptance probability of a worse solution. At high temperatures, the algorithm is more likely to accept worse solutions, allowing it to explore the search space more widely. As the temperature decreases, the algorithm becomes more conservative and less likely to accept worse solutions, converging toward the optimal solution.

Overall, Simulated Annealing is a powerful and widely used optimization method, especially for complex problems with many local optima, where traditional optimization techniques may struggle

3 UGR'16 Dataset

This dataset [6] is a collection of network flow traces gathered by monitoring the traffic in a real Internet Service Provider (ISP) network for five months. The data collection took place in two phases, utilizing Netflow sensors. The first capture occurred between March and June of 2016 under normal network usage conditions, with the goal of studying and modeling the typical behavior of network users and identifying anomalies such as SPAM campaigns. The flows were subsequently labeled as "background" (legitimate flows) or "anomalies" (non-legitimate flows) for use in the CALIBRATION portion of the dataset to train the models. The second capture took place between July and August of 2016, during which "controlled" (synthetic) attacks were launched to acquire a TEST dataset for validating anomaly detection algorithms. Twenty-five virtual machines were deployed within one of the ISP sub-networks, with five of these machines attacking the other twenty. This part forms the TEST section of the dataset, used to validate the trained models.

The researchers conducted four types of attacks during a twelve-day period, namely: Denial of Service (DoS), generated using hping3 tool; port scanning from one attacking machine to one victim machine (Scan11); port scanning from four attacking machines to four victim machines (Scan44); and Botnet traffic, produced by Neris software [13] (NerisBotnet). These attacks were carried out at various intervals, either according to a predetermined schedule or randomly and amidst actual background traffic.

This dataset has been used in many different works, so it is widely referenced in the cybersecurity literature [14–17]. Its main advantage is that it contains data collected from a real network, so it is very interesting to test and validate IDS approaches (or other methods). It also contains periodic or cyclostationary data, as the background traffic follows day/night and weekday/weekend patterns. The dataset can be downloaded from: https://nesg.ugr.es/nesg-ugr16/.

As stated, the dataset includes data from several months and has a global size of more than 200GB. However, we have considered a reduced set for the present study, that is the same used in previous works (for the sake of comparison). Namely, our dataset contains data from two weeks in the month of July, during which there were normal/background traffic and also synthetic attacks. Thus, we have a part devoted to CALIBRATION or training of the models, and another part, TEST, used for evaluating the performance of the obtained models, once the optimization process has finished.

4   MSNM Optimization

Multivariate Statistical Network Monitoring (MSNM) is the proposal of José Camacho et al. [4]. It is based on Multivariate Statistical Process Control (MSPC), an extension of SPC [18] for the consideration of multiple variables. However MSPC fails in the interpretation of relationships between variables, so the algorithm proposed by Camacho applies Principal Component Analysis (PCA) [5] PCA is a technique that operates on two-dimensional data, usually with mean-centering and sometimes auto-scaling. In these data, there are N observations and M variables. The goal of PCA is to identify the subspace of maximum variance in the M dimension by transforming the original variables, which may be correlated, into principal components (PCs). These PCs are uncorrelated variables, and they enable the reduction of data dimensionality without losing information. One of the advantages of PCA is its ability to handle a large amount of data, which is especially useful for network anomaly detection.

MSNM was initially an unsupervised system, in which only a dataset with no labels was provided. The great advantage of this type of system is that it does not require prior knowledge about the attacks to be detected. So, it is possible to detect any kind of anomaly, be it a known attack, an unknown attack, or a system failure.

Although the 'classic' MSNM algorithm is effective in many cases, certain anomalies require a more specific or supervised approach to be detected. To address this, a semi-supervised MSNM [7] aims to establish a hybrid method that combines the advantages of both systems. Researchers observed that certain attacks are more strongly influenced by particular variables than others. As a result, this methodology aims to replace the auto-scaling step of the algorithm by assigning weights to each of the 134 variables considered. However, this approach raises the issue of determining the optimal weight values to maximize the detection capacity of a particular attack or anomaly. Camacho et al. implemented a method to set the weights optimally, named Run-to-Run PLS (R2R-PLS) [8], being Partial Least Squares (PLS) optimization a multivariate regression technique, that makes MSNM a semi-supervised learning method.

Following the same idea, in the present work, different metaheuristics are used to optimize this set of weights, aiming to enhance the performance of MSNM in the detection of network traffic anomalies. We will consider especially Evolutionary Algorithms because they usually reach good solutions in many optimization problems, as well as PSO and SA. Thus, every individual in the EAs is an array of 134 values, each of them in [0,1]. These correspond to the associated weight to each variable, being '0' a non-considered variable and '1' the most relevant.

The approaches have been applied on the real dataset described in Sect. 3. The CALIBRATION part of UGR'16 dataset has been used for evolving the candidate solutions of the algorithms and fitting the MSNM with it. The TEST part has been used for evaluating the best final individual, or solution, in each case, without including these data for fitness computing (during optimization).

5   Experiments and Results

All the experiments have been carried out using the Global Optimization Toolbox [19] and MEDA Toolbox [20] for MATLAB. Since all the algorithms we have run include a stochastic component, 10 runs have been done for each configuration, calculating the average and standard deviation for the results. A different set of runs has been conducted for each type of attack since the optimal sets of weights are strongly dependent on it.

The considered metric (or quality measure) for the algorithms, as well as the fitness (evaluation function) of every individual in the EAs, is the *Area Under the Curve (AUC)* [21] obtained by semi-MSNM using the optimized weights. AUC is related with the ROC curve, which represents the proportion of true positives vs false positives in a classification system, so, the AUC is a value between 0 and 1. If it is 1, the solution would be perfect and the optimized MSNM would have detected all the anomalies in the dataset. If the value is 0, it indicates that no related anomaly has been detected. A value of 0.5 means a random classification.

5.1   Population Study

One of the most important parameters in an EA is the population size. So, before conducting all the experiments, we will study the best value for this parameter. To this end we have considered an EA approach using a Random population initialization for the genes of the individuals, an Auto Feasible Adaptive mutator [19] and Simulated Binary crossover, (SbX) [22]. Auto Adaptive Feasible Mutator generates genes with an adaptive direction with respect to the last successful or unsuccessful generation. The SbX crossover operator produces effective solutions for continuous search spaces and influences how the solutions/offspring are dispersed in comparison to those of their parents.

The results for the population study are shown in Fig. 2, which displays the results for the CALIBRATION (i.e. used in the evolution) 2a, and TEST (once evolution has finished) 2b data respectively.

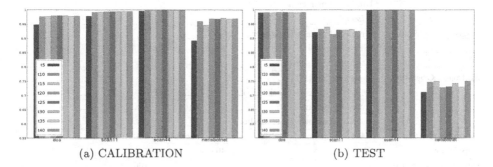

<table>
<tr><td>(a) CALIBRATION</td><td>(b) TEST</td></tr>
</table>

Fig. 2. Obtained fitness for each population size using CALIBRATION dataset (a) and TEST dataset (b).

Looking at Fig. 2 on the left-side graphs, it can be seen that for a population size equal to 40 the algorithm reaches the best solutions in three out of the four cases. This is confirmed in the right-side graphs, in which the results bring us to the same conclusion. It is important to note that NerisBot is the most difficult problem, so we consider these results with a higher relevance.

5.2   Experimental Setup

We have run the three algorithms described in Sect. 2. The configuration parameters for each algorithm are described in Table 1. It might be noticed the EA has been configured with several options, combining different ways to initialize the population (Binary and Random), different crossover operators (Arithmetic [23] and SbX [22]), and different mutation methods (Feasible Adaptive [19] and Gaussian [24]). The rest of the parameters, such as the selection method, population size, or replacement policy are fixed for all the configurations. Namely, the selection has been done using a probability Roulette Wheel method with scaled fitness [19] (a better individual has a bigger chance to be selected than a less fit one). The replacement is fixed to a generational model using elitism for best individual preservation. The stopping criteria was fixed to a number of generations, or iterations in the case of SA.

As it can be seen in Table 1 there are many configuration options in the EA case, so we have summarized all the combinations and assigned a name for each one (for the sake of clarity) in Table 2.

Table 1. Configuration parameters for EA, PSO, and SA algorithms.

Parameter	EA	PSO	SA
Initial Pop Generation	Random Generation $g_i \in [-10, 10]$		
	Uniform generation $g_i is 0 or 1$	-	-
Crossover Operator	Simulated Binary Crossover (SBX) [22] [19]	-	-
	Arithmetic crossover		
Mutation Operator	Gaussian [24] [19] (mutationgaussian method) [19]	-	-
	Auto Adaptative [19] (mutationadaptfeasible method)	-	-
		-	-
Stopping Criteria	100 Generations	100 Generations	1000 Iterations
Population size	40	40	-
Initial Temperature	-	-	100
Cooling function	-	-	Exponential $t_{i+1} = ti * 0,95^k$

Table 2. Acronyms of the different combinations for the EA configuration parameters.

Algorithm Label	Population Initialization	Mutation Operator	Crossover Operator
BAA	Binary	Auto Adaptative	Arithmetic
BAS	Binary	Auto Adaptative	SBX
BGA	Binary	Gaussian	Arithmetic
BGS	Binary	Gaussian	SBX
RAA	Random	Auto Adaptative	Arithmetic
RAS	Random	Auto Adaptative	SBX
RGA	Random	Gaussian	Arithmetic
RGS	Random	Gaussian	SBX

5.3 Variability and Evolution Study

The algorithms based on a population often presents the issue that the population does not have enough diversity or it is reduced along the iterations. As a consequence, the explored solutions do not cover the entire search space and the algorithm may get stuck on suboptimal solution.

To avoid this problem, a measure of the diversity of the population, *variability*, has been proposed and computed throughout all the executions. It is the average of the standard deviation calculated for each gene (variable). This measurement gives us idea of whether or not there is adequate diversity in the population. If the variability value is high, it means that all the standard deviations of each gene are high and therefore the population has good diversity. On the other hand, if it is low, the population has a low diversity. This measure can also be an indicator of whether the operators of each algorithm are providing enough diversity to the population.

To show this, we have represented a complete evolution of two of the Evolutionary Algorithm variants (RGS and BAS) in Fig. 3 including the minimum (Mn), mean (Av), and maximum (Mx) fitness values for one of the executions on the NerisBotnet problem, the hardest one. As it can be seen the behaviour of the algorithms regarding the fitness evolution is as desired, i.e. fitness maximum and average have a growing tendency. Fluctuations in average values are normal, given the stochastic component of EAs, which could lead to worse solutions/individuals in one generation. However they could drive to better ones in following generations, when combined with other individuals.

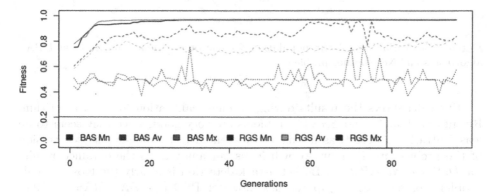

Fig. 3. The best, min, and average fitness of the population along the evolution for RGS and BAS algorithms for the Nerisbotnet problem.

Regarding the variability of the population, we have plotted it for the same algorithms and runs in Fig. 4. The results show how the variability in RGS remains high along the complete execution. In the case of BAS, it is high until generation 70 and the it starts to drop. This happens due to the Autoadaptive Mutation operator, which aims to add a high mutation component at the beginning of the run (more exploration is needed) and reduces this factor along generations (exploitation is more recommended).

5.4   Results

Firstly, we analyze the results after the optimization process for each EA combination, PSO, and SA. These are detailed in Table 3, including the average best fitness of the 10 runs at the end of the evolution, i.e. the best individual obtained considering the CALIBRATION dataset for the computation of its fitness. As it can be seen, all the AE combinations performs very well in almost all the problems/attacks, excepting BAA which is a step below the rest. RAS obtained the best results in almost all the cases, with PSO also obtaining very high values.

The main differences between the approaches can be seen in the case of NerisBotNet attack, which is quite complex to be detected for many optimizers[1].

---

[1] Indeed, is the semi-MSNM optimized by these algorithms which cannot detect these attacks.

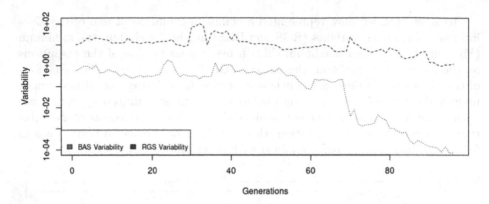

Fig. 4. The Variability of the population for one of the runs using RGS and BAS algorithms and Nerisbotnet problem.

Figure 5a shows the results during training/calibration for each algorithm. Results for *Scan11* and *Scan44* problems are very good, since they seem to be 'easy' attacks to detect, as it was also presented in the non-supervised version of the problem in [7]. Thus we will focus our analysis on the obtained results for *DoS* and *NerisBotNet*. Botnet attack detection is clearly the most difficult problem for most of the approaches, however PSO and BAS, RAS anf RGS obtain good detection values.

If we pay attention only to the EA configurations, SBX crossover is included in four of the best results for EAs (RAS, BAS, RGS, and BGS), so we can say that SBX crossover is a good option for this problem. It is important also to note that the results for the algorithms which include Auto Adaptative mutation, are also very good. So, combining these operators seem to be the best option.

Finally, the population initialization method is not as important as we could expect, since there seems not being remarkable differences between random and binary approaches.

Now, we will compare the best obtained solutions for each approach with two of the State-of-the-Art techniques, presented in [7], namely MSNM-AS (autoscaled) and MSNM-R2R-PLS (run to run). To conduct this comparison, we use the TEST partition of the dataset (new data for which none of the algorithms was trained). The results are displayed in Fig. 5b.

As it can be seen, these are remarkably worse than in CALIBRATION case, given that these are completely new data for all the methods. This is extremely noticeable in the case of NerisBotNet, where the maximum is reached around 74% by BAS, followed by RGS and MSNM-R2R-PLS around 72%. It is curious to see that one of the easy attacks to be detected during training (Scan11) is now much more difficult to be identified for all the algorithms. With regard to the comparison between our new approaches and those of the SotA, there are several EA configurations reaching better results than MSNM-R2R-PLS: most of them in the DoS and Scan11 attacks, and BAS and RGS in NerisBotNet case.

Table 3. Results for each EA combination, PSO and SA, including for each attack, the average and standard deviation for the CALIBRATION dataset. Best three values in each case are marked in boldface.

Approach	DoS	Scan11	Scan44	NerisBotNet
BAA	0.9529 ± 0.0017	0.9831 ± 0.0016	0.9970 ± 0.0002	0.8190 ± 0.0286
BAS	0.9788 ± 0.0021	**0.9944±0.0003**	**0.9989±0.0007**	0.9694±0.0025
BGA	0.9721 ± 0.0010	0.9922 ± 0.0005	0.9985 ± 0.0004	0.9415 ± 0.0079
BGS	**0.9797±0.0020**	0.9936 ± 0.0007	0.9973 ± 0.0004	**0.9704±0.0023**
RAA	0.9562 ± 0.0058	0.9868 ± 0.0017	0.9972 ± 0.0002	0.8316 ± 0.0033
RAS	**0.9799±0.0012**	**0.9941±0.0005**	**0.9989±0.0007**	**0.9703±0.0052**
RGA	0.9716 ± 0.0011	0.9923 ± 0.0005	0.9982 ± 0.0004	0.9387 ± 0.0063
RGS	0.9775 ± 0.0020	**0.9938±0.0007**	**0.9989±0.0007**	0.9681 + 0.0028
PSO	**0.9827±0.0048**	0.9744 ± 0.0034	**0.9996±0.00001**	**0.9735±0.0110**
SA	0.9325 ± 0.0052	0.9744 ± 0.0028	0.9961 ± 0.0003	0.7215 ± 0.0127

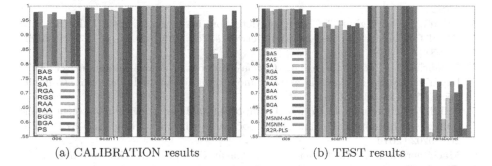

    (a) CALIBRATION results   (b) TEST results

Fig. 5. (a) Fitness comparison for each algorithm and each attack using CALIBRA-TION dataset. (b) Fitness comparison for each algorithm and each attack adding MSNM-AS and MSNM-R2R-PLS using TEST data.

6 Conclusions and Future Work

The purpose of this work is to enhance the performance of a cutting-edge Intrusion Detection System called Semi-Supervised MSNM, which identifies anomalies in data collected from real network traffic flows. To achieve this goal, we have applied a set of metehuristics based on different approaches of Evolutionary Algorithms (EAs), together with Particle Swarm Optimization (PSO) and Simulated Annealing methods. These methods are designed to optimize the weights utilized by the MSNM algorithm for detection.

Several experiments have been conducted to detect four different types of attacks: Denial of Service, Port Scan11, Port Scan44 and NerisBotNet. The results show that most of the EA configurations reach very good solutions in all the attacks, specially those using Auto-Adapted Mutation and SBX Crossovertogether, being even better than the original optimized version of the

MSNM algorithm: MSNM-R2R-PLS. PSO also performs very well in almost all the cases.

As future work, we will investigate the use of alternative operators that may be better suited for addressing highly challenging attacks. Additionally, we will intend to examine other EA models and assess the efficacy of other algorithms.

Acknowledgements. This work has been partially funded by projects PID2020-113462RB-I00, PID2020-115570GB-C22 and PID2020-115570GB-C21 granted by Ministerio Español de Economía y Competitividad; project TED2021-129938B-I0, granted by Ministerio Español de Ciencia e Innovación; as well as project A-TIC-608-UGR20 granted by Junta de Andalucía.

References

1. Anderson, J.P.: Computer security threat monitoring and surveillance. James P. Anderson Company, Technical Report (1980)
2. Hathaway, O.A., et al.: The law of cyber-attack. California law review, pp. 817–885 (2012)
3. Sabahi, F., Movaghar, A.: Intrusion detection: a survey. In: 2008 Third International Conference on Systems and Networks Communications, pp. 23–26. IEEE (2008)
4. Camacho, J., Pérez-Villegas, A., García-Teodoro, P., Maciá-Fernández, G.: Pca-based multivariate statistical network monitoring for anomaly detection. Comput. Secur. **59**, 118–137 (2016)
5. Jolliffe, I.T., Cadima, J.: Principal component analysis: a review and recent developments. Philosophical Trans. Roy. Soc. A Math. Phys. Eng. Sci. **374**(2065), 20150202 (2016)
6. Maciá-Fernández, G., Camacho, J., Magán-Carrión, R., García-Teodoro, P., Therón, R.: Ugr '16: a new dataset for the evaluation of cyclostationarity-based network idss. Comput. Secur. **73**, 411–424 (2018)
7. Camacho, J., Maciá-Fernández, G., Fuentes-García, N.M., Saccenti, E.: Semi-supervised multivariate statistical network monitoring for learning security threats. IEEE Trans. Inf. Forensics Secur. **14**(8), 2179–2189 (2019)
8. Camacho, J., Picó, J., Ferrer, A.: Self-tuning run to run optimization of fed-batch processes using unfold-pls. AIChE J. **53**(7), 1789–1804 (2007)
9. Eiben, A.E., Smith, J.E.: Introduction to evolutionary computing. Springer (2015)
10. Eberhart, R., Kennedy, J.: Particle swarm optimization. In: Proceedings of the IEEE International Conference on Neural Networks, 4, pp. 1942–1948 (1995)
11. He, Y., Ma, W.J., Zhang, J.P.: The parameters selection of pso algorithm influencing on performance of fault diagnosis. In: MATEC Web of Conferences, vol. 63, EDP Sciences (2016) 02019
12. Van Laarhoven, P.J., Aarts, E.H., van Laarhoven, P.J., Aarts, E.H.: Simulated annealing. Springer (1987)
13. Garcia, S., Grill, M., Stiborek, J., Zunino, A.: An empirical comparison of botnet detection methods. Comput. Secur. **45**, 100–123 (2014)
14. Rajagopal, S., Kundapur, P.P., Hareesha, K.S.: A stacking ensemble for network intrusion detection using heterogeneous datasets. Secur. Commun. Networks **2020**, 1–9 (2020)

15. Larriva-Novo, X., Vega-Barbas, M., Villagra, V.A., Rivera, D., Alvarez-Campana, M., Berrocal, J.: Efficient distributed preprocessing model for machine learning-based anomaly detection over large-scale cybersecurity datasets. Appl. Sci. **10**(10), 3430 (2020)
16. Fuentes García, N.M., et al.: Multivariate statistical network monitoring for network security based on principal component analysis (2021)
17. Magán-Carrión, R., Urda, D., Diaz-Cano, I., Dorronsoro, B.: Improving the reliability of network intrusion detection systems through dataset integration. IEEE Trans. Emerg. Top. Comput. **10**(4), 1717–1732 (2022)
18. Boardman, T.J.: The statistician who changed the world: W. edwards deming, 1900–1993. Am. Statistician **48**(3), 179–187 (1994)
19. Lofberg, J.: Yalmip: A toolbox for modeling and optimization in matlab. In,: IEEE international conference on robotics and automation (IEEE Cat. No. 04CH37508). IEEE **2004**, 284–289 (2004)
20. Camacho, J., Pérez-Villegas, A., Rodríguez-Gómez, R.A., Jiménez-Mañas, E.: Multivariate exploratory data analysis (meda) toolbox for matlab. Chemom. Intell. Lab. Syst. **143**, 49–57 (2015)
21. Hastie, T., Tibshirani, R., Friedman, J.H., Friedman, J.H.: The elements of statistical learning: data mining, inference, and prediction. Volume 2. Springer (2009)
22. Deb, K., Beyer, H.G.: Self-adaptive genetic algorithms with simulated binary crossover. Evol. Comput. **9**(2), 197–221 (2001)
23. Albadr, M.A., Tiun, S., Ayob, M., Al-Dhief, F.: Genetic algorithm based on natural selection theory for optimization problems. Symmetry **12**(11), 1758 (2020)
24. Lan, K.T., Lan, C.H.: Notes on the distinction of gaussian and cauchy mutations. In: 2008 Eighth International Conference on Intelligent Systems Design and Applications. Volume 1, IEEE (2008) 272–277

Iterative Graph Embedding
and Clustering

Artem Oborevich[2,3] and Ilya Makarov[1,2,3]($\boxtimes$)

[1] HSE University, Moscow, Russia
[2] Artificial Intelligence Research Institute (AIRI), Moscow, Russia
aoborevich@edu.hse.ru
[3] AI Center, NUST MISiS, Moscow, Russia
iamakarov@misis.ru

Abstract. Graph embedding can be seen as a transformation of any graph into low-dimensional vector space, where each vertex of the graph has a one-to-one correspondence with a vector in that space. The latest study in this field shows a particular interest in a slightly different approach of graph embedding, where each node is inclined to preserve a community membership to respect high-order proximity and community awareness. We investigate different options of solving both tasks jointly, so a practical solution to one problem could be shared to enhance a solution to another problem and vice-versa. We imply that many iterations of such transferring can be made to achieve better results in both problems simultaneously. As a result of our work, we introduce a model that outperforms traditional methods which consider problems of graph embedding and community detection separately.

Keywords: Network embedding · Community detection · Clustering

1 Introduction

Graph embedding enables us to apply standard machine learning methods, commonly applied in traditional machine learning tasks, such as text analysis, facial expression recognition [1,34], image classification and clustering [35], speech recognition [30,33], etc. Traditionally, this implies the construction of such vectors in a low-dimensional space, that with a given pair of nodes their proximity (distance) is transferred from the graph structure knowledge [2,22]. In other words, if two nodes have similar embedding, this should indicate the fact that the first node is reachable from the second within a few steps and vice versa (considering the undirected graph). The embedding space is sometimes called the feature space as semantically it should reflect abstract structural preferences of

The work of I. Makarov on Section 1 was supported by the Russian Science Foundation under grant 22-21-00227 and performed at National Research University Higher School of Economics (HSE University), Moscow, Russia.

I. Rojas et al. (Eds.): IWANN 2023, LNCS 14134, pp. 68–79, 2023.
https://doi.org/10.1007/978-3-031-43085-5_6

the network or graph. Previously, we have conducted multiple experiments with graph embeddings and feature engineering for network analysis [16–21,23,29], however, most of the models construct node embeddings not taking into account cluster structure.

One trade-off that comes with such representation is a necessity to encode rich graph semantics, whereas at the same time we want an embedding space to be small enough to allow us to train systems quickly and cheaply [8,31]. One of such semantics is communities, which are usually defined as groups of densely connected nodes. Recently, iterative methods have been acknowledged as a way to convert complex, non-linear network structures into formats more compatible with vector spaces. These methods use randomized walks as the main strategy to learn complex node relationships. While such methods perform well in preserving node proximity, they do not have any explicit knowledge of internal community structure. However, the most recent studies have proved the idea that clustering and community detection algorithms can be effectively incorporated into the iterative node embedding process. Moreover, their alliance leads to mutual benefits due to the fact that neighboring nodes should be detected within the same community and, oppositely, nodes within the same cluster should maintain near distance by definition.

In this paper, we investigate the real effectiveness of hybrid iterative graph embedding and clustering methods and show that their potential performance has not yet been reached extending results from [22].

2   Related Work

We survey some of the recent papers dedicated to graph embedding and community detection algorithms and observe that many studies bring different concepts, which make their parallel comparison especially intriguing.

2.1   Node Embedding

DeepWalk [27] performs multiple random walks on the graph yielding node embedding learned in an unsupervised manner. It applies the randomized path traversing technique to provide insights into localized structures within networks. The main concept of this algorithm is that each random walk is treated as a sentence, and, therefore, can be processed in a manner similar to the *Word2Vec* model.

There is a modification of the *DeepWalk* model called *Node2Vec* [9]. The main idea here is that you can tune your random walks using a set of two parameters p and q, which correspond to the probabilities the model switches between the local search (*BFS*) and the *DFS* respectively. In the case of a frequent *DFS*, the model learns about node types splitting (e.g. bridges or avenues on the map), whereas a frequent local search contributes to delimiting the neighborhoods and local communities.

2.2 Clustering

K-means [10,13,37] (or Lloyd's algorithm) is a popular method in cluster analysis, which shares idea of vector quantization to produce modeling of probability density functions [32] by the distribution of vectors. It minimizes within-cluster variances (squared Euclidean distances) using an iterative approach.

Gaussian Mixture Model (*GMM*) [11] is a popular statistically-based method [32] in clustering. More formally each cluster in it is mathematically represented as a parametric distribution, e.g., Gaussian one, and the entire dataset, therefore, can be modeled as a mixture of those distributions. The parameters are usually optimized using the Expectation-Maximization (EM) algorithm. Whereas *K-means* makes a hard assignment of each data point to a single cluster, the EM algorithm, in contrast, makes a soft assignment based on posterior probabilities.

2.3 Graph Embedding and Clustering

GEMSEC [28] is an extension over well-know techniques that considers both problems of node embedding and clustering simultaneously. More importantly, it shows that the community awareness, e.g. cluster centroids knowledge, can be effectively incorporated into previously described sequence-based models, similar to *Node2Vec*. Specifically, it adds *K-means* clustering cost regularization to the *Word2Vec* model to achieve superior accuracy and efficiency. It produces better results in the sense that each community becomes more tightly clustered, which makes nodes from different communities appear clearly separated.

There is an interesting study proposing a different way of graph embedding named community embedding (*ComE*) [3]. Their approach is mainly inspired by the *GMM*, and, therefore, they assume that the low-dimensional space that any community fits can be implemented by the multivariate Gaussian distribution. Based on this fact, they obtain community embedding and community detection together fitting *GMM*. They provide community embedding and community detection feedback to the node embedding and vice-versa to optimize their solution in a loop. Thus the authors of this work invite us to consider the problems of node embedding, community embedding, and community detection and node embedding holistically.

In another recent study, the authors investigate the effectiveness of community embedding algorithms [36] in real-world tasks and come to the conclusion that there is not enough evidence that the community embedding brings sufficient enhancement to the clustering without acquiring vasts of expensive resources (both time-wise and hardware-wise). Moreover, given enough resources, those algorithms require very accurate hyper-parameters tuning for each domain-specific task separately to achieve any sort of prevailing over the traditional clustering methods.

Cluster-GCN [6] is a modern algorithm for training deep and large Graph Convolutional Neural networks (*GCN*) [15] in semi-supervised manner. It tackles the problem of inefficient *SGD* on large networks by implementing a clustering internally to restrict further neighborhood search within dense areas. While this heuristic implementation leads to comparable accuracy with the previous state-of-the-art methods [4,5], it allows *Cluster-GCN* to surpass them in computational and memory efficiency [31].

2.4   Summary

While in general node embedding is an efficient way to embed nodes in moderately-scaled networks and, consequently, to depict them as it preserves node proximity by design, it does not necessarily leverage its community knowledge. Therefore it is commonly used in combination with traditional clustering algorithms as a bottom-level performance.

On the other hand, a major part of nowadays algorithms try to capitalize on solving node embedding and community detection problems jointly.

Table 1. Algorithms comparison

NAME	PROXIMITY			CLUSTERING
	FIRST-ORDER	SECOND-ORDER	HIGH-ORDER	
DEEPWALK	✓	✓	✗	✗
NODE2VEC	✓	✓	✗	✗
K-MEANS	✗	✗	✗	✓
GMM	✗	✗	✗	✓
GEMSEC	✓	✓	✓	✓
COME	✓	✓	✓	✓
CLUSTER-GCN	⋆	⋆	⋆	✓

Though *Cluster-GCN* does not necessarily provide any node embedding directly, their inside layers can be used to define feature space (node embedding) as well as shown in Table 1.

3   Selected Algorithm

Table 2. Notation in this paper

NOTATION	DESCRIPTION
$G(V,E)$	GRAPH G WITH A SET OF VERTICES V AND EDGES E
$\mathcal{L}$	LOSS FUNCTION
ϕ_i	NODE EMBEDDING OF v_i
C_i	SET OF CONTEXTS OF NODE v_i
ϕ'_j	i- TH CONTEXT EMBEDDING IN SET
$\mathbb{P}_n(\cdot)$	NEGATIVE SAMPLING PROBABILITY
ψ_k	MEAN VALUE OF CLUSTER k
σ	SIGMOID ACTIVATION FUNCTION

We formulate core notations of the paper in Table 2. Traditionally, the central focus of graph embedding algorithms is to preserve first-order proximity. Mathematically, such a relationship can be enforced by minimizing the following equation:

$$\mathcal{L}_1 = - \sum_{(v_i, v_j) \in V} \log \sigma(\phi_i^T \phi_j) \tag{1}$$

To preserve second-order proximity (i.e., neighbors within ζ steps), we introduce an extra embedding frequently named as a sharing context. Basically, it means a node context that is shared across its neighborhood (similarly to the *Word2Vec* model [26], where each node is defined by its context, e.g. surrounding words). Thus each node has a node embedding ϕ_i itself and a set of contexts C_i for other nodes. Then we define a function to measure how well node v_i generates its context $\phi_j' \in \Phi_i'$, using negative sampling [26] to make computations cheap:

$$\Delta_{ij} = log\sigma(\phi_j'^T \phi_i) + \sum_{l=1}^{m} \mathbb{E}_{v_l \sim \mathbb{P}_n(v_l)} \left[log\sigma(-\phi_l'^T \phi_i) \right] \tag{2}$$

Generally maximizing this equality enforces node v_i to generate such an embedding ϕ_i as to best fit its positive context ϕ_j', but not its negative samples ϕ_l' (m total). Then we can minimize the following objective to obtain the second-order proximity by:

$$\mathcal{L}_2 = - \sum_{v_i \in V} \sum_{v_j \in C_i} \Delta_{ij} \tag{3}$$

To reflect community structure (dense areas within the graph) we introduce $K - means$-loss as a regularization term. The main idea there is to minimize the sum of squared distances from all points to their cluster centers:

$$\mathcal{L}_3 = \alpha \sum_{i=1}^{m} \sum_{k=1}^{K} \mathbb{K}_{c_i=k} \|\phi_i - \psi_k\|^2 \tag{4}$$

However, we also want to take into account the inter-cluster distance measure as following:

$$\Delta_{inter} = \min_{i,j \in \{1,...,K\}} (\|\psi_i - \psi_j\|^2), \ i \neq j \tag{5}$$

We take only the minimum value as we want the smallest value to be maximized. Consequently, we derive our novel regularization term:

$$\mathcal{L}_3^* = \mathcal{L}_3 \frac{1}{\Delta_{inter}} \tag{6}$$

Thus, as a result we define:

$$\mathcal{L}(\Phi, \Phi', \Psi) = \mathcal{L}_1(\Phi) + \mathcal{L}_2(\Phi, \Phi') + \mathcal{L}_3^*(\Phi, \Psi) \tag{7}$$

Finally, we present our core Algorithm 1.

Algorithm 1. GEMSEC with inter-cluster regularization (**G-INTER**)

Data: $G = (V, E)$ - graph to be embedded.
N - number of samples per node.
w - context size.
l - length of sequence.
d - embedding dimension.
η - learning rate.
$t \leftarrow 0$.
$\Psi \leftarrow \Psi_0$.
for n in $\{1, \ldots, N\}$ **do**
    $\hat{V} = \mathbf{Shuffle}(V)$
    for v_i in $\hat{V}$ **do**
        $t \leftarrow t + 1$
        $\eta \leftarrow \mathbf{Update}\ (\eta, t)$
        Sample node using $Node2Vec\ (G, v_i, l)$
        $\phi_i, \Phi_i' \leftarrow \mathbf{Extract\ features}\ (v_i, w)$
        $Adam$ on Eq. 6 w.r.t. $\Psi \in \mathbb{R}^d$
        $Adam$ on Eq. 7 w.r.t. $\phi_i \in \mathbb{R}^d, \Phi_i' \in \mathbb{R}^d$
    end for
end for

---

4   Experiments

4.1   Data

We use the *CORA* dataset [24, 38] for testing hypothesis: this dataset consists of 2708 nodes distributed over 7 communities. *CiteSeer* [38] is another citation network that consists of 3327 nodes and 6 classes. Datasets statistics are shown in Table 3.

Table 3. Datasets

NAME	TYPE	CLASSES	NODES	EDGES
CORA	CITATION	7	2708	5429
CITESEER	CITATION	6	3327	4732

In Fig. 1, we can see an original *CORA* visualization according to its 7 classes. We may assume that this network structure is complicated enough for any problem on the table, especially in unsupervised learning. We use these node labels as masks for clustering evaluation.

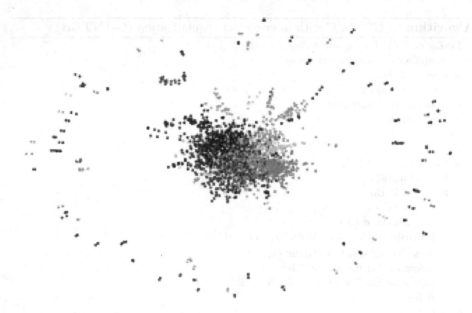

Fig. 1. *CORA* depicted with Fruchterman-Reingold force-directed algorithm [7]

4.2   Metrics

We compare model predictions with ground truth labels indifferently of absolute values of the labels: permutations of class labels result in the score value unchanged. As the main numerical metric, we introduce *Normalized Mutual Information* (*NMI*) score [25] to scale the results between 0 (no mutual information) and 1 (perfect correlation).

As a secondary metric we choose a harmonic mean score or better known as *F-score* [12], which is defined as follows:

$$F = \frac{(1 + \beta^2) \times recall \times precision}{(\beta^2 \times precision) \times recall}$$

We see no particular reason to favor precision or recall in our tests, thus we select $\beta = 1$ (F_1-*score*).

4.3   Results and Achievements

In this section, we evaluate and score graph embedding and clustering accuracy using previously discussed algorithms and metrics.

In Fig. 2, *Node2Vec* is applied on *Cora* graph with $\zeta = 20$ in $\mathbb{R}^d$, $d = 128$, $p = q = 1$.

In Fig. 3, we add *K-means* regularization (see Sect. 2.3) to the same model as in the previous test applied on *Cora* (see Fig. 2) with $\alpha = 0.5$.

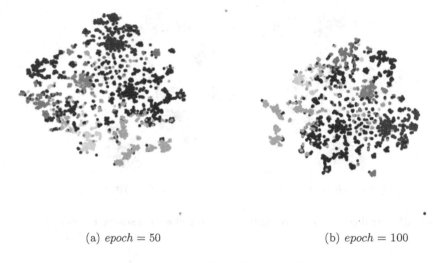

(a) *epoch* = 50   (b) *epoch* = 100

Fig. 2. *Node2Vec* embedding

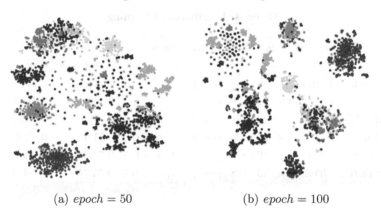

(a) *epoch* = 50   (b) *epoch* = 100

Fig. 3. *Node2Vec* embedding with *K-means* regularization

In Fig. 4, we add inter-cluster distance measure to the *K-means* loss function denominator (see Sect. 3) with no hyper-parameters tuning.

For each model we use adaptive learning rate (*lr*) with *Adam* optimizer [14] as follows:

$$lr = \begin{cases} 0.01, & 1 \leq epoch \leq 50 \\ 0.001, & 50 \leq epoch \leq 100 \end{cases}$$

The results of G-INTER algorithm show better quality compared to Node2vec and GEMSEC baselines as shown in Table 4.

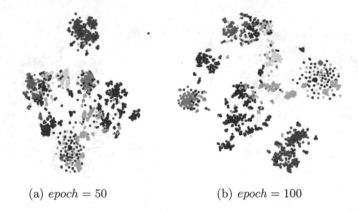

(a) *epoch* = 50   (b) *epoch* = 100

Fig. 4. *Node2Vec* embedding with *K-means* and inter-cluster distance regularization

Table 4. Unsupervised learning

NAME	DATASET	ζ	$\mathbb{R}^d$	PARAMETERS	ACCURACY	
					epoch = 50	*epoch* = 100
NODE2VEC	CORA	20	128	$p = q = 1$	0.7160	0.7210
NODE2VEC	CITESEER	20	128	$p = q = 1$	0.4690	0.4880
GEMSEC	CORA	20	128	$p = q = 1, \alpha = 0.5$	0.7200	0.7440
GEMSEC	CITESEER	20	128	$p = q = 1, \alpha = 0.5$	0.5170	0.5360
G-INTER*	CORA	20	128	$p = q = 1, \alpha = 0.5$	**0.7330**	**0.7560**
G-INTER*	CITESEER	20	128	$p = q = 1, \alpha = 0.5$	**0.5270**	**0.5470**

5   Conclusion

In our experiments, we prove the effectiveness of joint methods over node embedding algorithms with no particular community knowledge. Our extension of *GEMSEC* gives a slight edge in test results, which brings us to the idea that inter-cluster distance [33] can be an important area for further analysis. Comparing existing approaches for node embeddings, most of them do not incorporate cluster structure directly into optimization loss function, which we show as limitation and provide a base for further discussion.

Appendix

```
import torch

class KMeansClusteringLoss(torch.nn.Module):
    def __init__(self):
        super(KMeansClusteringLoss,self).__init__()

    def forward(self, encode_output, centroids):
        n = encode_output.shape[0]
        d = encode_output.shape[1]
        k = centroids.shape[0]

        z = encode_output.reshape(n,1,d)
        z = z.repeat(1,k,1)

        mu = centroids.reshape(1,k,d)
        mu = mu.repeat(n,1,1)

        inter = torch.cdist(centroids, centroids)
        tri_indices = torch.triu_indices(inter.shape[0],
                                  inter.shape[1], offset=1)
        min_inter = inter[tri_indices[0], tri_indices[1]].min()

        dist = (z-mu).norm(2,dim=2).reshape((n,k))
        dist = dist / min_inter
        loss = (dist.min(dim=1)[0]**2).mean()

        return alpha * (1/k * loss)
```

References

1. Antoniadis, P., Filntisis, P.P., Maragos, P.: Exploiting emotional dependencies with graph convolutional networks for facial expression recognition. In: 2021 16th IEEE International Conference on Automatic Face and Gesture Recognition (FG 2021), pp. 1–8. IEEE (2021)
2. Cai, H., Zheng, V.W., Chang, K.C.C.: A comprehensive survey of graph embedding: problems, techniques, and applications. IEEE Trans. Knowl. Data Eng. **30**(9), 1616–1637 (2018)
3. Cavallari, S., Zheng, V.W., Cai, H., Chang, K.C.C., Cambria, E.: Learning community embedding with community detection and node embedding on graphs. In: Proceedings of the 2017 ACM on Conference on Information and Knowledge Management, pp. 377–386 (2017)
4. Chen, J., Zhu, J., Song, L.: Stochastic training of graph convolutional networks with variance reduction. arXiv preprint arXiv:1710.10568 (2017)
5. Chen, J., Ma, T., Xiao, C.: FastGCN: fast learning with graph convolutional networks via importance sampling. In: Proceedings of International Conference on Learning Representations (ICLR) (2018). http://arxiv.org/abs/1801.10247'
6. Chiang, W.L., Liu, X., Si, S., Li, Y., Bengio, S., Hsieh, C.J.: Cluster-GCN: an efficient algorithm for training deep and large graph convolutional networks. In: Proceedings of the 25th ACM SIGKDD International Conference on Knowledge Discovery & Data Mining, pp. 257–266 (2019)
7. Gajdoš, P., Ježowicz, T., Uher, V., Dohnálek, P.: A parallel fruchterman-reingold algorithm optimized for fast visualization of large graphs and swarms of data. Swarm Evol. Comput. **26**, 56–63 (2016)

8. Grachev, A.M., Ignatov, D.I., Savchenko, A.V.: Neural networks compression for language modeling. In: Shankar, B.U., Ghosh, K., Mandal, D.P., Ray, S.S., Zhang, D., Pal, S.K. (eds.) PReMI 2017. LNCS, vol. 10597, pp. 351–357. Springer, Cham (2017). https://doi.org/10.1007/978-3-319-69900-4_44

9. Grover, A., Leskovec, J.: node2vec: scalable feature learning for networks. In: Proceedings of the 22nd ACM SIGKDD International Conference on Knowledge Discovery and Data Mining, pp. 855–864 (2016)

10. Hamerly, G., Elkan, C.: Learning the k in k-means. Adv. Neural Inf. Process. Syst. **16**, 281–288 (2004)

11. He, X., Cai, D., Shao, Y., Bao, H., Han, J.: Laplacian regularized gaussian mixture model for data clustering. IEEE Trans. Knowl. Data Eng. **23**(9), 1406–1418 (2010)

12. Hripcsak, G., Rothschild, A.S.: Agreement, the f-measure, and reliability in information retrieval. J. Am. Med. Inform. Assoc. **12**(3), 296–298 (2005)

13. Kanungo, T., Mount, D.M., Netanyahu, N.S., Piatko, C.D., Silverman, R., Wu, A.Y.: An efficient k-means clustering algorithm: analysis and implementation. IEEE Trans. Pattern Anal. Mach. Intell. **24**(7), 881–892 (2002)

14. Kingma, D.P., Ba, J.: Adam: A method for stochastic optimization. arXiv preprint arXiv:1412.6980 (2014)

15. Kipf, T.N., Welling, M.: Semi-supervised classification with graph convolutional networks. arXiv preprint arXiv:1609.02907 (2016)

16. Makarov, I.: Russian freight flights time prediction. In: 2019 2nd International Conference on Computer Applications & Information Security (ICCAIS), pp. 1–5. IEEE (2019)

17. Makarov, I., Gerasimova, O.: Link prediction regression for weighted co-authorship networks. In: Rojas, I., Joya, G., Catala, A. (eds.) IWANN 2019. LNCS, vol. 11507, pp. 667–677. Springer, Cham (2019). https://doi.org/10.1007/978-3-030-20518-8_55

18. Makarov, I., Gerasimova, O.: Predicting collaborations in co-authorship network. In: 2019 14th International Workshop on Semantic and Social Media Adaptation and Personalization (SMAP), pp. 1–6. IEEE (2019)

19. Makarov, I., Gerasimova, O., Sulimov, P., Zhukov, L.E.: Recommending co-authorship via network embeddings and feature engineering: the case of national research university higher school of economics. In: Proceedings of the 18th ACM/IEEE on Joint Conference on Digital Libraries, pp. 365–366. ACM (2018)

20. Makarov, I., Korovina, K., Kiselev, D.: JONNEE: joint network nodes and edges embedding. IEEE Access 1–14 (2021)

21. Makarov, I., Makarov, M., Kiselev, D.: Fusion of text and graph information for machine learning problems on networks. PeerJ Comput. Sci. **7**, e526 (2021)

22. Makarov, I., Oborevich, A.: Network embedding for cluster analysis. In: Proceedings of CINTI 2021, pp. 1–6. IEEE (2021)

23. Makarov, I., et al.: Temporal network embedding framework with causal anonymous walks representations. PeerJ Comput. Sci. **8**(e858), 1–27 (2022). https://doi.org/10.7717/peerj-cs.858

24. McCallum, A.: Cora dataset (2017). https://relational.fit.cvut.cz/dataset/CORA

25. McDaid, A.F., Greene, D., Hurley, N.: Normalized mutual information to evaluate overlapping community finding algorithms. arXiv preprint arXiv:1110.2515 (2011)

26. Mikolov, T., Sutskever, I., Chen, K., Corrado, G., Dean, J.: Distributed representations of words and phrases and their compositionality. arXiv preprint arXiv:1310.4546 (2013)

27. Perozzi, B., Al-Rfou, R., Skiena, S.: DeepWalk: online learning of social representations. In: Proceedings of the 20th ACM SIGKDD International Conference on Knowledge Discovery and Data Mining, pp. 701–710 (2014)
28. Rozemberczki, B., Davies, R., Sarkar, R., Sutton, C.: GEMSEC: graph embedding with self clustering. In: Proceedings of the 2019 IEEE/ACM International Conference on Advances in Social Networks Analysis and Mining 2019, pp. 65–72. ACM (2019)
29. Rustem, M.K., Makarov, I., Zhukov, L.E.: Predicting psychology attributes of a social network user. In: Proceedings of the Fourth Workshop on Experimental Economics and Machine Learning (EEML 2017), Dresden, Germany, 17–18 September 2017, pp. 1–7. CEUR WP (2017)
30. Savchenko, A.V.: Phonetic words decoding software in the problem of Russian speech recognition. Autom. Remote Control **74**, 1225–1232 (2013)
31. Savchenko, A.V.: Fast inference in convolutional neural networks based on sequential three-way decisions. Inf. Sci. **560**, 370–385 (2021)
32. Savchenko, A.V., Belova, N.S.: Statistical testing of segment homogeneity in classification of piecewise-regular objects. Int. J. Appl. Math. Comput. Sci. **25**(4), 915–925 (2015)
33. Savchenko, A.V., Savchenko, L.V.: Towards the creation of reliable voice control system based on a fuzzy approach. Pattern Recogn. Lett. **65**, 145–151 (2015)
34. Savchenko, A.V., Savchenko, L.V., Makarov, I.: Classifying emotions and engagement in online learning based on a single facial expression recognition neural network. IEEE Trans. Affect. Comput. **13**(4), 2132–2143 (2022)
35. Sokolova, A.D., Kharchevnikova, A.S., Savchenko, A.V.: Organizing multimedia data in video surveillance systems based on face verification with convolutional neural networks. In: van der Aalst, W.M.P., et al. (eds.) AIST 2017. LNCS, vol. 10716, pp. 223–230. Springer, Cham (2018). https://doi.org/10.1007/978-3-319-73013-4_20
36. Tandon, A., Albeshri, A., Thayananthan, V., Alhalabi, W., Radicchi, F., Fortunato, S.: Community detection in networks using graph embeddings. Phys. Rev. E **103**(2), 022316 (2021)
37. Yadav, J., Sharma, M.: A review of k-mean algorithm. Int. J. Eng. Trends Technol. **4**(7), 2972–2976 (2013)
38. Yang, Z., Cohen, W.W., Salakhutdinov, R.: Revisiting semi-supervised learning with graph embeddings (2016)

Boosting NSGA-II-Based Wrappers Speedup for High-Dimensional Data: Application to EEG Classification

Juan Carlos Gómez-López[1,2](✉), Daniel Castillo-Secilla[3], Dragi Kimovski[4], and Jesús González[1,2]

[1] Department of Computer Engineering, Automation and Robotics (ICAR), University of Granada, Granada, Spain
[2] Research Centre for Information and Communication Technologies (CITIC-UGR), University of Granada, Granada, Spain
{goloj,jesusgonzalez}@ugr.es
[3] Fujitsu Technology Solutions S.A., CoE Data Intelligence, Camino del Cerro de los Gamos, 1, Pozuelo de Alarcón, Madrid, Spain
daniel.castillosecilla@fujitsu.com
[4] Institute of Information Technology, University of Klagenfurt, Klagenfurt, Austria
dragi.kimovski@aau.at

Abstract. The considerable technological evolution during the last deca-des has made it possible to deal with biological datasets of increasing higher dimensionality, such as those used in BCI applications. Thus, techniques such as feature selection, which allow obtaining the underlying information of these datasets by removing those features considered redundant or noisy, have emerged. Over the years, wrapper approaches based on evolutionary algorithms have gained great relevance, as they have proven to be one of the best procedures to tackle this problem, with NSGA-II being one of the most used search strategies. Historically, these procedures have presented a well-known bottleneck in the evaluation method. However, a more significant bottleneck appears when dealing with high-dimensional datasets, which lies in the application of the NSGA-II's selection method to very large populations. For this reason, this paper aims to alleviate this problem and, consequently, develop a parallel strategy able to reach a superlinear speedup.

Keywords: High-dimensionality · Feature Selection · EEG Classification · Wrapper · Bottleneck · NSGA-II · Multi-population

1 Introduction

The growing technological evolution has led to a considerable increase in computational power, which has caused fields of study such as bioinformatics to experience a boom in the interest of countless researchers [13]. Bioinformatics focuses on studying biological datasets through computer engineering. Among

I. Rojas et al. (Eds.): IWANN 2023, LNCS 14134, pp. 80–91, 2023.
https://doi.org/10.1007/978-3-031-43085-5_7

the many datasets processed in different disciplines of bioinformatics are Electroencephalograms (EEGs), which are the data of interest in this paper. Briefly, EEGs, obtained through electrodes placed on the scalp in a non-invasive way, are the signals of study in Electroencephalography, a discipline framed within Brain-Computer Interface (BCI) applications.

One of the most recurrent problems when working with biological datasets is the *curse of dimensionality* [2]. The thousands or even tens of thousands of features that compose these datasets distort the final classification model, necessitating some kind of Feature Selection (FS) technique [10] that eliminates those considered noise or irrelevant. Over the years, different approaches to solving FS problems have been presented [4]. However, two have become the winning horses when this problem is addressed: wrapper and filter approaches. The former search for the best subset of features along with an induction algorithm that evaluates each possible solution to the problem. On the contrary, filter approaches select features using a preprocessing step, so they are independent of the induction algorithm used. Although filters are computationally less demanding than wrappers, they usually achieve lower-quality solutions than the latter, largely due to the bias introduced by preprocessing. In this context, wrappers perform the search without any prior bias, making them an effective FS technique for high-dimensional problems.

Essentially, a wrapper consists of two parts: a strategy that guides the search and an induction algorithm, in this case, framed in classification, that evaluates all the possible solutions to the problem. The Non-dominated Sorting Genetic Algorithm II (NSGA-II) has been used as the search strategy here, which evolves a population of individuals, or potential solutions, codifying different FSes. In addition, the evaluation of all the population individuals along generations is in charge of the k-Nearest Neighbors (k-NN) classifier. The choice of NSGA-II and k-NN was motivated by their good performance in FS problems [12].

Historically, wrappers based on Evolutionary Algorithms (EAs) have suffered from a well-known bottleneck caused by the evaluation method, i.e., the fitness function. Nevertheless, a new and even narrower bottleneck appears when applying NSGA-II to select features from high-dimensional datasets since, as detailed in [7], the population size for an EA used within a wrapper procedure for a given dataset should be at least the number of features of such a dataset, to escape from local optima. For this reason, the population will consist of thousands or even tens of thousands of individuals for high-dimensional problems, leading the selection in NSGA-II to become a bottleneck still more relevant than the one produced by the evaluation method. To the best of our knowledge, no study has addressed this issue yet, as most of them focus on optimizing the evaluation method. Thus, this paper aims to alleviate the NSGA-II selection time in a problem of EEG classification. Additionally, the procedure has been accelerated through a parallel strategy, obtaining a superlinear speedup.

After this introduction, Sect. 2 describes the datasets and execution platform used to run all the experiments, along with the details of the NSGA-II-based wrapper. Then, Sect. 3 exposes the new bottleneck of NSGA-II for high-

Table 1. Class distribution in subjects 104, 107, and 110 from the University of Essex database.

Subject	Dataset	Left hand	Right hand	Feet
104	Training	60	66	52
	Test	55	65	58
107	Training	56	57	65
	Test	58	58	62
110	Training	58	60	60
	Test	58	60	60

dimensional problems, which is alleviated in Sect. 4. Next, in Sect. 5, an efficient parallel island-based scheme is applied to the wrapper. Finally, Sect. 6 is focused on the quality of the solutions found, concluding the paper in Sect. 7.

2   Relevant Details

2.1   Datasets and Execution Platform Description

For the experiments, three datasets recorded at the BCI Laboratory of the University of Essex are used, which correspond to three human subjects coded as 104, 107, and 110. Specifically, these signals obtain SensoriMotor Rhythms (SMRs) from the sensorimotor cortex, which are modulated by the actual movement or Motor Imagery (MI). The process for recording the EEGs is described in [1], where the signals are acquired from 32 electrodes at 256 Hz with the BioSemi system [3] and processed using the MultiResolution Analysis (MRA) method. After that, a feature extraction step is carried out to build a final dataset of 356 signals of 3 600 features each, where the first 178 samples are used for training while the remaining 178 samples are kept for testing purposes. The datasets contain EEG signals that can belong to three different types of movements: left hand, right hand, and feet. Their class distribution is detailed in Table 1.

All the experiments have been run on a Personal Computer (PC) with two Intel Xeon Silver 4214 CPUs and a Quadro RTX 6000 Graphic Processing Unit (GPU). The operating system used is CentOS (v7.9.2009). PC energy consumption has been obtained by an Arduino-based energy metering system specially developed for the experiments, which captures active power, voltage, current, power factor, and energy consumption information each $200ms$.

2.2   NSGA-II-Based Wrapper Overview

Over the years, Python has become one of the most widely used languages by the Machine-Learning community [9]. This popularity is due to several reasons, such as ease of programming or the number of available libraries. For this reason, the wrapper presented in [7], which has already proved to work quite well analyzing

Table 2. Hyperparameter values used for the experiments.

Hyperparameter	Value
Crossover probability (P_c)	0.8
Mutation probability (P_m)	0.00028
Number of executions (N_e)	30
Number of generations (N_g)	500
Number of individuals (N_i)	3 840

several datasets, including the high-dimensional ones described above, has been implemented in Python to execute all the experiments performed in this paper. On the one hand, the *Distributed Evolutionary Algorithm in Python* (DEAP) library has been used to implement NSGA-II since it is one of the most widely used libraries for the scientific community in this kind of procedure. On the other hand, k-NN has been developed by using the *Scikit-learn* library. From now on, this wrapper implemented in Python will take the name of Single-Population Multi-Objective Evolutionary Wrapper (SPMOEW).

A significant aspect when executing any Machine-Learning procedure is the adjustment of its hyperparameters. In this sense, the value of the wrapper's hyperparameters can be found in Table 2. N_i has been set to 3 840 for a relevant DEAP restriction. It is important to remember that, as previously stated, N_i should be equal to the number of features of the dataset to be analyzed. Furthermore, to apply selection based on dominance in DEAP, the library requires that the population size be a multiple of 4 [6]. In contrast, P_m refers to the probability that each feature selected by individuals belonging to the offspring may undergo a mutation. With this in mind, the value of P_m has been adjusted according to the study performed in [8], where it was concluded that a suitable value for this hyperparameter could be $1/N_f$, with N_f being the number of features in the dataset. Finally, the value of k for k-NN has been adjusted to the square root of the number of samples in the dataset, as was done in [7].

3 Analysis of NSGA-II-Based Wrapper Bottlenecks for High-Dimensional Problems

As discussed in the introduction, the evaluation method is a well-known bottleneck in wrapper procedures. However, the selection procedure used within NSGA-II causes an even more relevant bottleneck when selecting features from high-dimensional datasets. This issue, due to the need for very large population sizes (of at least the number of features in the dataset [7]) to find appropriate solutions, has not been addressed yet, at least to the best of our knowledge.

The first question to be addressed is to know the exact percentage that the NSGA-II selection time represents with respect to the total execution time. In this regard, Fig. 1 depicts the percentage of the total runtime dedicated to

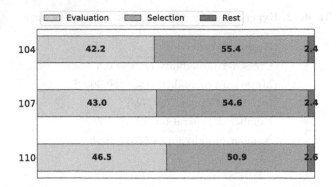

Fig. 1. Percentage of the total time dedicated to the evaluation, selection, and the rest of the functions of SPMOEW for each Essex EEG dataset.

the evaluation, selection, and the rest of the functions of SPMOEW for each Essex EEG dataset. It can be appreciated that the selection time becomes the main bottleneck of the procedure, whose percentage exceeds 50% of the total execution time, thus leaving the evaluation method as a secondary issue, which is rather surprising. Thus, the following section is centered on alleviating this new bottleneck.

4   Reducing the Selection Time

The NSGA-II's selection method is based on the dominance between the different individuals, which implies that all of them have to be compared to each other. In this sense, the number of comparisons needed is:

$$N_o \cdot \frac{N_i \cdot (N_i - 1)}{2} \tag{1}$$

where N_o is the number of objectives. Thus, the number of comparison increases quadratically as the population size grows. However, the mere fact of distributing individuals into subpopulations should divide the selection time among the number of subpopulations used. For this reason, a Multi-Population Multi-Objective Evolutionary Wrapper (MPMOEW) is proposed. The basic idea is to divide the SPMOEW's single population into different subpopulations, each evolving independently and exchanging individuals every few generations (migration). In this case, 25% of individuals of the Pareto front are exchanged every 20 generations following a uni-directional ring topology, as proposed in [11].

MPMOEW has been run with the same hyperparameters as SPMOEW (shown in Table 2), distributing the individuals into 1, 2, 4, 8, 16, and 24 subpopulations[1]. It is important to note that only subject 104 has been used in all the efficiency-related experiments. As the three Essex datasets have the same

---

[1] The last value of N_{sp} is set to 24 since that is the maximum number of CPU physical cores where the wrapper is executed.

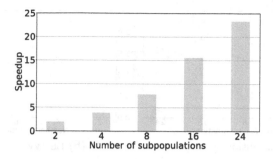

Fig. 2. Speedup achieved by the selection method when individuals are distributed into subpopulations for subject 104.

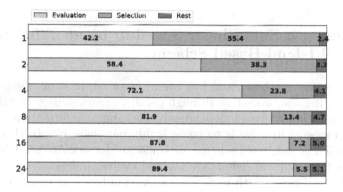

Fig. 3. Percentage of the total time spent on evaluation, selection, and other MPMOEW functions when individuals are distributed into subpopulations for subject 104.

number of features and samples, as well as a similar selection and evaluation time, analyzing just one of them is enough. However, all three datasets will be studied in Sect. 6 to check the quality of the solutions found by the wrapper.

Figure 2 shows how a linear speedup is achieved for the selection method when individuals are distributed into an increasing number of subpopulations for subject 104, as expected. Also, Fig. 3 illustrates the percentage of the total time spent on evaluation, selection, and other MPMOEW functions when individuals are distributed into subpopulations, showing that the selection time becomes practically negligible as the number of subpopulations increases, leaving the evaluation method as the only bottleneck of the procedure. Nonetheless, the percentage devoted to the remaining MPMOEW functions is also growing slightly, which is mainly due to the increase in the number of migrations. Finally, Fig. 4 shows the speedup in execution time and the savings in energy consumption achieved by MPMOEW when individuals are distributed into subpopulations for subject 104. It can be appreciated that, even though the wrapper is still running sequentially, a speedup of almost ×2.5 has been attained by the mere fact of distributing individuals into subpopulations, thanks to the decrease in the selection time.

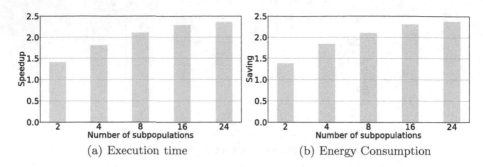

(a) Execution time   (b) Energy Consumption

Fig. 4. Speedup in execution time and savings in energy consumption achieved by MPMOEW when individuals are distributed into subpopulations for subject 104.

5   Obtaining a Superlinear Speedup by an Efficient Parallel Island-Based Scheme

Once the selection method issue has been mitigated, it is interesting to continue accelerating the procedure through parallel programming. In MPMOEW, every subpopulation evolves independently, exchanging information only in the migration process, which leads to being highly parallelizable. In this sense, an efficient parallel island-based scheme for Uniform Memory Access (UMA) architectures has been applied, where each island evolves independently in each of the CPU physical cores, with their communication being necessary only during the migration of individuals. The new procedure will be named Island-based Multi-Population Multi-Objective Evolutionary Wrapper (IMPMOEW).

The main objective is to ensure that all CPU cores are computing a task 100% of the time and that communication between them is asynchronous. This way, a CPU core will not remain idle waiting to receive data from another core. Figure 5 illustrates the developed scheme where each core is in charge of the complete evolution of a subpopulation. For this purpose, the parent process (P_0) creates as many processes as there are subpopulations, waiting for all the child processes to finish. A relevant issue here is the use of processes instead of threads. The Python interpreter uses the Global Interpreter Lock (GIL) [5], which allows the execution of only one thread simultaneously, so it is impossible to execute multiple threads that are managed by the same process as C++ does. In this context, real parallelism is provided in Python through the *Multiprocessing* library, where different processes are created to be executed in different cores at the same time. This type of parallelism is not as efficient as thread-based parallelism, but it is the only option at the moment. The two main drawbacks of process-based parallelism are the amount of memory needed and the overhead caused by the communication between processes. For the former little can be done, as it will depend on the number of individuals and the dataset size. Conversely, to overcome the second drawback, the code has been designed so that each subpopulation evolves independently, with communication between them being necessary to migrate individuals only.

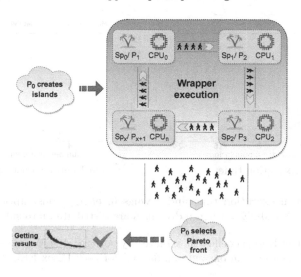

Fig. 5. IMPMOEW. *Island-based* scheme applied to MPMOEW for UMA architectures. Each subpopulation evolves independently on each of the CPU cores, exchanging individuals via asynchronous migrations.

Asynchronous migrations have been developed using the *Queue* class of the *Multiprocessing* library. These queues are responsible for communicating the different subpopulations in the migration process. At this moment, each subpopulation puts its migrants in its sending queue. By contrast, the reception of migrants has been simplified to prevent blocking the subpopulations by waiting for migrants. Each generation, the subpopulation checks if there are migrants in its receiving queue. If so, the migrants join the population, and then the selection of the best ones is applied. If there are no migrants in the queue, that generation proceeds as usual. This process will be repeated until the maximum number of established generations is completed or until convergence has been reached.

In this context, Fig. 6 illustrates the speedup in execution time and savings in energy consumption achieved by MPMOEW and IMPMOEW when individuals are distributed into subpopulations. The first point to note is that the speedup in execution time is higher than the savings in energy consumption because the greater the number of subpopulations, the more cores are used in parallel, which leads the energy consumption to grow accordingly. By contrast, the most relevant highlight is that a superlinear speedup in execution time is obtained. IMPMOEW aims to ensure that all the physical cores used are computing most of the time, so a speedup similar to the N_{sp} value should be obtained in this respect. If this is added to the fact that the selection time is divided by the number of subpopulations, this superlinear speedup is explained. Specifically, the speedup is around $\times 35$ with 24 subpopulations.

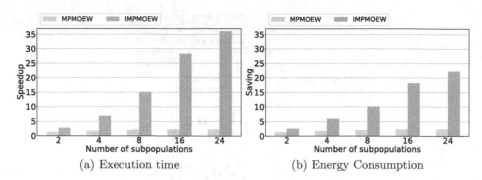

(a) Execution time   (b) Energy Consumption

Fig. 6. Speedup in execution time and savings in energy consumption achieved by MPMOEW and IMPMOEW when individuals are distributed into subpopulations.

Table 3. Mean test Kappa coefficient achieved by both SPMOEW (one subpopulation) and IMPMOEW (more than one subpopulation) for each Essex EEG dataset.

N_{sp}	Dataset		
	104	107	110
1	0.732 ± 0.022	0.562 ± 0.020	0.611 ± 0.021
2	0.731 ± 0.033	0.545 ± 0.027	0.601 ± 0.019
4	0.710 ± 0.035	0.540 ± 0.024	0.589 ± 0.037
8	0.725 ± 0.028	0.538 ± 0.025	0.597 ± 0.025
16	0.719 ± 0.025	0.548 ± 0.024	0.593 ± 0.024
24	0.710 ± 0.029	0.524 ± 0.030	0.593 ± 0.043

6   Statistical Study of the Quality of the Solutions Found by the Wrapper

Although the wrapper used as the basis for the development of SPMOEW has been proven to find quality solutions for the analyzed datasets, it should be verified that these solutions are statistically similar to those obtained when individuals are distributed into subpopulations. In this regard, Table 3 shows the mean test Kappa coefficient achieved by both SPMOEW (one subpopulation) and IMPMOEW (more than one subpopulation) for each Essex EEG dataset. It can be observed that, regardless of the wrapper used, similar quality solutions are achieved. However, even though the results obtained by using different values of N_{sp} seem similar for each dataset, this resemblance should be assessed statistically.

Accuracies obtained for subjects 104 and 107 follow a normal distribution, so their homoscedasticity should be tested. In this case, Bartlett's test has verified that all data are homoscedastic. Thus, the ANOVA test can be applied. In contrast, since not all the accuracies obtained for subject 110 follow a normal distribution, a Kruskal-Wallis non-parametric test has been applied. With this

Table 4. p-values obtained after comparing the accuracies when using different values of N_{sp} with the specific statistic test for each Essex EEG dataset.

Dataset	Test	p-value
104	ANOVA	0.16
107	ANOVA	0.14
110	Kruskal-Wallis	0.22

Table 5. Mean number of selected features achieved by both SPMOEW (one subpopulation) and IMPMOEW (more than one subpopulations) for each Essex EEG dataset.

N_{sp}	Dataset		
	104	107	110
1	24.132 ± 18.451	14.036 ± 7.910	13.667 ± 6.369
2	23.733 ± 13.830	14.400 ± 5.014	15.403 ± 8.089
4	21.333 ± 9.891	16.250 ± 14.489	17.467 ± 12.997
8	21.800 ± 13.383	13.867 ± 5.654	12.200 ± 5.588
16	19.933 ± 11.863	11.733 ± 4.234	11.533 ± 4.4992
24	20.362 ± 9.843	13.801 ± 6.348	18.133 ± 15.126

in mind, the p-values obtained for each dataset are shown in Table 4. As all the p-values are higher than a significance level $\alpha = 0.05$, it can be concluded that IMPMOEW obtains similar solutions to SPMOEW with 95% confidence.

Finally, Table 5 shows the mean number of features selected by both SPMOEW and IMPMOEW for each Essex EEG dataset. It can be seen how the wrapper can reduce the number of features considerably, regardless of the value of N_{sp}.

7 Conclusions

Although, a priori, it could be thought that the most computationally demanding part of an NSGA-II-based wrapper should be the evaluation of solutions, the analysis of such a wrapper applied to high-dimensional data has uncovered an even narrower bottleneck caused by the NSGA-II selection mechanism, whose computing time depends on the population size quadratically. Since the population size should be set to at least the number of features in the dataset, the selection method can even suppose more computing time than the evaluation of solutions, as exposed in Sect. 3. To address this problem, a multi-population approach has been adopted since, in this way, the selection time is divided by the number of subpopulations used. Additionally, as such kind of approach is highly parallelizable, an island-based scheme has been implemented so that each subpopulation evolves independently in each CPU physical core, with communication performed asynchronously and only for the migration process. In this way, a superlinear speedup in wrapper performance is attained while obtaining statistically similar solutions regardless of the number of subpopulations used.

It is important to emphasize that the wrapper procedure has been considerably accelerated without addressing the evaluation bottleneck, which demands about 90% of the total execution time when individuals are split among 24 subpopulations. Thus, there is still room to improve the wrapper performance, as a greater speedup could be obtained by focusing efforts on improving the performance of k-NN. The use of GPUs, distributed systems, or the implementation of k-NN in a more efficient language such as C/C++ are ideas to consider for the future.

Besides, migration hyperparameters have been set according to the guidelines of several studies, although none of them used the large population sizes required by the high-dimensional data processed in this paper. Thus, it is quite possible that these hyperparameter values can be fine-tuned to improve the solutions found. The in-depth study of their effect on wrappers applied to high-dimensional datasets is an open front where interesting research could be done.

Acknowledgment. This research has been funded by the Spanish Ministry of Science, Innovation, and Universities (grants PGC2018-098813-B-C31 and PID2022-137461NB-C31) and the ERDF fund. We would like to thank the BCI laboratory of the University of Essex, especially Dr. John Q. Gan, for allowing us to use their datasets.

References

1. Asensio-Cubero, J., Gan, J.Q., Palaniappan, R.: Multiresolution analysis over simple graphs for brain computer interfaces. J. Neural Eng. **10**(4), 21–26 (2013). http://doi.org/10.1088/1741-2560/10/4/046014
2. Bellman, R.E.: Adaptive Control Processes: A Guided Tour. Princeton University Press, Princeton (1961)
3. BioSemi: Biosemi System. https://www.biosemi.com/products.htm. Accessed 02 Dec 2019
4. Dash, M., Liu, H.: Feature selection for classification. Intell. Data Anal. **1**, 131–156 (1997). https://doi.org/10.1016/S1088-467X(97)00008-5
5. Karami, E.M.: Explaining the python global interpreter lock. https://www.obytes.com/blog/explaining-the-python-global-interpreter-lock. Accessed 24 Sept 2020
6. Fortin, F.A., De Rainville, F.M., Gardner, M.A., M. P., Gagn, C.: Tournament selection based on dominance. https://deap.readthedocs.io/en/master/api/tools.html#deap.tools.selTournamentDCD. Accessed 14 Sept 2020
7. González, J., Ortega, J., Damas, M., Martín-Smith, P., Gan, J.Q.: A new multiobjective wrapper method for feature selection - accuracy and stability analysis for BCI. Neurocomputing, 407–418 (2019). https://doi.org/10.1016/j.neucom.2019.01.017
8. Muhlenbein, H., Schlierkamp-Voosen, D.: Optimal interaction of mutation and crossover in the breeder genetic algorithm. In: Proceedings of the Fifth International Conference on Genetic Algorithms, p. 648. Citeseer, Morgan Kaufmann Publishers Inc., San Francisco (1993)

9. Pérez, F., Granger, B.E., Hunter, J.D.: Python: an ecosystem for scientific computing. Comput. Sci. Eng. **13**(2), 13–21 (2011). https://doi.org/10.1109/MCSE.2010.119
10. Tang, J., Alelyani, S., Liu, H.: Feature selection for classification: a review. Data Classif. Algorithms Appl. 37–64 (2014)
11. Tomassini, M.: Spatially Structured Evolutionary Algorithms: Artificial Evolution in Space and Time. Springer, Heidelberg (2006). https://doi.org/10.1007/3-540-29938-6
12. Xue, B., Zhang, M., Browne, W.N., Yao, X.: A survey on evolutionary computation approaches to feature selection. IEEE Trans. Evol. Comput. **20**, 606–626 (2016). http://doi.org/10.1109/TEVC.2015.2504420
13. Yin, Z., Lan, H., Tan, G., Lu, M., Vasilakos, A.V., Liu, W.: Computing platforms for big biological data analytics: perspectives and challenges. Comput. Struct. Biotechnol. J. **15**, 403–411 (2017). https://doi.org/10.1016/j.csbj.2017.07.004

Extending Drift Detection Methods to Identify When Exactly the Change Happened

Markus Vieth[✉][iD], Alexander Schulz[iD], and Barbara Hammer[iD]

Bielefeld University, Bielefeld, Germany
mvieth@techfak.uni-bielefeld.de

Abstract. Data changing, or drifting, over time is a major problem when using classical machine learning on data streams. One approach to deal with this is to detect changes and react accordingly, for example by retraining the model. Most existing drift detection methods only report that a drift has happened between two time windows, but not when exactly. In this paper, we present extensions for three popular methods, MMDDDM, HDDDM, and D3, to determine precisely when the drift happened, i.e. between which samples. One major advantage of our extensions is that no additional hyperparameters are required. In experiments, with an emphasis on high-dimensional, real-world datasets, we show that they successfully identify when the drifts happen, and in some cases even lead to fewer false positives and false negatives (undetected drifts), while making the methods only negligibly slower. In general, our extensions may enable a faster, more robust adaptation to changes in data streams.

Keywords: Data Streams · Drift Detection · Concept Drift

1 Introduction

When applying machine learning to data streams, one major problem are changes of the data over time, leading to a worsening of the model performance. Such changes are present in many different domains, such as in IoT [1] or in business applications [9]. Changes over time can be modelled in different ways, e.g. by covariate shift [15], assuming different marginal distributions between a training and test set, or possibly several changes of the former in the course of a data stream [5].

Approaches to deal with drift in data streams can be divided into active and passive [4]: passive approaches continuously learn on the new data when it becomes available, while active approaches aim to identify when an update of the

The project has been funded by the Ministry of Culture and Science of the Federal State North Rhine-Westphalia in the frame of the project RoSe in the AI-graduate-school https://dataninja.nrw.

I. Rojas et al. (Eds.): IWANN 2023, LNCS 14134, pp. 92–104, 2023.
https://doi.org/10.1007/978-3-031-43085-5_8

model is necessary by detecting changes, or drifts, in the data. Focusing on the active setting, many existing drift detection methods work with time windows [12] and only try to detect whether a drift has happened in between them, but do not identify when exactly in time, or between which consecutive data samples, the change occurred. Depending on the method, only a rough estimation is made. However, knowing the "when" is an advantage: if the drift is assumed to be earlier than it actually was, samples from the old data distribution are mixed into the samples from the new distribution, possibly leading to a worse detection of consecutive drifts and a poorer model performance. In the other direction, if the drift is assumed to have happened later than in reality, samples from the new distribution are incorrectly believed to come from the old distribution, again leading to a possibly worse model because of the smaller training set.

In this work, we consider unsupervised methods which is the more general setting, meaning that labels for the data are not required for drift detection. We also assume that drifts can occur between any two consecutive samples, and do not require that the data arrives in batches where all samples are guaranteed to come from the same concept/distribution.

We present extensions to the three popular drift detection methods D3, HDDDM, and MMDDDM [3,6,7], that enable a flexible and precise change point detection. These methods were chosen because they test for drifts in very different ways, and the basic idea of our extensions fits for all of them. We compare with the, to our knowledge, only existing method that identifies the precise change point of the drift, the Spectral Drift Detection Method [8]. We evaluate our extensions, with an emphasis on high-dimensional, real-world datasets, and show that in all cases and for each method our extension improves the localization of drift in time drastically. In some cases, this even improves the overall drift detection performance, as we demonstrate experimentally[1].

2 Problem Formalization

We assume a given data stream $D = \{x_0, x_1, \ldots x_t, \ldots\}$ that consists of unlabeled data observations over time t. We further assume that at specific and unknown time points t_i the data generation distribution $p_{t_i}(x)$ changes, such that $p_{t_i}(x) \neq p_{t_i-1}(x)$. Then we say that a drift occurred at time point t_i.

Many methods from literature [3,6,7] apply drift detection between time windows only, thereby analysing drift only at the given time point between the two time windows. More details are in the related work Sect. 3. We propose a method to augment these approaches by a fine-grained drift detection scheme.

3 Related Work

In the following, we recall the three popular drift detection methods D3, HDDDM, and MMDDDM that we utilize in our work and describe in how far

---

[1] https://github.com/mvieth/extending-drift-detection-methods.

they can detect the precise drift location. We also recall the change point detection approach SDDM that is designed to precisely detect the drift location.

The Discriminative Drift Detector (D3, [6]) uses a classifier to identify drift in the most recent data. It divides the window of the latest $w + w\rho$ samples into two parts, w older and $w\rho$ newer ones, and assigns different labels to each part (e.g. 0 and 1). If a classifier can accurately distinguish between the two parts, as measured by the AUC being larger than a given threshold, they likely come from different distributions. In this case a drift is detected and the old w samples are removed. Otherwise, only the oldest $w\rho$ samples are removed. In both cases, the remaining samples are shifted to the left and drift detection is applied again as soon as the window is filled up with new samples.

Accordingly, D3 can detect drifts at maximum every $w\rho$ time steps and not in between. Following the original article, we utilize a logistic regression model as the classifier (we use the implementation from scikit-learn [13]). D3 has three hyperparameters: the number of stored older samples w, of stored newer samples $w\rho$ as a ratio of the older ones, and the AUC threshold. We choose $\rho = 1$ to avoid class imbalance problems.

The Hellinger Distance Drift Detection Method (HDDDM, [3]) calculates for each feature k the Hellinger distance between histograms P_k and Q_k and takes the average over those w.r.t. the features, where P_k is based on the most recent batch of data and Q_k on the older ones:

$$\delta_H(t) = \frac{1}{d} \sum_{k=1}^{d} \sqrt{\sum_{i=1}^{b} \left(\sqrt{\frac{P_{i,k}}{\sum_{j=1}^{b} P_{j,k}}} - \sqrt{\frac{Q_{i,k}}{\sum_{j=1}^{b} Q_{j,k}}} \right)^2}$$

Subsequently, the difference between the last two measurements is computed as $\epsilon(t) = \delta_H(t) - \delta_H(t - 1)$ and drift is detected if $|\epsilon(t)| > \beta(t)$, where $\beta(t)$ is an adaptive threshold.

Consequently, HDDDM can detect drift only between batches of data. Hyperparameters constitute the batch size N, the number of bins b used in the histograms (default $b = \lfloor \sqrt{N} \rfloor$) and the choice of calculating $\beta(t)$ (two strategies are presented) as well as the according hyperparameter.

The Maximum Mean Discrepancy Drift Detection Method (MMDDDM, [7, 14]) uses the Maximum Mean Discrepancy (MMD) two sample test (here in its unbiased form) for drift detection:

$$MMD_u^2 = \frac{1}{m^2 - m} \sum_i^m \sum_{j,i \neq j}^m k(x_i, x_j) - \frac{2}{nm} \sum_i^m \sum_j^n k(x_i, y_j) + \frac{1}{n^2 - n} \sum_i^n \sum_{j,i \neq j}^n k(y_i, y_j)$$

where $x_1, \ldots, x_m$ constitute the first sample set and $y_1, \ldots, y_n$ the second. This involves computing a kernel k, usually a Gaussian kernel, on all pairs of stored samples. When the MMD is high enough, a drift is detected, and all samples from the old sample set deleted. MMDDDM has 3 hyperparameters: how many samples are used for the "after-the-split" sample set, the number of samples between each MMD test, and a threshold determining when a drift is detected. By default, we use the same value for the first two hyperparameters.

The Spectral Drift Detection Method (SDDM, [8]), first computes the similarity between all recent samples and compiles them in a kernel matrix. Then it applies a spectral clustering algorithm to this matrix and identifies change points by training a decision tree on the most important eigenvectors. The number of change points (or leaves in the decision tree) is determined via cross validation. This method has a rather large number of hyperparameters: the maximum number of stored samples, the minimum number of samples necessary before the stored data is tested for drifts, how often the method checks for drifts (e.g. every 50 samples), the number of eigenvectors used, the percentage of data used for testing in cross validation, the number of iterations in the cross validation, up to how many drift points (or different concepts) are considered possible in the data window during the cross validation, and the minimum number of samples each decision tree leaf must contain (this also determines how many samples must at least lie between two drifts). In [8], both a Gaussian kernel and a Moment Tree kernel are tested. In this paper, we only use the Gaussian kernel because this facilitates comparison with the MMDDDM, and because a Moment Tree kernel would require tuning additional hyperparameters. SDDM is a special case in that it can report the same drift multiple times, with slightly different location estimates.

4   Precise Localization of Change Points in Time

In this section, we describe our extensions to the drift detection methods D3, HDDDM, and MMDDDM. Important to note is that we only add additional steps when a drift was detected. Until that point, the methods are unchanged. What we do in general for all three methods, is going through all possible points where the drift could have happened, compute a measure or statistic describing how likely it is that the drift happened at that point, and choose the point with the highest value as the estimated drift point. Then, instead of the usual scheme of discarding old data as described above, we instead drop all samples received before our estimated drift point. This can even influence future detections in a positive way by making sure that no old samples, but all new samples are kept. As a side note, we do not have to account for the multiple-comparisons-problem since at this point we do not test whether a drift has happened, instead we know that this is the case and we only want to determine *when* it happened.

4.1   D3

For D3, we have the classifier that was used to determine whether there was a drift or not, which was trained to distinguish between old and new samples (before and after the drift). While some training samples are mislabeled (sample from before the drift labelled as "new" or sample from after the drift labelled as "old") unless the division into old and new data window fits perfectly by chance, we can still assume the classifier to be good enough since it passed the given AUC threshold. From the computation of the AUC we already have the probability of belonging to the new distribution for each of the recent, stored samples. Then we evaluate all possible split points: a naive idea would be to compute the mean probability of all samples newer and of all samples older than each tested split point, the split point with the highest difference of the means would be the estimated drift point. However, we need a correction for split points which have only very few newer (or older) samples, since the mean of these few samples may be by chance very high (or low, respectively), which would lead to an incorrectly estimated drift point. An option to solve this could be to only allow splits which have at least a certain number of points on each side. But this would introduce a new, undesirable hyperparameter (the minimum number of points), and could in some cases exclude the correct split point from consideration. Instead, we use the t-statistic from Welch's t-test to determine the optimal split point:

$$t = \frac{\bar{x}_n - \bar{x}_o}{\sqrt{\frac{s_n^2}{n_n} + \frac{s_o^2}{n_o}}}$$

where $\bar{x}_n, \bar{x}_o$ are the mean probabilities, s_n, s_o the corresponding standard deviations, and n_n, n_o the number of samples newer and older than the tested split point, respectively. Figure 1 shows an example of finding the best split point with the t-statistic.

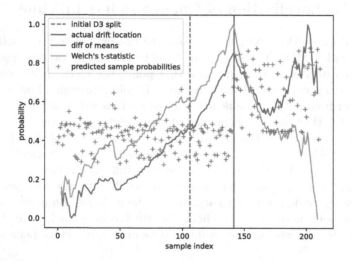

Fig. 1. D3 has detected a drift in the rialto dataset. All samples left of the red dashed line are labelled as old, all samples right of it as new. By looking at the predicted sample probabilities however, we can give a better estimate of the drift point. The naive difference of means would give a wrong estimate of the drift point, but Welch's t-statistic has its maximum at the actual drift point. (Color figure online)

Finally, instead of discarding all samples labelled as "old" (as in the original D3 algorithm), we discard all samples from before the estimated drift point. This means that we potentially keep samples mislabelled as "old" or drop samples mislabelled as "new".

4.2   HDDDM

For HDDDM, we have the two histograms, representing the data before and after the detected drift, respectively. For each stored sample, we determine whether it came from the old distribution, then we label it with 0, or from the new distribution, then we label it with 1. To do this, we check for each feature into which histogram bin the sample falls, and average the corresponding bin values for each histogram. Now we want to split this, time-wise, so that the one sample set contains as many zeros as possible, and the other as many ones as possible. At each possible split point, we run Fisher's exact test (one-sided), calculating the probability of having more samples with label 1 after the tested split point (and equally fewer samples with label 1 before). The split points with the lowest probability wins.

We can limit the number of possible split points we have to test: the drift could have happened within the samples used for the "new" histogram, or the drift could have happened within the samples that belonged to the "new" histogram in the last drift test, but now belong to the "old" histogram. This can be the case if not enough samples from the new concept had arrived yet for the last drift test. More concretely, if the "new" histogram contains n samples in each drift test, we only have to search for the true drift point within the $2n$ newest samples. If the drift had happened before that, it would have been detected in the previous drift test already. Like for D3, we discard all samples older than the estimated drift point.

4.3   MMDDDM

In the case of MMDDDM, the kernel matrix of the stored, recent samples has already been computed. Now we can simply compute the MMD for all possible splits, i.e. for all possible divisions along time into two sample sets. We expect the MMD to be highest when we split the recent samples at the actual drift point. Again, we have to account for the high uncertainty when one of the two sample sets (either the number of samples before or after the tested drift point) is very small. Then it can happen that the MMD is by chance higher than the MMD at the true drift point. We formulate a "corrected" MMD measure, similar to the previously used t-statistic:

$$MMD_t^2 = \frac{MMD_u^2}{\sqrt{\frac{s_x^2}{m(m-1)} + \frac{s_{xy}^2}{mn} + \frac{s_y^2}{n(n-1)}}}$$

where MMD_u^2 is the unbiased MMD as described above, and s_x, s_{xy}, and s_y are the standard deviations among $k(x,x)$, $k(x,y)$, and $k(y,y)$, respectively. An example of the precise localization of the drift point using the corrected MMD is presented in Fig. 2.

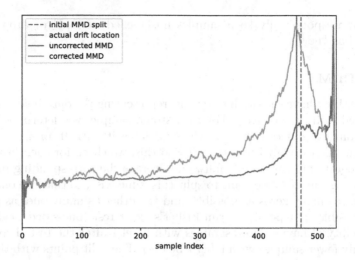

Fig. 2. MMDDDM has detected a drift in the mnist dataset. The red dashed line shows the initial division into pre- and post-drift sample sets. The true drift point is a few samples before that, where the corrected MMD is highest. In contrast, the uncorrected MMD has its maximum at the wrong split point. (Color figure online)

5   Experiments

In the following, we describe our experimental setup including the utilized data sets and evaluation scores. We detail our approach to select the best hyperparameters for each method and inspect the impact of our change point detection scheme on classical drift detection in Sect. 5.2. Finally, we present our main results for change point detection in Sect. 5.3.

5.1   Experimental Setup

We test all methods on data streams of length 1600, with 2 drifts introduced at known points. The first drift occurs between sample 380 and sample 700, the second drift between sample 900 and sample 1220, this way all three concepts have roughly equal length. The exact drift time points are random. Labels are not available to the drift detection methods. The samples are randomly chosen from the following datasets, with modifications for each concept:

- rialto bridge timelapse [11]. Each concept only contains samples from one class (one building). A drift means that the datastream switches to samples from another class/building. Each sample has 27 features.
- forest covertype [2]. With each drift, one of the 10 continuous features is shifted by 1 standard deviation, and one of the 44 binary features is inverted. Each sample has 54 features.
- mnist hand-written digits [10]. In each concept, the 28 by 28 images are shifted by 1 pixel up, down, left, or right, then downsampled to 14 by 14 images and

vectorized. Each datastream contains only samples from one class (one digit). Thus, each sample has 196 features.
- multi-label music emotions classification [16]. The three concepts correspond to the following emotions (in order): "amazed-suprised", "quiet-still", and "angry-aggresive" (sic, also called "angry-fearful" in the paper). These emotions are chosen because almost no songs are labelled with more than one of these emotions. Each sample is either a song labelled with the specific emotion of the concept (with probability of 66.7%), or a song labelled with none of the three mentioned emotions (probability 33.3%, noise to make the detection more difficult). Each sample has 72 features.

We consider the following metrics:

- false negatives: the number of drifts that have not been detected by the method. An equivalent metric would be the number of true positives, which is the number of actual drifts minus the number of false negatives
- false positives: the number of times the method has reported a drift but there was none
- location error: the number of samples between the actual drift point and the drift point estimated by the method. Since the method's estimate may be earlier or later than the actual drift point, the location error is the absolute value of this difference
- detection delay: the number of samples between the actual drift and when the method reported it. In an online scenario it is desirable to have a small delay, to e.g. update the model as soon as possible
- run time: how many seconds the method takes to run on one data stream. This should only be understood as a rough estimate since further code optimizations may make the method faster

To determine the number of true positives, false positives, and false negatives, we match each actual drift with the detection where the estimated drift location is closest to the actual drift. However, if they are more than 50 samples apart, we instead mark them as false negative and false positive, respectively, instead of a true positive with a very high location error. This cut-off empirically makes sense since no method estimates the drift location with such a high error.

The following methods and variations are compared: MMDDDM, HDDDM, and D3 in their original form (without our extensions), then *with* our extensions, and additionally MMDDDM and D3 without our extensions, but the drift test is run every time when a new data sample arrives, instead of once every few samples. Naively thinking, this could lead to a small location error and make our extensions unnecessary. Finally, we also compare to SDDM from the Related Works section.

5.2   Hyperparameter Selection and Drift Detection Results

In a first step we evaluate the mentioned approaches for classical drift detection in order to select the best hyperparameters for our subsequent evaluation of change point detection and to inspect possible consequences of our change point detection scheme for drift detection.

Figure 3 shows the (approximate) Pareto fronts, that is, all hyperparameter settings that do not lead to more false positives *and* more false negatives than some other hyperparameter setting.

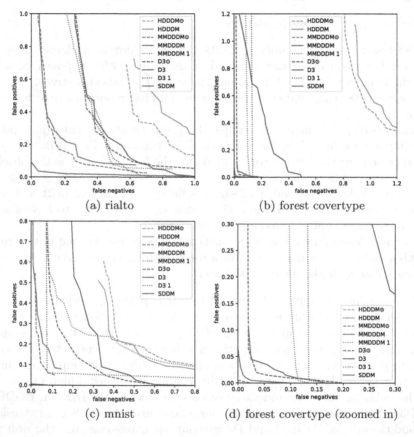

(a) rialto

(b) forest covertype

(c) mnist

(d) forest covertype (zoomed in)

Fig. 3. The (approximate) Pareto fronts of the different methods, i.e. for each method the hyperparameter settings that yield the best combinations of false positives and false negatives. A ⊙ means that the method is run with our extensions to localize the drift. A "1" after the method name means that the drift test is run every time a new sample is received.

It is visible that HDDDM does not perform well on the forest covertype dataset, this is because only 2 of the 54 features are changed with each drift, and the average of all feature-wise Hellinger distances does not pass the detection threshold for every actual drift. For some methods and datasets, our extensions lead to fewer false positives and negatives (e.g. D3 on forest covertype and mnist, HDDDM on rialto and forest covertype), while in other cases, there is no clear difference. On rialto and mnist, SDDM is the best method, and on forest covertype, MMDDDM is the best method (especially with drift localization).

5.3   Results for Change Point Detection

For a closer look, we choose for each method the hyperparameter setting which yields the lowest number of misdetections (false positives + false negatives). The results are presented in Table 1.

SDDM consistently achieves the lowest location error, and MMDDDM and D3 with our extensions are closely behind. Our extensions to HDDDM, MMD-DDM, and D3 greatly improve the location error on every dataset. Running the drift test for every received sample (MMDDDM 1 and D3 1) almost always makes the location error worse. This is because a drift will be detected earlier, so the drift location assumed by the method is always several samples before the actual drift location.

Consequently, the detection delay is almost always smaller when the drift test is run for every received sample. D3 has a rather high detection delay, while for the other methods, there is no clear trend recognizable.

Our extensions to HDDDM and MMDDDM make the methods only slightly slower. In the case of D3, our extension makes the method even slightly faster. This is because the optimal window size, which gives the lowest number of misdetections, is higher when our extension is used (versus when it is not used), thus the drift test with the costly training of the discriminator is performed less often. As it can be expected, running the drift test for every incoming sample is much slower. SDDM is also quite slow, it takes at least 5 times as long as HDDDM, MMDDDM, or D3 with localization. This may be improved with other hyperparameter settings (e.g. waiting for more samples before running the drift test again), but only at the cost of an increased detection delay. In our tests, the repeated fitting and evaluation of the decision trees in the cross-validation took by far the longest time in SDDM.

Table 1. A ⊙ means that the method is run with our extensions to localize the drift. A "1" after the method name means that the drift test is run every time a new sample is received. Shown are the means and standard deviations of 200 runs. The best value per dataset and metric is marked in bold.

	method	location error	detection delay	run time
rialto	HDDDM⊙	5.7 ± 7.4	76.6 ± 23.5	0.2 ± 0.1
	HDDDM	20.6 ± 11.6	77.3 ± 20.8	**0.1 ± 0.0**
	MMDDDM⊙	4.0 ± 6.3	67.4 ± 17.3	2.5 ± 0.5
	MMDDDM	17.2 ± 9.2	72.1 ± 15.1	1.9 ± 0.6
	MMDDDM 1	18.4 ± 9.5	**56.0 ± 13.7**	63.1 ± 37.1
	D3⊙	2.5 ± 4.3	95.4 ± 25.5	**0.1 ± 0.0**
	D3	19.2 ± 10.9	91.1 ± 18.7	**0.1 ± 0.0**
	D3 1	26.5 ± 10.7	65.5 ± 10.9	6.7 ± 0.7
	SDDM	**1.6 ± 1.9**	95.7 ± 37.0	30.8 ± 4.2
forest covertype	HDDDM⊙	1.4 ± 5.9	75.1 ± 23.9	0.3 ± 0.1
	HDDDM	20.2 ± 12.8	77.2 ± 22.9	**0.1 ± 0.0**
	MMDDDM⊙	0.5 ± 2.4	61.8 ± 16.1	2.6 ± 0.6
	MMDDDM	18.9 ± 8.8	65.4 ± 15.8	1.9 ± 0.4
	MMDDDM 1	24.6 ± 3.2	**41.4 ± 3.2**	66.6 ± 25.3
	D3⊙	1.3 ± 3.9	87.4 ± 25.1	3.5 ± 0.8
	D3	23.7 ± 10.6	102.8 ± 22.0	3.8 ± 0.7
	D3 1	19.7 ± 4.6	93.3 ± 4.6	273.5 ± 49.9
	SDDM	**0.2 ± 0.8**	60.1 ± 28.7	31.3 ± 4.4
mnist	HDDDM⊙	4.8 ± 8.3	71.2 ± 21.1	0.9 ± 0.1
	HDDDM	15.9 ± 9.6	67.5 ± 15.8	**0.5 ± 0.0**
	MMDDDM⊙	1.9 ± 3.0	64.8 ± 15.9	5.0 ± 1.1
	MMDDDM	16.4 ± 8.8	65.2 ± 13.7	4.5 ± 1.1
	MMDDDM 1	14.9 ± 7.2	**46.7 ± 11.8**	80.1 ± 36.2
	D3⊙	3.2 ± 6.2	106.3 ± 20.4	5.5 ± 0.8
	D3	17.9 ± 9.8	92.0 ± 16.7	6.3 ± 1.0
	D3 1	26.9 ± 7.4	80.1 ± 7.4	456.3 ± 68.2
	SDDM	**1.3 ± 1.4**	74.3 ± 26.2	25.3 ± 4.0
music	HDDDM⊙	2.9 ± 4.8	76.3 ± 18.5	0.4 ± 0.1
	HDDDM	17.4 ± 8.5	62.0 ± 15.3	**0.2 ± 0.0**
	MMDDDM⊙	1.9 ± 2.5	88.0 ± 16.4	1.3 ± 0.9
	MMDDDM	17.3 ± 8.0	71.6 ± 13.2	1.8 ± 0.2
	MMDDDM 1	23.3 ± 4.5	**52.7 ± 4.5**	30.8 ± 15.3
	D3⊙	3.4 ± 5.8	112.4 ± 24.6	2.8 ± 2.1
	D3	20.7 ± 10.8	103.5 ± 20.3	4.1 ± 0.5
	D3 1	24.2 ± 7.2	86.9 ± 7.2	312.4 ± 47.7
	SDDM	**1.7 ± 2.0**	173.3 ± 24.9	18.4 ± 2.6

6   Conclusion and Outlook

We presented extensions for three popular drift detection methods (MMDDDM, HDDDM, and D3), to find the exact drift location in time. We showed that these extensions lead to a much better estimate of the drift location, while having little effect on the run time of the method. In some cases, our extensions even led to fewer false positives and false negatives. This makes the drift detection methods with our extensions particularly useful for online scenarios.

Future work could include evaluations on additional datasets, and investigating how our extensions work with other drift detection methods.

References

1. Bifet, A., Gama, J.: IoT data stream analytics (2020)
2. Blackard, J.A., Dean, D.J.: Comparative accuracies of artificial neural networks and discriminant analysis in predicting forest cover types from cartographic variables. Comput. Electron. Agric. **24**(3), 131–151 (1999). https://doi.org/10.1016/S0168-1699(99)00046-0
3. Ditzler, G., Polikar, R.: Hellinger distance based drift detection for nonstationary environments. In: 2011 IEEE Symposium on Computational Intelligence in Dynamic and Uncertain Environments (CIDUE), pp. 41–48 (2011). https://doi.org/10.1109/CIDUE.2011.5948491
4. Elwell, R., Polikar, R.: Incremental learning of concept drift in nonstationary environments. IEEE Trans. Neural Netw. **22**(10), 1517–1531 (2011). https://doi.org/10.1109/TNN.2011.2160459
5. Gama, J., Žliobaitė, I., Bifet, A., Pechenizkiy, M., Bouchachia, A.: A survey on concept drift adaptation. ACM Comput. Surv. (CSUR) **46**(4), 1–37 (2014)
6. Gözüaçık, Ö., Büyükçakır, A., Bonab, H., Can, F.: Unsupervised Concept Drift Detection with a Discriminative Classifier. In: Proceedings of the 28th ACM International Conference on Information and Knowledge Management, pp. 2365–2368. ACM, Beijing China (2019). https://doi.org/10.1145/3357384.3358144
7. Gretton, A., Borgwardt, K., Rasch, M., Schölkopf, B., Smola, A.: A kernel method for the two-sample-problem. In: Advances in Neural Information Processing Systems. vol. 19. MIT Press (2006). https://proceedings.neurips.cc/paper/2006/hash/e9fb2eda3d9c55a0d89c98d6c54b5b3e-Abstract.html
8. Hinder, F., Artelt, A., Vaquet, V., Hammer, B.: Precise Change Point Detection using Spectral Drift Detection (2022). arXiv preprint arXiv:2205.06507
9. Klinkenberg, R.: Predicting phases in business cycles under concept drift. In: LLWA, pp. 3–10. Citeseer (2003)
10. Lecun, Y., Bottou, L., Bengio, Y., Haffner, P.: Gradient-based learning applied to document recognition. Proc. IEEE **86**(11), 2278–2324 (1998). https://doi.org/10.1109/5.726791
11. Losing, V., Hammer, B., Wersing, H.: KNN classifier with self adjusting memory for heterogeneous concept drift. In: 2016 IEEE 16th International Conference on Data Mining, pp. 291–300 (2016). https://doi.org/10.1109/ICDM.2016.0040
12. Lu, J., Liu, A., Dong, F., Gu, F., Gama, J., Zhang, G.: Learning under concept drift: a review. IEEE Trans. Knowl. Data Eng. **31**(12), 2346–2363 (2018)
13. Pedregosa, F., et al.: Scikit-learn: machine learning in python. J. Mach. Learn. Res. **12**(85), 2825–2830 (2011). https://jmlr.org/papers/v12/pedregosa11a.html

14. Rabanser, S., Günnemann, S., Lipton, Z.: Failing loudly: an empirical study of methods for detecting dataset shift. In: Advances in Neural Information Processing Systems. vol. 32 (2019). arxiv.org/abs/1810.11953

15. Sugiyama, M., Krauledat, M., Müller, K.R.: Covariate shift adaptation by importance weighted cross validation. J. Mach. Learn. Res. **8**(5), 985–1005 (2007)

16. Trohidis, K., Tsoumakas, G., Kalliris, G., Vlahavas, I.: Multi-label classification of music by emotion. EURASIP J. Audio Speech Music Process. **2011**(1), 1–9 (2011). https://doi.org/10.1186/1687-4722-2011-426793. Publisher: SpringerOpen

Pedestrian Multi-object Tracking Algorithm Based on Attention Feature Fusion

Yan Zhou$^{(\boxtimes)}$, Zhennan Du, and Dongli Wang

Xiangtan University, Xiangtan 411105, Hunan, China
{yanzhou,wangdl}@xtu.edu.cn

Abstract. Multi-Object Tracking (MOT) is a challenging research area in computer vision with significant practical applications. With the advent of deep neural networks, significant progress has been made in MOT, and Qdtrack has become a widely used algorithm due to its relatively simple structure and high performance. However, accurate target tracking in complex scenes with mutual occlusion, motion blur and complex backgrounds is still a significant challenge. To address the problem of low target tracking accuracy in complex scenes, this paper proposes an end-to-end multi-object tracking algorithm based on attention feature fusion. First, a new lightweight attention module is introduced into the backbone network, which enhances the ability of the network to capture key information and locate targets without increasing the computational complexity. Second, the feature pyramid structure is improved to reduce the loss of features in the fusion process and improve the feature expression ability of the model. Finally, the Intersection over Union (IoU) of the original model is optimised and the regression ability of the bounding box is improved using polyloss to optimise the cross-entropy loss. Experimental results on the MOT16 and MOT17 benchmarks show that the proposed algorithm effectively improves the accuracy and robustness of multi-object pedestrian tracking compared to other algorithms, and has better tracking performance.

Keywords: deep learning · multi-object tracking · attentional mechanism · feature pyramid

1 Introduction

The development of object tracking can be divided into two categories: single object tracking and multi-object tracking. While single object tracking has seen significant breakthroughs in recent years, multi-object tracking remains an area of continuous development. The purpose of multi-object tracking is to continuously track the trajectory of multiple objects in a video frame. This technology is at the heart of many intelligent applications, including automotive driving, security surveillance, robotics and military security. Video-based multi-object tracking has a wide range of applications and significant research implications.

I. Rojas et al. (Eds.): IWANN 2023, LNCS 14134, pp. 105–118, 2023.
https://doi.org/10.1007/978-3-031-43085-5_9

Previous multi-object tracking methods involved providing the initial frame and then predicting the position of the object in subsequent frames. However, this method was unable to handle the frequent appearance and disappearance of objects, making it unsuitable for current multi-object tracking application scenarios. However, with the rapid development of deep learning, deep neural network models have become increasingly accurate due to their powerful feature extraction and representation capabilities. Deep learning algorithms have been widely validated in computer vision and have become one of the mainstream models of artificial intelligence systems due to their outstanding performance. As a result, object detection algorithms based on deep learning have been continuously improved, and many algorithms with good performance have been proposed.

Tracking by Detection (TBD) has gradually become the mainstream framework for multi-object tracking due to its leading tracking performance. Although the model is complex and the computational cost is high, some Joint Detection and Tracking (JDT) algorithms have emerged in recent years, which provide a single-stage tracking algorithm that is jointly learned by multiple modules. Despite the superior progress made in multi-object tracking, this task remains challenging due to the complexity of the scene. Several difficulties persist in multi-object tracking, including morphological change, scale change, occlusion and disappearance, and image blur. Addressing these difficulties is necessary to improve the performance of detectors and data association algorithms and to effectively distinguish objects from the background.

The main contributions of this paper are as follows: (1) The SA (Shuffle Attention) lightweight module is introduced into the backbone network, which improves the feature extraction capability with almost no additional computational parameters. (2) Based on the feature pyramid network (FPN), a feature pyramid structure with multi-scale feature fusion is proposed to improve the performance of the model. (3) IoU (Intersection over Union) is replaced with to improve the regression accuracy of bbox and polyLoss is introduced to optimise the cross-entropy loss.

2 Related Work

Object detection and data association are two key components of multi-object tracking. Object detection estimates the object's boundary box and uses data association to obtain the object's identity tracking information.

2.1 Object Detection in Multi-object Tracking

Deep learning is a powerful tool in computer vision that can extract deep semantic and visual features for various tasks, including object classification, detection and segmentation. In particular, object detection plays a fundamental role in multi-object tracking, a hot research topic in computer vision with broad applications in areas such as automotive driving, security surveillance, robotics, and military security. In recent years, the application of deep learning to object detection has significantly improved the accuracy of tracking algorithms. Kim et al. were the first to apply object detection to multi-object tracking using deep features extracted from pre-trained convolutional neural networks (CNN) and a multi-hypothesis tracking algorithm. Since then, more advanced object detectors have been developed, including one-stage and two-stage object detection methods. One-stage object detectors, such as the YOLO series, RetinaNet and CenterNet, directly output classification and location information without generating candidate regions. Two-stage object detectors, such as the Faster R-CNN series, generate candidate regions separately and then classify them.

2.2 Data Association

Data association plays a key role in multi-object tracking, where the similarity between trajectories and detection frames is calculated and matched. Similarity measurement is a critical step in data association, which is based on various factors such as motion, appearance and location information. A popular approach is the Simple Online and Realtime Tracking (SORT) proposed by Alex et al [1]. This method combines motion and position information by predicting the position of the tracked box using a Kalman filter and calculating the Intersection over Union (IOU) between the detection frame and the prediction frame as the similarity. In contrast, appearance information is more accurate for measuring similarity in long-range matching. To address this challenge, Alex et al. proposed DeepSORT [2], which extracts appearance features using an independent Re-ID model. After the object is blocked for a period of time, re-identification can be used to extract the cosine similarity of the appearance features. The matching strategy is the next critical step in data association, where the ID is assigned to the object based on the calculated similarity. The Hungarian or Greedy algorithm is commonly used for this purpose. MOTDT first matches using appearance similarity and then uses IOU similarity to match tracks that were not previously matched. QuasiDense [3] converts appearance similarity into probability to complete the matching by two-way softmax operation and nearest neighbour search. This matching strategy is adopted in our proposed method.

3 Model Algorithm

The overall network structure of the model in this paper is shown in Fig. 1. The model is an improvement of the QuasiDense Track [3] approach and the backbone network is based on Resnet-50 by introducing the lightweight attention module of SA-Net corresponding to the SResnet module in Fig. 1, which is used to extract image feature information. PGRE is a new feature fusion module improved for this paper. The model takes two image frames in parallel and uses region proposals generated by RPN (Region Proposal Network) to learn the instance similarity using quasi-dense matching. The data association part converts the appearance similarity into probability through a bidirectional softmax operation and uses nearest neighbour search to achieve matching.

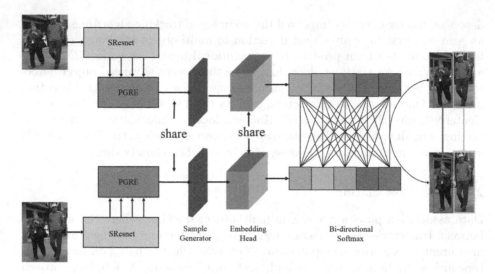

Fig. 1. Overall network structure.

3.1 SA Lightweight Module

The attention mechanism has been widely applied in computer vision tasks, including channel attention and spatial attention. In the convolutional neural network, the channel dimension and the scale space of the image are crucial. Channel attention highlights the significant channels and suppresses the less relevant ones to improve the performance of the network. Spatial attention, on the other hand, focuses on task-relevant feature points or regions in the image, improving the efficiency of the network. However, the traditional model lacks the ability to learn the importance of different channels and spatial features, leading to suboptimal performance. To address this issue, this paper introduces the attention mechanism in the backbone network to capture the relationships between different channels and spaces and learn the importance of different features. The lightweight module SA-Net (Shuffle attention) used in this study combines the two attention mechanisms with minimal additional computation. It first uses a series of convolutional methods to extract the channel features and reduce the computational cost. Then, the features of channel attention and spatial attention are computed separately and concatenated within the group. The feature graph with the same size as the input is merged to obtain the final result. Finally, the group is rearranged using the Channel Shuffle [4] operation to allow information interaction between different groups. The structure of SA-Net is shown in Fig. 2. Where $x \in R^{c \times h \times w}$. c, h, w indicate the channel number, spatial height, and width, respectively. $W \in R^{c/2g \times 1 \times 1}$, $b \in R^{c/2g \times 1 \times 1}$.

Table 1 shows a comparison of the performance of different attention modules on the same ResNet network using the ImageNet dataset. The top-1 accuracy indicates the ability of the model to correctly predict the most similar category. It can be seen that after adding the SA-Net module, there is almost no increase

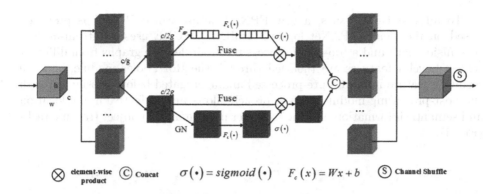

Fig. 2. The structure of SA-Net.

in computational complexity while achieving better results. By introducing the SA-Net module into the ResNet trunk, with a depth of convolutional neural network, the model gains extremely light cross-modal interaction information, which enhances its ability to accurately track objects over a long period of time.

Table 1. .

methods	Param	GFLOPs	TOP-1 Accuracy(%)
ResNet	25.557M	4.122	76.384
+SENet	28.088M	4.130	77.462
+CBAM	28.090M	4.139	77.626
+ECA-Net	25.557M	4.127	77.480
+SA-Net	25.557M	4.125	77.724

3.2 Feature Fusion Module

In a deep convolutional neural network, features are propagated from convolutional features to subsequent layers. However, this can result in ignoring location and detail information, leading to poor performance for small objects or those with occlusions. To address this, the feature pyramid structure (FPN) has been proposed, which integrates features of different scales to improve model performance in object recognition and multi-object tracking.

In quasi-dense similarity learning, RoI (Region of interest) feature vectors are extracted using a lightweight embedding head, and positive and negative GT values are discriminated based on IoU values. However, FPN has its limitations, such as the lack of location information and the uneven use of semantic and location information for large and small suggestions.

To address these issues, a new FPN structure called PGRE, is proposed based on the ideas of PANet [6] and GroIE [7]. PGRE uses a bottom-up feature fusion path and a top-down process to obtain feature graphs from different scales. Level 4 features are then fed through the RoI Pooler module, and the resulting feature maps are pre-processed and aggregated before passing through the post-processing module. This approach allows more effective use of location and semantic information, leading to better performance in object tracking tasks (Fig. 3).

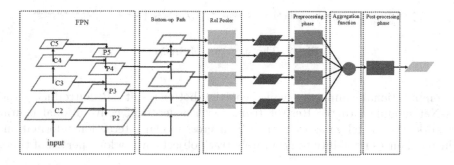

Fig. 3. PGRE network structure.

3.3 α - CIoU

The Intersection over Union (IoU) [8] is a critical metric used in tracking tasks to measure the similarity between predicted and real boundary boxes. However, the standard IoU has limitations in accurately reflecting the degree of overlap between boxes, particularly with respect to their distance. This insensitivity can lead to inaccurate tracking, as shown in Fig. 4, where the IoU values are equal but the degree of overlap varies significantly. Furthermore, when there is no overlap between the predicted and real boxes, the IoU is equal to 0. This results in a gradient of 0 in the optimisation loss function, which prevents successful optimisation. To address these issues, this paper proposes the use of α - CIoU [9] as the loss function for regression training. α - CIoU takes into account the overlap area, center point distance, aspect ratio, and regularization terms in IoU, making it more sensitive to distance. These regularisation terms help to improve convergence speed, final tracking performance and overall model robustness. Therefore, accurate bounding box regression training is critical for successful tracking.

$$IoU = \frac{X \cap X^{kt}}{X \cup X^{kt}} \tag{1}$$

$$LOSS = 1 - IoU \tag{2}$$

$$L_{\alpha-CIoU} = 1 - IoU^{\alpha} + \frac{\rho^{2\alpha}\left(x, x^{kt}\right)}{d^{2\alpha}} + (\beta \nu)^{\alpha} \tag{3}$$

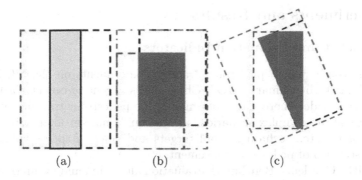

(a)   (b)   (c)

Fig. 4. The prediction results of different qualities with equal IoU

where $X^{kt} = \left(a^{kt}, b^{kt}, l^{kt}, n^{kt}\right)$ is the ground-truth,$X = (a, b, l, n)$, x and x^{kt} denote the central points of X and X^{kt}, $\rho(\bullet)$ is the Euclidean distance, and d is the diagonal length of the smallest enclosing box covering the two boxes, β is a positive trade-off parameter, and ν measures the consistency of aspect ratio,$\nu = \frac{4}{\pi^2}\left(\arctan\frac{l^{kt}}{n^{kt}} - \arctan\frac{l}{n}\right)^2$, $\beta = \frac{\nu}{(1-IoU)+\nu}$, α is the hyperparameter is usually set to 2 or 3

3.4   Cross-Entropy Loss as PolyLoss

In general, a good loss function can take a more flexible form and should be tailored to different tasks and data sets. Inspired by Taylor Expansion, some researchers proposed a new framework for understanding and designing loss function – PolyLoss [10], which regarded and designed loss function as a linear combination of a polynomial function. They believed that the commonly used classification loss function, such as cross-entropy loss, decomposed into a series of weighted polynomial bases. They can be decomposed into the form of $\sum_{j-1}^{\infty} \lambda_j (1 - p_m)^j$, Where $\lambda_j \in Q^+$ is the polynomial coefficient, p_m is the predicted probability of the target class tag. Each polynomial basis $(1 - p_m)^j$ is weighted by the corresponding polynomial coefficient $\lambda_j \in Q^+$, this enables polyLoss to easily adjust different polynomial bases. The Taylor expansion of cross-entropy loss based on $(1 - p_m)^j$ is as follows.

$$L_{CE} = -\log(P_m) = \sum_{j=1}^{\infty} \frac{1}{j}(1 - P_m)^j = (1 - P_m) + \frac{1}{2}(1 - P_m)^2... \quad (4)$$

When $\lambda_j = \frac{1}{j}$, polyloss is equivalent to cross-entropy loss, but this coefficient may not be optimal. Experiments show that simply adjusting the single polynomial coefficients is sufficient to achieve better performance than cross-entropy loss. The specific calculation formula is as follows.

$$L_{Poly-1} = (1 + \varepsilon_1)(1 - p_m) + \frac{1}{2}(1 - p_m)^2 + ... = -\log(p_m) + \varepsilon_1(1 - p_m) \quad (5)$$

4   Experiments and Results

4.1   Data Set and Evaluation Indicators

This paper conduct multiple object tracking experiments on the MOT16 and MOT17 datasets benchmark. MOT Challenge is a generic object tracking the complex scene video more data sets mainly for pedestrian tracking, including a series of real life complex scenarios, such as buses market monitoring, mobile video, recorder, etc., including small targets and goals of sports video, bright and dark scenario of pedestrian movement, etc.

The MOT Challenge Benchmark evaluation algorithm and evaluation indicator package contains the following: Multiple Object Tracking Accuracy (MOTA) and ID F1 Score (IDF1) are the most general metrics in the MOT task. MOTA is computed based on FP, FN, and IDs. IDF1 evaluates the identity preservation ability and focuses more on the association's performance. IDs are the number of identity exchanges of all tracking targets during the tracking of video sequences. FP represents the number of false positives in the entire video sequence, the hypothetical boundary box that cannot match the boundary box of the real trajectory in the video sequence is called false positive. FN represents the number of false negatives in the whole video sequence, and the boundary box that cannot match the real trajectory of the assumed boundary box in the video sequence is called false negatives. MT is the number of tracking tracks that at least 80% of the video frames of each target can be correctly tracked in the tracking process. ML is the number of tracking tracks that at most 20% of the video frames of each target can be correctly tracked in the tracking process.

4.2   Experimental Environment

This paper is based on pyTorch-1.7 and the MMTracking source framework provided by the Chinese University of Hong Kong. The experiments are performed on Ubuntu 16.04 system. Training was performed on a NVIDIA GeForce RTX 3090 graphics card server. The software environment used was PyTorch 1.70, Python 3.7 and CUDA 11.1. The model in this paper adopts the transfer learning method. When training the model, we first load the pre-training weights of ResNet50 on ImageNet and the pre-training on coco-person. We use stochastic gradient descent (SGD) to optimise the loss function with an initial learning rate of 0.005, and the batch size was set to 4. A total of 24 epochs were trained. Using the multiscale training enhancement method, the input image size was set to 1088×1088.

4.3   Ablation Experiment

Ablation experiments are a commonly used experimental method in the field of deep learning and are mainly used to analyze the influence of different network branches on the entire model. To further analyze the influence of the SA attention module, PGRE module, $\alpha - CIoU$, PolyLoss. We conducted the ablation experiment. The specific results are shown in Table 2, where "$\sqrt{}$" represents the improved method. The experimental data show that the improved strategy can improve the tracking accuracy in different degrees, which proves the effectiveness of the improved method.

Table 2. Comparison of the algorithm in this paper with a baseline on MOT17. ↑ means higher is better.

Method	SA-Net	PGRE	$\alpha - CIoU$	PolyLoss	MOTA↑	IDF1↑
baseline					67.1	64.3
Improve 1	$\sqrt{}$				67.7	65.1
Improve 2	$\sqrt{}$	$\sqrt{}$			68.4	66.0
Improve 3	$\sqrt{}$	$\sqrt{}$	$\sqrt{}$		68.8	66.3
Improve 4	$\sqrt{}$	$\sqrt{}$	$\sqrt{}$	$\sqrt{}$	69.2	66.7

4.4 Comparison of Different Algorithms

To further verify the effectiveness of this algorithm and scientific paper, we compared it with some mainstream multi-object tracking algorithms. The comparison results are shown in Table 3 and Table 4. It can be seen from the table that our algorithm is obviously superior to other models in MOTA, IDF1, MT, and FN indicators, and has strong competitiveness in ML, FP and IDs indicators. MOTA and IDF1 are multi-object tracking accuracy and identity identification F1 value, respectively, which are the main evaluation indices reflecting the overall detection and tracking performance of the model. The algorithm in this paper is superior to other models in MOTA and IDF1, which indicates that it has better overall performance in the tracking process robustness. The attention module and feature fusion modules proposed by the algorithm in this chapter greatly improve the overall detection performance of the network and help the network to better mine the location and tracking information of the target, thus achieving obvious effects on the MT and FN indicators. However, by improving the network detection performance, the omission of targets is reduced, which inevitably leads to some increase in FP. It is proved that our algorithm is robust and has certain significance for pedestrian multi-object tracking.

4.5 Visualization

To see the tracking results more directly, we visualized the tracking effects of several different scenes from M0T16 and MOT17.

Figure 5 shows the video recorded under the front lens of a busy shopping mall camera and the elevated viewing platform of a pedestrian street at night. In daily life, interactive actions are always performed between targets, which will cause occlusion phenomenon. Meanwhile, the pedestrian targets are in a moving

Table 3. Comparison of the algorithm in this paper with a baseline on MOT16. ↑ means higher is better, ↓ means lower is better.

Method	MOTA↑	IDF1↑	MT↑	ML↓	FP↓	FN↓	IDs↓
Lif-T [11]	61.3	64.7	27.0%	34.0%	–	–	–
DeepSORT [2]	61.4	62.2	32.8%	18.2%	12852	56668	781
Tube-TK [12]	64.0	59.4	33.5%	19.4%	10962	53626	1117
JDE [13]	64.4	55.8	35.4%	20.0%	–	–	1544
CNNMTT [14]	65.2	62.2	32.4%	21.3%	6578	55896	946
CTracker [15]	67.6	57.2	32.9%	23.1%	8934	48305	5529
TraDeS [16]	70.1	64.7	37.3%	20.0%	8091	45210	1144
ours	70.1	67.3	41.6%	19.7%	8781	44703	1038

Table 4. Comparison of the algorithm in this paper with a baseline on MOT17. ↑ means higher is better, ↓ means lower is better.

Method	MOTA↑	IDF1↑	MT↑	ML↓	FP↓	FN↓	IDs↓
Tracktor [17]	54.4	56.1	25.7%	29.8%	44109	210774	2574
DeepSORT [2]	60.3	61.2	31.5%	20.3%	36111	185301	2442
Lif-T [11]	60.5	65.6	26.9%	28.8%	14966	206619	1189
Tube-TK [12]	63.0	58.6	31.2%	19.9%	27060	177483	4137
MOTR [18]	65.1	66.4	33.0%	25.2%	45486	149307	2049
CTracker [15]	66.6	57.4	32.2%	24.2%	22284	160491	5529
CenterTrack [19]	67.8	64.7	34.6%	24.6%	18498	160332	3039
TraDeS [16]	69.1	63.9	36.4%	21.5%	20892	150060	4833
ours	69.2	66.7	40.7%	21.4%	25621	145367	3042

state in the video frame, and the size is also different. As can be seen from the comparison diagram of tracking effects before and after improvement (the part in red circle), compared with before improvement, the method in this paper has improved the cases of missing tracking caused by objects with small scale and occlusion, and is more suitable for actual application scenarios.

As shown in Fig. 6, these are the 384 frame, the 393 frame, and the 403 frame of the Street View video taken on the mobile platform. Above are the results of the baseline visualization, and below are the results of the algorithms in this paper. Both algorithms successfully track people in the video at frame 384 (baseline ID is 141, algorithm ID in this paper is 131). In frame 393, the target is about to pass another target for occlusion. At frame 403, the baseline target ID has changed to 155 due to occlusion, and the algorithm in this chapter still maintains the same ID for successful tracking.

Figure 7 is the video taken at the intersection in the moving bus, which are frames 651, 666, and 694, respectively. Above are the baseline tracking results,

(a) baseline   (b) ours

(c) baseline   (d) ours

Fig. 5. Different scenes miss objects visualizations.

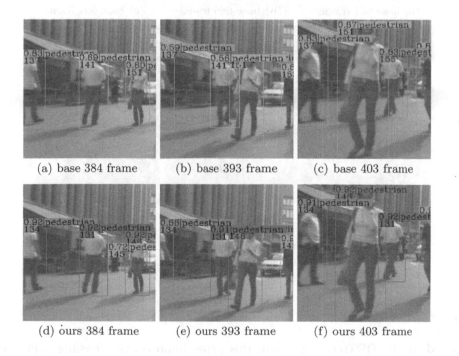

(a) base 384 frame   (b) base 393 frame   (c) base 403 frame

(d) ours 384 frame   (e) ours 393 frame   (f) ours 403 frame

Fig. 6. Tracking comparison visualization of Street View shot on mobile platform.

and below are the results of the methodology in this paper. In frame 651, we all tracked the object (the baseline ID is 666, and the algorithm ID in this chapter is 706). In frame 666, the baseline lost the target due to angle changes and occlusion. However, we continued tracking and also detected pedestrians with background acquaintances at a greater distance. At frame 694, the baseline tracks the target again, but the ID is changed to 960. However, the algorithm in this chapter continues to track both pedestrians and the ID remains unchanged. The visualization results of several scenarios show that the algorithm proposed in this chapter can effectively improve the tracking robustness.

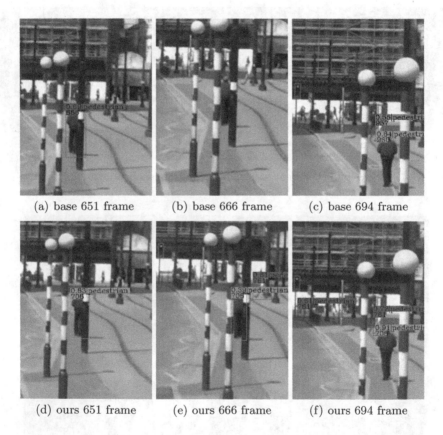

(a) base 651 frame   (b) base 666 frame   (c) base 694 frame

(d) ours 651 frame   (e) ours 666 frame   (f) ours 694 frame

Fig. 7. Video taken from the bus at the intersection.

5   Conclusion

Based on the QDTrack algorithm, this paper improves the tracking ability of the model in the face of occlusion, motion, scale change, and other problems by improving the backbone network, feature fusion module, IoU and Loss. Experimental results show that the proposed model achieves good results on MOT17 pedestrian multi-object tracking data set, and effectively improves the accuracy and robustness of the model. To improve the real-time performance of the model is the next research direction.

Conflicts of interest. The authors state that they have no conflicting financial interests or personal connections that may have influenced the work reported in this paper.

References

1. Bewley, A., Ge, Z., Ott, L., et al.: Simple online and realtime tracking. In: 2016 IEEE International Conference on Image Processing (ICIP), pp. 3464–3468. IEEE (2016)
2. Wojke, N., Bewley, A., Paulus, D.: Simple online and realtime tracking with a deep association metric. In: 2017 IEEE International Conference on Image Processing (ICIP), pp. 3645–3649. IEEE (2017)
3. Pang, J., Qiu, L., Li, X., et al.: Quasi-dense similarity learning for multiple object tracking. In: Proceedings of the IEEE/CVF Conference on Computer Vision and Pattern Recognition, pp. 164–173 (2021)
4. Ma, N., Zhang, X., Zheng, H.T., et al.: ShuffleNet V2: practical guidelines for efficient CNN architecture design. In: Proceedings of the European Conference on Computer Vision (ECCV), pp. 116–131 (2018)
5. Zhang, Q.L., Yang, Y.B.: SA-Net: shuffle attention for deep convolutional neural networks. In: ICASSP 2021–2021 IEEE International Conference on Acoustics, Speech and Signal Processing (ICASSP), pp. 2235–2239. IEEE (2021)
6. Liu, S., Qi, L., Qin, H., et al.: Path aggregation network for instance segmentation. In: Proceedings of the IEEE Conference on Computer Vision and Pattern Recognition, pp. 8759–8768 (2018)
7. Rossi, L., Karimi, A., Prati, A.: A novel region of interest extraction layer for instance segmentation. In: 2020 25th International Conference on Pattern Recognition (ICPR), pp. 2203–2209. IEEE (2021)
8. Rezatofighi, H., Tsoi, N., Gwak, J.Y., et al.: Generalized intersection over union: a metric and a loss for bounding box regression. In: Proceedings of the IEEE/CVF Conference on Computer Vision and Pattern Recognition, pp. 658–666 (2019)
9. He, J., Erfani, S., Ma, X., et al.: α-IoU: a family of power intersection over union losses for bounding box regression. Adv. Neural Inf. Process. Syst. **34**, 20230–20242 (2021)
10. Leng, Z., Tan, M., Liu, C., et al.: PolyLoss: A polynomial expansion perspective of classification loss functions. arXiv preprint arXiv:2204.12511 (2022)
11. Hornakova, A., Henschel, R., Rosenhahn, B., et al.: Lifted disjoint paths with application in multiple object tracking. In: International Conference on Machine Learning, pp. 4364–4375. PMLR (2020)
12. Pang, B., Li, Y., Zhang, Y., et al.: TubeTK: adopting tubes to track multi-object in a one-step training model. In: Proceedings of the IEEE/CVF Conference on Computer Vision and Pattern Recognition, pp. 6308–6318 (2020)
13. Wang, Z., Zheng, L., Liu, Y., Li, Y., Wang, S.: Towards Real-Time Multi-Object Tracking. In: Vedaldi, A., Bischof, H., Brox, T., Frahm, J.-M. (eds.) ECCV 2020. LNCS, vol. 12356, pp. 107–122. Springer, Cham (2020). https://doi.org/10.1007/978-3-030-58621-8_7

14. Mahmoudi, N., Ahadi, S.M., Rahmati, M.: Multi-target tracking using CNN-based features: CNNMTT. Multimedia Tools Appl. **78**, 7077–7096 (2019)
15. Peng, J., et al.: Chained-tracker: chaining paired attentive regression results for end-to-end joint multiple-object detection and tracking. In: Vedaldi, A., Bischof, H., Brox, T., Frahm, J.-M. (eds.) ECCV 2020. LNCS, vol. 12349, pp. 145–161. Springer, Cham (2020). https://doi.org/10.1007/978-3-030-58548-8_9
16. Wu, J., Cao, J., Song, L., et al.: Track to detect and segment: an online multi-object tracker. In: Proceedings of the IEEE/CVF Conference on Computer Vision and Pattern Recognition, pp. 12352–12361 (2021)
17. Bergmann, P., Meinhardt, T., Leal-Taixe, L.: Tracking without bells and whistles. In: Proceedings of the IEEE/CVF International Conference on Computer Vision, pp. 941–951 (2019)
18. Zeng, F., Dong, B., Zhang, Y., Wang, T., Zhang, X., Wei, Y.: MOTR: end-to-end multiple-object tracking with transformer. In: Avidan, S., Brostow, G., Cissé, M., Farinella, G.M., Hassner, T. (eds.) Computer Vision-ECCV 2022. ECCV 2022. Lecture Notes in Computer Science. vol. 13687. Springer, Cham (2022). https://doi.org/10.1007/978-3-031-19812-0_38
19. Zhou, X., Koltun, V., Krähenbühl, P.: Tracking objects as points. In: Vedaldi, A., Bischof, H., Brox, T., Frahm, J.-M. (eds.) ECCV 2020. LNCS, vol. 12349, pp. 474–490. Springer, Cham (2020). https://doi.org/10.1007/978-3-030-58548-8_28

Fairness-Enhancing Ensemble Classification in Water Distribution Networks

Janine Strotherm$^{(\boxtimes)}$ (ID) and Barbara Hammer (ID)

Center for Cognitive Interaction Technology, Bielefeld University, Bielefeld, Germany
{jstrotherm,bhammer}@techfak.uni-bielefeld.de

Abstract. As relevant examples such as the future criminal detection software [1] show, fairness of AI-based and social domain affecting decision support tools constitutes an important area of research. In this contribution, we investigate the applications of AI to socioeconomically relevant infrastructures such as those of water distribution networks (WDNs), where fairness issues have yet to gain a foothold. To establish the notion of fairness in this domain, we propose an appropriate definition of protected groups and group fairness in WDNs as an extension of existing definitions. We demonstrate that typical methods for the detection of leakages in WDNs are unfair in this sense. Further, we thus propose a remedy to increase the fairness which can be applied even to non-differentiable ensemble classification methods as used in this context.

Keywords: Fairness · Disparate Impact · Equal Opportunity · Leakage Detection in Water Distribution Networks

1 Introduction

Due to the increasing usage of AI-based decision making systems in socially relevant fields of application the question of *fair decision making* gained importance in recent years (cf. [1,6]). Fairness is hereby related to the several (protected) groups or individuals which are affected by the algorithmic decision making and characterized by *sensitive features* such as gender or ethnicity. Most algorithms on which these tools are based on rely on data which can be biased with respect to the questions of fairness without intention, resulting in skewed models. Also the algorithm itself can discriminate protected groups or individuals without explicitly aiming to do so due to an undesirable algorithmic bias (cf. [11,12]).

Several definitions of fairness have been discussed (cf. [3–5,11,12,15]). From a legal perspective, one distinguishes between *disparate treatment* and *disparate impact* (cf. [3]). While disparate treatment occurs whenever a group or an individual is intentionally treated differently because of their membership in a protected class, disparate impact is a consequence of indirect discrimination happening despite "seemingly neutral policy" (cf. [12]). From a scientific viewpoint, the variety of fairness notions is much larger where many popular approaches focus mainly on (binary) classification tasks (cf. [4,11,12]).

I. Rojas et al. (Eds.): IWANN 2023, LNCS 14134, pp. 119–133, 2023.
https://doi.org/10.1007/978-3-031-43085-5_10

Besides the definition of fairness, the problem arises how to enhance fairness in well known AI methods while maintaining a reasonable overall performance. Approaches can hereby be grouped in pre-processing, in-processing and post-processing techniques (cf. [12]), whereas we will focus on in-processing methods.

The question of fairness becomes especially relevant when the decisions of a machine learning (ML) model take impact on socioeconomic infrastructure, such as water distribution networks (WDNs). To the best of our knowledge, the question of fairness has not been approached within this domain. We address the important problem of leakage detection in WDNs and investigate in how far typical models treat different groups of consumers of the WDN (in)equally. We hereby focus on *group fairness*, which in contrast to *individual fairness* focuses on treating different groups among the WDN equally instead of treating similar individuals similarly (cf. [11]).

To come up with a first approach to improve fairness in such a domain of high social and ethical relevance, based on [15], we consider the empirical covariance between the sensitive features and the model's prediction as a proxy for the fairness measure. The algorithms can handle even multiple non-binary sensitive features and satisfy both the concept of disparate treatment *and* disparate impact simultaneously (cf. [15]). Then, we extend [15] by (a) giving explicit generalized definitions of well-known fairness measures even for multiple non-binary sensitive features, (b) modifying their idea to any possibly non-convex classification model including arbitrary ensemble classifier, and (c) presenting a method to handle the problem of potential non-differentiability of AI algorithms.

The rest of the work is structured as followed: In Sect. 2, we introduce two definitions of group fairness for multiple non-binary sensitive features. Afterwards, in Sect. 3, we present a standard methodology to detect leakages in WDNs and investigate whether the resulting model makes fair decisions with respect to the previously defined notions of fairness. Then, in Sect. 4, we propose and evaluate several adaptations to this methodology that enhance fairness. Finally, our findings are summarized and discussed in Sect. 5.

The implementation of our results can be found on GitHub[1].

2   Fairness in Machine Learning

Let $(\Omega, \mathcal{F}, \mathbb{P})$ be a probability space of interest, consisting of the sample space $\Omega \neq \emptyset$, a σ-field $\mathcal{F}$ and a probability measure $\mathbb{P}$. Moreover, let $\hat{Y} : \Omega \to \mathcal{Y}$ be some binary classifier, i.e., $\mathcal{Y} = \{0, 1\}$, being trained to model some true labels $Y : \Omega \to \mathcal{Y}$. Usually, $\hat{Y}$ can be written as some model $f : \mathcal{X} \to \{0, 1\}$, applied to the features $X : \Omega \to \mathcal{X}$, i.e., $\hat{Y} = f(X)$ holds. In recent years, the interest towards the question of such classifier $\hat{Y}$ being fair with respect to some additional, sensitive feature $S : \Omega \to \mathcal{S}$ has risen. Mostly, $\mathcal{S} = \{0, 1\}$ gives binary information about the membership or non-membership of a protected class, such as some certain gender or ethnicity (cf. [12]). While the majority of

---

[1] https://github.com/jstrotherm/FairnessInWDNS.

the literature focuses on a single binary sensitive feature S (cf. [4,11,12]), in this work, we generalize the understanding of fairness to multiple binary sensitive features $S_1, ..., S_K$ that model a single non-binary or K different (non-)binary sensitive feature(s).

Within this work, we will focus on group fairness. Assuming that all of the following conditional probabilities exist, one well-known notion of group fairness based on the predictor $\hat{Y}$ and the binary sensitive feature S is called disparate impact (DI), requiring that

$$\frac{\mathbb{P}(\hat{Y} = 1 \mid S = 0)}{\mathbb{P}(\hat{Y} = 1 \mid S = 1)} \geq 1 - \epsilon$$

is satisfied for some given value $\epsilon \in [0, 1]$, assuming that $\{S = 0\}$ is the protected group and that the nominator is smaller than the denominator (cf. [12]). The disparate impact notion is "designed to mathematically represent the legal notion of *disparate impact*" (cf. [12]). It assures that the relative amount of positive predictions within the protected group $\{S = 0\}$ deviates at most $100\epsilon\%$ from the relative amount of positive predictions within the non-protected group $\{S = 1\}$.

We generalize this definition to multiple binary sensitive features by

$$\text{DI} := \min_{k_1, k_2 \in \{1, ..., K\}} \frac{\mathbb{P}(\hat{Y} = 1 \mid S_{k_1} = 1)}{\mathbb{P}(\hat{Y} = 1 \mid S_{k_2} = 1)} \geq 1 - \epsilon. \tag{1}$$

Criticism of the disparate impact score DI could be the missing dependence on the true label Y (cf. [8]). We thus introduce another notion of group fairness, called equal opportunity (EO). In standard definition, equal opportunity holds whenever

$$\left| \mathbb{P}(\hat{Y} = 1 \mid S = 0, Y = 1) - \mathbb{P}(\hat{Y} = 1 \mid S = 1, Y = 1) \right| \leq \epsilon$$

is satisfied for some given value $\epsilon \in [0, 1]$ (cf. [11,12]). Equal opportunity ensures the true positive rates (TPRs) among protected and non-protected groups to differ at most $100\epsilon\%$.

Similarly, we generalize this definition to multiple binary sensitive features:

$$\text{EO} := \max_{\substack{k_1, k_2 \\ \in \{1, ..., K\}}} \left| \mathbb{P}(\hat{Y} = 1 \mid S_{k_1} = 1, Y = 1) - \mathbb{P}(\hat{Y} = 1 \mid S_{k_2} = 1, Y = 1) \right| \leq \epsilon.$$

$$\tag{2}$$

Remark 1. Our generalized notions of fairness go hand in hand with the conventional ones: In the conventional definitions, a single binary random variable S gives information about the membership of a protected group $\{S = 0\}$ or the membership of a non-protected group $\{S = 1\}$. Our definition handles the existence of K different groups without defining which of the groups are protected in advance. By defining the protected group $\{S = 0\}$ as group 1 and the non-protected group $\{S = 1\}$ as group 2 and the random variables S_k giving information about the membership ($S_k = 1$) or non-membership ($S_k = 0$) of group k for $k = 1, 2$, the conventional definitions and our definitions (cf. Eq. (1) and (2)) coincide, because $\{S = 0\} = \{S_1 = 1\}$ and $\{S = 1\} = \{S_2 = 1\}$ holds.

3   Leakage Detection in Water Distribution Networks

A key challenge in the domain of WDNs is to detect leakages. In this task, Ω corresponds to possible states of a WDN, given by time-dependant demands of the end users of the D nodes in the network. We assume that among those, d nodes are provided with sensors (usually, $D \gg d$), which deliver pressure measurements $p(t) \in \mathbb{R}^d$ for different times $t \in \mathbb{R}$ and which can be used for the task at hand. As we usually measure pressure values within fixed time intervals $\delta \in \mathbb{R}_+$, we introduce the notation $t_i := t_0 + i\delta$, where t_0 is some fixed reference point with respect to time.

3.1   Methodology

There are several methodologies that make use of pressure measurements to approach the problem of leakage detection using ML, i.e., by training a classifier $\hat{Y} \in \{0,1\} = \mathcal{Y}$ that predicts the true state of the WDN $Y \in \mathcal{Y}$ with respect to the question whether a leak is active (1) or not (0). One standard approach comes in two steps: In first instance, so called *virtual sensors* are trained, i.e., regression models being able to predict the pressure at a given node $j \in \{1, ..., d\}$ and some time $t_i \in \mathbb{R}$, based on measured pressure at the remaining nodes $\hat{j} \neq j$ and over some discrete time interval of size $T_r + 1 \in \mathbb{N}$. Subsequently, these virtual sensors are used to compute *pressure residuals* of measured and predicted pressure to train an ensemble classifier that is able to predict whether a leakage is present in the WDN at the time of the used residual (cf. [9]).

Virtual Sensors. The virtual sensors $f_j^r : \mathbb{R}^{d_r} \to \mathbb{R}$ for each node $j \in \{1, ..., d\}$ and $d_r := d - 1$ are linear regression models trained on leakage free training data $\mathcal{D}_j^r = \{(\overline{p}_{\neq j}(t_i), p_j(t_i)) \in \mathbb{R}^{d_r} \times \mathbb{R} \mid i = 0, ..., n_r\}$. More precisely, $y(t_i) = 0 \in \mathcal{Y}$ holds for all realisations $i = 0, ..., n_r$ of Y and the inputs are given by the rolling means $\overline{p}_{\neq j}(t_i) := (T_r + 1)^{-1} \sum_{\iota=0}^{T_r} p_{\neq j}(t_i - \iota\delta)$ at all nodes except the node j, which is the only preprocessing required for the training pipeline (cf. [2]).

Ensemble Leakage Detection. Standard leakage detection methods rely on the residuals $r_j(t_i) := |p_j(t_i) - f_j^r(\overline{p}_{\neq j}(t_i))| \in \mathbb{R}_+$ we obtain from the true pressure measurements $p(t_i) \in \mathbb{R}^d$ and the virtual sensor predictions $f_j^r(\overline{p}_{\neq j}(t_i)) \in \mathbb{R}$ for each sensor node $j \in \{1, ...d\}$ and (possibly unseen) times $t_i \in \mathbb{R}$ (cf. [9]).

A simple detection method performing good on standard benchmarks is the *threshold-based ensemble classification* introduced by [2]: Without any further training, we can define a classifier $f_j^c : \mathbb{R}_+ \to \mathcal{Y}$ by

$$f_j^c(r_j(t_i)) = f_j^c(r_j(t_i), \theta_j) := \mathbb{1}_{\{r_j(t_i) > \theta_j\}}$$

for each sensor node $j \in \{1, ...d\}$ and a node-dependant hyperparameter $\theta_j \in \mathbb{R}_+$. We easily obtain an ensemble classifier $f^c : \mathcal{X} \to \mathcal{Y}$, called the H-method, with

Hyperparameter $\Theta := (\theta_j)_{j=1,...,d} \in \mathcal{X}$ for $\mathcal{X} := \mathbb{R}_+^{d_c}$ and $d_c := d$ that predicts whether there is a leakage present in the WDN at time $t_i \in \mathbb{R}$ or not, defined by

$$f^c(r(t_i)) = f^c(r(t_i), \Theta) := \mathbb{1}_{\{\sum_{j=1}^{d_c} f_j^c(r_j(t_i)) \geq 1\}}. \qquad (3)$$

Evaluation. We evaluate the H-method in terms of general performance, measured by accuracy (ACC), and in terms of fairness, measured by disparate impact as well as equal opportunity score DI and EO, respectively (cf. Eq. (1) and (2)) in Sect. 3.3, after introducing the application domain and data set.

3.2   Application Domain and Data Set

One key contribution of this work is to introduce the notion of fairness in the application domain of WDNs. The WDN considered is *Hanoi* (cf. [13]) displayed in Fig. 1. It consists of 32 nodes and 34 links.

To evaluate the H-method presented in Sect. 3.1 on Hanoi, we generate pressure measurements with a time window of $\delta = 10$min. using the atmn toolbox (cf. [14]). The pressure is simulated at the sensor nodes displayed in Fig. 1 and for different leakage scenarios, which differ in the leakage location and size. As the WDN is relatively small, we are able to simulate a leakage at each node in the network and for three different diameters $d \in \{5, 10, 15\}$cm. For the preprocessing according to Sect. 3.1, we choose $T_r = 2$ such as [2] do.

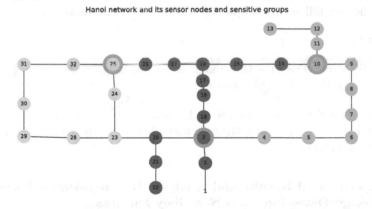

Hanoi network and its sensor nodes and sensitive groups

Fig. 1. The Hanoi WDN, its sensor nodes (IDs 3, 10 and 25) and the protected groups, each highlighted in another color (group 1 on the left side in light shade, group 2 in the middle in dark shade, group 3 on the right side in middle shade). The sensor nodes are colored in the same color of the protected group they belong to and highlighted with a grey circle.

The question arises how the leakage detection is related to fairness. Knowing that each node of the network corresponds to a group of consumers, a natural question is whether these local groups benefit from the WDN and its related services in equal degree. To ensure that the algorithms that will be presented in Sect. 4.1 scale to larger WDNs, we do not consider single nodes but groups of nodes in the WDN as protected groups in terms of fairness. Then, for all $k = 1, ..., K$, we define the sensitive feature $S_k \in \{0, 1\}$ to give answer to the question whether (1) or whether not (0) a leakage is active in such a protected group k. In terms of equal service one would expect an equally good detection of leakages independent on the leakage location, i.e., the protected group. For Hanoi, we work with $K = 3$ different groups, also displayed in Fig. 1.

Given *this* definition of sensitive features S_k for $k = 1, ..., K$ in the WDN, we obtain the following important results with regard to the notions of fairness.

Lemma 1 (Equivalence of disparate impact and equal opportunity in WDNs). *Let S_k be the sensitive feature describing whether a leakage is active in the protected group k of the WDN for each $k = 1, ..., K$. Moreover, let $\epsilon, \tilde{\epsilon} \in [0, 1]$ and define $\max_k := \max_{k \in \{1, ..., K\}} \mathbb{P}(\hat{Y} = 1 \mid S_k = 1)$.*

1. *If disparate impact holds with ϵ, equal opportunity holds with $\tilde{\epsilon} = \epsilon \max_k$.*
2. *If equal opportunity holds with $\tilde{\epsilon}$, disparate impact holds with $\epsilon = \tilde{\epsilon}(\max_k)^{-1}$.*

Proof. First of all, note that for any $k \in \{1, ..., K\}$ and $\omega \in \Omega$, $S_k(\omega) = 1$ implies $Y(\omega) = 1$ by definition of the sensitive feature S_k. Therefore, $\{Y = 0, S_k = 1\}$ is empty. Subsequently, we obtain $\{Y = 1, S_k = 1\} = \{S_k = 1\}$ and thus, $\mathbb{P}(\hat{Y} = 1 \mid Y = 1, S_k = 1) = \mathbb{P}(\hat{Y} = 1 \mid S_k = 1)$.

Secondly, we also define $\min_k := \min_{k \in \{1, ..., K\}} \mathbb{P}(\hat{Y} = 1 \mid S_k = 1)$. We then obtain $\mathrm{DI} = \frac{\min_k}{\max_k}$ and, together with the first observation, $\mathrm{EO} = \max_k - \min_k$. Now the rest follows by simple equivalent transformations.

Corollary 1. *Given the setting of Lemma 1,*

1. *$\mathrm{EO} = \tilde{\mathrm{EO}}$ for $\tilde{\mathrm{EO}} := (1 - \mathrm{DI}) \cdot \max_k$ and*
2. *$\mathrm{DI} = \tilde{\mathrm{DI}}$ for $\tilde{\mathrm{DI}} := 1 - \frac{\mathrm{EO}}{\max_k}$ holds.*

Proof. This is a direct consequence of Lemma 1, where we choose $\epsilon := 1 - \mathrm{DI}$ in setting 1 and $\tilde{\epsilon} := \mathrm{EO}$ in setting 2, and where we can work with equalities instead of estimations.

3.3 Experimental Results and Analysis: Residual-Based Ensemble Leakage Detection Does Not Obey Fairness

In Table 1, the results of the H-method presented in Sect. 3.1 are shown. The hyperparameter $\Theta \in \mathcal{X} = \mathbb{R}_+^{d_c}$ is chosen manually per diameter d such that the test accuracy is close to maximal. On the one hand, we see that the method in general performs better the larger the leakage size is (measured in ACC), as larger leakages are associated with larger pressure drops. Note that the method is capable of detecting even small leakages with high accuracy in larger (and therefore, more realistic) WDNs (cf. [2]).

Table 1. Results of the H-method with $\max_k = \max_k \mathbb{P}(\hat{Y} = 1 \mid S_k = 1)$ and $\min_k = \min_k \mathbb{P}(\hat{Y} = 1 \mid S_k = 1)$ according to (the proof of) Lemma 1. Moreover, disparate impact score DI and $\tilde{\text{DI}}$ as well as equal opportunity score EO and $\tilde{\text{EO}}$ according to Eq. (1), (2) and Corollary 1.2, 1.1, respectively.

d	ACC	$\max_k$	$\min_k$	DI	EO	$\tilde{\text{DI}}$	$\tilde{\text{EO}}$
5	0.6223	0.8468	0.4880	0.5763	0.3558	0.5763	0.3588
10	0.7998	0.9983	0.6372	0.6383	0.3611	0.6383	0.3611
15	0.8837	1.0000	0.6402	0.6402	0.3598	0.6402	0.3598

On the other hand, we see that the method is unfair in terms of disparate impact score DI, where a value of 0.8 or larger is desirable (cf. [15]), and equal opportunity score EO. However, the experimental evaluation confirms the mathematical findings of Corollary 1 by comparing the corresponding columns in Table 1. This also justifies that in our setting, the usage of one of the two measures is sufficient. Therefore, from now on, we work with disparate impact score DI only.

4 Fairness-Enhancing Leakage Detection in Water Distribution Networks

Motivated by the result that the standard leakage detection method presented in Sect. 3.1 does not satisfy the notions of fairness, as another main contribution of this work, we modify this H-method to enhance fairness as introduced in Sect. 2. Given virtual sensors f_j^r for all $j = 1, ..., d$ and resulting residuals $r(t_i) \in \mathcal{X} = \mathbb{R}_+^{d_c}$ (cf. Sect. 3.1) as well as labels $y(t_i) \in \mathcal{Y} = \{0, 1\}$ for times $t_i \in \mathbb{R}$, we can turn the choice of the hyperparameter $\Theta := (\theta_j)_{j=1,...,d} \in \mathcal{X}$ of the ensemble classifier f^c (cf. Eq. (3)) into an optimization problem (OP) with corresponding function space $\mathcal{H} := \{f^c : \mathcal{X} \to \mathcal{Y}, \ r \mapsto f^c(r, \Theta) \mid \Theta \in \mathcal{X}\}$. In the following section, we therefore present different, in contrast to the H-method optimization-based, methods that aim at optimizing the parameter $\Theta \in \mathcal{X}$ in order to obtain an optimal ensemble classifier $f^c(\cdot, \Theta_{\text{opt}}) \in \mathcal{H}$. These methods on the one hand are further baselines, where treating the modelling problem as an OP enables us to optimize the result of the H-method itself without fairness considerations. On the other hand, we consider fairness-enhancing methods, where the parameter $\Theta \in \mathcal{X}$ needs to be optimized such that the resulting ensemble classifier is as accurate and fair on the given training data as possible.

4.1 Methodology

The following methods define training algorithms based on labeled training data $\mathcal{D}^c = \{(r(t_i), y(t_i)) \in \mathcal{X} \times \mathcal{Y} \mid i = 1, ..., n_c\}^2$ for an $n_c > n_r$, which also holds

---

[2] In practise, we train and test the (ensemble) classifier(s) on unseen data for times $i \geq n_r + 1$. However, for the sake of readability, we choose the indices $i = 1, ..., n_c$ instead of $i = n_r + 1, ..., n_c$ here.

data based on leaky states of the WDN. For simplicity, we omit the dependence of all (loss) functions on the training data.

Optimizing Loss with Fairness Constraints. In general, a learning problem can be phrased as an OP, where the objective is to minimize some suitable loss function $L : \mathcal{X} \to \mathbb{R}$ with respect to the parameter $\Theta \in \mathcal{X}$, i.e.,

$$\left\{ \min_{\Theta \in \mathcal{X}} \; L(\Theta). \right. \tag{4}$$

The advantage of redefining the choice of hyperparameters Θ (H-method) as an OP is that we can now extend this OP by fairness constraints, which can be given by side constraints $C_k : \mathcal{X} \to \mathbb{R}$ of the underlying OP:

$$\begin{cases} \min_{\Theta \in \mathcal{X}} & L(\Theta), \\ \text{s.t.} & C_k(\Theta) \geq 0 \; \forall k = 1, ..., \hat{K}. \end{cases} \tag{5}$$

Choice of Loss Functions. In view of the notions of fairness (cf. Sect. 2), an intuitive and by means of linearity easily to differentiate loss function is given by the difference of the false positive rate (FPR) and the TPR, i.e., $L_1(\Theta) := - \text{TPR}(\Theta) + \text{FPR}(\Theta)$. Another classical evaluation score which we can use as a loss function is the accuracy $L_2(\Theta) := - \text{ACC}(\Theta)$.

Choice of Fairness Constraints. In terms of fairness constraints, [15] introduce the covariance between a single binary sensitive feature and the signed distance of a feature vector and the decision boundary of a convex margin-based classifier as a proxy for fairness measurements. We adapt this idea to our setting by considering the covariance of each sensitive feature and replacing the signed distance by the prediction of the ensemble classifier $\hat{Y} = f^c(X, \Theta)$. Using that $\hat{y}(t_i) = f^c(r(t_i), \Theta)$ holds for all realisations $i = 1, ..., n_c$, for all sensitive features S_k for $k = 1, ..., K$, the *empirical* covariance is given by

$$\text{Cov}_{\text{emp.}}(S_k, \hat{Y}) = \frac{1}{n_c} \sum_{i=1}^{n_c} (s_k(t_i) - \overline{s_k}) \cdot f^c(r(t_i), \Theta). \tag{6}$$

Remark 2. The usage of the (empirical) covariance as a proxy for fairness is based on the idea that fairness of a machine learning model $\hat{Y}$ can be interpreted as the assumption of $\hat{Y}$ being independent of the sensitive feature S (cf. [11]), or in our case, each of the sensitive features S_k for $k = 1, ..., K$. As independence of two random variables implies their covariance being equal to zero, the latter can be interpreted as a necessary condition for fairness. Therefore, we consider the covariance of each sensitive feature S_k for $k = 1, ..., K$ and the prediction of the ensemble classifier $\hat{Y} = f^c(X, \Theta)$, which by linearity is given by

$$\text{Cov}(S_k, \hat{Y}) = \mathbb{E}((S_k - \mathbb{E}(S_k)) \cdot (\hat{Y} - \mathbb{E}(\hat{Y})) = \mathbb{E}((S_k - \mathbb{E}(S_k)) \cdot \hat{Y}).$$

However, as the probability measure $\mathbb{P}(S_k, \hat{Y})^{-1}$ on $\mathcal{Y} \times \mathcal{Y}$ is unknown, we replace it by its empirical approximation $\frac{1}{n_c} \sum_{i=1}^{n_c} \delta_{(s_k(t_i), \hat{y}(t_i))}$ and obtain the empirical covariance (6). As in practise, an exact value of zero will rarely be achieved, enforcing the empirical covariance to be small is a reasonable fairness proxy.

Assuming that a comparatively high (empirical) covariance (in either positive or negative direction) between a sensitive feature S_k for $k \in \{1, ..., K\}$ and the model's prediction $\hat{Y} = f^c(X, \Theta)$ implies a significant difference in the relative amount of positive predictions in contrast to the remaining sensitive features leads to the idea of constraining the absolute value of the (empirical) covariance as a side constraint in the above considered OP (cf. Eq. (4)) (cf. [15]).

Motivated by that, we require $\text{Cov}_{\text{emp.}}(S_k, \hat{Y}) \leq c$ and $\text{Cov}_{\text{emp.}}(S_k, \hat{Y}) \geq -c$ or, equivalently formulated in standard form,

$$C_k(\Theta) := c - \text{Cov}_{\text{emp.}}(S_k, \hat{Y}) \geq 0 \text{ and } C_k(\Theta) := c + \text{Cov}_{\text{emp.}}(S_k, \hat{Y}) \geq 0$$

to hold for all $k = 1, ..., K$ (i.e., $\hat{K} = 2K$ in Eq. (5)). Hereby, the hyperparameter $c \in [0, \infty)$ regulates how much the covariance's absolute value is bounded and therefore, the desired fairness (cf. Remark 2).

Table 2. Overview of the proposed methods.

Method	Loss	Constraints
T-F-PR	L_1	-
T-F-PR+F	L_1	emp. Cov
ACC	L_2	-
ACC+F	L_2	emp. Cov

Explicit Methods. The resulting methods as a combination of used loss function with or without the fairness-enhancing side constraint (cf. OP (4) or (5)) deliver two baseline and two fairness-enhancing ensemble leakage detection algorithms, summarized in Table 2.

Differentiable Approximation of the Learning Problems. Loss function and side constraint (cf. Eq. (6)) clearly depend on the model's prediction $\hat{Y} = f^c(X, \Theta)$ resp. $y(t_i) = f^c(r(t_i), \Theta)$ for all $i = 1, ..., n_c$. However, in view of the model's definition (cf. Eq. (3)), f^c is not differentiable with respect to Θ. To make $\hat{Y} = f^c(X, \cdot)$ differentiable, we approximate each indicator function $\mathbb{1}_{\{v>0\}}$ by the sigmoid function $\text{sgd}_b(v) - (1 + \exp^{-bv})^{-1}$ with hyperparameter $b \in \mathbb{R}_+$. All in all, we obtain a differentiable OP by replacing the ensemble classifier's prediction $f^c(r(t_i), \Theta)$ (cf. Eq. (3)) by

$$\hat{f}^c(r(t_i), \Theta) := \text{sgd}_b \left(\sum_{j=1}^d \text{sgd}_b(r_j(t_i) - \theta_j) - T \right) \tag{7}$$

for all $i = 1, ..., n_c$, where we replace the threshold 1 of the exact ensemble classifier f^c by a hyperparameter $T \in [0, 1]$ to handle the insecurity of the sigmoid function around zero. Then, by expressing the losses $L_1 = -\text{TPR} + \text{FPR}$ and $L_2 = -\text{ACC}$ by

$$L_1(\Theta) = -\frac{\sum_{i=1}^{n_c} y(t_i) \cdot f^c(r(t_i), \Theta)}{\sum_{i=1}^{n_c} y(t_i)} + \frac{\sum_{i=1}^{n_c}(1 - y(t_i)) \cdot f^c(r(t_i), \Theta)}{\sum_{i=1}^{n_c}(1 - y(t_i))},$$

$$L_2(\Theta) = \frac{\sum_{i=1}^{n_c} y(t_i) \cdot f^c(r(t_i), \Theta) + \sum_{i=1}^{n_c}(1 - y(t_i)) \cdot (1 - f^c(r(t_i), \Theta))}{n_c},$$

their approximated versions using $\hat{f}^c$ instead of f^c will be differentiable with respect to Θ as well, and so is the empirical covariance (cf. Eq. (6)) when using $\hat{f}^c$ instead of f^c. The resulting approximated OPs can therefore be optimized with a gradient-based optimization technique.

Optimizing Fairness with Accuracy Constraints. Instead of optimizing some loss function L under some fairness side constraints, [15] suggest to optimize a fairness proxy under loss constraints. They use the covariance as a proxy while constraining the training loss by some percentage of the optimal loss obtained when training without fairness considerations. As a variation, we use the disparate impact score DI directly as a loss function and the accuracy ACC for the constraint. The resulting DI+ACC-method is therefore given by

$$
\begin{cases}
\min_{\Theta \in \mathcal{X}} & -\operatorname{DI}(\Theta), \\
\text{s.t.} & \operatorname{ACC}(\Theta) \geq (1 - \lambda) \operatorname{ACC}_{\text{opt.}}.
\end{cases} \tag{8}
$$

The hyperparameter $\lambda \in [0, 1]$ hereby regulates how much the obtained accuracy $\operatorname{ACC}(\Theta)$ is allowed to differ from the optimal accuracy $\operatorname{ACC}_{\text{opt.}}$ received in the ACC-method (cf. Table 2).

In contrast to the methods proposed in Sect. 4.1, we like to test the OP (8) as a non-differentiable OP, which therefore requires a non-gradient-based optimization technique.

Evaluation. We evaluate all presented methods, i.e., the standard H-method (Sect. 3.1), the optimization-based baselines T-F-PR- and ACC-method as well as the fairness-enhancing T-F-PR+F-, ACC+F- (cf. Sect. 4.1 and Table 2) and DI+ACC-method (cf. Sect. 4.1), as we did in Sect. 3.1.

4.2   Experimental Results and Analysis

Based on the pressure measurements in the Hanoi WDN as introduced in Sect. 3.2 and the resulting residuals, we test all six methods introduced in Sect. 3.1 (H-method) and Sect. 4.1 (T-F-PR-, ACC-, T-F-PR+F, ACC+F, DI+ACC-method, also see Evaluation in Sect. 4.1) per diameter d in practise.

Setup. *H-method* We use the H-method presented in Sect. 3.1 and tested in Sect. 3.3 as a baseline. Subsequently, we use the hyperparameter found here as an initial parameter $\Theta_0 \in \mathcal{X}$ for the remaining optimization-based methods.

Gradient-Based Methods. While the T-F-PR- and the ACC-method are used as another baseline, the remaining methods are fairness-enhancing methods. The magnitude of fairness can be regulated by a hyperparameter: The T-F-PR+F- and ACC+F-method ensure fairness by bounding the empirical covariance of each sensitive feature and the models approximated prediction (cf. Eq. (6) and (7)) by the hyperparameter $c \in [0, \infty)$. In addition, for all these methods, we choose $b = 100$ and $T = 0.8$.

Non-Gradient-Based Method. In contrast, the DI+ACC-method regulates fairness by different choices of the hyperparameter $\lambda \in [0, 1]$ that controls how much loss in accuracy is allowed while increasing fairness.

Transforming Constraint OPs in Non-Constraint OPs. For all OPs, we use the log-barrier method (cf. [10]) to transform the constrained OP into a non-constrained one and tune the regularization hyperparameter per method.

Optimization Techniques. For the differentiable OPs (T-F-PR-, ACC-, T-F-PR+F- and ACC+F-method), we use BFGS (cf. [10]) to find the optimal parameter $\Theta_{opt.} \in \mathcal{X}$. For the non-differentiable OP (DI+ACC-method), we use Downhill-Simplex-Search, also known as Nelder-Mead (cf. [7]). Each method is trained per diameter d on 40% of the data and evaluated on the remaining data.

Results and Analysis. *Increasing Fairness.* In Fig. 2, we see the performance of each ensemble classifier measured in accuracy and disparate impact score. For the fairness-enhancing methods, we test different hyperparameters, causing error bars for these methods. We start with a hyperparameter c and λ that causes an accuracy of 0.5 and disparate impact score of 1.0 whenever possible and in- resp. decrease the hyperparameter by 0.01 until the disparate impact score of the fairness-enhancing model achieves the disparate impact score of the corresponding baseline (T-F-PR for T-F-PR+F and ACC for ACC+F and DI+ACC). The height of the bars with error bars correspond to the mean accuracy and disparate impact score, achieved by each method over all hyperparameters tested. The error bars themselves reach from the lowest to the largest score of the two scores considered.

We see that for all fairness-enhancing methods and all leakage sizes, the fairness-enhancing methods on average increase fairness, measured by disparate impact score, while on average, mostly decreasing accuracy by only some small percent-

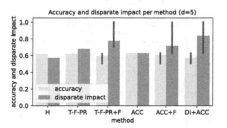

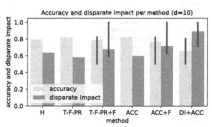

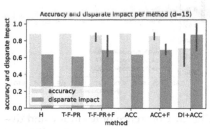

Fig. 2. Accuracy and disparate impact score per method and leakage diameter.

age compared to their corresponding baselines. For $d = 5$, all fairness-enhancing methods allow a large range of fairness improvement at cost of a small range in accuracy, which is only due to the relatively poor accuracy of the leakage detector in this scenario in general. For the other diameters, the ranges of fairness and accuracy are similarly large. Thus, one can say that fairness and overall performance are mutually dependent to about the same extent.

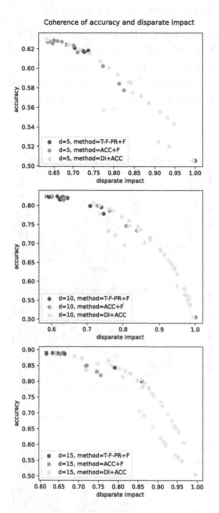

Fig. 3. Coherence of accuracy and disparate impact score for the different fairness-enhancing methods and different leakage sizes. The cross data points visualize the disparate impact score and accuracy of the non-fairness-enhancing baselines (T-F-PR, dark blue, for T-F-PR+F and ACC, light blue, for ACC+F and DI+ACC). (Color figure online)

The Coherence of Fairness and Overall Performance. A more detailed visualization of how fairness is related to the overall performance of the model can be found in Fig. 3. For each tested hyperparameter c and λ, respectively, depending on what fairness-enhancing method was used, the obtained disparate impact score is plotted together with the observed accuracy. For better readability, we split these observations by the leakage sizes tested.

The characteristic curve that can be observed in all subimages is the so-called pareto front, visualizing that the increase in fairness is accompanied by the reduction in accuracy score and vice versa. A desired disparate impact score of about 0.8 can be achieve by a decrease of accuracy by approximately 0.03–0.05 points below the optimal accuracy obtained. Hereby, both fairness and overall performance can be influenced by the fairness hyperparameters c and λ, respectively. Deciding which choice of fairness hyperparameter is optimal is a difficult task that depends on the extent of the decisions of the underlying ML model as well as legal requirements. Regarding legal requirements, by not using the sensitive features for the decision making of the algorithm, the methods presented can satisfy the legal definition of disparate treatment *and* disparate impact (depending on the hyperparameter chosen) simultaneously.

Another observation is that the largest accuracies of the fairness-enhancing methods are approximately as good as the accuracy of their baseline methods while achieving equal or better fairness results. In opposite direction, perfect fairness of 1.0 can be achieved at a cost of the worst possible accuracy of 0.5. While for the covariance-based algorithms (T-F-PR+F- and ACC+F-method), the jump in disparate impact and accuracy score is rather abrupt when reaching the extreme of (1.0, 0.5), the method relying on the optimization of fairness while constraining on the accuracy (DI+ACC-method) allows more fine-grained variations in both scores. This is due to the fact that the hyperparameter λ regulates the accuracy and not the fairness measure. The accuracy constraint is less sensitive to the log barrier method than the covariance constraints, since a too small choice of the hyperparameter c quickly sets all punishment terms to infinity and thus outputs the trivial solution.

The Influence of the Hyperparameters on Fairness and Overall Performance. In Fig. 4, we show how the hyperparameters are related to disparate impact and accuracy score. Each of the two scores is plotted against the used hyperparameter for all fairness-enhancing methods and leakage diameters tested.

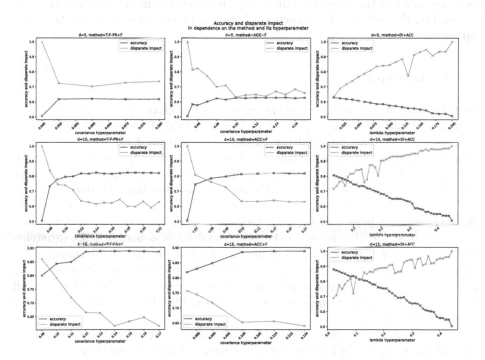

Fig. 4. Coherence of accuracy, disparate impact and the training hyperparameter.

For the T-F-PR+F- and the ACC+F-method, the decrease of the hyperparameter c is accompanied by the improvement of the fairness measure as well as the decrease of the performance measure. This can be explained by the intuition described before: A high empirical covariance of a sensitive feature and the prediction of the approximated ensemble model means that the relative number of positive predictions within the related group differs significantly from the relative number of positive predictions within a group with small covariance. Thus, the more the covariance is constrained, the less such extreme differences in the relative number of positive predictions across groups occur, leading to a better fairness score. In the case of disparate impact, therefore, a (better) higher score at the expense of a (worse) lower overall performance - compared to the overall performance that occurs in the unconstrained case or for a looser constraint, that is a larger bound c, - appears.

In contrast, for the DI+ACC method the increase of the hyperparameter λ is accompanied by the improvement of the fairness measure as well as the decrease of the performance measure due to the fact that a higher hyperparameter λ allows a larger deviation of the optimal accuracy score. Thus, a worse accuracy is penalized less so that the fairness measure can be optimized to a larger extend.

Non Optimality. Last but not least, note that the non-optimal solutions and the local jumps recognized in Fig. 3 and 4, respectively, can be explained by the non-convexity of the objective functions. Therefore, the found solutions strongly depend on the initialized parameter Θ_0 and might not correspond to the global optimum.

5   Conclusion

In this work, we introduced the notion of fairness in an application domain of high social and ethical relevance, namely in the field of water distribution networks (WDNs). This required the extension of fairness definitions for a single binary sensitive feature to single non-binary or multiple, possibly even non-binary, sensitive features. We then investigated on the fairness issue in the area of leakage detection within WDNs. We showed that standard approaches are not fair in the context of different groups related to the locality within the network. As a remedy, we presented methods that increase fairness of the ensemble classification model with respect to the introduced fairness notion while satisfying the legal notions of disparate treatment and disparate impact simultaneously. We empirically demonstrated that fairness and overall performance of the model are interdependent and the use of hyperparameters provides the ability to trade off fairness and overall performance. However, this trade off lies in the responsibility of the policy maker.

To allow more fine-grained steps between improving fairness and decreasing overall performance in the presented covariance-based approaches, next steps would be to swap loss function and constraint to achieve similar results as in the approach with accuracy constraint. Moreover, the notion of fairness within the water domain is still in its beginning and extensions to more complex WDNs as well as more powerful ML algorithms is essential.

Acknowledgments. We gratefully acknowledge funding from the European Research Council (ERC) under the ERC Synergy Grant Water-Futures (Grant agreement No. 951424).

References

1. Angwin, J., Larson, J., Mattu, S., Lauren Kirchner, L.: Machine Bias - There's software used across the country to predict future criminals. And it's biased against blacks. ProPublica (2016)
2. Artelt, A., Vrachimis, S., Eliades, D., Polycarpou, M., Hammer, B.: One Explanation to Rule them All - Ensemble Consistent Explanations. arXiv preprint arXiv:2205.08974 (2022)
3. Barocas, S., Hardt, M., Narayanan, A.: Fairness and Machine Learning: Limitations and Opportunities. fairmlbook.org (2019). www.fairmlbook.org
4. Castelnovo, A., Crupi, R., Greco, G., Regoli, D., Penco, I.G., Cosentini, A.C.: A clarification of the nuances in the fairness metrics landscape. Sci. Rep. **12**(1), 4209 (2022)
5. Dwork, C., Hardt, M., Pitassi, T., Reingold, O., Zemel, R.: Fairness through awareness. In: Proceedings of the 3rd Innovations in Theoretical Computer Science Conference, pp. 214–226 (2012)
6. European Commission and Directorate-General for Communications Networks, Content and Technology: Ethics guidelines for trustworthy AI. Publications Office (2019). https://doi.org/10.2759/346720
7. Gao, F., Han, L.: Implementing the Nelder-Mead simplex algorithm with adaptive parameters. Comput. Optim. Appl. **51**(1), 259–277 (2012)
8. Hardt, M., Price, E., Srebro, N.: Equality of opportunity in supervised learning. In: Advances in Neural Information Processing Systems, vol. 29 (2016)
9. Isermann, R.: Fault-Diagnosis Systems. Springer, Heidelberg (2006). https://doi.org/10.1007/3-540-30368-5
10. Jorge Nocedal, S.J.W.: Numerical Optimization, vol. 2. Springer, New York (2006)
11. Mehrabi, N., Morstatter, F., Saxena, N., Lerman, K., Galstyan, A.: A survey on bias and fairness in machine learning. ACM Comput. Surv. (CSUR) **54**(6), 1–35 (2021)
12. Pessach, D., Shmueli, E.: A review on fairness in machine learning. ACM Comput. Surv. (CSUR) **55**(3), 1–44 (2022)
13. Santos-Ruiz, I., López-Estrada, F.R., Puig, V., Valencia-Palomo, G., Hernández, H.R.: Pressure sensor placement for leak localization in water distribution networks using information theory. Sensors **22**(2), 443 (2022)
14. Vaquet, J.: Automation Toolbox for Machine learning in water Networks (2023). www.pypi.org/project/atmn/
15. Zafar, M.B., Valera, I., Rogriguez, M.G., Gummadi, K.P.: Fairness constraints: mechanisms for fair classification. In: Artificial Intelligence and Statistics, pp. 962–970. PMLR (2017)

Measuring Fairness with Biased Data: A Case Study on the Effects of Unsupervised Data in Fairness Evaluation

Sarah Schröder[✉][iD], Alexander Schulz[iD], Ivan Tarakanov, Robert Feldhans[iD], and Barbara Hammer[iD]

CITEC, Bielefeld University, Inspiration 1, 33619 Bielefeld, Germany
{saschroeder,aschulz,itarakanov,
rfeldhans,bhammer}@techfak.uni-bielefeld.de

Abstract. Evaluating fairness in language models has become an important topic, including different types of measurements for specific models, but also fundamental questions such as the impact of pre-training biases in fine-tuned models. Ultimately, many rely on a data based evaluation using one of the few larger datasets for this purpose. We investigate one of them, the BIOS dataset [1], that has been employed in several studies. It is an entirely unsupervised dataset, in the sense that it is scraped from the web and automatically labeled. We investigate this dataset in depth and expose a variety of flaws such as out-of-domain samples or falsely labeled samples, which particularly affect the biases measured on this dataset. We consider a subset that we review, relabel and clean, then reproduce fairness experiments from the literature both on the original and cleaned subset and demonstrate, that biases are strongly affected by the mentioned problems.

Keywords: Large Language Models · Fairness · Text Corpora

1 Introduction

With the advent and success of deep learning models, the datasets that are used to training these models have grown drastically, containing billions of training instances [2] most recently. In context of Large Language Models (LLMs) it is known since several years that these models tend to incorporate biases from their training datasets and methods to measure and remove such biases have been investigated [3,4]. It has also been shown that even filtered datasets can contain problematic content [5]. One common approach to investigate bias in LLMs is to analyse their behaviour w.r.t. specifically for that designed bias evaluation datasets. Such a dataset has been proposed by De-Arteaga et al.: Bias in Bios [1]. More precisely, they published a crawler, that collects short biographies from the internet and automatically derives a job label and gender label. Naturally, other

I. Rojas et al. (Eds.): IWANN 2023, LNCS 14134, pp. 134–145, 2023.
https://doi.org/10.1007/978-3-031-43085-5_11

works used the dataset [7–9]. For instance, Steed et al. [8] did rather fundamental work by investigating the bias transfer hypothesis, i.e. to what extent does bias from pre-trained language models persist after fine-tuning and affect the downstream task. However, since none of these works provide the actual dataset, but only reuse the code for crawling the dataset from the web, reproducibility of their works and comparability with potential follow-up work is strictly limited. Moreover, when recreating the BIOS dataset ourselves and investigating it, we found several flaws that arise as a result from the unsupervised/self-supervised fashion in which the dataset is scraped. This includes issues such as out-of-domain samples and falsely labeled samples. Since the bias measure used by De-Arteaga et al. [1] specifically compares true positive rates per gender, any systematic issue with the ground truth labels may impact their bias measures significantly. In addition, when investigating fundamental questions such as the bias transfer hypothesis, using a high-quality (supervised) dataset appears to be the better choice. This paper offers the following contributions: (i) We analyze the BIOS dataset and document quality issues together with several sources for it; (ii) we propose criteria for quality control and labeling and apply these to a subset of the data; (iii) utilizing the reviewed subset, we reproduce the bias experiments by De-Arteaga et al. and show how the dataset quality affects bias measurements; (iv) we publish code to reconstruct our reviewed subset to allow fellow researchers to reproduce our work[1].

Specifically, in Sect. 3 we summarize issues that we found in the dataset. In Sect. 4 we introduce our quality criteria, describe the review process and the results of our review. We reproduce the bias experiments in Sect. 5.

2 Foundations

In this section, we detail the central concepts of the study [1], which we will build upon in our own work. On a high level, the authors collect data of descriptions including persons name and an occupation, train a classifier to predict the binary gender from each description and evaluate how much these predictions are biased for each occupation. Details on the data set and gender bias quantification is given in the following.

Dataset. De-Arteaga et al. [1] acquired the Bios dataset from the Common Crawl[2] by searching for strings of the form "<name> is a(n) <description> <title>", where <name> presents a name-like pattern (a sequence of capitalized words) and <title> an occupation. Each biography was labeled according to the identified name, title and had a gender assigned based on the most frequent pronouns (male or female). They chose twenty-eight occupations from the BLS Standard Occupation Classification system[3] that were most commonly found in a subset of the Common Crawl. We use the author's code to crawl the dataset

---

[1] https://github.com/HammerLabML/MeasuringFairnessWithBiasedData.
[2] https://commoncrawl.org/.
[3] https://www.bls.gov/soc/.

and apply their pre-processing, as the dataset itself has not been published. For our dataset review, we use a subset of the data containing only ten classes. This review process itself will be explained in Sect. 4.

Explicit Gender Indicators. Just like De-Arteaga et al. [1] we consider two versions of biographies, first the raw texts (with explicit gender indicators), second we employ the "gender scrubbing" from [1], which deletes the subject's first name and the words *he, she, her, his, him, hers, himself, herself, mr, mrs, and ms.*

Gender GAP. In [1] the true positive rate $\text{TPR}_{g,y}$ is recorded for predicting the correct occupation y, with gender $g \in \{\text{male}, \text{female}\}$. Then, they compute $\text{GAP}_{g,y} = \text{TPR}_{g,y} - \text{TPR}_{\sim g,y}$, where $\sim g$ refers to $\{\text{male}, \text{female}\} \setminus g$. Finally, the authors inspect $\text{GAP}_{\text{female},y}$ over $\pi_{\text{female},y}$, where $\pi_{g,y}$ denotes the percentage of people with gender g in occupation y and observe a strong correlation. Noteworthy, this bias measure corresponds to Equalized Odds [10], which is more prominently used in the literature.

3   Dataset Analysis

In this section we perform an in-depth analysis of the utilized data set and report incorrect labeling regarding the name detection, occupation detection and gender scrubbing. We point to several reasons for these issues and present examples.

Reproducibility. In order to allow researchers to reproduce their work, the authors published the code used to create the dataset. However, it must be noted that this code might produce an incomplete version of the potential dataset if server calls to the Common Crawl fail. Consequently, there is no guarantee that different works using the BIOS dataset actually used the exact same data.

3.1   Quality Issues

Other than that, due to the automated approach of data acquisition and labeling without any human supervision, there is no guarantee for the quality of the dataset. By reviewing parts of the dataset, we identified a number of issues. To some extent these issues could be considered normal or unavoidable in a real world dataset, maybe even valuable as training data in terms of producing robust models. However, users of the dataset need to be aware of these things to make informed decision in their experiments or model deployment. Furthermore, when evaluating either performance, robustness or fairness we need a reliable test set. Especially false or insufficient labels (or false pronouns for fairness evaluation) would immediately impact evaluation results.

Name Detection. Since the name-like pattern was identified by a sequence capitalized words, any other capitalized word could falsely be identified as (part of) the name. Furthermore, some names might not represent persons, but places. For example, "Abu Dhabi is a model for modernization in the Islamic world [...]" was selected as a biography with name "Abu Dhabi" and occupation "model" or in "Photographer <first name> <last name> [...]", photographer was interpreted as the first name, followed by a middle and last name. However, the same biography existed without the leading "Photographer", where the name was correctly identified. When removing duplicates from the dataset, these biographies were considered unique, since the names were different. Overall, there are potential issues if the name detection is faulty:

1. random texts are falsely considered a biography (in combination with ambiguous occupation words, e.g. model)
2. false names will affect the gender-scrubbing (removing too much information or leaving a part of the name)
3. failure to remove duplicates

Ambiguous Occupation Terms. One of the twenty-eight titles is *model*, which is frequently used in another sense as the occupation. Consequently, there is a larger number of samples, which were falsely labeled as *model* or are not biographies at all. Even for less ambiguous occupations there are cases, where these terms are in another context or refer to another kind of occupation, such as *software architects* being labeled as *architects*, while there also exists the class *software engineer*. Examples are:

1. "Abu Dhabi is a model for modernization in the Islamic world [...]" (model)
2. "Jason Varitek is a true model for a catcher doing his homework [...]" (model)
3. "Jenny Simons is a naughty nurse [...]" (nurse)
4. "Bonas Khanal is a software architect who works primarily on the Microsoft stack [...]" (architect)

Label is Not Correct or Complete. The pre-processing assumes biographies starting with a sequence of the form "<name> is a(n) <description> <title>", automatically assigning the title as label without checking the rest of the sentence or further context. However, there are numerous biographies starting with patterns such as "<name> is a(n) <description> <title1>, <title2> and <title3>" or "<name> is a(n) <description> <title1> turned <title2>", where only one of many titles or the wrong (old) title was labeled. Another issue are biographies starting with something like "<name> is a(n) <title>'s wife/husband", where the presumed label refers to a third person and the actual occupation is not known. Such samples pose contradictions to the classifier and hence would be at higher risk of being misclassified or contribute to a sub-optimal classifier. (apart from this, there are cases where a former occupation is mentioned in the first sentence, but the actual occupation will be mentioned in the second sentence.)

Text Length/Quality/Style. The criteria for identifying biographies all consider the first sentence (or part of the first sentence). It is simply assumed that the following text will be a biography referring to the before mentioned person. However, there are several cases that violate this assumption:

- the following text is (stylistically) not a biography (could be an advertisement, book/movie summary or complete unrelated to the first sentence)
- the following text is extremely short/has no actual content (could be as short as "read more..." or "check out his/her Instagram..."
- the following text (partially) refers to another person or multiple other persons (hence the detected gender could be false)
- the following biography refers only to a person's hobbies, personal life or misfortune/illness rather than their professional background

Insufficient Gender-Scrubbing. During the dataset pre-processing, a gender-scrubbed version of each biography is created as explained in Sect. 2. While it might be impossible to fully remove gender indications (these might be arbitrarily subtle) there remain some obvious gender indicators left out by this approach. Some example phrases found after gender-scrubbing are: "_ is [...] the only male psychologist [...]", "As the first son [...]" and "[...] is a loving father [...]". For terms as *mother*, *father* or *husband* it is not trivial for an automated approach to decide whether these refer to the protagonist or a third person, but simply removing or replacing such terms with a neutral term would be an easy measure to reduce the number of gender indicators.

On another note, in-consequent gender-scrubbing might actually be harmful for people with anti-stereotypical biographies (in terms of gender) or in homosexual relationships. For instance, given a biography containing "[he] currently lives in New York City with [his] husband and their two dogs", a language model trained on hetero-normative texts might be able to infer the protagonists gender as male given the pronouns. But after removing the pronouns, the most likely inference would be female. Hence, any gender bias in favor of or against women, could also apply for men in homosexual relationships and vice-versa. Similar to relationships any context anti-stereotypical for the protagonist's gender could act as a clue for the opposite gender and have similar effects. While trade-off between individual and group fairness or between different sensible attributes may not be trivial to solve, awareness of these effects is critical when investigating fairness issues or mitigating biases.

4   Dataset Review

In this section, we present our data set review, that consists of, first, an automatic filtering for valid descriptions, and second, of a subsequent manual review to check our defined quality criteria.

4.1   Selecting a Subset for Review

To make the review process feasible, we used a subset of about 10% of the original dataset. Specifically, we selected the classes *dentist, architect, nurse, surgeon, physician, psychologist, attorney, photographer, teacher, journalist*. These are the ten largest classes after excluding the largest class *professor* and account for approximately half of the dataset. In addition we identified samples that were potentially mislabeled or invalid (i.e. no biography or unsuitable for the classification task) by applying a keyword search to the biography content, identifying wrong names and samples with multiple titles or no valid title.

Titles. Regarding the titles, we took the first sentence of each biography and tested if each of the titles used in the crawler were present in that sentence. Titles were only counted if followed by a space or punctuation. Specifically, in cases such as "[...] <title>'s husband/wife" the title did not contribute to the number of titles. If there were no titles or more than one title found, the sample was added to the review set. Furthermore, we searched the first sentences for phrases such as *turned*, *retired* or *former*, which often indicated a change of occupations and hence potential mislabeling.

Names. During our dataset analysis we found a number of capitalized words that were falsely identified as a name or part of the name: *Dad, Mom, Brother, Sister, If, The, His, Her, Is, Share, What, Why, Who, Where, Would*. Hence if one of these words was part of a name, the sample was added to the review subset.

Content. We also identified a number of "biographies", which were actually reviews or summaries of books or movies for instance. To filter these, we checked the biographies for phrases such as 'written by', 'watch the trailer', 'imdb', 'this movie/book', 'throughout this video' or terms and phrases that were more likely used in fiction than actual biographies (e.g. 'paranormal', 'superhero', 'fetish', 'vampire', 'with a secret', 'what she is not expecting', 'blowjob', 'vicious killer'). In addition, we filtered samples, where the URI contained certain keywords, such that we could expect the webpages to contain either movie reviews, fan fiction/wikis or adult content. Lastly, very short texts (less than 20 words) were also added to the review set.

4.2   Quality Criteria

As a baseline for the dataset review, we determined the following criteria to assign titles and determine the validity of samples.

Titles. Any current occupation mentioned in the first sentence, was labeled as a raw title with the exact wording as used in the sentence, regardless if they matched the predefined ten classes or not. Former occupations were not labeled.

Invalid Samples. To count as valid, a sample had to meet the following criteria:

- the text is a biography in the broadest sense
- the biography refers to one person whose occupation is stated
- the biography content refers to the person's professional background at some point

The first point particularly excludes other texts such as movie review or fan fiction if the focus of the text is the fiction or the content is too unrealistic.

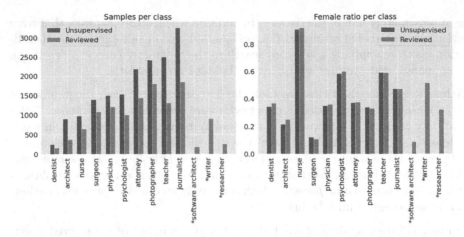

Fig. 1. Number of samples and gender ratio per class, using the automatically assigned titles (unsupervised) and the correct titles (reviewed) including three additional titles.

Samples, which describe the biography of a fictitious person, but are still a biography and meet the other criteria can be accepted. Regarding the second point, we specifically check (or correct) the name, verify that the mentioned title(s) and the following text refers to this person. Biographies that mentioned multiple persons or such as "<name> is a <title>'s wife/husband" where marked invalid. Additionally, we verified that the following text (that is used for classification) contained some references to the persons career or education. Additionally, hobbies, family or other personal matters may be mentioned, but we require some information of the professional background, even if the title cannot be concluded unambiguously.

4.3   Merging Raw Titles

The titles labeled during review were merely raw titles that do not necessarily fit the predefined classes. Hence, after review, we matched the raw titles with these classes. Those titles that did not fit any of the classes were grouped into other occupations categories in order to identify potential other classes. Specifically, we found a significant number of samples that fit into the categories *writer,* *software architect* and *researcher.* Raw titles, which could not be grouped with other occupations, and groups that did not contain too many samples were merged into the *other* category. For our experiments with the reviewed titles, the samples were assigned any matching titles (including the ten chosen classes and the new classes *writer, software architect, researcher*). The *other* category was not assigned. Samples that only had raw titles from the *other* category did not get a title assigned and hence were not used.

4.4   Review Results

The results of our review are listed in Table 1. Due to the automated selection of samples for review, the number of samples per class varies, in particular for the class *dentist*. For most classes, only around 60% to 80% of reviewed samples were found valid. In case of the *teacher* class, only half of all samples were valid. Of the valid samples, the majority was correctly labeled, except for the class *architect*, where a substantial amount of samples actually refers to *software architects*. Around 10% to 25% (40% of *physicians*) had in fact multiple labels.

Table 1. Statistics of the reviewed dataset. The percentage of valid samples refers to the number of reviewed samples, other percentages to the resulting number of valid samples.

	architect	surgeon	dentist	teacher	psychologist
samples (reviewed)	901	1404	249	2501	1536
% valid	0.663	0.767	0.662	0.481	0.656
% one title	0.864	0.909	0.878	0.817	0.751
% wrong title	0.374	0.001	0	0.023	0.029
% multiple titles	0.135	0.09	0.121	0.182	0.248
	nurse	photographer	physician	attorney	journalist
samples (reviewed)	982	2420	1504	2182	3261
% valid	0.684	0.729	0.803	0.668	0.587
% one title	0.751	0.927	0.599	0.918	0.827
% wrong title	0.034	0	0.004	0.034	0.046
% multiple titles	0.248	0.072	0.4	0.081	0.172

Figure 1 shows the number of samples and gender ratio of our subset before and after review. Although a significant number of samples is invalid, the impact on the gender ratios before/after review is very low.

5   Implications for Bias Evaluation

Experimental Setup. To assess the impact of the dataset quality on the results of [1], we reproduce the experiments with our reviewed subset. Thereby we distinguish three cases: First we use the complete subset as if never reviewed (*unsupervised*). For the other two cases, we use the review results, i.e. remove invalid samples and use the corrected labels, once modeled as a multi-label classification (*reviewed-multi*), secondly only using single-label samples and hence modeling the problem as a single-label classification (*reviewed-single*). Furthermore, for each of these versions, we conduct one experiment with the *raw* biographies (including gender indicators) and *scrubbed* biographies (after removing gender indicators). For each case, we train our BERT classifier as described below. For the *reviewed-multi* cases we include the three additional classes (*writer, researcher, software architect*) in the *reviewed-single* case only *software architect* as the other two classes were underrepresented in the single-label samples. We use 5-fold cross validation and report mean and standard deviation for all scores.

Semantic Representation. While in [1] the semantic representation have been bag-of-words, fastText word embeddings and a trained deep network, we utilize BERT [6] as a state-of-the-art model that is still feasible to train and show that we do get comparable results on the unsupervised subset. We use a *SequenceClassification* model from Hugginface[4] with the pre-trained weights of the *bert-base-uncased* model. Since this implementation only supports single-label classification, we used a custom classification head for multi-label classification, consisting of a dropout layer ($p = 0.3$) and a linear layer mapping from the embeddings dimensions to the number of classes. As optimizer we used *AdamW* with a learning rate of $1e-5$. The loss for single-label classification was the cross entropy loss and BCE loss for multi-label classification. We trained each model for 2 epochs with a batch size of 8.

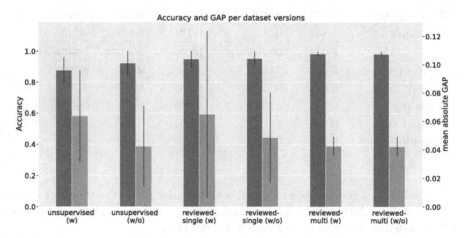

Fig. 2. Accuracy and GAP scores (mean absolute of class-wise GAPs) for the different dataset versions. Standard deviations for the GAP scores are mostly high.

5.1   Results

Figure 2 shows the performance and overall GAP score of our occupation classifier trained on the different versions of the dataset. We observe higher accuracies than De-Arteaga et al., which is not surprising as we have a lower number of classes. On the reviewed dataset, accuracies are higher compared to the unsupervised dataset and in the multi-label case, we report the highest accuracies. In the

---

[4] https://huggingface.co/.

unsupervised and *reviewed-single* case GAP scores are rather unstable across the different folds. On average, gender scrubbing reduces the GAP in the single label cases. In the *reviewed-multi* case we see a different behavior: Even for *raw* texts, we report one of the lowest GAP scores. Furthermore, gender scrubbing does not influence the GAPs and in both cases, we report low standard deviations.

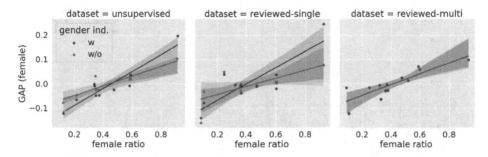

Fig. 3. GAP_{female} vs. π_{female} for all occupations with/without gender indicators. Correlation coefficients: unsupervised-w 0.93, unsupervised-wo 0.89, reviewed-single-w 0.85, reviewed-single-wo 0.67 (p $= 0.023$), reviewed-multi-w 0.85, reviewed-multi-wo 0.86. If not specified, p-values are smaller than 0.01.

Regarding Fig. 3 we can reproduce the results from De-Arteaga et al. on the unsupervised dataset. The range of GAP scores is smaller, which is not surprising as the most extreme occupations from their work are not part of our dataset. Other than that, we observe an even stronger correlation of GAP scores and gender ratio, but similar differences between the cases with/without gender indicators. On the reviewed datasets, we report smaller but still significant correlation, although in the *reviewed-single* case with gender scrubbing, we report a p-value larger than 0.01. In the multi-label case, we report the same GAPs and hence same Pearson correlation and slope for raw and scrubbed texts. This contradicts the findings from De-Arteaga et al. and our findings for the other two dataset versions, where gender-scrubbing did reduce both GAP scores and the correlation with the gender ratio.

Additionally, we visualize the class-wise GAPs for the *raw* dataset in Fig. 4. This shows different behaviors for the classes. In some cases, GAPs decrease on the reviewed datasets, in others they increase or switch signs (only if the GAP is comparably low). In the *reviewed-multi* case, the two most extreme GAPs (for *nurse* and *software engineer*) are reduced significantly, which mostly explains the overall lower GAPs reported in Fig. 2 and 3. For most classes we report high standard deviations in the *unsupervised* and *reviewed-single* case, but consistently low standard deviations for multi-label classification. This particularly shows in the classes with extreme gender ratios, where the minority gender is more likely to be over- or underrepresented in the train-split due to the comparably small size of our dataset. However, this highlights the importance of choosing sufficiently large test sets or using cross-validation with such imbalanced classes. The changes regarding the *architect* class can be directly explained by our review. By filtering the (predominantly male) software architects, the gender GAP shifts in favor of women, which can be explained by the changed gender ratio after review.

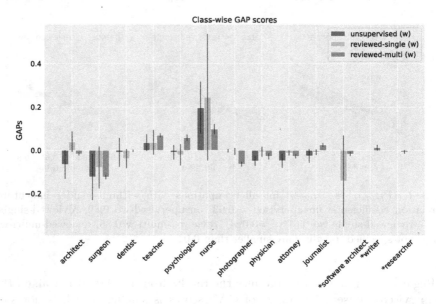

Fig. 4. GAP scores per class. Positive values indicate a higher TPR for women than for men.

6   Conclusion and Discussion

During our analysis of the BIOS dataset, we found a number of problems, such as mislabeled samples or samples that were not biographies at all, due to the simple and automated approach of data acquisition and labeling. Some of these issues can be solved by using a more thorough approach to identify potential biographies and label them, like checking for multiple occupations and modelling the classification task as multi-label classification. Our dataset review showed that a large percentage of samples were indeed invalid, had additional titles or in some cases even the wrong title assigned. Unsurprisingly, reviewing the dataset and modelling it correctly, lead to an improvement in performance. More importantly, we found that the review also reduced the overall bias, although there were trade-offs between the classes. The general correlation between gender ratios in the training data and gender GAPs persist in the reviewed dataset. This is not surprising as it is beneficial for model performance to optimize performance for over represented groups. As explained earlier, the gender-scrubbing was not too thorough and we can expect modern language models to be able to pick up more subtle hints on a person's gender or find other factors correlating with gender.

Hence, we cannot expect this measure to fully remove biases. Nevertheless, the overall reduced gender GAPs on the reviewed data with multi-label classification shows that data quality cannot be neglected, neither in terms of performance nor fairness. The fact that gender-scrubbing had no effect on biases in the *reviewed-multi* case is concerning as it shows that the method itself is flawed and that the data quality/problem modelling had a significant influence on bias and the effect of this debiasing measure. Hence, it is questionable how reliable other bias experiments on the unsupervised BIOS dataset are. While we appreciate works that produce downstream tasks applicable for bias evaluation, we emphasize the need of proper dataset quality or, at the very least, documentation of datasets that give fellow researchers the necessary insights into a dataset to decide its usability for their research.

References

1. De-Arteaga, M., et al.: Bias in bios: a case study of semantic representation bias in a high-stakes setting. In: Proceedings of the Conference on Fairness, Accountability, and Transparency, pp. 120–128 (2019)
2. Schuhmann, C., et al.: LAION-5B: an open large-scale dataset for training next generation image-text models. arXiv preprint arXiv:2210.08402 (2022)
3. Bolukbasi, T., Chang, K.W., Zou, J.Y., Saligrama, V., Kalai, A.T.: Man is to computer programmer as woman is to homemaker? Debiasing word embeddings. In: Advances in Neural Information Processing Systems, vol. 29 (2016)
4. Schroeder, S., Schulz, A., Kenneweg, P., Hammer, B.: So can we use intrinsic bias measures or not? In: Proceedings of the 12th International Conference on Pattern Recognition Applications and Methods, Setúbal, Portugal, pp. 403–410. SCITEPRESS - Science and Technology Publications (2023)
5. Birhane, A., Prabhu, V. U., Kahembwe, E.: Multimodal datasets: misogyny, pornography, and malignant stereotypes. arXiv preprint arXiv:2110.01963 (2021)
6. Devlin, J., Chang, M. W., Lee, K., Toutanova, K.: Bert: pre-training of deep bidirectional transformers for language understanding. arXiv preprint arXiv:1810.04805 (2018)
7. Jourdan, F., Kaninku, T.T., Asher, N., Loubes, J.-M., Risser, L.: How optimal transport can tackle gender biases in multi-class neural network classifiers for job recommendations. Algorithms **16**(3), 174 (2023). https://doi.org/10.3390/a16030174
8. Steed, R., Panda, S., Kobren, A., Wick, M.: Upstream mitigation is not all you need: testing the bias transfer hypothesis in pre-trained language models. In: Proceedings of the 60th Annual Meeting of the Association for Computational Linguistics (Volume 1: Long Papers), pp. 3524–3542 (2022)
9. Webster, K., et al.: Measuring and reducing gendered correlations in pre-trained models. arXiv preprint arXiv:2010.06032 (2020)
10. Hardt, M., Price, E., Srebro, N.: Equality of opportunity in supervised learning. In: Advances in Neural Information Processing Systems, vol. 29 (2016)

Advances in Artificial Neural Networks

Observe Locally, Classify Globally: Using GNNs to Identify Sparse Matrix Structure

Khaled Abdelaal(✉)📵 and Richard Veras📵

University of Oklahoma, Norman, OK 73019, USA
{khaled.abdelaal,richard.m.veras}@ou.edu

Abstract. The performance of sparse matrix computation highly depends on the matching of the matrix format with the underlying structure of the data being computed on. Different sparse matrix formats are suitable for different structures of data. Therefore, the first challenge is identifying the matrix structure before the computation to match it with an appropriate data format. The second challenge is to avoid reading the entire dataset before classifying it. This can be done by identifying the matrix structure through samples and their features. Yet, it is possible that global features cannot be determined from a sampling set and must instead be inferred from local features. To address these challenges, we develop a framework that generates sparse matrix structure classifiers using graph convolutional networks. The framework can also be extended to other matrix structures using user-provided generators. The approach achieves 97% classification accuracy on a set of representative sparse matrix shapes.

Keywords: Sparse Matrix · Graph Neural Networks · Classification

1 Introduction

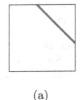

	Diagonal	Random	Rand+Diag	Kronecker
Diagonal	**500**	0	0	0
Random	0	**490**	10	0
Rand+Diag	0	40	**460**	0
Kronecker	0	20	0	**480**

(a)   (b)   (c)

Fig. 1. Efficacy of our classifier framework at determining structure when the data is permuted. (a) is an off-diagonal matrix, (b) is a re-labelled variant of (a), and (c) is the confusion matrix for the classifier framework on re-labelled matrices similar to (b). By using GCNs our approach is invariant to node labelling and achieves around 97% accuracy.

© The Author(s), under exclusive license to Springer Nature Switzerland AG 2023
I. Rojas et al. (Eds.): IWANN 2023, LNCS 14134, pp. 149–161, 2023.
https://doi.org/10.1007/978-3-031-43085-5_12

Sparse matrices represent a fundamental building block used throughout the field of scientific computing in applications, such as graph analytics, machine learning, fluid mechanics, and finite element analysis [6,13]. Such matrices appear as operands in numerous fundamental computational kernels such as sparse matrix-vector multiplication (SpMV), Cholesky factorization, LU factorization, sparse matrix-dense matrix multiplication, and matricized tensor times Khatri-Rao product (MTTKRP) among others. Building efficient algorithms for this class of kernels mainly depends on the storage format used for the sparse matrix as observed in different studies [1,4]. A variety of such formats are proposed in literature [9,10]. Hence, it is crucial to identify the structure of the matrix to choose the ideal sparse format, and eventually tailor the algorithm to that format to optimize the workload performance. However, identifying the structure of the matrix is not always trivial. Figure 1a shows a spy plot of an off-diagonal sparse matrix, and Fig. 1b shows the same matrix, with some of the original row indices and column indices re-labelled. It is less obvious for the latter figure to provide an insight of the original structure of the non-zeros within the matrix. Additionally, in case of huge sparse matrices, we might only have access to samples of the matrix. This could be because of computational or storage restrictions, or missing data. In these two cases (re-labelling and sub-sampling), we need efficient techniques to recognize the shape of the input matrix.

To tackle mentioned issues, we propose a framework to identify sparse matrices structures, using graph neural networks. Figure 1c shows the confusion matrix for the proposed framework using four sample classes on re-labelled variants. The framework design is modular, allowing users to easily augment it with new structures generators or feature sets. The main contributions of this paper are as follows:

- Proposing a novel, modular Graph Neural Network framework to accurately predict the shapes of sparse matrices, including partial samples, and re-labelled variants of original matrices.
- Presenting a new balanced synthetic dataset for structured sparse matrices.
- Providing a performance analysis of graph-level classification on sparse matrices, using different feature sets.
- Introducing two new compact and efficient feature sets for matrices as graphs, namely: Linear and Exponential Binned One-Hot Degree Encoding.

The rest of this paper is organized as follows: Sect. 2 introduces the necessary background, Sect. 3 details the design of the proposed framework, Sect. 4 discusses the evaluation and results of the framework, while Sect. 5 describes related work. Finally, Sect. 6 summarizes the findings of the paper.

2   Background

2.1   Graph Neural Networks

Graph neural networks (GNNs) [17] are a class of deep learning models that operate on graphs or networks. Unlike traditional neural networks that operate on structured data such as images or sequences, GNNs can handle arbitrary graph structures with varying node and edge attributes, enabling them to learn powerful representations of graph-structured data. The key idea behind GNNs is to iteratively update node embeddings by aggregating information from the embeddings of their neighbors through the "graph convolution" operation. By stacking multiple layers of graph convolution and non-linear activation functions, GNNs can learn hierarchical representations of the graph that capture both local and global information.

2.2   Structured Matrices

Several common structures are observed in sparse matrices, such as:

Diagonal all non-zeros are located on the main or a secondary diagonal. This structure represents a 1D mesh and commonly appears in various scientific and engineering applications.

Random the non-zero elements are randomly distributed across the matrix, with variable density. Such matrices have no specific identifiable structure.

Kronecker Graphs [11] are a class of synthetic graphs that have been widely used to model real-world networks, and are generated by recursively applying the Kronecker product of a small base graph with itself. Let A and B be two matrices. Then, their Kronecker product $A \otimes B$ is given by

$$A \otimes B = \begin{pmatrix} a_{11}B & \cdots & a_{1n}B \\ \vdots & \ddots & \vdots \\ a_{m1}B & \cdots & a_{mn}B \end{pmatrix} \tag{1}$$

where a_{ij} are the entries of A. We use these three classes of structures, and a combination of them, as a representative set that can be combined to form more complex relationships [16,20]. Our framework is not limited to only these structures, and they serve as an example to evaluate its performance.

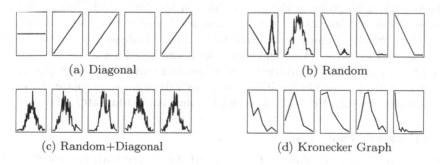

(a) Diagonal   (b) Random

(c) Random+Diagonal   (d) Kronecker Graph

Fig. 2. Global Degree Distribution for samples in each matrix (graph) class studied in this paper. In our approach we classify the shape based on local views from sampled data.

Degree as a Representative Node Feature. Figure 2 illustrates that one can accurately distinguish between the different classes based on the degree distribution of the representative graph. For example, for Diagonal matrices (Fig. 2a), the degree for all nodes is low, and is either constant or linear across all nodes. Kronecker graphs follow a power-law degree distribution, with only a few nodes having many connections (high degree) and most of the nodes having relatively few connections (low degree). However, only the local per-node degree view may be immediately available, and not the global graph view. An example of such a case is only having a sample of the graph and not the entire graph due to storage or computational limitations. The power of GNNs can be leveraged to carry out the required task: the prediction of the sample matrix structure.

3   Framework Description

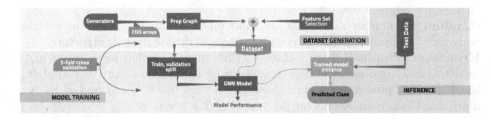

Fig. 3. High-Level overview of the framework. It consists of three main phases: **dataset generation**, where the synthetic sparse matrices are generated, prepared as graphs, and have feature set attached. Then, the GNN **model training** using 5-fold cross validation to capture the model performance, and then generate a trained model instances, that is used later in the **inference** phase.

The goal of the proposed framework is to predict the structure of the input sparse matrix through its classification into one of the configured target classes. We use diagonal, random, diagonal+random, and Kronecker graph as examples of these classes to evaluate the performance of the framework. New structures can be seamlessly integrated. Figure 3 shows a high-level overview of the proposed framework. It consists of three main stages: Dataset generation, Model Training, and Inference. A synthetic dataset is generated using different generators for different shapes of matrices, which are then represented as graphs. In the training phase, we use GNN with 5-fold cross validation to evaluate the model performance. Finally, the trained model instance is used for later inference.

3.1   Dataset Generation

We generate a balanced dataset of 40K synthetic sparse matrices, covering the four sample classes through individual generators. Each of the generators returns a Coordinate (COO) representation for the matrix, excluding the actual non-zero values. The COO representation is then used as the adjacency list to build the graph representation.

3.2   Feature Set Selection

A per-node feature vector is necessary for the graph neural network to classify matrices. Node degree can be calculated for rows/columns in input matrices from their graph representation.

One-Hot Degree Encoding uses a number of features equal to the maximum degree $+ 1$. A limitation of this encoding is that it requires the knowledge of the maximum degree in the entire dataset before training. Also, the required storage is proportional to the maximum degree recorded in the dataset, which increases memory requirements and reduces maximum possible batch size during training. Moreover, it poses complications during inference if the input matrix has a degree greater than the maximum degree in the training set. In our dataset, the maximum training set degree is 7710, so the length of one-hot encoding feature vector per node is 7711, limiting the maximum batch size on GPU to only 1 graph.

Local Degree Profile (LDP) [3] captures the local structural information of nodes in their immediate neighborhood. LDP is calculated for each node as a five-feature vector: the node degree, the minimum degree of its neighbors, the maximum degree of its neighbors, the mean degree for its neighbors, and the standard deviation of the degrees of each neighbors. LDP features are easy to compute for any given graph. Additionally, the number of features per node is fixed, regardless of the used training data. LDP incurs low storage overhead.

Linear Binned One-Hot Degree Encoding. (LBOH) We implement a modified version of one-hot encoding, to address its limitations. LBOH works by having a fixed number of buckets for representing one-hot degrees. Buckets ranges are designed as follows: a set individual sequential buckets from 0 t α (inclusive) where α is a small integer (less than 10). Then, we add a set of buckets with fixed step β from α: $(\alpha + \beta), (\alpha + 2\beta), ..., (\alpha + k\beta)$ where $(\alpha + k\beta)$ is the maximum degree threshold. Any degree greater than $(\alpha + k\beta)$ is mapped to the final bucket.

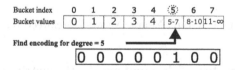

Fig. 4. An example of finding the linear binned one-hot degree encoding for a node with degree $= 5$, where the parameters for the encoding scheme are $\alpha = 5$, $\beta = 3$, and $k = 2$. Degree 5 is mapped to its associated bucket (5 to 7), then the bucket index (5) is represented using one-hot encoding (1 at the position where the value 5 exists, 0 otherwise).

Figure 4 shows an example of LBOH encoding. As opposed to One-Hot Encoding, LBOH provides a fixed number of features regardless of the maximum degree in the training dataset. At the inference stage, only the values of α, β, and k are needed.

Exponential Binned One-Hot Degree Encoding (EBOH). The main difference between EBOH and LBOH is the kind of step between buckets ranges. Instead of a linear step in LBOH, EBOH uses an exponential step to cover more degree values with a small number of features. First, the value of α is chosen such that $1 \leq \alpha \leq 3$. Then for the buckets, a sequential one-to-one mapping is performed for values 0 through 2^α. For the following buckets, the upper bound (inclusive) is $2^{\alpha+i}$ where $i \in [1, k]$ and $k \in \mathbb{N}^*$.

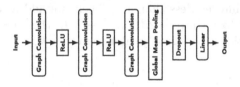

Fig. 5. An example of finding the exponential binned one-hot degree encoding for a node with degree = 7880, where the parameters for the encoding scheme are $\alpha = 2$ and $k = 3$. Degree 7880 is mapped to its associated bucket (33 to ∞), then the bucket index (8) is represented using one-hot encoding (1 at the position where the value 8 exists, 0 otherwise).

Figure 5 shows an example of EBOH encoding. EBOH encoding still provides the benefit of having the number of features independent of the maximum degree in the training set.

3.3   The Graph Neural Network Architecture

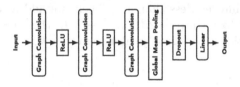

Fig. 6. Graph Neural Network architecture.

To identify the structure of the input matrix, the matrix is viewed as the adjacency list of a graph, enabling the use of machine learning methods designed for graph data. GNNs provide additional benefits such as allowing the use of matrices (graphs) of arbitrary sizes as input, Also, GNNs are agnostic to node ordering. This powerful property enables re-labelling or permuting nodes in a graph representing a sparse matrix, while maintaining accurate predictions. The machine learning task of interest is graph-level prediction since a single label (class) is needed for an entire graph (matrix). The GNN architecture is shown in Fig. 6. The hidden layers are three graph convolution layers and one linear (output) layer. Graph convolution is an operation where node embeddings are iteratively generated as the aggregations from the node neighborhoods. This operation is used to capture complex features of the graph. The first convolution layer aggregates information from the local neighborhood of each node. This operation is repeated in subsequent convolution layers in order to propagate information to increasingly larger neighborhoods. By the end of three convolution layers, the model has learned a hierarchical representation of the graph, where the features at each layer capture increasingly complex structural patterns. The learned representation so far is "node embeddings". Then, learned node embeddings are reduced into a single graph embedding using a global mean pooling operation (called readout layer). Samples are randomly dropped out to reduce overfitting. Finally, a linear classifier is applied to the graph embedding.

```
def generateDiagRandom(size, threshold=2):
    """ A function to generate a Diag+Random square matrix """
    tuples = [(x,y) for x in range(size) for y in range(size) if (random
        .randint(0,10) <= threshold or x == y)]
    # seperate tuples into two lists: the row array and the column array
    coo_rep = list(map(list, zip(*tuples)))
    return coo_rep[0], coo_rep[1], [size, size]
```

```
def process(self):
    catMap = [...., {
        # Number of instances to generate for this class
        'num_iter':10000,
        # Name of the generator function
        'generator':generateDiagRandom,
        # A string list of required generator function param
        'gen_params':['random.randint(MIN_DIM_SIZE,MAX_DIM_SIZE)'] }]
```

Fig. 7. The two steps needed to add a new class to the classifier framework. First (top), create a new generator function in the generators file, and second (bottom), add a dictionary entry to catMap list in the process method of the dataset class.

Modularity. New shapes of matrices can be easily integrated in our framework. To achieve this, two steps are needed as shown in Fig. 7: (1) write a generator for that new shape, and (2) add an entry to the categories (shapes) map in the dataset class for this shape, containing the number of dataset instances to generate, the name of the generator function, and the different required parameters. The generator is required to return the COO representation excluding values, and the matrix dimensions. After generating the new dataset instances for this class, one does not need to re-train the entire model again. Transfer learning [21] can be used to replace the last layer of the trained model with a new layer that has the appropriate number of outputs, after introducing the new shape(s). Then, the weights of all previous layers are frozen and only the new layer is trained. Another aspect of modularity in our framework is the ability to seamlessly attach different feature sets. Feature sets are only computed when the graph is queried. To implement a new feature set, a modification to the **get** method of the dataset is needed. This method first reads in the graph file from disk, calculates the new feature set, and attaches it to the graph.

4 Analysis

We run a set of experiments to evaluate the accuracy of our approach in detecting structures, using the different feature representation discussed in Sect. 3.2.

4.1 Evaluation

Experimental Setup. Table 1 shows the experimental setup and learning parameters used in the experiments. We use PyTorch Geometric [8] for the GNN.

Evaluation Metrics. To evaluate the prediction accuracy of the framework, four derived metrics are used: accuracy, precision, recall, and F1-score. We report per-class and overall accuracy and F1 score numbers, since the latter is the harmonic mean of precision and recall. Using both accuracy and F1-score helps provide a more comprehensive evaluation of the framework's performance. Accuracy gives an overall view of how well the classifier is performing, while the F1-score provides insights into its ability to correctly classify positive instances.

Table 1. Experimental setup and Training parameters used in the experiments. * Batch size used for traditional one-hot encoding is 1.

Component	Specification	Parameter	Value
GPU	NVIDIA RTX A6000	Optimizer	Adam
GPU Memory	48 GB GDDR6	Learning Rate	0.01
CUDA Version	11.8	Error Criterion	Cross Entropy
Main Memory	64 GB DDR4	Batch Size	256*
Operating System	Ubuntu 22.04	Cross Validation Folds	5

4.2 Results

Table 2. Performance of the classifier for different degree representations

	Degree Representation							
	One-Hot Encoding		LDP		LBOH		EBOH	
	Accuracy	F1 Score	Accuracy	F1 Score	Accuracy	F1 Score	Accuracy	F1 Score
Diagonal	1.0	0.90	1.0	0.97	1.0	1.0	1.0	1.0
Random	0.90	0.91	0.64	0.76	0.92	0.95	0.95	0.96
Random+Diagonal	0.86	0.99	0.98	0.83	0.96	0.94	0.97	0.96
Kronecker	0.90	0.94	0.90	0.93	0.98	0.97	0.96	0.98
Overall	0.90	0.90	0.88	0.88	0.97	0.97	0.97	0.98

Classification Performance. Table 2 shows the accuracy and F1 score for the classifier using the different degree representations discussed in Sect. 3.2.

Performance results show that both LBOH, and EBOH provide high prediction accuracy of around 97% and a F1 score of around 98%. On the other hand, traditional one-hot encoding exhibits a lower accuracy of around 90%. One-Hot Encoding requires a significantly large number of features per node (7711), limiting the training batch size on the A6000 GPU to only one graph. This forces the optimizer to adjust the neural network weights very frequently, hence hurting the overall accuracy. Using LDP as a feature set exhibits variant model performance across folds depending on the validation set being used. In some folds, LDP provides high accuracy of around 97% to 98% similar to EBOH. In other folds, LDP fails to converge to an acceptable loss value, and ends up with an accuracy of around 74% on the last few epochs. This performance variance across folds deems LDP unfit for the purposes of our application. It significantly fails in two classes: Random and Kronecker. It predicts Random matrices as Random+Diagonal for more than 32.5% of the instances. This is likely due to the prevalence of the local degree neighbor summary features (the last four LDP features) instead of focusing on the node degree. This eventually results in failing to discover the global hierarchical structures in the matrix. LDP still shows perfect accuracy in case of diagonal matrices since almost all nodes in the matrix's graph have the same degree. LDP prediction quality for Kronecker graphs is also lower than other evaluated feature sets (around 81% in some folds) for the same reasons.

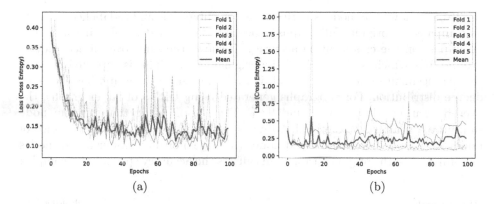

(a)   (b)

Fig. 8. Cross Entropy Loss across different folds in 5-fold cross validation training using (a) EBOH, and (b) LDP feature set.

Fig. 8a demonstrates the validation loss across the 5 different folds for EBOH. It shows almost no variance in the loss across the different folds, indicating the stability of the model's performance across folds. On the other hand, Fig. 8b shows the validation loss for LDP and illustrates that the loss does not converge in 2 out of 5 folds.

Classifying Sub-samples and Re-labelled Subgraphs. To test the efficacy of GNN on both aspects, we generate 200 new matrices: 50 for each of the four classes. For each of them, we generate 10 subgraphs and 10 re-labelled variants. To generate the subgraphs, we use uniform random node sampling (URNS) [12]: nodes are randomly selected with uniform probability, as well as the edges connecting the selected nodes. Re-labelling of a graph G simply renames the nodes V of the graph, and produces a new graph G' with the same size and degree distribution of the original graph G. Figure 9 shows an example of both URNS and random re-labelling.

Table 3. Accuracy comparison for node sampling, node re-labelling, and original graphs using EBOH feature set.

Class	Node Sampling	Node Re-labelling	Original Graphs
Diagonal	1	1	1
Random	0.83	0.98	0.98
Random+Diagonal	0.92	0.92	0.92
Kronecker	0.94	0.96	0.96
Overall	**0.92**	**0.97**	**0.97**

Table 3 shows the model's performance on subgraphs and re-labelled variants as compared to original full graphs. The table shows that re-labelling node has no impact on the classification accuracy; it shows the same overall accuracy of around 97% which is observed for the original graphs. This is expected because the arrangement of nodes in a graph is irrelevant, since the graph has the same degree distribution. For subgraphs generated using URNS of larger graphs, the overall accuracy drops to around 92%. The reason being that random node sampling can alter the degree distribution of the graph. The random choice of nodes can result in either isolated nodes (no edges) or much lower degree nodes as compared to the original graph. This affects the accuracy specially for complex

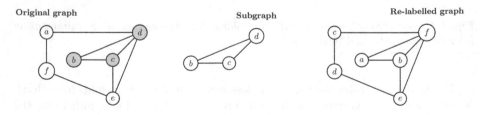

Fig. 9. Example of generating a random sub-sample and a re-labelled variant of a graph. The original graph (left) contains six nodes. Using URNS, a random subgraph (middle) of three nodes is generated. A re-labelled variant (right) is generated using a random 1:1 mapping between the original and new node labels.

shapes such as random and Kronecker graphs. One way to reduce the accuracy loss for samples is to use a more sophisticated graph sampling technique rather than randomly selecting nodes or edges.

5   Related Work

Prediction on Sparse Matrices. Several studies investigated the use of machine learning to predict the optimal sparse format for SpMV on CPU and GPU [2,14,15,18,19,23]. Our framework does not directly predict the best sparse format, instead, we only predict the structure of the input matrix. This allows de-coupling the sparsity pattern from the sparse format, following the argument adopted by AlphaSparse [7] since our framework also allows the seamless integration of new classes. Existing techniques collect a set of features from each matrix such as: the number of diagonals, the ratio of true diagonals to total diagonals, the (maximum) number of non-zeros per row, the variation of the number of nonzeros per row, the ratio of nonzeros in DIA and ELL data structures, and a factor or power-law distribution. We only need to calculate one feature per node: its degree. Also, [23] uses a CNN approach to treat matrices as images, and in order to fix the size of the matrix, they normalize input matrices into fixed size blocks, losing partial matrix information in the process. In contrast, our approach handles arbitrary sizes of matrices, without losing precision, leveraging the power of Graph Neural Networks. We can optionally sample large matrices and maintain high prediction accuracy. An additional benefit to our framework is that it is order in-variant, since matrices are represented as graphs.

Graph Representation for Learning. Representing non-attribute graphs is an open problem [5]. Common approaches employ graph properties such as node degree, more specifically a one-hot encoding of the degree [22]. One-hot encoding suffers from numerous limitations (Sect. 3.2). LDP [3] provides a compact representation for graph using five features per node. Although the computation of such feature vector is efficient, using LDP results in unreliable model performance for our task (Sect. 4.2). Both our representations (LBOH and EBOH) outperform one-hot encoding and LDP while addressing their shortcomings.

6   Summary

In this paper, we proposed a GNN based framework to classify structured sparse matrices. We introduced two novel non-attribute graph representations based on node degrees: LBOH, and EBOH. We evaluated the efficacy of our framework on a synthetic, balanced dataset of matrices that we generated containing random matrices from four sample classes: diagonal, random, random+diagonal, and Kronecker graphs. Performance results demonstrate a high classification accuracy of 97% for the framework when using our feature sets: LBOH and EBOH. They also show high accuracy of 92% and 97% on random node subsamples and re-labelled variants respectively. Our framework is modular, allowing the inclusion of additional classes with minimal user effort. Future endeavors target the automatic generation of the optimal sparse data format and algorithm for sparse matrix kernels, using the obtained prediction results from the current framework.

References

1. Bell, N., Garland, M.: Implementing sparse matrix-vector multiplication on throughput-oriented processors. In: Proceedings of the Conference on High Performance Computing Networking, Storage and Analysis, pp. 1–11 (2009)
2. Benatia, A., Ji, W., Wang, Y., Shi, F.: Sparse matrix format selection with multiclass SVM for SPMV on GPU. In: 2016 45th International Conference on Parallel Processing (ICPP), pp. 496–505 (2016)
3. Cai, C., Wang, Y.: A simple yet effective baseline for non-attribute graph classification. arXiv preprint arXiv:1811.03508 (2018)
4. Choi, J.W., Singh, A., Vuduc, R.W.: Model-driven autotuning of sparse matrix-vector multiply on GPUs. Association for Computing Machinery, New York (2010)
5. Cui, H., Lu, Z., Li, P., Yang, C.: On positional and structural node features for graph neural networks on non-attributed graphs. Association for Computing Machinery, New York (2022)
6. Davis, T.A., Hu, Y.: The university of Florida sparse matrix collection, vol. 38, no. 1 (2011)
7. Du, Z., Li, J., Wang, Y., Li, X., Tan, G., Sun, N.: Alphasparse: generating high performance SPMV codes directly from sparse matrices. In: Proceedings of the International Conference on High Performance Computing, Networking, Storage and Analysis, SC 2022. IEEE Press (2022)
8. Fey, M., Lenssen, J.E.: Fast graph representation learning with PyTorch Geometric. In: ICLR Workshop on Representation Learning on Graphs and Manifolds (2019)
9. Filippone, S., Cardellini, V., Barbieri, D., Fanfarillo, A.: Sparse matrix-vector multiplication on GPGPUs, vol. 43, no. 4 (2017)
10. Langr, D., Tvrdík, P.: Evaluation criteria for sparse matrix storage formats. IEEE Trans. Parallel Distrib. Syst. 27(2), 428–440 (2016)
11. Leskovec, J., Chakrabarti, D., Kleinberg, J., Faloutsos, C., Ghahramani, Z.: Kronecker graphs: an approach to modeling networks. J. Mach. Learn. Res. 11(33), 985–1042 (2010)
12. Leskovec, J., Faloutsos, C.: Sampling from large graphs. Association for Computing Machinery, New York (2006)
13. Leskovec, J., Krevl, A.: SNAP Datasets: Stanford large network dataset collection (2014). https://snap.stanford.edu/data
14. Li, J., Tan, G., Chen, M., Sun, N.: SMAT: an input adaptive auto-tuner for sparse matrix-vector multiplication, vol. 48, no. 6 (2013)
15. Li, K., Yang, W., Li, K.: Performance analysis and optimization for SPMV on GPU using probabilistic modeling. IEEE Trans. Parallel Distrib. Syst. 26(1), 196–205 (2015)
16. Puschel, M., et al.: Spiral: code generation for DSP transforms. Proc. IEEE 93(2), 232–275 (2005)
17. Scarselli, F., Gori, M., Tsoi, A.C., Hagenbuchner, M., Monfardini, G.: The graph neural network model. IEEE Trans. Neural Networks 20(1), 61–80 (2009)

18. Su, B.Y., Keutzer, K.: clSpMV: a cross-platform OpenCL SpMV framework on GPUs. Association for Computing Machinery, New York (2012)
19. Tan, G., Liu, J., Li, J.: Design and implementation of adaptive SPMV library for multicore and many-core architecture, vol. 44, no. 4 (2018)
20. Van Loan, C.: Computational frameworks for the fast Fourier transform. SIAM (1992)
21. Weiss, K., Khoshgoftaar, T.M., Wang, D.: A survey of transfer learning. J. Big Data 3(1), 1–40 (2016)
22. Xu, K., Hu, W., Leskovec, J., Jegelka, S.: How powerful are graph neural networks? In: International Conference on Learning Representations (2019)
23. Zhao, Y., Li, J., Liao, C., Shen, X.: Bridging the gap between deep learning and sparse matrix format selection. Association for Computing Machinery, New York (2018)

Unsupervised Topic Modeling with BERTopic for Coarse and Fine-Grained News Classification

Mohamad Al Sayed[1], Adrian M.P. Brașoveanu[1,2(✉)], Lyndon J.B. Nixon[1,2], and Arno Scharl[2,3]

[1] Modul Technology GmbH, Am Kahlenberg 1, 1190 Vienna, Austria
{sayed,brasoveanu,nixon}@modultech.eu
[2] Modul University Vienna, Am Kahlenberg 1, Vienna, Austria
{adrian.brasoveanu,lyndon.nixon,arno.scharl}@modul.ac.at
[3] webLyzard technology, Liechtensteinstrasse 41/26, 1090 Vienna, Austria
scharl@weblyzard.com

Abstract. Transformer models have achieved state-of-the-art results for news classification tasks, but remain difficult to modify to yield the desired class probabilities in a multi-class setting. Using a neural topic model to create dense topic clusters helps with generating these class probabilities. The presented work uses the BERTopic clustered embeddings model as a preprocessor to eliminate documents that do not belong to any distinct cluster or topic. By combining the resulting embeddings with a Sentence Transformer fine-tuned with SetFit, we obtain a prompt-free framework that demonstrates competitive performance even with few-shot labeled data. Our findings show that incorporating BERTopic in the preprocessing stage leads to a notable improvement in the classification accuracy of news documents. Furthermore, our method outperforms hybrid approaches that combine text and images for news document classification.

Keywords: News Classification · Topic Modeling · BERTopic · Sentence Transformers · SetFit

1 Introduction

Discovering the most relevant content is a key challenge not only for web search, but also for personalization, targeted advertising, or media analysis tasks. Organizations can increase their revenue if their content pipelines satisfy and engage their target audiences. In light of the recent advances in generative models, the importance of fine-grained news classification has increased considerably, as large repositories of text are used to train Large Language Models (LLMs) for a wide variety of downstream tasks [25].

LLMs like BERT [4] and RoBERTa [13] have achieved accurate results for classification problems, but are often not ideal for fine-grained classification.

I. Rojas et al. (Eds.): IWANN 2023, LNCS 14134, pp. 162–174, 2023.
https://doi.org/10.1007/978-3-031-43085-5_13

Leaving aside the computational costs, a major challenge is the overfitting of such models in resource scarcity situations (e.g., smaller datasets, fewer training examples per class) [5,20]. Significant changes to the architecture (e.g., the addition of new layers or modules) are also needed to generate the desired probabilities for each class in multi-class settings. Considering these issues, we have searched for a solution that is easier to train with a lower number of examples, while also offering good results for multi-class classification. Another set of important demands was that any new model should be interpretable, work for both coarse-grained and fine-grained datasets, and offer good performance in multi-label settings as well (e.g., by assigning multiple categories to a single document). These requirements led us to examine topic models, as they naturally lend themselves to such multi-label settings since they can help surface the various topics present in a document. The early generation of topic models like Latent Dirichlet Allocation (LDA) discarded semantic relations within the textual entities due to their bag-of-words representations. Neural topic models have introduced clustered embeddings as a way to contextualize semantic attributes [19]. BERTopic [7], a recent neural topic model, builds upon the idea of clustered embeddings and delivers better accuracy than the earlier models. The topic clusters generated by this model collect documents with similar semantics. They are easy to understand, explain, interpret, or even modify by human users. The results of the BERTopic clustering can then further be used by modern classifiers based on Transformer models like RoBERTa or SetFit [21], and this will lead to significant improvements compared to the baselines. BERTopic also supports multiple languages, which makes it ideal for building large multilingual classification systems.

Presenting a method to build superior multi-class classifiers by using BERTopic as a preprocessor for classifiers built on RoBERTa or SetFit, the remainder of this paper is structured as follows: Sect. 2 reviews the current literature. Section 3 describes the text classification pipeline. Section 4 then presents the results of our experiments. The concluding Sect. 5 summarizes the results and presents future research directions.

2   Related Work

Text classification is one of the oldest tasks in computer science, but until recently there have been very few good surveys about it. Li and his colleagues wrote a comprehensive survey [10] that covers 60 years of developments for this task, including traditional methods that include feature extraction (e.g., bag of words, TF-IDF), early ML classifiers (e.g., Naive Bayes, Support Vector Machines, Random Forest, etc.), and the recent Deep Learning (DL) developments (e.g., Convolutional Neural Networks, Transformers, Graph Convolutional Networks). Another recent survey [14] is focused only on the DL progress for this task. This survey mentions RoBERTa [13] and XLNet [23] as the most promising DL models for text classification.

Besides text classification, LLMs have been demonstrated to successfully solve a wide variety of NLP tasks from named entity extraction to summarization. A recent survey about LLMs [25] presents the most promising models, the training methodology, and the tasks they were trained for. To navigate the Transformer landscape today, we need a taxonomy [11], and possibly an overview of its formalization [16], especially if we plan to change the training method or the various encoder and decoder modules. BERT and RoBERTa are included in the pre-trained encoder branch of the Transformer taxonomy, together with BERT and RoBERTa. Depending on the use case, the models that build on foundational models, like BERT or RoBERTa, might also be included in the application branch.

Zong and colleagues [26] examine the best models available for text classification and analyze the cost/quality compromise. They also include some recent models like GPT-3 [2] or T5 [17], and provide an analysis of the cost of accessing and training such models under a set of fixed constraints. They conclude that the selection of a model depends on a variety of factors, including cost (e.g., training cost, annotation cost, or inference cost) and desired accuracy.

Instead of using RoBERTa directly, many of the recent classification papers that use LLMs tend to focus on pre-processing techniques or on fine-tuning existing models. Pre-processing can also use embeddings from other BERT models. A recent model based on topic modeling and hierarchical clustering is called BERTopic [7]. The main idea of the model is to extract a coherent topic representation from a collection of documents using clustering. The underlying clustering algorithm is HDBSCAN, but it can also be replaced with k-Means [6]. Such a change typically results in fewer outliers and better performance on short texts. BERTopic variants have been successfully used for analyzing media coverage and people's opinion on distance learning during COVID-19 pandemic [8] or for analyzing the links between the topics included in companies' financial disclosures and their risk ratings [9].

3   Method

Our initial approach towards the news classification problem was to examine websites that already categorize their content feed with different classes (e.g., entertainment, sports, politics, etc.). By analyzing their URLs, we discovered that it was possible to identify the classes directly from the URL patterns (e.g., URLs that contain the string www.euronews.com/news/sport will be categorized as Sports). We initially built a large silver standard for multiple languages with documents crawled from the web. After an initial set of experiments, we decided to switch to datasets collected from well-known sources (e.g., New York Times, Guardian, or BBC) for English. This decision was taken due to a desire to increase the reproducibility and interpretability of our method. Besides these two aspects, the search for a new solution was driven by the idea of cost-effectiveness, as it was important to find new models that had lower operational costs (e.g., training and adaptability). The main problem with the models we used based on

RoBERTa was the fact that it was difficult to output the desired probabilities for each class in a multi-class settings environment without significant architectural changes. This was mainly because these early models were not really able to correctly interpret the semantics of the documents. Another class of models seemed to be better suited for such a task: topic models, as they already did a good job of selecting the most important topics from a document. A particularly recent development, neural topic models that use clustered embeddings like BERTopic [1], drew our attention. The rest of the section describes how we managed to use BERTopic as a good preprocessor for multi-class classifiers based on RoBERTa and SetFit and significantly improve their performance in the process.

3.1 Problem Definition

We define the new classification problem as a multi-class classification problem in which the classifier needs to predict the category with the highest probability of being associated with a document.

We consider $X \in \mathbb{R}^{n \times d}$ the documents from each dataset considered. We define $Y \in \{1, 2, \ldots, k\}^n$ to be the vector of class labels and θ to be the model's parameters. "n" refers to the number of instances/documents in our dataset. And "d" is used to indicate the number of features or dimensions in each instance of the dataset. In the context of document classification, "d" refers to the number of distinct words or tokens considered in our analysis. The main goal is to learn the following function $f_\theta : X \to Y$. The objective function is designed to map the documents to their correct label.

3.2 Preprocessing with BERTopic

In contrast to classic large language models like BERT and RoBERTa which were designed to solve a wide variety of general NLP tasks, BERTopic[1] [7] was specifically created with topic modeling in mind. BERTopic is a neural topic model that uses class-based TF-IDF vectorization. It takes document collections as input, not just sentences or single documents. The output of a BERTopic prediction is a set of topics that summarize the main themes from the document collection it was fed. These topics are represented by clusters that are relatively easy to understand and evaluate by humans. Figure 1 showcases how various BERTopic clusters are built.

The generation of these clusters begins with the transformation of documents into embeddings using sentence-transformer models. These dense vector representations of the documents are then used in the UMAP algorithm for dimensionality reduction, which transforms the high-dimensional document embeddings into lower-dimensional embeddings. This step is followed by HDBSCAN, a density-based clustering algorithm, which groups similar lower-dimensional embeddings into clusters, hence creating topics. The topics are then represented by a set of representative words generated based on class-based TF-IDF scores.

---

[1] https://maartengr.github.io/BERTopic/index.html.

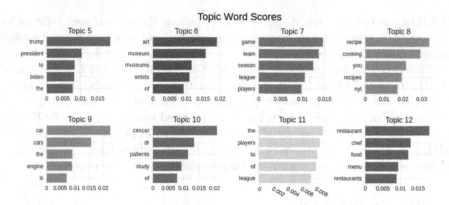

Fig. 1. Clusters created with BERTopic for N24 dataset.

We use BERTopic to eliminate noisy samples from the initial training datasets. BERTopic provides a topic label for each document based on their embeddings which is a dense vector representation. By default, BERTopic assigns a "−1" label to the documents that do not belong to any obvious cluster or topic. $g : X \to Z$, where $Z \in 1, 2, \ldots, m, -1^n$ is a vector of topic labels, and m represents the number of distinct topics. Eliminating topic "−1" can be represented by the transformation function that takes the initial dataset as input such as $T(X, Z) = X'$, where $X' \in \mathbb{R}^{n' \times d}$, and n' is the number of remaining documents after processing (Fig. 2).

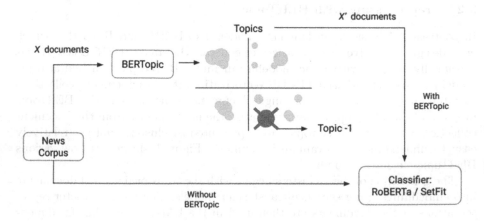

Fig. 2. Data Pipeline

3.3 Improving Transformer Classifiers

The output of the cluster embeddings is then fed into Transformer models based on RoBERTa or SetFit. We assume that the resulting models will converge better during the optimization process when we eliminate the noisy samples from the initial training datasets. The assumption is based on the idea that the Transformer classifier can focus on relevant patterns in the training data instead of being distracted by noisy ones.

Let $L_\theta(X, Y)$ denote the loss function for the model with parameters θ on input data X and true labels Y. We reflect on our assumption with the following:

$$L_\theta(X', Y') \leq L_\theta(X, Y) \tag{1}$$

where $Y' \in 1, 2, \ldots, k^{n'}$ is the vector of class labels corresponding to the remaining samples in X'.

We then perform a comparison between the models' accuracy with and without the use of the preprocessing method with BERTopic. Our goal is to show that the classification accuracy with the preprocessing technique is higher than the one produced by the models trained on the initial training data. We reflect this by the following:

$$\text{Accuracy}(f_\theta(X')) \geq \text{Accuracy}(f_\theta(X)) \tag{2}$$

4 Experiments

In this section, we evaluate the effectiveness of BERTopic preprocessing technique for Transformers on three public datasets using two models (RoBERTa-base and SetFit) for the news classification task.

4.1 Datasets

We have decided to test our approach on both coarse and fine-grained datasets, as it was important to understand if the method works for both cases.

N24News[2] [22] is a fine-grained multimodal news dataset built using the New York Times API. The dataset contains news, pictures, and videos published between 2010 and 2020. The dataset contains 61218 real news documents with text or images that belong to 24 classes. We have only used the abstract and body partitions of the dataset for our experiments, as our experiments are focused on classifying text documents instead of multimodal classification.

The coarse-grained Guardian News[3] dataset collects around 150000 news articles (2000 articles per month) split into 5 classes. The articles were published by The Guardian between January 2016 and June 2022.

---

[2]https://github.com/billywzh717/N24News.
[3]https://www.kaggle.com/datasets/adityakharosekar2/guardian-news-articles.

The coarse-grained BBC News[4] dataset was published by Kaggle in 2018 and collects articles published by the news division of the BBC. It contains 2225 real news documents that belong to 5 classes.

Table 1 summarizes the different categories included in the three datasets we have used for experiments. As can easily be seen, there is a lot of overlap between the two datasets. N24News dataset is more fine-grained compared to the other two, but on the other hand, it also contains some of the categories already included in them.

Table 1. Categories and their counts for various datasets.

Dataset	Type	Count	Categories
N24News	Fine	24	Art & Design, Automobiles, Books, Dance, Economy, Education, Fashion, Food, Global Business, Health, Media, Movies, Music, Opinion, Real Estate, Science, Sports, Style, Technology, Television, Theater, Travel, Well, Your Money
Guardian	Coarse	5	Business, Culture, Politics, Sport, Technology
BBC News	Coarse	5	Business, Entertainment, Politics, Sport, Tech

4.2 Models

The same experimental settings and hyperparameters were used for all models to ensure a fair comparison. We trained the models with a learning rate of 1e-5, and a batch size of 16. We employed a maximum sequence length of 512 tokens and used early stopping with patience of two epochs to avoid overfitting.

For the RoBERTa model, we selected the hyperparameters based on best practices established in the existing research literature. Extensive experimentation with hyperparameters has been carried out in numerous previous studies (e.g., [12] or [24]), and we leveraged these findings to optimize the performance of our RoBERTa model.

Regarding the BERTopic model, we adhered to the default parameters provided by the library, with a single exception: we adjusted the size of the topic clusters. This modification was made after a series of tests with varying cluster sizes, and it was determined that the optimal cluster size was 120.

We report accuracy as the primary evaluation metric. We provide the number of training and testing samples for each experiment to understand the impact of the preprocessing method on the size of the dataset. Since the end results may be influenced not only by the choice of the preprocessing technique but also by the selected classifier, we used two baseline classifiers to verify our results.

RoBERTa is the original classifier that obtained good performance for shorter texts, and also a wide number of downstream tasks at the time it was released

---

[4]https://www.kaggle.com/c/learn-ai-bbc.

(2019). RoBERTa was trained on 5 large-scale datasets totaling over 160GB of text and obtained good results for GLUE, SQuAD, and RACE. For our experiments, we used the RoBERTa$_{BASE}$.

The second model is the recent SetFit[5] [21], an efficient few-shot model introduced in 2022. Few-shot learners were introduced as a solution for scenarios in which labeled resources are scarce. Similar to RoBERTa which is essentially a fine-tuned BERT, SetFit (Sentence Transformer Fine-tuning) is a fine-tuned version of the Sentence Transformers (ST)[6] [18]. STs use Siamese networks for generating sentence embeddings for classification. SetFit is trained in two steps: i) an initial step generates sentence pairs and fine-tunes a pretrained ST using a contrastive training approach to learn common and rare attributes for various classes; ii) a logistic regression model called classification head is then trained on the resulting sentence embeddings and their class labels. The resulting model is efficient in the sense that it is much faster at inference and its training requires smaller sample sizes than similar models. The number of labeled examples per class is small (e.g., 8 or 16 samples per class) and the number of parameters is also relatively small compared to other models (e.g., 110 million parameters for SetFit compared to 175 billion parameters for GPT-3). The main benefits of using SetFit are the prompt-free design (e.g., no need to use handcrafted prompts, as a small set of examples is enough), multilingual ability, small size, and a significant reduction in the cost and time of training (e.g., fewer samples, smaller datasets, no need to use pretrained LLMs).

4.3   Results and Discussion

The first phase of our experiments was focused on understanding which embeddings would work the best for the preprocessing phase. Since categories can be thought of as clusters, we have tested different sentence embeddings models[7] from GloVe, to all-mpnet-base-v2 and BERTopic. GloVe [15] was built by Stanford scientists. We used both word and document versions of GloVe. The all-mpnet-base-v2 model is a sentence Transformer built by Microsoft. Due to eliminating the "topic-1" and using various embeddings, the number of training and testing samples varies across the runs, as can easily be seen in Table 2. The best runs used BERTopic and RoBERTa, but also had a lower number of training and testing samples than we would have expected.

The second phase of our experiments is presented in Table 3. Here we evaluated the baselines and the best models on multiple datasets, including N24News abstract and body, Guardian News and BBC News. The same observation about the number of training and testing samples applies to this set of experiments. As it can easily be seen, by removing the problematic cases, the BERTopic with RoBERTa classifier obtains the best scores.

---

[5]https://github.com/huggingface/setfit.

[6]https://www.sbert.net/.

[7]The pre-trained models were initially selected from this list: https://www.sbert.net/docs/pretrained_models.html.

What is important to notice is that regardless of the experiment, the models that included BERTopic embeddings have performed better. This suggests that the method is relatively stable and robust. Percentage-wise, the gains obtained when using SetFit classifiers with BERTopic preprocessors are impressive, and such classifiers do obtain the best scores for the BBC News dataset.

The best models for the N24 datasets for both cases (abstract and body) used RoBERTa classifiers. The N24News dataset is also the dataset with the largest number of categories. The results for the two partitions suggest that this method might be a good fit for fine-grained classifiers.

By combining BERTopic's clustered embeddings with a Sentence Transformer fine-tuned with SetFit, we obtain a prompt-free framework that demonstrates competitive performance even with few-shot labeled data. While results obtained with classic Transformers like RoBERTa might still end up being better, the training samples will be much larger and less cost-effective.

While the unsupervised topic modeling with BERTopic can be applied for both coarse and fine-grained news classification, results suggest that the choice of the classifier (e.g., RoBERTa, SetFit, etc.) should be taken depending on the type of dataset (e.g., RoBERTa may be a better fit for coarse classification).

Finally, an important aspect to understand that is not really evidenced through the evaluation is the fact that the topic clusters produced by BERTopic can easily be inspected via the BERTopic library, as it has already been evidenced through Fig. 1. The fact that such topic clusters are interpretable (e.g., we can easily see which words are included in each cluster) is perhaps one of the best arguments for using this neural topic model.

Table 2. N24News evaluation with various embeddings. Numbers from the #Training and #Testing columns represent counts. Bold values indicate the best run.

Corpus	Model	#Training	#Testing	Accuracy
N24News (Abstract)	RoBERTa	48974	12244	78.77
N24News (Abstract)	Glove Word	16868	4218	83.01
N24News (Abstract)	RoBERTa w BERTopic	27445	6862	**85.34**
N24News (Body)	RoBERTa	46838	11710	87.65
N24News (Body)	Glove Word	23620	5905	89.8
N24News (Body)	Glove Doc	25440	6360	89.88
N24News (Body)	all-mpnet-base-v2	24639	6160	89.78
N24News (Body)	RoBERTa w BERTopic	23743	5936	**90.54**

Table 3. Evaluation on multiple datasets. Numbers from the #Training and #Testing columns represent counts. Bold values indicate the best run for a RoBERTa or SetFit model.

Corpus	Model	#Training	#Testing	Accuracy
N24News (Abstract)	RoBERTa	48974	12244	78.77
N24News (Abstract)	RoBERTa w BERTopic	27445	6862	**85.34**
N24News (Abstract)	SetFit	2880	5380	64.6
N24News (Abstract)	SetFit w BERTopic	2880	6641	**80.6**
N24News (Body)	RoBERTa	46838	11710	87.65
N24News (Body)	RoBERTa w BERTopic	23743	5936	**90.54**
N24News (Body)	SetFit	1440	5773	76.5
N24News (Body)	SetFit w BERTopic	1440	5936	**87.3**
Guardian News	RoBERTa	41976	10494	96.26
Guardian News	RoBERTa w BERTopic	23936	5984	**96.66**
Guardian News	SetFit	600	4510	93.3
Guardian News	SetFit w BERTopic	600	5984	**95.3**
BBC News	RoBERTa	1780	445	72.32
BBC News	RoBERTa w BERTopic	1555	389	**75**
BBC News	SetFit	100	397	65.5
BBC News	SetFit w BERTopic	100	49	**75.5**

5   Conclusion

This paper presents a method to preprocess documents with BERTopic to help increase the accuracy of Transformer classifiers like RoBERTa or SetFit. The chosen approach not only manages to outperform hybrid models presented in the literature on certain datasets like N24, but also is well-suited for a range of multi-class classification tasks. Since the resulting framework is prompt-free, an upcoming publication will investigate how this approach compares with prompt-based classifiers [3].

Future work will also focus on correctly classifying documents in multi-label settings, and on visual means to explore and better understand the results. This will help advance work in fields as diverse as affective knowledge classification, story detection, or news recommendation. Another important research direction will be multilingualism, as our current experiments have been solely focused on English. The current solution uses a human-in-the-loop to verify the content of the clusters, but future versions might increasingly rely on automated verification methods.

Acknowledgements. The research presented in this paper has been partially conducted within the GENTIO project funded by the Austrian Federal Ministry for Climate Action, Environment, Energy, Mobility and Technology (BMK) via the ICT of the Future Program (GA No. 873992). Adrian M.P. Brasoveanu has been partially funded by the Vienna Science and Technology Fund (WWTF) [10.47379/ICT20096].

References

1. Abdelrazek, A., et al.: Topic modeling algorithms and applications: a survey. Inf. Syst. **112**, 102131 (2023). https://doi.org/10.1016/j.is.2022.102131
2. Brown, T.B., et al.: Language models are few-shot learners. In: Larochelle, H. et al. (eds.) Advances in Neural Information Processing Systems 33: Annual Conference on Neural Information Processing Systems 2020, NeurIPS 2020 (December), pp. 6–12. Virtual (2020). https://arxiv.org/abs/2005.14165
3. Clavié, B., et al.: Large language models in the workplace: a case study on prompt engineering for job type classification. In: Mètais, E., Meziane, F., Sugumaran, V., Manning, W., Reiff-Marganiec, S. (eds.) NLDB 2023. LNCS, vol. 13913, pp. 3–17. Springer, Cham (2023). https://doi.org/10.1007/978-3-031-35320-8_1
4. Devlin, J., et al.: BERT: pre-training of deep bidirectional transformers for language understanding. In: Burstein, J., Doran, C., Solorio, T. (eds.) Proceedings of the 2019 Conference of the North American Chapter of the Association for Computational Linguistics: Human Language Technologies, NAACL-HLT 2019, Minneapolis, MN, USA, 2–7 June 2019, Volume 1 (Long and Short Papers), pp. 4171–4186. Association for Computational Linguistics (2019). https://doi.org/10.18653/v1/n19-1423
5. Ferrell, B.J., et al.: Attention-based models for classifying small data sets using community-engaged research protocols: classification system development and validation pilot study. JMIR Formative Res. **6**(9), e32460 (2022). https://doi.org/10.2196/32460
6. de Groot, M., Aliannejadi, M., Haas, M.R.: Experiments on generalizability of BERTopic on multi-domain short text. CoRR abs/2212.08459 (2022). https://doi.org/10.48550/arXiv.2212.08459
7. Grootendorst, M.: BERTopic: neural topic modeling with a class based TF-IDF procedure. CoRR abs/2203.05794 (2022). https://doi.org/10.48550/arXiv.2203.05794
8. Hristova, G., Netov, N.: Media coverage and public perception of distance learning during the COVID-19 pandemic: a topic modeling approach based on BERTopic. In: Tsumoto, S., et al. (eds.) IEEE International Conference on Big Data, Big Data 2022, Osaka, Japan, 17–20 December 2022, pp. 2259–2264. IEEE (2022). https://doi.org/10.1109/BigData55660.2022.10020466
9. Kim, M.G., Kim, K.S., Lee, K.C.: Analyzing the effects of topics underlying companies' financial disclosures about risk factors on prediction of ESG risk ratings: emphasis on BERTopic. In: Tsumoto, S., et al. (eds.) IEEE International Conference on Big Data, Big Data 2022, Osaka, Japan, 17–20 December 2022, pp. 4520–4527. IEEE (2022). https://doi.org/10.1109/BigData55660.2022.10021110
10. Li, Q., et al.: A survey on text classification: from traditional to deep learning. ACM Trans. Intell. Syst. Technol. **13**(2), 31:1–31:41 (2022). https://doi.org/10.1145/3495162
11. Lin, T., et al.: A survey of transformers. CoRR abs/2106.04554 (2021). https://arxiv.org/abs/2106.04554
12. Liu, X., Wang, C.: An empirical study on hyperparameter optimization for fine-tuning pre-trained language models. In: Zong, C., et al. (eds.) Proceedings of the

59th Annual Meeting of the Association for Computational Linguistics and the 11th International Joint Conference on Natural Language Processing, ACL/IJCNLP 2021, (Volume 1: Long Papers), Virtual Event, 1–6 August 2021, pp. 2286–2300. Association for Computational Linguistics (2021). https://doi.org/10.18653/v1/2021.acl-long.178

13. Liu, Y., et al.: RoBERTa: a robustly optimized BERT pretraining approach. CoRR abs/1907.11692 (2019). http://arxiv.org/abs/1907.11692

14. Minaee, S., et al.: Deep learning-based text classification: a comprehensive review. ACM Comput. Surv. **54**(3), 62:1–62:40 (2022). https://doi.org/10.1145/3439726

15. Pennington, J., Socher, R., Manning, C.D.: Glove: global vectors for word representation. In: Moschitti, A., Pang, B., Daelemans, W. (eds.) Proceedings of the 2014 Conference on Empirical Methods in Natural Language Processing, EMNLP 2014, 25–29 October 2014, Doha, Qatar, A Meeting of SIGDAT, a Special Interest Group of the ACL, pp. 1532–1543. ACL (2014). https://doi.org/10.3115/v1/d14-1162

16. Phuong, M., Hutter, M.: Formal algorithms for transformers. CoRR abs/2207.09238 (2022). https://doi.org/10.48550/arXiv.2207.09238

17. Raffel, C., et al.: Exploring the limits of transfer learning with a unified text-to-text transformer. J. Mach. Learn. Res. 21, **140**(1–140), 67 (2020). http://jmlr.org/papers/v21/20-074.html

18. Reimers, N., Gurevych, I.: Sentence-BERT: sentence embeddings using Siamese BERT-networks. In: Inui, K., et al. (eds.) Proceedings of the 2019 Conference on Empirical Methods in Natural Language Processing and the 9th International Joint Conference on Natural Language Processing, EMNLPIJCNLP 2019, Hong Kong, China, 3–7 November 2019, pp. 3980–3990. Association for Computational Linguistics (2019). https://doi.org/10.18653/v1/D19-1410

19. Sia, S., Dalmia, A., Mielke, S.J.: Tired of topic models? Clusters of pretrained word embeddings make for fast and good topics too! In: Webber, B., et al. (eds.) Proceedings of the 2020 Conference on Empirical Methods in Natural Language Processing, EMNLP 2020, Online, 16–20 November 2020, pp. 1728–1736. Association for Computational Linguistics (2020). https://doi.org/10.18653/v1/2020.emnlp-main.135

20. Tirumala, K., et al.: Memorization without overfitting: analyzing the training dynamics of large language models. CoRR abs/2205.10770 (2022). https://doi.org/10.48550/arXiv.2205.10770, arXiv:2205.10770

21. Tunstall, L., et al.: Efficient few-shot learning without prompts. CoRR abs/2209.11055 (2022). https://doi.org/10.48550/arXiv.2209.11055

22. Wang, Z., et al.: N24News: a new dataset for multimodal news classification. In: Calzolari, N., et al. (eds.) Proceedings of the Thirteenth Language Resources and Evaluation Conference, LREC 2022, Marseille, France, 20–25 June 2022, pp. 6768–6775. European Language Resources Association (2022). https://aclanthology.org/2022.lrec-1.729

23. Yang, Z., et al.: XLNet: generalized autoregressive pretraining for language understanding. In: Wallach, H.M., et al. (eds.) Advances in Neural Information Processing Systems 32: Annual Conference on Neural Information Processing Systems 2019, NeurIPS 2019(December), Vancouver, BC, Canada, pp. 8–14, pp. 5754–5764 (2019). https://proceedings.neurips.cc/paper/2019/hash/dc6a7e655d7e5840e66733e9ee67cc69-Abstract.html

24. Yaseen, T.B., et al.: JUST-BLUE at SemEval-2021 Task 1: predicting lexical complexity using BERT and RoBERTa pre-trained language models. In: Palmer, A.,

et al. (eds.) Proceedings of the 15th International Workshop on Semantic Evaluation, SemEval@ACL/IJCNLP 2021, Virtual Event/Bangkok, Thailand, 5–6 August 2021, pp. 661–666. Association for Computational Linguistics (2021). https://doi.org/10.18653/v1/2021.semeval-1.85

25. Zhou, C., et al.: A comprehensive survey on pretrained foundation models: a history from BERT to ChatGPT. arXiv preprint arXiv:2302.09419 (2023). https://doi.org/10.48550/arXiv.2302.09419

26. Zong, S., et al.: Which model shall I choose? Cost/quality trade-offs for text classification tasks. CoRR abs/2301.07006 (2023). https://doi.org/10.48550/arXiv.2301.07006

Double Transfer Learning to Detect Lithium-Ion Batteries on X-Ray Images

David Rohrschneider[ID], Nermeen Abou Baker[(✉)][ID], and Uwe Handmann[ID]

Computer Science Institute, Ruhr West University of Applied Sciences,
Luetzowstrasse 5, 46236 Bottrop, Germany
Nermeen.Baker@hs-ruhrwest.de

Abstract. With the soaring popularity of electronic gadgets, Lithium-Ion Batteries (LIB) have witnessed a remarkable surge. The inspiration behind this study arises from the urgent need to automate the identification of batteries in diverse contexts, such as electronic waste recycling facilities or security screening at airports. Ultimately, it strives to minimize health hazards associated with battery recycling by enabling more accurate sorting with minimal human involvement. In this paper, we applied double transfer learning to eight cutting-edge object detectors, unlocking the potential of X-Ray images in recognizing and categorizing electronic mobile devices (EMD) along with their embedded Lithium-Ion batteries (LIB).

Keywords: Benchmark · Object Detection · Transfer Learning · Double transfer Learning · X-Ray Images · Lithium-Ion Batteries

1 Introduction

1.1 Problem Statement

The presence of such substances is noticeable in X-Ray images by darker color at the location of the battery. Based on this fact, there are a couple of applications that can benefit from detecting and sorting EMD and LIB. The usage of EMD has increased significantly in recent years, and it can be concluded that the number of disposals of these devices is also increasing. A study by [1] reported that, before sorting and recycling batteries, each electronic device must first be classified as to whether it contains batteries or not. In this process, there is a proportion of devices where removing batteries is more difficult because they have been glued or welded together. This results in considerable manual time and expenses, and can also pose an increased health risk to personnel, e.g. due to damaged LIBs. Recent research has already inspected that it is possible to classify electronic devices by model on RGB images, capturing each device from a top view and using a convolutional neural network for classification [2].

D. Rohrschneider and N.A. Baker—Equal contribution.

I. Rojas et al. (Eds.): IWANN 2023, LNCS 14134, pp. 175–188, 2023.
https://doi.org/10.1007/978-3-031-43085-5_14

Recognizing the model series can deliver exact information on the number, type, and location of batteries inside the device, but it requires the neural network to already know all possible device models found in the recycling facility, which is why direct inspection of EMD internals using X-Ray images may be more applicable. Another application is the security inspection of passenger baggage of LIB to prevent potential risks of heating and burning during the flight [3].

1.2   Research Gap

Some of the main challenges that cause low recycling rates of Lithium are barriers for sorting, disassembly, and pre-treatment steps, evoked by diversity and non-standardization of LIBs [4]. Being able to detect and classify EMD and LIB on X-Ray images using artificial intelligence, the electronic waste could be processed more efficiently to improve the economical viability of recycling electronic waste. To ensure the passengers' safety within an aircraft for example in the USA, the Federal Aviation Administration states that EMD should be kept in carry-on baggage and, if stored in checked baggage, should be packed to protect them from any outer damage [5]. Utilizing state-of-the-art (SOTA) object detection methods in this use case could enhance the speed and quality of baggage inspection. To our knowledge, none of the previous works tested You Only Look Once (YOLO)v7 [6], YOLOv8 [7] or vision transformer models in detecting EMD and LIB on X-Ray images.

Section 2 investigates related work and the SOTA of deep learning approaches to object detection. Section 3 describes the datasets preparation and model implementation in detail. Then, in Sect. 4 model performance is compared, and the impact of employing double transfer learning is analyzed. Finally, the experimental results are discussed, and future prospects are suggested.

2   State of the Art

2.1   Related Work

In addition to our previous work, which will be described in Sect. 2.2, two other publications have addressed the detection of EMD or batteries on X-Ray images.

The work by [8] introduced the HiXray dataset, which will also be briefly presented in Sect. 2.3, along with the Lateral Inhibition Module. With the module being detached from deep learning models and backbone structures, it can be integrated into existing architectures to enhance a model's performance. The HiXray dataset was tested using three different types of deep learning models: Single Stage Detector (SSD), Fully convolutional one-stage object detection (Fcos), and YOLOv5s [9]. The results show that the suggested module improves the mean Average Precision (mAP) of each of the existing models by an average of 1.5% and achieves the best MAP of 96.8% when combined with YOLOv5.

The authors of [1] deal with the detection and classification of batteries in the context of automating electronic waste recycling. For the experiment, 532 objects of electronic waste were imaged with a Computed Tomography scanner to visualize the batteries within, regardless of external dirt or damage to the object. Each of the 943 batteries was manually labeled, each assigned to one of six battery types, including prismatic and pouch lithium-ion batteries among others. The YOLOv2 model with a LightNet backbone was used for the detection and evaluated by the precision, recall, and the F1-score. The precision of just classifying whether or not an electrical device can be seen on an X-Ray image is 89% and the precision for detecting batteries within the image is 82% for the pouch LIB class and 75% for the cylindrical LIB class.

2.2 Motivation

In our previous work [10], we tested the detection of EMD and LIB using YOLOv5m [9] and the HiXray dataset. Multiple transfers of weights are utilized to detect the EMD and LIB, achieving a precision of 0.94 for detecting EMD and 0.935 for detecting LIB. It is also noted that the performance significantly improves when applying a second knowledge transfer for LIB detection in X-Ray images. Building upon this finding and utilizing the same database, this study incorporates the transfer of weights from the EMD detection task to the LIB detection task. However, the previous work was limited to YOLOv5m only and therefore was bound to the performance of a 1-Stage detector model, which was designed for real-time object detection tasks [9].

2.3 Public X-Ray Datasets

X-Ray image datasets related to object detection include, SIXray [11], OPIXray [12], PIDray [13], HiXray [8], and GDXray Baggage [14]. Solving our problem requires a dataset with appropriate object classes to detect EMD and LIB in X-Ray images. Furthermore, it should have a sufficient number of samples since the batteries contained in EMD appear smaller in relation to surrounding objects and are more difficult to identify. To present an overview of recent X-Ray datasets for object recognition, Table 1 compares their numbers of images and classes" for improved phrasing and readability.

2.4 Methods of Object Recognition

In the past, object recognition models were traditionally categorized into two types: two-stage and one-stage models. In 2020, a new technique called "End-to-End Object Detection with Transformers (DETR)" was introduced [15]. By utilizing the transformers' self-attention mechanism, these models can put objects in a global context of the image and create relationships between them, which helps in finding the final bounding box and classification decisions. Recent SOTA examples of object detection models employing the transformer architecture are the Swin transformer [16] or DINO models [17].

This work tested 8 object detection models, as shown in Table 2.

Table 1. Public X-Ray datasets (MD classes are written in bold)

Dataset	Number of images	Classes
SIXray	1,059,231	Gun, Knife, Wrench, Pliers, Scissors, Hammer
OPIXray	8,885	Folding Knife, Straight Knife, Scissors, Utility Knife, Multi-tool Knife
PIDray	47,677	Gun, Bullet, Knife, Wrench, Pliers, **Powerbank**, Baton, Lighter, Sprayer, Hammer, Scissors, Handcuffs
HiXray	45,364	**Portable Charger 1**, **Portable Charger 2**, **Mobile Phone**, **Laptop**, **Tablet**, Cosmetic, Water, Nonmetallic Lighter
GDXray Baggage	8,150	Handgun, Razor Blade, Shuriken, Pen Case, Clip, Spring, Door Key, Knife

Table 2. Comparison of models with pre-trained weights

Technique	2-Stage	1-Stage					Transformer	
Model	Faster R-CNN [18]	SSD Lite [19]	YOLO v5m	YOLO v7 W6	YOLO v8	Efficient Det D1 [20]	Casc. Mask R-CNN	DINO 4scale
Backbone	Res Net50	Mobile Net v2	YOLO v5	E-ELAN	Darknet -53	Efficient Net-b1 BiFPN	Swin-S	Swin-L
#M Parameters	29.162	4.475	21.2	70.4	25.9	6.6	107	218
COCO mAP	29.3	29.1	45.4	54.9	50.2	38.4	51.9	58.0

3 Materials and Methods

3.1 Selection of Models and Dataset

Models. To compare the impact of transfer learning on the detection of LIB on X-Ray images among the three object detection strategies presented in Sect. 2.4, at least one SOTA model was selected for each approach for the experiment. Considering the limited computing time and capacity available in practical scenarios, the models are selected to achieve a trade-off between computational cost and prediction accuracy. To perform double transfer learning, it is necessary to have access to pre-trained weights from a large and high-quality dataset that is publicly available.

Dataset. To choose a suitable database for this benchmark, the datasets compared in Table 1 were evaluated based on the number of samples and the EMD classes present. The HiXray dataset was selected among them, because it consists of more than 45,000 images in total and 5 different classes of EMD, while the other X-Ray datasets have, at most, one EMD class. It is not publicly available but can be obtained for academic purposes upon request. The set comes with images saved in JPG format and has an average resolution of 1,200 × 900 px, but they vary from image to image. The images are divided into two sets, with 36,295 assigned to the training set and 9,069 to the test set, resulting in a ratio of approximately 4:1. The bounding boxes for each image were manually annotated for the eight different object classes. The descriptions of these bounding boxes are provided in separate text files that correspond to the respective images. On average, there are 2.27 classes per image, indicating that multiple objects can be assigned to the eight object classes [8].

3.2 Dataset Preparation

The HiXray dataset consists of five EMD classes, namely portable charger1, portable charger2, mobile phone, laptop, and tablet, "as well as three additional classes: water, cosmetic, and non-metallic lighter". Since this work focuses on the detection of EMD and batteries, retaining these three classes might adversely affect the results, thus, they are initially excluded from the label files. For the first dataset shown in Fig. 1(a), 12,000 annotated samples and 2,000 unannotated samples are used for the training split, while 3,000 annotated samples and 500 unannotated samples are used for the testing split, all derived from the original HiXray dataset. Noting that the "Mobile Phone" class has the biggest number of occurrences, which poses a challenge in achieving a balanced distribution.

The second dataset was manually derived as a subset of the remaining images for the battery contained in the EMD. As previously indicated in our previous work [10], the three different LIB classes, namely 'Prismatic LIB,' 'Cylindrical LIB,' and 'Pouch LIB,' were found across different EMD classes, regardless of

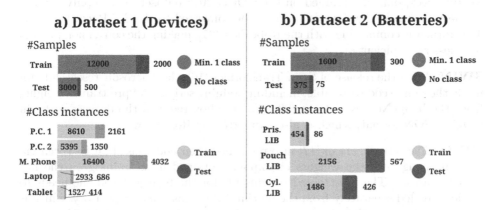

Fig. 1. Distribution of samples and class instances per dataset

their brand or model series. The 'Pouch LIB' appears in 'Mobile Phone', 'Laptop' and 'Tablet', 'Prismatic LIB' could only be found inside 'Portable charger1' and 'Cylindrical LIB' was found to appear in 'Portable charger2', as well as in some instances of the 'Laptop' class. As a result, 300 samples containing only the class 'Portable Charger2' and three times 300 samples containing only the classes 'Mobile Phone', 'Laptop' and 'Tablet' are created, along with 400 samples containing only the class 'Portable Charger1' since the prismatic LIB appears only once per instance. Another 300 samples without any annotations are added too. The testing set is created likewise, making up a total of 1,900 training and 450 testing samples, which then are manually annotated by drawing a bounding box around each battery instance and discarding the EMD classes. To increase the number of training samples, two augmentations are applied per image, including horizontal flipping and cropping with a zoom rate of 0–50%. The entire process is implemented using Roboflow [21] and the results in the train-test-split are visualized in Fig. 1(b).

3.3   Setting up the Environment

For the implementation, Google Colab Pro+ is used. To train and evaluate the chosen models with pre-trained weights and the custom dataset, we utilized the official code repositories published by the authors. These repositories offer prepared scripts, which are used to perform the training using equal hyperparameters and the evaluation with equal metrics described in the following sections.

YOLO. As stated in Table 2, three models from the YOLO family were tested in this study: YOLOv5m [9], YOLOv7-W6 [6] and YOLOv8m [7]. Since the dataset format is the same for all three models, our two datasets were exported only once using RoboFlow [21].

TensorFlow Object Detection API. To prepare Faster R-CNN [18], SSD Lite [19] and EfficientDet D1 [20] for the training and evaluation in Google Colab Pro+, the TensorFlow Object Detection API [22] is used. The original repository provides a wide range of object detection models and pre-trained weight checkpoint files trained on the COCO 2017 dataset. To simplify the use of the framework, the dataset needs to be converted to the TF-Record format. This format is compatible with the RoboFlow [21], making the conversion process seamless and efficient.

SWIN. With the release of Swin Transformer [16], the Microsoft research team made the source code available, including training scripts and pre-trained weights from the ImageNet-1K dataset [23]. The toolbox requires the dataset to be in COCO-JSON format, which is also supported by RoboFlow.

DINO. The researchers of [17] also published their code based on PyTorch via GitHub along with the weights checkpoint obtained from training on the MS COCO dataset. The pre-trained weights file for the Swin-L backbone needs to be downloaded separately from the official Swin transformer repository, which is also used above. DINO also excepts the dataset to be in COCO-JSON format, which had already been exported before.

Hyperparameters. The entire experiment was conducted using one Google Colab Pro+ notebook per framework. This setup ensures that all tests are conducted on a Tesla T4 GPU with 15 GB of RAM on two Intel(R) Xeon(R) CPUs @ 2.30 GHz sharing 52 GB of RAM. To ensure a fair comparison of the tested object detection models, the hyperparameters are fixed before starting the experiments and then passed to each configuration file. The input image size is set to 640x640 Pixels, the optimizer used was the Stochastic Gradient Descent (SGD), the initial learning rate was set to 0.01 with cosine decay, the confidence threshold during training to 0.001, and Intersection over Union (IoU)-threshold was set to 0.7 for the anchor-based models. Momentum is used with a β value of 0.937, along with the three warm-up epochs for learning rate and momentum term. The number of training epochs is fixed at 20 for fine-tuning on the first and 30 for fine-tuning on the second dataset while early stopping was used to ensure generalization. Considering the varying sizes of models and datasets, as well as the restricted resources in Google Colab Pro+, the batch size for each training was determined based on the capacity of GPU memory. Details on these will be described in Sect. 4.3.

4 Evaluation

4.1 Evaluation Metrics

The performance evaluation of each object detection model was conducted using the original repositories and the COCO API package [24]. The COCO API provides access to commonly used metrics, including mAP@0.5 and mAP@0.5:0.95 per category. The numbers following the @-symbol denote the IoU threshold used to calculate the mAP. The notation 0.5:0.95 indicates that the average mAP is calculated for each IoU threshold between 0.5 and 0.95, with a step size of 0.05. Therefore, the mAP@0.5:0.95 is a more critical metric as it measures the accuracy of predicted bounding box coordinates in relation to the ground truth, often resulting in lower values compared to mAP@0.5. The early stopping method was utilized, and the epoch number at which the highest evaluation

Transfer Learning Flow

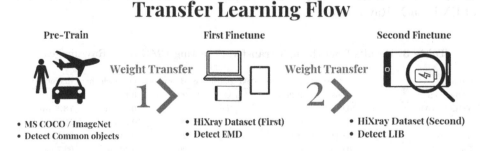

Fig. 2. Visual summary of the finetuning process applied to each model. The two weight transfers are colored red. (Color figure online)

score was achieved, was recorded. Additionally, the inference time was measured in milliseconds, and the mean epoch time was measured in hours. These metrics were compared by considering the batch sizes used for each training.

4.2 Evaluation Strategy

After setting up all the frameworks and datasets, the pre-trained models are fine-tuned on the first dataset for 20 epochs. At this point, the first knowledge transfer from the general task of detecting common objects within the MS COCO dataset to the specific task of detecting EMD on X-Ray images is performed, as can also be seen in Fig. 2.1. The COCO evaluation metrics are calculated on the testing split after each training epoch and displayed in the console. Additionally, a checkpoint file is created after each epoch, and the early stopping method is employed to select the checkpoint state with the highest average mAP across all classes for the second training, detecting three classes of batteries on X-Ray images. For this, the second dataset, which was introduced in Sect. 3.2, is uploaded into the notebooks. Next, the training is started using the same hyperparameters, the best checkpoint file from the previous training, and setting the epochs to 30. By doing so, the second knowledge transfer takes place by using the weights from the task of detecting EMD on X-Ray images and finetuning each model to predict bounding boxes for the batteries inside of the EMD. This transfer of weights is pointed out in Fig. 2.2. Once again, the evaluation metrics are computed after each epoch and the training is stopped when there are no significant improvements in the average mAP across all classes. Following each training iteration in the experiment, an additional evaluation is conducted on the testing set using a batch size of 1 to obtain mAP values for a valid model comparison. Furthermore, the final training speed is calculated by dividing the elapsed time taken to reach the best checkpoint by the number of epochs. Finally, if the framework does not display the inference time per image, an individual image inference is performed to measure the time in milliseconds per image.

4.3 Benchmark Results

Detection of EMD. Table 3 displays the evaluation results for the detection of EMD on X-Ray images.

Table 3. Results from the first transfer: Detecting EMD on X-Ray images

Model	mAP@ 0.5	mAP@ 0.5:0.95	Batch size	Best epoch	Ø Epoch time	Inference time
YOLOv5m	0.968	**0.605**	32	20	0.13 h	17 ms
YOLOv7-W6	0.969	0.598	16	20	**0.078 h**	**13.6 ms**
YOLOv8m	0.973	0.585	32	20	0.28 h	15 ms
SSD Lite	0.941	0.471	16	20	0.135 h	35.1 ms
EfficientDet D1	0.939	0.425	8	20	0.407 h	68.7 ms
Faster R-CNN	0.96	0.59	4	20	0.262 h	110.6 ms
Cas. Mask R-CNN (Swin-S)	0.959	0.581	4	20	2.33 h	57 ms
DINO (Swin-L)	**0.976**	0.562	2	8	3.73 h	700 ms

As shown in Table 3, the models achieved a minimum mAP@0.5 of 0.939 when predicting EMD on X-Ray images using their pre-trained weights. The DINO model with a Swin-L backbone performs the best out of the eight models, with a value of 0.976, and thus having an average miss-prediction rate of 2.4%. DINO achieves its best evaluation results at epoch 8 of 20 and attains the highest mAP@0.5 among all the experiments. On the other hand side, DINO can only be trained with a batch size of 2, resulting in the longest average epoch time of 3.73 h and the slowest inference speed of 700 ms per image. This happens due to its big number of parameters, which were compared in Table 2. Similarly, the Cascade Mask R-CNN model with Swin-S transformer backbone has the second-longest average epoch time of 2.33 hours but is more than 12 times faster than DINO when it comes to a single image inference. In contrast to this, the 2-Stage detector Faster R-CNN with a ResNet50 backbone shows up nearly 10 times smaller average epoch time and higher mAP scores, but an inference speed almost twice as long. Among the 1-Stage detectors, the YOLO models, despite having fewer parameters, can compete with the larger vision transformer models in terms of mAP results. Each of the three tested YOLO models achieved higher mAP@0.5:0.95, could be trained with larger batch size, and exhibited faster epoch and inference times compared to the transformer models. In particular, the recently published model YOLOv8m could achieve a mAP@0.5 of 0.973, exhibiting only 0.31% lower accuracy and more than 46 times faster when performing a single image inference compared to the DINO vision transformer. Furthermore, the YOLO models surpass the other 1-Stage detectors (SSD-Lite and EfficientDet-D1) in terms of mAP@0.5, mAP@0.5:0.95, and inference time.

Detection of LIB. Table 4 shows the evaluation results for the detection of LIB on X-Ray images.

Table 4. Results from the second transfer: Detecting Batteries on X-Ray images

Model	mAP@ 0.5	mAP@ 0.5:0.95	Batch size	Best epoch	Ø Epoch time	Inference time
YOLOv5m	0.928	0.746	32	30	**0.021 h**	18 ms
YOLOv7-W6	0.932	0.731	32	30	0.033 h	**13.5 ms**
YOLOv8m	0.94	**0.753**	32	24	0.06 h	14.8 ms
SSD Lite	0.827	0.516	16	30	0.074 h	30 ms
EfficientDet D1	0.86	0.547	8	30	0.207 h	50 ms
Faster R-CNN	0.863	0.551	4	30	0.198 h	109 ms
Cas. Mask R-CNN (Swin-S)	0.944	0.72	8	20	0.245 h	67 ms
DINO (Swin-L)	**0.947**	0.727	2	3	2.82 h	660 ms

Using the weights from the first training to train the models on detecting the three LIB classes in X-Ray images, the results are noted in Table 4. It is worth mentioning that the second dataset contains fewer samples than the first, as previously shown in Fig. 1. Therefore, the batch size could be raised in the cases of YOLOv7-W6 and Cascade Mask R-CNN with Swin-S backbone. A similar pattern to the results in Table 3 can be observed in the current table. Starting with the DINO model, it is able to achieve the highest mAP@0.5 score with 0.947 at the early epoch of 3, which is also the highest value among all models. Following closely, the second vision transformer model, Cascade Mask R-CNN with Swin-S backbone, achieves a mAP@0.5 value of 0.944 at the 20th epoch. Interestingly, its average epoch time is less than one-tenth of DINO's average epoch time, likely due to the fact that the batch size in the second training is twice as large as in the first training. The 2-Stage model Faster R-CNN remains the second slowest model when it comes to the inference per image speed and is, along with SSD-Lite and EfficientDet-D1, almost 10% less precise than the YOLO and vision transformer models in terms of both mAP categories. Among the YOLO models, YOLOv8m achieves the highest mAP@0.5 of 0.94 and outperforms all other models with a mAP@0.5:0.95 of 0.753. On the other hand, it has the slowest average epoch time compared to YOLOv7-W6 and YOLOv5m in both experiments but is still faster than the remaining five models. Moreover, the mAP@0.5 values, particularly for SSD-Lite, EfficientDet-D1, and Faster R-CNN, are lower than in the first training with the task of detecting EMD on X-Ray images and the mAP@0.5:0.95 values are higher. One possible reason for this is the presence of smaller and more occluded object instances, such as the cylindrical LIBs. Figure 3 emphasizes this aspect in a side-by-side comparison of three original images from the dataset.

Prismatic LIB   Pouch LIB   Cylindrical LIB

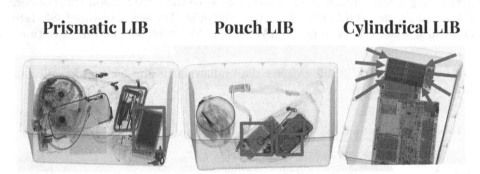

Fig. 3. Side-by-side comparison of the three different battery instances (Marked by red rectangles and arrows). (Color figure online)

To further investigate this phenomenon, Table 5 provides a closer examination of the mAP values, along with the average image proportion of a bounding box per category.

It is noticeable, that the cylindrical LIB is not predicted as accurately as the other classes in terms of mAP@0.5:0.95 for all models. Table 5 provides additional details on why the three models mentioned earlier achieve significantly lower mAP results. SSD Lite, EfficientDet-D1, and Faster R-CNN achieve a maximum score of 0.753 for mAP@0.5 and 0.361 for mAP@0.5:0.95. Since the cylindrical LIB is approximately one-third the size of the other two classes, meaning that

Table 5. Comparison of bounding box proportion and mAP per class

	Pouch LIB		Prismatic LIB		Cylindrical LIB	
Øbox/image	1.69%		1.68%		0.57%	
	mAP@ 0.5	mAP@ 0.5:0.95	mAP@ 0.5	mAP@ 0.5:0.95	mAP@ 0.5	mAP@ 0.5:0.95
YOLOv5m	0.966	**0.816**	0.901	0.799	0.917	0.624
YOLOv7-W6	**0.975**	0.809	0.894	0.785	0.926	0.6
YOLOv8m	0.962	0.806	0.922	0.814	**0.936**	**0.638**
SSD Lite	0.91	0.595	0.904	0.622	0.667	0.33
EfficientDet D1	0.929	0.642	0.948	0.696	0.703	0.303
Faster R-CNN	0.923	0.627	0.912	0.665	0.753	0.361
Cas. Mask R-CNN (Swin-S)	0.964	0.767	0.935	0.793	0.934	0.601
DINO (Swin-L)	0.958	0.78	**0.969**	**0.815**	0.913	0.586
Average	0.948	0.73	0.923	0.749	0.844	0.505

a single instance's bounding box makes up only 0.57% of the image it is found in, there could be a close relationship to the models detecting them with less precision. However, the other models can overcome this obstacle, with YOLOv8m reaching a maximum score of 0.936 mAP@0.5.

4.4   Double Transfer Learning

To address the question of how well the double transfer learning applies to the task of detecting EMD and LIB on X-Ray images, the loss function trend of all the 8 individual trainings is visualized in Fig. 4. The loss values are normalized to a range between 0 and 1 for each model using Min-Max-Normalization, which involves taking the minimum and maximum values from the combined set of EMD and LIB training losses. The loss graphs displayed in blue correspond to the first training to detect EMD, while the red graphs represent the second training to detect LIB. This benchmark evaluates a total of eight different models, with each model having one blue and one red graph, resulting in a total of 16 loss graphs displayed in Fig. 4.

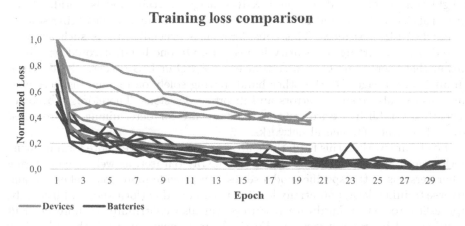

Fig. 4. Comparison of normalized training loss between detecting devices and batteries (Color figure online)

The graphic demonstrates that during the first training, the majority of loss values are higher compared to the second training after transferring weights for the second time. Additionally, it can be observed that in most of the red graphs, the initial loss value in epoch 1 is lower. This suggests that reusing weights from the first training on EMD leads to lower losses and, consequently, higher prediction accuracy values at the early stage of the second training. This gradual introduction allows the model to learn to detect more specific and smaller object categories, such as the LIB within the EMD. This is particularly useful for vision transformer models, as they typically require a substantial amount of data to achieve performance comparable to convolutional neural networks [25]. Transfer learning, or in this case, double transfer learning, can help to reduce the amount of data needed for object recognition from a new context.

5 Summary and Outlook

In summary, it is possible to outperform the SOTA in detecting EMD and LIB on X-Ray images using pre-trained models, regardless of whether a 1-Stage, 2-Stage, or vision transformer model is employed. The results in Sect. 4.3 demonstrate that, with the utilization of double transfer learning, a mAP@0.5 value of 0.947 after the third training epoch in the case of DINO could be achieved. This can be a significant advantage if the training process of a complex model is computationally expensive and there are limited resources available for the task of detecting batteries whether in a recycling facility or at an airport security control. The use of multiple weight transfer, particularly in the case of vision transformers such as Swin or DINO, shows promise as a method to address the challenge of requiring extensive data for training these types of models. Furthermore, it has become evident, that the YOLO models, especially YOLOv8m, can achieve nearly equal performance with a tenth of the number of parameters compared to DINO, resulting in a more than 45 times faster single image inference speed. Although our previous work [10] demonstrated the positive impact of double weight transfer on LIB detection in X-Ray images, certain models, notably SSD-Lite, EfficientDet-D1, and Faster R-CNN, faced challenges in detecting smaller cylindrical LIB instances. This consideration is crucial in tasks such as automated battery sorting or security inspections. On one hand, it could be a limitation that there is no guarantee that every EMD or LIB is correctly identified on an X-Ray image. On the other hand, relying solely on manual inspection by humans can also result in errors and incorrect decisions. Therefore, it becomes crucial to establish precise criteria in the future for determining the required level of accuracy for neural networks.

For future work in this field, there are several avenues to explore for improving performance. Firstly, utilizing the entire HiXray dataset, as well as considering other versions of the specific model series, can provide more comprehensive and diverse training data, potentially leading to enhanced performance. Additionally, upgrading to better hardware resources can also contribute to more efficient training and inference processes. Furthermore, a new dataset with real X-Ray images of electronic waste can be created, which can be beneficial for the context of automating battery sorting in recycling facilities. Object recognition using machine learning models could also act as a part of an EMD or LIB detection pipeline, in which multiple sensors like serial number detectors or tools for weight and chemical composition analysis are combined to improve performance and stability. Lastly, the classification of smartphone model series using deep learning was also covered in previous research [2], and finding an appropriate combination of these techniques could result in more efficient solutions, as well as a lower health risk for the human being.

References

1. Sterkens, W., Diaz Romero, D., Goedemé, T., Dewulf, W., Peeters, J.R.: Detection and recognition of batteries on x-ray images of waste electrical and electronic equipment using deep learning. Resour. Conserv. Recycl. **168**, 105246 (2021)
2. Abou Baker, N., Stehr, J., Handmann, U.: Transfer learning approach towards a smarter recycling. In: Pimenidis, E., Angelov, P., Jayne, C., Papaleonidas, A., Aydin, M. (eds.) ICANN 2022. LNCS, vol. 13529, pp. 685–696. Springer, Cham (2022). https://doi.org/10.1007/978-3-031-15919-0_57
3. Dangerous goods. https://www.easa.europa.eu/en/domains/passengers/dangerous-goods. Accessed 10 Feb 2023
4. Ma, X., Azhari, L., Wang, Y.: Li-ion battery recycling challenges. Chem **7**(11), 2843–2847 (2021)
5. Lithium batteries in baggage. https://www.faa.gov/newsroom/lithium-batteries-baggage. Accessed 22 July 2022
6. Wang, C.-Y., Bochkovskiy, A., Liao, H.-Y.M.: YOLOv7: trainable bag-of-freebies sets new state-of-the-art for real-time object detectors. arXiv preprint arXiv:2207.02696 (2022)
7. Jocher, G., Chaurasia, A., Qiu, J.: YOLO by Ultralytics (2023). https://ultralytics.com/. Accessed 06 Feb 2023
8. Tao, R., et al.: Towards real-world x-ray security inspection: a high-quality benchmark and lateral inhibition module for prohibited items detection. In: 2021 IEEE/CVF International Conference on Computer Vision (ICCV), pp. 10903–10912 (2021)
9. YOLOv5 and Vision AI. https://ultralytics.com/. Accessed 22 July 2022
10. Abou Baker, N., Rohrschneider, D., Handmann, U.: Battery detection of xray images using transfer learning. In: The 30th European Symposium on Artificial Neural Networks (ESANN 2022), (Bruges, Belgium), pp. 241–246 (2022)
11. Miao, C., et al.: Sixray: a large-scale security inspection x-ray benchmark for prohibited item discovery in overlapping images. in: 2019 IEEE/CVF Conference on Computer Vision and Pattern Recognition (CVPR), pp. 2114–2123 (2019)
12. Wei, Y., Tao, R., Wu, Z., Ma, Y., Zhang, L., Liu, X.: Occluded prohibited items detection: an x-ray security inspection benchmark and de-occlusion attention module. CoRR, abs/2004.08656 (2020)
13. Wang, B., Zhang, L., Wen, L., Liu, X., Wu, Y.: Towards real-world prohibited item detection: a large-scale x-ray benchmark. In: 2021 IEEE/CVF International Conference on Computer Vision (ICCV), pp. 5392–5401 (2021)
14. Mery, D., et al.: Gdxray: the database of x-ray images for nondestructive testing. J. Nondestr. Eval. **34**, 1–12 (2015)

15. Carion, N., Massa, F., Synnaeve, G., Usunier, N., Kirillov, A., Zagoruyko, S.: End-to-end object detection with transformers. In: Vedaldi, A., Bischof, H., Brox, T., Frahm, J.-M. (eds.) ECCV 2020. LNCS, vol. 12346, pp. 213–229. Springer, Cham (2020). https://doi.org/10.1007/978-3-030-58452-8_13

16. Liu, Z., et al.: Swin transformer: hierarchical vision transformer using shifted windows. arXiv preprint arXiv:2103.14030 (2021)

17. Zhang, H., et al.: Dino: Detr with improved denoising anchor boxes for end-to-end object detection (2022)

18. Ren, S., He, K., Girshick, R., Sun, J.: Faster R-CNN: towards real-time object detection with region proposal networks. In: Advances in Neural Information Processing Systems, vol. 28 (2015)

19. Liu, W., et al.: SSD: single shot multibox detector. In: Leibe, B., Matas, J., Sebe, N., Welling, M. (eds.) ECCV 2016. LNCS, vol. 9905, pp. 21–37. Springer, Cham (2016). https://doi.org/10.1007/978-3-319-46448-0_2

20. Tan, M., Pang, R., Le, Q.V.: Efficientdet: scalable and efficient object detection. In: 2020 IEEE/CVF Conference on Computer Vision and Pattern Recognition (CVPR), pp. 10778–10787 (2020)

21. RoboFlow. https://roboflow.com/. Accessed 22 July 2022

22. Huang, J., et al.: Speed/accuracy trade-offs for modern convolutional object detectors. In: Proceedings of the IEEE Conference on Computer Vision and Pattern Recognition, pp. 7310–7311 (2017)

23. Deng, J., Dong, W., Socher, R., Li, L.-J., Li, K., Fei-Fei, L.: Imagenet: a large-scale hierarchical image database. In: 2009 IEEE Conference on Computer Vision and Pattern Recognition, pp. 248–255 (2009)

24. Lin, T.-Y., et al.: Microsoft COCO: common objects in context. In: Fleet, D., Pajdla, T., Schiele, B., Tuytelaars, T. (eds.) ECCV 2014. LNCS, vol. 8693, pp. 740–755. Springer, Cham (2014). https://doi.org/10.1007/978-3-319-10602-1_48

25. Beal, J., Kim, E., Tzeng, E., Park, D.H., Zhai, A., Kislyuk, D.: Toward transformer-based object detection. CoRR, vol. abs/2012.09958 (2020)

MLFEN: Multi-scale Long-Distance Feature Extraction Network

Yuhua Wang$^{(\boxtimes)}$ and Yuhao Lian

Chongqing University of Posts and Telecommunications, Chongqing 400065, China
curley0619@gmail.com, lianyuhao@ieee.org

Abstract. Hyperspectral image fusion frequently leverages panchromatic and multispectral images. Although remote sensing images exhibit multi-scale features, prior research has predominantly focused on local feature extraction using convolutional approaches, thereby neglecting long-range dependencies among image elements. To overcome this limitation, we introduce a network for multi-scale long-distance feature extraction that incorporates an encoder-decoder structure with skip connections and a multi-layer perceptron block with attention mechanisms. By capturing features from multiple scales and distant locations within the image, the proposed network improves the performance of image fusion. Our experimental findings demonstrate that the proposed network achieves state-of-the-art performance in image fusion tasks.

Keywords: Image Fusion · Pansharpening · Multilayer perceptron · Encoder - decoder

1 Introduction

Pansharpening is a technique that fuses panchromatic (PAN) and multispectral (MS) images acquired simultaneously over the same geographic area. This process is a specific problem of data fusion, as it aims to combine the spatial detail provided by the PAN image (which does not exist in the MS image) with the spectral detail of the MS image to obtain a high-resolution multispectral image. Nowadays, both PAN and MS images can be acquired from commercial satellites such as IKONOS and GeoEye. Pansharpening is currently the optimal solution for obtaining high-resolution images since physical limitations prevent this goal from being achieved through a single sensor. Additionally, pansharpening is used as a preprocessing step for many upstream remote sensing tasks such as change detection, target recognition, and scene interpretation to enhance the images and achieve better results.

Pansharpening consists of two main approaches: component substitution (CS) and multi-scale analysis (MRA). CS methods rely on using the PAN image to replace components, and include techniques such as intensity-hue-saturation (IHS) [1], principal component analysis (PCA) [2], and Gram-Schmidt (GS) [3]

© The Author(s), under exclusive license to Springer Nature Switzerland AG 2023
I. Rojas et al. (Eds.): IWANN 2023, LNCS 14134, pp. 189–199, 2023.
https://doi.org/10.1007/978-3-031-43085-5_15

spectral sharpening. MRA methods involve injecting spatial details by resampling the MS bands using a multi-resolution decomposition of the PAN image. Several forms of MRA can be used to extract spatial details, including DWT [4], UDWT [5], ATWT [6], LP [7].

With the widespread application of deep neural networks in image processing, researchers have begun to explore ways to use deep learning to solve image fusion. Some well-known methods include PNN [8] based on convolutional neural networks, which adopts the architecture previously proposed for image super-resolution [10]. Another method is PanNet [9], which focuses on preserving spatial and spectral information.

The aforementioned methods treat pansharpening as a simple image regression problem and do not consider the deep features of the images. The main contributions and nobility of this paper are summarized as follows:

- This paper presents a novel approach to incorporate the extraction of long-range point relationships into deep learning frameworks. The approach employs fully connected layers to extract the relationships between distant modules and introduces attention modules to address redundant information introduced by the fully connected layers.
- Our proposed method incorporates multiscale information of remote sensing images by leveraging an encoder-decoder structure to extract these features. To fuse shallow and deep features, we use skip connections. Experimental results demonstrate that our approach, MLFEN, outperforms other existing methods, achieving state-of-the-art performance.

This paper is structured as follows: firstly, a review of prior work on remote sensing image fusion is presented. The second section focuses on discussing fully connected networks, encoder-decoder structures, and their influence on the proposed network architecture. In the third section, the problem is formalized, and the MLFEN architecture is elaborated in detail. The fourth section reports on experimental evaluations of the MLFEN model, including comparisons with existing methods, as well as a thorough ablation study. Finally, the proposed MLFEN model is summarized in the last section, which shows a significant improvement in image fusion accuracy and substantial advancements in both quantitative analysis metrics and visual analysis.

2   Related Work

In recent decades, numerous methods for pansharpening have been proposed, with the most commonly used being component substitution techniques such as IHS [1], PCA [2], and Brovey [11]. These methods are direct and fast, but they often succeed in only approximating the spatial resolution of the HRMS image contained in the PAN image, at the cost of introducing spectral distortions. To address this problem, more complex techniques have been proposed, such as adaptive methods and dependency-based methods. In multi-resolution methods, the PAN image and the LRMS image are decomposed, for example using

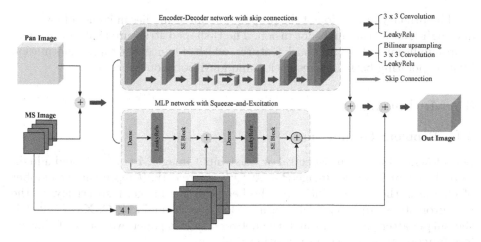

Fig. 1. Details of the structure of MLFEN.

wavelets or Laplacian pyramids, and then fused. Other model-based methods encode beliefs about the relationships between PAN, HRMS, and LRMS images in a regularized objective function, and then treat the fusion problem as an image restoration optimization problem. Many of these algorithms have achieved good results. In this study, we selected the best-performing methods among these for experimental comparison.

MLP-Mixer. The MLP Block proposed in this article is mainly derived from a recent computer vision framework called MLP-Mixer, which was introduced by the Google ViT team. MLP-Mixer replaces the convolutional operations in traditional CNNs and the self-attention mechanism in Transformers with multilayer perceptron (MLP) operations.

MLP-Mixer mainly consists of three parts: Per-patch Fully-connected, Mixer Layer, and Classifier. The Classifier part adopts the traditional Global Average Pooling (GAP)+Fully-connected layer (FC)+Softmax method. The Fully-connected layer, unlike the convolutional layer, cannot extract information between local regions. To solve this problem, MLP-Mixer converts the input image into a 2D table through Per-patch Fully-connected, making it easier to fuse information between local regions in the subsequent stages.

Encoder Decoder Structure. The encoder-decoder architecture proposed in this paper draws inspiration from the Unet model in the field of image segmentation. Unet is an extension of the Fully Convolutional Network (FCN) that addresses the challenge of extracting both contextual and positional information. Its U-shaped architecture consists of a feature extraction network and a feature fusion network. The feature extraction network encodes the input image to obtain local features that are subsequently transformed into abstract semantic features. The feature fusion network then utilizes these abstract features to recover the original image size.

The primary innovation of the proposed architecture lies in its use of low-level and high-level feature maps for fusion. The U-shaped structure facilitates a more thorough fusion of front and back features, which enhances the target information by incorporating both high-resolution and low-resolution information.

3   Methodology

3.1   Problem Formulation

Let a low-resolution multispectral or noisy input image $\mathbf{M} \in \mathbb{R}^{C,w,h}$, and a high-resolution panchromatic image $\mathbf{P} \in \mathbb{R}^{c,W,H}$, where c, W, H represent the number of channels, the image width and the height of the image, respectively. In the panchromatic sharpening problem, $w \ll W$, $h \ll H$, and $c \ll C$. $\mathbf{X} \in \mathbb{R}^{C,W,H}$ is the output after passing through the network. In this paper, we use the following loss function to solve the optimization problem:

$$Loss = L(\mathbf{M,X,P}) \tag{1}$$

The variable L represents a unique loss function for each task. In this formula, the optimization problem incorporates both multispectral image $\mathbf{M}$ and panchromatic image $\mathbf{P}$. This structure efficiently integrates multiscale spatial details and semantic features. In the following section, a novel architecture named Multiscale Long-distance Feature Extraction Network is proposed for various image fusion tasks.

For the purpose of quantitatively validating the results, both reference-based and reference-free standards were employed in this study. Reference-based standards encompassed Spectral Angle Mapper (SAM [16]), Spatial Correlation Coefficient (SCC [17]), Peak Signal-to-Noise Ratio (PSNR [18]), and the relative dimensionless global error in synthesis (ERGAS [15]). In the context of full-color sharpening, there exist various criteria for assessing the performance of the method on the original spatial resolution image in the absence of reference data. These standards comprise the spectral quality index (D_λ), the spatial quality index (D_S), and the no-reference spectral and spatial joint quality index (QNR [15]). Two of the more important formulas are shown in Eqs. 2, 3:

$$SAM(I_i, J_i) = arccos(\frac{< I_i, J_i >}{\|I_i\|\|J_i\|}) \tag{2}$$

In the Eq. 2, I_n represents the pixel vector of the n-band MS image, J_n represents the predicted vector of n band HRMS, $< \cdot, \cdot >$ denotes the inner product, and $\|\cdot\|$ denotes the magnitude. The SAM global value for the entire image is calculated by taking the average of individual measurements over all pixels. The optimal value for SAM is 0, indicating perfect spectral alignment between the fused image and the reference image.

$$ERGAS = \frac{100}{R} \sqrt{\frac{1}{N} \sum_{i=1}^{n} (\frac{RMSE(I_i, J_i)}{\mu(I_i)})^2} \tag{3}$$

Equation (3) is *Erreur Relative Globale Adimensionnelle de Synthèse* (ERGAS). In the Eq. 3, RMSE is defined as root mean square error, which will not be described in this paper because of its common use. μ represents the average of the image (the average over pixels), which is ideally 0.

3.2 MLFEN Architecture

By studying the PanNet network architecture, this study further investigates and builds upon its success. Like other pan-sharpening methods, our deep network is designed to preserve both spectral and spatial information simultaneously. Given the potential correlations between distant blocks in remote sensing images, we incorporate a Multi-Layer Perceptron (MLP) module to extract long-range feature relations. To prevent the introduction of redundant information, an attention module is utilized to extract important information within the MLP module. Additionally, to account for the multi-scale nature of remote sensing images, we employ an encoder-decoder structure with skip connections to extract and fuse multi-scale features. We will discuss each of these topics in more detail below.

The overall architecture, as illustrated in Fig. 1, consists primarily of an encoder-decoder structure with skip connections, as well as two network structures featuring Multi-Layer Perceptron (MLP) blocks with attention mechanisms. The attention block, depicted in Fig. 2, employs attention mechanisms in the channel dimension. Prior to applying the Squeeze-and-Excitation (SE) attention mechanism, the feature map assigns equal importance to each channel. Subsequently, the SE block assigns different weights to different channels, as represented by the different colors in the figure. This results in different levels of importance for each feature channel, allowing the neural network to prioritize channels with higher weight values.

MLP Block with Attention Mechanism. The lower part in Fig. 1 uses MLP blocks to extract the relationship features between distant pixels of the image and prevents the redundancy of information through the attention mechanism. The attention mechanism shown in Fig. 2 uses another new neural network to obtain the importance of each channel of the feature map by means of automatic learning, and then assigns a weight value to each feature using this importance, so that the neural network can focus on certain feature channels. The channels of feature maps that are useful for the current task are promoted, and the feature channels that are not useful for the current task are suppressed. Among them, Squeeze compreszes the two-dimensional features ($H * W$) of each channel to 1 real number by global average pooling, and the feature map is transformed from $[c, w, h] =¿ [c, 1, 1]$. excitation generates a weight value for each feature channel. In this paper, two fully connected layers are used to build the correlation between channels, and the number of output weight values is the same as the number of input feature maps. $[c, 1, 1] =¿ [c, 1, 1]$. Finally, Scale is used to weight the previously obtained normalized weights to the features of each channel. Multiplication is used in this paper, multiplying the weight coefficients channel by channel. $[c, w, h] * [c, 1, 1] => [c, w, h]$.

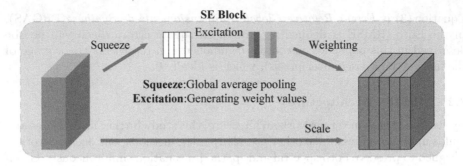

Fig. 2. Specific architecture of SE attention block.

The core idea of attention is to automatically learn feature weights according to loss through a fully connected network, rather than directly according to the numerical assignment of feature channels, so that the weights of effective feature channels are large. Of course, the SE attention mechanism inevitably increases some parameters and computational complexity, which is considered acceptable in this paper. The benefit is that it has more nonlinearities and can better fit the complex correlations between channels.

Encoder-Decoder Architecture with Skip Connections. As shown in Fig. 1, the encoder-decoder network structure is similar to Unet and can extract features of images at multiple scales. The multi-scale features represent the semantic features of the image from low to high level, and the semantic features are used to guide the parameter estimation of the decoder. Skip connections are used to combine shallow features and deep features to extract multi-scale features of the image.

4   Experiments

We conducted several experiments using data acquired from the IKONOS satellite. To minimize the objective function in Eq. 1, we employed the adaptive momentum stochastic optimization method Adam. For our experiments, we extracted PAN/LRMS/HRMS patch pairs of size 64×64. We randomly divided these patches into training and validation sets, with a ratio of 90% and 10%, respectively. We compared the performance of six commonly used pan-sharpening methods, namely IHS [1], Brovey [11], GSA [12], MTF-GLP-HPM [13], PanNet [9], and PNN [8]. Multiple parameter settings were tested for each method, and the optimal configuration was selected based on performance metrics.

4.1   Dataset

IKONOS is a high resolution, optical remote sensing commercial satellite, including land, agriculture, forestry, environmental protection and other departments to provide HD satellite images. The resolution of panchromatic images and multi-spectral images used in this paper is 4m and 1m, respectively. Here, four different image scenes are selected for experiments, including agricultural, urban, forest, or a mixture of these scenes.

4.2   Results

Table 1 presents the results of the test images in terms of quantitative metrics, indicating that MLFEN outperforms existing methods across all metrics. Additionally, visually, the HR-MS images reconstructed by MLFEN exhibit superior spectral and spatial information quality. Figure 3 displays the RGB image of the fused HR-MS image, with detailed image comparisons provided after zooming in. Our analysis reveals that while PanNet, and PNN obtained clearer edges in the fused image, they also suffered from significant spectral distortion, which led to discrepancies in color and reference images. In contrast, MLFEN retains the spectral information while also generating spatial details that are similar to the reference data. This reduces the error of the fused image, and we evaluate the performance of the fused image using the original resolution image.

Table 1. Quantitative evaluation results of pansharpening in different image scenes. ↓ indicates the lower the better, ↑ indicates the higher the better, and bold indicates the optimal value.

	IHS	Brovery	GSA	MGH	PanNet	PNN	MLFEN
PSNR↑	25.4087	25.2994	26.4543	26.8807	28.4330	28.5427	**32.2875**
SAM↓	0.1796	0.1451	0.212	0.1993	0.1473	0.1419	**0.1106**
SCC↑	0.8831	0.8914	0.9189	0.928	0.9483	0.9554	**0.9654**
ERGAS↓	7.8865	7.9374	6.885	6.4427	5.1673	4.8567	**4.4947**
D_λ↓	0.1047	0.0606	0.1604	**0.0734**	0.0784	0.1063	0.0751
D_s↓	0.1564	0.1485	0.1374	0.0889	0.0757	0.0795	**0.0722**
QNR↑	0.7553	0.7999	0.7243	0.8442	0.8519	0.8226	**0.8581**

In Table 1, where MGH stands for MTF-GLP-HPM. Table 1 shows that MTF-GLP-HPM produces the minimum value of D_λ, indicating that it provides the most similar reconstructed spectrum to the LR-MS image's spectrum. Meanwhile, all other indicator values are optimal for MLFEN, implying its superiority in terms of spatial details over other methods, and its strong performance in spectral preservation. Moreover, MLFEN's superiority in the QNR index over other methods suggests its ability to balance the trade-off between spectral resolution and spatial resolution effectively, resulting in a superior QNR value.

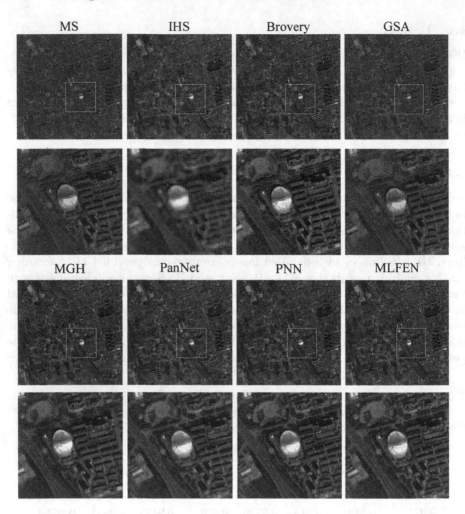

Fig. 3. Image fusion result. First row: RGB image of the pansharpening MS image. Second row: magnified RGB image. (Color figure online)

Here, only PSNR values in the training process of deep learning methods are compared. As depicted in Fig. 4, the PSNR values of all networks exhibit an increasing trend as the number of epochs increases. However, the PSNR values eventually reach a peak and do not increase indefinitely. The results demonstrate that the proposed MLFEN architecture achieves the highest PSNR value and outperforms other compared models, indicating its effectiveness in enhancing image fusion quality.

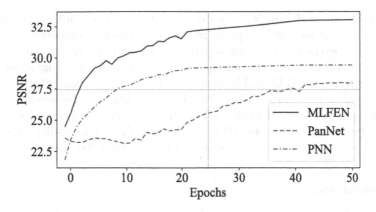

Fig. 4. Variation of PSNR for each network structure.

4.3   Ablation Experiment

In order to investigate the role of certain modules within the MLFEN framework, a series of ablation experiments were conducted in this study. The experiments involved 4 different configurations, and the results are presented in Table 2.

MLP Block. The inclusion of MLP blocks in the network architecture improves the ability to extract long-range point relationships, which is particularly relevant for remote sensing images. In the first and second experiments, we have demonstrated that the addition of MLP blocks leads to significant improvements in various performance indicators, while the removal of this module results in a deterioration of these indicators. This suggests that the MLP block plays a crucial role in capturing distant point relationships.

Table 2. The results of ablation experiments.

Configuration	MLP	Encoder-Decoder	SE	PSNR↑	SCC↑	SAM↓	ERGAS↓
I	×	×	×	29.9282	0.9603	0.1296	4.7315
II	√	×	×	30.1919	0.9617	0.1194	4.7202
III	√	×	√	30.6091	0.9650	0.1067	4.5010
IV	×	√	×	32.0280	0.9635	0.1121	4.5100
MLFEN	√	√	√	**32.2875**	**0.9654**	**0.1106**	**4.4947**

SE Block. In experiments two and three, we investigated the impact of integrating SE blocks with attention mechanisms to enhance the network's performance. The SE block processes redundant information in images introduced by the MLP block using attention methods, thereby improving network performance. The results presented in Table 2 demonstrate a significant improvement in the performance of various indicators when SE blocks are included.

Encoder-Decoder Moudle. This paper considers the multi-scale nature of remote sensing images and proposes an encoder-decoder architecture with skip connections to extract multi-scale features. To evaluate its effectiveness, we constructed a neural network with an encoder-decoder block and conducted experiments as presented in Table 2. The results demonstrate a significant improvement in peak signal-to-noise ratio (PSNR) with the inclusion of this module, indicating its efficacy in noise reduction. While the performance of other indicators is slightly inferior to that of the MLP block used alone, the results support the rationale behind using an encoder-decoder block with skip connections.

5   Conclusion

This article proposes a novel network architecture for image fusion that enables effective extraction of long-range pixel relationships and multi-scale features. The network structure is specifically designed for remote sensing images, allowing for the utilization of distant point relationships and spatial and semantic features at different scales. The proposed method demonstrates superior performance in the task of pansharpening image fusion, as evidenced by objective quantitative metrics and visual inspection.

Acknowledgements. I would like to express my deepest gratitude to my thesis advisor Shuyue Luo, for their guidance, support, and patience throughout the research process. Their insightful feedback and expertise have been invaluable in shaping this thesis. I would like to thank my colleagues Yuhao Lian, for their helpful discussions, encouragement, and assistance in various stages of my research. Their insights and ideas have greatly contributed to the development of this thesis. My gratitude also goes to my family and friends for their unwavering support, encouragement, and love. Their emotional support has helped me to stay motivated and focused during the difficult times of my research. Lastly, I would like to thank the participants of my study, without whom this research would not have been possible. Their willingness to participate and share their experiences and insights have been crucial to the success of this study. Thank you all for your invaluable support and contributions.

References

1. Chavez Jr., P.S., Sides, S.C., Anderson, J.A.: Comparison of three different methods to merge multiresolution and multispectral data: Landsat TM and SPOT panchromatic. Photogramm. Eng. Remote Sens. **57**(3), 295–303 (1991)
2. Shettigara, V.K.: A generalized component substitution technique for spatial enhancement of multispectral images using a higher resolution data set. Photogramm. Eng. Remote Sens. **58**(5), 561–567 (1992)
3. Laben, C.A., Brower, B.V.: Process for enhancing the spatial resolution of multispectral imagery using pan-sharpening. U.S. Patent 6 011 875 (2000)
4. Mallat, S.: A theory for multiresolution signal decomposition: the wavelet representation. IEEE Trans. Pattern Anal. Mach. Intell. **11**(7), 674–693 (1989)

5. Nason, G.P., Silverman, B.W.: The stationary wavelet transform and some statistical applications. In: Antoniadis, A., Oppenheim, G. (eds.) Wavelets and Statistics. LNS, vol. 103, pp. 281–299. Springer, New York (1995). https://doi.org/10.1007/978-1-4612-2544-7_17

6. Shensa, M.J.: The discrete wavelet transform: wedding the á trous and Mallat algorithm. IEEE Trans. Sig. Process. **40**(10), 2464–2482 (1992)

7. Burt, P.J., Adelson, E.H.: The Laplacian pyramid as a compact image code. IEEE Trans. Commun. vol. **COM-31**(4), 532–540 (1983)

8. Masi, G., Cozzolino, D., Verdoliva, L., Scarpa, G.: Pan-sharpening by convolutional neural networks. Remote Sens. **8**(7), 594 (2016)

9. Yang, J., Fu, X., Hu, Y., Huang, Y., Ding, X., Paisley, J.: PanNet: a deep network architecture for pan-sharpening. In: 2017 IEEE International Conference on Computer Vision (ICCV) (2017)

10. Dong, C., Loy, C.C., He, K., Tang, X.: Image super-resolution using deep convolutional networks. IEEE Trans. Pattern Anal. Mach. Intell. **38**(2), 295–307 (2016)

11. Gillespie, A.R., Kahle, A.B., Walker, R.E.: Color enhancement of highly correlated images. ii. Channel ratio and chromaticity transformation techniques. Remote Sens. Environ. **22**(3), 343–365 (1987)

12. Li, X., Zhang, Y., Gao, Y., Yue, S.: Using guided filtering to improve gram-Schmidt based pansharpening method for GeoEye-1 satellite images. In: Proceedings of the 4th International Conference on Information Systems and Computing Technology (2016)

13. Shen, K., Yang, X., Lolli, S., Vivone, G.: A continual learning-guided training framework for pansharpening. ISPRS J. Photogramm. Remote. Sens. **196**, 45–57 (2023)

14. Wald, L.: Data Fusion: "Definitions and Architectures-Fusion of Images of Different Spatial Resolutions." Les Presses de l' École des Mines, Paris, France (2002)

15. Alparone, L., et al.: Multispectral and panchromatic data fusion assessment without reference. Photogramm. Eng. Remote Sens. **74**(2), 193–200 (2008)

16. Yuhas, R.H., Goetz, A.F.H., Boardman, J.W.: Discrimination among semi-arid landscape endmembers using the Spectral Angle Mapper (SAM) algorithm. In: Proceedings of Summaries 3rd Annual JPL Airborne Geoscience Workshop, pp. 147–149 (1992)

17. Vallejos, R., Pérez, J., Ellison, A.M., Richardson, A.D.: A spatial concordance correlation coefficient with an application to image analysis. Spat. Stat. **40**(100405), 100405 (2020)

18. Chaudhary, N., Mahajan, R.: Spectrum sensing using 16-QAM and 32-QAM modulation techniques at different signal-to-noise ratio: a performance analysis. IAES Int. J. Artif. Intell. (IJ-AI) **12**(2), 966 (2023)

Comparison of Fourier Bases and Asymmetric Network Bases in the Bio-Inspired Networks

Naohiro Ishii[1]([⊠]), Kazunori Iwata[2], Yuji Iwahori[3], and Tokuro Matsuo[1]

[1] Advanced Institute of Industrial Technology, Tokyo 140-0011, Japan
nishii@acm.org, matsuo@aiit.ac.jp
[2] Aichi University, Nagoya 453-8777, Japan
kazunori@aichi-u.ac.jp
[3] Chubu University, Kasugai 487-0027, Japan
iwahori@isc.chubu.ac.jp

Abstract. Machine learning, deep learning and neural networks are extensively developed in many fields, in which neural network architectures have shown a variety of applications. However, there is a need for explainable fundamentals in complex neural networks. In this paper, it is shown that bio-inspired networks are useful for the explanation of network functions. First, the asymmetric network is created based on the bio-inspired retinal network. They have orthogonal bases which correspond to the Fourier bases. Second, the classification performance of the asymmetric network is compared to the conventional symmetric network. Further, the asymmetric network is extended to the layered networks, which generate higher dimensional orthogonal bases. Their replacement operation is shown to be useful in the classification. These higher dimensional bases preserve the independence of patterns in their layered networks. Finally, it is shown that the sparse codes made of the higher dimensional bases are applied to the classification of real-world data.

Keywords: asymmetric and symmetric networks · generation of orthogonal bases · classification performance of networks · replacement of bases · independence in extended layered network

1 Introduction

Recently, there has been a great deal of excitement and interesting in deep neural networks, because they have achieved breakthrough results in areas as machine learning, computer vision, neural computations and artificial intelligence [1, 2]. Their networks are expected to be transparent, understandable and explainable in their successive processing in the multilayered structures [2]. Pseudo orthogonal bases perform generalization capability in neural network learning [3]. In the network developments, orthogonality is a fundamental topic in learning and neural networks [9]. In this paper, it is shown that bio-inspired network generates useful bases for the independence of features classification. First, the asymmetric network with nonlinear functions is created based on the bio-inspired retinal network. The asymmetric network and the Fourier bases are compared in their structures, and they show the same classification performance. Second the extended layered asymmetric networks are derived, which are also based on the model of the brain cortex network. Then, the higher dimensional orthogonal bases are generated in the layered networks. In the extended layered networks, the replacement of bases shows better classification performance. Further, the higher dimensional bases preserve the independence of patterns and useful for the classification as the sparse coding [4]. Their sparse codes are applied to the classification of the real-world data [10].

2 Bio-inspired Neural Networks

2.1 Background of Asymmetric Neural Networks Based on the Bio-inspired Network

In the biological neural networks, the structure of the network, is closely related to the functions of the network. Naka et al. [7] presented a simplified, but essential networks of catfish inner retina as shown in Fig. 1.

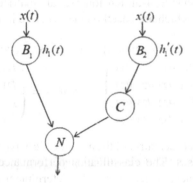

Fig. 1. Asymmetric network with linear and squaring nonlinear pathways

The asymmetric neural network is extracted from the catfish retinal network [7]. The asymmetric structure network with a quadratic nonlinearity is shown in Fig. 1, which composes of the pathway from the bipolar cell B to the amacrine cell N and that from the bipolar cell B, via the amacrine cell C to the N [7, 8]. It is shown that N cell response is realized by a linear filter, which is composed of a differentiation filter followed by a low-pass filter. Thus, the asymmetric network in Fig. 1 is composed of a linear pathway and a nonlinear pathway with the cell C, which works as a squaring function.

2.2 Model of Asymmetric and Symmetric Networks

Models of the asymmetric and symmetric networks, both of which are the bio-inspired networks [7, 9] are shown in Fig. 2(a) and (b), respectively, in which impulse response functions of cells are shown in $h_1(t)$ and $h'_1(t)$. The ( )^2 shows a square operation. The symmetric model called energy model is proposed [5] in the visual system.

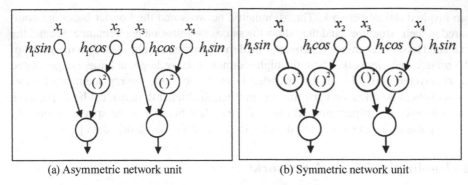

(a) Asymmetric network unit   (b) Symmetric network unit

Fig. 2. Asymmetric unit and symmetric unit in the 1st layer of the network

3 Fourier Bases and Asymmetric Network Bases

The asymmetric network bases are described in $\{sin\,x_1, cos^2x_2, cos\,x_3, sin^2x_4\}$, while the Fourier bases are described in $\{sin\,x_1, cos\,2x_2, cos\,x_3, sin\,2x_4\}$. For the comparison of the classification performance, 4-dimensional square matrices (*Asym.*) and (*Fourier*) are considered as follows, which are based on the structure in Fig. 2.

$$\{\sin x_1, \cos 2x_2, \cos x_3, \sin 2x_4\} \; \{\sin x_1, \cos^2 x_2, \cos x_3, \sin^2 x_4\}$$

$$(Asym.) = \begin{pmatrix} a_{11} & a_{12} & a_{13} & a_{14} \\ a_{21} & a_{22} & a_{23} & a_{24} \\ a_{31} & a_{32} & a_{33} & a_{34} \\ a_{41} & a_{42} & a_{43} & a_{44} \end{pmatrix} \; (Fourier) = \begin{pmatrix} f_{11} & f_{12} & f_{13} & f_{14} \\ f_{21} & f_{22} & f_{23} & f_{24} \\ f_{31} & f_{32} & f_{33} & f_{34} \\ f_{41} & f_{42} & f_{43} & f_{44} \end{pmatrix} \quad (1)$$

where (*Asym.*) shows the output matrix of the asymmetric network bases, while (*Fourier*) shows that of Fourier bases. The classification performance is evaluated using the independence or dependence, which imply the determinant to be non-zero or zero, respectively. The determinants of (*Asym.*) and (*Fourier*) are described in $\|Asym.\|$ and $\|Fourier\|$, respectively.

3.1 Relation Between Fourier and Asymmetric Networks Bases

We assume the first 4-dimensional input as the $X_1 = (x_{11}\,x_{12}\,x_{13}\;x_{14})$ which is a component of the total input $X = [X_1\,X_2\,X_3\,X_4]$. The X is described in 4-dimensional input matrix. The output in Fig. 3 becomes for the input $\{x_1, x_2, x_3, x_4\}$, in which h_1 to be 1 for the simplicity. We assume the input x to be 0 or 1. Then, we set $sin1 = a$ and $cos1 = b$.

$$X = \begin{bmatrix} x_{11} & \cdots & x_{14} \\ \vdots & \ddots & \vdots \\ x_{41} & \cdots & x_{44} \end{bmatrix} \; \left(\equiv \begin{bmatrix} 1 & 0 & 0 & 1 \\ 0 & 1 & 1 & 0 \\ 0 & 0 & 1 & 1 \\ 1 & 1 & 0 & 0 \end{bmatrix} \right) \quad (2)$$

The determinant with asymmetric net bases, $\{sin(x_1),\; cos^2(x_2),\; cos(x_3),\; sin^2(x_4)\}$.

is shown in Eq. (3).

$$\|Asym.\| \, for \, Eq. \, (2) = \begin{Vmatrix} a & 1 & 1 & a^2 \\ 0 & b^2 & b & 0 \\ a & 1 & b & a^2 \\ 0 & b^2 & 1 & 0 \end{Vmatrix} \tag{3}$$

Similarly, the determinant with Fourier bases, $\{sin(x_1), \, cos(2x_2), \, cos(x_3), \, sin(2x_4)\}$ is shown in Eq. (4).

$$\|Fourier\| \, for \, Eq. \, (2) = \begin{Vmatrix} a & 1 & 1 & 2ab \\ 0 & 2b^2 - 1 & b & 0 \\ a & 1 & b & 2ab \\ 0 & 2b^2 - 1 & 1 & 0 \end{Vmatrix} \tag{4}$$

Lemma 1. Under the assumption that the input x is 0 or 1, Fourier basis $sin \, 2x_4$ and asymmetric network basis $sin^2 x_4$ are same in the determinants, $\|Asym.\|$ and $\|Fourier\|$ of the matrices (*Asym.*) and (*Fourier*) in Eq. (1).

This is proved as follows. The Fourier basis $sin \, 2x_4$ is in the 4th column in (*Asym.*). Since the Fourier basis, $sin \, 2x_4 = 2 \, sin \, x_4 \cdot cos \, x_4 = 2ab$ or 0 in the 4th column,

The matrix (*Asym.'*) is defined, in which the 4th column composes of 0 or 1 by setting $2ab \bullet \|Asym.'\|$ (which is equal to $\|Asym.\|$). Since the asymmetric basis $sin^2 x_4 = a^2$ holds, in the matrix (*Fourier*), the matrix (*Fourier'*) is defined, in which the 4-th column composes of 0 or 1 by setting $a^2 \bullet \|Fourier'\|$ (which is equal to $\|Fourier\|$. Thus, the 4th column is same in both $\|Asym.'\|$ and $\|Fourier'\|$.

Lemma 2. The relation between the Fourier basis $cos \, 2x_2$ and the asymmetric network bas $cos^2 x_2$ holds as $cos \, 2x_2 = 2 \, cos^2 x_2 - 1$.

Using Lemma 2, the basis relation in the 2nd column satisfies $(1/2) \times cos2x_2 + 1 = cos^2 x_2$. Using this relation and Eq. (1), the determinants equation is derived in the following.

Theorem 3. Between $\|Fourier\|$ and $\|Asym.\|$, the following determinants equation is derived.

$$\{sin \, x_1, cos \, 2x_2, cos \, x_3, sin \, 2x_4\} \quad \{sin \, x_1, cos \, 2x_2, cos \, x_3, sin \, 2x_4\} \quad \left\{ sin \, x_1, cos^2 x_2, cos \, x_3, sin^2 x_4 \right\}$$

$$\left(\frac{1}{2}\right) \begin{Vmatrix} f_{11} & f_{12} & f_{13} & f_{14} \\ f_{21} & f_{22} & f_{23} & f_{24} \\ f_{31} & f_{32} & f_{33} & f_{34} \\ f_{41} & f_{42} & f_{43} & f_{44} \end{Vmatrix} + \begin{Vmatrix} f_{11} & 1 & f_{13} & f_{14} \\ f_{21} & 1 & f_{23} & f_{24} \\ f_{31} & 1 & f_{33} & f_{34} \\ f_{41} & 1 & f_{43} & f_{44} \end{Vmatrix} = \xi \cdot \begin{Vmatrix} a_{11} & a_{12} & a_{13} & a_{14} \\ a_{21} & a_{22} & a_{23} & a_{24} \\ a_{31} & a_{32} & a_{33} & a_{34} \\ a_{41} & a_{42} & a_{43} & a_{44} \end{Vmatrix} \tag{5}$$

where ξ is a constant. Equation (5) is described shortly using determinant notation as

$$\|Fourier\| + \|B\| = \xi \|Asym.\| \tag{6}$$

where the additional determinant, $\|B\|$ is the determinant with replaced in the 2nd column by all one in the $\|Fourier\|$.

The determinant of the $\|Fourier\|$ for the given input matrix, X is shown in Eq. (2) using the determinants of sub-matrices, $[T_1]$ and $[T_2]$, which is derived from the expansion of the 2nd column of the Fourier matrix in Eq. (1), (Fourier)). Fourier basis, $cos\,2x$ in the 2nd column consists of $(2b^2-1)$ or 1. Then, Eq. (7) is obtained.

$$\|Fourier\| = (a)(ab)\left\{\left(\pm\left(2b^2 - 1\right)\right)[T_1] + (\pm 1)[T_2]\right\} \tag{7}$$

where (a) in Eq. (7) is from the 1st column on the basis $sin\,x_1$ and (ab) in Eq. (4) is from the 4th column on the basis $sin\,2x_4$. $\lceil T_1 \rceil$ shows the sum of the determinant of the cofactor matrix with respect to $(2b^2-1)$ in the 2nd column, while $[T_2]$ shows that of the cofactor matrix with respect to 1 in the 2nd column. Similarly, the determinant of the asymmetric network for the given input matrix, X, $\|Asym.\|$ is shown in Eq. (8).

$$\|Asym.\| = (a)\left(a^2\right)\left\{(\pm\left(b^2\right)[T_1] + (\pm 1)[T_2]\right\} \tag{8}$$

where (a) in Eq. (8) is from the 1st column on the basis $sin\,x_1$ and (a^2) in Eq. (8) is from the 4th column on the basis $sin^2 x_4$. Note here that $\lceil T_1 \rceil$ and $[T_2]$ are same in Eq. (7) and Eq. (8), since the cofactor matrices are same for the same given X. The relation of determinants between Fourier bases and asymmetric network bases are shown in the following theorem.

Theorem 4. $\|Asym.\| \neq 0$ holds if and only if $\|Fourier\| \neq 0$.

This is proved by taking the contrapositive. Firstly, we prove a sufficient condition that if $\|Fourier\| \neq 0 holds$, then, $\|Asym.\| \neq 0$ holds. By its contrapositive, if $\|Asym.\|=0$ holds, $b^2[T_1] + 1[T_2] = 0$ from Eq. (4), where $(\pm)$ in Eq. (4) is assumed to be included in $[T_1]$ and $[T_2]$. From the equation, $b^2[T_1] + 1[T_2] = 0$, $b^2[T_1] = -1[T_2]$ holds. If $[T_1] \neq 0$ holds, the left side $b^2[T_1]$ shows more than 2 order with respect to b, while the right side $-1[T_2]$ becomes 1st order with respect to b. Then, $b^2[T_1] \neq -1[T_2]$ holds. This contradicts the assumption $b^2[T_1] = -1[T_2]$. Thus, $[T_1] = 0$ is satisfied. From the assumption, $b^2[T_1] + 1[T_2] = 0$, also $[T_2] = 0$ is satisfied. Since the determinant, $\|Fourier\|$ consists of $(2b^2 - 1))[T_1] + 1[T_2]$, which becomes 0 by $[T_1] = 0$ and $[T_2]=0$ by the same in the $[T_1]$ and $[T_2]$ used in the $\|Asym.\|$.

A necessary condition that if $\|Asym.\| \neq 0$ holds, then, $\|Fourier\| \neq 0$ is satisfied, is also proved by applying the contrapositive.

We define here the classification performance to be the number of the determinant of the input matrix with non-zero value for the independence.

Corollary 5. $\|Asym.\| = 0$ holds if and only if $\|Fourier\| = 0$.

Corollary 6. The classification performance of the asymmetric network is equal to that of the Fourier performance.

Corollary 7. Determinants $[T_1]$ and $[T_2]$ are same in $\|Fourier\|$ and $\|B\|$.

Corollary 7 implies that $[T_1]$ and $[T_2]$ in Eq. (5) take the same values. Then, there are no conflicts between $\|Fourier\|$ and $\|B\|$. Experiments for the verification of theorem 4, corollaries 5, 6 and 7 are performed generating 4-dimensional input matrix of random patterns with 0 and 1. Determinants of $\|Fourier\|$ and $\|Asym.\|$ are computed as shown in Fig. 3. Theorem 4 and corollary 5 are experimented by generating 4-dimensional {0.1} patterns. Figure 3 shows the Fourier and the asymmetric networks create the same results.

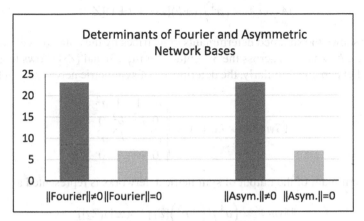

Fig. 3. Comparison of determinants of Fourier and asymmetric network bases

3.2 Independence for Classification in the Network

Independence for classification in the network is important also with learning. The network performance for classification is evaluated using mean squared loss function, *Loss* in Eq. (9) between the inputs and outputs of the network.

$$Loss = \frac{1}{m}\|XW - Z\|^2 \tag{9}$$

where X and Z are the inputs and output matrices of the network, respectively. W is the connection weight matrix between the inputs and outputs. The optimization of the loss function in Eq. (9) is realized by the derivative of Eq. (10) with respect to W

$$\frac{\partial}{\partial W}(\frac{1}{M}\|XW - Z\|^2) = \frac{1}{M}X^T(XW - Z) = 0 \tag{10}$$

By the differentiation in Eq. (10),

$$W = (X^TX)^{-1}X^TZ \tag{11}$$

is obtained. From Eq. (11), a sufficient condition for the existence of the weights is the input matrix, X to be non-singular, i.d., the X consists of independence input vectors. For the classification of input vectors and the learning by the modified weight ΔW in the network, the sufficient condition implies the determinant of the matrix to be non-zero.

4   Evaluation in the Asymmetric and the Symmetric Networks

The determinant of the outputs of asymmetric network is represented as

$$\|Asym.\| = \left(a^3\right)\{(\pm b)[Z_1] + (\pm 1)[Z_2]\} \tag{12}$$

where $[Z_1]$ shows the summed determinants of matrices by the cofactor, with respect to the variable $\pm b$ expansion across the 3rd column in Eq. (3) and $[Z_2]$ shows those by the cofactor ± 1 expansion. Similarly, the determinant of symmetric networks is in Eq. (13).

$$\|Sym.\| \text{ for Eq.(3)} = \begin{Vmatrix} a^2 & 1 & 1 & a^2 \\ 0 & b^2 & b^2 & 0 \\ a^2 & 1 & b^2 & a^2 \\ 0 & b^2 & 1 & 0 \end{Vmatrix} \tag{13}$$

The determinant of the output of symmetrical networks is represented as

$$\|Sym.\| = \left(a^4\right)\{\left(\pm b^2\right)[Z_1] + (\pm 1)[Z_2]\} \tag{14}$$

Note here the determinant $\{[Z_1] + (\pm 1)[Z_2]\}$ is same in both Eqs. (13) and (14).

4.1   Conditions of Independence of $\|Asym.\|$ and $\|Sym.\|$

Independence of data plays an important role for the classification of their data. The independence in the asymmetric networks and the symmetric ones are evaluated based on the $\|Asym.\|$ and $\|Sym.\|$

Theorem 8.   Under the condition $[Z_1] \neq 0$, the determinant of asymmetric networks, $\|Asym.\| \neq 0$ holds.

This is proved as follows. The first term, $(\pm b)[Z_1]$ in the determinant of asymmetric networks in Eq. (12), have the odd order exponentiation of the variable b, while the second term $(\pm 1)[Z_2]$ have the even order exponentiation of the variable b. Thus, these terms do not generate the summation to be zero.

Lemma 9.   The determinants $[Z_1]$ and $[Z_2]$ in the asymmetric and symmetric network are described in a quadratic polynomial of variable b in the following.

$$[Z_1] = mb^2 + l \quad \text{and} \quad [Z_2] = kb^2 + n \tag{15}$$

where m, l, k and n are numerical coefficients.

This is proved from the definition of the matrix. Since $[Z_1]$ and $[Z_2]$ are computed by the cofactor of expansion across the 3rd column in the 4-demensional matrix, variable b^2 exists only in the 2nd column of the matrix. Since only one variable in the 2nd column is used in the each expansion term from the definition of the determinant, the expansion terms consist of the quadratic polynomial in Eq. (15).

Theorem 10. $\|Asym.\| = 0$ holds if and only if only if $m = 0$, $l = 0$, $k = 0$ and $n = 0$. Similarly, $\|Sym.\| = 0$ holds if and only if $m = 0$, $(l + k) = 0$ and $n = 0$.

Corollary 11. When $\|\text{Sym.}\| = 0$ holds, the parameters $l = -k \neq 0$ holds. Then, if $l = -k \neq 0$ holds, $\|\text{Asym.}\| \neq 0$ is satisfied.

Theorem 12. When $\|\text{Sym.}\| \neq 0$ holds, $\|\text{Asym.}\| \neq 0$ is also satisfied.

This is proved by the contradiction of the statement; if $\|Asym.\| = 0$ holds, then.

$\|\text{Sym.}\|=0$ holds. From theorems, the performances are compared in the next theorem.

Theorem 13. The performance of the classification of $|Asym.|$ include that of $|Sym.|$. This shows $|Asym.|$ is superior to $|Sym.|$ in the classification performance.

4.2 Input Patterns for Independence in Asymmetric and Symmetric Networks

Definition 1. Symmetric patterns X is defined to have the following rows $\boxed{1}$ or $\boxed{2}$ in the 4-dimensional matrix.

$$\boxed{1}\, X_i = (x_{i_1} x_{i_2} x_{i_3} x_{i_4}) \ \text{and}\ X_j = (x_{j_1} x_{j_2} x_{j_3} x_{j_4}),\ \text{in which}\ x_{j_1} = x_{i_4}, x_{j_2} = x_{i_3},$$
$$x_{j_3} = x_{i_2}\ \text{and}\ x_{j_4} = x_{i_1}\ \text{are satisfied.} \tag{16}$$

$$\boxed{2}\, X_k = (x_{k_1} x_{k_2} x_{k_3} x_{k_4}) = (x_{k_4} x_{k_3} x_{k_2} x_{k_1}) \tag{17}$$

(Example: the matrix with $\{0.1\}$ in Eq. (3) is a symmetric pattern). The 1st and 2nd rows show from $\boxed{2}$ in Definition 1, while the 3rd and 4th rows show from $\boxed{1}$ in Definition 1. Experimental results using symmetric patterns are shown in Fig. 4.

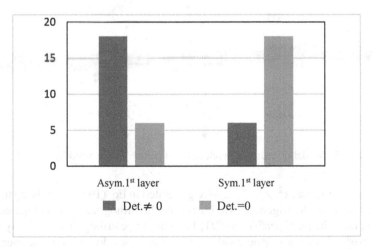

Fig. 4. Experimental results of independent and dependent ratio for symmetric patterns

5 Generation of Orthogonality in Asymmetric Layered Networks

5.1 Realization of the Asymmetric Layered Networks

Figure 5 shows a model of the V1 followed by the MT area in the cortex [6] after the retinal network, which is the higher order layer followed by the asymmetric neural networks unit in Fig. 2(a). Further,*Gabsin* and *Gabcos* in Fig. 5 show Gabor filters. Then, the half-wave rectification is approximated in the following equation.

$$f(x) = \frac{1}{1 + e^{-\eta(x-\theta)}} \tag{18}$$

By Taylor expansion of Eq. (18) at $x = \theta$, the Eq. (19) is derived as follows,

$$f(x)_{x=\theta} = f(\theta) + f'(\theta)(x - \theta) + \frac{1}{2!}f''(\theta)(x - \theta)^2 + \ldots$$

$$= \frac{1}{2} + \frac{\eta}{4}(x - \theta) + \frac{1}{2!}\left(-\frac{\eta^2}{4} + \frac{\eta^2 e^{-\eta\theta}}{2}\right)(x - \theta)^2 + \cdots \tag{19}$$

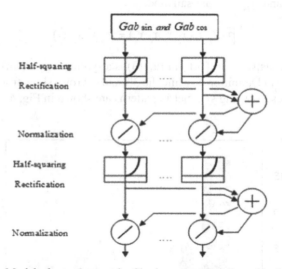

Fig. 5. Model of neural network of brain cortex V1 followed by MT [6]

The nonlinear terms, $x^2, x^3, x^4, \ldots$. Are generated in Eq. (19). By applying the power reducing formula in the trigonometric functions, the orthogonality is computed. When n and m are odd, the pair$\{(sin^n t),(cos^m t)\}$ becomes to be orthogonal. Similarly, the pair $\{(sin^n t), (cos^m t)\}$becomes orthogonal in case of n to be odd and m to be even, and in case of n to be even and m to be odd, pair}is not orthogonal.

5.2 Replacement of the Orthogonal Basis in the 2nd Layered Network

In the 1st step, the replace operation $sin^2 x_4 \cdot cos^2 x_2$ is performed with $sin^2 x_4$ in the asymmetric 2nd layer network in Fig. 7. In the 2nd step, the basis $cos x_3$ is replaced with

$\{cos\,x_3 \cdot cos^2 x_2\}$ in the asymmetric 2nd layer network. Similarly, in the symmetric 2nd layer network, $cos^2 x_3$ is replaced with $\{cos^2 x_3 \cdot cos^2 x_2\}$. As the 3rd step, two bases, $cos^2 x_2$ and $sin^2 x_4$ are replaced with $\{cos^2 x_2 \cdot sin^2 x_4\}$ and $\{sin^2 x_4 \cdot sin\,x_1\}$, respectively in the asymmetric 2nd layer network. Similar 3rd operation is performed in the symmetric 2nd layer network. Thus, all the determinants of patterns are non-zero as shown in Fig. 6.

As the real data problem, the part of the Reuters collections [10] was used, which consists of 'Cocoa data', Copper data' and 'Cpi data'. We used 16 reduced words from the original data of Cocoa, Copper and Cpi data, which are normalized. These reduced data, is transformed to the array in the asymmetric network. The correct classification among 'Cocoa', 'Copper' and 'Cpi' data groups are realized using linear discrimination (black and gray bars) followed by the higher dimensional bases(brown bars) in Fig. 7.

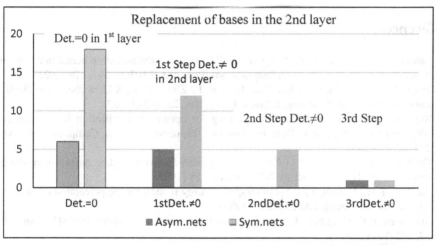

Fig. 6. Generation of independence by bases replacements in the 2nd layer

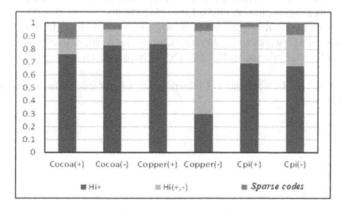

Fig. 7. Higher dimensional bases classify correctly Cocoa, Copper and Cpi classes

6 Conclusion

Studies of machine learning, artificial intelligence and neural networks have been developed greatly. In this paper, starting from the bio-inspired neural networks, it is shown that the asymmetrical network with nonlinear functions have characteristics of orthogonal basis, which are compared to the Fourier bases. To improve the classification performance, the extended asymmetric layered networks are developed in which the higher dimensional bases are created. The replacement of the higher dimensional bases is proposed which improves greatly the classification performance. Further, they are used the sparce codes for the classification of real world data. These higher dimensional orthogonal bases are expected to create the synthetic features and to generate the integrated features in the higher layered networks.

References

1. Samek, W., Lapusckn, S., Anders, C.J., Muller, K.-R.: Explaining deep neural networks and beyond: a review of methods and applications. Proc. IEEE **103**(3), 247–278 (2021)
2. Peng, X., Li, Y., Tsang, I.W., Zhu, H., Lv, J., Zhou, J. T.,:XAI beyond classification: interpretable neural clustering. J. Mach. Learn. Res. **23**, 1–28 (2022)
3. Sugiyama, M., Ogawa, H.: Active learning for optimal generalization in trigonometric polynomial models. IEICE Tran. Fundamental. Electron. Commun. Comp. Sci. **E84-A**(9), 2319–2329 (2001)
4. Olshausen, B.A., Field, D.J.: Sparse coding with an overcomplete basis set: a strategy employed by V1? Vision. Res. **37**(23), 3311–3325 (1997)
5. Adelson, E.H., Bergen, J.R.: Spatiotemporal energy models for the perception of motion. J. Optical Soc. of America **A**, 284–298 (1985)
6. Simonceli, E.P., Heeger, D.J.: a model of neuronal responses in visual area MT. Vision Res. **38**, 743–761 (1996)
7. Naka, K-I., Sakai, H.M., Ishii, N.: Generation of transformation of second order nonlinearity in catfish retina, Ann. Biomed. Eng **16**, 53–64 (1988)
8. Sakai, H, M., Naka, K-I.: Dissection of the neuron network in the catfish inner retina. I. Trans. Ganglion Cells J. Neurophysiol. **60**(5), 1549–1567 (1988)
9. Ishii, N., Deguchi, T., Kawaguchi, M., Sasaki, H., Matsuo, T.: Orthogonal properties of asymmetric neural networks with gabor filters. In: Hybrid Artificial Intelligent Systems, HAIS 2019, LNCS 11738, Springer, pp.589–601 (2019). https://doi.org/10.1007/978-3-030-29859-3_50
10. Reuters-21578 Text Categorization Collection. https://kdd.ics.edu/databases/reuters21578/reuters21578.html

Temporal Dynamics of Drowsiness Detection Using LSTM-Based Models

Rafael Silva[1,2](✉) (iD), Lourenço Abrunhosa Rodrigues[3,4] (iD),
André Lourenço[3,4] (iD), and Hugo Plácido da Silva[1,2] (iD)

[1] Department of Bioengineering (DBE), Instituto Superior Técnico (IST),
Av. Rovisco Pais 1, 1049-001 Lisboa, Portugal
`rafael.c.silva@tecnico.ulisboa.pt`
[2] Instituto de Telecomunicações (IT), Av. Rovisco Pais 1,
Torre Norte—Piso 10, 1049-001 Lisboa, Portugal
`hsilva@lx.it.pt`
[3] Instituto Superior de Engenharia de Lisboa (ISEL), 1600-312 Lisboa, Portugal
[4] CardioID Technologies LDA, Lisboa, Portugal
`{lar,arl}@cardio-id.com`

Abstract. Different LSTM-based models were tested for binary drowsiness detection using the ULg Multimodality Drowsiness Database (DROZY). The dataset contains physiological signals and behavioral measures collected from participants during different experimental conditions designed to induce varying levels of drowsiness. The LSTM models were trained using a sequential approach using the inter-beat intervals, where they were exposed to increasing levels of drowsiness over time. The performance of the models was evaluated in terms of accuracy, precision, recall, F1-score, and AUC. The results showed that the stacked bidirectional LSTM model achieved the highest performance with an accuracy of 0.873, precision of 0.825, recall of 0.793, F1-score of 0.808, and AUC of 0.918. These findings suggest that LSTM-based models can learn to capture the temporal dynamics of drowsiness and make accurate predictions based on the current and previous levels of drowsiness.

Keywords: Drowsiness Detection · LSTM · Recurrent Neural Networks · ECG · HRV

1 Introduction

Driving while drowsy is a major cause of road accidents and fatalities [3]. Public awareness for this fact has increased in recent years, with an effort in driver drowsiness detection research and even policies made to stimulate the mitigation of this problem [4]. While there are several potential mechanisms to detect driver impairment caused by drowsiness, monitoring of steering wheel dynamics and vehicle lateral position being some of the already implemented strategies [5], monitoring of physiological information is of special interest given the privileged point of view it provides onto the physiological process of drowsiness onset.

I. Rojas et al. (Eds.): IWANN 2023, LNCS 14134, pp. 211–220, 2023.
https://doi.org/10.1007/978-3-031-43085-5_17

With the evolution of practical and minimally intrusive wearables with biometric sensors incorporated, such as photoplethysmography (PPG) enabled smart watches and wrist bands, as well as fully off-the-person alternatives like the CardioWheel [9], a steering wheel with conductive material patches that allows the opportunistic acquisition of high-quality electrocardiogram (ECG) without changing in any way the driver-vehicle interaction, in-vehicle continuous monitoring of physiological signals is more feasible than ever.

These signals, PPG and ECG, can be used to derive the time series of instantaneous heart rate, by measuring the successive time intervals between heartbeats, the inter-beat intervals (IBIs), whose analysis, denominated Heart Rate Variability (HRV) analysis, provides insightful information about the autonomic nervous system (ANS) regulation that is closely related with the alertness/drowsiness state of an individual [1,17]

One of the most interesting, but also difficult, dimensions of this problem is to capture the temporal evolution of HRV characteristics with the onset of a drowsy state. This is a continuous process through which the sympathetic and parasympathetic systems must negotiate the urge to sleep and the need to allocate cognitive resources to the driving task. A naive classification scheme that tries to classify drowsiness based only on a single time window observation will encounter a large confusion region when the driver is in this transition phase. However, by including information on the continuous evolution of this process, the system can learn to identify that the individual is accumulating signs of drowsiness onset, and so correctly identify the beginning of impairment.

This work explores the use of Long-Short Term Memory (LSTM) models trained sequentially with IBI data acquired in individuals with increasing levels of sleep deprivation. By training the models with ordered data segments we expect the models to learn the temporal structure inherent to the drowsiness process.

2   Background

2.1   Driver Drowsiness Detection

Driver drowsiness has been researched through various avenues. Different classes of information and sensor types have been evaluated for this classification task. Halin *et al.* [5] provide a good review of these works, identifying vehicle, environment and driver related sensors in 56 papers. Vehicle sensors include the measurement of steering wheel movements, car lateral position, and acceleration patterns. These are justified by being directly impacted by the drowsiness level that a driver experiences. This driving performance impact is used as an input for drowsiness detection algorithms in [13,20], but a point of criticism to this approach is that it is only able to warn users of their impairment when it starts impacting their control of the vehicle. Environment indicators instead focus on traffic levels, road geometry, and factors like weather and luminosity, which are important dimensions to consider, as they provide information about a task complexity level, as explored in the iDreams project [18]. Finally, a set of driver-centric metrics can be used. Halin and colleagues further divide this category into two: behavioral and physiological. While the first includes indicators like body position, facial expression, gaze, and Percentage of Eye closure (PERCLOS) that can suffer from the same limitation as vehicle indicators, the latter provides insight into internal physiological processes that start to change before any concrete manifestation of drowsiness appears.

From the physiological signals included in this last set of indicators, heart rate monitoring is the easiest to implement, given the penetration of smartwatches and wristbands as everyday fashionable accessories, and the possibilities for off-the-person ECG acquisition. Moreover, drowsiness levels can be quantified using the Karolinska Sleepiness Scale (KSS), which is a subjective self-report scale that rates the level of sleepiness on a scale ranging from 1 (extremely alert) to 9 (very sleepy).

2.2   Heart Rate Variability

HRV is the analysis of the successive inter-beat intervals. The sequence of IBIs is obtained by locating the R-peaks in an ECG signal, or the systolic peaks in a PPG, and computing the time differences between them. The analysis can be done in three different domains: time, frequency and non-linear [15].

The time domain corresponds to direct statistical analysis of the IBI time series, taking the average, standard deviation and percentage of successive differences above a certain threshold. The frequency domain, on the other hand, first evenly re-samples the IBI sequence and computes a frequency transform of that data. The power present in different bands of the IBI spectrum is commonly assessed using the high-frequency band (0.15–0.4 Hz), the low-frequency band (0.04–0.15 Hz), and the very low-frequency band (0.0033–0.04 Hz). The ratio between low and high frequencies is regarded as a good measurement between the sympathetic and parasympathetic nervous systems [1], which in turn are linked to the alertness/drowsiness level of an individual. Finally, the non-linear domain measures the recurrence and self-similarity properties of the IBI sequence, with common metrics being the axis of the Pointcaré plot and entropy.

2.3   Long-Short Term Memory Models

A Long Short-Term Memory (LSTM) is a type of Recurrent Neural Network that processes sequential data (Fig. 1). It consists of four components: an input gate, a forget gate, an output gate, and a cell state [19]. These components work together to allow the LSTM cell to store and manipulate information over time:

- Input gate I_t: controls the flow of information into the cell state
- Forget gate F_t: controls the flow of information out of the cell state
- Output gate O_t: controls the flow of information from the cell state to the output
- Cell state C_t: stores information over time and is updated at each time step by a combination of the current input, the previous cell state, and the output of the input and forget gates

Each gate consists of fully connected layers with sigmoid activation functions, and training results in each gate "learning" how to modify the information encoded in the internal cell state given the input sequence evolution.

LSTM-based models are able to effectively process time series data because they have the capacity to maintain a "memory" of past observations and use this information to predict future outcomes and identify sequential patterns. This ability has been successfully applied in several domains including natural language processing, speech recognition, time series forecasting, and signal classification.

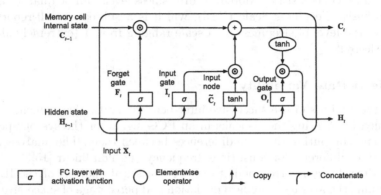

Fig. 1. Schematic representation of an LSTM model. Extracted from [19].

3 Related Work

Previous works attempted to establish models capable of distinguishing alert from drowsy periods of driving from HRV features. Both Oliveira and Silveira [12,16] trained traditional models as support vector machines, gradient boosting machines, and random forests, to obtain high accuracy scores (¿80%) for alert samples and lower scores (∼50%) for drowsy ones. Rodrigues [14] confirmed this trend, showing that the transition phase labels, provided through the self-assessment scale of Karolinska Sleepiness Scale (KSS) [8], was a major source of confusion for the models. All these works treat samples as independent to their position in time, a point that, revised, could clear the confusion created by alert-to-drowsy transition phases by allowing identification of the beginning and continuation of such phases.

In a work by Iwamoto et al. [7], a Deep Learning architecture is used to build a driver drowsiness model based on an LSTM and an Auto-Encoder (AE) that receives as input sequences thirty IBI values, using the LSTM components to learn HRV features and the AE to compress those features and best reconstruct the original sequence of IBI. By training the model only with alert time segments, Reconstruction error could be used to detect the "outlier" drowsy epochs. The good results (AUC = 0.88) obtained with LSTM-generated HRV features provide validation for our proposed methodology.

4   Proposed Approach

4.1   LSTM-Based Models

To evaluate the best LSTM-based approach for time-series classification for the drowsiness detection problem, a systematic approach is presented with different network architectures (Fig. 2). These include:

- Vanilla LSTMs: consist of a single LSTM cell.
- Stacked LSTMs: consist of multiple LSTM layers, with the output of one layer serving as the input to the next layer.
- Bidirectional LSTMs: process the input sequence in two directions, going both forward and backward.
- Stacked Bidirectional LSTMs: the input is processed in both directions by multiple LSTM cells.

To adapt the structure of LSTM networks to the classification problem, the output of the LSTM is fed to a dense layer with a single node and sigmoid activation function.

In this study, the IBIs sequence of the ECG recordings was selected as the source of HRV measurement, and the different LSTM networks proposed were tested. Furthermore, the KSS values were converted to binary labels, where KSS = 6 served as the threshold between alert and drowsy states, as proposed in [11].

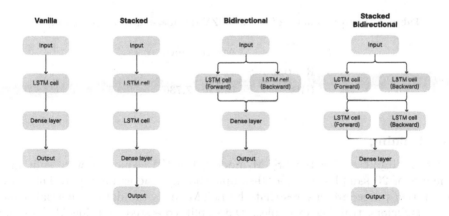

Fig. 2. Different LSTM-based network architectures for classification.

4.2   Data

The ULg Multimodality Drowsiness Database (DROZY) [10] is a comprehensive dataset designed to enhance the development of drowsiness detection algorithms. The dataset comprises data from 14 young and healthy participants (11 females and 3 males), who underwent three 10-min psychomotor vigilance tests

(PVTs) while experiencing increasing levels of sleep deprivation due to prolonged wakefulness. The dataset contains a total of 42 samples, with data for each subject and each PVT. The data involves self-reported drowsiness levels using the KSS, PVT stimulus times and corresponding reaction times, and Polysomnography (PSG) signals acquired at 512 Hz including ECG, Electroencephalography (EEG), Electrooculography (EOG), and Electromyography (EMG). Additionally, the dataset also includes near-infrared (NIR) intensity videos, 2D and 3D annotations of face landmarks for selected frames, and interpolation coefficients and indices for missing frames.

4.3   Preprocessing

The BioSPPy toolbox [2] was utilized to preprocess the ECG signals. A high-pass Finite Impulse Response (FIR) filter with a cut-off frequency of 0.5 Hz was applied to all ECG signals to improve the signal quality. To detect R-peaks in the dataset, the algorithm proposed by Hamilton [6] was employed. The IBI sequence was extracted from each recording by calculating the difference between consecutive R-peaks and converting it to milliseconds (ms). Additionally, to ensure that the extracted IBIs were within acceptable physiological reference values, only those below 1200 ms were selected. The number of IBIs extracted from the recordings is presented in Table 1. The number of ECG recordings (that were used does not correspond to the number of acquisitions as some of them were corrupted.

Table 1. Data Summary of the DROZY database after preprocessing.

	ALERT	DROWSY
RECORDINGS	22	13
IBIs	14,890	7,736

4.4   Training

To fix the input size of the LSTM models, the IBI sequences were split into segments of 20 samples using a 20-sample sliding window of step 1. This data preparation procedure ensures that the LSTM reads the data sequentially and it has sufficient training examples. Data split consisted of using 70% of the recordings for training, 10% for validation and 20% for testing. The LSTM-based models were built with LSTM cells of 200 units. Additionally, to further assess the generalization ability of the best-performing model, a 5-fold cross-validation procedure was applied to the training data (80%) and then evaluated on the test data (20%).

The training procedure was performed using Tensorflow version 2 in a Python 3 environment. The loss function was set as the binary cross-entropy and data were fed into the models using batches of 32 samples, and the Adam optimizer algorithm was used with a learning rate of 0.0001 for a total of 50 epochs.

5   Results and Discussion

After training, the test set was used to evaluate the classification performance of the different models, using standard binary-classification performance metrics (Table 2).

The results presented indicate that the LSTM-based models performed well in detecting drowsiness levels using the RR-interval sequence of ECG recordings. The Stacked Bidirectional LSTM model achieved the highest accuracy (87.3%), precision (82.5%), recall (79.3%), F1-Score (77.1%), and AUC (91.8%) values, indicating its superior performance in drowsiness detection in comparison to the remaining models. Figure 3 shows the training and ROC curves for this model. Table 3 also presents the result from the 5-fold cross-validation procedure, with a slight decrease in the classification performance.

The training curve shows a consistent decrease in both the training and validation loss, which indicates that the model learned effectively from the training data and did not overfit the training set. However, there are some spikes of losses that go up to 0.8, which suggests that the model struggled to generalize on certain data points during training. Despite these spikes, the model's overall performance is quite strong, as evidenced by the ROC curve's high AUC value of 0.918.

The vanilla LSTM model performed the worst among the models tested, with an F1-Score of 65.2%. These results suggest that the vanilla LSTM model struggled to learn the relevant features of the input data and was not effective in distinguishing between drowsy and non-drowsy states. The Stacked LSTM and Bidirectional LSTM models, with F1-scores of 75% and 77.1%, respectively, achieved better performance than the vanilla LSTM model, but still fell short of the performance of the Stacked Bidirectional LSTM model.

Given the training procedure chosen, i.e., exposing the models to increasing levels of drowsiness, the models were able to capture the temporal dynamics of drowsiness in IBI sequences and make accurate predictions based on the current and previous levels of drowsiness.

Table 2. Classification performance metrics of the different LSTM-based models using a single train-test split.

LSTM MODEL	ACCURACY	PRECISION	RECALL	F1-SCORE	AUC
Vanilla	0.661	0.500	0.938	0.652	0.580
Stacked	0.841	0.799	0.707	0.750	0.872
Bidirectional	0.851	0.803	0.742	0.771	0.895
Stacked Bidirectional	0.873	0.825	0.793	0.808	0.918

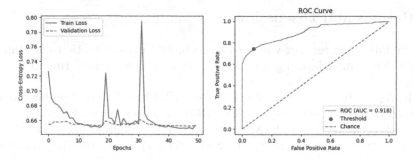

Fig. 3. Training curve (**left**) and ROC curve (**right**) of the Stacked Bidirectional LSTM model. The chosen threshold for binary classification is presented in the ROC curve.

Table 3. Classification performance metrics of the stacked bidirectional model using 5-fold cross-validation.

LSTM MODEL	ACCURACY	PRECISION	RECALL	F1-SCORE	AUC
Stacked Bidirectional	0.862	0.814	0.768	0.790	0.895

6   Conclusions and Future Work

This study provides valuable insights into the effectiveness of LSTM-based models in detecting drowsiness levels using RR-interval sequences of ECG recordings. The results demonstrate that the Stacked Bidirectional LSTM model outperforms the other models tested in terms of accuracy, precision, recall, F1-Score, and AUC values. The training and ROC curves for this model indicate that it learned effectively from the training data and achieved a high level of generalization, although there were some spikes in the losses that suggest room for improvement.

The proposed methodology involved simplifying the ordinal scale (KSS) into a binary one (alert/drowsy), which yielded satisfactory results. This simplification was necessary due to the high level of discretization in the KSS scale and its subjective nature, which compromises the label accuracy. Moreover, this approach aligns with the requirements of a real-time monitoring system, where users are alerted in case of drowsiness. In the future, a potential step could be the development of a 3-class model that incorporates an intermediate level of drowsiness, as suggested in the literature [11].

Overall, the findings suggest that LSTM-based models can effectively capture the temporal dynamics of drowsiness and make accurate predictions based on the current and previous levels of drowsiness. This has important implications in safety-critical domains such as transportation, where real-time drowsiness detection systems based on physiological signals could potentially prevent accidents. Further research is needed to validate the findings of this study on larger and more diverse datasets and to explore the potential of incorporating additional features to improve model performance.

One future step for improving the performance of the models could be to implement transfer learning for the DROZY database, given its relatively small size compared to other ECG databases. Another approach could be to incorporate spectral information into the IBI data for the drowsiness detection problem, as this information may be more closely related to the activity of the ANS. Additionally, exploring other models (e.g., transformers) or combining the features from multiple models could be another approach for improving performance.

Funding. This work was partially funded by the Instituto de Telecomunicações (IT), by the Fundação para a Ciência e Tecnologia (FCT) under the project UIDB/50008/2020 and under the Scientific Employment Stimulus Individual Call grant 2022.04901.CEECIND, and by the European Regional Development Fund (FEDER) through the Operational Competitiveness and Internationalization Programme (COMPETE 2020), and by National Funds (OE) through the FCT under the LISBOA-01-0247-FEDER-069918 "CardioLeather".

References

1. Baharav, A., et al.: Fluctuations in autonomic nervous activity during sleep displayed by power spectrum analysis of heart rate variability. Neurology **45**, 1183–1187 (1995)
2. Carreiras, C., et al.: BioSPPy: Biosignal Processing in Python (2015). https://github.com/PIA-Group/BioSPPy/
3. Commission, E.: Road safety thematic report - fatigue. Technical report, European Commission, Directorate General for Transport (2021)
4. European Parliament, Council of the European Union: Regulation (EU) 2019/2144 of the European Parliament and of the Council, November 2019
5. Halin, A., Verly, J.G., Droogenbroeck, M.V.: Survey and synthesis of state of the art in driver monitoring. Sensors **21** (2021). https://doi.org/10.3390/s21165558
6. Hamilton, P.: Open source ECG analysis. In: Computers in Cardiology, pp. 101–104, September 2002. https://doi.org/10.1109/CIC.2002.1166717, iSSN 0276-6547
7. Iwamoto, H., Hori, K., Fujiwara, K., Kano, M.: Real-driving-implementable drowsy driving detection method using heart rate variability based on long short-term memory and autoencoder. IFAC-PapersOnLine **54**, 526–531 (2021). https://doi.org/10.1016/j.ifacol.2021.10.310
8. Kaida, K., et al.: Validation of the Karolinska sleepiness scale against performance and EEG variables. Clin. Neurophysiol. Off. J. Int. Federation Clin. Neurophysiol. **117**, 1574–1581 (2006). https://doi.org/10.1016/j.clinph.2006.03.011
9. Lourenço, A., Alves, A.P., Carreiras, C., Duarte, R.P., Fred, A.: Cardiowheel: ECG biometrics on the steering wheel. Mach. Learn. Knowl. Discov. Databases, 267–270 (2015)
10. Massoz, Q., Langohr, T., Francois, C., Verly, J.G.: The ULg multimodality drowsiness database (called drozy) and examples of use. In: Proceedings of the 2016 IEEE Winter Conference on Applications of Computer Vision (WACV 2016), Lake Placid, NY (2016)
11. Oliveira, L., Cardoso, J.S., Lourenço, A., Ahlström, C.: Driver drowsiness detection: a comparison between intrusive and non-intrusive signal acquisition methods. In: 2018 7th European Workshop on Visual Information Processing (EUVIP), pp. 1–6 (2018). https://doi.org/10.1109/EUVIP.2018.8611704

12. Oliveira, L., Cardoso, J.S., Lourenço, A., Ahlström, C.: Driver drowsiness detection: a comparison between intrusive and non-intrusive signal acquisition methods. In: Proceedings - European Workshop on Visual Information Processing, EUVIP 2018-November (2019). https://doi.org/10.1109/EUVIP.2018.8611704
13. Philip, P., et al.: Effect of fatigue on performance measured by a driving simulator in automobile drivers. J. Psychosomatic Res. **55**, 197–200 (2003). https://doi.org/10.1016/S0022-3999(02)00496-8
14. Rodrigues, L.: Driver Drowsiness Detection with Peripheral Cardiac Signals. Master's thesis, Instituto Superior Técnico, Lisbon, Portugal (2021)
15. Shaffer, F., Ginsberg, J.P.: An overview of heart rate variability metrics and norms. Front. Publ. Health **5**, 258 (2017). https://doi.org/10.3389/fpubh.2017.00258, https://www.ncbi.nlm.nih.gov/pmc/articles/PMC5624990/
16. Silveira, C.S., Cardoso, J.S., Lourenço, A.L., Ahlström, C.: Importance of subject-dependent classification and imbalanced distributions in driver sleepiness detection in realistic conditions. IET Intell. Transp. Syst. **13** (2019). https://doi.org/10.1049/iet-its.2018.5284
17. Sztajzel, J.: Heart rate variability: a noninvasive electrocardiographic method to measure the autonomic nervous system. Swiss Med. Wkly **134**, 514–522 (2004)
18. Talbot, R., et al.: Framework for operational design of experimental work in i-dreams. Deliverable 3.1 of the EC H2020 project iDREAMS (2020)
19. Zhang, A., Lipton, Z.C., Li, M., Smola, A.J.: Dive into deep learning. CoRR abs/2106.11342 (2021). https://arxiv.org/abs/2106.11342
20. Čulík, K., Kalašová, A., Štefancová, V.: Evaluation of driver's reaction time measured in driving simulator. Sensors **22** (2022). https://doi.org/10.3390/s22093542

Fine-Tuned SqueezeNet Lightweight Model for Classifying Surface Defects in Hot-Rolled Steel

Francisco López de la Rosa[1,2(✉)], José Luis Gómez-Sirvent[1,3],
Lidia María Belmonte[1,2], Rafael Morales[1,2],
and Antonio Fernández-Caballero[1,3]

[1] Instituto de Investigación en Informática de Albacete (I3A), Universidad de Castilla-La Mancha, Avenida de España s/n, 02071 Albacete, Spain
francisco.lopezrosa@uclm.es
[2] Department of Electrical, Electronic, Automatic and Communications Engineering, Universidad de Castilla-La Mancha, Avenida de España s/n, 02071 Albacete, Spain
[3] Departamento de Sistemas Informáticos, Universidad de Castilla-La Mancha, Avenida de España s/n, 02071 Albacete, Spain

Abstract. The advent of powerful artificial intelligence-based tools is opening up new opportunities to improve the efficiency of processes in the manufacturing industry. One of those processes is visual inspection, where deep learning approaches, particularly convolutional neural networks (CNNs), have achieved human-level defect classification performance. The problem in using CNNs is that training takes a long time and they are computationally expensive. However, in recent years, lightweight models have emerged as an alternative. The most representative lightweight model is SqueezeNet, which can compete with deep CNNs in terms of classification performance with few trainable parameters. In this paper, a SqueezeNet model is used to classify surface defects on hot-rolled steel plates in a public surface defect database. The model is fine-tuned using a grid search algorithm and evaluated by 5-fold cross-validation, achieving an F1-score of 0.96984.

Keywords: Lightweight CNN model · SqueezeNet · Defect classification · Automated inspection systems · Industry 4.0

1 Introduction

Industry 4.0 has arrived on the industrial scene. Many different industries have embraced its concepts and changed the way they operate to achieve greater efficiency in their processes [14]. The consolidation of Industry 4.0 is made possible by the advancement of artificial intelligence (AI) and explainable artificial intelligence (XAI) [1]. One of the most widespread applications of AI in industry is in visual inspection systems. Deep learning (DL) techniques, which harness the wealth of information collected throughout the manufacturing chain, are being used to detect and classify defects early in the manufacturing process [4,5,12], preventing cost overruns in subsequent operations and saving time.

I. Rojas et al. (Eds.): IWANN 2023, LNCS 14134, pp. 221–233, 2023.
https://doi.org/10.1007/978-3-031-43085-5_18

Among the DL techniques, convolutional neural networks (CNNs) are the most suitable for classification problems when sufficient image data is available. Probably the main disadvantage of using CNNs, besides the possibility of not having enough images to train them, is that these models are very time and resource consuming [18]. Fortunately, there is an alternative to the deep and heavy CNN models that have been developed in recent years: the lightweight models. Lightweight CNN models combine high classification performance (close to the state of the art of heavyweight models) with much lower time consumption than deep CNN models. These characteristics make them suitable for use even in real-time applications, contributing to greater efficiency in inspection processes.

There are several studies based on CNNs that have worked with the Northeastern University (NEU) surface defect database [17] to classify the surface defects of the hot rolled steel that make up the database. For example, a SqueezeNet model [6] was used to classify the defects of the dataset even under noisy conditions [3]. A very recent paper [2] developed a CNN (DSTEELNet) to classify the defects from the NEU dataset after applying sophisticated data augmentation techniques. Another work [19] showed a self-designed CNN that was able to achieve good classification results after applying data augmentation techniques. Finally, a hybrid method that merged a CNN and a transformer [9] outperformed other state-of-the-art CNN models such as ResNet and GoogleNet.

The present paper also uses the NEU surface defect database to fine-tune a given topology of the lightweight SqueezeNet model for hot-rolled steel surface defect classification. The model is optimized through 5-fold cross-validation along with a grid search algorithm to ensure the selection of optimal hyperparameters for the model. To the best of our knowledge, no previous approach has attempted to classify defects using the lightweight SqueezeNet model configuration presented in this paper. Furthermore, the validation and evaluation of the model is done in a different way.

The remainder of the paper is organized as follows. Section 2 collects information about the NEU database and describes the implementation of the SqueezeNet lightweight model. Section 3 shows and discusses the results obtained by the proposed approach. Finally, Sect. 4 provides a set of conclusions and suggestions for future work.

2   Materials and Methods

This section is divided into two subsections. In the first one, a thorough description of the data used in this work is presented, while the second subsection deals with the proposed methodology used to classify the hot-rolled steel surface defects that make up the NEU dataset.

2.1   The NEU Dataset

The NEU dataset contains up to six different classes of defects that typically occur during hot rolling of steel (*inclusion, crazing, patches, pitted surface,*

Table 1. Description of the defect classes

Defect class	Description
Inclusion	Remains of impurities often pressed into the steel during the rolling process
Crazing	Defects of this class appear as cracks along the surface of the material
Patches	Marks of different shades that appear all over the surface of the metal
Pitted surface	A particular type of corrosion that occurs in a circular pattern and penetrates into the interior of the material [11]
Scratches	Wear marks caused by the rollers during the rolling process
Rolled-in scale	A rolled-in scale defect is the result of the introduction of mill scale into the metal during the rolling process

scratches and *rolled-in scale*). Each of the classes contains 300 grayscale images whose size is 200 × 200 pixels. Therefore, the dataset consists of 1800 images in total. Table 1 describes the main features of the different classes of the dataset.

Figure 1 illustrates the six different defect classes that make up the NEU surface defects dataset.

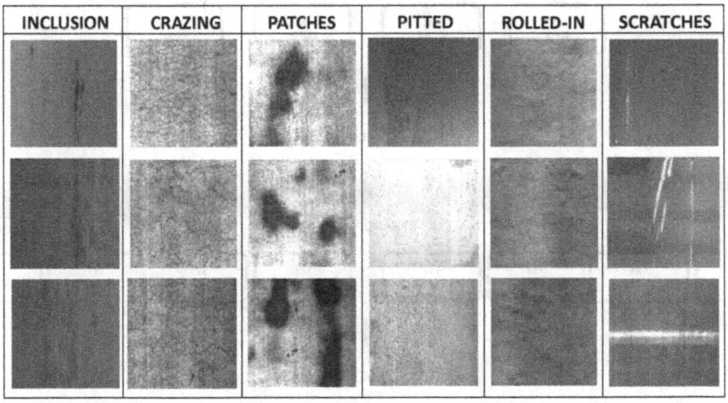

Fig. 1. Samples of defect classes in the NEU dataset

Although the defects that make up each class have the same origin, they can be very different from each other due to several aspects (material changes, different illumination, different defect arrangement). Moreover, some of the classes have similar features. These two peculiarities make the classification tasks studied in this work more difficult. In addition, 300 images of each class may be a little short of what is needed to perform defect classification using CNN. However, the authors of this paper aim to show how far a fine-tuned lightweight model such as SqueezeNet can go without the need for data augmentation.

2.2   SqueezeNet Lightweight Model

This section is divided into two subsections. The first describes the fundamentals of SqueezeNet and its basic architecture, while the second details the SqueezeNet-based approach used to perform defect classification.

SqueezeNet Architecture. SqueezeNet has completely changed the paradigm of CNNs. The trend in CNNs has been to develop deeper and deeper networks with millions of trainable parameters. These deep CNN models achieve high classification scores and are well suited for a wide range of applications. However, their heavy weights make them unsuitable for some tasks that require immediate response or when available resources are insufficient to run these models. SqueezeNet combines good classification performance with a relatively small number of trainable parameters (50 times less than AlexNet [8]), which makes this model suitable for more modest equipment and classification tasks requiring fast response. The main design innovation introduced by SqueezeNet is the fire module (see Fig. 2).

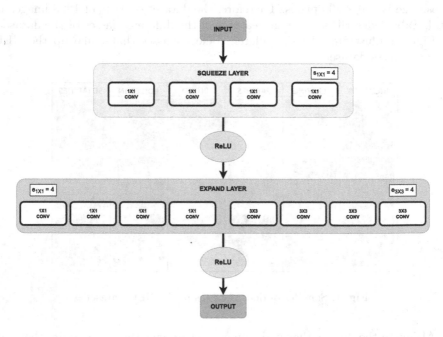

Fig. 2. SqueezeNet fire module

The fire module consists of a layer composed of 1×1 kernels (squeeze layer), followed by a layer combining 1×1 and 3×3 kernels (expand layer). The expand layer is preceded and followed by a rectifier linear unit (ReLU) activation. The fire modules are sized according to the following three design parameters:

- $s_{1\times1}$: number of 1×1 kernels in the squeeze layer
- $e_{1\times1}$: number of 1×1 kernels in the expand layer
- $e_{3\times3}$: number of 3×3 kernels in the expand layer

There is a design rule that must be followed to reduce the number of input channels and thus the number of trainable parameters: $s_{1\times1}$ must be less than the sum of $e_{1\times1}$ and $e_{3\times3}$. Figure 3 shows the three typical models of SqueezeNet. The base model is the one on the left. The other two versions are obtained by adding a simple (center) or complex (right) bypass. The authors of this research chose the baseline model for the experiments because it is the lightest model. Finally, Table 2 collects all the details of the SqueezeNet model as designed for this particular work. The code of the baseline model has of course been modified to suit the specifics of this particular application, i.e. the number of classes.

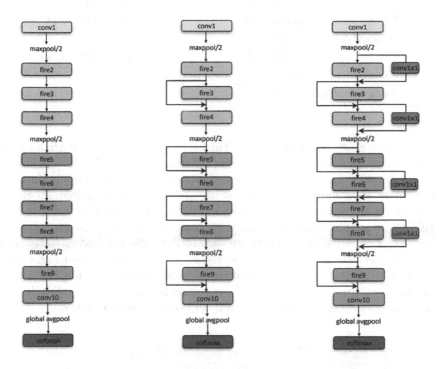

Fig. 3. SqueezeNet versions. Left: baseline model. Center: Simple bypass. Right: Complex bypass [6]

SqueezeNet-Based Approach. The different steps involved in training the SqueezeNet lightweight model are described next. These steps, which are detailed below, are data preparation, hyperparameter tuning, and evaluation.

Table 2. Proposed SqueezeNet architecture

Layer	Output size	Stride/Filter size	$s_{1\times1}$	$e_{1\times1}$	$e_{3\times3}$
Input	400×400				
Conv 1	200×200×32	7×7/(×32)			
MaxPool 1	100×100×32	2×2/2			
Fire 2	100×100×32		8	16	16
Fire 3	100×100×32		8	16	16
Fire 4	100×100×64		16	32	32
MaxPool 4	50×50×64	2×2/2			
Fire 5	50×50×64		16	32	32
Fire 6	50×50×128		32	64	64
Fire 7	50×50×128		32	64	64
Fire 8	50×50×256		64	128	128
MaxPool 8	25×25×256	2×2/2			
Fire 9	25×25×256		64	128	128
Conv 10	25×25×6	1×1/1(×6)			
AveragePool 10	1×1×6				

– *Data preparation.* As mentioned in the previous sections, there are a total of 1800 images from six different classes (300 images per class). Since the images are of the same format and size, and there is no intention to apply data augmentation techniques, no preprocessing operations are required. Therefore, the images from the original dataset are divided into different subsets. First, 10% of the images from each class (180 images in total) are used to create the test set. The remaining images are sent to five different folders (324 images each) to perform a 5-fold cross-validation. In this process (see Fig. 4), four of the five folders are used to train the SqueezeNet model, while the remaining folder is used for validation. Because this is a 5-fold cross-validation, the process is repeated five times to cover all possible combinations.

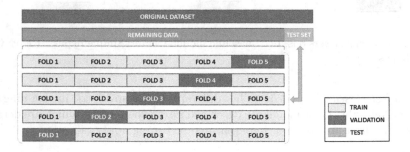

Fig. 4. 5-fold cross-validation

– *Hyperparameter tuning.* To determine the optimal hyperparameter settings, each of the data combinations from the previous step was introduced into a grid search algorithm [10]. The grid search algorithm generates a model for each possible combination of parameters, resulting in multiple iterations. The optimal combination of hyperparameters is then determined based on a performance metric. In this case, two different hyperparameters were considered: optimizer and learning rate (LR) (see Table 3). The optimizer tries to reduce the model loss by adjusting the weight of the neurons and the LR. In this research, adaptive moment estimation (ADAM) [7] and stochastic gradient descent (SGD) [15] are the optimizers to be tested. The learning rate determines how much the model weights change as a function of the error in the previous iteration. In this case, this hyperparameter takes values from 10^{-1} to 10^{-6} on a logarithmic scale. Each of the different data combinations from the previous steps are therefore introduced into the grid search to find the optimal value for the LR and the best performing optimizer.

Table 3. Grid search hyperparameters considered for SqueezeNet tuning

Hyperparameters	Values
Optimizer	ADAM, SGD
Learning rate	$10^{-1}, 10^{-2}, 10^{-3}, 10^{-4}, 10^{-5}, 10^{-6}$

– *Model evaluation.* Finally, the test set is used to evaluate the performance of the different models generated by the grid search algorithm. A performance metric must be used to evaluate the performance of the model on the test set data. In this case, the authors have chosen one of the most complete performance metrics available, the F1-score. The model with the highest F1-score and the lowest validation loss will be selected as the optimal model. The F1-score can be defined as the harmonic mean between precision (P) and recall (R) and is computed as:

$$\text{F1-score} = \frac{2 \times P \times R}{P + R} \tag{1}$$

where P expresses the proportion of successful positive predictions and is computed as:

$$P = \frac{TP}{TP + FP} \tag{2}$$

and R represents the ratio of relevant predictions and is determined as:

$$R = \frac{TP}{TP + FN} \tag{3}$$

where TP means true positives, FP stands for false positives and FN represents the false negatives.

3   Results and Discussion

Each experiment in this paper was performed on a workstation computer with the following hardware specifications: 3.80 GHz Intel i7-10700KF v8 CPU, 32 GB RAM, and NVIDIA GeForce RTX 2070 SUPER 8 GB. The Keras (https://keras.io/) and TensorFlow (https://www.tensorflow.org) libraries were used to train the models, with Pycharm (https://www.jetbrains.com/pycharm/) as the integrated development environment (IDE) and Python 3.8 (https://www.python.org/) as the programming language.

The model was trained using a batch size of 1 for 100 consecutive epochs and selecting categorical cross-entropy as the loss function. Note, however, that early stopping conditions were introduced to avoid excessive overfitting. Thus, if the validation loss did not improve for 10 consecutive epochs, the training of the model was automatically stopped. With these training conditions, each of the combinations from the 5-fold cross-validation was trained for each set of hyperparameters of the grid search algorithm. Table 4 shows the results of the grid search algorithm.

Table 4. Results of the SqueezeNet model grid search

Iteration	Optimizer	LR	F1-score$\pm\sigma$	Validation loss$\pm\sigma$
1	ADAM	10^{-1}	0.04762 ± 0.00000	2.25767 ± 1.60761
2	SGD	10^{-1}	0.04762 ± 0.00000	2.10058 ± 0.00006
3	ADAM	10^{-2}	0.04762 ± 0.00000	1.79175 ± 0.00000
4	SGD	10^{-2}	0.19638 ± 0.11839	0.64420 ± 1.97706
5	ADAM	10^{-3}	0.04762 ± 0.00000	1.79175 ± 0.00000
6	SGD	10^{-3}	0.79179 ± 0.31543	0.16847 ± 0.12652
7	**ADAM**	10^{-4}	$\mathbf{0.96985 \pm 0.01646}$	$\mathbf{0.09264 \pm 0.04687}$
8	SGD	10^{-4}	0.93373 ± 0.10054	0.23980 ± 0.10069
9	ADAM	10^{-5}	0.94147 ± 0.05539	0.14873 ± 0.02904
10	SGD	10^{-5}	0.09366 ± 0.15878	1.82672 ± 0.26398
11	ADAM	10^{-6}	0.83891 ± 0.03120	0.48055 ± 0.13085
12	SGD	10^{-6}	0.04189 ± 0.04364	1.88787 ± 0.00656

Note that the calculated F1 score and validation loss correspond to the mean F1 score and mean validation loss along with its corresponding standard deviation for each of the 5-fold cross-validation scenarios for each tuple of grid search hyperparameters, ensuring good generalisation of the model and proper evaluation of the results. Table 4 shows that depending on the LR, ADAM will outperform SGD or vice versa.. In terms of LR, values between 10^{-4} and 10^{-5} give the best results. The best hyperparameter combination, highlighted in the table, is the one with ADAM as the optimizer and an LR of 10^{-4}, which gives a mean F1-score of 0.96984 and a mean validation loss of 0.09264.

Figure 5 and Fig. 6 shows the learning curves and the confusion matrix of the best performing model, respectively. Looking at Fig. 5, the training of the model looks correct as there is no excessive overfitting and both the accuracy curve and the loss curve of the training and validation sets converge smoothly. On the other hand, the confusion matrix of the model (see Fig. 6) shows that the classification performance of the model is excellent, as shown by the F1-scores displayed in the principal diagonal.

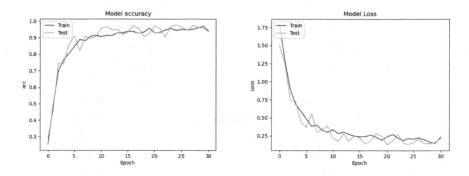

Fig. 5. SqueezeNet accuracy and loss curves

ACTUAL VALUES

		Inclusion	Crazing	Patches	Pitted	Rolled-in	Scratches
	Inclusion	1.00	0	0	0	0	0
	Crazing	0	0.91	0	0.09	0	0,01
PREDICTED VALUES	Patches	0.01	0	0.09	0	0	0
	Pitted	0	0.01	0.01	0.94	0.03	0
	Rolled-in	0.01	0	0	0	0.99	0
	Scratches	0	0.01	0	0.01	0	0.99

Fig. 6. Confusion matrix of the SqueezeNet model

The results obtained with our approach are comparable to the other studies mentioned above that have used the NEU database to detect surface defects in hot-rolled steel as shown in Table 5. First, we compare a SqueezeNet model similar to ours [3] for defect classification on the NEU dataset. The results are impressive, even reaching 100% accuracy when the SqueezeNet model is supported by a low-level feature fine-tuning algorithm and an MRF module, clearly outperforming the SqueezeNet method presented in our work. However, the implementation of a cross-validation method, as introduced in our work, significantly improves the reliability of the obtained results. Therefore, in some cases, it is worth using a smaller percentage of data for training and losing some classification performance in order to obtain fully reliable results.

Table 5. Comparison with related works

Reference	CNN model	F1-score	Accuracy
[3]	SqueezeNet based methodology	—	0.997
	SqueezeNet+ELF	—	1
	SqueezeNet+ELF+MRF	—	1
[2]	VGG-16	0.884	—
	VGG-19	0.894	—
	ResNet50	0.920	—
	DSTEELNet	0.956	—
	MobileNet	0.920	—
	Yolov5	0.836	—
	Yolov5-SE	0.886	—
[19]	CNN own design	—	0.991
	CNN own design no augmentation	—	0.883
[9]	GoogleNet	—	0.972
	ResNet18	—	0.981
	CNN-T	—	0.989
	MobileNetV2	—	0.986
	VIT	—	0.961
Ours	**SqueezeNet based methodology**	**0.970**	**0.970**

Second, let us look at DSTEELNet [2], an approach developed and compared with other state-of-the-art CNN models to classify defects in the NEU dataset. The results obtained with all the CNN models studied (F1 scores ranging from 0.836 for Yolov5 to 0.956 for DSTEELNet) are slightly worse than those obtained with the SqueezeNet method used in our work, even with data augmentation techniques. Moreover, they do not include cross-validation, so the results are a priori less reliable than those of our work.

The next paper [19] proposes a self-developed CNN model to classify the defects in the NEU surface database. Their approach achieves an impressive result of 0.991 accuracy when applying data augmentation techniques, but on the original dataset they achieve only 0.883 accuracy, much lower than that obtained by our approach on the same data. Finally, the authors of a recent paper [9] compare the T-CNN model with other state-of-the-art models. The results obtained by these models are similar to those obtained by our method, which is able to compete with these heavy CNN models in terms of classification performance. Another aspect worth mentioning is that, unlike our work, data augmentation is applied. Moreover, no cross-validation method is included, so the results obtained would be less reliable than in our work.

4 Conclusions and Future Work

In this work, the implementation of a lightweight model for the classification of surface defects on hot-rolled steel plates was investigated. First, the authors worked with the data from the original dataset to prepare them for consistent model training and evaluation. Then, the basics of the SqueezeNet lightweight model were clearly explained and the proposed architecture was presented. Then, the training strategy and its steps were detailed. In this case, a grid search algorithm that plays with the optimizer and the LR was used to fine-tune the model, while 5-fold cross-validation was used to perform a comprehensive evaluation of the model using the correct performance metrics.

Finally, the results obtained with the proposed approach were presented, focusing on the learning curves and the confusion matrix of the best performing model, which obtained a mean F1-score of 0.96984 and a validation loss of 0.09264. These results are impressive for such a light model and comparable to those obtained by similar works, considering that no data augmentation was applied. Since reliability plays an important role in any manufacturing industry, the use of validation strategies such as cross-validation is highly recommended. The proposed SqueezeNet method is not only very good in terms of classification performance, but also fully reliable thanks to the use of 5-fold cross-validation.

In terms of future work, there are several possibilities for further research. First, the lightweight SqueezeNet model could be applied to other domains. Another possibility is to apply novel data augmentation techniques in order to make a direct comparison with the results obtained in other works [13]. Finally, the authors would like to develop a methodology to deal with the detection of unknown defects [16], which is a major problem in any manufacturing industry.

Acknowledgments. This work was partially supported by iRel40, a European co-funded innovation project that has been granted by the ECSEL Joint Undertaking (JU) (grant number 876659). The funding of the project comes from the Horizon 2020 research programme and participating countries. National funding is provided by Germany, including the Free States of Saxony and Thuringia, Austria, Belgium, Finland, France, Italy, the Netherlands, Slovakia, Spain, Sweden, and Turkey. Grant PCI2020-112001 funded by MCIN/AEI/10.13039/ 501100011033 and by the "European Union NextGenerationEU/PRTR". Grant 2022-GRIN-34436 funded by Universidad de Castilla-La Mancha and by "ERDF A way of making Europe". Grant 2023-PRED-21291 funded by Universidad de Castilla-La Mancha and by "ESF Investing in your future". Grant BES-2021-097834 funded by MCIN/AEI/10.13039/501100011033 and by "ESF Investing in your future".

References

1. Ahmed, I., Jeon, G., Piccialli, F.: From artificial intelligence to explainable artificial intelligence in industry 4.0: a survey on what, how, and where. IEEE Trans. Industr. Inf. **18**(8), 5031–5042 (2022). https://doi.org/10.1109/TII.2022.3146552

2. Ahmed, K.R.: DSTEELNET: a real-time parallel dilated CNN with atrous spatial pyramid pooling for detecting and classifying defects in surface steel strips. Sensors **23**(1), 544 (2023). https://doi.org/10.3390/s23010544
3. Fu, G., et al.: A deep-learning-based approach for fast and robust steel surface defects classification. Opt. Lasers Eng. **121**, 397–405 (2019). https://doi.org/10.1016/j.optlaseng.2019.05.005
4. Gómez-Sirvent, J.L., López de la Rosa, F., Sánchez-Reolid, R., Morales, R., Fernández-Caballero, A.: Defect classification on semiconductor wafers using fisher vector and visual vocabularies coding. Measurement **202**, 111872 (2022). https://doi.org/10.1016/j.measurement.2022.111872
5. Gómez-Sirvent, J.L., de la Rosa, F.L., Sánchez-Reolid, R., Fernández-Caballero, A., Morales, R.: Optimal feature selection for defect classification in semiconductor wafers. IEEE Trans. Semicond. Manuf. **35**(2), 324–331 (2022). https://doi.org/10.1109/TSM.2022.3146849
6. Iandola, F.N., Han, S., Moskewicz, M.W., Ashraf, K., Dally, W.J., Keutzer, K.: SqueezeNet: AlexNet-level accuracy with 50x fewer parameters and<0.5 MB model size. arXiv:1602.07360 (2016). https://doi.org/10.48550/arXiv.1602.07360
7. Kingma, D.P., Ba, J.: Adam: a method for stochastic optimization. arXiv:1412.6980 (2014). https://doi.org/10.48550/arXiv.1412.6980
8. Krizhevsky, A., Sutskever, I., Hinton, G.E.: ImageNet classification with deep convolutional neural networks. Commun. ACM **60**(6), 84–90 (2017). https://doi.org/10.1145/3065386
9. Li, S., Wu, C., Xiong, N.: Hybrid architecture based on CNN and Transformer for strip steel surface defect classification. Electronics **11**(8), 1200 (2022). https://doi.org/10.3390/electronics11081200
10. Liashchynskyi, P., Liashchynskyi, P.: Grid search, random search, genetic algorithm: a big comparison for NAS. ArXiv:1912.06059 (2019). https://doi.org/10.48550/arXiv.1912.06059
11. Lv, X., Duan, F., Jiang, J.J., Fu, X., Gan, L.: Deep metallic surface defect detection: the new benchmark and detection network. Sensors **20**(6), 1562 (2020). https://doi.org/10.3390/s20061562
12. López de la Rosa, F., Gómez-Sirvent, J.L., Morales, R., Sánchez-Reolid, R., Fernández-Caballero, A.: A deep residual neural network for semiconductor defect classification in imbalanced scanning electron microscope datasets. Appl. Soft Comput. **131**, 109743 (2022). https://doi.org/10.1016/j.asoc.2022.109743
13. López de la Rosa, F., Gómez-Sirvent, J.L., Sánchez-Reolid, R., Morales, R., Fernández-Caballero, A.: Geometric transformation-based data augmentation on defect classification of segmented images of semiconductor materials using a resnet50 convolutional neural network. Expert Syst. Appl. **206**, 117731 (2022). https://doi.org/10.1016/j.eswa.2022.117731
14. Psarommatis, F., Prouvost, S., May, G., Kiritsis, D.: Product quality improvement policies in industry 4.0: characteristics, enabling factors, barriers, and evolution toward zero defect manufacturing. Front. Comput. Sci. **2** (2020). https://doi.org/10.3389/fcomp.2020.00026
15. Robbins, H., Monro, S.: A stochastic approximation method. Ann. Math. Stat. 400–407 (1951). https://doi.org/10.1214/aoms/1177729586
16. López de la Rosa, F., Gómez-Sirvent, J.L., Kofler, C., Morales, R., Fernández-Caballero, A.: Detection of unknown defects in semiconductor materials from a hybrid deep and machine learning approach. In: Ferrández Vicente, J.M., Álvarez-Sánchez, J.R., de la Paz López, F., Adeli, H. (eds.) IWINAC 2022, vol. 13259. pp. 356–365. Springer, Cham (2022). https://doi.org/10.1007/978-3-031-06527-9_35

17. Song, K., Yan, Y.: A noise robust method based on completed local binary patterns for hot-rolled steel strip surface defects. Appl. Surf. Sci. **285**, 858–864 (2013). https://doi.org/10.1016/j.apsusc.2013.09.002
18. Sánchez-Reolid, R., et al.: Artificial neural networks to assess emotional states from brain-computer interface. Electronics **7**(12) (2018). https://doi.org/10.3390/electronics7120384
19. Yi, L., Li, G., Jiang, M.: An end-to-end steel strip surface defects recognition system based on convolutional neural networks. Steel Res. Int. **88**(2), 1600068 (2017). https://doi.org/10.1002/srin.201600068

Shot Boundary Detection
with Augmented Annotations

Miguel Jose Esteve Brotons[1], Jorge Carmona Blanco[1],
Francisco Javier Lucendo[1], and José García-Rodríguez[2]([✉])

[1] Telefónica I+D, Madrid, Spain
{miguel.estevebrotons,jorge.carmonablanco,
javier.lucendodegregorio}@telefonica.com
[2] Computers Technology Department, University of Alicante, Alicante, Spain
jgarcia@dtic.ua.es

Abstract. In recent years, deep learning approaches have been considered to provide state-of-the-art results in shot boundary detection. These approaches revolve around the need for large annotated datasets. The quality of the annotations is crucial to the robustness of the algorithm. Having graphical tools to verify the correct annotation of the original datasets, as well as the correct generation of synthetic datasets is a must. In this paper we propose a framework that allow the visual inspection of the datasets, incorporating the option of editing the annotations manually, as well as annotations from other algorithms, generating a set of augmented annotations. In addition, we benchmark the performance of TransNet in three scenarios, 1) using the datasets with their original annotations, 2) using automatically generated annotations, and 3) using the combination of the previous annotations, as augmented annotations. We conclude that the usage of augmented annotations significantly improves the network results.

Keywords: Shot boundary detection · augmented annotations · visual inspection

1 Introduction

Shot boundary detection is a problem that has been largely studied for several years. The problem consists of doing a first temporal segmentation of the video into shots. There is no formal definition of a shot, but we can say that a shot is a group of frames that share a high degree of similarity to consider that they form part of the same scene. Following the approach, a scene could be considered a group of shots, all with the same semantics. Over the years, a lot of papers and studies have been published with algorithms to solve the problem of adequately segmenting the video into shots. One group of recent algorithms is based on supervised deep learning techniques, where a given algorithm is trained based on a dataset of videos properly annotated with the shots. One of the most recent and relevant algorithm is TransNet V2 [11], trained and tested on the datasets BBC [2], RAI [3], ClipShots [13] and IACC3 [1] datasets.

© The Author(s), under exclusive license to Springer Nature Switzerland AG 2023
I. Rojas et al. (Eds.): IWANN 2023, LNCS 14134, pp. 234–250, 2023.
https://doi.org/10.1007/978-3-031-43085-5_19

We have reviewed the last innovations in the field of shot boundary detection, using deep learning techniques, and reviewed the datasets being utilized in those recent research activities. After gaining access to the BBC, RAI, Clipshots, and IACC3 datasets, and performing the first completely random and manual visual inspection, we were surprised to verify that in many videos, there were a high number of not annotated transitions. This led us to question whether the annotations were being made on shots or on scenes. However, we came to the conclusion that in most cases it could have been due to the intrinsic difficulty of maintaining a single criterion on the part of the human annotator. Due to the high number of datasets and videos to be inspected, we considered it necessary to develop a tool that would allow us to select a dataset programmatically, list all its videos, and by selecting the video, view the annotations, both abrupt and synthetic.

After developing the visual inspection tool, we observed that a significant number of videos, especially in the RAI and Clipshots datasets, contained a high number of false negatives, that is, non-annotated transitions. We wondered how we could improve state of the art in the shot detection task if the datasets we started with did not give us enough confidence in the quality of their annotations. That is why we decided to analyze in detail the mechanism for generating a dataset for the task in question, and the impact that starting with datasets whose annotations were not wholly accurate could have. Then we set out to investigate how using annotations obtained from the predictions of a shot detector available on the internet like pyscenedetect would influence the final result. The objective would be to compare the results with a reference base model, using the original annotations, with those obtained with pyscene, and finally, with a combined set of annotations, which we call augmented, as the union of the original annotations and those obtained with pyscene.

We generated the dataset that feeds the model synthetically containing abrupt and gradual transitions. To do this, we use video segments from the orig inal datasets that do not contain transitions, and applying a blending algorithm, we generate the synthetic transition. If the original dataset contains many false negatives, we will take segments that already contain a transition to generate a new one. If it contains false positives, we could generate a synthetic transition between two segments that belong to the same shot.

Some novel aspects of our work include the following:

1. A fully detailed preprocessing pipeline for training and evaluating a supervised algorithm for shot detection with several modes to generate the data to be utilized for training, evaluation, and testing.
2. A GUI SBD data framework to run all preprocessing steps required to generate the training, evaluation, and test datasets
3. As part of such GUI, we include video and mosaic players allowing to select any publicly available dataset and visualize any of the videos and their annotations, which serve as a useful inspection tool. We also include the option to manually edit any annotations provided and add/remove new annotations.

4. Integration with pySceneDetect [4] in order to augment and further improve the original annotations.
5. Comparative results of using annotations in three scenarios: using the annotations provided in the aforementioned datasets, which we call original annotations, using annotations obtained with pyscene, which we abbreviated as "cd" annotations, and those obtained as a result of combining both, which we call augmented annotations and we abbreviate as "reanots".

The remaining of the paper is organized as follows: Sect. 2 revised related works. In Sect. 3 we present our general approach to improve video annotations while Sect. 4 describes the use of Pyscene. Section 5 describe our experimentation to validate the proposal and we finish with our conclusions and further work.

2   Related Work

In this chapter we review the main contributions to the shot boundary detection problem based on a supervised model, reviewing the solution proposed in the design of the data processing pattern for the model in question. We find that almost all the articles focus on detailing the architecture of the solution, the datasets used and the results of the experimentation, without specifying in detail how the data processing was designed. In most cases, the datasets are also not publicly available and neither is the source code. Therefore, what we do is describe the processing scheme based on our own interpretation of what is exposed in the contributions.

In [2], a deep learning paradigm using visual and textual features is presented, stating that when the textual features are not directly available, they can be obtained with speech recognition. Deep-learned features are then used together with a spectral clustering algorithm to segment the video. The deep learning part is based on a Siamese network with two branches that share the same architecture and weights. Each branch takes as input the middle frame of two different shots. Branch outputs are shot visual and textual descriptors that are concatenated and given to a top network that computes a similarity function. The siamese network is pre-trained with ImageNet dataset [9] plus Places dataset [17]. Finally, the network is fine-tuned using the BBC dataset [11], which contains around 4900 shots and 670 scenes. Ten videos are used for training and one for tests. As most of the shots labeled belong to different scenes, the author artificially generates shot pairs belonging to the same scene to balance the training set per batch. Therefore the dataset for the model presented in this paper consists of two input frames from the same or different shots and with a label, 0 or 1 if both shots belong to a different or same scene.

In [15], the approach for shot boundary detection presented consists of frame similarity with adaptive threshold and a two-round bisection-based comparison of segments. At the end of a candidate segment selection process, cut and gradual transition frames are selected by comparing the five top classes of the candidate frames after being classified using a CNN model similar to the ImageNet challenge winning model. The five classes are called TAG. The assumption is that frames in the same shot tend to have similar TAG while frames in different shots do not.

3D convolutional networks have been proposed to tackle some video-related tasks, like in [7] and [8]. Compared to 2D ConvNet, 3D ConvNet can model temporal information better owing to 3D convolution and 3D pooling operations. In 3D ConvNets, convolution and pooling operations are performed spatio temporally while in 2D ConvNets they are done only spatially. Previous to 3D ConvNet there were some trials to learn temporal information using 2D convnets, like [10], where although the temporal stream network takes multiple frames as input, because of the 2D convolutions, after the first convolution layer, temporal information is collapsed completely. The same happens with fusion models in [8], where most of the networks lose their temporal input signal after the first convolution layer. Only the slow fusion model in [8] uses 3D convolutions.

In [16] a more deep analysis of 3D ConvNet is done, playing with two types of architectures: homogeneous temporal depth and varying temporal depth. The paper introduces C3D network, which represents state-of-the-art in action recognition using UCF101 dataset [8].

DeepSBD is presented in [6], which consists of a 3D ConvNet inspired in C3D [16] that allows to the classification of a segment of 16 frames as input in either gradual, sharp, or no-transition. After a 3D Convnet, an SVM is utilized for classification, followed by a merging and post-processing step. The merging step merges consecutive segments with the same labels, and the post-processing step just reduces false positives, by estimating color histograms of the first and end frame and measuring with Bhattacharyya distance between these histograms. Therefore it handles the problem of shot boundary detection as a classification problem, where each sample is composed of a video segment of length 16 and the corresponding label between 3 possible candidates. They generate their own dataset in two portions: first portion, SBD_syn, generated synthetically using image composition models, with 220,339 sharp and gradual segments, and a second portion, SBD_BT, containing 4,427 no transition segments, acting as hardnegatives. They also used TRECVID [1] release with just 18.027 total transitions.

In [5] a fully convolution network is proposed. The network takes 10 frames as input, runs them through four layers of 3D convolutions, each followed by a ReLU, and finally classifies if the two center frames come from the same shot if there is an ongoing transition. With this approach, the dataset is generated with two types of samples: 10 frames with or without a transition in the frame index 6. The transitions are generated synthetically, including dissolves, fade-in fade-out, and wipes. Non-transition frames are added including flashes. The network is able to detect if the middle center of the sample has a transition or not, but not the type of transition nor its duration.

In [13] shot detection is performed in a cascade approach with first initial filtering using adaptive thresholds, which produces a set of transition candidates, then those candidates are further fed into a strong cut transition detector to filter out false cut transitions. The cut transition detector is tested with four approaches based on measuring similarity: siamese, feature concatenation, image concatenation, and C3D Convnet. The remaining are fed into a gradual transition detector based on a single shot detector but using default segments instead default boxes.

In [12] shot detection is done using Dilated Convolutions with different dilation rates for the time dimension, in order to augment the receptive field without increasing the number of network parameters. Dilated Convolutions outputs are concatenated in the channel dimension forming a Dilated DCNN layer. Then, multiple Dilated DCNN layers are stacked with spatial max pooling to form a Stacked DDCNN block. The network proposed then consists of multiple SDD-CNN blocks, very next block operating on smaller resolution but a greater channel dimension. The network is trained using 3000 videos in TRECVID IACC3 dataset with automatic creation of transitions. Evaluation was done using 100 IACC.3 [1] videos different from training set and testing done using RAI [2].

A second version of the network provided in [12] is presented in [11], which improve by adding Convolution Kernel Factorization, Frame Similarities as Features and Multiple Classification Heads.

3   Approach

3.1   Overview

In this chapter, based on [11], we describe a general view of the pipeline to pre-process data for the task at hand: supervised video shot boundary detection. In Fig. 1, we show the functional blocks involved in the complete data handling process and the interfaces between those blocks. For the common understanding, we define available blocks and reference points between functional blocks, R_j.

The functional blocks are listed as follow:

1. Consolidate: the process by which the different format of the different original datasets is normalized to a common notation syntax.
2. Create: the process by which, based on the consolidated annotations, datasets are generated based on video segments of different types: transition, non-transition, and test segments. Transitions segments are video segments that contain at least one transition frame. Non-transition segments are segments that do not have any transition; therefore, all frames on the segment are expected to be of the same shot. Test segments are segments that may or not contain transitions. Segments are generated and held in memory and then written in tensorflow TFRecords. We have then three different record types according to the type of segments they contain.
3. Input pipeline: the process that reads the TFRecords, parses it and generates the samples to feed the model. For each record, there is a specific pipeline, so we have a native pipeline, for transition segments, a synthetic pipeline for non-transition segments, and a test pipeline, for test segments.
4. Model: the model accepts video shots of length 100.

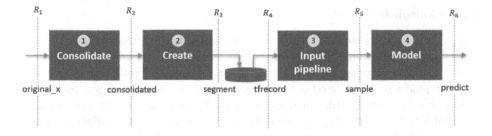

Fig. 1. General overview of the processing pipeline.

The interfaces between blocks are detailed as follow:

- R_1: Raw data in its original form as provided by data supplies. These are datasets in their original form. We name those datasets as originals.
- R_2: Consolidated files. These are *.txt files, one per original, containing a comma separated video path in each line and its respective annotations files already normalized and a standard syntax.
- R_3: Video segments, of different types, that may or not contain transitions.
- R_4: TFRecords, that store the video segments in disk, in the binary format defined in Tensorflow.
- R_5: Video samples to feed the model. Those video samples are of length 100, with one or more transitions per sample, and its corresponding annotations.

3.2 Dataset Generation Options

We define four different modes to create the TFRecords datasets. The modes are enumerated from 0 to 4. These are model 0 or static online, mode 1 or dynamic online, mode 3 or dynamic offline, and mode 4 offline or assisted. Static versus dynamic resides in how the original datasets are assigned to the different types of TFRecords: without transitions, with transitions, test. Online versus offline difference between the modes depends if the creation process is done as part of the reading pipeline (online) or prior to the pipeline (offline). In modes 0 and 1, the process of creating synthetic samples, as well as the temporal augmentation for native samples is done as part of the pipeline, that is, online. In mode 3, the process of creating synthetic samples, as well as the time augmentation for native samples, is done as part of the dataset creation process. Thus, TFRecords generated in mode 3 already contain segments with synthetically generated transitions as well as time augmentation. In this mode, the pipeline does not perform synthetic generation or temporary augmentation, which results in improved training times. Mode 4 is dynamic online using a configuration file that defines which original datasets are used, as well as the weight of each of them.

3.3   Consolidate

Each dataset has its own folder structure. Common to all datasets is that we have videos and annotations. Each of these datasets has its own directory structure, and the annotations are provided in different formats. Therefore, these datasets are not ready to be utilized directly but require some form of processing. We name the datasets in this form as "originals", as these are datasets in their original form. The first step we follow is what we named as consolidation step. In the consolidation step, the different originals are scripted to a common directory structure and the same annotation syntax.

At the output of the consolidation process, we get in a ROOT_CONSOLI-DATED folder defined by configuration, a directory where there are as many consolidated .txt files as originals, each containing per line a video path and its annotation file path, separated by commas. Two annotations files are generated, following the next syntax: {video_name.orig}.txt and {video_name}.reanot.txt. At the end of this process, both contain exactly the same annotations. However, reanot files are generated to allow a further step to change the provided annotations. The overview of the process is shown in Fig. 2.

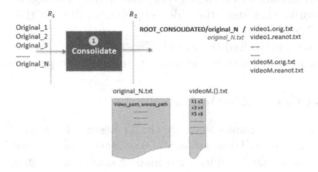

Fig. 2. General overview of the consolidate process. Note that two annotation files are generated, videoM.orig.txt and videoM.reanot.txt, to allow reanotations in further processing

3.4   Create

Once the original datasets are normalized, the next phase is related to generating the samples to feed our model. The samples must consist of a sequence of frames and labels. The network in a 3D Convolution based model requires a fixed length, so in our particular case, we will feed the model with videos of length 100, with labels consisting of a 1-D tensor where all values are zero except where there is a transition frame, where the value will be 1. In order to generate those samples, we take into account several considerations: first, transition samples can be of two types: cuts and gradual. Cuts are those transitions corresponding to abrupt transitions in between shots, where there is only one frame transition.Gradual are those transitions that consist of more than one frame, and typically correspond to video production effects like zoom-in, zoom-out, fade, or dissolves. In addition, transitions can be native or synthetic. Native transitions are those that come from annotations in the original datasets, and synthetic transitions are those that come from synthetically generated transitions using any synthetic method available.

In between the samples that feed the model and the original videos we generate segments, and those segments are stored in TensorFlow TFRecords [14]. When generating TFRecords, we define three types of TFRecords: non-transition TFRecords, transition TFRecords, and test TFRecords. Non-transition TFRecords contain segments of default duration equal to 100 frames. Those segments are taken from original videos and do not contain any transition. These non-transition TFRecords will be processed by a data pipeline that will generate synthetic transitions, cut or graduals. Non-transition TFRecords will contain just segments, and no labels, as labels will be generated at the time that samples are generated during the data preprocessing pipeline. Transition TFRecords contains segments of default length 160, with a transition always placed in the middle of the segment. These transition TFRecords will be processed by a specific pipeline that will generate a transition sample, moving the position of the transition and therefore emulating a temporal augmentation pattern. Test TFRecords will contain frames with their corresponding label as transition or non-transition frame.

3.5 Input Pipeline

The input pipeline is the process in the preprocessing workflow that takes samples from the dataset and generates samples that feed the model during training, evaluation, and testing. The pipeline is based on TensorFlow data API and takes as input the TensorFlow examples saved in TFRecords during the create step.

As there are three types of TFRecords, non-transition, transition, and test, we need three different input processing pipelines for each TFRecord type. Three different pipelines are required to handle each type of TFRecords. Non-transitions TFRecords are handled by the Synthetic pipeline. Transition TFRecords are handled by the Native pipeline and test TFRecords are handled by the test pipeline. The native pipeline always takes transitions TFRecords (transition segments), then as per the TensorFlow pattern, parses the TFRecords, converts from segment to shot, and does segment augmentation. The conversion from segment to shot consists in translating a 160 video segment with the transition placed in the center, to a 100 frames shot duration. The synthetic pipeline always takes non-transition TFRecords (non-transition segments). Then, as per the TensorFlow pattern, parse the TFRecords, convert from segment to shot, then does shot augmentation, and finally contact shots.

Synthetic Shots. The entry to this function is two shots of any length. Synthetic shots are generated by mixing two different shots using some form of an algorithm. Two algorithms are available: basic and advanced. For basic synthetic, let X_1, X_2 be the two shots to concatenate, and $alpha$ be a transition vector, then $\text{Synthetic}[X_1, X_2] = X_1 * alpha + X_2 * (1 - alpha)$. A transition_boundary is generated by randomly selecting an index in the range of shot_len. A dissolve_len is generated by randomly selecting a value in between a min_transition_len and a max_transition_len provided as parameters, such that $[0 < \text{min\_transition\_len} < \text{maximum\_transition\_len} < \text{shot\_len}]$. A dissolve kernel vector is generated of length dissolve_len, with values uniformly spaced between 1.0 and 0.0. The dissolve vector is centered in the transition_boundary index. The transition vector "alpha" is generated from the dissolve vector by padding with one on the left and zeros on the right, in case padding was required. The overview of the synthetic shot generation is shown in Fig. 3.

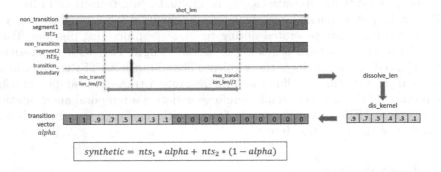

Fig. 3. General overview of synthetic shot generation.

3.6   Model Network

The whole approach described here is for a model, as the one proposed by [11] and [12], that expects as input a sequence of shot_len frames, with a label that will have 1's in all index frames that corresponds with a transition frame. Thus, a sample shot with no transitions will have as a label a vector with all zeros. A sample shot with a single cut transition at index i will have as a label a vector with all zeros except index i where there will be a one. A sample shot with a gradual transition at index i to $i + N$ will have as label a vector with all zeros, except index positions from i till $i + N$ where will be 1. A label with multiple consecutive 1 s corresponds with a multiple hot vector representation. Still, a gradual transition can be represented with a single 1 just in the middle of the transition. In such cases, this label corresponds with a one-hot representation of the transition. So, the model is intended to support two heads, a first head named onehot, and a second head named manyhot. The overview model is shown in Fig. 4.

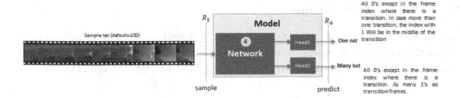

Fig. 4. Model overview with input and labels.

4 Annotations with Pyscene

As part of our framework, we provide the option to annotate all videos in an original dataset with annotations obtained using PySceneDetect [4]. PySceneDetect is a command line application and a python library that allows for shot detection following three methods: "Content Aware Detector", "Adaptive Content Detector" and "Threshold Detector". The overview of PySceneDetect annotations generations is illustrated in Fig. 5.

We allow generating cut annotations following the same schema that annotations generated with the Consolidate option, but in specific annotation files depending on the method. The annotation files generated using PySeceneDetect, are saved with name {video_name}.{method}.txt, where the method is "cd" for Content-Aware, "ad" for Adaptive Content" and "td" for Threshold Detector.

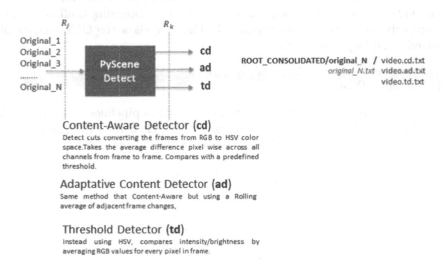

Fig. 5. Anotations using pySceneDetect.

4.1 Manual Annotations

One of the functions we implement in our framework is to manually annotate videos with cut annotations or with gradual annotations, and also remove annotations. The manually annotated shots are added to the *.reanot.txt files.

4.2 Merging Annotations

In addition, annotations generated using PySceneDetector can also be merged on top of the original annotations and integrated in the reannot.txt files. We allow the option to merge file per file or the complete dataset at once. Also, we enable undoing the merge operations. With undo, the *.reannot.txt is returned as a copy of *.orig.txt annotations.

4.3 Inspect Annotations

In any supervised algorithm problem, inspecting the annotations of the datasets is a crucial step in order to ensure that annotations are correctly generated. As part of the framework we develop a tool that allows selecting any of the originals, then listing all videos and their annotations, selecting one of the videos and its annotation file, and for that tuple, videoannots, then displaying data as total shots, total cuts, total graduals, total frames, and the list of shots. Then, we allow opening a video widget and a mosaic widget. In the video widget we can play the video, and in the mosaic widget, we can see a mosaic representation of the video, with the annotations. This allows for inspecting if all transitions are properly annotated or not visually. In Fig. 6 we show the GUI components involved when inspecting dataset annotations.

4.4 Inspect Pipeline

We also develop a tool to inspect each of the three pipelines developed. The tool allows to select a TFRecord once generated, then runs the pipeline that corresponds to the selected TFRecord according to its type.

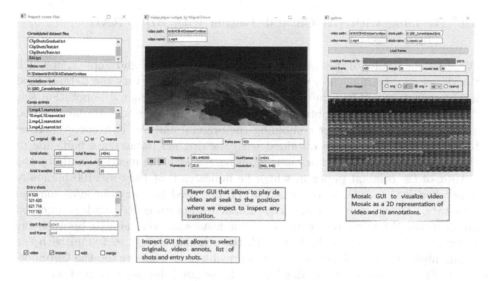

Fig. 6. GUI for inspect annotations

5 Experiments

We want to check if using a set of annotations obtained with PyScene and augmented with the original dataset improves the obtained metrics obtained using only the original dataset. As a base model, we use Transnet. We decided to carry out different types of experiments: a first type where we train and validate with IACC3, and test with a different dataset: RAI, BBC, and ClipShotsTest. In the original Transnet paper, the author uses IACC3 to train and validate and uses RAI to test, which helps us to verify that we reproduce and even improve the results obtained by the latter. Then we move to a second type of experiment, where we train, eval, and test with a split of the same dataset. In all experiments, we use 20 epochs, with batch size 4 for train and evaluation. We employ Adam optimizer with a learning rate 0.001. We employ Transnet with a slight modification with regard to the original paper. In the original paper, the author says to utilize a last dense layer of dimension 2, and softmax output. Instead, we employ a last dense layer of dimension 1, and a sigmoid activation function.

5.1 Training with IACC3 and Testing with BBC, RAI, ClipsShotsTest

We show the results of the first set of experiments in Fig. 7 and Table 1, in which the dataset for training and validation is IACC3, and we vary the dataset used for testing. In graphics, the term "orig" refers to the original annotation set, "cd" to the annotations obtained with PyScene and "reanot" the augmented annotations, obtained after combining the original and the PyScene. When we test with RAI, the results obtained with the original dataset are bad, due to the poor quality of the annotations. This agrees with the mention made by the author of the original paper in which he indicates that they use 100 manually annotated RAI videos. In our case, we do not need to carry out such manual annotation effort since we can verify similar results using RAI with augmented annotations. When we test with BBC, the results are very similar between the three sets of annotations. In fact, with the original, we get 92.9, and with the augmented one, 92.4. The worst results are obtained when we use the ClipShotsTest. Similar values are obtained with the original dataset 64.6 and with the augmented one, 62.2. The rational behind is as follows: when we test with RAI orig we get a bad result, 0.27, however, when we test with RAI generated from pySceneDetect, the result improves to 0.897. When we test with RAI augmented, we get 0.87. In the case of testing with the original ClipShotsTest, we get 0.65. When we use ClipShotsTest from pySceneDetect, we get 0.4387. This is evidence that ClipShotsTest is composed of videos with complex segmentation in shots. That is why PySceneDetect does not perform well and consequently, the result with ClipShotsTest using augmented annotations does not improve either. We conclude that given a poorly annotated dataset, the use of augmented annotations will improve the result as long as the automatically assisted annotation gets a good score as well, as is the case with RAI. With regards ClipShots, If we had used the entire dataset (ClipShotsTrain and ClipShotsGradual) and not only ClipShotsTests, we would very possibly have obtained better results. We leave this verification for future experiments.

Table 1. Best F1 score IACC3 orig with regards annotations

dataset/annotations	orig	cd	reanot
BBC	0.9286	0.8522	0.9241
ClipShotTest	0.6457	0.4387	0.6218
RAI	0.2651	0.8975	0.8665

5.2   Training and Testing with Dataset Splits

We obtain the F1score with each dataset for the second set of experiments, carrying out a training, validation, and test split. Table 2 shows each dataset's train eval test split. Then, Fig. 8 shows the results of each trained model on the corresponding test subset with each of the annotations. The best results obtained are with original BBC, reaching 98.3, followed by augmented RAI, with 97.4, augmented IACC3 with 94.4, and original Clipshots with 88.4.

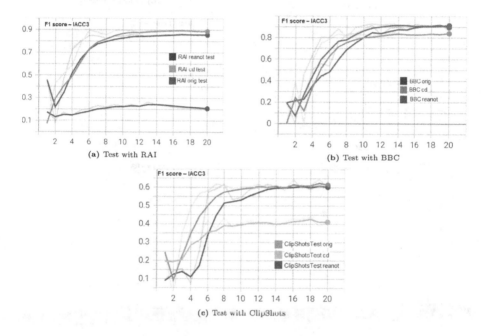

Fig. 7. F1 score training with IACC3 orig

Table 2. Splits for IACC3, RAI, BBC and ClipShots.

split/dataset	IACC3	RAI	BBC	ClipShotsTrain	ClipShotsGradual	ClipShotsTest
train	0.9	0.6	0.6	0.9	0.9	–
eval	0.05	0.2	0.2	0.1	0.1	–
test	0.05	0.2	0.2	–	–	1.0

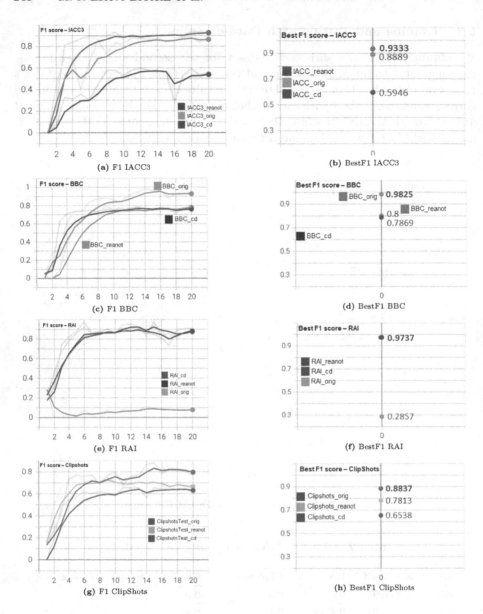

Fig. 8. F1 score training with each dataset

6   Conclusions

Despite being a widely researched task in recent years, video shot detection is still subject to possible improvements, especially in video production for broadcast and streaming. Accurate shot detection is only a first step in video time segmentation, so accuracy in this detection is critical to ensure equally accurate detection at a semantic level in the hierarchy.

Despite the simplicity of the approach, and as far as we know, our contribution presents a first comparative evaluation of shot detection metrics considering the datasets referenced in most studies and the potential improvement that could be obtained using augmented annotations from the programmatic generation carried out with a conservative algorithm such as PySceneDetect. We concluded that it is not necessary to manually annotate a dataset for testing but that we can build on the annotations already made and augment them to evaluate new models without compromising the reliability of the result.

Acknowledgment. We would first like to thanks Telefonica I+D for supporting the Industrial Phd of Miguel Esteve Brotons. We would like to thank 'A way of making Europe" European Regional Development Fund (ERDF) and MCIN/AEI/10.13039/501100011033 for supporting this work under the TED2021-130890B (CHAN-TWIN) research Project funded by MCIN/AEI /10.13039/501100011033 and European Union NextGenerationEU/ PRTR, and AICARE project (grant SPID202200X139779IV0). Also the HORIZON-MSCA-2021-SE-0 action number: 101086387, REMARKABLE, Rural Environmental Monitoring via ultra wide-ARea networKs And distriButed federated Learning. Finally, we also would like to thank Nvidia for their generous hardware donations that made these experiments possible.

References

1. Awad, G., et al.: Trecvid 2017: Evaluating ad-hoc and instance video search, events detection, video captioning and hyperlinking. In: TREC Video Retrieval Evaluation (2017)
2. Baraldi, L., Grana, C., Cucchiara, R.: A deep SIAMESE network for scene detection in broadcast videos. In: Proceedings of the 23rd ACM international conference on Multimedia. ACM (2015). https://doi.org/10.1145/2733373.2806316
3. Baraldi, L., Grana, C., Cucchiara, R.: Shot and scene detection via hierarchical clustering for re-using broadcast video. In: Azzopardi, G., Petkov, N. (eds.) CAIP 2015. LNCS, vol. 9256, pp. 801–811. Springer, Cham (2015). https://doi.org/10.1007/978-3-319-23192-1_67
4. Brandon Castellano. https://scenedetect.com/en/latest/
5. Gygli, M.: Ridiculously fast shot boundary detection with fully convolutional neural networks (2017). https://arxiv.org/abs/1705.08214, https://doi.org/10.48550/ARXIV.1705.08214
6. Hassanien, A., Elgharib, M., Selim, A., Bae, S.H., Hefeeda, M., Matusik, W.: Large-scale, fast and accurate shot boundary detection through spatio-temporal convolutional neural networks (2017). https://arxiv.org/abs/1705.03281, https://doi.org/10.48550/ARXIV.1705.03281
7. Ji, S., Wei, X., Yang, M., Kai, Yu.: 3D convolutional neural networks for human action recognition. IEEE Trans. Pattern Anal. Mach. Intell. **35**(1), 221–231 (2013). https://doi.org/10.1109/TPAMI.2012.59

8. Karpathy, A., Toderici, G., Shetty, S., Leung, T., Sukthankar, R., Fei-Fei, L.: Large-scale video classification with convolutional neural networks. In: 2014 IEEE Conference on Computer Vision and Pattern Recognition, pp. 1725–1732 (2014). https://doi.org/10.1109/CVPR.2014.223

9. Krizhevsky, A., Sutskever, I., Hinton, G.E.: Imagenet classification with deep convolutional neural networks. In: Pereira, F., Burges, C.J., Bottou, L., Weinberger, K.Q. (eds.) Advances in Neural Information Processing Systems, vol. 25. Curran Associates Inc. (2012). https://proceedings.neurips.cc/paper/2012/file/c399862d3b9d6b76c8436e924a68c45b-Paper.pdf

10. Simonyan, K., Zisserman, A.: Two-stream convolutional networks for action recognition in videos (2014). https://arxiv.org/abs/1406.2199, https://doi.org/10.48550/ARXIV.1406.2199

11. Souček, T., Lokoč, J.: Transnet v2: an effective deep network architecture for fast shot transition detection (2020). https://arxiv.org/abs/2008.04838, https://doi.org/10.48550/ARXIV.2008.04838

12. Souček, T., Moravec, J., Lokoč, J.: Transnet: a deep network for fast detection of common shot transitions (2019). https://arxiv.org/abs/1906.03363, https://doi.org/10.48550/ARXIV.1906.03363

13. Tang, S., Feng, L., Kuang, Z., Chen, Y., Zhang, W.: Fast video shot transition localization with deep structured models (2018). https://arxiv.org/abs/1808.04234, https://doi.org/10.48550/ARXIV.1808.04234

14. Tensorflow. TFRecord and tf.train.Example. https://www.tensorflow.org/tutorials/load_data/tfrecord

15. Tong, W., Song, L., Yang, X., Qu, H., Xie, R.: CNN-based shot boundary detection and video annotation. In: 2015 IEEE International Symposium on Broadband Multimedia Systems and Broadcasting, pp. 1–5 (2015). https://doi.org/10.1109/BMSB.2015.7177222

16. Tran, D., Bourdev, L., Fergus, R., Torresani, L., Paluri, M.: Learning spatiotemporal features with 3D convolutional networks (2014). https://arxiv.org/abs/1412.0767, https://doi.org/10.48550/ARXIV.1412.0767

17. Zhou, B., Lapedriza, A., Xiao, J., Torralba, A., Oliva, A.: Learning deep features for scene recognition using places database. In: Ghahramani, Z., Welling, M., Cortes, C., Lawrence, N., Weinberger, K.Q. (eds.) Advances in Neural Information Processing Systems, vol. 27. Curran Associates Inc. (2014). https://proceedings.neurips.cc/paper/2014/file/3fe94a002317b5f9259f82690aeea4cd-Paper.pdf

3D Human Body Models: Parametric and Generative Methods Review

Nahuel Emiliano Garcia-D'Urso[(✉)] [ID], Pablo Ramon Guevara [ID],
Jorge Azorin-Lopez [ID], and Andres Fuster-Guillo

Department of Computing Technology, University of Alicante, Alicante, Spain
{nahuel.garcia,pguevara,jazorin,fuster}@ua.es

Abstract. This paper provides an overview of the current state-of-the-art in the field of 3D human body model estimation, reconstruction, and generation in computer vision. The paper focuses on the most widely used parametric and generative methods and their applications. The paper highlights the use of different input data formats, including 2D images, videos, and 3D scans of the human body in various fields such as medicine, film industry, video game industry, extended reality, and clothing. The field of 3D human body recovery has seen a significant advancement with the development of parametric models. These approaches use a set of parameters to represent body shape and pose and are widely used for reconstructing 3D human body. Some approaches emphasize body deformations, while others concentrate on shape and pose optimization or the separation of body shape into identity-specific and pose-dependent components, among other aspects. The advancements in the field have led to improved accuracy and stability in representing human body shapes and poses. On the other hand, in recent years various generative methods have been developed to generate 3D models of the human body. Variational Autoencoder (VAEs) and Generative Adversarial Networks (GANs) are two commonly used types of neural networks for this purpose. These methods can generate 3D human body models by learning the distribution of the data.

Keywords: Parametric · Generative · 3D Model · Human Body

1 Introduction

Over the past few years, there has been a significant increase in the use of parametric models and generative models in computer vision. These parametric and generative human body models have been used to solve wide range of problems, from image synthesis to human body animation. This review paper aims to provide an overview of the current state-of-art in the field of 3D human body model reconstruction, generation, and acquisition. We will focus on the most widely used methods and their applications, as well as the challenges and future directions of this research area.

I. Rojas et al. (Eds.): IWANN 2023, LNCS 14134, pp. 251–262, 2023.
https://doi.org/10.1007/978-3-031-43085-5_20

The human body is one of the most complex structures to model and animate, parametric models such as [4] and [5] have been proposed to address this challenge. Parametrics are mathematical methods based on a set of parameters to define the body shape and pose, making it possible to generate realistic 3D human body models.

On the other hand, generative models are a class of machine learning methods that are trained to generate new, previously unseen samples that are similar to the ones it was trained on. These methods learn the underlying probability distributions of the data to be able to generate new instances of the data [38, 39, 46, 49, 50]. These methods enable the creation of rapid and accurate 3D human body models including clothing and texture. Additionally, they do not have the issue of estimating pose parameters, which is a challenging task due to the representation of 3D rotations.

A wide variety of input data formats are used by the different methods that address the reconstruction, generation and acquisition of 3D body models. The use of 2D images has become popular for model creation. Some methods use a single image [16, 18, 22, 26, 27] while others use multiple images [24] with different viewpoints. Additionally, videos [28] and 3D human models obtained with RGBD cameras [25] have also been used as inputs. Generative methods are also known to use sometimes as input data descriptive text of the 3D model to be generated [53, 54]. These various input methods have proven to be effective in generating realistic and accurate 3D human body models, and they have been utilized in a wide range of applications.

There are many areas that can greatly benefit from using these models. Some of the most significant applications include:

- Medicine: Human body models are valuable in the study of anatomy [1] and for patient monitoring.
- Film industry: Human body models can be used to capture motion data and render high-quality CGI humans.
- Video game industry: Human body models can be used to create realistic animations and interactions between characters [2].
- Extended reality: Human body models can be used to capture user input in virtual reality as well as rendering realistic characters.
- Clothing: These models can be used for fitting virtual clothes [3] and creating realistic images of clothing products.

Our main goal is to provide a comprehensive overview of the current state-of-the-art in the field of parametric and generative techniques for estimating, predicting or reconstructing 3D models of the human body. In order to achieve this, we have conducted extensive research and analysis, drawing upon various databases including Scopus and Web of Science. We focus on recent developments and advancements in the field, and aim to provide a thorough analysis of the strengths and limitations of different methods. Additionally, we aim to provide a review of the current generative methods, as many other research studies focus mainly on parametric models [6, 7]

This paper is organized as follow: in Sect. 2, we will review the most widely used parametric models and their specifications. We will discuss their strengths, limitations, and the most common usage where they have been employed. In Sect. 3 we will review the most widely used generative models. We will discuss their architecture, training methods, and the most common usage where they have been employed. In Sect. 4, we will discuss the current challenges and future directions of this research area. We will highlight the open problems and the areas where further research is needed.

2 Parametric Models

The use of parametric models to represent and generate 3D human body shapes and poses has garnered increasing attention in recent years. These 3D models of the human body are based on a set of parameters that can be adjusted to change its pose and shape. This allows for not only the reconstruction of a human body but also the creation of new and unique bodies. Parametric models of the human body have been widely used in various applications, such as animation, virtual reality, and garment simulation.

One of the earliest and most referenced works in the field of parametric models for representing and generating 3D human body shapes and poses is [4]. This paper presents a method for creating a parametric model of the human body using a statistical model that represents the variations in body shape and pose. The parametric model can be adjusted to match the pose of the human in a registered 3D scan of the human body by interpolating from a set of example scans with similar joint angles. To model body shape variation across different people, the paper proposes to morph a generic template shape into 250 scans of different humans in the same pose. The variability of human shape is captured by performing principal component analysis (PCA) over the displacements of the template points. However, it's important to note that this method does not model how body shape changes with pose. The model is primarily used for hole-filling of scans, and it served as a foundation for many of the parametric models that have been developed since then.

Another significant parametric model is the SCAPE (Shape Completion and Animation of People) model [5]. SCAPE models body deformations caused by both shape and pose as triangle deformations. The model utilizes a linear regression approach to determine the pose deformation from adjacent joint angles. Although this parametric model aims to represent the muscle deformations caused by body movements, it does not capture the difference between specific muscle activities, such as a flexed bicep and a relaxed one, making it insufficiently expressive for certain applications. Additionally, tissue perturbations caused by motion (e.g., fat wiggling) are not represented in the model. Despite these limitations, SCAPE has the advantage of requiring minimal human intervention, with only a small set of markers required to be placed on the scans as a starting point for registration. This makes the model easily adaptable to other datasets.

An extension of this work is the parametric model of BlendSCAPE [9], which addresses the registration process of the human body by optimizing over both shape and pose simultaneously. This model uses a blending technique, where the deformation of a triangle is represented as a linear combination of several parts that influence the triangle. However, it is important to note that the Blend-SCAPE body model formulation is not compatible with standard graphics packages and may not be suitable for distribution in common graphics applications. This limitation should be considered when assessing the suitability of this model for a specific application [8].

SymmetricSCAPE [10] is a variant of the SCAPE and BlendSCAPE models that leverages the inherent symmetry present in the human body. By incorporating symmetry constraints during the parametric modeling process, SymmetricSCAPE improves the accuracy and stability of the resulting model. Using symmetry constraints, SymmetricSCAPE results in a more robust, and accurate parametric representation of the 3D human body in comparison to traditional models that do not take symmetry into consideration.

The SMPL (Skinned Multi-Person Linear Model) [11] model is a widely-used parametric model for 3D human body shape and pose representation. This model separates the body shape and pose into a set of parameters, similar to the approach used in the SCAPE model. Unlike SCAPE, SMPL uses a vertex-based skinning approach that includes shape and pose corrective blend shapes, which allows for greater flexibility in representing the body shape and pose. Additionally, SMPL has the capability to be integrated with various graphics applications, such as Autodeks Maya, Blender or Unity [35–37].

In recent years, there have been several improvements to the SMPL model, such as SMPL-H [12] which incorporates an articulated hand pose and shape parameters. Further advancements have been made through SMPL-X [13], which includes hand and face pose and shape parameters.

The recent STAR [14] (Sparse Trained Articulated Human Body Regressor) model improve the limitations of SMPL. An issue commonly recognized in linear blends skinning is the reduction of the mesh volume around the joints, STAR addresses this issue by learning corrective blend shapes. Generally speaking STAR learn how to deform the 3D model of the human body in a realistic manner. Additionally, the parametric model used in STAR reduces the number of parameters required, by 20% compared to SMPL.

Recent research has explored various ways of formulating parametric models. One such model is presented in [15] which utilizes a bone-level skinning approach. This method is distinct from traditional techniques in which the scale of bones is established before creating the template, as opposed to after. The process begins by determining the lengths and angles of bones to establish the skeleton, then incorporates identity-specific variations to the surface. These elements are combined through Linear Blend Skinning (LBS) to produce the final animation.

2.1   Parametric Model Estimation

Recent studies have demonstrated the effectiveness of Convolutional Neural Networks (CNNs) in accurately estimating a person's pose [55]. However, with the advancement of parametric models of the human body, researchers have been able to reconstruct both the pose and shape of human bodies from various sources such as images, videos, 3D models, and others. The availability of both synthetic and real datasets of 3D human body models has greatly impacted this field. In this section we will review works addressing the recovery of the 3D model of the human body by estimating parametric model components.

The paper [16] is the first to achieve the goal of reconstructing the 3D human body from a single image. It uses the SMPL parametric model of the human body to estimate the 3D human pose and shape. The approach demonstrates its effectiveness in reconstructing the human body from single images. To estimate the pose, they employ a CNN-based method named DeepCut [17]. The shape of the human body is predicted by minimizing an objective function that measures the difference between the projected 3D model joints and the detected 2D joints.

The paper [18] proposed a 3D deformation model for capturing faces, hands, and bodies of human subjects. The method uses a combination of multiples cameras and depth sensors to accurately capture the subject. The resulting model is a combination of different models. They used SMPL for the body, FaceWarehouse [19] for the face and a rigged hand mesh with 16 joints for the hand. In [20] presents the first method to capture the 3d total motion of a person. This method takes as input a single image or a monocular video and outputs the parameters of the 3D human body model [18].

The approach described in [21] is a combination between a regression deep neural network that predicts the model parameters and an iterative optimization routine to fit the SMPL model to a set of 2D keypoints. Low-resolution images are a major challenge in 3D human shape and pose estimation. This is because the information necessary for accurate 3D human body model reconstruction is often lost or reduced when an image is downsampled. However, recent advances in deep learning techniques have enabled the development of models that can effectively handle low-resolution images. One such work is the paper [22].

The estimation of multiple 3D human body models using 2D images as input is a challenging task in the area of computer vision and graphics. The paper [23] presents a method for addressing this problem through a monocular, one-stage network regression approach. The authors propose a deep neural network architecture that takes a single RGB image as input and outputs the parameters of multiple SMPL models, allowing for the estimation of multiple human bodies in a single image.

Estimating parametric models to dress 3D human bodies or obtaining anthropometric measurements has gained considerable interest in recent years. One such paper is [24] which presents a Multi-Garment Network (MGN) that infers the garments, human shape, and canonical pose from eight images. Another example is [25] proposes a pipeline for fitting a SMPL template to 3D body scans, and then obtaining anthropometric measurements from the fitted template.

The paper [26] presents a method for monocular 3D human mesh recovery that differs from most existing methods because neural networks are trained using synthetic data, instead of real data. The neural network takes as input a single 2D image and outputs the parameters of a SMPL-X [13] model. This approach offers the advantage of being able to generate unlimited amounts of training data and the ability to control various aspects of the synthetic data, such as lighting conditions, poses, and body shapes. The results of the paper demonstrate the effectiveness of the synthetic training approach in improving the accuracy of monocular 3D human mesh recovery, such as.

BodyNet [38] is a direct neural network that predicts volumetric body shape from a single image. It incorporates 3D loss, multi-view re-projection loss and intermediate supervision of 2D pose, 2D body part segmentation and 3D pose to achieve performance improvement. It outperforms recent approaches and also enables volumetric body-part segmentation [16,56].

The work reported in [45] is a learning-based model that infers the 3D shape of people from 1–8 frames of monocular video in less than 10 s with a reconstruction accuracy of 5 mm. The model predicts the parameters of a statistical body model and instance displacements to add clothing and hair to the shape. It predicts using both bottom-up and top-down streams to allow information to flow in both directions and is learned only from synthetic 3D data. The model can reconstruct shapes from a single image with an accuracy of 6mm.

To summarize the information, a Table 1 compiles the input methods, the approach used, the generated output, and the relevant application.

Table 1. Summary of parametric methods, including the inputs used, outputs generated, and real-world applications

Input	Method	Output	Application
2D image	[16]	SMPL	Recovery
2D + 3D images	[18]	Adam	Recovery
2D image	[20]	Adam	Recovery
2D image	[21]	SMPL	Recovery
Low resolution 2D images	[22]	SMPL	Recovery
2D image	[23]	SMPL	Recovery
2D images	[24]	SMPL	Recovery + Clothing
3D scan	[25]	SMPL	Recovery + Clothing
2D image	[26]	SMPL-X	Recovery
2D image	[38]	SMPL	Recovery
2D images	[45]	SMPL	Recovery + Clothing

3   Generative Models

Models capable of generating visual content have experienced a significant surge in popularity.

Autoencoders are a type of neural network that can be used to generate new data by encoding and decoding existing data. The architecture of an autoencoder typically consists of an encoder and a decoder. The encoder maps the input data to a lower-dimensional representation, called the latent code, and the decoder maps the latent code back to the original space.

Generative Adversarial Networks (GANs) consist of two neural networks, a generator and a discriminator, that work together to generate new data. The generator network takes a random noise as input and generates new data, while the discriminator network takes as input generated (fake) data or real data and returns a probability that it is real. The generator and discriminator are trained in an adversarial way, where the generator tries to generate data that can fool the discriminator, and the discriminator tries to correctly classify if that data is real or fake.

Diffusion-based methods use a flow-based model, where the flow is a series of invertible transformations, to generate high-resolution data from low-resolution data. The process involves adding noise to the low-resolution data and then flowing it through the invertible transformations to generate an intermediate high-resolution representation. The noise is then removed from the intermediate representation to obtain the final high-resolution data.

Style transfer is a technique in computer vision and image processing that aims to transfer the artistic style of one image to another image while preserving the content of the target image. The goal is to generate a new image that has the same content as the target image but with the style of the source image. This technique can be used in the context of human body generation by transferring data such as pose, shape or body features from a source image or video to another.

3.1   3D Human Model Generation Using Generative Models

When it comes to generating 3D human models, there are two main approaches. The first approach is to use a general generator and guide it to generate human models. The second approach is to use a generator that has been specifically designed to generate human models from the start.

Human Specific. SiCloPe [39] models clothed human bodies using deep generative models. It can reconstruct a complete and textured 3D model of a person wearing clothes from a single input picture. It uses a silhouette-based representation that combines 2D silhouettes and 3D joints of a body pose to describe the complex shape variations of clothed people. It synthesizes consistent silhouettes and feeds them into a deep visual hull algorithm for 3D shape prediction and uses a conditional generative adversarial network to infer the texture of the subject's back view.

PIFu [40] is a highly effective implicit representation that locally aligns pixels of 2D images with their corresponding 3D object. The method can infer 3D surface and texture from a single image or multiple input images. It can handle intricate shapes and their variations and deformations, and can produce high-resolution surfaces including largely unseen regions such as the back of a person. It extends naturally to arbitrary number of views and is memory efficient, spatially aligned with the input image and can handle arbitrary topology. PIFuHD [41] builds on top of PIFu with an additional module and applies it to the task of human digitalization.

Tex2Shape [42] is a simple method to infer detailed full human body shape from a single photograph. It turns shape regression into an aligned image-to-image translation problem and estimates detailed normal and vector displacement maps from partial texture maps of the visible region. The results feature details even on parts that are occluded in the input image and the model generalizes well to real-world photographs.

HumanMeshNet [43] regresses a template mesh's vertices and receives regularization from 3D skeletal locations in a multi-branch, multi-task framework. It focuses on implicitly learning the mesh representation and is a novel model for 3D human body reconstruction from a monocular image.

DeepHuman [44] is image-guided 3D human reconstruction network that leverages a dense semantic representation and fuses different scales of image features into the 3D space. The visible surface details are refined through a normal refinement network. The method outperforms state-of-the-art approaches in 3D human model estimation from a single image.

HumanGen [47] is a 3D human generation scheme with detailed geometry and 360∘ realistic free-view rendering. The scheme marries the 3D human generation with various priors from the 2D generator and 3D reconstructor of humans through the design of an "anchor image." The authors adopt a pronged design to disentangle the generation of geometry and appearance and use an anchor image to adapt a 3D reconstructor for fine-grained details synthesis and propose a two-stage generation scheme for geometry and appearance.

HumanNeRF [48] describes a neural representation for high-fidelity free-view synthesis of dynamic humans. It uses an aggregated pixel-alignment feature with a pose embedded non-rigid deformation field and raw HumanNeRF can already produce reasonable rendering on sparse video inputs. The approach is improved with in-hour scene-specific fine-tuning and appearance blending. The authors show that this approach is effective in synthesizing photorealistic free-view humans with sparse camera view inputs.

General. The CoCosNet [51] (and CoCosNet v2 [52]) paper introduces full-resolution correspondence learning for cross-domain image translation. It uses a hierarchical strategy that employs the correspondence from coarse to fine levels and utilizes the ConvGRU module to refine the current correspondence. The result is a highly efficient and effective approach for exemplar-based image translation that outperforms state-of-the-art literature (Table 2).

Table 2. Summary of generative methods, including the inputs used, outputs generated, and real-world applications.

Input	Method	Output	Application
2D image	[38]	3D Human Body Model	Generation
2D image	[39]	3D Human Body Model + Clothes	Generation + Clothing
2D image	[40]	3D Human Body Model + Clothes	Generation + Clothing
2D image	[41]	3D Human Body Model + Clothes	Generation + Clothing
2D image	[42]	3D Human Body Model + Clothes	Generation
2D image	[43]	3D Human Body Model	Generation
2D image	[44]	3D Human Body Model	Generation + Clothing
2D image	[47]	3D Human Body Model	Generation
Video	[48]	3D Human Body Model	Generation
2D image	[52]	2D image	Generation

4 Conclusions

In conclusion, this paper provides an overview of the recent developments in the field of 3D human body recovery, focusing on both parametric and generative methods. The recovery of parametric models from 2D images or 3D models is an active area of research, with various techniques proposed [16, 18, 22, 24–28, 38, 45]. The predominant parametric method for 3D human body reconstruction has been SMPL and its variants. These methods are trained on real-world data and provide a high level of accuracy in reconstructing human shapes and poses. The release of different real [25, 29–31] and synthetic [32–34] dataset has greatly contributed to the development of 3D human body reconstruction methods.

Regarding to the generative methods, there are two main approaches. One is to use a general generator and guide it towards generating human models, while the other approach is to use a generator specifically designed to create 3D human body models from the start. In this paper, we focused on the latter approach. Based on our research, we have observed that the most commonly used input form are RGB images.

Acknowledgment. This work was supported by the Spanish State Research Agency (AEI) under grant PID2020-119144RB-I00 funded by MCIN/AEI/10.13039/501100011033, and has also been developed with the support of valgrAI - Valencian Graduate School and Research Network of Artificial Intelligence and the Generalitat Valenciana, co-funded by the European Union.

References

1. Smith, C., Tollemache, N., Covill, D., Johnston, M.: Take away body parts! An investigation into the use of 3D-printed anatomical models in undergraduate anatomy education. Anat. Sci. Educd. **11**, 44–53 (2018). https://anatomypubs. onlinelibrary.wiley.com/doi/abs/10.1002/ase.1718

2. Starke, S., Zhao, Y., Zinno, F., Komura, T.: Neural animation layering for synthesizing martial arts movements. ACM Trans. Graph. **40** (2021). https://doi.org/10. 1145/3450626.3459881

3. Apeagyei, P., et al.: Application of 3D body scanning technology to human measurement for clothing Fit. Int. J. Dig. Content Technol. Appl. **4**, 58–68 (2010)

4. Allen, B., Curless, B., Popović, Z.: The space of human body shapes: reconstruction and parameterization from range scans. In: ACM SIGGRAPH 2003 Papers, SIGGRAPH 2003, pp. 587–594 (2003)

5. Anguelov, D., Srinivasan, P., Koller, D., Thrun, S., Rodgers, J., Davis, J.: SCAPE: shape completion and animation of people. ACM Trans. Graph **24**, 408–416 (2005)

6. Muhammad, Z., Huang, Z., Khan, R.: A review of 3D human body pose estimation and mesh recovery. Digit. Signal Processi. Rev. J. **128**, 103628 (2022)

7. Brunton, A., Salazar, A., Bolkart, T., Wuhrer, S.: Review of statistical shape spaces for 3D data with comparative analysis for human faces. Comput. Vis. Image Underst. **128**, 1–17 (2014)

8. Mahmood, N., Ghorbani, N., Troje, N., Pons-Moll, G., Black, M.: AMASS: archive of motion capture as surface shapes. CoRR. abs/1904.03278 (2019). http://arxiv. org/abs/1904.03278

9. Hirshberg, D.A., Loper, M., Rachlin, E., Black, M.J.: Coregistration: simultaneous alignment and modeling of articulated 3D shape. In: Fitzgibbon, A., Lazebnik, S., Perona, P., Sato, Y., Schmid, C. (eds.) ECCV 2012. LNCS, vol. 7577, pp. 242–255. Springer, Heidelberg (2012). https://doi.org/10.1007/978-3-642-33783-3_18

10. Chen, Y., Song, Z., Xu, W., Martin, R., Cheng, Z.: Parametric 3D modeling of a symmetric human body. Comput. Graphics **81**, 52–60 (2019)

11. Loper, M., Mahmood, N., Romero, J., Pons-Moll, G., Black, M.: SMPL: a skinned multi-person linear model (2015)

12. Romero, J., Tzionas, D., Black, M.: Embodied hands. ACM Tran. Graphics **36**, 1–17 (2017). https://doi.org/10.1145

13. Pavlakos, G., et al.: Expressive body capture: 3D hands, face, and body from a single image. In: Proceedings IEEE Conference on Computer Vision And Pattern Recognition (CVPR) (2019)

14. Osman, A.A.A., Bolkart, T., Black, M.J.: STAR: sparse trained articulated human body regressor. In: Vedaldi, A., Bischof, H., Brox, T., Frahm, J.-M. (eds.) ECCV 2020. LNCS, vol. 12351, pp. 598–613. Springer, Cham (2020). https://doi.org/10. 1007/978-3-030-58539-6_36 https://star.is.tue.mpg.de

15. Wang, H., Güler, R.A., Kokkinos, I., Papandreou, G., Zafeiriou, S.: BLSM: a bone-level skinned model of the human mesh. In: Vedaldi, A., Bischof, H., Brox, T., Frahm, J.-M. (eds.) ECCV 2020. LNCS, vol. 12350, pp. 1–17. Springer, Cham (2020). https://doi.org/10.1007/978-3-030-58558-7_1

16. Bogo, F., Kanazawa, A., Lassner, C., Gehler, P., Romero, J., Black, M.: Keep it SMPL: automatic estimation of 3d human pose and shape from a single image (2016). https://arxiv.org/abs/1607.08128

17. Pishchulin, L., et al.: DeepCut: Joint subset partition and labeling for multi person pose estimation (2015). https://arxiv.org/abs/1511.06645

18. Joo, H., Simon, T., Sheikh, Y.: Total capture: a 3D deformation model for tracking faces, hands, and bodies (2018). https://arxiv.org/abs/1801.01615
19. Cao, C., Weng, Y., Zhou, S., Tong, Y., Zhou, K.: FaceWarehouse: a 3D facial expression database for visual computing. IEEE Trans. Vis. Comput. Graph. **20**, 413–425 (2014)
20. Xiang, D., Joo, H., Sheikh, Y.: Monocular total capture: posing face, body, and hands in the wild (2018). https://arxiv.org/abs/1812.01598
21. Kolotouros, N., Pavlakos, G., Black, M., Daniilidis, K.: Learning to reconstruct 3D human pose and shape via model-fitting in the loop. In: ICCV (2019)
22. Xu, X., Chen, H., Moreno-Noguer, F., Jeni, L., Torre, F.: 3D Human Shape and Pose from a Single Low-Resolution Image with Self-Supervised Learning (2020). https://arxiv.org/abs/2007.13666
23. Sun, Y., Bao, Q., Liu, W., Fu, Y., Black, M., Mei, T.: Monocular, one-stage, regression of multiple 3D people (2020). https://arxiv.org/abs/2008.12272
24. Bhatnagar, B., Tiwari, G., Theobalt, C., Pons-Moll, G.: Multi-garment net: learning to dress 3D people from images (2019). https://arxiv.org/abs/1908.06903
25. Yan, S., Wirta, J., Kämäräinen, J.: Anthropometric clothing measurements from 3D body scans (2019). https://arxiv.org/abs/1911.00694
26. Sun, Y., et al.: Synthetic training for monocular human mesh recovery (2020), https://arxiv.org/abs/2010.14036
27. Pavlakos, G., Zhu, L., Zhou, X., Daniilidis, K.: Learning to estimate 3D human pose and shape from a single color image (2018). https://arxiv.org/abs/1805.04092
28. Kanazawa, A., Zhang, J., Felsen, P., Malik, J.: Learning 3D Human Dynamics from Video (2019)
29. Ionescu, C., Papava, D., Olaru, V., Sminchisescu, C.: Human3.6M: large scale datasets and predictive methods for 3D human sensing in natural environments. IEEE Trans. Pattern Anal. Mach. Intell. **36**, 1325–1339 (2014)
30. Sigal, L., Balan, A., Black, M.: HumanEva: synchronized video and motion capture dataset and baseline algorithm for evaluation of articulated human motion. Int. J. Comput. Vision **87**, 4–27 (2010)
31. Mahmood, N., Ghorbani, N., Troje, N., Pons-Moll, G., Black, M.: AMASS: archive of motion capture as surface shapes (2019). https://arxiv.org/abs/1904.03278
32. Patel, P., Huang, C., Tesch, J., Hoffmann, D., Tripathi, S., Black, M.: AGORA: avatars in geography optimized for regression analysis (2021). https://arxiv.org/abs/2104.14643
33. Varol, G., et al.: Learning from synthetic humans. In: 2017 IEEE Conference On Computer Vision And Pattern Recognition (CVPR) (2017)
34. Yu, T., Zheng, Z., Guo, K., Liu, P., Dai, Q., Liu, Y.: Function4D: real-time human volumetric capture from very sparse consumer RGBD sensors. In: IEEE Conference On Computer Vision And Pattern Recognition (CVPR2021) (2021)
35. Autodesk, INC. Maya (2019). https://autodesk.com/maya
36. Community, B.: Blender - a 3D Modelling and Rendering Package. Blender Foundation (2018). http://www.blender.org
37. Haas, J.: A History of the Unity Game Engine. Worcester Polytechnic Institute (2014)
38. Varol, G., et al.: BodyNet: volumetric inference of 3D human body shapes (2018). https://arxiv.org/abs/1804.04875
39. Natsume, R., et al.: SiCloPe: silhouette-based clothed people (2019). https://arxiv.org/abs/1901.00049

40. Saito, S., Huang, Z., Natsume, R., Morishima, S., Kanazawa, A., Li, H.: PIFu: pixel-aligned implicit function for high-resolution clothed human digitization (2019). https://arxiv.org/abs/1905.05172

41. Saito, S., Simon, T., Saragih, J., Joo, H.: PIFuHD: multi-level pixel-aligned implicit function for high-resolution 3D human digitization (2020). https://arxiv.org/abs/2004.00452

42. Alldieck, T., Pons-Moll, G., Theobalt, C., Magnor, M.: Tex2Shape: detailed full human body geometry from a single image (2019). https://arxiv.org/abs/1904.08645

43. Moon, G., Lee, K.: I2L-MeshNet: image-to-lixel prediction network for accurate 3D human pose and mesh estimation from a single RGB image (2020). https://arxiv.org/abs/2008.03713

44. Zheng, Z., Yu, T., Wei, Y., Dai, Q., Liu, Y.: DeepHuman: 3D human reconstruction from a single image (2019). https://arxiv.org/abs/1903.06473

45. Alldieck, T., Magnor, M., Bhatnagar, B., Theobalt, C., Pons-Moll, G.: Learning to reconstruct people in clothing from a single RGB camera (2019). https://arxiv.org/abs/1903.05885

46. Sarkar, K., Liu, L., Golyanik, V., Theobalt, C.: HumanGAN: a generative model of humans images (2021). https://arxiv.org/abs/2103.06902

47. Jiang, S., Jiang, H., Wang, Z., Luo, H., Chen, W., Xu, L.: HumanGen: generating human radiance fields with explicit priors (2022). https://arxiv.org/abs/2212.05321

48. Weng, C., Curless, B., Srinivasan, P., Barron, J., Kemelmacher-Shlizerman, I.: HumanNeRF: free-viewpoint rendering of moving people from monocular video (2022). https://arxiv.org/abs/2201.04127

49. Yang, C., et al.: BodyGAN: general-purpose controllable neural human body generation. In: 2022 IEEE/CVF Conference On Computer Vision And Pattern Recognition (CVPR), pp. 7723–7732 (2022)

50. Isola, P., Zhu, J., Zhou, T., Efros, A.: Image-to-image translation with conditional adversarial networks (2016). https://arxiv.org/abs/1611.07004

51. Zhang, P., Zhang, B., Chen, D., Yuan, L., Wen, F.: Cross-domain correspondence learning for exemplar-based image translation (2020). https://arxiv.org/abs/2004.05571

52. Zhou, X., et al.: CoCosNet v2: full-resolution correspondence learning for image translation (2020). https://arxiv.org/abs/2012.02047

53. Poole, B., Jain, A., Barron, J., Mildenhall, B.: DreamFusion: text-to-3D using 2D diffusion (2022)

54. Lin, C., et al.: Magic3D: high-resolution text-to-3D content creation. ArXiv Preprint ArXiv:2211.10440 (2022)

55. Tian, Y., Zhang, H., Liu, Y., Wang, L.: Recovering 3D human mesh from monocular images: a survey (2022). https://arxiv.org/abs/2203.01923

56. Kanazawa, A., Black, M., Jacobs, D., Malik, J.: End-to-end recovery of human shape and pose (2017). https://arxiv.org/abs/1712.06584

Deep Learning Recommendation System for Stock Market Investments

Michal Parzyszek[1] and Stanislaw Osowski[1,2(✉)]

[1] Faculty of Electrical Engineering, Warsaw University of Technology, Warsaw, Poland
stanislaw.osowski@pw.edu.pl
[2] Electronic Faculty, Military University of Technology, Warsaw, Poland

Abstract. The paper proposes and compares two models for creating a recommendation system in the stock market, based on convolutional neural networks (CNN). The first model encodes the values of the time series of the stock exchange quotations into multiple technical indicators incorporating them in the form of an image. These indicators are defined on the closing prices of the stock quotes from the previous day The second model introduces some modifications of the previous approach by changing the definitions of the technical indicators. The numerical experiments have shown its improved performance. Both models are generated from the one-dimensional stock market data and saved as images. The CNN neural network uses these images in the training and testing phases. The numerical experiments aimed at maximizing profit from the investments have been performed on the stock data of the six largest companies listed on the Warsaw Stock Exchange. The recommendations for companies were classified in the form of three classes (Buy, Sell, Hold). The numerical results for the proposed methods are presented and compared with other investment methods typically used in the stock market.

Keywords: data analysis · artificial intelligence · CNN · time series analysis · stock market prediction

1 Introduction

Stock market forecasting is still a challenging problem for stock analysts. Nowadays machine learning models applying deep learning techniques can effectively predict the time series, including forecasting stock index prices [1, 2]. The highest efficiency is due to solutions applying the recurrent neural networks (RNN) organized in the form of long short-term memory (LSTM) [3–5] or convolutional neural networks (CNN) [6].

In this study, we propose an approach that converts 1-D financial time series into 2-D images, which are applied as the input attributes to the deep convolutional neural network (CNN) for the trading system. The CNN architectures are regarded now as the most efficient image processing tools able to solve the classification or regression problem with the highest accuracy [7, 9]. The important property of such a solution is an association in one common structure of two tasks: generation and selection of the diagnostic features and undertaking the final decision of classification or regression.

I. Rojas et al. (Eds.): IWANN 2023, LNCS 14134, pp. 263–275, 2023.
https://doi.org/10.1007/978-3-031-43085-5_21

The first step in our system is to convert the time series into a 2-D image. To develop such a representation, 16 different technical indicator variables with various parameter settings are defined. They are presented in the form of the image corresponding to the actual (real) values of the days' time series. The x-axis of the image represents the data of 16 days for 16 technical indicators, represented in the y-axis. As a result, 16 × 16 pixels sized images are created and served as input to the deep convolutional neural network. The CNN is learned on the training data and the trained network is verified on data not used in the learning stage.

The proposed approach can be divided into a few main steps: dataset extraction and transformation to the technical indicators, labeling data, creation of the image based on the technical parameter values, CNN analysis, and financial evaluation phase. The aim is to determine the best network model to indicate points in the time series to buy, sell, or hold shares considering the trend of the company's stock price. The numerical experiments aimed at maximizing profit from the investment have been performed on the stock data of the six largest companies listed on the Warsaw Stock Exchange. The company's data were classified into three classes (Buy, Sell, Hold). The numerical results for the above methods are presented and compared with several investment methods popularly used in the stock market.

2 Database

The numerical experiments regarding the recommendation for stock market investments are performed for 6 Polish leading companies. They include PKO BP Bank (PKO bank Polski), CD Project S.A. (CDR), LW Lubelski Wegiel Bogdanka S.A. (LWB), KGHM Corporate (KGH), Orange Polska (OPL) and LPP S.A. (LPP). All companies are among the largest companies in the Polish market and are members of the WIG20 index. The stock market data considered in the experiments were collected from January 1, 2010, to November 12, 2019. The above-mentioned companies were chosen because they are representatives of different trends in the market in the considered period. For example, CDR represented a strong rising uptrend, LWB – a strong downtrend, and the other four – a horizontal trend.

Figure 1 presents the changes in daily prices of these companies within the considered time. Three colors are used to show the division of data into a learning set (blue), validation (green), and only testing (red). The small uncolored part of data at the beginning of the set (until the dotted line) has been used only for the creation and validation of the technical indicators of the stock market. The selected companies show many different aspects of the trend changes and may be treated as representative of other companies in the stock.

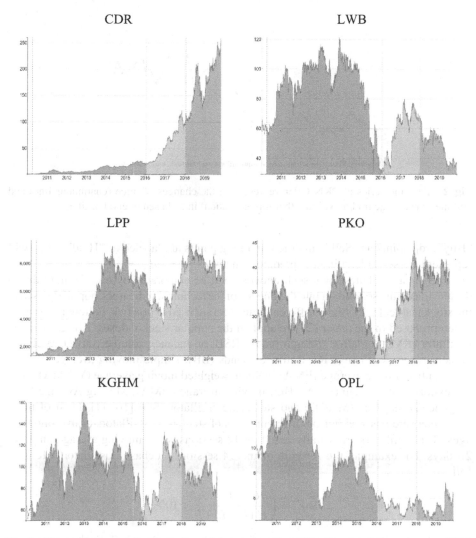

Fig. 1. The graphs presenting the prices of the companies within the period from the year 2010 to 2019. (Color figure online)

3   Technical Indicators of the Stock Market

Stock market prices are characterized by abrupt changes in time. This can be seen well in the charts shown in Fig. 1. The price changes are well associated with the changes in trading volume on the market. Such association for PKN Orlen is presented in Fig. 2

When developing the automatic system for recommending the decision on the stock market (Buy, Sell, or Hold) the important point is to prepare a proper form of input data. This was done manually through the analysis of stock market behavior in the past. The extracted time series were manually marked for each day based on the daily close prices, considering the top and bottom points in a sliding window. Bottom points are labeled as

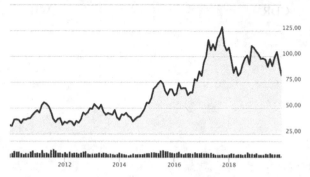

Fig. 2. The time series of PKN Orlen representing the changes of price (continuous line) and volume of exchange trading volume (bar representation) in a chosen interval of time

"Buy", top points as "Sell", and the remaining points are labeled as "Hold". They will represent classes in learning the prediction network.

The input attributes for a such network will be elaborated by transforming the 1-dimensional time series data into the images of the technical indicators, popularly used in the stock market [1, 2]. Many different indicators are used in practice. Following the approach proposed in the paper [2], we applied the popular stock indexes: Klinger volume oscillator (KVO), relative strength index (RSI), on balance volume (OBV), Williams percent range (Williams%R), Bollinger bands (BB), simple moving average (SMA), exponential moving average (EMA), volume weighted moving average (VWMA), double exponential average (DEA), Hull moving average (HMA), Moving Average Convergence/Divergence (MACD) and stochastic oscillator (SO) [10, 11]. Most of these parameters are based either on different types of stochastic oscillators or moving averages. The oscillators are usually based on 14 sessions while moving average on 12 or 26 days. For example, the formula of the 14 sessions stochastic oscillator %K is as follows

$$\%K = \left(\frac{C - L14}{H14 - L14} \right) \times 100 \tag{1}$$

where %K represents the current value of the stochastic indicator, C – is the most recent closing price, L14 – is the lowest price traded of the 14 previous trading sessions, and H14 – the highest price traded in the same 14-day period.

Williams's percent range is another indicator of the oscillator type (similar to %K) that compares a stock's closing price to the high-low range over a specific period (typically 14 days). It is defined as follows

$$\%R = \left(\frac{H14 - C}{H14 - L14} \right) \times 100 \tag{2}$$

where the parameters in the expression have the same meaning as in the definition of the previous indicator.

A simple moving average (SMA) indicator is an arithmetic moving average calculated by adding recent prices and then dividing that figure by the number of periods in the calculation average. The exponential moving average (EMA) is a weighted moving average that gives more weighting or importance to recent price data. The typical formula for the day t and n representing the number of days in EMA is following

$$EMA(t) = price(t) \times \frac{2}{n+1} + EMA(t-1) \times \left(1 - \frac{2}{n+1}\right) \tag{3}$$

Moving average convergence/divergence is calculated by subtracting the 26-period exponential moving average (EMA) from the 12-period EMA.

The relative Strength Index is a momentum indicator identifying the situation when a stock is overbought or oversold. It is defined by the formula

$$RSI = 100 - \frac{100}{1 + U/D} \tag{4}$$

where U represents the average number of upward-moving days over 14 trading days and D is the average number of downward-moving days. RSI above 70 indicates an overbought situation, and an RSI below 30 is oversold. For investors, it reveals the higher than usual probability that the stock is overbought and is due for a downturn.

On-balance volume OBV is a technical indicator of momentum, using volume changes to make price predictions. For the point t, it is simply defined by the formula

$$OBV(t) = OBV(t-1) + \begin{cases} volume & if\ close(t) > close(t-1) \\ 0 & if\ close(t) = close(t-1) \\ -volume & if\ close(t) < close(t-1) \end{cases} \tag{5}$$

A Bollinger Band is an indicator defined by a set of trendlines, plotted two standard deviations (positively and negatively) away from a simple moving average of a security's price defined as (high + low + close)/3. It is composed of three lines: a simple moving average (middle band), and an upper and lower band. The upper and lower bands are typically 2 standard deviations ± from a 20-day simple moving average (center line).

4 Deep Models for Stock Market Prediction

The proposed recommendation system for the stock market is based on the application of convolutional neural networks (CNN) and deep learning [7, 8]. For each day the set of technical indicators described in the previous section is calculated based on the past days. The image will be created by their values for 16 succeeding days. Similarly, to the proposition presented in [2] the following technical indicators are used in this step.

1. Klinger Volume Oscillator based on the 34- and 55-period EMAs of volume force.
2. The relative Strength Index measures the speed and magnitude of a security's recent price changes to evaluate overvalued or undervalued conditions in the price of the security.
3. On Balance Volume as an indicator using volume flow to predict changes in stock price.
4. William's Percent Range %R.

5. Bollinger Bands -returning 2 parameters.
6. Simple Moving Average of 45 days.
7. Exponential Moving Average of 26 days.
8. Exponential Moving Average of 12 days.
9. Volume Weighted Moving Average for 5 days.
10. Double Exponential Moving Average for 6 days.
11. Hull Moving Average for 5 days.
12. Moving Average Convergence/Divergence (12,26,9) returning 2 parameters: the MACD representing the difference of EMAs with characteristics for 12 and 26 days and the second exponential average of the MACD series for 9 days.
13. The stochastic Oscillator returns 2 values: fast stochastic indicator %K and slow stochastic indicator %D representing a 3-period moving average of %K.

The total number of indicators defined in this way is 16. Gathering them for 16 succeeding days we get the 16x16 image representing the day. Each day is associated with the proper recommendation of the class (Buy, Sell, or Hold).

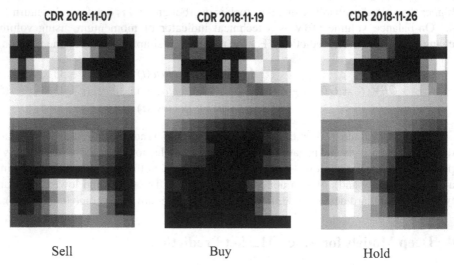

Fig. 3. The typical images representing three classes (Sell, Buy, and Hold) for the company CDR and three days 2018-11-07, 2018-11-19, and 2018-11-26 for a standard definition of the technical indicators (the first system).

By repeating this procedure for all days under consideration we can create the input database used in learning the first proposed recommendation system.

Figure 3 presents the typical images representing these three classes for the company CDR. The second recommendation system proposed by us presents some modifications in the calculation of the moving averages. In the first system, the moving average indicators have been calculated for all 16 days forming the image separately with the same periods.

In the second modified system, the averages are calculated only for the actual day, however, with different periods. The starting length was 20 days, and the next 19, 18 up to 5 (16 values). The following changes of parameters (from 6 to 11) in the previously defined set have been defined in this way.

6. Moving Average of open price 20–5 days.
7. Moving Average of close price 20–5 days.
8. Exponential Moving Average 20–5 days.
9. Volume Weighted Moving Average 20–5 days.
10. Double Exponential Moving Average 20–5 days.
11. Hull Moving Average 20–5 days.

The other parameters were left unchanged. Figure 4 presents the images representing the same three classes for the company CDR after such modifications.

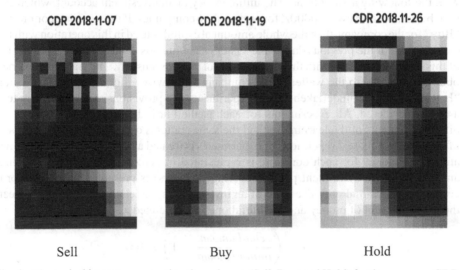

Fig. 4. The typical images representing three classes (Sell, Buy, and Hold) for the company CDR and three days 2018-11-07, 2018-11-19, and 2018-11-26 a modified way of definition of the technical indicators (second system).

The significant differences in images representing the same days are visible. This is especially well seen in the case of the Sell and Buy classes.

5   Results of Numerical Experiments

The images created by the technical indicators described in the previous section create the input data for the CNN network. The CNN network was built from scratch using a Python environment. The final structure of CNN applied in experiments is as follows.

'data' Image Input 16 × 16.
'conv1' Convolution 52 14 × 14, filter 3 × 3, stride 1 × 1 and padding [0 0].
'relu' ReLU.
'conv2' Convolution 52 12 × 12, filter 3 × 3, stride 1 × 1 and padding [0 0].
'relu' ReLU.
'pool2' MaxPooling 52 6 × 6 filter 2 × 2 stride 2 × 2.
'drop' Dropout 52 6 × 6.
'flatten' Fully connected layer 1872.
'drop' Dropout 50%
'flatten' Fully connected layer 156.
'drop' Dropout 50%
'output' classification 3 classes.

The loss function for softmax classifier was defined as cross-entropy and the applied learning algorithm was ADAM [9]. The numerical experiments have been performed with the following assumptions. The initial money in an investment account, which can then be used to invest was 10000zl for each of the 6 companies. If the system recommends "Buy" for the company, then the whole amount of capital is used in this operation with the closing price of the previous day. If the system recommends "Sell" then all stock shares of this company are sold with the closing price of the previous day. The amount of money obtained is added to the wallet of the company. Otherwise, after the recommendation "Hold" no action is undertaken. In the experiments, the provisions, and transaction fees are not considered. All experiments for each method have been repeated 3 times with different contents of the learning set and the same contents of testing data. The average of the results of these 3 repetitions of experiments is treated as the final score. The initial amount of money for each company is represented in percentage as 100%. The final amount (after the investment period covering 1.5 years) is presented in relative form concerning the initial one. The yield of an investment is the relative difference between the actual amount of money and the initial one and is defined as follows

$$yield = \left(\frac{actual\ amount}{initial\ amount} - 1 \right) \times 100 \tag{6}$$

The positive value of yield means profit, and the negative one – the loss.

Table 1. The results of the first system for testing data and all companies obtained in three repeated runs (all values expressed in %)

Run	CDR	LWB	PKO	KGH	LPP	OPL	Mean ±std
1	286.53	52.63	99.84	92.03	91.62	73.95	116.10 ±85.19
2	179.15	69.24	98.60	81.84	94.07	84.65	101.26 ±39.51
3	245.60	50.94	83.57	72.78	88.88	85.14	104.48 ±70.49
Mean	231.09	57.60	94.00	82.22	91.52	81.25	106.28 ±62.48
Average yield	131.09	−42.40	−4.00	−17.78	−8.48	−18.75	6.61 ±62.41

Table 1 presents the results of the first system in the form of the final amount of money for each company (percentage value) obtained in 3 repetitions of the learning/testing procedure. In each run different set of data was used in learning and testing phases. The last row shows the yield. The numerical results, presented in percentage, correspond to only testing data (different in each run) not taking part in the learning and validation phases.

Table 2 presents the corresponding results of the modified (second) system obtained for the same testing data, also in 3 runs. The total average value of yield for all companies is positive in both solustions of the predicting system. In the first case the global profit is equal to 6.61%. However, the positive value was obtained thanks to the very good results of one company (CDR), while the rest companies observed losses.The modified system has generated total profit equal to 10.24%. It is much higher than in the first system. Moreover, for each company (except CDR) the system has obtained significant improvement of results. It means more balanced solution of the prediction system.

Table 2. The results of the second (modified) system for testing data and all companies obtained in three repeated runs (all values expressed in %)

Run	CDR	LWB	PKO	KGH	LPP	OPL	Mean ±std
1	225.92	77.08	91.90	96.54	80.91	112.00	114.06 ±56.18
2	176.20	77.13	98.55	80.96	108.70	84.19	104.29 ±37.17
3	211.82	100.79	99.00	68.75	95.31	99.33	112.50 ±50.13
Mean	204.64	85.00	96.48	82.08	94.76	98.50	110.24 ±46.71
Average yield	104.64	−15.00	−3.52	−17.92	−5.24	−1.50	10.24 ±46.71

The interesting is also the plot of actions recommended in the process of investments. Figure 5 presents such plots for OPL company at the application of the first and second systems. The blue curve presents the actual prices of shares, the red point – is the system recommendation to sell, and the green – is the recommendation to buy. In most cases, they are in good agreement with the trend of prices.

However, some recommendations were not quite optimal. In the first system we can recognize 17 sell recommendations and 8 buy recommendations representing such cases. The corresponding numbers in the modified system are 12 (sell) and 6 (buy).

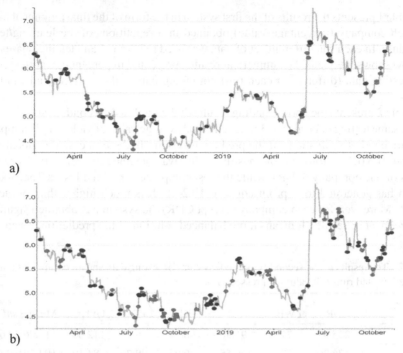

a)

b)

Fig. 5. The plots of recommendations taken in the process of trading of OPL company by the first (a) and second (b) systems. The blue curve presents the actual prices of shares, the red point recommendation to sell, and the green – recommendation to buy. (Color figure online)

6 Comparison to Other Recommendation Methods Used in the Stock Market

To assess the quality of our results, we have compared them with the other methods of recommendation, which are used in the stock market. The most conservative is the "Buy and Hold" approach [10, 11]. It is a passive investment strategy in which an investor buys stocks and holds them for a long period regardless of fluctuations in the market. An investor has no concern for short-term price movements and values of technical indicators. The general opinion is that this method is a good choice for individuals seeking healthy long-term returns.

The other method, sometimes used by investors, is the so-called recurrent plot. A recurrence plot is a visualization (image) of a square matrix X, in which the matrix elements correspond to those times at which a state of a dynamical system recurs (columns and rows correspond then to a certain pair of times). Each element $X(i,j)$ of this matrix is defined in such a way, that $x(i)$ corresponds to the particular point of the time series and $x(j)$ to the point, which is sufficiently close to the value of $x(i)$. As a result, the recurrent plot fulfills the same role as the images created by our methodology based on technical indicators.

The third method used in this comparison is a direct application of the moving average convergence/divergence (MACD). MACD is calculated here by subtracting the long-term EMA (26 periods) from the short-term EMA (12 periods). If the MACD is above the signal line, the histogram will be above the MACD's baseline or zero line. If the MACD is below its signal line, the histogram will be below the MACD's baseline. In this way, it is possible to identify when bullish or bearish momentum is high and possibly overbought/oversold.

Table 3. The comparative results of investments based on different recommendations. The best results for the company are denoted in bold.

Method	CDE	LWB	PKO	KGH	LPP	OPL	Mean yield
Buy and hold	218.25	53.99	88.55	85.22	87.67	**100.70**	5.74
Recurrence plot	139.98	81.74	98.85	86.20	**102.02**	76.87	−2.39
Random	143.55	63.30	**98.68**	86.26	92.46	94.66	−3.52
MACD	114.98	59.44	71.78	**116.34**	92.21	91.17	−9.01
Proposed system 1	**231.09**	57.60	94.00	82.22	91.52	81.25	6.28
Proposed system 2	204.65	**85.00**	96.48	82.08	94.76	98.50	**10.24**

The next comparative method is the application of throwing a dice (random choice). The probability of taking particular action corresponds now to the percentage of the classes (Buy, Sell, Hold) in the training data.

The comparative analyses have been performed for the same learning and testing data as used in the previous experiments. Table 3 presents the average relative values (expressed in percentage) of the final prices of the companies, as well as the average yield for all compared methods.

Figure 6 depicts the results of yield for all compared methods in a graphical form. It is seen that both systems proposed by us have generated the highest profit. The modified way of defining the technical indicators was unbeatable. From the rest methods only the most conservative "Buy and Hold" has brought the total positive yield. The rest of the approaches resulted in a loss (negative values of yield).

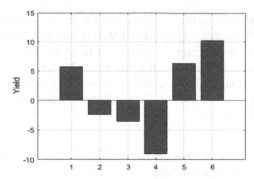

Fig. 6. The average values of yields for all six companies generated in the trading process by different methods of system recommendation. The succeeding numbers in horizontal axis represent the following methods of recommendation: 1 - Buy and Hold, 2 – recurrence plot, 3- random, 4 - MACD, 5 - the first proposed system based on technical indicators, and 6 - the second system applying the modified way of defining technical indicators.

7 Conclusions

The paper has proposed and investigated two automatic recommendation systems for stock market investment. They apply the selected sets of technical indicators of the stock organized in the form of an image. These images are delivered as input attributes to the CNN network, responsible for generating three classes of recommendation: Buy, Sell, and Hold. The results of numerical experiments have shown the advantage of such an approach over the typical methods used by investors in the stock market.

Although the proposed approach is promising, more emphasis should be paid to some other aspects of system creation. First, more indicators may be used in creating the optimal set of technical parameters converted to the image. Moreover, more attention should be paid to selecting the proper image size applied to CNN as well as the optimized structure of CNN architecture. The interesting direction is to use an ensemble of different types of CNN networks and integrate their predictions into a common final decision. In future work more stock companies will be used to create more data for the deep learning models.

References

1. Li, X., Kang, Y., Li, F.: Forecasting with time series imaging. arXiv:1904.08064 (2020). doi.org/10.48550/arXiv:1904.08064
2. Sezer, O., Ozbayoglu, M.: Algorithmic financial trading with deep convolutional neural networks: time series to image conversion approach. Appl. Soft Comput. **70**, 525–538 (2018). https://doi.org/10.1016/j.asoc.2018.04.024
3. Fischer, T., Krauss, C.: Deep learning with long short-term memory networks for financial market predictions. Eur. J. Oper. Res. **259**(2), 689–702 (2017). https://doi.org/10.1016/j.ejor.2017.11.054
4. Moghar, A., Hamiche, M.: Stock market prediction using LSTM recurrent neural network. Procedia Comput. Sci. **170**, 1168–1173 (2020). https://doi.org/10.1016/j.procs.2020.03.049

5. Osowski, S.: Prediction of air pollution using LSTM. In: International Work-Conference on Artificial Neural Networks, Madera (2021). https://doi.org/10.1007/978-3-030-85099-917
6. Chen, J.F., Chen, W.L., Huang, C.P., Huang, S.H., Chen, A.P.: Financial time-series data analysis using deep convolutional neural networks. In: 7th International Conference on Cloud Computing and Big Data (CCBD), IEEE (2016). https://doi.org/10.1109/CCBD.2016.027
7. Goodfellow, I., Bengio, Y., Courville, A.: Deep Learning. MIT Press (2016)
8. Canziani, A., Paszke, A., Culurciello, E.: An analysis of deep neural network models for practical applications. arXiv:1605.07678 (2016). https://doi.org/10.48550/arXiv.1605.07678
9. Kingma, D.P., Ba, J.L.: Adam, a method for stochastic optimization, arXiv:1412.6980v0 (2014). https://doi.org/10.48550/arXiv.1412.6980
10. B. Shannon, Technical analysis using multiple timeframes, LifeVest, Colorado, USA (2008)
11. Murphy, J.J.: Technical analysis of the financial markets: a comprehensive guide to trading methods and applications, Penguin Group (USA) Inc. (1999)

Minimal Optimal Region Generation for Enhanced Object Detection in Aerial Images Using Super-Resolution and Convolutional Neural Networks

Iván García-Aguilar[1]($\boxtimes$) (iD), Lipika Deka[2](iD), Rafael Marcos Luque-Baena[1](iD), Enrique Domínguez[1](iD), and Ezequiel López-Rubio[1](iD)

[1] Department of Computer Languages and Computer Science, University of Málaga, Bulevar Louis Pasteur, 35, 29071 Málaga, Spain
ivangarcia@lcc.uma.es
[2] The De Montfort University Interdisciplinary Group in Intelligent Transport Systems (DIGITS). De Montfort University, The Gateway, Leicester LE1 9BH, UK

Abstract. In recent years, object detection has experienced impressive progress in several fields. However, identifying objects in aerial images remains a complex undertaking due to specific challenges, including the presence of small objects or tightly clustered objects. This paper proposes a novel methodology for enhancing object detection in aerial imagery by combining super-resolution and convolutional neural networks (CNNs). We begin by binarizing the grey zone of the image to detect roads and other regions of interest using the You Only Look Once (YOLO) model. Next, we generate minimal optimal regions. For every one of them, we apply Super-resolution (SR) to improve the number of pixels, generating a new image on which to re-infer. We then apply a CNN to these regions to detect objects more accurately. Our results show that the proposed methodology increases mean average precision in the Unmanned Aerial Vehicle Benchmark Object Detection and Tracking Dataset (UAVDT).

Keywords: Object detection · Small scale · Super-resolution · Convolutional neural networks

1 Introduction

Currently, detecting small objects in images is challenging in computer vision, particularly when using convolutional neural networks (CNNs). While CNNs have achieved state-of-the-art results in various object detection tasks, their performance is often compromised when detecting small objects due to their limited receptive field and feature resolution. This problem is exacerbated by small objects often having low contrast and may be occluded or partially visible, making their detection even more difficult.

Feature Pyramid Networks (FPN) have recently become popular in state-of-the-art detectors due to their effectiveness in detecting objects of different

I. Rojas et al. (Eds.): IWANN 2023, LNCS 14134, pp. 276–287, 2023.
https://doi.org/10.1007/978-3-031-43085-5_22

scales. However, detecting small objects with low resolution and dense distribution in complex scenes still presents a challenge. Kyungseo Min et al. propose a new feature pyramid architecture called Attentional Feature Pyramid Network (AFPN) [1] to improve small object detection. It consists of three components: Dynamic Texture Attention, Foreground-Aware Co-Attention, and Detail Context Attention. Another popular approach is to use attention mechanisms to focus on informative regions of the input image. This can improve the detection of small objects by allowing the network to focus selectively on regions likely to contain small objects. Lian et al. propose a small object detection method in traffic scenes based on attention feature fusion [2]. Wang et al. [3] propose a method to improve the performance of small object detection in UAV aerial images. The proposed method focuses on improving the data and network structure aspects. In addition to these approaches, other recent works have proposed various other techniques for small object detection, including context-aware methods [4,5], and multi-scale detection strategies [6,7].

Concerning Super-Resolution, recent advances in computer vision have led to the development of sophisticated algorithms and models. There are several techniques for achieving super-resolution in images, and some of the most popular methods include FSRCNN [8], or VDSR [9], among others. The Fast Super-Resolution Convolutional Neural Network (FSRCNN) is a deep learning model that uses a lightweight architecture to achieve fast and accurate super-resolution. The Very Deep Super-Resolution (VDSR) model is a deep convolutional neural network that was proposed for single-image super-resolution.

Generative Adversarial Networks (GANs) have emerged as a powerful tool for addressing image super-resolution problems. Wang et al. propose the Super-Resolution Generative Adversarial Network (SRGAN) [10]. It is based on a Residual-in-Residual Dense Block (RRDB) without batch normalization as the basic building unit. This method adopts the relativistic GAN approach for the discriminator to predict relative realness and improves the perceptual loss by utilizing pre-activation features to provide more substantial supervision for brightness consistency and texture recovery. They extend the powerful ESRGAN to a practical restoration application named Real-ESRGAN [11]. A U-Net discriminator with spectral normalization is employed, which increases the discriminator's capability and stabilizes the training dynamics.

In this work, we propose an efficient methodology to improve the detection of small objects without retraining. Our methodology is based on combining the YOLOv7 object detection model [12] with the Real-ESRGAN super-resolution model [11]. The proposed approach enables the detection of small objects with improved accuracy by first applying super-resolution to the input image and then using the enhanced image for object detection. A set of experiments have been performed using the UAVDT Dataset [13].

The paper is structured as follows. Section 2 presents the methodology of the proposal. The results are detailed in Sect. 3. Finally, some conclusions are provided in Sect. 4.

2 Methodology

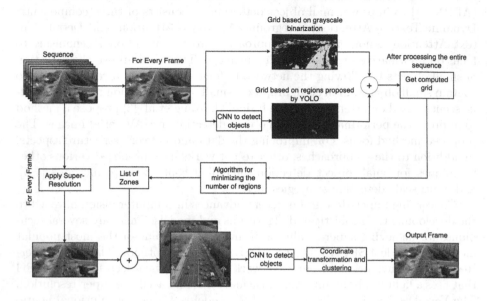

Fig. 1. Workflow of the proposed technique.

This section details the proposed methodology to improve vehicle detection in aerial sequences as shown in Fig. 1. Our proposal begins with an unannotated video sequence, converted into a set of frames that will be processed in the next step. More specifically, we have:

$$D = \{(\mathbf{Y}_l) \mid l \in \{1, ..., N\}\} \tag{1}$$

where D is the unlabeled dataset, and $\mathbf{Y}_l$ represents each extracted low-resolution frame. Also, N stands for the total number of frames that compose the sequence. The first step involves processing each image $\mathbf{Y}_l$ extracted from the sequence and performing binarization on it to get a binary mask. For this purpose, the color space is changed from RGB (Red, Green, Blue) to HSV (Hue Saturation Value). In the HSV color space, each pixel is represented as a 3-dimensional vector, where the first dimension represents the hue, the second dimension is the saturation, and the third dimension refers to the overall brightness of the pixel. Saturation represents the intensity of color in a pixel, with 0 indicating a pure grey tone and 100 indicating a pure color tone. The threshold to define where pixels are selected is defined as follows:

$$threshold = max\_val \frac{th2}{100} \tag{2}$$

where *max_val* is the maximum saturation value in the image, and *th2* is a threshold value representing the percentage of saturation below which a pixel is considered part of the background. In our experiments, *th2* has been set to 10% of color saturation, as it has been observed that this value produces good results in object detection in images. That is, the objects of interest usually have higher color saturation than the background, so saturation-based binarization effectively separates objects from the background. After that, a binary mask is created using the threshold value:

$$\text{mask} = \begin{cases} 0, & \text{if } saturation < threshold \\ 1, & \text{in other cases} \end{cases} \tag{3}$$

The resulting binary image is stored in a matrix denoted as *binarized_grid*. Next, each image is processed using the YOLOv7 object detection model[1] $\mathcal{G}$ for object detection, whose input is an image $\mathbf{Y}$, and yields a set *ROI* of proposed regions of interest (ROIs):

$$W = \mathcal{G}(\mathbf{Y}) \tag{4}$$

$$ROI = \{(\alpha_i, \beta_i, \gamma_i, \delta_i) \mid i \in \{1, ..., N\}\} \tag{5}$$

where N stands for the number of proposed regions, the coordinates of the top left corner of the i-th detection within $\mathbf{Y}$ are noted $(\alpha_i, \beta_i) \in \mathbb{R}^2$, and the coordinates of the bottom right corner of the i-th detection are $(\gamma_i, \delta_i) \in \mathbb{R}^2$.

For each ROI, a variable *grid_rpn* is incremented by 1 for each pixel within the bounding box. After processing the entire sequence, both *binarized_grid* and *grid_rpn* are normalized. Then *computed_grid* is obtained by multiplying both grids by respective weights w_1 and w_2, where $w_1 = 0.4$ and $w_2 = 0.6$. This gives higher priority to the ROIs proposed by YOLOv7. The formula for *computed_grid* is as follows:

$$computed\_grid = w_1 \cdot binarized\_grid + w_2 \cdot grid\_rpn \tag{6}$$

Once *computed_grid* is obtained, the minimum optimal set of regions of interest is calculated, which covers a minimum coverage percentage. The algorithm generates all possible bounding boxes with a size acceptable to the YOLOv7 network and calculates the number of 1's (foreground pixels) within each box. Both are represented in Fig. 2.

Next, the proposed algorithm establishes the zones that maximize the number of 1's until they cover the minimum coverage percentage of the 1's in *computed_grid*. The steps for the algorithm are as follows:

---

[1] https://github.com/WongKinYiu/yolov7.

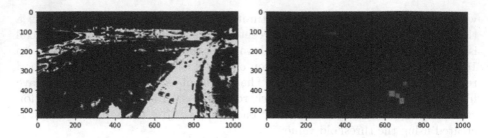

Fig. 2. Grid based on grayscale binarization and regions proposed by YOLO.

1. Set a minimum coverage threshold.
2. Divide the matrix into zones that meet the coverage threshold.
3. Determine the number of obtained zones.
4. Calculate the amount of coverage within a given zone.
5. Identify the best location for a zone based on the obtained coverage.
6. Generate the zones based on the coverage threshold and best locations. An example is represented in Fig. 3.
7. Calculate the coverage at a given location based on a set of points and a specified radius.

Once the minimum optimal zones are obtained, they are extracted from each image in the sequence, and super-resolution is applied. $\mathcal{F}$ is the applied super-resolution model, which receives as input $\mathbf{Y}_{LR}$, a low-resolution image and processes it to obtain $\mathbf{Y}_{HR}$, an image with a higher number of pixels and an enhanced quality.

Fig. 3. Example of generated zones.

For our proposal, we have selected the Real-ESRGAN[2] super-resolution model. This model has several versions available for use. Each determines a particular Z upscaling factor ($X2$, $X3$, $X4$). Since the selected upscaling factor represented by the variable Z is $X2$, the low-resolution image $\mathbf{Y}_{LR}$ will be processed, obtaining a new one denoted as $\mathbf{Y}_{HR}$ with the double of pixels on each

---

[2] https://github.com/xinntao/Real-ESRGAN.

spatial dimension. The resulting sub-images have an optimal size for the selected object detection model.

The next step is to apply the object detection network to set a new list of detections, leading to step six of the workflow:

$$W_{HR} = \mathcal{G}\left(\mathbf{Y}_i\right) \tag{7}$$

The next step is to translate the locations of each element in W_{HR} from the coordinate system of $\mathbf{Y}_i$ to the coordinate system of $\mathbf{Y}_{LR}$. The equation that translates a point $\hat{\mathbf{C}}_i$ expressed in the coordinate system of $\mathbf{Y}_i$ to the same point $\mathbf{C}_i$ expressed in the coordinate system of $\mathbf{Y}_{LR}$ is the following:

$$\mathbf{C}_i = (\chi_i, \psi_i) + \frac{1}{Z} \cdot \hat{\mathbf{C}}_i \tag{8}$$

The list of elements with the translated coordinates is denoted as W_{HRT}.

$$W_{HRT} = \{(\alpha_{k,l}, \beta_{k,l}, \gamma_{k,l}, \delta_{k,l}, \lambda_{k,l}, \rho_{k,l}) \mid l \in \{1, ..., N_k\}\} \tag{9}$$

$$(\alpha_{k,l}, \beta_{k,l}) = \mathbf{y}_k + \tfrac{1}{Z}\left(\hat{\alpha}_{k,l}, \hat{\beta}_{k,l}\right) \quad (\gamma_{k,l}, \delta_{k,l}) = \mathbf{y}_k + \tfrac{1}{Z}\left(\hat{\gamma}_{k,l}, \hat{\delta}_{k,l}\right) \tag{10}$$

$$\lambda_{k,l} = \hat{\lambda}_{k,l} \quad \rho_{k,l} = \hat{\rho}_{k,l} \tag{11}$$

Finally, the detections stored in the W_{HRT} list are evaluated to identify whether an element has been detected for the first time or already corresponded to a known detection in W_{LR}. To do so, the intersection-on-union metric is applied for each pair of detections $\mathbf{d}_j, \mathbf{d}_k \in W_{HRT}$:

$$IOU = \frac{Area\left(\mathbf{d}_j \cap \mathbf{d}_k\right)}{Area\left(\mathbf{d}_j \cup \mathbf{d}_k\right)} \tag{12}$$

A pair of detections $\mathbf{d}_j$ and $\mathbf{d}_k$ are associated with the same object if and only if $IOU > \theta$, where the threshold θ must be adjusted. Then, the final set of detections W'_{HRT} is obtained by collecting the remaining detections after removing duplicates. The κ clustering forms a list with the detections obtained for each element. The detection with the highest score for each element is selected, and the bounding box is adjusted based on the detections set for the same element.

3   Experiments

This section evaluates the proposed methodology on several sequences that require object detection. This task can be difficult due to the wide range of sizes of the elements contained in it. The following subsections detail the dataset used, the chosen object detection model, the super-resolution model and the testing process for evaluation.

3.1   Dataset

The UAVDT (Unmanned Aerial Vehicle Detection and Tracking) dataset has been used to evaluate the proposal. The dataset comprises over 80,000 high-resolution images captured from various unmanned aerial vehicles (UAVs) and ground-based cameras. The images cover multiple scenarios, such as urban and rural areas, highways, and intersections. One of the main reasons for the popularity of the UAVDT dataset in this context is the significant variation in the sizes of the objects, particularly the vehicles, in the images. This variation poses a significant challenge to detection. Additionally, the dataset provides annotations for object detection, which enables the evaluation of the performance of the proposed methodology.

Among all the available sequences in the UAVDT dataset, we have evaluated five sequences in our research. These sequences were selected based on their diversity in terms of object sizes and background complexity, enabling us to assess our proposal's robustness and accuracy.

3.2   Object Detection

In this research, the Yolov7 object detection model was selected as the core component of the proposed object detection. Yolov7 is a state-of-the-art object detection model that has achieved superior performance in terms of accuracy and speed. The model is based on a deep convolutional neural network that can detect objects in real-time. One of the advantages of using a pre-trained Yolov7 model is that it is pre-trained with the Common Objects in Context (COCO) dataset [14], which contains over 330,000 images with over 2.5 million object instances. The COCO dataset is large and diverse, including objects of different sizes, shapes, and orientations, making it an ideal dataset for pre-training object detection models. Pre-training with COCO enables the model to learn general object detection features that can be fine-tuned for specific tasks, such as detecting objects in aerial images.

Although Yolov7 is a fast and accurate object detection model, it has limitations when detecting small objects. This is because the model divides the input image into a grid and assigns each cell a bounding box set. The size of these bounding boxes is fixed and cannot be adjusted, which makes it difficult to detect small objects accurately. Despite these limitations, it was found that Yolov7 was a suitable model for the proposed methodology. A comparison has been made between the following methods:

- Original Model (RAW): The direct application of the unmodified raw object detection model.
- Proposed: The presented methodology is based on the re-inference of super-resolved optimal areas.

Original Frame

Upscaled Zone   Super-Resolved Zone

Fig. 4. Differences between upscaling and applying Super-Resolution with a scaling factor of x2 on a vehicle extracted from the UAVDT - Sequence M0301.

3.3   Super-Resolution

Real-ESRGAN was used as the model for applying super-resolution in the proposed methodology. Real-ESRGAN is a deep learning model for image super-resolution based on the Enhanced Super-Resolution Generative Adversarial Network (ESRGAN). The model was trained on a large dataset of high-resolution images to learn the mapping between low-resolution and high-resolution images. Real-ESRGAN has demonstrated state-of-the-art performance in a range of super-resolution tasks, including image upscaling and restoration. One of the main advantages is its ability to generate high-quality, high-resolution images with fine details and textures. The model achieves this using a deep neural network architecture with multiple layers that can capture complex image features and patterns. Real-ESRGAN also incorporates a perceptual loss function that ensures the generated images are visually similar to the original images.

Another advantage of Real-ESRGAN is that it allows for applying an upscaling factor specific to the dataset. This means that the model can adapt this parameter according to the characteristics of the aerial images used in the proposed methodology, which may have different resolution requirements than other image datasets. This customization ensures that the generated high-resolution images are optimized for the specific dataset. Figure 4 shows the differences between the original image, direct upscaling and super-resolution applications with a scaling factor of x2 according to a vehicle detected in the frame. Comparing the two extracted areas, it can be observed that the super-resolved image exhibits a significant improvement in image quality and sharpness.

3.4 Experimental Setups

During the experimental testing phase, we selected a hyper-parameter set specified in Table 1. The model Yolo v7 was selected to perform the object detection. This model accepts an input image with a fixed size (640 × 640 px). An upscaling factor of 2 was selected to apply Super-Resolution. For this reason, the areas extracted from the original input image have a size of 320 × 320 px. Once the Super-Resolution was applied to one of these, the new super-resolved sub-image had the perfect input size according to the object detection model. The Threshold has been set to 80% to cover the possible zones where the model could identify objects. The selected classes to perform the evaluation have been defined according to the predominance of the elements contained in the sequences of the UAVDT Dataset.

Table 1. Selected values of the hyperparameters.

Parameter	Value
Minimum score to select a detection	0.8
Upscaling Factor to apply Super-Resolution	2
Window size to generate new areas	320 × 320
Minimum Threshold to cover the possible regions	80%
Evaluated classes	Car

3.5 Results

During the experimentation phase, mean Average Precision (mAP) is used to determine the effectiveness of the presented proposal from a quantitative point of view. The mAP is a widely used measure of accuracy in object detection and image segmentation tasks. It measures how well an algorithm can localize objects in an image or sequence of images, taking into account precision. To evaluate the performance of our methodology using mAP, we used the Common Objects in Context (COCO) evaluator. The evaluation was performed according to different levels of intersection over union (IoU) and object sizes within the sequence.

Table 2 summarizes the mean Average Precision for the RAW model and our proposed methodology across five different sequences. The results demonstrate a significant improvement in mAP for our proposed methodology compared to the RAW model. Specifically, our methodology achieves a median detection rate of 38.3% for sequence M1305, compared to only 15.6% for the RAW model. This improvement in detection accuracy is not limited to larger objects but extends to smaller and medium-sized objects. These results suggest that our proposed methodology provides more reliable detections through re-inference and detects new objects that the RAW model previously missed. These results highlight the

Table 2. Mean average precision (mAP) was obtained from diverse sequences of the UAVDT Dataset using the Yolo v7 model. A detection is considered valid when its score exceeds the set threshold of 80%. The best results are marked in **bold**.

		Mean Average Precision (mAP) - Yolov7 - Score > 80%				
		IoU=0.50:0.95|area=all	IoU>0.50|area=all	IoU>0.75|area=all	IoU=0.50:0.95|area=Small	IoU>0.50|area=Medium
Seq. M0301	RAW	0.113	0.252	0.060	0.114	0.109
	OURS	**0.223**	**0.466**	**0.167**	**0.218**	**0.294**
Seq. M0604	RAW	0.324	0.468	0.387	0.077	0.497
	OURS	**0.417**	**0.639**	**0.481**	**0.196**	**0.569**
Seq. M1301	RAW	0.074	0.116	0.089	0.030	0.291
	OURS	**0.154**	**0.295**	**0.132**	**0.131**	**0.282**
Seq. M1305	RAW	0.156	0.229	0.199	0.081	0.413
	OURS	**0.383**	**0.638**	**0.422**	**0.338**	**0.549**
Seq. M1401	RAW	0.183	0.244	0.229	0.045	0.491
	OURS	**0.323**	**0.481**	**0.377**	**0.199**	**0.603**

Fig. 5. An example is applied to frame 317 of Sequence M0301. The left side shows the Raw model results, while the right side shows the detections after applying the presented proposal using Yolo v7.

Fig. 6. An example is applied to frame 8 of Sequence M1301. The left side shows the Raw model results, while the right side shows the detections after applying the presented proposal using Yolo v7.

effectiveness of our proposed approach for object detection and demonstrate its potential to improve the accuracy of existing models.

Figures 5 and 6 provide qualitative data on the performance of our proposed methodology. These figures demonstrate how our methodology outperforms the RAW model regarding the number of objects identified without requiring any retraining of the object detection model. Our methodology detects significantly more objects than the RAW model, even in cases where the objects are small or partially occluded. Overall, these qualitative results provide further evidence of the effectiveness of our proposed methodology for object detection tasks.

4   Conclusions

We presented an approach to improve object detection accuracy without retraining the object detection model. Our proposed methodology involves selecting optimal minimum zones for re-inference after applying super-resolution to the input images. The results of our study demonstrate that this approach effectively improves the mAP for object detection across a range of scenarios. Specifically, our methodology significantly improves mAP compared to the RAW model, with an average increase of up to 22.7% across test sequences evaluated. Moreover, our approach detected a more significant number of objects, including small and medium-sized objects, without requiring retraining of the object detection model. Our study also provides some insights into future research directions. One potential avenue for future research is to explore the impact of different minimum zone sizes on the performance of our methodology and the application of different upscaling factors using the Super-resolution model according to a heuristic.

Acknowledgments. This work is partially supported by the Autonomous Government of Andalusia (Spain) under project UMA20-FEDERJA-108, project name Detection, characterization and prognosis value of the non-obstructive coronary disease with deep learning. It includes funds from the European Regional Development Fund (ERDF). It is also partially supported by the University of Málaga (Spain) under grants B1-2019_01, project name Anomaly detection on roads by moving cameras; B1-2019_02, project name Self-Organizing Neural Systems for Non-Stationary Environments; B1-2021_20, project name Detection of coronary stenosis using deep learning applied to coronary angiography, and B4-2022, project name Intelligent Clinical Decision Support System for Non-Obstructive Coronary Artery Disease in Coronarographies. Iván García-Aguilar is funded by a scholarship from the Autonomous Government of Andalusia (Spain) under the Young Employment operative program [grant number SNGJ5Y6-15]. The authors thankfully acknowledge the computer resources, technical expertise and assistance provided by the SCBI (Supercomputing and Bioinformatics) center of the University of Málaga. They also gratefully acknowledge the support of NVIDIA Corporation with the donation of a RTX A6000 GPU with 48Gb. The authors also thankfully acknowledge the grant of the Universidad de Málaga and the Instituto de Investigación Biomédica de Málaga y Plataforma en Nanomedicina-IBIMA Plataforma BIONAND.

References

1. Min, K., Lee, G.-H., Lee, S.-W.: Attentional feature pyramid network for small object detection. Neural Networks **155**, 439–450 (2022). ISSN: 0893–6080. https://doi.org/10.1016/j.neunet.2022.08.029

2. Lian, J., et al.: Small object detection in traffic scenes based on attention feature fusion. Sensors **21**(9), 3031 (2021). https://doi.org/10.3390/s21093031

3. Wang, X., Zhu, D., Yan, Y.: Towards efficient detection for small objects via attention-guided detection network and data augmentation. Sensors **22**(19), 7663 (2022). https://doi.org/10.3390/s22197663

4. Guan, L., Wu, Y., Zhao, J.: SCAN: semantic context aware network for accurate small object detection. Int. J. Comput. Intell. Syst. **11**(1), 951 (2018). https://doi.org/10.2991/ijcis.11.1.72

5. Gong, Y., et al.: Context-aware convolutional neural network for object detection in VHR remote sensing imagery. IEEE Trans. Geosci. Remote Sens. **58**(1), 34–44 (2020). https://doi.org/10.1109/TGRS.2019.2930246

6. Liu, B., et al.: Small object detection using multi-scale feature fusion and attention. In: 2022 41st Chinese Control Conference (CCC), pp. 7246–7251 (2022). https://doi.org/10.23919/CCC55666.2022.9902202

7. Li, L., Li, B., Zhou, H.: Lightweight multi-scale network for small object detection. PeerJ Comput. Sci. **8**, e1145 (2022). https://doi.org/10.7717/peerj-cs.1145

8. Dong, C., Loy, C.C., Tang, X.: Accelerating the super-resolution convolutional neural network (2016). arXiv: 1608.00367 [cs.CV]

9. Kim, J., Lee, J.K., Lee, K.M.: Accurate image super-resolution using very deep convolutional networks (2016). arXiv: 1511.04587 [cs.CV]

10. Wang, X., et al.: ESRGAN: enhanced super-resolution generative adversarial networks (2018). arXiv: 1809.00219 [cs.CV]

11. Wang, X., et al.: Real-ESRGAN: training real-world blind super-resolution with pure synthetic data (2021). arXiv: 2107.10833 [eess.IV]

12. Wang, C.-Y., Bochkovskiy, A., Liao, H.-Y.M.: YOLOv7: trainable bag-of-freebies sets new state-of-the-art for real-time object detectors (2022). arXiv: 2207.02696 [cs.CV]

13. Du, D., et al.: The unmanned aerial vehicle benchmark: object detection and tracking (2018). arXiv: 1804.00518 [cs.CV]

14. Lin, T.-Y., et al.: Microsoft COCO: common objects in context (2015). arXiv: 1405.0312 [cs.CV]

Long-Term Hail Risk Assessment
with Deep Neural Networks

Mikhail Mozikov[1,3](✉), Ivan Lukyanenko[2], Ilya Makarov[3,4],
Alexander Bulkin[1,5,7], and Yury Maximov[6]

[1] Skolkovo Institute of Science and Technology, Moscow, Russia
mozikov.mb@phystech.edu
[2] Moscow Institute of Physics and Technologies, Moscow, Russia
lukianenko.ia@phystech.edu
[3] Artificial Intelligence Research Institute (AIRI), Moscow, Russia
[4] AI Center, NUST MISiS, Moscow, Russia
iamakarov@misis.ru
[5] International Center for Corporate Data Analysis, Grenoble, France
a.bulkin@iccda.io
[6] Los Alamos National Laboratory Los Alamos, New Mexico, USA
yury@lanl.gov
[7] Moscow State University, Moscow, Russia

Abstract. Hail risk assessment is crucial for businesses, particularly in
the agricultural and insurance sectors, as it helps estimate and mitigate
potential losses. Although significant attention has been given to short-
term hail forecasting, the lack of research on climatological-scale hail risk
estimation adds to the overall complexity of this task. Hail events are
rare and localized, making their prediction a long-term open challenge.

One approach to address this challenge is to develop a model that
classifies vertical profiles of meteorological variables as favorable for hail
formation while neglecting important spatial and temporal information.
The main advantages of this approach lie in its computational efficiency
and scalability. A more advanced strategy involves combining convolutional layers and recurrent neural network blocks to process geospatial
and temporal data, respectively.

This study compares the effectiveness of these two approaches and
introduces a model suitable for forecasting changes in hail frequency.

Keywords: Climate modeling · Hail · Machine learning · Deep
Learning

1 Introduction

According to Verisk's 2021 report [29], losses due to hail in 2020 reached $14.2
billion in the USA. Insurance companies are especially vulnerable to hail events.
Urban sprawl and population growth in large cities, such as Dallas/Fort Worth,
Texas; St. Louis, Missouri; Chicago, Illinois; and Denver, Colorado, have made
property damage from hail events more likely. It is worth noting that climate

I. Rojas et al. (Eds.): IWANN 2023, LNCS 14134, pp. 288–301, 2023.
https://doi.org/10.1007/978-3-031-43085-5_23

change increases both the impact and complexity of research in this area. The model designed to estimate hail risk can potentially improve modern economic risk models, helping to reduce costs for vulnerable businesses.

Hailstones are formed when raindrops are carried upward by thunderstorm updrafts into icy atmosphere areas and freeze. Hailstones grow by colliding with liquid water drops that freeze on their surface [16]. Hail falls when the thunderstorm's updraft can no longer support the weight of the hailstone because either the stone becomes large enough or the updraft weakens. Most hail growth occurs at a temperature of approximately $-10°C$ to $-25°C$ [10]. Time is an essential factor in creating large hailstones, with studies showing that they spend 10–15 min or more in the growth regions of storms [2].

Hail is an extreme event, with its main complexity arising from being very local in time and space compared to the resolution of available climate models. The hail/no hail ratio in the experimental setting is around 1%. Short-term (up to 24 h) and nowcasting (up to 2 h) hail forecasts exist, utilising various meteorological models combined with machine learning approaches [3].

However, short-term forecasts differ from models for estimating hail frequencies on a climatological scale due to differences in available data. A statistical approach taken by the model in [23] works with the joint distribution of atmospheric indices and provides a map of the current hail probability distribution around the globe. It is robust and accurate for plain terrains of Europe and the USA but disregards landscape influence.

This study aims to develop a machine-learning model capable of working with various climatic models and climate change scenarios. The model should estimate changes in hail probability for a given area using a combination of CNN and LSTM neural networks. This approach enables the model to capture the spatial [13,18–20,25] and temporal [11,17,21,24,27] data structure. In contrast, the baseline is a structure-oblivious machine learning method dismissing spatial and temporal components.

The task is to identify favorable hail development conditions (classification) and evaluate the frequency of hail events in a given area, relying on climate models included in Coupled Model Intercomparison Project (CMIP5) [28].

The following are our main contributions:

– we have developed a distinct neural network architecture specifically designed to predict alterations in hail frequency for the upcoming decades using CMIP data.
– we have conducted several experiments and compared the results with a baseline model. Our findings reveal that our approach approximates the natural yearly cycle of hail frequencies (averaged between 2010 to 2015) and performs better than the baseline model.

Overall, our research provides an innovative solution to forecast hail frequency changes and improve predictions for future weather patterns.

The rest of the work is organized as follows. Section 2 provides information on the experimental setup, including the data used, the quality metrics, the learning settings, and the problem setup. Section 3 introduces our neural network approach to hail forecasting – HailNet. We explain the architecture of our network and discuss why we chose this particular type. Additionally, we present the results of our experiments, comparing the performance of HailNet and a baseline solution based on gradient boosting with the model introduced in [23].

2 Related Work

This section begins by providing an overview of existing approaches to meteorological and climate forecasts. Additionally, it discusses methods used to work with geospatial data more broadly. In meteorology and agricultural applications, machine learning approaches are increasingly being integrated into conventional routine [5] or for predicting specific weather phenomena like heavy rains [7]. Machine learning is also utilized in estimating crop water demand using on-farm sensors and public weather information in meteorological science applications to agriculture. In a climate study, the primary focus is designing or improving global climate models with machine learning techniques [1]. Another machine learning application [4] uses satellite and climate data to predict wheat yield in Australia.

The article [30] outlines the fundamental concept of why the application of machine learning in weather forecasts (meteorology) and climate forecasts appear to be closely related, yet substantially different. The key distinguishing factor between the two fields is the significant difference in lead times. Meteorology typically focuses on predicting weather patterns in the short-term, ranging from a few hours to a few days ahead. In contrast, climate forecasting is concerned with long-term trends and patterns, ranging from several months to decades ahead. Due to these distinct differences in timescales, the methods and models used for machine learning applications must be carefully chosen and adapted accordingly to ensure accurate and reliable results.

Let us consider the part of the meteorological field that works with hail prediction. In the study [3], the authors process output of a convection-allowing ensemble system and produce a 24-h forecast as a result. However, convection-allowing ensembles are available in only a few countries. Study [23], based on the joint distributions of atmosphere parameters, produces a map of the current hail probability distribution around the globe. It is the closest work to the proposed problem statement, but its aim is to create a current map based only on historical data. It was shown that the model has good generalization ability in Europe and the USA, except for mountain areas.

Our study relates to geospatial time series analysis. Some approaches to visualize such data are described in [14]. In the initial steps of research, visualization can help geoscientists to gain a better understanding of geospatial time series. The article [8] elaborates on AI-based techniques for 3D point clouds and geospatial data as generic components of geospatial AI.

3   Experiment Setting

This section provides information regarding the data sources and preprocessing steps used in the hail forecasting project, along with the problem statement and performance metrics.

3.1   Data

The data required to perform hail forecasts are geospatial time series. Such data consist of values of a quantity obtained at successive times with equal intervals, and spatial relations are involved. It is worth noting that hailstone formation is associated with cumulonimbus clouds, which, unlike others, have an extended vertical structure that can exceed 18 km in some cases [15]. It is important to address this by using data from different isobaric surfaces. Thus, the data structure of the geospatial time series expands to three dimensions.

The dataset for model training and evaluation includes three sources:

- ERA5 dataset [12] consists of historical data and is used to derive feature space for training and evaluation of the model performance.
- CMIP5 MRI-CGCM3 [31] consists of climate projections and is used to derive feature space for forecasting.
- Storm Events Database [22] is used to collect information about target – presence or absence of hail event on a given day.

The ERA5 hourly data on single levels by European Centre for Medium-Range Weather Forecasts (ECMWF) were used to derive surface climate variables. To address the third dimension of data structure, variables from ERA5 hourly data on pressure levels were collected.

The dataset used in our study covers the period from 2010 to 2021, with a time resolution of one hour and a spatial resolution of $0.25° \times 0.25°$ latitude and longitude. The target variable is binary, indicating whether hail was present (1) or not (0), and was obtained from The National Oceanic and Atmospheric Administration's (NOAA) Storm Events Database. This database contains information about storms and other significant weather events that have caused loss of life, injuries, property damage, and disruptions to commerce in the United States. We extracted the start and end times and the coordinates of hail events from this database.

It is worth noting that the Storm Events Database may not be entirely homogeneous on a large timescale due to changes in observing techniques and variations in data sources. However, such inhomogeneity is almost negligible within the last decade. Table 1 presents a complete list of variables used in this study.

The Coupled Model Intercomparison Project Phase 5 (CMIP5) [28] includes climate models developed by different universities under assumptions about climate change scenarios. Representative Concentration Pathway 8.5 (RCP 8.5) was selected, as it reflects the modeled climate changes that could occur if emissions continue to rise throughout the 21st century. This scenario is considered the "worst case" option in CMIP5.

Hail events are typically associated with local weather patterns. Using data with the highest possible spatial resolution is necessary to prevent such information from being lost in data with low spatial density. The dataset available from the Meteorological Research Institute (MRI-CGCM3) in CMIP5 has one of the highest spatial resolutions ($1° \times 1°$), which meets this criterion. Therefore, we used it as the source of future climate projections.

Table 1. Data Description.

Dataset	variables	Description	Time span
ERA5 on pressure levels	U-component of wind	The eastward component of the wind vector	2010–2021
	V-component of wind	The northward component of the wind vector	
	Specific humidity	The mass of water vapour per kilogram of moist air	
	Temperature	The temperature in the atmosphere on the given isobaric surface	
ERA5 on single level	Surface pressure	The pressure (force per unit area) of the atmosphere at the surface of land, sea and inland water	2010–2021
Storm Events Database	Hail event	Indicator of presence of hail in a given area and time	2010–2021
CMIP5 MRI-CGCM3 RCP8.5	U-component of wind	The eastward component of the wind vector	2022–2050
	V-component of wind	The northward component of the wind vector	
	Specific humidity	The mass of water vapour per kilogram of moist air	
	Temperature	The temperature in the atmosphere on the given isobaric surface	
	Surface pressure	The pressure (force per unit area) of the atmosphere at the surface of land, sea and inland water	

To address the spatial and temporal differences between ERA5 and CMIP5 data, we preprocessed all data to the resolutions of the least dense dataset, CMIP5 MRI-CGCM3 [31].

In addition, we needed to address the issue of imbalanced target data during preprocessing. To address this problem, we applied the synthetic minority oversampling technique (SMOTE) [6]. After this step, the ratio of hail to no hail events increased from 1:100 to 1:1.

Overall, these preprocessing steps allowed us better to match the spatial and temporal dimensions of the data and to account for imbalanced target data.

3.2   Problem Statement

Let us define our problem and assign formal designations.

$n \in \mathbb{N}$ – number of labeled grid cells;

$X_{i,j,k} \in \mathbb{R}^{n \times \text{lat} \times \text{long}}$ – a tensor of climate variables at one timestamp;

$\widetilde{X}_{t,i,j,k}$ – a time-series of climate variables corresponding to one day;

$f(w, X)$ – predicting model.

We aim to solve the time series classification problem for every point on the grid of the target region. Our objects are time series of tensors denoted by $X_{i,j,k}$ with a 6-h period. The targets are lat × long matrices, where zeros indicate no hail, and ones represent hail events.

The output of the function $f(w, X)$ is a 2-dimensional probability pseudo-distribution, with every value being in the range of 0 to 1. However, the sum of all matrix values is not necessarily equal to 1, as favorable conditions for hail formation may occur at multiple points on the same day. Therefore, we define the optimization problem of minimizing the Mean Squared Error between the pseudo-distributions output and the target grids to find the best parameters for our model.

$$w^* = \underset{w}{\text{argmin}} \frac{1}{N} \sum_{t=1}^{N} \|f(w, X_t) - Y_t\|^2 \tag{1}$$

The schematic pipeline of the desired model is presented in Fig. 1.

3.3   Quality Metrics

We utilize precision (2) and recall (3), which are two fundamental metrics, and additionally, we incorporate RMSE (4), which is a metric specific to the domain:

$$Precision = \frac{TP}{TP + FP}, \tag{2}$$

$$Recall = \frac{TP}{TP + FN}, \tag{3}$$

where:

TP - true positives of model predictions;

FP - false positives of model predictions;

FN - false negative of model predictions.

Regarding domain-specific metrics, it is crucial to ensure that the model can accurately capture the annual periodic structure of hail occurrence. This can be achieved with a particular metric introduced in [23]. The metric is a basic root-mean-squared error ($RMSE_{om}$) that calculates the difference between the normalized observed and modeled annual cycle of monthly mean hail frequency across the target region.

$$RMSE_{om} = \sqrt{\sum_{i=1}^{12} (o_m - m_m)^2} \tag{4}$$

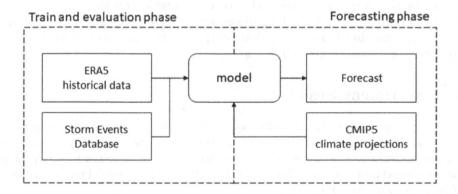

Fig. 1. The schematic pipeline of the desired model.

where:

o_m - normalized observed annual cycle of monthly mean hail frequency;

m_m - normalized modeled annual cycle of monthly mean hail frequency;

Normalization is essential in the context of hail observation datasets as it addresses the issue of partial inhomogeneity. Although this inhomogeneity is minor within the period of 2010–2021, normalization remains crucial to enhance the model's stability and aligns with the methodology of the original study [23] for comparison purposes. We can classify the target data into two categories: 1) "hail" and 0) "no hail". Given the objective of our study, it is imperative to prioritize the recall metric in our model evaluation. The cost of false negatives (when hail frequency is underestimated) is anticipated to outweigh the cost of false positives. Therefore, the recall metric, which measures the model's ability to detect true positives, should be the focus of our evaluation.

4   Neural Network Approach (HailNet)

We developed a model capable of capturing the spatial and temporal structure of climate variables by introducing a neural network with an architecture displayed in Fig. 2, which we will refer to as HailNet further in the paper. The neural network comprised two steps of feature extraction:

– Convolutional (Conv2d) in parallel with dense layers
– Recurrent layers (LSTM)

The first step involved Convolutional (Conv2d) and dense layers, which enabled the gathering of the spatial structure of data. After that, the concatenation of the Dense and Convolutional layers' outputs mapped each time series climate tensor to an embedding space [26]. Objects in embedding space had fewer dimensions and were specified with essential spatial information extracted from raw climate tensors. Subsequently, we observed time series of elements from embedding space. Using the LSTM network (second step), HailNet extracted temporal features from time series. The output of the last LSTM unit proceeded to the Dense layers with a sigmoid activation function on the output. As an output of HailNet, we obtained a 2-dimensional grid with a shape of lat × long with values ranging from 0 to 1. We interpreted this grid as a 2-dimensional probability pseudo-distribution.

5   Experiment Results

In this section, we describe a baseline approach to the problem of hail forecasting. Later, we compare the performance of the HailNet to the baseline model and to a model introduced in [23]. In the last part, we present a sample of hail frequency forecast based on the CMIP5 MRI-CGCM3 (RCP 8.5) [31] model for Texas, part of New Mexico, and Oklahoma.

5.1 Baseline Approach Description

At the outset of our experiment, we developed a simple baseline solution to which we could refer. Within this framework, we grouped all features into a one-day time scale and assumed the absence of temporal and spatial structure within it. This assumption was based on the fact that hail events are typically quite local, and observing a 24-h period should yield all of the features corresponding to a single event.

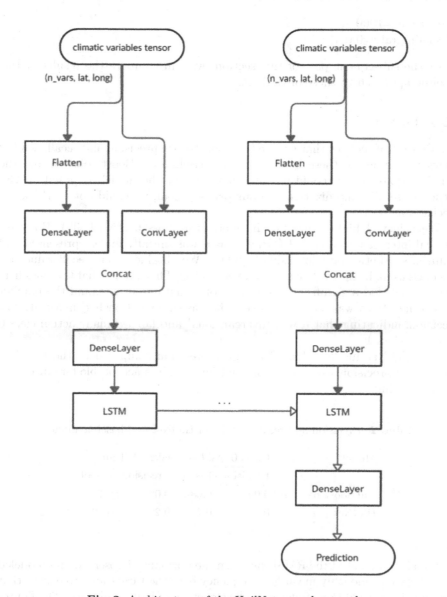

Fig. 2. Architecture of the HailNet neural network.

Notably, ignoring spatial structure prevents the model from incorporating information on associated synoptic processes. However, due to the uncertainty of whether data with a resolution of $1° \times 1°$ adequately preserves spatial features corresponding to these processes, ignoring spatial structure was deemed acceptable.

To create the baseline model, we employed the CatBoost classifier [9]. The model hyperparameters were optimized using a grid search, and the most suitable parameters were determined to be:

- iterations=1000,
- learning_rate=0.01.

In the subsequent part of this section, we will compare the results of this baseline approach to those of HailNet.

5.2   HailNet

The first set of metrics that we need to consider are precision and recall. Table 2 shows the values of these metrics for the baseline CatBoost solution and the HailNet. However, we could not obtain values for the model proposed in [23]. Although recall maximization was our primary goal, we could not overlook the precision metric.

We achieved high values for the metrics in class 0 ("No hail"). However, the real interest lies in class 1 ("Hail"), as it was initially underrepresented and required synthetic sampling using SMOTE. We obtained good recall values for this class using both baseline and HailNet models. This implies that the classifiers correctly labeled a significant proportion of actual hail events, and the number of false negatives was small. However, the baseline solution had markedly low precision, indicating that it was "overcautious" and labeled many actual class 0 events as class 1.

The HailNet model achieved a higher precision compared to the baseline. Although a precision score of 0.2 may not be high, it is acceptable for our experiment's setting.

Table 2. Precision and recall metrics of HailNet and baseline model.

Model	Class 0: No hail		Class 1: Hail	
	Precision	Recall	Precision	Recall
CatBoost baseline	0.99	0.68	0.02	0.77
HailNet	0.98	0.76	0.20	0.79

We also calculated the RMSE between the normalized observed and modeled annual cycle of monthly mean hail frequency over the target area to ensure that our model captured the annual-periodic structure of hail occurrence. The figures of the corresponding curves are shown in Fig. 3. We observed that the HailNet approximated all months well, except for May, which requires further research.

The baseline model achieved a score of 0.19 for $RMSE_{om}$, which is close to the values reported in [23]. In the best case, the scores from the reference ranged from approximately 0.18 to 0.20 for the model considered optimal based on the number of features used. The HailNet attained a score of 0.16 and slightly outperformed the model proposed in [23] in terms of the $RMSE_{om}$ metric.

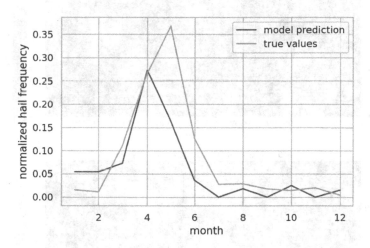

Fig. 3. Normalized observed and modeled annual cycle of monthly mean hail frequency.

5.3   CMIP-Based Forecast

Using data from the CMIP5 MRI-CGCM3 model (scenario RCP 8.5) [31], we have produced a hail frequency forecast. The frequency of hail for Texas, part of New Mexico, and Oklahoma for the years 2025 and 2030 is presented in Fig. 4. The heat maps of hail frequency reveal a distinct peak in Northern Texas that persists over time. This peak is likely associated with the leeward side of mountains situated to the West. However, the extent and direction of this influence require further investigation.

The pattern in the plain area is more complex. Although hail frequency at some points may be comparable to the Northern peak, there is no consistent geographical location. Figure 5 illustrates the difference in hail frequency between the years 2030 and 2025. We observe an upward trend in the leeward area and high variability in the frequency difference in the plain area.

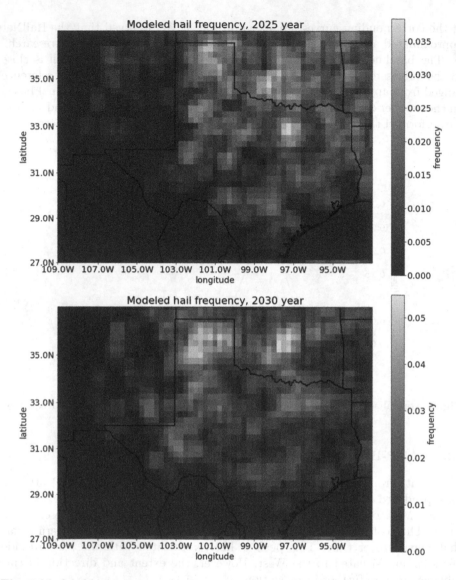

Fig. 4. Modeled hail frequency for 2025 and 2030 years based on CMIP5 MRI-CGCM3 model (RCP 8.5).

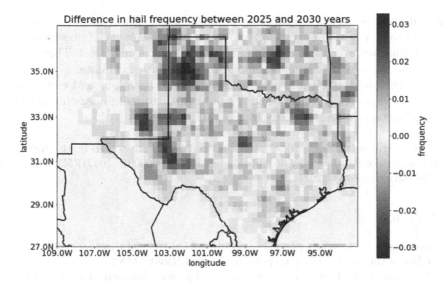

Fig. 5. Difference in hail frequency between 2030 and 2025 years.

6   Conclusion

In this paper, we introduced a new model – HailNet, designed to produce climatological hail forecasts based on CMIP data. It slightly outperforms the model, designed for a close purpose [23].

Still, there are limitations to the model. The model is tied to the locations it has been trained on, meaning that to use it for different regions, retraining and adaptation to the target dataset are required. Additionally, the model tends to be overly cautious, sometimes labeling non-hail events as hail. This is reflected in a high recall metric but a low precision metric, indicating that while the model can accurately identify most hail events, it also generates a significant number of false positives.

In the final part of the paper, we presented a sample of hail forecast for Texas, part of New Mexico, and Oklahoma and performed a brief analysis of it. Keeping limitations in mind, our model contributes to the hail risk estimation field. It may help to more accurately estimate the risk proposed by changing hail frequency and reduce the costs for dependent businesses and governmental organizations such as agricultural, insurance companies, and the green energetics sector (considering solar panels installation). The outputs of HailNet can be used to improve modern economic risk models.

Hail forecasting is a challenging task. One of the main reasons for this is the sparsity of the target. Moreover, the complexity grows when scientists do not have a proper dataset of hail observations, which is the case for almost all countries except the USA and some parts of the European Union. However, even with the available datasets, researchers should work with caution, considering that a single dataset (as with Storm Events Database) may be compiled from different sources, which introduces inhomogeneity to data that could be quite severe in some cases. Observation techniques and the quality of observations, in general, improve over time, creating an artificial increasing trend. These factors need to be taken into account when working with historical data.

Nevertheless, the re-equipment of meteorological stations and expansion of station coverage will certainly yield a great benefit to hail forecasting in particular and meteorological forecasts in general.

Acknowledgments. The work of Mikhail Mozikov was supported by the Analytical Center under the RF Government (subsidy agreement 000000D730321P5Q0002, Grant No. 70-2021-00145 02.11.2021).

References

1. Brenowitz, N.D., et al.: Machine learning climate model dynamics: offline versus online performance. arXiv preprint arXiv:2011.03081 (2020)
2. Browning, K.A., Foote, G.: Airflow and hail growth in supercell storms and some implications for hail suppression. Q. J. R. Meteorol. Soc. **102**(433), 499–533 (1976)
3. Burke, A., Snook, N., Gagne, D.J., II., McCorkle, S., McGovern, A.: Calibration of machine learning-based probabilistic hail predictions for operational forecasting. Weather Forecast. **35**(1), 149–168 (2020)
4. Cai, Y., et al.: Integrating satellite and climate data to predict wheat yield in Australia using machine learning approaches. Agric. For. Meteorol. **274**, 144–159 (2019)
5. Chase, R.J., Harrison, D.R., Burke, A., Lackmann, G.M., McGovern, A.: A machine learning tutorial for operational meteorology, Part I: traditional machine learning. arXiv preprint arXiv:2204.07492 (2022)
6. Chawla, N.V., Bowyer, K.W., Hall, L.O., Kegelmeyer, W.P.: Smote: synthetic minority over-sampling technique. J. Artif. Intell. Res. **16**, 321–357 (2002)
7. Choi, C., Kim, J., Kim, J., Kim, D., Bae, Y., Kim, H.S.: Development of heavy rain damage prediction model using machine learning based on big data. In: Advances in Meteorology 2018 (2018)
8. Döllner, J.: Geospatial artificial intelligence: potentials of machine learning for 3d point clouds and geospatial digital twins. PFG-J. Photogrammetry Remote Sens. Geoinf. Sci. **88**(1), 15–24 (2020)
9. Dorogush, A.V., Ershov, V., Gulin, A.: Catboost: gradient boosting with categorical features support. arXiv preprint arXiv:1810.11363 (2018)
10. Foote, G.B.: A study of hail growth utilizing observed storm conditions. J. Appl. Meteorol. Climatol. **23**(1), 84–101 (1984)
11. Grachev, A.M., Ignatov, D.I., Savchenko, A.V.: Neural networks compression for language modeling. In: Shankar, B.U., Ghosh, K., Mandal, D.P., Ray, S.S., Zhang, D., Pal, S.K. (eds.) Neural networks compression for language modeling. LNCS, vol. 10597, pp. 351–357. Springer, Cham (2017). https://doi.org/10.1007/978-3-319-69900-4_44
12. Hersbach, H., et al.: The era5 global reanalysis. Q. J. R. Meteorol. Soc. **146**(730), 1999–2049 (2020)
13. Korinevskaya, A., Makarov, I.: Fast depth map super-resolution using deep neural network. In: IEEE International Symposium on Mixed and Augmented Reality (ISMAR 2018), New York, USA, pp. 117–122. IEEE (2018)

14. Köthur, P., Sips, M., Unger, A., Kuhlmann, J., Dransch, D.: Interactive visual summaries for detection and assessment of spatiotemporal patterns in geospatial time series. Inf. Vis. **13**(3), 283–298 (2014)
15. Krauss, T.W., Sinkevich, A.A., Veremey, N.E., Dovgalyuk, Y.A., Stepanenko, V.D.: Study of the development of an extremely high cumulonimbus cloud (Andhra Pradesh, India, September 28, 2004). Russ. Meteorol. Hydrol. **32**(1), 19–27 (2007)
16. Laboratory, N.S.S.: Nssl's hail research (2020). https://www.nssl.noaa.gov/education/svrwx101/hail
17. Makarov, I., Bakhanova, M., Nikolenko, S., Gerasimova, O.: Self-supervised recurrent depth estimation with attention mechanisms. PeerJ Comput. Sci. **8**(e865), 1–25 (2022)
18. Makarov, I., Gerasimova, O.: Link prediction regression for weighted co-authorship networks. In: Rojas, I., Joya, G., Catala, A. (eds.) IWANN 2019. LNCS, vol. 11507, pp. 667–677. Springer, Cham (2019). https://doi.org/10.1007/978-3-030-20518-8_55
19. Makarov, I., Gerasimova, O.: Predicting collaborations in co-authorship network. In: Proceedings of the 14th IEEE International Workshop on Semantic and Social Media Adaptation and Personalization (SMAP 2019), New York, USA, 09–10 June 2019, pp. 1–6. Cyprus University of Technology. IEEE (2019)
20. Makarov, I., Korinevskaya, A., Aliev, V.: Fast semi-dense depth map estimation. In: Proceedings of the ACM Workshop on Multimedia for Real Estate Tech, New York, USA, 11 June 2018, pp. 18–21. University of Tokyo, ACM (2018)
21. Makarov, I., et al.: Temporal network embedding framework with causal anonymous walks representations. PeerJ Comput. Sci. **8**(e858), 1–27 (2022)
22. NOAA: National Oceanic and Atmospheric Administration. Storm events database (2018). https://www.ncdc.noaa.gov/stormevents/
23. Prein, A.F., Holland, G.J.: Global estimates of damaging hail hazard. Weather Clim. Extremes **22**, 10–23 (2018)
24. Savchenko, A.V.: Phonetic words decoding software in the problem of Russian speech recognition. Autom. Remote. Control. **74**, 1225–1232 (2013)
25. Savchenko, A.V.: Fast inference in convolutional neural networks based on sequential three-way decisions. Inf. Sci. **560**, 370–385 (2021)
26. Savchenko, A.V., Belova, N.S.: Unconstrained face identification using maximum likelihood of distances between deep off-the-shelf features. Expert Syst. Appl. **108**, 170–182 (2018)
27. Savchenko, A., Khokhlova, Y.I.: About neural-network algorithms application in viseme classification problem with face video in audiovisual speech recognition systems. Opt. Mem. Neural Netw. **23**(1), 34–42 (2014)
28. Taylor, K.E., Stouffer, R.J., Meehl, G.A.: An overview of cmip5 and the experiment design. Bull. Am. Meteor. Soc. **93**(4), 485–498 (2012)
29. The hail hazard and its impact on property insurance (2020). https://www.verisk.com/insurance/capabilities/weather-risk/hail-and-severe-thunderstorm-risk/
30. Watson-Parris, D.: Machine learning for weather and climate are worlds apart. Phil. Trans. R. Soc. A **379**(2194), 20200098 (2021)
31. Yukimoto, S., et al.: A new global climate model of the meteorological research institute: MRI-CGCM3-model description and basic performance. Meteorol. J. **90**, 23–64 (2012)

Video Scene Segmentation Based on Triplet Loss Ranking

Miguel Jose Esteve Brotons[1], Jorge Carmona Blanco[1],
Francisco Javier Lucendo[1], and José García-Rodríguez[2]($\boxtimes$)

[1] Telefónica I+D, Madrid, Spain
{miguel.estevebrotons,jorge.carmonablanco,
javier.lucendodegregorio}@telefonica.com
[2] Computers Technology Department, University of Alicante, Alicante, Spain
jgarcia@dtic.ua.es

Abstract. Scene segmentation is the task of segmenting the video in groups of frames with a high degree of semantic similarity. In this paper, we contribute to the task of video scene segmentation with the creation of a novel dataset for temporal scene segmentation. In addition, we propose the combination of two deep models to classify whether two video frames belong to the same or a different scene. The first model consists of a triplet network that is composed of 3 instances of the same 2D convolutional network. These instances correspond to a multi-scale net that performs frame embedding efficiently based on their similarity. We feed this network with an efficient triplet sampling algorithm. The second model is responsible for classifying whether these embeddings correspond to frames from different scenes by fine-tuning a siamese network.

Keywords: Shot boundary detection · scene segmentation · keyframes · triplet network · multi-scale network · siamese network

1 Introduction

Video technology is growing at a very high rate. Thanks to these advances, the volume of video content available for users' consumption is increasing exponentially, especially on Video-on-Demand (VoD) platforms. Some streaming platforms such as YouTube store more than 300 h of video content per minute. This data growth raises the problem of data management. The number of users who consume it and, therefore, the revenues generated by these platforms have also increased. According to Statista [21], the number of users in the VoD segment is expected to reach 1640 million by 2027. In addition, use cases for improving the user experience are beginning to arise in the ecosystem of streaming platforms. Some of these use cases can be solved with machine learning techniques. Because of this, video analysis and processing is one of the most active areas in video streaming and the video entertainment industry, including fields such as classification or stream segmentation. Segmentation in shots as well as in scenes

© The Author(s), under exclusive license to Springer Nature Switzerland AG 2023
I. Rojas et al. (Eds.): IWANN 2023, LNCS 14134, pp. 302–315, 2023.
https://doi.org/10.1007/978-3-031-43085-5_24

plays a fundamental role as the first step in extracting features from video assets. However, its practical application is lagging behind due to the complexity of the methods.

To start analyzing a sequence, we should look at the video decomposition hierarchy, shown in Fig. 1. A video can be decomposed into different sets of frames based on its content: scenes consist of a consecutive series of shots representing a high-level concept or story; shots are defined as a consecutive series of frames that are taken with a single camera, representing a continuous action in time and space; and keyframes are the most representative frames of the shot to which they belong.

Therefore, to enable a detailed search within the VoD content available to the user, it is necessary to perform scene detection, which consists of automatically segmenting an input video into meaningful and story-telling parts without any help. This scene segmentation task requires a good representation of the video content, specifically of the frames that compose it. Recently, Convolutional Neural Networks [12] have shown their powerful abilities in image representation, generating amazing results in image processing tasks such as classification or object detection.

The contribution of this work is two-fold: on one hand we create a novel dataset for temporal scene segmentation composed of keyframes of shots belonging to the same and different scenes. On the other hand, we propose the combination of two deep models to classify video frames in scenes.

The remainder of this paper is organized as follows: Sect. 2 briefly reviews the work related to scene segmentation. Section 3 addresses an overview of the deep learning model proposed. Then, Sect. 4 introduces a new dataset created and used for our approach. Section 5 describes the multi-scale and the siamese network architectures developed. Section 6 details the evaluation metrics used. Section 7 presents some techniques for optimizing model training. Section 8 shows training curves and evaluation metrics. And Sect. 9 analyzes the results obtained and summarizes the conclusions and possible next steps.

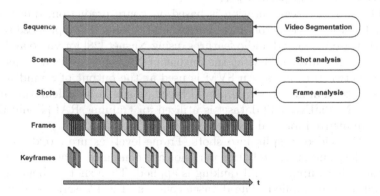

Fig. 1. Hierarchy of video decomposition.

2   Related Work

Most scene detection algorithms seek to find similarities between the content of adjacent shots. In [2], they extract visual features through CNNs and textual features thanks to a variant of the bag-of-words approach, obtaining descriptors clustered using k-means and the cosine distance between the words. To obtain these representations, they concatenate a parameter indicating the center frame of the shot, providing the network temporal information between the two shots. Then a siamese network processes the set of frames of two different shots, extracts their features, and concatenates the similarity scores to build a similarity matrix, which is then given to a spectral clustering algorithm to obtain the final scene boundaries.

In [17], the video sequence is treated as multi-modal data that contains different high-level elements. This approach does not take into account a single global representation of the shot but extracts and combines features from different elements. It processes keyframes and audio to extract features of places, cast detected instances, actions, and speech details deleting the background sound. Then they formulate scene segmentation as a binary classification problem on shot boundaries using a net consisting of two branches with temporal convolution layers. Each branch embeds the shots before and after the boundary respectively. Finally, they calculate the difference between each shot representation.

The scene segmentation task appears to be a bimodal problem, however, we will work only with visual cues. In [22], although they also use a bimodal approach, we observe how it performs shot clustering based on visual appearance. Depending on its length, they extract a set of keyframes from each shot. Each key-frame is represented as a visual descriptor defined as the concatenation of the HSV color histograms and the low-level features of ResNet50 [10] CNN architecture. In order to determine the similarity between the two shots, they train a triplet-based network architecture. The network's input is a set of three images, called triplet, which is fed into three identical deep neural networks that share architecture and parameters.

In [6], the scene segmentation is based on genre prediction, differentiating between 4 genres: action, romance, horror, and drama. They propose an architecture that use video and audio features, using Nasnet [28] for video and Soundnet for audio features. LSTM at the output of Nasnet is utilized to model the transition between framers. An SVM is used at the output of SoundNet [1]. A transition detection method is employed based on similarity calculation between shots genre. MediaEval [20] dataset is utilized for training. RAI [4] and BBC [3] datasets are utilized for eval and testing.

In [5], the video is split into shots. Frame-level features, textual features from subtitles, and start-end time per shot are combined. The inter-shot similarity matrix is computed, and ranking is applied. Then shot sequence regrouping algorithm was applied. The dataset consists of 2 TV series - 5 seasons of Game of Thrones and 3 seasons of Breaking Bad, manually annotated in scenes. Evaluation metrics used are Coverage [24], Purity Clustering, WindowDiff, and Pk.

In [8] based on [9], the authors propose a self-supervised shot contrastive learning approach to learn a shot representation that maximizes the similarity between nearby shots compared to randomly selected shots. The self-supervised part consists of an encoder network for visual and audio modalities and momentum contrastive learning to contrast the similarity of the embedded shots. Finally, the problem is formulated as a binary classification problem of determining if a shot boundary is also a scene boundary or not. An MLP with 3 Fully Connected layers is utilized as a classifier. [14] follows the same approach than [8] and contributes using additional pretext tasks. In [27], another self-supervised model is proposed with a Siamese Network of Encoders in the self-supervised part, and a MLP/Bi-LSTM for classification.

3   Overview

Based on Deep Ranking CNN [25], the idea of our model is to learn image similarity by learning distributed embedding representation of data points in a way that in the high dimensional vector space, contextually similar data points are projected in the nearby region whereas dissimilar data points are projected far away from each other. So, we can define the similarity between images I and Q as the Euclidean Distance between f(I) and f(Q) in the image embedding space. We decided to replace the Euclidean Distance with the Cosine Similarity (Eq. 1) since the latter is bounded in the range [0,1].

$$S(f(I), f(Q)) = \frac{f(I) \cdot f(Q)}{\|f(I)\|\|f(Q)\|} \tag{1}$$

For this, we build a set of triplets, where each triplet $T_i = (A_i, P_i, N_i)$ is composed of an anchor image A_i, a positive image P_i (which is similar to the anchor image), and a negative image N_i (which is dissimilar to the anchor image). A triplet characterizes the relative similarity relationship for the three images. Then we can define the following loss function (Eq. 2) for a triplet T_i:

$$L(A_i, P_i, N_i) = max\{0, S(f(A_i), f(N_i)) - S(f(A_i), f(P_i)) + g\} \tag{2}$$

where g is a parameter that regularizes the similarity gap between the two image pairs. The basic idea is to formulate a loss function that increases the similarity between A_i and P_i, and decreases the similarity between A_i and N_i by a gap, as shown in Fig. 2.

In this model, the most crucial component is to learn an image embedding function $f(.)$. Once this model is able to represent each frame correctly, we use its architecture and parameters to fine-tune a siamese network. This network is responsible of determine if two different keyframes belong to the same scene by generating a good representation of two frames from different shots and classifying those representations with a fully connected shallow network.

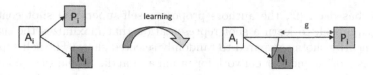

Fig. 2. Triplet loss performance.

4 Dataset

To carry out a scene understanding, and be able to extract visual features according to the visual and conceptual similarity between frames of two different shots, we build the MovieClips dataset, consisting of a set of triplets, where the positive image belongs to the same scene as the query image, and the negative image belongs to a different scene within the same movie. Then, for the classification task, we split each triplet sample into two different samples, each of them composed of two images and a binary label that defines whether both frames belong to different scenes.

4.1 Data Collection

Movieclips [23] is a YouTube channel that collects the best moments, scenes, and dialogues from many movies, making it one of the largest collections of movie clips on the web. Then, in order to build an appropriate dataset for our model, each video is filtered and classified into different movies by performing regular expression matching operations on the video title. Once the video is downloaded, Shot Boundary Detection is performed using PySceneDetect [7]. Finally, we extract the center frame of each shot and run a triplet sampling algorithm over all the extracted frames in order to have the data available for training our triplet-based network. The final dataset comprises more than 13K movie clips, translating to more than 500 h of video content, classified into more than 1.3K movies. From these videos, we extract more than 618K frames.

4.2 Extracting Frames

It is possible that the video contains an outro (sequence at the end of the video) which should be omitted as it contains frames that do not belong to the scene. An algorithm has been developed that identifies the outro of the video based on the similarity of the last frame given a set of possible options. The video can also contain a letterbox/pillarbox which is intended to preserve the aspect ratio of the content. Since these black bars do not contain useful information to determine the similarity between the content of the frames, an algorithm capable of detecting and removing these black bars has been implemented as described below.

Outro Detection. The Movieclips channel presents 5 different outros in its videos (although it may also have none). The algorithm aims to identify the video's outro by measuring the similarity between the last frame of the video and the frame of each of the possible outro's ends and comparing this similarity value with a certain threshold. Once the outro is detected, the sbd annotation will be modified, indicating $EndSceneInstant = EndVideoInstant - OutroDuration$.

Comparing an image with noisy variations of itself, we concluded that Multi-Scale Structural Similarity [26] is less sensitive to noise, generating more optimal values for slight visual changes in the image.

Letterbox/Pillarbox Detection. The aspect ratio of the video content collected in Movieclips is very varied, so an algorithm capable of detecting black bars dynamically is needed. For this purpose, an analysis of the intensity changes in both axes along the video is made. This analysis consists of the following:

1. For each center frame of the shot, after being converted to grayscale and normalized to the range [0,1], the derivatives are calculated using the Scharr function [16], which is more accurate than the standard Sobel function for kernel size 3×3.
2. To correctly determine the edge points, an adaptive thresholding binarization algorithm [15] is applied to both derivatives.
3. Normalize the sum of the center frames derivatives.
4. The letterbox/pillarbox detection is based on the number of edge points in the mean derivatives frames of the corresponding axis, horizontal changes for letterboxes and vertical changes for pillarboxes. Taking into account that the presence of letterbox/pillarbox is symmetrical with respect to the corresponding axis, it is convenient to sum the number of edge points on both sides when comparing with the threshold. Then, the method considers the existence of letterbox/pillarbox if the percentage of edge points in each pair of rows/columns exceeds 97.5%.

4.3   Triplet Sampling

The choice of a triplet selection strategy is also crucial for the model to learn visual similarity. If we feed the network with triplets composed of random images, there is a chance it may learn via easy samples early on training, hit a local minima, and fail in optimum convergence. In this paper, we employ an offline sampling scheme, which consists of choosing a Soft Positive and a Hard Negative for each Anchor. Soft Positive means that it is easy to find similarities between images of the same class and Hard Negative means that it is difficult to find dissimilarities between images of different classes. As our dataset is composed of frames from different scenes, we assume that each scene corresponds to a class and that the possible hard negative classes will be the rest of the scenes belonging to the same movie. The idea is to calculate the cosine similarity between each Anchor and each possible Positive/Negative.

We use convolutional ResNet50 CNN architecture pre-trained on the ImageNet dataset (ILSVRC 2012) [19] as feature extractor, generating a batch of features B_i for each scene i, considering each video as a unique and independent scene. Each batch of features B_i is encoded as a $n \times d$ tensor, where n is the number of frames that make up the scene and d is the dimension of the output feature vectors. Once all feature batches from a movie are generated, cosine similarity-based triplet mining begins.

First, we need to find the most similar scene within the same movie. For this purpose, a new feature batch M is computed which contains the average feature vector of each scene by making a weighted average of all the key-frame feature vectors of the shots that make up the scene. From M, which is a $s \times d$ tensor, where s is the total number of videos within a movie, we can compute a batch pairwise similarity (Eq. 3) between each feature vector of both batches, where B_i/B_j is the feature batch of $class_i/class_j$, and $\|B_i\|/\|B_j\|$ includes the Euclidean norm of each feature vector in B_i/B_j encoding a $n_i/n_j \times 1$ tensor. This computation results in a $n_i \times n_j$ tensor, where each value x, y indicates the similarity between the feature vector x of B_i and the feature vector y of B_j.

$$batch\_parwise\_similarity(B_i, B_j) = \frac{B_i \cdot B_j^t}{\|B_i\| \cdot \|B_j\|^t} \tag{3}$$

Then, with Eq. 4, we find the hardest class, where H encodes the indexes of the most similar class in a 1-dimensional tensor of size s.

$$H = argmax[batch\_parwise\_similarity(M, M)] \tag{4}$$

Finally, we find for each anchor image, its correct positive and negative images. To this end, the batch pairwise similarity between the corresponding batches is computed as shown in Eq. 5 and Eq. 6, where B_i is the feature batch of class i (anchors) and B_j is the feature batch of the most similar class to the class i ($j = H_i$).

$$positives(B_i) = argmax[batch\_parwise\_similarity(B_i, B_i)] \tag{5}$$

$$negatives(B_i) = argmax[batch\_parwise\_similarity(B_i, B_j)] \tag{6}$$

5   Network Architecture

A triplet-based architecture is proposed for the model loss function (Eq. 2). This network takes three different image batches as input. The first batch contains the query images, the second one the corresponding positive images, and the third one the corresponding negative images. Each of them is fed independently into three identical deep neural networks with shared architecture and weights. Then, the deep neural network computes the embedding of each image E_i: $f(I_i) \in R^d$, where d is the dimension of the feature embedding.

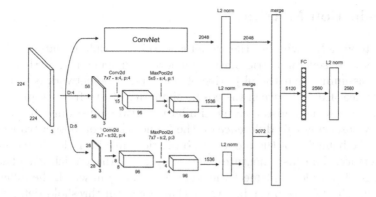

Fig. 3. Multi-scale network architecture.

We design a multi-scale deep neural architecture, shown in Fig. 3. The ConvNet corresponds to the ResNet50 convolutional layers encodes strong invariance and captures the image semantics. The other two parts of the network takes downsampled images and use shallower network architecture to capture the visual appearance as result of having less invariance. Finally, we normalize the embeddings from the three parts, and combine them in a fully-connected layer. Although ResNet50 achieves very good performance for image classification, the strong invariance encoded in its architecture can be harmful for fine-grained image similarity tasks. ReLU activation function is used after each convolutional and linear layer.

Then, we implement a siamese network composed of two branches, each of the them corresponds to the multi-scale network defined above. A fully-connected shallow network takes the joint representations of both shots and learns how to weight the components to determine, based on the similarity between them, whether they belong to the same scene. This network process an input of 5120 (2×2560) components through 3 hidden layers using ReLu activation function after each of them. Finally, an output layer with a single neuron and a Sigmoid activation function, which seems to be better than Softmax for Binary Classification tasks, make the prediction based on the correlation between both frames.

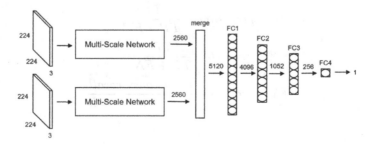

Fig. 4. Siamese network architecture.

6   Evaluation Metrics

In order to visualize whether the model is learning to differentiate the frames of different scenes, several metrics have been developed depending on their purpose.

First, assuming that the objective of the multi-scale network is to maximize the difference (Eq. 7) between the cosine similarity of two frame representations from the same scene and the cosine similarity of two embeddings from different scenes, by observing this difference on the frames that compose a triplet we can evaluate the behavior of the model. Two metrics have been developed based on this difference. The first measures the accuracy of the model, calculating the percentage of samples that meet the condition (Eq. 8) in which the difference of both cosine similarities must be higher than a certain threshold defined by the margin of the loss function. The second metric simply shows the average value of this difference for each epoch, trying to verify that this value increases as the network learns to discriminate the elements of the triplet.

$$S(f(A_i), f(P_i)) - S(f(A_i), f(N_i)) \tag{7}$$

$$S(f(A_i), f(N_i)) + g <= S(f(A_i), f(P_i)) \tag{8}$$

We have also designed two measures of accuracy for the classification network. The first is based on the percentage of correctly predicted labels and the other based on the F1 score of the model. This last metric takes the well-detected scene changes (TP, label $= 1$, pred $= 1$), the number of frame pairs that are grouped in the same scene but belong to different scenes, and therefore the scene change is not detected (FN, label $= 1$, pred $= 0$), and the pairs that belong to the same scene but the algorithm indicates that the similarity between them is not enough to classify them correctly (FP, label $= 0$, pred $= 1$). Then we compute three different metrics:

- Precision (Eq. 9): shows the hit rate of detected scene changes.
- Recall (Eq. 10): indicates the percentage of scene changes detected with respect to the ground truth.
- F1-score (Eq. 11): Combine both precision and recall metrics.

$$precision = \frac{TP}{TP + FP} \tag{9}$$

$$recall = \frac{TP}{TP + FN} \tag{10}$$

$$F1\_score = 2 \times \frac{precision \times recall}{precision + recall} \tag{11}$$

7 Training

Training a deep neural network usually needs a large amount of data. This data can be loaded from disk (loading data during training) or from memory (loading data before training). For the first case, the dataset contains the image paths and those images are loaded and transformed at each pass through the network, which considerably increases the training time. For the second case, the images are loaded and stored in memory with the transformations already applied, which leads to a significant reduction in training time, but also a large memory consumption and the time taken to load the dataset.

On the other hand, we use the pre-trained ResNet50 weights on ImageNet in the ConvNet of the multi-scale network to reduce the number of epochs at which the model converges and, consequently, the time spent in training. For this triplet based model, we employ the stochastic gradient descent with Nesterov momentum algorithm. The momentum algorithm [11] converges faster than traditional stochastic gradient methods. Back-propagation scheme is used to compute the gradient. To avoid overfitting, dropout with keeping probability 0.6 is applied to all the linear layers. The loss function used is the one explained in Sect. 3.

The siamese network follows the same stochastic method as the previous network (SGD). However, since this model is intended to solve a binary classification problem, the Binary Cross Entropy Loss [18] will be used as loss function. It is a criterion that measures the binary cross entropy between the target and the input probabilities. Since the multi-scale network is in charge of generating a good representation of the corresponding frames, we freeze its parameters during the classification training, fine-tuning the model and learning to find the differences between the two representations. Dropout with probability 0.6 is also applied to all the linear layers.

8 Experiments

We train both models using a subset of our MovieClips dataset. This subset is composed of 20 different movies (176 videos), generating a total data volume of 3253 frames. Then, applying our triplet sampling method on this subset, we will obtain 3253 samples to train the triplet-based model and, since each triplet is broken down into two different duplets for the siamese network, 6506 samples to learn the classification task. After shuffling the samples, the subset is split into training and validation at a ratio of 0.8 and 0.2 respectively, distributing the triplets/duplets in batches of size 8. Data preprocessing consists of resizing the samples to 224×224 using bilinear interpolation, followed by central cropping and finally rescaling pixel values to $[0.0, 1.0]$.

Several trainings have been performed trying with different optimization algorithms. The Adam algorithm, for example, is not able to minimize the loss function for our case. However, using SGD with a momentum factor of 0.8 and adjusting the learning rate, we managed to find suitable coefficients that minimize the cost function in a reasonable number of epochs.

First, to learn the frame embedding function we train the triplet-based model by setting the learning rate to 0.01. The learning curves for the first 100 epochs are shown in Fig. 5. Although we are able to minimize the loss function for the training subset, the model is not able to generalize correctly, leading to an overfitting problem. This problem can arise as a consequence of using a small and not very varied dataset or an overly complex model which generates an output with an excessive amount of features. Considering that we have only used 1% of the MovieClips dataset, it is possible that by training the model with all the samples we may mitigate this error. In addition, we can also add regularization, implement data augmentation techniques by creating modified copies of the existing data or reduce the number of features by applying PCA [13].

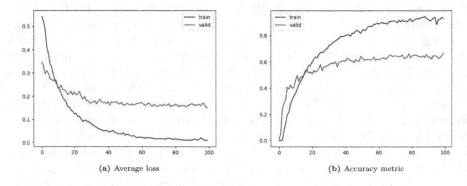

(a) Average loss   (b) Accuracy metric

Fig. 5. Multi-scale network learning curves.

On the other hand, we apply transfer learning by loading the trained weights at epoch 50 of the multi-scale network on each of the siamese network branches and perform fine-tuning on the fully-connected layers. For this purpose, we have employed decay over a learning rate initialized to 0.001, which will be divided by 10 every 15 epochs. Although the frame representation is not optimal, the

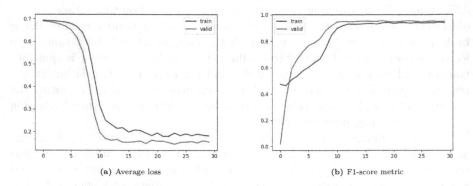

(a) Average loss   (b) F1-score metric

Fig. 6. Siamese network learning curves.

classification model seems to be able to distinguish whether two frames belong to different scenes. This can be seen in Fig. 6 where, after training for 30 epochs, the loss function of both subsets converges to around 0.2 and the f1-score rises to 0.9, indicating that both the precision and recall of the model are high.

9 Conclusions

To perform a scene segmentation algorithm, it is necessary to analyze the relationship between the content of adjacent shots. This relationship is multi-modal, since visual and audio signals play an important role. The built dataset is a good option to train a model to analyze the visual and semantic similarity between the most representative frames of multiple shots. For this reason, it is important to use a key-frame extraction strategy, so that each of the dataset samples contains relevant information of the shot to which it belongs. Then, we introduce two deep learning models to analyze and classify this similarity. The training experiments show that the model in charge of creating frame feature embeddings suffers from overfitting. However, this problem can be solved in future trainings. On the other hand, the siamese network appears to be effective and to achieve good frame classification performance based on their visual features. All of the studies in this paper show that scene analysis is a difficult but important topic that deserves research.

Acknowledgment. We would first like to thanks Telefonica I+D for supporting the Industrial Phd of Miguel Esteve Brotons. We would like to thank 'A way of making Europe" European Regional Development Fund (ERDF) and MCIN/AEI/10.13039/501100011033 for supporting this work under the TED2021-130890B (CHAN-TWIN) research Project funded by MCIN/AEI /10.13039/501100011033 and European Union NextGenerationEU/ PRTR, and AICARE project (grant SPID202200X139779IV0). Also the HORIZON-MSCA-2021-SE-0 action number: 101086387, REMARKABLE, Rural Environmental Monitoring via ultra wide-ARea networKs And distriButed federated Learning. Finally, we also would like to thank Nvidia for their generous hardware donations that made these experiments possible.

References

1. Aytar, Y., Vondrick, C., Torralba, A.: Learning sound representations from unlabeled video, Soundnet (2016)
2. Baraldi, L., Grana, C., Cucchiara, R.: A deep siamese network for scene detection in broadcast videos. In: Proceedings of the 23rd ACM International Conference on Multimedia, MM '15, pp. 1199–1202, New York, NY, USA (2015). Association for Computing Machinery
3. Baraldi, L., Grana, C., Cucchiara, R.: A deep Siamese network for scene detection in broadcast videos. In: Proceedings of the 23rd ACM International Conference on Multimedia, ACM (2015)

4. Baraldi, L., Grana, C., Cucchiara, R.: Shot and scene detection via hierarchical clustering for re-using broadcast video. In: Azzopardi, G., Petkov, N. (eds.) CAIP 2015. LNCS, vol. 9256, pp. 801–811. Springer, Cham (2015). https://doi.org/10.1007/978-3-319-23192-1_67

5. Berhe, A., Guinaudeau, C., Barras, C.: Video scene segmentation of tv series using multi-modal neural features (2019)

6. Bouyahi, M., Benayed, Y.: Video scenes segmentation based on multimodal genre prediction. Procedia Comput. Sci. **176**,10–21 (2020)

7. Castellano, B.: PySceneDetect 2014–2022. https://github.com/Breakthrough/PySceneDetect

8. Chen, S., Nie, X., Fan, D., Zhang, D., Bhat, V., Hamid, R.: Shot contrastive self-supervised learning for scene boundary detection (2021)

9. He, K., Fan, H., Wu, Y., Xie, S., Girshick, R.: Momentum contrast for unsupervised visual representation learning (2020)

10. He, K., Zhang, X., Ren, S., Sun, J.: Deep residual learning for image recognition. In: 2016 IEEE Conference on Computer Vision and Pattern Recognition (CVPR), pp. 770–778 (2016)

11. Shun ichi Amari: Backpropagation and stochastic gradient descent method. Neurocomputing **5**(4), 185–196 (1993)

12. Krizhevsky, A., Sutskever, I., Hinton, G.E.: ImageNet classification with deep convolutional neural networks. In: Pereira, F., Burges, C.J., Bottou, L., Weinberger, K.Q., editors, Advances in Neural Information Processing Systems, vol. 25. Curran Associates Inc (2012)

13. Maćkiewicz, A., Ratajczak, W.: Principal components analysis (PCA). Comput. Geosci. **19**(3), 303–342 (1993)

14. Mun, J., et al.: Boundary-aware self-supervised learning for video scene segmentation (2022)

15. OpenCV. Image Thresholding (2023). https://docs.opencv.org/4.x/d7/d4d/tutorial_py_thresholding.html

16. OpenCV. Sobel Derivatives (2023). https://docs.opencv.org/3.4/d2/d2c/tutorial_sobel_derivatives.html

17. Rao, A., et al.: A local-to-global approach to multi-modal movie scene segmentation. In: Proceedings of the IEEE/CVF Conference on Computer Vision and Pattern Recognition, pp. 10146–10155 (2020)

18. Ruby, U., Yendapalli, V.: Binary cross entropy with deep learning technique for image classification. Int. J. Adv. Trends Comput. Sci. Eng. **9**, 10 (2020)

19. Russakovsky, O., et al.: ImageNet large scale visual recognition challenge. Int. J. Comput. Vis. (IJCV) **115**(3), 211–252 (2015)

20. Sivaraman, K., Somappa, G.: MovieScope: movie trailer classification using deep neural networks (2017)

21. Statista. Digital media - video on demand worldwide (2023). https://www.statista.com/outlook/dmo/digital-media/video-on-demand/worldwide#revenue

22. Tapu, R., Mocanu, B., Zaharia, T.: DEEP-AD: a multimodal temporal video segmentation framework for online video advertising. IEEE Access **8**, 99582–99597 (2020)

23. Rotten Tomatoes. @movieclips, 2006–2023. https://www.youtube.com/@MOVIECLIPS

24. Vendrig, J., Worring, M.: Systematic evaluation of logical story unit segmentation. Multimedia, IEEE Trans. **4**, 492–499 (2003)

25. Wang, J., et al.: Learning fine-grained image similarity with deep ranking. In: 2014 IEEE Conference on Computer Vision and Pattern Recognition, pp. 1386–1393 (2014)
26. Wang, Z., Simoncelli, E.P., Bovik, A.C.: Multiscale structural similarity for image quality assessment. In: The Thrity-Seventh Asilomar Conference on Signals, Systems & Computers, vol. 2, pp. 1398–1402 (2003)
27. Wu, H., et al.: Scene consistency representation learning for video scene segmentation (2022)
28. Zoph, B., Vasudevan, V., Shlens, J., Le, Q.V.: Learning transferable architectures for scalable image recognition (2018)

A Model for Classifying Emergency Events Based on Social Media Multimodal Data

ZhenHua Wu(iD), Liangyu Chen(iD), and YuanTao Song$^{(\boxtimes)}$(iD)

University of Chinese Academy of Sciences, Beijing 100043, China
{wuzhenhua21,chenliangyu21}@mails.ucas.ac.cn,
songyuantao@ucas.ac.cn

Abstract. Social media has emerged as a crucial source of information for emergency management. However, the diverse range of data types, including textual and visual information, presents a significant challenge for scholars seeking to analyze this information effectively. In this paper, we propose a novel multimodal model that employs cross-attention mechanisms to effectively integrate textual and visual information. The model is further enhanced with attention-based pooling layer, whole word masking, and RandAugment image data enhancement techniques, which are leveraged to classify contingencies in social media tweets. Empirical evaluation on the CrisisMMD dataset demonstrates that our model outperforms multiple existing baseline approaches for informative tasks and humanitarian action classification tasks. These results affirm the effectiveness of the model in integrating features from multiple modalities and demonstrate its superior generalization capabilities.

Keywords: Multimodality · Attention mechanism · Emergencies classification

1 Introduction

In the current age of flourishing social media platforms, an enormous quantity of images and texts are being uploaded every second to social media platforms worldwide. Social media platforms generate tweets at a high velocity such that it is no longer arduous to acquire real-time data on a large scale. However, these tweets contain a substantial amount of content that is irrelevant to the topic, thus automating the processing and analysis of social media big data remains a challenge in emergency situations. In the context of emergency situational awareness, it is imperative to efficiently sort through the vast amount of social media data to extract relevant information that can aid in emergency management, which has significant value in achieving timely response and decision-making.

Alam et al. [3] developed a CrisisMMD dataset that comprises approximately 20, 000 manually labeled image-text pairs. Meanwhile, Ofli et al. [15] conducted a study on this dataset, demonstrating that models incorporating both text and image modalities outperform their unimodal counterparts. Notably, Fig. 1 showcases a tweet regarding the Iran earthquake within the dataset, together with its associated image. Apparently, the

I. Rojas et al. (Eds.): IWANN 2023, LNCS 14134, pp. 316–327, 2023.
https://doi.org/10.1007/978-3-031-43085-5_25

Fig. 1. Text of original tweet: "Iran's earthquake exposes political rifts and ineffective governance."

text alone cannot accurately infer the extent of the damages resulting from the earthquake. Nonetheless, a more intuitive comprehension of the damages can be achieved through analysis of the tweet's attached image.

As previously indicated, a distinct correspondence exists between the visual and textual content. In instances where the textual data is insufficient to explicate the context, analysis of the corresponding images facilitates the derivation of additional correlation information. This approach enhances the model's capacity to render accurate inferences and judgments.

In short, we propose a novel emergency classification model, termed EMC-SMMD (Emergency Classification Model for Social Media Multimodal Data), on the Crisis-MMD dataset. A cross-attention mechanism is introduced to enhance the interaction between the different modal features. We also introduce an attention-based pooling layer, thereby improving the receptive field of the image model. In addition, whole word masking technique is employed to preserve the contextual information of words, while RandAugment [5] is applied during model finc-tuning to improve the generalization ability.

Based on experimental findings, our model outperforms the existing unimodal and multimodal baseline method in terms of enhancing the interaction of significant information in both textual and visual feature and ensuring accurate inference.

2   Related Work

2.1   Application of AI in Emergency Response

The unprecedented growth of the Internet has led to an exponential surge in the amount of breaking news-related information available on social media platforms. Effectively navigating and monitoring data has become increasingly challenging. In recent years, there has been significant research activity focused on emergency analysis, which has yielded several proposed methods. However, these techniques rely on manual feature extraction, which is often plagued by the subjectivity of the individuals involved and necessitates extensive pre-processing. Consequently, it is challenging to obtain accurate and comprehensive representation features of social events. The advent of deep learning has revolutionized the field of AI in social media [7].

The current research on emergency detection in social media mostly concentrated on textual data. Notably, Kumar et al. [10] proposed a tweets monitoring system for providing first action following a critical incident, while Shekhar et al. [18] introduced a system that can estimate the severity of damage to facilities and the level of suffering experienced by victims.

In contrast to prior research, we center on the integration of textual and visual modalities to enhance the accuracy of emergent event classification in the domains of social media.

2.2 Multimodal Learning

In the domain of multimodal learning, neural networks are commonly employed to fuse complementary information extracted from multiple modalities of a given phenomenon [1,17]. This fusion strategy has proven to be effective in several applications [12]. Nonetheless, the utilization of current multimodal learning frameworks in the context of breaking news remains limited. To address this gap, several recent studies have proposed novel multimodal fusion approaches. For instance, Nie et al. [14] proposed an end-to-end multiscale fusion model and embedded semantic information into hash codes for cross-modal retrieval tasks, while Wu et al. [20] used three different sub-networks to extract features, and then deeply fused them by overlaying a common attention layer.

The majority of existing research focuses on constructing image and text feature extractors while neglecting the importance of cross-modal feature interactions. We employ a cross-attention mechanism for model fusion, which enables more profound information interaction across different modalities than mere concatenation. Moreover, our proposed model introduces an attention-based pooling layer that effectively conveys meaningful information and enhances the receptive field of the feature extraction model.

3   Methodology

We will elaborate our model in 6 parts: we use VGG19 and ALBERT [11] to extract image feature maps and text embeddings, respectively, in the first two parts; the third part comprises cross-attention methods that fuse features; the fourth part applies attention-based pooling layer replacing the Max-pooling layer of VGG19 to enhance the actual receptive field; and the last two parts comprise whole-word masking and RandAugment. We apply these techniques to improve the generalization ability of our model.

3.1   Image Model

We employ a transfer learning methodology to process the data in image modalities, which has been demonstrated as a promising technique in the field of visual recognition tasks [16]. The model, VGG19, is a deep convolutional neural network that has

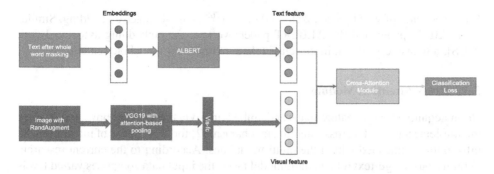

Fig. 2. EMC-SMMD model structure overview

demonstrated remarkable efficacy in image feature extraction. It leverage a deep archi-tecture that incorporates multiple convolutional and pooling layers. And we employ transfer learning by utilizing the pre-trained weights of the VGG19 model on ImageNet as the initialization parameters. Moreover, we modify the Softmax classification layer to account for 2 and 5 classes that are relevant to the informative task and the humanitarian action task, respectively.

For each input image v_i:

$$f_i = VGG19(v_i) \tag{1}$$

f_i is the vectorized form of the feature map in the VGG19, which dimension $D_f = W \times H \times C$. Of course, W, H, C represents height, width and the number of channels of the feature map, respectively.

3.2 Text Model

ALBERT, a lightweight pre-trained natural language processing model, has been intro-duced as a variant of BERT. Its model structure comprises a multi-layer Transformer encoder, which is utilized for encoding input sentences. This model's salient charac-teristics include improved efficiency and reduced number of parameters. In this study, ALBERT has been employed as a text feature extraction module to obtain pre-trained word embeddings.

Based on the trade-off between model performance and computational resources utilized, we choose ALBERT as textual processing model. And it has undergone pre-training on tweet data pertaining to emergency events on Wiki dataset.

Given a text content T, we model T as a sequence of words $T = t_1, t_2, ..., t_n$, where n denotes the number of words in the text. The input ALBERT network yields the trans-formed feature representation as $e_1, e_2, ..., e_n$, where e_i corresponds to the transformed characteristic of t_i. The representation of the word e_i is obtained from the pre-trained model ALBERT by converting. For input text t_i, we have:

$$e_i = ALBERT(t_i) \tag{2}$$

t_i is a sequence of word-piece tokens and $e_i \in R^{768}$ is the sentence embedding. Similar to the BERT paper and the ALBERT paper, we take the embedding associated with [CLS], a trainable vector akin to a class token, to represent the whole sentence.

3.3   Cross-Attention Module

Upon acquiring image feature maps and embeddings via the models above, we propose the implementation of a cross-attention mechanism [2] for the purpose of integrating the information enmeshed within the multi modalities. According to the current research, in numerous image-text related multimodal tasks, the input data comprises varied levels of noise or erroneous information. Such scenarios may lead to the transfer of adverse information that may reduce the model's performance during training. However, using a cross-attention mechanism can mitigate the influence of negative information.

We presents a cross-attention module, which leverages a confidence-based blocking mechanism to prevent negative features from being processed. The partially blocked features obtained from visual and textual modalities will turn to a self-attention layer that subsequently filters the information and determines which features are relevant for propagation to the next layer.

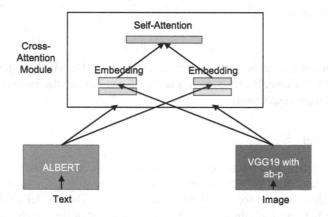

Fig. 3. Cross-attention module structure

The self-attention layer uses the fully-connected layer to project the image feature mapping to a fixed dimension K (the value of K is determined based on the experimental results, and in this model, the optimal K value is chosen as 100). The feature representation obtained after the operation is $\hat{f}_i$. Similarly, the embedding of sentences is similarly mapped and projected to a fixed dimension to obtain $\hat{e}_i$, this can be expressed as follows:

$$\hat{f}_i = F(W_v^T f_i + b_v) \tag{3}$$

$$\hat{e}_i = F(W_e^T e_i + b_e) \tag{4}$$

We use ReLU as the activation function F.

Without attention mechanism, the integration of features derived from diverse modalities poses a significant challenge as it could potentially undermine the efficacy of the model when confronted with spurious information in any given modality. This novel cross-attention mechanism relies entirely on the text embedding e_i to generate the visual attention mask A_{mv_i}, and solely on the image embedding v_i to generate the textual attention mask A_{me_i}. So we have:

$$A_{mv_i} = \sigma(W_v'^T[f_i \mid e_i] + b_v') \qquad (5)$$

$$A_{me_i} = \sigma(W_e'^T[f_i \mid e_i] + b_e') \qquad (6)$$

We use the Sigmoid function as σ, and | is concatenation.

Finally, the resultant amalgamation of embeddings will be fed into a fully connected network comprised of two layers. Self-attention is incorporated into the fully-connected network comprised of two layers, and the Softmax cross-entropy loss function is employed to execute the classification process.

3.4 Attention-Based Pooling Layer

The application of attention mechanisms has become a popular and powerful method to enhance the performance of neural networks. One such critical use case of the mechanism is to augment the effective receptive field of the network, which is essential for capturing long-range dependencies in the input data. Luo et al. [13] contend that CNN models are highly localized, where the contribution of various pixels within the receptive field is not uniform, and the effective receptive field is a Gaussian distribution that accounts for only a fraction of the theoretical receptive field. The authors demonstrate that the effective receptive field scales at the $\sqrt{n}$ level as the number of network layers n increases, and the ratio of the effective receptive field will decrease at the $\frac{1}{\sqrt{n}}$ level. Furthermore, the receptive field grows during the model training process. Araujo et al. [4] computed the receptive field of the neural network and concluded that the theoretical receptive field of the original VGG19 model is 260×260. To improve the receptive field, we propose an attention-based pooling layer to substitute the original model's last max-pooling layer. This proposed layer is similar to a single Transformer block, which weights the input features to better participate in the classification decision.

Fig. 4. Attention-Based Pooling

At the end of the backbone network, the feature vectors produced by said network undergo a process of weighting and aggregation via the implementation of an attention-based pooling layer, illustrated in Fig. 4. Weighting of the vectors is determined based on the degree of similarity between pixel regions and the [CLS] vector. The resulting d-dimensional vectors, which have undergone aggregation, are subsequently summed with the [CLS] vectors before being processed through a feed-forward network (FFN). We introduce one-dimensional convolution for visual features to perform local information aggregation and dimensionality reduction. Pooling operator is implemented through a Softmax layer. We employ a single-headed attention model to prevent dilution of attention across multiple channels. By converting the [CLS] vector to a matrix, the mechanism implemented is capable of focusing on category labels. While the replacement with attention-based pooling will increase computational usage, we balance it by reducing the learning rate and employing a lower batch size. As per relevant receptive field calculations, the improved VGG19 model offers a field of 324×324, a 55.3% increase compared to the original model.

3.5 Whole Word Masking

Whole word masking is a prevalent method in NLP tasks, which involves predicting the complete sequence of a word rather than just its constituent characters. This technique conceals some words during language preprocessing, allowing the model to predict these hidden words based on contextual information. Compared to random masking, whole word masking offers a superior advantage in preserving contextual information of words, thereby enabling the model to better learn the features of language and enhance the prediction accuracy for unknown words. The implementation comprises two parts: word embeddings for encoding the text's words and the Transformer architecture for modeling the contextual relationships among the words.

3.6 Image Data Enhancement

RandAugment is a data enhancement method, which includes color transformations, geometric transformations, filter transformation, image distortion and pixel control.

These involves selecting parameters within a specific range base on the application scenario. The technique offers an advantage in that it allows for a diverse set of training images to be generated through random transformations. This is beneficial for enhancing the robustness and generalization performance of the model, while also mitigating the risk of overfitting. And we propose to apply it during model fine-tuning.

4   Experimental Setup

4.1   Introduction to the Dataset

We employ the CrisisMMD dataset, a multimodal dataset containing tweets related to seven natural disasters in 2017. Our study exclusively concentrates on distinguishing between informational and not informational content, categorizing humanitarian actions.

Informative vs. Not Informative: Determining whether tweets collected during a critical incident can be used for humanitarian action purposes.

Humanitarian Action: Categorizing tweets from test set into one of the following 5 categories:

- Infrastructure and utility damage
- Rescue volunteering or donation effort
- Affected individuals (injured, killed, reconnected, etc.)
- Other relevant information (relevant news broadcasts, etc.)
- Not humanitarian

In light of manual annotation of text and images separately, tweets and images of the same emergent event may have different labels. We, therefore, focus on data where a given task has identical labels for both text and images. Such filtering inevitably leads to a paucity of samples for some categories, thus distorting the overall label distribution and potentially impeding the model's ability to learn adequate semantic features. To address this issue, we combine semantically similar or related categories. The end result is a subset of five categories for the humanitarian act tasks.

Given that tweets on Twitter can have up to four appended images, the CrisisMMD dataset contains several tweets that possess multiple images. It was imperative to ensure the absence of tweet text duplication. Therefore, for tweets with multiple images, we put them all in the training set. The experimental setup employed a 70%, 15%, and 15% data split ratio for the training, validation, and test sets respectively (Table 1).

Table 1. List of categories and data splitting for different tasks.

Task	Train(70%)		Val(15%)		Test(15%)		Total	
	Text	Image	Text	Image	Text	Image	Text	Image
Informative Task	8,293	9,601	1,573	1,573	1,534	1,534	11,400	12,708
Humanitarian Action Task	5,263	6,126	998	998	955	955	7,216	8,079

4.2 Data Preprocessing

Apparently, the textual content in dataset is filled with a variety of noise. So we reduce it by removing deactivated words, non-ASCII characters, numbers, and URL links from the text. In addition, punctuation marks are replaced with spaces.

We employ an approach for processing images included in tweets, which involves scaling the pixel values between the range of 0 to 1 and augmenting the training images using RandAugment. And we normalize each channel of the images.

4.3 Baselines

In this paper, we present a comparative evaluation of our proposed model against some existing unimodal and multimodal classification techniques. Specifically, we assess the performance of our model by comparing with VGG19, ResNet18 [8] and BERT, after subjecting them to fine-tuning in accordance with the actual training set structure.

To evaluate various multimodal baseline techniques, we focus on recently introduced multimodal fusion methods that have been utilized for classification purposes.

Of note, two of the methods assessed have previously been implemented in the Crisis-MMD dataset.

- SCBD: The proposed method employs DenseNet and BERT to extract features. Cross-attention techniques are utilized in modal fusion, and random shared embedding (SSE) serves as a regularization approach to mitigate overfitting.
- COBRA [19]: This article presents a Contrastive Bimodal Representation Algorithm for joint cross-modal embedding. They utilize a set of loss functions that incorporate multimodal fusion, as well as a contrast algorithm to enable the models to jointly maintain the relationships between distinct internal cross-modal data samples.
- MMBT [9]: The study proposes a model under scrutiny incorporates two modalities and is designed to learn from labeled data with the aid of a Transformer-based architecture.
- CBP [6]: The study introduces a concise approach for Bilinear Pooling in the context of visual question answering.

4.4 Evaluation Metrics

We follow the data partitioning and preprocessing methods of previous research work on CrisisMMD dataset, using Accuracy, Precision, Recall, and F1-Score as the evaluation metrics for model performance.

4.5 Parameter Settings

For the multimodal embedding of emergency, we extract image features using VGG19 pre-trained on ImageNet, and the image embedding has a dimension of $7\times7\times512$. The text features are extracted using pre-trained ALBERT, and according to the original setting in the paper, the text embedding has a dimension of 768. We use standard Adam optimizer with a total of 100 epochs, with an initial learning rate of 2×10^{-3} and a small batch setting of 64. In addition, when the number of accuracy stopping boosting rounds on the validation set reaches 20, the initial learning rate will be reduced by a multiplier of 0.1 and training will continue.

In the training phase, for images, we apply RandAugment to get more training images by randomly flipping and rotating them. And we randomly mask a complete word with a 30% probability when rotating the text.

5   Experimental Results

5.1 Comparison of Performance

Table 2 presents the experimental results of EMC-SMMD, which has exhibited notable advantages compared to independent models such as VGG19, ResNet18 and BERT, for both informative classification tasks and humanitarian tweet classification tasks. And it has also outperformed several multimodal baseline methods, namely CBP, COBRA and MMBT. When compared to the selected SCBD approach, our proposed model exhibits an overall superior performance in informative classification, and shows competitive performance in humanitarian tweet classification. In comparison to the SSE technique used by SCBD, the whole word masking and RandAugment we used are more generalizable and can be applied to data enhancement in downstream tasks. Relevant ablation experiments have been conducted in Sect. 5.2 to investigate the impact of these modules (cross-attention mechanism, attention-based pooling layer, whole word masking and RandAugment) on the model's performance.

Table 2. Different Models' Performance on CrisisMMD.

Model	Informative Task				Humanitarian Categorization Task			
	Accuracy	Precision	Recall	F1-Score	Accuracy	Precision	Recall	F1-Score
VGG19	0.803	0.807	0.794	0.801	0.771	0.767	0.753	0.760
ResNet18	0.836	0.834	0.827	0.830	0.819	0.812	0.803	0.807
BERT	0.849	0.846	0.851	0.848	0.857	0.853	0.848	0.850
SCBD	0.893	0.881	0.894	0.887	**0.904**	**0.897**	0.883	**0.890**
COBRA	0.906	0.898	0.893	0.895	0.884	0.881	0.877	0.879
MMBT	0.825	0.817	0.801	0.809	0.858	0.854	0.847	0.850
CBP	0.881	0.876	0.862	0.869	0.863	0.870	0.833	0.851
EMC-SMMD(ours)	**0.915**	**0.906**	**0.897**	**0.901**	0.893	0.895	**0.886**	**0.890**

5.2 Ablation Study

In the ablation experiments, we examine the various modules: cross-attention mechanism, attention-based pooling layer, whole word masking, and RandAugment. All experiments in this section were performed in the informative task test set. The experimental results are shown in Table 3.

Table 3. Ablation Study for Informative Task.

Model	Test Set			
	Accuracy	Precision	Recall	F1-Score
EMC-SMMD(ours)	**0.915**	**0.906**	**0.897**	**0.901**
-Cross Attention	0.890	0.883	0.875	0.879
-Attention-based Pooling	0.903	0.894	0.897	0.895
-Whole Word Masking	0.905	0.899	0.893	0.896
-RandAugment	0.898	0.892	0.889	0.890

The experimental findings presented demonstrate the significance of the cross-attention mechanism in multimodal fusion. Removal of the cross-attention mechanism and using only simple concate operations led to a notable decrease in accuracy from

91.5% to 89.0%. Furthermore, utilizing the attention-based pooling layer was found to be more effective than before, as the model's accuracy dropped to 90.3% after using Max-pooling. Moreover, the study highlights the importance of employing whole word masking and RandAugment to enhance the model's generalization. The results indicate that omitting the whole word masking during the pre-training phase caused a decline in accuracy from 91.5% to 90.5%, while the exclusion of RandAugment in the fine-tuning phase resulted in an accuracy drop to 89.8%. These findings suggest that targeted data enhancement applied to the input text and image data during model training is an effective technique for enhancing the model's performance.

6   Conclusions and Future Work

In this paper, a multimodal attention-based model is presented to enhance the performance and generalization of emergent event classification task. Our model leverages the complementary information embedded in text and images, and integrates their features to emphasize the relevant information, eliminate the interference of irrelevant information, and foster cross-modal interactions. Empirical evaluations on the CrisisMMD dataset demonstrate that our model can selectively attend to informative features in both modalities, resulting in an improved joint multimodal representation.

For future work, given the limited size of the CrisisMMD dataset, we focus to apply our method to other larger, multimodal data containing more social events for emergency detection, not only for natural disaster-related datasets. Exploring the balance between model performance and computational resources spent is also an important research direction to get the model into practical application.

References

1. Abavisani, M., Joze, H.R.V., Patel, V.M.: Improving the performance of unimodal dynamic hand-gesture recognition with multimodal training. In: Proceedings of the IEEE/CVF Conference on Computer Vision and Pattern Recognition, pp. 1165–1174 (2019)
2. Abavisani, M., Wu, L., Hu, S., Tetreault, J., Jaimes, A.: Multimodal categorization of crisis events in social media. In: Proceedings of the IEEE/CVF Conference on Computer Vision and Pattern Recognition, pp. 14679–14689 (2020)
3. Alam, F., Ofli, F., Imran, M.: CrisisMMD: multimodal twitter datasets from natural disasters. In: Proceedings of the International AAAI Conference on Web and Social Media, vol. 12 (2018)
4. Araujo, A., Norris, W., Sim, J.: Computing receptive fields of convolutional neural networks. Distill (2019). https://doi.org/10.23915/distill.00021, https://distill.pub/2019/computing-receptive-fields
5. Cubuk, E.D., Zoph, B., Shlens, J., Le, Q.V.: RandAugment: practical automated data augmentation with a reduced search space. In: Proceedings of the IEEE/CVF Conference on Computer Vision and Pattern Recognition Workshops, pp. 702–703 (2020)
6. Fukui, A., Park, D.H., Yang, D., Rohrbach, A., Darrell, T., Rohrbach, M.: Multimodal compact bilinear pooling for visual question answering and visual grounding. arXiv preprint arXiv:1606.01847

7. Guntuku, S.C., Preotiuc-Pietro, D., Eichstaedt, J.C., Ungar, L.H.: What twitter profile and posted images reveal about depression and anxiety. In: Proceedings of the International AAAI Conference on Web and Social Media, vol. 13, pp. 236–246 (2019)
8. He, K., Zhang, X., Ren, S., Sun, J.: Deep residual learning for image recognition. In: Proceedings of the IEEE Conference on Computer Vision and Pattern Recognition, pp. 770–778 (2016)
9. Kiela, D., Bhooshan, S., Firooz, H., Perez, E., Testuggine, D.: Supervised multimodal bitransformers for classifying images and text. arXiv preprint arXiv:1909.02950 (2019)
10. Kumar, S., Barbier, G., Abbasi, M., Liu, H.: TweetTracker: an analysis tool for humanitarian and disaster relief. In: Proceedings of the International AAAI Conference on Web and Social Media, vol. 5, pp. 661–662 (2011)
11. Lan, Z., Chen, M., Goodman, S., Gimpel, K., Sharma, P., Soricut, R.: ALBERT: a lite BERT for self-supervised learning of language representations. arXiv preprint arXiv:1909.11942 (2019)
12. Lee, K.H., Chen, X., Hua, G., Hu, H., He, X.: Stacked cross attention for image-text matching. In: Proceedings of the European Conference on Computer Vision (ECCV), pp. 201–216 (2018)
13. Luo, W., Li, Y., Urtasun, R., Zemel, R.: Understanding the effective receptive field in deep convolutional neural networks. In: Advances in Neural Information Processing Systems, vol. 29 (2016)
14. Nie, X., Wang, B., Li, J., Hao, F., Jian, M., Yin, Y.: Deep multiscale fusion hashing for cross-modal retrieval. IEEE Trans. Circ. Syst. Video Technol. $31(1)$, 401–410 (2020)
15. Ofli, F., Alam, F., Imran, M.: Analysis of social media data using multimodal deep learning for disaster response. arXiv preprint arXiv:2004.11838 (2020)
16. Ozbulak, G., Aytar, Y., Ekenel, H.K.: How transferable are CNN-based features for age and gender classification? In: 2016 International Conference of the Biometrics Special Interest Group (BIOSIG), pp. 1–6. IEEE (2016)
17. Perera, P., Abavisani, M., Patel, V.M.: In2l: Unsupervised multi-image-to-image translation using generative adversarial networks. In: 2018 24th International Conference on Pattern Recognition (ICPR), pp. 140–146. IEEE (2018)
18. Shekhar, H., Setty, S.: Disaster analysis through tweets. In: 2015 International Conference on Advances in Computing, Communications and Informatics (ICACCI), pp. 1719–1723. IEEE (2015)
19. Udandarao, V., Maiti, A., Srivatsav, D., Vyalla, S.R., Yin, Y., Shah, R.R.: COBRA: contrastive bi-modal representation algorithm. arXiv preprint arXiv:2005.03687 (2020)
20. Wu, Y., Zhan, P., Zhang, Y., Wang, L., Xu, Z.: Multimodal fusion with co-attention networks for fake news detection. In: Findings of the association for computational linguistics: ACL-IJCNLP 2021, pp. 2560–2569 (2021)

A Performance Evaluation of Lightweight Deep Learning Approaches for Bird Recognition

Dmitrij Teterja[1(✉)], Jose Garcia-Rodriguez[1], Jorge Azorin-Lopez[1],
Esther Sebastian-Gonzalez[2], Srdjan Krco[3], Dejan Drajic[3],
and Dejan Vukobratovic[4]

[1] Department of Computer Science and Technology, University of Alicante, Alicante,
Spain
{dteterja,jgarcia}@dtic.ua.es, jazorin@ua.es
[2] Department of Ecology, University of Alicante, Alicante, Spain
esther.sebastian@ua.es
[3] DunavNet, Novi Sad, Serbia
{srdjan.krco,dejan.drajic}@dunavnet.eu
[4] Department of Power, Electronic and Communication Engineering,
Faculty of Technical Sciences, University of Novi Sad, Novi Sad, Serbia
dejanv@uns.ac.rs

Abstract. Reliable identification of bird species is a critical task for
many applications, such as conservation biology, biodiversity assessments, and monitoring bird populations. However, identifying birds in
the wild by visual observation can be time-consuming and prone to
errors. There is a growing need for efficient and accurate bird recognition methods that can help researchers and conservationists identify
bird species quickly and reliably. In this paper, we present a comparative
analysis of the performance of state-of-the-art deep convolutional neural
networks on a significantly sized bird dataset. Our goal is to develop a
more accurate and efficient bird recognition method that can be deployed
on edge computing devices. The results show that lightweight networks
as EfficientNetB0 provide a great accuracy (more than 97%) and low
time of response with a small demand for technological resources. Our
findings could provide a reliable means of identifying bird species in the
wild, which is essential for many conservation and management efforts.

Keywords: bird recognition · deep neural networks · edge computing

1 Introduction

Many biologists, including ornithologists, agree that birds play a vital role in our
ecosystem. They provide several ecological functions such as pesticide control,
pollination, seed dispersal, fertilization of the soil, and plant reproduction [7].
According to the Organisation for Economic Co-operation and Development
(OECD), biodiversity, which refers to the presence of various animal species in

© The Author(s), under exclusive license to Springer Nature Switzerland AG 2023
I. Rojas et al. (Eds.): IWANN 2023, LNCS 14134, pp. 328–339, 2023.
https://doi.org/10.1007/978-3-031-43085-5_26

an ecosystem, is crucial for sustaining life, food provisioning, water purification, flood and drought control, and nutrient cycling. Additionally, it plays a role in climate regulation [1]. These services are essential for supporting human well-being and economic prosperity, which is why many countries actively participate in promoting biodiversity conservation and sustainable use [2].

Climate change has led to a decline in the ability of many native species to provide ecosystem services at their previous levels. This situation opens up opportunities to explore future relationships between biodiversity, ecological integrity, and ecosystem services [3]. Numerous countries are actively engaged in promoting biodiversity conservation and sustainable use [2] while others are studying the impacts of climate change on biodiversity, ecosystems, and ecosystem services, as well as the implications for natural resource management [4].

The use of automated systems enables the possibility of bird image classification. This advancement opens up opportunities for ornithologists to utilize highly accurate bird classification and predictions of spotted species using various modalities, such as sounds [21], visuals [23] or a combination of both [26] for monitoring and preservation activities [8].

Bird recognition poses a crucial yet challenging task due to the variations in their appearances under different weather and environmental conditions [9]. Each situation requires individual review and treatment, adding to the complexity. However, advancements in computer vision, driven by the efforts of the research community, have led to improvements in the recognition task [27]. In our study, we considered previous works on human recognition methods that utilized neural networks and trajectory analysis [28], as well as studies focused on group activity description and recognition through trajectory analysis [29]. Additionally, we explored a predictive method for early recognition of global human behavior [30].

There are at least two factors that highlight the non-trivial nature of bird identification. First, there is a similarity among different species, including their forms, as well as similarities in image and sound backgrounds with other animals. Secondly, the limited experience of bird watchers can degrade the accuracy of species recognition [12]. These reasons have prompted researchers to explore more advanced systems that can study bird species with minimal human intervention, in an automated manner.

Due to advancements in fast and effective machine learning training techniques, there is an ongoing search for the most efficient method of bird species recognition. Researchers are seeking an efficient model that reduces training time and can be deployed on mobile or edge-computing devices. It is not only bird image recognition that benefits from fast models, but also behavior recognition [5]. When selecting the optimal solution for bird species classification, it is important to consider various parameters that contribute to better efficiency, with smaller values indicating improved performance. These parameters include hyperparameters such as the number of convolutional layers, number of sizes of kernels, activation function, pooling size (if applicable), number of dense layers, connectivity pattern, number of neurons, weight regularization, dropout (if applicable), batch size, learning rule, and learning rate [15]. Furthermore, these hyperparameters can be adjusted and optimized for a specific model using a technique known as a hyperparameter tuning [16].

In this study, our objective is to evaluate and compare deep neural network methods optimized for bird recognition on mobile or edge-computing devices using images captured in natural environments [5]. We computed and compared state-of-the-art methods using various metrics and employed transfer learning techniques and classifiers. As a result, we achieved an f1-score accuracy of 97.6369%.

Considering the deployment environment of our proposed system, it is crucial to utilize lightweight devices that can operate in the wild environment and accurately recognize birds' behavior. This study is being conducted with the aim of addressing this requirement.

The structure of the paper is as follows: Sect. 2, Dataset Selection, describes the data selection process, including its sources, processing methods, and rationale. Section 3, Model Selection, provides an overview of the researched methods' architecture, highlighting their strengths and weaknesses. Section 4, Experiments and Results, presents the conducted experiments and their outcomes, with a focus on the applicability to mobile and edge-computing systems in the wild. Section 5, Conclusion, discusses the current and long-term implications of the experimental results for our research.

2 Dataset Selection

In this section, we will discuss our image data selection process. Selecting suitable data for high-accuracy species recognition posed several challenges. While numerous bird datasets are freely available online, not all provide the necessary quality for our purposes. Quality concerns include low-resolution images, significant background components, and birds in positions that may hinder species differentiation. However, the most significant challenge was curating a balanced dataset with appropriately chosen training, testing, and validation sets.

There are several popular publicly available datasets commonly used for bird recognition. These include:

1. CUB-200-2011 (Caltech - UCSD Birds - 200 - 2011): This dataset consists of 11,788 images categorized into 200 bird species. It comprises 5,994 training images and 5,794 test images [22,23,26].
2. SSW60 (Sapsucker Woods 60 Audiovisual Dataset): This dataset contains 31,221 images, 5,400 videos, and 3,861 audio sequences across 60 bird classes. It is used for multi-modal recognition [17].
3. Birds 500 Species dataset: This dataset consists of 500 bird species, with 80,085 training images, 2,500 test images, and 2,500 validation images [5,22]).

In this study, we utilized images from the Birds 500 Species dataset [10]. The dataset is organized into 500 sub-directories, each representing a different bird species. It is divided into three subsets: training set, test set, and validation set. The images have a resolution of 224 × 224 pixels with 3 color channels (RGB). A sample of these images is shown in Fig. 1.

Fig. 1. Random selection of bird images from the Birds 500 Species dataset.

3 Model Selection

In this section, we present the model architecture designed for testing with images captured in the wild environment on mobile or edge-computing devices.

Our study employs various CNN models, namely ResNet50, VGG16, MobileNetV2, EfficientNetB0, Xception, InceptionV3, DenseNet121, and NASNetMobile. Each model underwent transfer learning and was trained to accommodate the Birds 500 Species dataset. Our objective was to identify the most efficient yet high-performing methods suitable for edge-computing devices, such as Raspberry Pi or Nvidia Jetson Nano.

– *ResNet50.* The ResNet50 model is a variant of the Residual Network architecture, featuring 50 convolutional layers and 16 residual units. It incorporates skip connections to bypass layers, mitigating the issue of vanishing gradients caused by layer stacking [19].
– *MobileNetV2.* The MobileNetV2 model is a CNN specifically designed for mobile devices. It utilizes an inverted residual block structure called MBConv Block. The architecture begins with a fully convolutional layer with 32 filters, followed by 19 residual bottleneck layers [14].
– *NASNetMobile.* NASNet is a convolutional neural network (CN N) architecture developed using reinforcement learning to optimize neural network cells. The mobile version, NASNet-Mobile, comprises 12 cells and has approximately 5.3 million trainable parameters (variables). It performs approximately 564 million multiply-accumulate operations and requires around 23 megabytes (MB) of memory [5].
– *EfficientNetV2B0.* This CNN architecture is a more efficient variant compared to EfficientNet, achieved by employing a new inverted residual structure called MBConv in the initial layer. Although it requires less memory compared to

some other architectures, the extensive use of parameters can lead to increased memory consumption [5].

- *DenseNet121.* This CNN architecture has been successful in achieving high accuracy and efficient training. It consists of 121 layers with trainable weights and incorporates dense connections. These connections address the vanishing-gradient problem and enhance feature propagation while effectively reducing the number of model parameters [18].
- *InceptionV3.* This architecture shows improved performance on classification benchmarks by using mini-networks with 8×8 convolutions, two Inception modules, and a concatenated output filter size of 2048 per tile. These design choices enhance network stability and robustness to variations [20].
- *Xception.* The Xception architecture is an extension of the Inception model, employing depthwise separable convolutions. With 126 layers, it effectively extracts key features using 36 convolutional layers. Average pooling layers replace the fully connected layers, reducing the parameter count. The output layer utilizes SoftMax activation for classification [11].
- *VGG16.* The VGG16 model is pretrained on ImageNet and serves as the initial parameter values for feature extraction, tailored to the dataset specifics. It consists of 3 convolutional layers and 3 fully connected layers [24].

4   Experiments and Results

In this section, we present our experimental results using the following approach. The bird data was split into 80,085 training images, 2,500 test images, and 2,500 validation images. The models were trained using mini-batch gradient descent with a batch size of 32 and a learning rate of 0.0001. To prevent overfitting, we employed five-fold cross-validation and early stopping for efficient hyperparameter optimization. Early stopping was triggered if the model's validation loss did not improve for 5 consecutive epochs. The evaluations were conducted on a system with a 12th Gen Intel Core i5 12400F @2.5 GHz, 8GB RAM, NVidia GeForce RTX 3060 12GB graphics card, running on Microsoft Windows 11 Build 22000.

4.1   Resource Limitations

In order to assess the suitability of convolutional neural networks (CNNs) for conducting the bird recognition task on mobile/edge-computing devices, it is essential to consider the resource limitations of these devices. The following are the key factors to be taken into account:

- RAM Size: Mobile/edge devices typically have limited RAM compared to desktop or server systems. The available RAM size affects the model's memory consumption during training and inference. Insufficient RAM can lead to out-of-memory errors or performance degradation. The RAM size of the devices on which the CNNs are intended to run should be known and considered while evaluating their performance.

- Hard Drive/SSD Size: The storage capacity of mobile/edge devices plays a crucial role in storing model files, datasets, and any additional resources required by the CNNs. The size of the CNN models and associated data should not exceed the available storage capacity of the device.
- Maximum Image Processing Time: Mobile/edge devices often have limited computational power compared to high-end desktops or servers. The processing time required for image classification using CNNs can vary based on the complexity of the model and the size of input images. It is important to determine the maximum acceptable image processing time for the bird recognition task on these devices to ensure real-time or near-real-time performance.
- Power Consumption: Mobile/edge devices are typically powered by batteries, and power consumption is a crucial factor to consider. CNN models with high computational requirements may drain the battery quickly, reducing the device's usability. Models that strike a balance between accuracy and computational efficiency should be preferred for energy-constrained environments.
- Network Size: The number of mobile/edge devices connected to a network refers to the network's scale, which is determined by the three primary nodes of edge computing: the device edge, local edge, and the cloud.
- Computational Compatibility: Mobile/edge devices may have specific hardware architectures, such as ARM processors or dedicated neural network accelerators. The CNN models should be compatible with the target device's hardware to ensure efficient execution and optimal performance.

By understanding and taking into account these resource limitations, we can evaluate the CNNs' performance and determine which network is best-suited for conducting the bird recognition task on mobile/edge-computing devices. The selected CNN should provide a balance between accuracy and resource efficiency, ensuring reliable and efficient bird classification within the given resource constraints.

4.2   Results

In this chapter, we conducted a review of various studies to identify models suitable for mobile or edge-computing devices. We analyzed 8 models and evaluated their performance metrics to select the most appropriate candidate for bird species recognition deployment. To expedite the training process, we trained all models for an equal number of epochs. The outcomes were represented in four graphs, illustrating accuracy in relation to model size, number of parameters, CPU time per inference step, and GPU time per inference step. Following the application of transfer learning, we trained these 8 models using our new bird dataset.

As a result, we present the measurements in Table 1 and four figures: Fig. 2, Fig. 3, Fig. 4, and Fig. 5.

We evaluated the fitted models using a five-fold non-shuffled cross-validation and measured their performance in terms of accuracy f1-score. The program initially reads all dataset images from a local disk into a matrix structure and then downloads pre-trained models from a remote repository to the local disk. Training on the bird dataset was performed for each model using the Adam optimizer with a learning rate of 0.0001 and categorical cross-entropy loss function. The maximum number of training epochs for all methods was set to 100. The methods reached convergence at different epochs: MobileNetV2 at epoch 46, NASNetMobile at epoch 91, InceptionV3 and Xception at epoch 63, EfficientNetB0 at epoch 54, ResNet50 at epoch 38, DenseNet121 at epoch 70, and VGG16 at epoch 54.

Table 1. Model size, CPU time, GPU time and f1-score accuracy comparison with accuracy after transfer learning.

Model	Size (MB)	Top-1 Acc (%)	Top-5 Acc- (%)	Param-eters (Mill)	Time (ms) per step CPU	Time (ms) per step GPU	Acc (f1-score)	Macro avg (f1-score)	Weigh-ted avg (f1-score)	Placement (size, param, CPU, GPU, acc)
MobileNetV2	14	71.3	90.1	3.5	25.9	3.8	94.4965	94.4340	94.49	1/1/1/1/5
NASNetMobile	23	74.4	91.9	5.3	27.0	6.7	92.5860	92.5109	92.58	2/2/2/6/8
InceptionV3	92	77.9	93.7	23.9	42.2	6.9	93.3258	93.2629	93.33	6/6/3/7/7
EfficientNetB0	29	77.1	93.3	5.3	46.0	4.9	97.6369	97.6075	97.63	3/3/4/4/1
ResNet50	98	74.9	92.1	25.6	58.2	4.6	96.5256	96.4640	96.52	7/7/5/3/2
DenseNet121	33	75.0	92.3	8.1	77.1	5.4	96.3664	96.3042	96.36	4/4/7/5/3
Xception	88	79.0	94.5	22.9	109.4	8.1	94.3872	94.3274	94.38	5/5/8/8/6
VGG16	528	71.3	90.1	138.4	69.5	4.2	94.9741	94.9236	94.97	8/8/6/2/4

Table 1 presents 8 selected models based on their performance. The table includes the size of each model (in MB), Top-1 and Top-5 accuracy (in percentage), number of parameters (in millions), CPU and GPU time per inference step. The subsequent metrics in the table pertain to the post-transfer learning phase, including f-1 score accuracy, f-1 macro average, f-1 weighted average. The last column indicates the rankings of the models in terms of size, parameters, CPU and GPU time per cycle, and accuracy. To interpret the table, let's consider the first model, MobileNetV2. According to the "Placement" column, it ranks first in terms of size and number of parameters. It is also the most CPU- and GPU-efficient. However, in terms of accuracy, it ranks third, with ResNet50 and VGG16 surpassing it in performance.

4.3   Analysis of the Results

In this section, we provide an in-depth analysis of the results, focusing on a comprehensive comparison of various parameters.

Results of the Comparison Model Size and Accuracy. In the first graph (Fig. 2), EfficientNetB0 achieves the highest accuracy of 97.6469%, followed by ResNet50 with 96.5256% accuracy. MobileNetV2 is the most compact model, while NAS-NetMobile ranks poorly in accuracy despite being small. VGG16 stands out as the largest model at 528MB. Xception and DenseNet121 offer a balance between size and accuracy, suitable for a general-purpose bird recognition system.

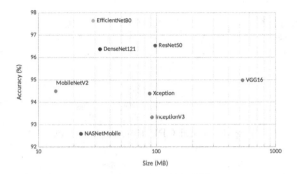

Fig. 2. Model Size vs Accuracy.

Results of the Comparison Parameters vs Accuracy. The second graph (Fig. 3) depicts the relationship between model parameters and accuracy. The situation here is similar to the previous graph, where EfficientNetB0 performed the best in terms of accuracy. Xception and DenseNet121 are a compromise between accuracy and the number of parameters, while MobileNetV2 is shown to have the smallest number of parameters among all eight models.

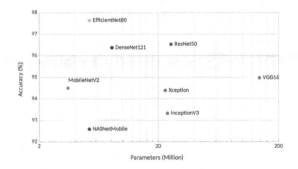

Fig. 3. Model Parameters number vs Accuracy.

Results of the Comparison CPU Time vs Accuracy. The third graph (Fig. 4) represents the accuracy of CPU-only systems. It is evident that MobileNetV2 is the fastest but not the least accurate model. However, considering our objectives, we prioritize precision and processing times for CPU-only systems. In this regard, EfficientNetB0 stands out as the most precise model, offering shorter-than-average execution times. Therefore, we identify EfficientNetB0 as our top choice for mobile or edge-computing devices without a GPU.

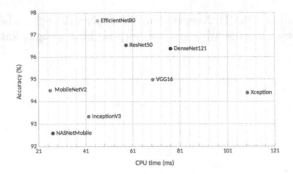

Fig. 4. Model CPU time vs Accuracy.

Results of the Comparison GPU Time vs Accuracy. The final graph (Fig. 5) compares GPU-enabled models in terms of computational times and accuracy. Computational times range from around 4 milliseconds for MobileNetV2 to over 8 milliseconds for Xception. ResNet50, DenseNet121, and EfficientNetB0 performed well in accuracy, with VGG16 slightly less so. However, MobileNetV2 stands out as the top choice for mobile or edge-computing systems due to its speed, especially in low-resource environments. It surpasses ResNet50, Efficient-NetB0, DenseNet121, and even VGG16 while maintaining a 94.4965% F1-score accuracy.

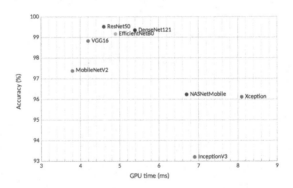

Fig. 5. Model GPU time vs Accuracy.

Our measurements reveal that the models with smaller sizes exhibited shorter computational times on both CPU and GPU systems. Columns 8, 9, and 10 (refer to Table 1) display the accuracy percentages sorted by model size. In this context, "Acc (f1-score)" refers to the f1 accuracy of a model after transfer learning, "Macro avg (f1-score)" represents the f1 macro average of a model after transfer learning, and "Weighted avg (f1-score)" indicates the f1 weighted average of a model after transfer learning. The column "Placement" denotes the ranking of accuracy achieved by each trained network after transfer learning with our dataset.

5   Conclusion

After evaluating various deep learning models for bird recognition on mobile or edge-computing devices, we have concluded that MobileNetV2 is the most suitable model where scarcity of resources is vital, while EfficientNetB0 stands out for its accuracy. MobileNetV2 offers faster inference times, lower power consumption, and a smaller size. However, EfficientNetB0 provides the highest accuracy, making it a comparable alternative for complex natural environments. If a balance is needed, then in CPU systems VGG16 mode will be the preference, while in GPU-enabled systems ResNet50 is the preference. The choice of model depends on factors such as memory availability, architecture type, machine learning acceleration availability, dataset parameters, number of output classes, etc. The final model selection should be based on actual system configuration and test results. In our future work, we will validate and test our models using real-world bird images to recognize bird species and behaviors in the wild. This research has practical implications for developing efficient and accurate bird identification tools, and we plan to extend our work to other domains beyond bird recognition.

Acknowledgment. We would like to express our gratitude to the "A way of making Europe" European Regional Development Fund (ERDF) and MCIN/AEI/10.13039/ 501100011033 for their support of this work under the "CHAN-TWIN" project (grant TED2021-130890B-C21) and AICARE project (grant SPID202200X139779IV0). We are also thankful for the funding received from the HORIZON-MSCA-2021-SE-0 action number: 101086387, REMARKABLE (Rural Environmental Monitoring via ultra wide-ARea networKs And distriButed federated Learning). Additionally, we extend our appreciation to Nvidia for their generous hardware donations, which made these experiments possible.

References

1. Upton, S.: Biodiversity and ecosystems (2014). Organisation for Economic Co-operation and Development. https://www.oecd.org/env/resources/OECD-work-on-biodiversity-and-ecosystems.pdf
2. Organisation for economic co-operation and development, results of the survey on the coefficients applied to RIO marker data when reporting to the un conventions on climate change and biodiversity, DCD/DAC/STAT(2020) 41/REV2
3. Ohsawa, T.: Idea paper: how are ecosystem services related to biodiversity and ecological integrity in each site under climate change? First published: 10 March 2022, Volume 37, Issue 4. https://doi.org/10.1111/1440-1703.12302
4. Weiskopf, S.R., et al.: Climate change effects on biodiversity, ecosystems, ecosystem services, and natural resource management in the United States, science of the total environment, vol. 733, 137782 (2020). ISSN 0048–9697. https://doi.org/10.1016/j.scitotenv.2020.137782

5. Jakariaa, A., Ferdinandus Pardedea, H.: Comparison of classification of birds using lightweight deep convolutional neural networks. Jurnal Elektronika dan Telekomunikasi (JET), Vol. 22, No. 2, December 2022, pp. 87–94. Accredited by KEMDIKBUDRISTEK, Decree No: 158/E/KPT/2021. https://doi.org/10.55981/jet.503

6. Tabur, Ali, M., Ayvaz, Y.: Ecological Importance of Birds (2010)

7. Belaire, J.A., et al.: Urban residents' perceptions of birds in the neighborhood: Biodiversity, cultural ecosystem services, and disservices. The Condor. **117**, 192–202. (2015). https://doi.org/10.1650/CONDOR-14-128.1

8. Narasimhan, R., et al.: Simultaneous segmentation and classification of bird song using CNN. In: 2017 IEEE International Conference on Acoustics, Speech and Signal Processing (ICASSP), New Orleans, LA, USA, pp. 146–150 (2017). https://doi.org/10.1109/ICASSP.2017.7952135

9. Jancovic, P., Köküer, M.: Bird species recognition using unsupervised modeling of individual vocalization elements. IEEE/ACM Trans. Audio, Speech, Lang. Process. **27**(5), 932–947 (2019). https://doi.org/10.1109/TASLP.2019.2904790

10. Gerry, "BIRDS 510 SPECIES- IMAGE CLASSIFICATION", views - 280689, downloads - 40432. https://www.kaggle.com/datasets/gpiosenka/100-bird-species

11. Sungsiri, A., et al.: The classification of edible-nest swiftlets using deep learning. In: 2022 6th International Conference on Information Technology (InCIT), Nonthaburi, Thailand, pp. 404–409 (2022). https://doi.org/10.1109/InCIT56086.2022.10067665

12. Al-Showarah, S., Qbailat, S.: Birds identification system using deep learning. Int. J. Adv. Comput. Sci. Appl. (IJACSA) **12**(4) (2021)

13. Huang, Y.P., Basanta, H.: Bird image retrieval and recognition using a deep learning platform. IEEE Access **7**, 66980–66989 (2019). https://doi.org/10.1109/ACCESS.2019.2918274

14. Chin, T.-W., Ding, R., Zhang, C., Marculescu, D.: Towards efficient model compression via learned global ranking (2019). https://doi.org/10.48550/arXiv.1904.12368

15. Zhu, W., et al.: Evolutionary convolutional neural networks using ABC. In: ICMLC '19: Proceedings of the 2019 11th International Conference on Machine Learning and Computing, 156–162. https://doi.org/10.1145/3318299.3318301

16. Yang, L., et al.: On hyperparameter optimization of machine learning algorithms: theory and practice (2020). Department of Electrical and Computer Engineering, The University of Western Ontario, arXiv:2007.15745v3 [cs.LG] 5 Oct 2022

17. Van Horn, G., et al.: Exploring fine-grained audiovisual categorization with the SSW60 dataset (2022). https://doi.org/10.48550/arXiv.2207.10664

18. Huang, G., et al.: Densely connected convolutional networks (2017). https://doi.org/10.48550/arXiv.1608.06993

19. Xiang, W., et al.: Birds detection in natural scenes based on improved faster RCNN. Appl. Sci. **12**, 6094 2022. https://doi.org/10.3390/app 12126094

20. Rahman, Md. M., et al.: Recognition of local birds of bangladesh using MobileNet and Inception-v3. Int. J. Adv. Comput. Sci. Appl. **11** (2020). https://doi.org/10.14569/ijacsa.2020.0110840

21. Bold, N., et al.: Bird species classification with audio-visual data using CNN and multiple kernel learning. In: 2019 International Conference on Cyberworlds (CW), Kyoto, Japan, pp. 85–88 (2019). https://doi.org/10.1109/CW.2019.00022

22. Liu, H., et al.: TransIFC: invariant cues-aware feature concentration learning for efficient fine-grained bird image classification. IEEE Trans. Multimedia (2023). https://doi.org/10.1109/TMM.2023.3238548

23. Nawaz, S., et al.: Are these birds similar: learning branched networks for fine-grained representations. In: 2019 International Conference on Image and Vision Computing New Zealand (IVCNZ), Dunedin, New Zealand, pp. 1–5 (2019). https://doi.org/10.1109/IVCNZ48456.2019.8960960
24. Wu, P., et al.: Classification of birds based on weighted fusion model. In: 2021 7th International Conference on Big Data Computing and Communications (BigCom) 90–97 (2021)
25. Alswaitti, M., et al.: Effective classification of birds' species based on transfer learning. Int. J. Electrical Comput. Eng. (IJECE), Vol. 12, No. 4, pp. 4172 4184 (2022). ISSN: 2088–8708, https://doi.org/10.11591/ijece.v12i4.pp4172-4184
26. Bold, N., et al.: Cross-domain deep feature combination for bird species classification with audio-visual data. In: 2019 Volume E102.D(10), pp. 2033–2042 (2019). https://doi.org/10.1587/transinf.2018EDP7383
27. Huang, Y.P., Basanta, H.: Recognition of endemic bird species using deep learning models. IEEE Access 9, 102975–102984 (2021). https://doi.org/10.1109/ACCESS.2021.3098532
28. Azorín-López, J., Saval-Calvo, M., Fuster-Guilló, A., García-Rodríguez, J.: Human behaviour recognition based on trajectory analysis using neural networks. In: The 2013 International Joint Conference on Neural Networks (IJCNN), Dallas, TX, USA, pp. 1–7 (2013). https://doi.org/10.1109/IJCNN.2013.6706724
29. Azorin-Lopez, J., Saval-Calvo, M., Fuster-Guillo, A., Garcia-Rodriguez, J., Cazorla, M., Signes-Pont, M.T.: Group activity description and recognition based on trajectory analysis and neural networks. In: 2016 International Joint Conference on Neural Networks (IJCNN), Vancouver, BC, Canada, 2016, pp. 1585–1592 (2016). https://doi.org/10.1109/IJCNN.2016.7727387
30. Azorin-Lopez, J., Saval-Calvo, M., Fuster-Guillo, A., Garcia-Rodriguez, J.: A novel prediction method for early recognition of global human behaviour in image sequences. Neural Process. Lett. 43(2), 363–387 (2015). https://doi.org/10.1007/s11063-015-9412-y

Classification of Subjectively Evaluated Images with Self-Enforcing Networks Using Reference Types, and a Cue Validity Factor

Christina Klüver$^{(\boxtimes)}$ ⓘ and Jürgen Klüver

CoBASC Research Group, Essen, Germany
cobasc@rebask.de

Abstract. Learning and classifying images with neural networks that are evaluated by humans according to subjective criteria is a major challenge. In this case, the ambient lights for luxury cars were scratched or manipulated under laboratory conditions to check the homogeneous radiation of the light. The intact and defective ambient lights were inserted in a special system and recorded with a built-in camera and at a later point in time evaluated. Using a Self-Enforcing Network (SEN), forming reference types for each flowless and fault type, and determining a so-called cue validity factor is a promising approach to correctly cluster and classify subjectively labeled images. We demonstrate the advantage of SEN based on the classification of ambient lighting, compared to previous approaches with Convolutional Neural Networks (CNN) and Deep Belief Networks (DBN).

Keywords: Self-Enforcing Networks (SEN) · Reference Types · Cue Validity Factor

1 Introduction

Ambient lights have been used in luxury vehicles for years, with the aim of improving driver orientation and comfort [5, 11, 15]. Experiments have been conducted to find a suitable position in vehicles based on individual perceptions (e.g., [8, 14]).

The concept and technology of ambient lighting has changed considerably over the years. Currently, various aspects are being investigated, especially the importance for automated driving assistance [2, 3, 7, 16].

Few studies, on the other hand, are related to the evaluation of ambient lighting or light guides from the point of view of homogeneity distribution [1, 12]. Defects can hardly be noticed [1] but have importance for the perceived quality of the interior [12]. This poses a major challenge for inspection in production.

The light emission in the interior of a vehicle should be pleasant for the driver – again, a subjective perception. In addition to aesthetics, ambient lighting should help with orientation in the car, but not be distracting while driving.

During the production process of ambient lighting, the light guides are subjected to manual quality inspection by human evaluators. The classification of the light guides into

© The Author(s), under exclusive license to Springer Nature Switzerland AG 2023
I. Rojas et al. (Eds.): IWANN 2023, LNCS 14134, pp. 340–352, 2023.
https://doi.org/10.1007/978-3-031-43085-5_27

the categories "homogeneous" or "non-homogeneous" can be subjective and depends on many factors such as experience, degree of fatigue or environmental influences.

In [1, p. 662], the problem of detecting "invisible" inhomogeneity is mentioned. Fig. 1 shows two such ambient lights, where a layman can hardly decide which light guide produces a homogeneous light and which one does not.

Fig. 1. Light guides for car interior lighting. Flawless light guide (upper one) and faulty light guide (lower one). (Images by Mentor GmbH & Co. Präzisions-Bauteile KG)

The project was supported by the company Mentor GmbH & Co. Präzisions-Bauteile KG, under whose guidance the light guides were manipulated, since in the production process most of the lights are of course completely intact.

Several light guides were clamped into a specially developed system with a camera to ensure that the light guide was always in the same position in the image.

A total of 201 images of light guides with a size of 4112 x 188 pixels are available: 82 images are labeled as free of defects and 119 images of possible production defects such as scratches, defective spots, too dark light emission, too yellow light emission and combinations thereof. These images were divided into different classes.

Since the perception and human evaluation of ambient lighting is subjective and depends on the personal condition, an automated analysis is obvious. For the production process the company demanded a small amount of training data and a fast adaptation. For this task we have tested different neural networks, as Convolutional (CNN) [13], Deep Belief (DBN) [2], and in this contribution, a Self-Enforcing Network (SEN).

The challenge is that, to our knowledge, there are no open datasets and no comparative studies available in which neural networks are used to evaluate the homogeneity distribution of vehicle lighting during the production process.

In the following, we will present the modeling concept with a SEN in detail and then briefly outline the results obtained with CNN and DBN.

2 The Self-Enforcing Network (SEN)

The Self-Enforcing Network (SEN) is a deterministic, two-layered, and self-organized learning neural network that acquires and orders knowledge according to cognitive theory learning models [10].

The characteristic features of the *SEN algorithm* and *SEN tool* are a) a semantic matrix, b) a cue validity factor (cvf), c) the transformation of the data from the semantic matrix into the weight matrix by the specific learning rule, and d) various visualizations.

a) *The semantic matrix* is the basis of the learning process. It contains the essential attributes (features) and their degree of membership to an object. In this case, the data consist of the images of the intact and defective ambient lights, and are normalized in the interval of $(-1, 1)$.

b) For each feature, *a cue validity factor* (cvf) is set when building the model for highlighting essential attributes. The cue validity factor influences the strength of an attribute's effect on activation by the network. The cvf is set manually or as presented here, an image is selected, and its characteristics automatically determine the cvf.

c) The *Self-Enforcing Rule* (SER) i.e., the learning rule used in SEN transforms the values of the semantic matrix v_{sm} into a weight value between attribute and object w_{ao} with the learning rate c and the cue validity factor (cvf_a):

$$w_{ao} = c * v_{sm} * cvf_a \tag{1}$$

Accordingly (Eq. 1), the special feature of SEN is that the weight matrix is not randomly generated.

Several activation functions are available; for this problem, the enforcing activation function (EAF) is used:

$$a_j = \sum_{i=1}^{n} \frac{w_{ij} * a_i}{1 + |w_{ij} * a_i|} \tag{2}$$

d) After the learning process, the results are visualized. In the *map visualization*, the data are ordered by their similarity: the more similar the data are, the closer they are displayed to each other.

When new data are presented as input vectors, additional visualizations are activated. In the *SEN-visualization*, the activation values indicate the strength of similarity of the input data to the data in the semantic matrix. The higher the final activation values, the more similar the input data are to the learned data.

The calculated Euclidean distances are also computed, with similarity indicated by the smallest distance (see below). In the case of image processing, only the distances are considered.

For many applications it is useful to form so-called *reference types* (e.g., [17]), which contain characteristic properties of the problem. For the case described here, a reference type consists of different images of light guides which are particularly representative for a class like defect-free or defective light guides. Since in SEN the weight matrix is not randomly generated, it is possible to see exactly what SEN is learning. For illustration purposes, Fig. 2 shows three original images with flowless light guides on the left and the images learned in SEN and subsequently exported on the right.

Fig. 2. Original (left) and learned (right) images representing flowless light guides.

A reference type is automatically formed in SEN by selecting these images (Fig. 3).

Fig. 3. Reference type for flowless light guides

The reference type consists of the calculated average values of the three images.

Fig. 4 presents the SEN *tool*; the map visualization shows the three images and the formed reference type for the class of flowless light guides. In addition, there are three images showing faulty light guides, and a new one as an input vector.

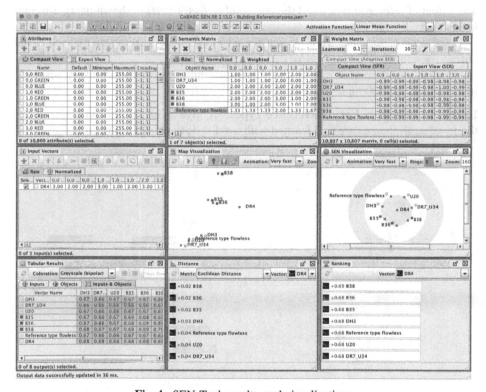

Fig. 4. SEN-Tool, results, and visualizations

For the illustration, the images are scaled to $120 \times 30 \times 3$, resulting in 10.800 input neurons for each RGB value. With a learning rate $c = 0.1$, 10 iterations, and the linear mean function the learning process is after 36 ms completed.

An important note must not be missing: the formation of reference types cannot be done by automatic averaging over all images of a class. Only if the data are close to each other in the map visualization, they are sufficiently similar to form a reference type. Accordingly, only images B36 and B35 should be selected to represent a class for defective light guides; B38 apparently has some other characteristics.

By adding a new image (DR4) as an input vector (circle), the similarity to the already learned images (red cross) is both visualized as an approximation and calculated. The highest activation value of the objects is represented by the "ranking", the smallest distance by the "distance". In this case, image B38 (faulty) with the closest match is indicated, but also very similar to DH3, which is labeled as flowless.

In general, building reference types can considerably reduce the amount of data to be learned without significant loss of features, provided that the data are carefully selected. In the next section, this advantage is substantiated based on the results.

3   Results

The effect of reference types and the cue validity factor (cvf) is shown using different levels of modeling.

The same parameters were chosen for all models: the learning rate $c = 0.1$, the enforcing activation function (EAF), and 1 learning step.

Images were imported in grayscale with a resolution of 180×35 pixels, resulting in a number of 6.300 input neurons.

Learn All Images to Form Reference Types. To give an idea of the heterogeneity of the images, all images are first learned with a cvf = 1; the result is shown in Fig. 5.

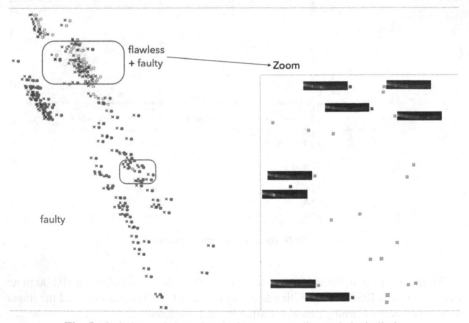

Fig. 5. Ordering and clustering the images according to their similarity

As can be seen in Fig. 5, there is no clear differentiation between flowless and faulty light guides. In particular, in the outlined cluster, there is a strong mixture between the labeled images - this may, of course, be a result of subjective evaluation.

In the next modelling phase, the reference types are formed.

Determine the Reference Types. Figure 5 already shows a certain order of the images: some are clustered very close to each other. Images that lie centrally within a cluster are particularly suitable for the formation of the reference types, and accordingly these were selected for further processing. In total 31 images were selected: 7 for two classes containing defect-free light guides (3 bright and 4 darker) and 24 images for 4 classes containing images with different defects. The remaining 170 images were imported as input vectors. The obtained result of the classification of the input images is shown in Fig. 6.

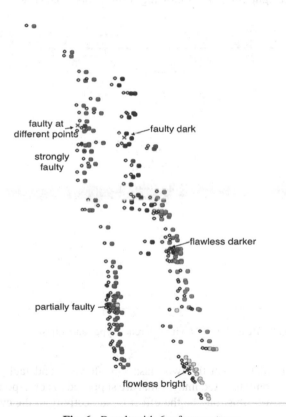

Fig. 6. Result with 6 reference types

The result has been significantly improved, but a clear separation between the patterns is still not given. Some of the images labeled as *faulty dark* are very close to the *flowless darker* light guides. In the next modeling step, the cvf was changed.

Change the Cue Validity Factor Automatically. In image processing with more than 6.000 input neurons, it is not practical to change the cvf manually [18]. Instead, an import of an image with an intact light guide is done as a grayscale image, where the cvf is 0 at black areas and approaches 1 with increasing brightness.

The effects of this procedure on all images are exemplified in Fig. 7.

Fig. 7. Effect of the cvf on images. On the left a flowless, on the right a faulty light guide

By forming the reference types and the automatically modified cvf by an image containing an intact light guide, the following result is obtained (Fig. 8):

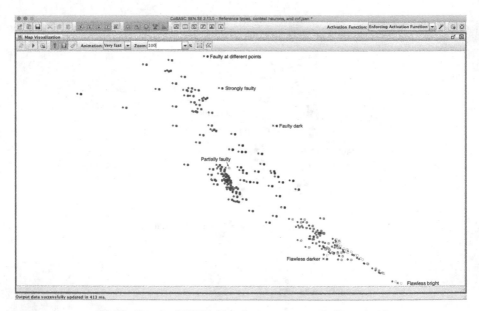

Fig. 8. Result of SEN with reference types and adjusted cvf

SEN can clearly distinguish the two classes of flowless and faulty. Within the 6 individual classes are only isolated images that are supposed to correspond to a different reference type. However, this may be the effect of subjectivity in the evaluation of the images.

The results also show how heterogeneous the defect classes are. Since these are artificial manipulations, a future step will be to examine only those light guides that exhibit defects in the real production process.

In the following section, the results by CNN and DBN are briefly presented in comparison.

4 Convolutional and Deep Belief Networks

In previous experiments, the established Convolutional Neural Networks (CNN) [13] and Deep Belief Networks (DBN) [2] were used to solve the problem. In the following, the respective architecture as well as the most important results are summarized.

4.1 Convolutional Neural Networks (CNN)

The image of a light guide is presented to the network in RGB colors; at a resolution of 4112×188 pixels, this results in 773.056 pixels.

By using the convolutional and pooling layers, the amount of data is reduced, and the network contains a total of "only" 120.122 connection weights to be adjusted for all layers. Fig. 9 shows the architecture of the CNN.

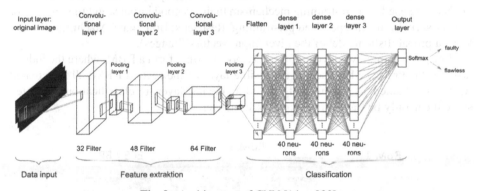

Fig. 9. Architecture of CNN [14, p.330]

The convolutional and pooling layers are used for feature extraction and to reduce the data for the actual learning process of the supervised learning network. The network is trained with the well-known *backpropagation rule* (gradient descent) with the error computation:

$$\delta_j = \begin{cases} f'(net_j)(t_j - o_j) \text{if j is an output neuron} \\ f'(net_j) \sum_k \delta_k w_{ik} \text{ else} \end{cases} \tag{3}$$

For the experiments, 55% (112 images) were selected for training and 45% (89 images) for testing. The parameters are: learning rate 0.001, the activation function ReLU with a batch size of 3. The Softmax function is used for classification in the two classes *faulty* or *flowless*.

The CNN achieves a recognition rate of 97.8 % in the best case (Table 1).

Table 1. Result of CNN

n = 89	CNN	flawless	faulty	
predicted	flawless	37	2	94.87%
	faulty	4	46	92.00%
		90.24%	96.83%	**97.80%**

The architecture and results for 2 classes with a Deep Belief Network are shown below.

4.2 Deep Belief Networks (DBN)

To reduce computational load, while keeping the RGB values, the images are scaled to 180 x 35 x 3, resulting in 18.900 input neurons. To ensure comparability with CNN, the same split of training (55%) and test (45%) data was maintained.

In this case, DBN were trained in addition to the fully connected neural networks. DBNs were used as a pre-training mechanism to find a good initial weighting matrix for the network. In the unsupervised pre-training, Restricted Boltzmann Machines (RBMs) learn a probabilistic model of the given input vectors (images).

For pre-training, two layers of the network at a time form a RBM, where the hidden layer of the first RBM is used as the visible layer of the second RBM; all the created RBMs are stacked, as shown in Fig. 10. Only the last layer for classification is omitted since it can only be trained in a supervised manner.

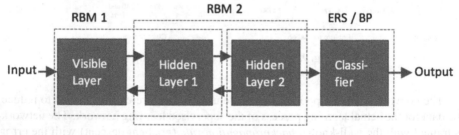

Fig. 10. Architecture of DBN [2, p. 29]

The learning rules used are the *backpropagation* (BP, see Eq. 3) and the *enforcing-rule supervised* (ERS). The learning rule introduced by [10] is a simplified alternative to BP that is not based on the gradient descent method. Furthermore, the weighting values may only lie in an interval between [−1, 1].

The variations ERS_2 and ERS_2-DL (for deep learning) is based on Eq. 4, where c is the equivalent of the learning rate:

$$\Delta w_{ij} = c \cdot \left| \left(1 - |w_{ij}| \right) \right| \cdot \delta_j \cdot o_i \qquad (4)$$

The error is computed for ERS_2 as

$$\delta_j = \begin{cases} (t_j - o_j) \text{if j is an output neuron} \\ \sum_k \delta_k w_{ik} \text{ else} \end{cases} \tag{5}$$

and the modification for deep learning ERS_2-DL is

$$\delta'_j = 2d \cdot \delta_j / n_i \tag{6}$$

where d is the number of layers and n_i the number' of neurons in the layer directly above the layer of neuron j.

The best architecture for ERS_2-DL is 18.900-500-500-2, and for BP 18900-500-2, with a learning rate 0.01, the logistic activation function, and a batch size of 2. The results are shown in Table 2.

Table 2. Results of DBN

predicted	**ERS2-DL**	**flawless**	**faulty**	100.00%
	flawless	34	0	
	faulty	2	54	96.43%
		94.44%	100.00%	**97.78%**
predicted	**BP**	**flawless**	**faulty**	80.00%
	flawless	36	9	
	faulty	0	45	100.00%
		100.00%	83.33%	**90.00%**

The result of ERS is comparable to that of CNN, while the network with the BP learning rule detects all the faulty light guides but cannot identify the flowless ones as well.

In summary, based on the results so far, it is clear where the advantages of SEN lie: Less data are needed to form the reference types than for training with other methods. In addition, a gray scale of the images is sufficient, which significantly reduces the number of input neurons.

Due to the visualizations and comprehensibility of the image arrangement, the heterogeneity of the data can be seen immediately. Adjustment of the reference types is thus easy and allows without effort to increase or decrease the number of reference types as needed.

Classification with 4 Classes with DBN and SEN. To demonstrate this, the 6 reference types in the SEN were reduced to 4 to allow comparison to the results with a DBN.

Since SEN is a self-organized learning network and thus has no targets, the quality of the model requires that the individual vectors are manually assigned to the classes accordingly to the labels. Table 3 shows the distribution of the images.

Table 3. Image assignment to four classes

		flawless	yellow	faulty	dark
DBN (n = 111)	Training data set	46	9	47	9
(n = 90)	Test data set	36	8	39	7
SEN (n = 4)	Reference types	1	1	1	1
(n = 170)	Test data set	64	10	68	28

Table 4. Classification results

	predicted	flawless	yellow	faulty	dark	Total
n = 90	ERS$_2$	36	6	32	6	88.89%
n = 90	BP	36	6	32	3	85.56%
n = 170	SEN	63	5	59	18	**93.00%**

The results with a DBN trained with the BP and ERS$_2$ learning rules [2] and with SEN are summarized in Table 4.

The classification quality becomes worse when there are multiple classes, as has been shown in other studies when subjective influences play a role (e.g. [4, 4]). In SEN, 93% of the images are correctly assigned with a higher number of test data. Even in SEN, the assignment to the given four classes is not entirely successful. Figures 6 and 8 already show that some images, e.g., for the flowless bright and flowless darker light guides, are not clearly separated. This is a reference to the subjective perception of the evaluator and the resulting label of light homogeneity.

SEN achieves better results because the weight values cannot be modified in a possible "wrong" direction as in the case of supervised-learning networks by the specification of the target vectors. Due to the learning rule in SEN, the images are transformed directly into the weight matrix and since false or subjective labels do not play a role in the learning process, there is less misclassification. Determining the cue validity factor (cvf) is also central to the effectiveness of SEN.

5   Conclusion

In this paper, it was shown that SEN, a self-organized learning network, can achieve a clear separation between flowless and faulty light guides, although the evaluation of the images was under subjective influences. However, this is only achieved by forming multiple reference types for different defect classes and using the cvf.

The formation of reference types should be done using the map visualization only after the learning process in SEN has been completed with some example images to find suitable candidates for averaging. Subsequently, the selection of an image that unambiguously represents a defect-free light guide provides reinforcement of these characteristic features by the cvf. This can, for example, eliminate the subjective assessment of brightness when experts determine which images best represent a light guide that emits a homogeneous light.

References

1. Blankenbach, K., Hertlein, F., Hoffmann, S.: Advances in automotive interior lighting concerning new LED approach and optical performance. J. Soc. Inf. Display **28**, 655–667 (2020)
2. Braun, G., Brokamp, M., Klüver, C.: Training subjective perception biased images of vehicle ambient lights with deep belief networks using backpropagation- and enforcing-rules supervised. In: Rutkowski, L., Scherer, R., Korytkowski, M., Pedrycz, W., Tadeusiewicz, R., Zurada, J.M. (eds) Artificial Intelligence and Soft Computing. ICAISC 2022. Lecture Notes in Computer Science, vol 13588. Springer, Cham. (2023). https://doi.org/10.1007/978-3-031-23492-7_3
3. Capallera, M., et al.: A contextual multimodal system for increasing situation awareness and takeover quality in conditionally automated driving. IEEE Access **11**, 5746–5771 (2023). https://doi.org/10.1109/ACCESS.2023.3236814
4. Diederichs, F., et al.: Improving driver performance and experience in assisted and automated driving with visual cues in the steering wheel. IEEE Trans. Intell. Transp. Syst. **23**(5), 4843–4852 (2022). https://doi.org/10.1109/TITS.2022.3162522
5. Fernandez, V., Chavez, J., Kemper, G.: Device to evaluate cleanliness of fiber optic connectors using image processing and neural networks. Int. J. Electr. Comput. Eng. (IJECE) **11**(4), 3093–3105 (2021)
6. Flannagan, M.J., Devonshire, J.M.: Effects of automotive interior lighting on driver vision. Leukos **9**(1), 9–23 (2012)
7. Hassib, M., Braun, M., Pfleging, B., Alt, F.: Detecting and influencing driver emotions using psycho-physiological sensors and ambient light. In: Lamas, D., Loizides, F., Nacke, L., Petrie, H., Winckler, M., Zaphiris P. (eds) Human-Computer Interaction – INTERACT. Lecture Notes in Computer Science, vol 11746. Springer, Cham (2019). https://doi.org/10.1007/978-3-030-29381-9_43
8. Khanh, T., Erkan, A., Röckl, P.: Adaptive lighting technology of future vehicle generations. ATZ Worldwide **125**, 18–23 (2023). https://doi.org/10.1007/s38311-022-1453-z
9. Kim, T., Kim, Y., Jeon, H., Choi, C.-S., Suk, H.-J.: Emotional response to in car dynamic lighting. Int. J. Autom. Technol. **22**(4), 1035–1043 (2021)
10. Klüver, C., Klüver, J.: New learning rules for three-layered feed-forward neural networks based on a general learning schema. In: Madani K. (Ed.) Proceedings of ANNIIP: International Workshop on Artificial Neural Networks and Intelligent Information Processing. Portugal: Scitepress, 2014, pp. 27–36 (2014)
11. Klüver, C., Klüver, J.: Self-organized learning by self-enforcing networks. In: Rojas, I., Joya, G., Cabestany, J. (Eds.): IWANN 2013, LNCS 7902, Berlin Heidelberg: Springer, Part I, pp. 518–529, (2013). https://doi.org/10.1007/978-3-642-38679-4_52
12. Schellinger, S., Franzke, D., Klinger, K., Lemmer, U.: Advantages of ambient interior lighting for drivers contrast vision. In: Proceedings SPIE 6198, Photonics in the Automobile II, 61980J (2006)

13. Stylidis, K., Woxlin, A., Siljefalk, L., Heimersson, E., S¨oderberg, R.: Understanding light. A study on the perceived quality of car exterior lighting and interior illumination. Procedia CIRP **93**, 1340–1345 (2020)
14. Thiemermann, S., Braun, G., Klüver, C.: Homogeneity testing of LED light guides by neural networks. In Klüver, C, Klüver, J. (Eds.): New algorithms for practical problems: variations on artificial intelligence and artificial life (pp. 325-339). Wiesbaden: Springer Fachmedien Wiesbaden (2021). (In German)
15. Weirich, C., Lin, Y., Khanh, T.Q.: Evidence for human-centric in-vehicle lighting: part 1. Appl. Sci. **12**, 552 (2022)
16. Winklbauer, M., Bayersdorfer, B., Lang, J.: Evocative lighting design for premium interiors. ATZ Worldwide **117**, 32–35 (2015)
17. Zhang, H., Lee, S.: Drowsiness prevention system in automatic driving-effects of light position on comfortable and unconscious wakefulness during driving. Intell. Hum. Syst. Integr. **69**, 8–16 (2023)
18. Zinkhan D., Eiermann S., Klüver C., Klüver J.: Decision Support Systems for Air Traffic Control with Self-enforcing Networks Based on Weather Forecast and Reference Types for the Direction of Operation. In: Rojas I., Joya G., Catala A. (eds) Advances in Computational Intelligence. IWANN 2021. Lecture Notes in Computer Science, vol. 12862. Springer, Cham. pp. 404 – 415, (2021). https://doi.org/10.1007/978-3-030-85099-9_33
19. Zurmaar, B.: Image recognition of traffic signs with self-enforcing networks. In: Klüver, C., Klüver, J. (eds) New algorithms for practical problems. Springer Vieweg, Wiesbaden. (2021). (In German)

Advanced Home-Based Diabetes Monitoring System: Initial Real-World Experiences

David Díaz Jiménez[1]([✉]) [iD], José Luis López Ruiz[1] [iD],
Alicia Montoro Lendínez[1] [iD], Jesús González Lama[2,3] [iD],
and Macarena Espinilla Estévez[1] [iD]

[1] Department of Computer Science, University of Jaén, Jaén, Spain
ddjimene@ujaen.es
[2] Cabra Clinical Management Unit, Sur de Córdoba Health Management Area,
Cordoba, Spain
[3] Maimonides Biomedical Research Institute of Cordoba (IMIBIC),
Reina Sofia University Hospital, University of Cordoba, Cordoba, Spain

Abstract. Diabetes is a disease that requires monitoring of healthy habits for its treatment. A recent study suggests that sensor-based activity recognition approaches are a suitable tool for monitoring such habits that are established between clinical personnel and the patient through a therapeutic contract. To date, there is no fully described system architecture for implementing a sensor-based activity recognition approach to monitor healthy habits. In this paper, we present the advanced architecture of a system for monitoring healthy habits of multiple diabetes patients in their homes. The presented system, called AI2EPD, features a complex architecture that encompasses from the sensor devices deployed in each patient's home to persistence in the central server and visualisation in the technical and clinical interface. The system has been deployed in the municipality of Cabra in Córdoba (Spain) in collaboration with the Cabra Health Centre. The approach for sensor installation in the patient's home, as well as the issues and challenges encountered during system deployment, are presented in this section.

Keywords: sensor-based activity recognition · monitor healthy habits · diabetes patients · complex architectures with sensors

1 Introduction

Individuals diagnosed with diabetes face an increased risk of developing severe health complications, leading to high healthcare costs, decreased quality of life,

This result has been partially supported through the Spanish Government by the project PID2021-127275OB-I00, FEDER "Una manera de hacer Europa", the European Union's Horizon 2020 research and innovation programme under grant agreement No 8571 and the Consejería de Salud y Familias de la Junta de Andalucía for the project with code AP-0233-2019.

and increased mortality [5]. As such, optimising diabetes management is a crucial public health priority [10]. According to the International Diabetes Federation's (IDF) diabetes atlas, there were an estimated 451 million people worldwide between the ages of 18 and 99 with diabetes in 2017 [4]. The IDF predicts that by 2045, this number will rise to 693 million. Lifestyle modifications such as maintaining a healthy diet, engaging in regular physical activity, not smoking, and maintaining a healthy weight are essential in the fight against diabetes [12].

Another crucial aspect of managing diabetes is patient involvement, with recommendations tailored to individual needs and therapeutic objectives agreed upon by the patient and healthcare provider. Individualised interventions have shown long-term beneficial clinical outcomes in patients with diabetes [7,14].

In this context, human activity recognition has advanced significantly in recent years, enabling a wide range of applications such as healthcare, smart homes and manufacturing. There are two main types of human activity recognition systems: video-based and sensor-based [1,9]. The former use cameras to capture images or videos of people's behaviour, while the latter rely on body or environmental sensors to detect details of movement or traces of activity. Due to the privacy concerns posed by cameras in personal spaces, sensor-based systems are more commonly used to monitor everyday activities [15]. In addition, the proliferation of smart devices and the Internet of Things has made it possible to embed sensors in a variety of objects, such as phones, watches, cars, walls and furniture, providing a ubiquitous and non-intrusive record of human motion information [8,11].

In the literature, there is a new study that propose sensor-based activity recognition approaches to empower patients with diabetes to improve healthy habits due to the fact that these approaches have the ability to record and analyse people's behaviours [14]. To this end, a therapeutic contract is proposed between the clinician and each patient with diabetes, agreeing on the healthy habits that each patient should have. These habits include, among others, the number of hours the patient should sleep or the time the medication should be taken.

A home activity monitoring system (Beprevent) was initially proposed for the treatment of patients with type 2 diabetes mellitus [14]. Beprevent [6], was a multipurpose home activity control system that provides information about people's routines through a mobile application; its target population is mainly made up of elderly, dependent, chronically or mentally ill people or people with some kind of physical and/or mental disability who live alone. For this purpose, it provides a central device with a power cable and 5 adhesive labels, which are attached to everyday objects. Through a mobile app, the routines under study are reported with these 5 labels.

Due to the major challenges addressed inter by this extension for patients with diabetes in a clinical setting, this extension was not ultimately carried out. Therefore, there is currently no real system with a detailed description of its architecture that implements a sensor-based activity recognition approach to monitor healthy habits in the therapeutic contract prescribed by the clinician.

Due to this gap, this paper presents the advanced architecture of such system, entitled AI2EPD, to monitor the behaviours of multiple patients at home, following the protocol and the experimental design proposed in their study *Feasibility of an Activity Control System in Patients with Diabetes: A Study Protocol of a Randomized Controlled Trial* [14]. Thus, this work presents the proposal of the integral technical architecture that includes the devices with sensors used to collect the information, the data collection in each home, the persistence in the database and, finally, the visualisation of the data from both the technical role and the clinical role.

The proposed system has been implemented in collaboration with the Cabra Health Centre and deployed in the municipality of Cabra in Córdoba, Spain. This paper also presents the approach used for sensor installation, as well as the challenges and issues encountered during system deployment.

This contribution is structured as follows: Sect. 2 reviews the importance of the therapeutic contract in the diabetic context. Section 3 presents the set of sensor devices to monitor therapeutic objectives. Section 4 presents the architecture of AI2EPD to monitor the behaviours of multiple patients at home. Section 5 presents the deployment experience of the system. Finally, Sect. 6 presents the conclusions.

2   Therapeutic Contract to Monitor Diabetic Patients

Diabetes mellitus is a chronic condition that poses a significant burden on healthcare systems worldwide. Patients with this condition face a higher risk of developing severe health problems that could be life-threatening. The management of diabetes requires significant lifestyle modifications, including regular physical activity, healthy diet, and blood glucose monitoring. Despite the availability of effective treatments, patients with diabetes often struggle to adhere to their prescribed treatment regimen. Non-adherence can result in poor health outcomes, increased healthcare costs, and reduced quality of life.

To address this issue, the study presented in [14] aimed to evaluate the feasibility of a multipurpose activity control solution for home activity (home activity control system) in improving treatment adherence in diabetic patients. The system was designed to provide real-time information on the activities of daily living carried out outside the home, enabling clinicians to monitor patients' behaviour and adjust their treatment regimen accordingly. By incorporating patient data into treatment plans, clinicians can develop tailored interventions that address each patient's unique needs, improving the likelihood of adherence to therapeutic objectives.

The therapeutic objectives for each patient are established in a therapeutic contract, which outlines the activities to be performed and their frequency. The contract also sets the frequency of follow-up visits to the healthcare centre to monitor the patient's condition. The therapeutic contract is reviewed at each follow-up visit, and any necessary adjustments are made in collaboration with the patient.

The importance of the therapeutic contract lies in its ability to enhance communication between healthcare providers and patients, promote collaboration and active patient participation in their own healthcare, and improve treatment adherence. By establishing clear and measurable goals and specifying the activities required to achieve them, the therapeutic contract helps ensure that patients receive appropriate and personalised treatment.

This paper presents the advanced architecture to monitor the therapeutic contract of diabetic patients. In collaboration with the Cabra health centre, the activities that will be included for monitoring and their frequency are as follows:

- Physical activity: walk at least 30 min in the morning and 30 min in the evening every day.
- Rest: the patient should sleep for a minimum of 5 h each night.
- Self-care:
  - Tooth brushing: at least 3 times a day after main meals.
  - Shower: once a day.
- Nutrition: eat at least X times a day.
- Medication: take at least X times every X hours.

The X variables will be configured in a personalised way between the clinician and the patient.

3   Sensor Devices to Monitor Therapeutic Objectives

The prototype presented here is composed of a variety of sensors and devices, as shown in Fig. 1. These include intelligent nodes, among which central nodes are distinguished, responsible for managing the sensors, and anchor nodes, which determine the location.

The following section outlines the necessary sensors required to carry out the monitoring of the activities previously established. In order to ensure accurate measurement, specific devices have been selected for each activity.

For physical activity monitoring, the use of an activity bracelet, as well as a sensor for opening and closing the main door, has been chosen. In this way, the time dedicated to performing physical exercises, as well as access to the dwelling, can be precisely recorded.

For monitoring rest, a motion and light sensor will be employed to supervise sleep patterns, thus enabling analysis of sleep habits.

Regarding tooth brushing activity, the location of the toothbrush will be taken into consideration to select the appropriate sensor. If the toothbrush is located within a cabinet, an opening and closing sensor will be used. Conversely, if it is located in an accessible place, a motion sensor will be employed.

In terms of shower monitoring, a combination of a temperature and humidity sensor, as well as a motion sensor, will be used to collect information about shower duration and water temperature.

For food consumption, a motion sensor will be used to record meal times. This will allow for monitoring of feeding schedules.

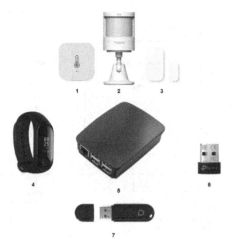

Fig. 1. Sensors and devices. **1** Temperature and humidity sensor. **2** Motion and light sensor. **3** Open and close sensor. **4** Activity tracker. **5** Nodes (Raspberry Pi). **6** Bluetooth adapter. **7** Conbee 2.

Lastly, for medication activity, an opening and closing sensor has been chosen. It will be placed on the medication container to record the moment when medication is taken.

4 Architecture of AI2EPD System

The architecture of the system, consisting of a home layer and a server layer, is illustrated in Fig. 2.

4.1 Home Layer

The home layer is composed of each of the homes that will be monitored.

Within each home, there is a Raspberry Pi configured to act as a central node in an IoT device network. The central node hosts the MQTT (MQ Telemetry Transport) broker[1], which is a subscription and publication messaging protocol used for communication between IoT devices. The MQTT broker allows connected devices in the IoT network to efficiently and reliably send and receive messages. Through this protocol, the rest of the nodes in the network in that home, called anchors, will send the information related to the RSSI of the activity tracker in that home.

The central node of each home also collects data from the activity tracker, as the node is linked to the tracker. In this sense, the information collected from the activity tracker is published on a topic, to be processed later in the MQTT client hosted on the same node. Furthermore, the Raspberry Pi is configured

---

[1] https://mqtt.org/.

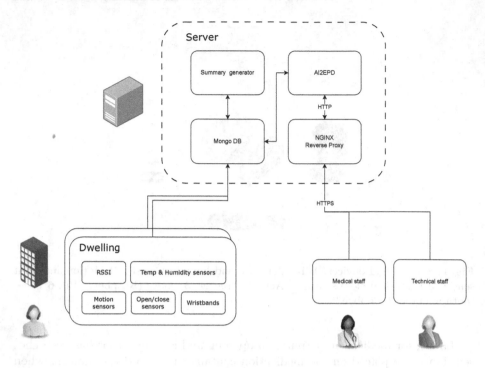

Fig. 2. System architecture.

with the installation of Home Assistant [3], an open-source home automation platform.

The Home Assistant platform is necessary to manage the previously described sensors from the central node, which in combination with the Zigbee module, allows it to act as a Zigbee coordinator. The information received through this protocol is dumped to topics, where the MQTT client will process them.

Once the client receives information from the different topics, it inserts the information into the database collection associated with the assigned home.

4.2   Server Layer

General Description. This layer is responsible for carrying out all intelligent processing while providing information of interest to both technical and healthcare personnel.

The process begins when the home layer sends data to the server. This data is stored in a specific database on the server, which also stores the therapeutic contracts that have been previously defined for each patient.

With the data streams collected at the patient's home and therapeutic contract, compliance summaries are generated. These summaries are generated using automated algorithms and processes that analyse the data and compare it with the defined therapeutic contracts, using the methodologies proposed in [2, 13].

To begin with, the location of the monitored patient within the home is determined. This is done by using the Received Signal Strength Indication (RSSI) between the anchors and the patient's activity tracker. Based on the strength, the patient's location in the environment can be determined approximately [2]. In addition, the combination of RSSI data with data from sensors distributed throughout the home allows inference of the patient's activities, such as walking, showering, sleeping, among others [13].

Once the compliance summaries have been generated, the inferred information is stored in the same database. Subsequently, healthcare personnel can access the dashboard and consult the compliance summaries to evaluate treatment progress and make informed decisions about the treatment plan for each patient.

Reverse Proxy. Data security is a crucial issue in any context, especially in the field of healthcare. For this reason, special emphasis has been placed on implementing measures that ensure adequate protection against potential threats.

One of the measures adopted to guarantee data security is the use of a reverse-proxy[2]. This method involves using a web server, specifically NGINX, that sits between the application and clients. The main function of the reverse proxy is to provide HTTPS(HyperText Transfer Protocol Secure) to clients, which encrypts communications, serves the static content of the application, and offers mitigation's against certain types of attacks, such as DDoS (Distributed Denial of Service).

The use of a reverse-proxy helps to increase the level of application security by providing an additional layer of protection.

Database. In a monitoring system, it is common to generate large volumes of data of different types and formats due to the large number of sensors and devices involved. These data may be of a heterogeneous nature, which makes their storage and management using traditional databases difficult.

In this regard, a non-relational database, specifically MongoDB[3], has been chosen as the solution for storing and managing the data generated by the system. MongoDB is a document-oriented database that allows for the storage of unstructured or semi-structured data, making it ideal for storing heterogeneous data generated by various sensors and devices.

Web System. In order to ease the management of the system and allow healthcare professionals to monitor patients, a web application has been developed

---

[2] https://www.cloudflare.com/learning/cdn/glossary/reverse-proxy/.

[3] https://www.mongodb.com/.

using Django, a high-level web development framework in Python. This application has been specifically designed to meet the needs of technical and healthcare personnel responsible for managing the system.

The web application has an authentication system using credentials, which ensures that only authorised users can access the different functions of the system. Each user has different types of access to the system's functions, depending on their role. This allows for greater security and more precise control over who can access what information and functions within the system.

As a result of this application, technical staff are able to effectively manage the system, while healthcare staff can oversee and monitor the status of patients.

- Technical view. The application designed for technical staff enables them to carry out important tasks for the proper functioning of the platform. Firstly, they can register and edit sensors, activity bracelets, and homes, among other devices and elements necessary for patient monitoring. In addition, it provides the possibility of consulting the status of the monitored homes, which allows for detecting possible faults in the devices and solving them quickly. It also makes it possible to retrieve the information recorded by the sensors and devices in the homes and visualise it through the use of graphs, Fig. 3, which enables analysis and decision-making by the technical staff.
- Clinical view. The application for healthcare staff consists of a main page that provides information on patients, medical staff and compliance with the therapeutic contract. In addition, it allows healthcare professionals to consult summaries indicating the degree of compliance with the activities established in each patient's individual therapeutic contracts, including details on the objective of each activity and the percentage of compliance within the planned schedules, Fig. 4. The application also allows the creation, modification and consultation of therapeutic contracts, including information on the date of entry into force, the author of the modification and the number of modifications made. Finally, it provides the option of downloading in CSV format the activities carried out by patients in a given date range for subsequent analysis.

5   Real-World Implementation: Issues and Challenges

The home-based patient monitoring system for diabetes proposed in this contribution has been deployed in a patient's home in Cabra de Córdoba (Spain) in collaboration with the Cabra Health Centre.

This section outlines the approach used for sensor installation in the patient's home, along with the issues and challenges encountered during system deployment.

5.1   Home Deployment Approach

The process of deploying a system in a real-world setting is a critical activity that requires a well-defined series of steps.

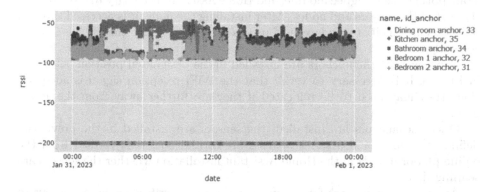

Fig. 3. RSSI of the anchors installed in the home.

Compliance with the therapeutic contract by day
Date of summary: 05/01/2023

Patient:

Contract id: 1

Activity	Objective	Done	Schedule	Progress	Percentage
Physical activity	150 minutes	202 minutes	●		100%
Rest	420 minutes	375 minutes	●		89%
Self-care: toothbrushing	2 times	2 times	●		100%
Self-care: shower	1 time	1 time	●		100%
Meals	3 times	3 times	●		100%
Medication	2 times	2 times	●		100%

Fig. 4. Summary.

Prior to deploying the system in the patient's home, a thorough interview is conducted to gather detailed information about the patient's habits in relation to their daily activities. Details are collected about the activities performed, approximate schedules, and the most common locations where these activities are carried out. In addition, a detailed analysis of the spatial distribution in the home is conducted to determine the most suitable locations for the sensors and devices.

During this stage, it is important to verify whether the home has access to the internet. If not, a mobile access point with a SIM card is provided for data transmission. Once the initial interview has been conducted and appropriate locations for the sensors and devices have been determined, the system is deployed in the home.

During the initial deployment of the system, a technical team travels to the patient's home to perform the installation. First, the central node, as shown in Fig. 5a, is installed, which contains the MQTT (Message Queuing Telemetry Transport) broker, Zigbee module, and Home Assistant necessary to interconnect the rest of the sensors and nodes. After installing the node, it is accessed through SSH (Secure Shell) to update it if necessary and to verify that it has started correctly.

Subsequently, the anchors are installed in the designated position, as shown in Fig. 5b. It is necessary to verify that the WiFi reception signal is adequate since they may need to be relocated if they are further away from the access point.

Once the anchors are installed, the sensors are installed in the previously defined locations, and a series of tests are carried out. It is verified from the online platform or from the Home Assistant installation whether the sensors are sending data.

As an additional check for the motion sensors, a verification is carried out to determine if the range and vision of the sensor are adequate and only detect the desired motion patterns. If not, the parameters are adjusted from the platform.

As the last step of the system deployment, the activity bracelet is linked, and it is verified that the anchors experience a variation in the RSSI signal strength as they move closer or further away from them.

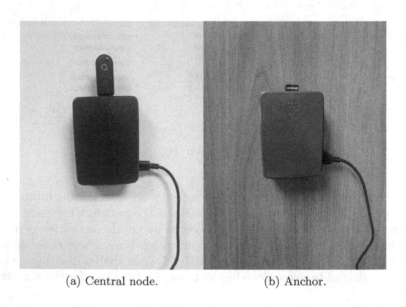

(a) Central node.   (b) Anchor.

Fig. 5. Nodes installed in the home.

Once the installation process described above is completed, the system will be properly configured and ready for use in the patient's home. This way, the patient can continue with their usual daily activities while the system constantly monitors their therapeutic goals.

5.2   Issues and Challenges

The deployment of a patient's home monitoring system brought to light a number of issues and challenges that are presented in this section.

Firstly, it was observed that some sensors and devices stopped operating correctly spontaneously due to a variety of factors, such as power loss and sensor disconnection. In the event of power outages, the devices have automatic reset functions, although occasionally they need to be manually disconnected and reconnected. Sensor disconnection, on the other hand, requires in-person reconnection.

Secondly, the energy cost associated with the operation of the system is a concern for some patients, even though the energy consumption is low and results in low operational costs. Therefore, it is important to evaluate compensatory measures for patients who may be affected financially.

Finally, remote updating of the smart nodes can be a complex process that affects the system's operation. During critical updates that involve changes to the project's code as well as third-party services and applications integrated with the system, remote updating of the smart nodes can be a complex process that affects the system's operation. In such cases, a team of technical experts, including some members present in the patient's home, is necessary to assist with the updates. It is crucial to carefully plan and communicate the update process to avoid any disruptions to the system's operation and ensure prompt resolution of any issues that may arise.

6   Conclusions

A fully defined system architecture for implementing a sensor-based activity recognition approach to monitor healthy habits in diabetic context. In this study, we have presented the advanced architecture of the AI2EPD system, which is used to monitor the healthy habits of multiple diabetes patients in their homes. This complex architecture encompasses devices with sensors deployed in each patient's home to data persistence in the central server and visualisation in the technical and clinical interface. Additionally, this system has been implemented in collaboration with the Cabra Health Centre in Córdoba, Spain. This section has outlined the methodology employed for the installation of the sensors in the patient's home, along with the problems and challenges encountered during the deployment of the system.

The proposed system represents a significant advancement in the management of diabetes by providing clinicians with accurate and up-to-date information on patient behaviour. By enabling real-time monitoring of patient activity, the system can facilitate early intervention and prevent complications associated with non-adherence. Further research is needed to assess the long-term effectiveness of the system and its impact on patient behaviour and treatment adherence.

References

1. Abdallah, Z.S., Gaber, M.M., Srinivasan, B., Krishnaswamy, S.: Activity recognition with evolving data streams: a review. ACM Comput. Surv. (CSUR) **51**(4), 1–36 (2018)
2. Albin-Rodriguez, A.P., De-La-Fuente-Robles, Y.M., Lopez-Ruiz, J.L., Verdejo-Espinosa, A., Espinilla Estévez, M.: UJAmI location: a fuzzy indoor location system for the elderly. Int. J. Environ. Res. Public Health **18**(16), 8326 (2021)
3. Assistant, H.: Home assistant (2023). https://www.home-assistant.io/
4. Atlas, I.D.: Global estimates of diabetes prevalence for 2013 and projections for 2035. Diabetes Res. Clin. Pract. **103**(2), 137–49 (2014)
5. Baena-Díez, J.M., et al.: Risk of cause-specific death in individuals with diabetes: a competing risks analysis. Diabetes Care **39**(11), 1987–1995 (2016)
6. Beprevent: Objetos inteligentes, discretos observadores. https://beprevent.es/
7. Dambha-Miller, H., Feldman, A.L., Kinmonth, A.L., Griffin, S.J.: Association between primary care practitioner empathy and risk of cardiovascular events and all-cause mortality among patients with type 2 diabetes: a population-based prospective cohort study. Ann. Family Med. **17**(4), 311–318 (2019)
8. Hassan, M.M., Uddin, M.Z., Mohamed, A., Almogren, A.: A robust human activity recognition system using smartphone sensors and deep learning. Futur. Gener. Comput. Syst. **81**, 307–313 (2018)
9. Herath, S., Harandi, M., Porikli, F.: Going deeper into action recognition: a survey. Image Vis. Comput. **60**, 4–21 (2017)
10. Hills, A.P., Misra, A., Gill, J.M., Byrne, N.M., Soares, M.J., Ramachandran, A., Palaniappan, L., Street, S.J., Jayawardena, R., Khunti, K., et al.: Public health and health systems: implications for the prevention and management of type 2 diabetes in South Asia. Lancet Diabetes Endocrinol. **6**(12), 992–1002 (2018)
11. Jiang, W., Yin, Z.: Human activity recognition using wearable sensors by deep convolutional neural networks. In: Proceedings of the 23rd ACM International Conference on Multimedia, pp. 1307–1310 (2015)
12. Lazarou, C., Panagiotakos, D., Matalas, A.L.: The role of diet in prevention and management of type 2 diabetes: implications for public health. Crit. Rev. Food Sci. Nutr. **52**(5), 382–389 (2012)
13. Lopez-Medina, M., Espinilla, M., Cleland, I., Nugent, C., Medina, J.: Fuzzy cloud-fog computing approach application for human activity recognition in smart homes. J. Intell. Fuzzy Syst. **38**(1), 709–721 (2020)
14. Montagut-Martínez, P., et al.: Feasibility of an activity control system in patients with diabetes: a study protocol of a randomised controlled trial. Diabetes, Metabolic Syndrome and Obesity: Targets and Therapy, pp. 2683–2691 (2022)
15. Zhang, M., Sawchuk, A.A.: USC-HAD: a daily activity dataset for ubiquitous activity recognition using wearable sensors. In: Proceedings of the 2012 ACM conference on Ubiquitous Computing, pp. 1036–1043 (2012)

On Comparing Early and Late Fusion Methods

Luis Manuel Pereira, Addisson Salazar[(✉)], and Luis Vergara

Institute of Telecommunications and Multimedia Applications,
Universitat Politècnica de València, Valencia, Spain
asalazar@dcom.upv.es

Abstract. This paper presents a theoretical comparison of early and late fusion methods. An initial discussion on the conditions to apply early or late (soft or hard) fusion is introduced. The analysis show that, if large training sets are available, early fusion will be the best option. If training sets are limited we must do late fusion, either soft or hard. In this latter case, the complications inherent in optimally estimating the fusion function could be avoided in exchange for lower performance. The paper also includes a comparative review of the fusion state of the art methods with the following divisions: early sensor-level fusion; early feature-level fusion; late score-level fusion (late soft fusion); and late decision-level fusion (late hard fusion). The main strengths and weaknesses of the methods are discussed.

Keywords: data fusion · early fusion · late fusion · hard late fusion · decision fusion

1 Introduction

Nowadays, information is growing exponentially in complexity, volume, variety and veracity. We can extract and derive valuable information from data to learn about the nature of things. Rather than base decisions or conclusions on personal experiences or intuitions, we are more likely to do so and feel more confident about it [1, 2].

Recent advances in the development of fast and reliable analytical techniques, and advanced devices combined with multivariate statistical approaches, have produced results that are more informative. In general, processes that require parameter estimation from multiple sources can benefit from data or information fusion in a context of multiple classifiers. Data fusion aims to combine the information received from the real world to make the results more heterogeneous, informative, and synthetic than the original ones. In this way, the aim is to increase the reliability of the classification and the quality of the extracted information [3, 4].

This paper presents a theoretical comparison of early and late fusion methods. The methods are called "early fusion" or "late fusion", respectively, depending on whether they are located before or after the classification. An initial discussion on the conditions to apply early or late (soft or hard) fusion is introduced. Besides, the paper also includes a comparative review of the fusion state of the art methods with the following divisions: early sensor-level fusion; early feature-level fusion; late score-level fusion (late soft fusion); and late decision-level fusion (late hard fusion).

© The Author(s), under exclusive license to Springer Nature Switzerland AG 2023
I. Rojas et al. (Eds.): IWANN 2023, LNCS 14134, pp. 365–378, 2023.
https://doi.org/10.1007/978-3-031-43085-5_29

2 Early Fusion/Late Fusion Comparison

2.1 Early Fusion

Let us assume for simplicity the scenario of two classes ($k = 1, 2$), and two sets of features $\mathbf{x}_1$, $\mathbf{x}_2$, from two different modalities that we will assume conditionally independent, i.e.: $p(\mathbf{x}_1, \mathbf{x}_2/k) = p(\mathbf{x}_1/k)p(\mathbf{x}_2/k)$. Suppose we perform early fusion, choosing the class that maximizes the "a posteriori" probability given the two sets of features $\mathbf{x}_1$, $\mathbf{x}_2$ i.e., the decision rule will be:

$$P(k = 1/\mathbf{x}_1, \mathbf{x}_2) \underset{\substack{< \\ k=2}}{\overset{\substack{k=1 \\ >}}{}} P(k = 2/\mathbf{x}_1, \mathbf{x}_2) \;\Leftrightarrow\; P(k = 1/\mathbf{x}_1, \mathbf{x}_2) \underset{\substack{< \\ k=2}}{\overset{\substack{k=1 \\ >}}{}} 0.5 \qquad (1)$$

where we have taken into account that $P(k = 1/\mathbf{x}_1, \mathbf{x}_2) + P(k = 2/\mathbf{x}_1, \mathbf{x}_2) = 1$. ... Note that rule (1) implies minimization of the probability of error if we assume that the costs of being wrong in the decision are symmetric and normalized to 1. That is, considering that we had exact knowledge of $P(k = 1/\mathbf{x}_1, \mathbf{x}_2)$ no other fusion rule allows us to reduce the probability of error more than (1).

2.2 Late Soft Fusion

Suppose now that we perform soft late fusion. For this purpose we generate an a posteriori probability (score) separately for each modality: $s_1 = P(k = 1/\mathbf{x}_1)$ and $s_2 = P(k = 1/\mathbf{x}_2)$. Note that when it is fulfilled that $P(k = 1/\mathbf{x}_1) + P(k = 2/\mathbf{x}_1) = 1$ and $P(k = 1/\mathbf{x}_2) + P(k = 2/\mathbf{x}_2) = 1$, It is enough to consider the scores s_1 and s_2 defined. Then we generate a new score s fusing s_1 and s_2 through a certain fusion function $s = f(s_1, s_2)$ $0 \le s \le 1$ and the new decision rule will be:

$$s \underset{\substack{< \\ k=2}}{\overset{\substack{k=1 \\ >}}{}} 0.5 \qquad (2)$$

Note that $s = f(s_1, s_2) = f(P(k = 1/\mathbf{x}_1), P(k = 1/\mathbf{x}_2))$ is ultimately a function of multivariate random variables $\mathbf{x}_1$, $\mathbf{x}_2$, which will generally be different from $P(k = 1/\mathbf{x}_1, \mathbf{x}_2)$ in (1) and therefore the rule (2) can never achieve a probability of error lower than the rule (1), assuming knowledge of $P(k = 1/\mathbf{x}_1, \mathbf{x}_2)$. For instance, let us analyze the case of the fusion method based on α-integration [5–7]. First, we apply Bayes' rule:

$$s_1 = P(k = 1/\mathbf{x}_1) = \frac{p(\mathbf{x}_1/k = 1)P_1}{p(\mathbf{x}_1)}; \quad s_2 = P(k = 1/\mathbf{x}_2) = \frac{p(\mathbf{x}_2/k = 1)P_1}{p(\mathbf{x}_2)} \qquad (3)$$

We will now consider the α-integration of both scores to obtain s s

$$s = \left(w_1 s_1^{\frac{1-\alpha}{2}} + w_2 s_2^{\frac{1-\alpha}{2}} \right)^{\frac{2}{1-\alpha}} = \left(w_1 \left(\frac{p(\mathbf{x}_1/k=1)P_1}{p(\mathbf{x}_1)} \right)^{\frac{1-\alpha}{2}} + w_2 \left(\frac{p(\mathbf{x}_2/k=1)P_1}{p(\mathbf{x}_2)} \right)^{\frac{1-\alpha}{2}} \right)^{\frac{2}{1-\alpha}}$$

$$= \left(\frac{w_1(p(\mathbf{x}_1/k=1)p(\mathbf{x}_2)P_1)^{\frac{1-\alpha}{2}} + w_2(p(\mathbf{x}_2/k=1)p(\mathbf{x}_1)P_1)^{\frac{1-\alpha}{2}}}{(p(\mathbf{x}_1)p(\mathbf{x}_2))^{\frac{1-\alpha}{2}}} \right)^{\frac{2}{1-\alpha}} \tag{4}$$

$$= \frac{\left(\left(p(\mathbf{x}_1/k=1)p(\mathbf{x}_2)w_1^{\frac{2}{1-\alpha}} \right)^{\frac{1-\alpha}{2}} + w_2\left(p(\mathbf{x}_2/k=1)p(\mathbf{x}_1)w_2^{\frac{2}{1-\alpha}} \right)^{\frac{1-\alpha}{2}} \right)^{\frac{2}{1-\alpha}}}{p(\mathbf{x}_1)p(\mathbf{x}_2)} P_1$$

On the other hand, the score generated in rule (1) can be written as:

$$P(k = 1/\mathbf{x}_1, \mathbf{x}_2) = \frac{p(\mathbf{x}_1, \mathbf{x}_2/k=1)P_1}{p(\mathbf{x}_1, \mathbf{x}_2)} = \frac{p(\mathbf{x}_1/k=1)p(\mathbf{x}_2/k=1)}{p(\mathbf{x}_1, \mathbf{x}_2)}P_1 =$$

$$= \frac{p(\mathbf{x}_1/k=1)p(\mathbf{x}_2/k=1)}{p(\mathbf{x}_1)p(\mathbf{x}_2/\mathbf{x}_1)}P_1 \tag{5}$$

where we have taken into account the assumed conditional independence between the characteristics of the two different modalities $\mathbf{x}_1$, $\mathbf{x}_2$. Comparing (4) and (5) we can conclude an interpretation of the α-integration, as an attempt to approximate the optimal score in (5). On the one side, in relation to the denominator, when operating the two modalities through separate channels before merging, it is not taken into account that $\mathbf{x}_1$ and $\mathbf{x}_2$ are (unconditionally) dependent. On the other hand, we can see that the parameters α, w_1 and w_2, which are estimated by minimizing a certain cost function (e.g. error probability), should tend to be adjusted so that $p(\mathbf{x}_2)w_1^{\frac{2}{1-\alpha}} \simeq p(\mathbf{x}_2/k=1)$ and $p(\mathbf{x}_1)w_2^{\frac{2}{1-\alpha}} \simeq p(\mathbf{x}_1/k=1)$, in order to obtain the minimum probability of error of the rule (1). In any case, it is clear that, assuming perfect knowledge of late soft fusion can never provide a lower probability of error than early fusion. However, we must also consider that the use of particular classifiers are conditioned to the requirements of the size of the training datasets, and thus, conditioning the fusion of the classifiers. The estimation of the sample size for training to obtain an estimated error is a complex task, see for instance the proxy learning curve for the Bayes classifier [8].

2.3 Late Hard Fusion

In this case, first, the binary decisions of each modality are generated and from them, a decision fusion rule is established to generate the final decision. Let us denominate d_1, d_2 the binary variables corresponding, respectively, to the binary decisions of each modality, i.e. We can write:

$$d_1 = u(s_1 - 0.5); \quad d_2 = u(s_2 - 0.5) \tag{6}$$

where $u(x) = \begin{cases} 1 & x > 0 \\ 0 & x < 0 \end{cases}$ is the step function. As we can see, the binary variables are equal to 1 if we choose class 1 and equal to 0 if we choose class 2. The final decision will be the result of applying a fusion rule to d_1, d_2. In the case of two modalities, we have only two possible reasonable rules:

$$d = g_1(d_1, d_2) = \begin{cases} 1 & \text{si } d_1 = 1, \ d_2 = 1 \\ 0 & \text{all other cases} \end{cases} \qquad d = g_2(d_1, d_2) = \begin{cases} 0 & \text{si } d_1 = 0, \ d_2 = 0 \\ 1 & \text{all other cases} \end{cases}$$
(7)

It seems natural that hard late fusion, in exchange for its simplicity, leads to worse performance than soft late fusion since binarization prior to fusion entails the loss of information. Somewhat heuristically, we can reason as follows. Suppose that the late soft fusion function $f(s_1, s_2)$ is optimized to minimize a certain cost function of the binary classifier, such as the probability of error. If we take into account that d_1, d_2 are a function of the scores s_1, s_2, we can think that ultimately the functions $g_1(d_1, d_2)$ and $g_2(d_1, d_2)$ in (7) are functions of s_1, s_2, which will, in general, be different from the optimal function used in the late soft fusion and which will, therefore, not minimize the chosen cost function. To be more specific, let's consider again the integration. It is easy to check that:

$$\alpha \to \infty \ \Rightarrow \ s \to \min(s_1, s_2) \, \alpha \to -\infty \ \Rightarrow \ s \to \max(s_1, s_2) \qquad (8)$$

But the following fusion rules are equivalents.

$$\min(s_1, s_2) \mathop{\gtrless}_{k=2}^{k=1} 0.5 \ \Leftrightarrow \ g_1(d_1, d_2) \quad \max(s_1, s_2) \mathop{\gtrless}_{k=2}^{k=1} 0.5 \ \Leftrightarrow \ g_2(d_1, d_2) \qquad (9)$$

That is, we can interpret the late hard fusion as a late soft fusion for a particular choice of integration parameters of α-integration, which will not, in general, be those that minimize the cost function and will therefore lead to worse (or, at best, equal) performances than those achievable with the late soft fusion.

In short, we conclude that if the "a posteriori" probability or score $P(k = 1/\mathbf{x}_1, \mathbf{x}_2)$ is known precisely, the late fusion will not be able to outperform the early fusion. On the other side, the late hard fusion will not be able to outperform the late soft fusion either, as we have justified above.

The exact knowledge of $P(k = 1/\mathbf{x}_1, \mathbf{x}_2)$ is only strictly possible if we have an infinite training set and we assume statistical consistency, i.e. $\hat{P}(k = 1/\mathbf{x}_1, \mathbf{x}_2) \to^{N \to \infty} P(k = 1/\mathbf{x}_1, \mathbf{x}_2)$, where N the size of the training set of each class (for simplicity we assume the same for both classes). From a practical point of view, accurate estimation implies having sufficiently large training sizes for each class. Unfortunately, it is not easy to determine in general what the adequate minimum size is. This outside the scope of this paper. However, there are recent works that provide certain criteria in this regard.

3   A Data Fusion Methods Review

In this section, we will review the existing data fusion methods. We will also present these methods' strengths, weaknesses, and related works. It is valid to clarify that due to the length limitation of this document, we will only refer to the most cited works.

3.1   Early Sensor-Level Fusion

The process of combining sensor data or data derived from disparate sources to reduce the uncertainty in the resulting information compared to what would be possible if these sources were used individually is known as sensor fusion (Fig. 1).

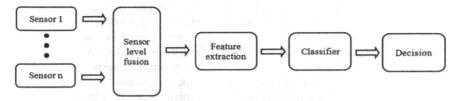

Fig. 1. Early sensor-level fusion scheme.

At the sensor level, there are several data fusion methods. We can find Kalman Filter, Bayesian Inference, Fuzzy Logic, and Artificial Neural Networks (ANN) among them. Below we present a table where we will find the weaknesses and strengths of each of these methods and related works (Table 1).

Table 1. Comparison between early sensor-level fusion methods and related works.

Sensor-level fusion methods	Strengths	Weaknesses	Works
Kalman Filter	It uses the minimum squares method to recursively estimate the state at time k, which is linear, unbiased, and of minimum variance. They are computationally efficient since processing the entire sequence of old observations with each new observation is unnecessary	It requires the initial conditions of the mean and variance of the state vector. It requires a broad knowledge of probabilities, especially the subject of Gaussian conditionality in random variables	[9–12]
Bayesian Inference	It is a recursive technique and can compute probabilities and posterior probabilities for multiple hypotheses. It offers a convenient setup for various models, such as hierarchical models and missing data problems	The probability distribution of the states must be known a priori. It often comes with a high computational cost, especially in models with many parameters	[13–15]

(continued)

Table 1. (*continued*)

Sensor-level fusion methods	Strengths	Weaknesses	Works
Fuzzy Logic	Accurate results in non-linear and challenging to model processes. It is based on logical sets and reasoning that are easy to understand and, therefore, to use Provides a simple mechanism for reasoning with vague, ambiguous, or imprecise information	Extensive validation and verification of fuzzy algorithms are necessary Defining precise fuzzy sets or membership functions requires time and effort Fuzzy control systems depend on human experience and knowledge	[16, 17]
Artificial Neural Networks	It is self-learning and can execute tasks that a linear program cannot and is able to process unorganized data. Its structure is adaptive in nature When an element of the neural network slows down, it can continue without problems, thanks to its parallel characteristics and is efficient of data noise, separating only the necessary information	Require prior training to operate, a large amount of data to achieve adequate efficiency and a lot of processing time for large neural networks Require specific hardware equipment to operate due to their computational complexity	[18–21]

3.2 Early Feature-Level Fusion

These methods or techniques aim to reduce the feature vector's dimension while preserving as much information as possible (Fig. 2).

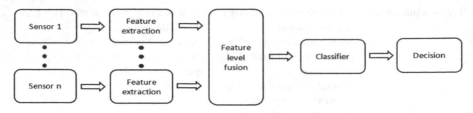

Fig. 2. Early feature-level fusion scheme.

Among the feature-level fusion methods, we can find Principal Component Analysis (PCA), Singular Value Decomposition (SVD), Multidimensional Scaling (MDS), and

Deep Learning. Below, we will detail their strengths and weaknesses and the related works that have been most cited (Table 2).

Table 2. Comparison between early feature-level fusion methods and related works.

Feature-level fusion methods	Strengths	Weaknesses	Works
Principal Component Analysis	Reduces the complexity of the data and identifies the most important features The method works quite fast, although for large data sets it requires relatively high computational complexity and memory requirements After the transformation, the variance of the data is preserved	It is necessary to choose the correct number of principal components needed for the dataset to avoid some loss of information Before applying the method, it is necessary to normalize the data; otherwise, it is difficult to find the optimal principal components	[4, 22, 23, 24]
Singular Value Decomposition	Simplifies the data and removes noise improving the results of the algorithm where SVD is used	It only makes use of a single data set, and by default, the resulting dimension reduction cannot incorporate any additional information that may be relevant Transformed data can be difficult to understand	[25–27]
Multidimensional Scaling	The solutions are relatively accurate, and the algorithm consumes little computer time	It does not allow quantifying the level of quality of the result. Since it is based on the relationship between dimensions or factors, evaluating this relationship in numbers is tough The interval scale condition can only sometimes be met in the data	[28–30]

(*continued*)

Table 2. (*continued*)

Feature-level fusion methods	Strengths	Weaknesses	Works
Deep Learning	It assists in trend and pattern detection and does not need human assistance, i.e., it makes its own decisions. It can handle many multidimensional data and constantly improves the algorithm to achieve more accurate results	Requires a large amount of data for training, which is time-consuming and computationally complex. Therefore, more powerful computers are needed to work	[31–33]

3.3 Late Score-Level Fusion (Late Soft Fusion)

Information from different detectors or classifiers is often referred to as scoring [34]. Score-level fusion is usually preferred because it offers the best compensation in terms of information content and ease of fusion [35, 36] (Fig. 3).

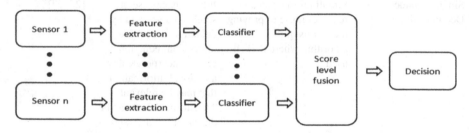

Fig. 3. Late score-level fusion scheme.

According to the literature, the Sum Rule, Likelihood Radius (LR), Support Vector Machines (SVM), and Alpha Integration are widely used score-level fusion methods. The following is a description of the strengths and weaknesses of these methods, as well as related work (Table 3).

Table 3. Comparison between late score-level fusion methods and related works.

Score-level fusion methods	Strengths	Weaknesses	Works
Likelihood Radio	Is able to handle discrete values in the score distribution It does not involve the normalization of the score vector but the transformation of its respective likelihood radio If the densities of the scores are accurate, an optimum level is reached at any desired value of false accept rate (FAR)	Requires detailed modeling of score distributions It is complex to implement due to the estimation of densities and is computationally complicated It is very time consuming as it involves a large amount of training samples Requires a high knowledge of statistical techniques	[37, 39]
Sum Rule	It does not require training samples No sample distribution modeling is required Does not require knowledge of statistics	Requires estimation of normalized parameter and weights vector and Its accuracy is rarely consistent Involves transformation of the score vector It requires that match scores be of the same nature It assumes comparable scales and strengths for input match scores	[40–43]
Support Vector Machines (SVM)	When there is a distinct margin of separation, it works incredibly well and It is effective in high-dimensional spaces In situations where there are more dimensions than samples, it is effective	When the data set is very large, it requires a lot of training time Does not work well when there is a lot of noise It is not a probabilistic model, so we cannot explain the classification in terms of probability	[44, 45]

(continued)

Table 3. (*continued*)

Score-level fusion methods	Strengths	Weaknesses	Works
Alpha Integration	It integrates many classic fusion operators and classifiers optimizing fusion parameters and achieving more results that are accurate	Optimizing the parameters is done by the gradient method, which may not converge to the global optimum. This method would inherit the weaknesses of the optimization method used	[5–7]

3.4   Late Decision-Level Fusion (Late Hard Fusion)

Decision-level fusion aims to combine the decisions made by different classifiers to reach a common consensus and obtain a more accurate decision (Fig. 4).

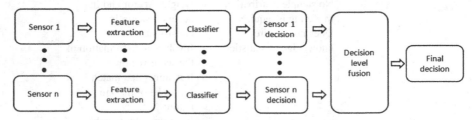

Fig. 4.   Late decision-level fusion scheme.

Majority Voting, Bagging, and Boosting are the most commonly used decision-level fusion methods. We present below the works related to these methods and a comparison taking into account their strengths and weaknesses (Table 4).

Table 4. Comparison between late decision-level fusion methods and related works.

Decision-level fusion methods	Strengths	Weaknesses	Works
Majority Voting	Since this method is based on the linear combination of multiple detection algorithms, errors or misclassifications of one model do not affect the result. The excellent performance of the others can compensate for the poor performance of one clarifier. It allows the results to be more robust and prone to overfitting	If different detection algorithms have significantly different levels of efficiency, the result may be erratic It should also be noted that the computational complexity could be high	[46, 47]
Bagging	Reduces variance and, in many cases, improves the accuracy of some predictors Increased stability and eliminates the problem of overfitting	It introduces a loss of interpretability of the model; it may experience biases when the proper procedure is ignored It can be computationally expensive	[48–50]
Boosting	Reduces variance and bias. It can generate a combined model minimizing errors as it avoids the drawbacks of single models Weights those classifiers with better performance on training data	It does not help to solve the overfitting problem; on the contrary, it may increase it It can be computationally expensive	[49–51]

4 Conclusions

A theoretical comparison of early and late fusion methods has been presented. An initial discussion on the conditions to apply early or late fusion has been introduced. From this, we can conclude that, if large training sets are available, early fusion will be the best option. The option of using dimensionality reduction methods on the fused feature vector should be considered, as well as the use of data augmentation techniques. If training sets are limited we must do late fusion, either soft or hard. In this latter case, we will avoid the complications inherent in optimally estimating the fusion function in exchange for lower performance. In any case, determining the best option in a specific application context requires a strong experimental component. In addition, a review of the fusion state of the art methods have been included. A comparative analysis of the main strengths and weaknesses of the methods has been provided. The advantages of early sensor-level fusion lie in simplicity, as the data to be fused are derived directly from the sensors. However, this approach may not be right when relationships are only possible at a more abstract level. The benefits of early feature-level fusion are found in the detection of correlated feature values. On the other hand, late score-level fusion offers advantages of performance, intuitiveness and simplicity because of the rules used. Finally, late decision-level fusion allows us to take advantage of the redundancy of a set of independent classifiers to achieve greater robustness by combining the single classifier results. There are several lines of research open from this work, e.g., an analysis of the different kinds of fusion when posteriori probabilities are unknown and have to be estimated; include in comparison of the methods, other factors such as computational complexity for real-time implementation.

References

1. Usa, H., Escamilla-Ambrosio, P.J., Mort, N.: Hybrid kalman filter-fuzzy logic adaptive multisensor data fusion architectures (2003)
2. Vergara, L., Soriano, A., Safont, G., Salazar, A.: On the fusion of non-independent detectors. Digit. Signal Process. **50**, 24–33 (2016)
3. Salazar, A., Safont, G., Vergara, L., Letters, E.V.-P.R.: Pattern recognition techniques for provenance classification of archaeological ceramics using ultrasounds. Pattern Recogn. Lett. **135**, 441–450 (2020)
4. Jolliffe, I.T., Cadima, J.: Principal component analysis: a review and recent developments. Philos. Trans. R. Soc. A Math. Phys. Eng. Sci. **374**(2065), 20150202 (2016)
5. Soriano, A., Vergara, L., Ahmed, B., Salazar, A.: Fusion of scores in a detection context based on Alpha integration. Neural Comput. **27**(9), 1983–2010 (2015)
6. Safont, G., Salazar, A., Vergara, L.: Multiclass Alpha integration of scores from multiple classifiers. Neural Comput. **31**(4), 806–825 (2019)
7. Safont, G., Salazar, A., Vergara, L.: Vector score alpha integration for classifier late fusion. Pattern Recogn. Lett. **136**, 48–55 (2020)
8. Salazar, A., Vergara, L., Vidal, E.: A proxy learning curve for the Bayes Classifier. Pattern Recogn. **136**(109240), 1–14 (2023)
9. Kalman, R.E.: A new approach to linear filtering and prediction problems. J. Basic Eng. **82**(1), 35–45 (1960)
10. Kalman, R.E., Bucy, R.S.: New results in linear filtering and prediction theory. J. Basic Eng. **83**(1), 95–108 (1961)
11. Moon, S., Park, Y., Ko, D.W., Suh, I.H.: Multiple kinect sensor fusion for human skeleton tracking using Kalman filtering (2016)
12. Yazdkhasti, S., Sasiadek, J.Z.: Multi sensor fusion based on adaptive Kalman filtering. In: Dołęga, B., Głębocki, R., Kordos, D., Żugaj, M. (eds.) Advances in Aerospace Guidance, Navigation and Control, pp. 317–333. Springer, Cham (2018). https://doi.org/10.1007/978-3-319-65283-2_17
13. Dempster, A.P.: A generalization of Bayesian inference. J. Roy. Stat. Soc.: Ser. B (Methodol.) **30**(2), 205–232 (1968)
14. Coninx, A., et al.: Bayesian sensor fusion with fast and low power stochastic circuits. In: 2016 IEEE International Conference on Rebooting Computing, ICRC 2016 - November (2016)
15. Coué, C., Fraichard, T., Bessiere, P., Mazer, E.: Multi-sensor data fusion using Bayesian programming: an automotive application. In:IEEE International Conference on Intelligent Robots and Systems, vol. 1, pp. 141–146 (2002)

16. Amin, M., Akhoundi, A., Valavi, E.: Multi-sensor fuzzy data fusion using sensors with different characteristics (2010)
17. Stover, J.A., Hall, D.L., Gibson, R.E.: A fuzzy-logic architecture for autonomous multisensor data fusion. IEEE Trans. Industr. Electron. **43**(3), 403–410 (1996)
18. McCulloch, W.S., Pitts, W.: A logical calculus of the ideas immanent in nervous activity. Bull. Math. Biophys. **5**(4), 115–133 (1943)
19. Quan, Y., Zhou, M.C., Luo, Z.: On-line robust identification of tool-wear via multi-sensor neural-network fusion. Eng. Appl. Artif. Intell. **11**(6), 717–722 (1998)
20. Lee, J., Steele, C.M., Chau, T.: Swallow segmentation with artificial neural networks and multi-sensor fusion. Med. Eng. Phys. **31**(9), 1049–1055 (2009)
21. Kańtoch, E.: Human activity recognition for physical rehabilitation using wearable sensors fusion and artificial neural networks. Comput. Cardiol. **2010**(44), 1–4 (2017)
22. Hotelling, H.: Analysis of a complex of statistical variables into principal components. J. Educ. Psychol. **24**(6), 417–441 (1933)
23. Abdi, H., Williams, L.J.: Principal component analysis. Wiley Interdiscip. Rev. Comput. Stat. **2**(4), 433–459 (2010)
24. Hasan, M.M., Islam, N., Rahman, M.M.: Gastrointestinal polyp detection through a fusion of contourlet transform and neural features. J. King Saud Univ. – Comput. Inf. Sci. **34**(3), 526–533 (2022)
25. Nasir, H., Stanković, V., Marshall, S.: Singular value decomposition based fusion for super-resolution image reconstruction. Signal Process Image Commun. **27**(2), 180–191 (2012)
26. Zhao, X., Ye, B.: Singular value decomposition packet and its application to extraction of weak fault feature. Mech. Syst. Signal Process **70–71**, 73–86 (2016)
27. Zhu, H., He, Z., Wei, J., Wang, J., Zhou, H.: Bearing fault feature extraction and fault diagnosis method based on feature fusion, vol. 21, no. 7, p. 2524 (2021)
28. Ye, X., Gao, W., Yan, Y., Osadciw, L.A.: Multiple tests for wind turbine fault detection and score fusion using two- level multidimensional scaling (MDS), vol. 7704, pp. 70–77 (2010)
29. Tian, G.Y., Taylor, D.: Colour image retrieval using virtual reality. In: Proceedings of the International Conference on Information Visualisation, vol. 2000-July, pp. 221–225 (2000)
30. Choo, J., Bohn, S., Nakamura, G.C., White, A.M., Park, H.: Heterogeneous data fusion via space alignment using nonmetric multidimensional scaling. In: Proceedings West Mark Ed Association Conference, pp. 177–188 (2012)
31. He, Q.-H., et al: Deep learning and radiomic feature-based blending ensemble classifier for malignancy risk prediction in cystic renal lesions (2023)
32. Wang, K., Xu, C., Li, G., Zhang, Y., Zheng, Y., Sun, C.: Combining convolutional neural networks and self-attention for fundus diseases identification (2023)
33. Jing, J., Wu, H., Sun, J., Fang, X., Zhang, H.: Multimodal fake news detection via progressive fusion networks. Inf. Process Manag. **60**(1) (2023)
34. Salazar, A., Safont, G., Soriano, A., Vergara, L.: Automatic credit card fraud detection based on non-linear signal processing. In: ICCST, pp 207–2012 (2012). Article no. 6393560
35. Salazar, A., Safont, G., Vergara, L.: Surrogate techniques for testing fraud detection algorithms in credit card operations. ICCST **2014**(6986987), 124–129 (2014)
36. Vergara, L., Salazar, A., Belda, J., Safont, G., Moral, S., Iglesias, S.: Signal processing on graphs for improving automatic credit card fraud detection. In: ICCST, pp. 1–6, Spain (2017)
37. Nanni, L., Lumini, A., Brahnam, S.: Likelihood ratio-based features for a trained biometric score fusion. Expert Syst. Appl. **38**(1), 58–63 (2011)
38. Zafar, R., et al.: Prediction of human brain activity using likelihood ratio based score fusion. IEEE Access **5**, 13010–13019 (2017)
39. Salazar, A., Safont, G., Rodriguez, A., Vergara, L.: Combination of multiple detectors for credit card fraud detection. In: ISSPIT, pp. 138–143, Limassol, Cyprus (2016)

40. Hube, J.P.: Neyman-Pearson biometric score fusion as an extension of the sum rule, vol. 6539, pp. 200–208 (2007)
41. Hammouche, R., Attia, A., Akhrouf, S.: Score level fusion of major and minor finger knuckle patterns based symmetric sum-based rules for person authentication. Evolving Syst. **13**(3), 469–483 (2022)
42. Kittler, J., Hatef, M., Duin, R.P.W., Matas, J.: On combining classifiers. IEEE Trans. Pattern Anal. Mach. Intell. **20**(3), 226–239 (1998)
43. Eskandari, M., Toygar, Ö.: Fusion of face and iris biometrics using local and global feature extraction methods. Signal Image Video Process **8**(6), 995–1006 (2014)
44. Fierrez-Aguilar, J., et al.: A comparative evaluation of fusion strategies for multimodal biometric verification. Lect. Notes Comput. Sci. **2688**, 830–837 (2003)
45. Arigbabu, O.A., et al.: Integration of multiple soft biometrics for human identification. Pattern Recogn. Lett. **68**, 278–287 (2015)
46. Ruta, D., Gabrys, B.: Classifier selection for majority voting. Inf. Fusion **6**(1), 63–81 (2005)
47. Jimenez, L.O., Morales-Morell, A.: Classification of hyperdimensional data based on feature and decision fusion approaches using projection pursuit, majority voting, and neural networks. IEEE Geosci. Remote. Sens. **37**(3 I), 1360–1366 (1999)
48. Breiman, L.: Bagging predictors. Mach. Learn. **24**(2), 123–140 (1996)
49. Ćwiklińska-Jurkowska, M.: Boosting, bagging and fixed fusion methods performance for aiding diagnosis. Biocybern. Biomed. Eng. **32**(2), 17–31 (2012)
50. Dietterich, T.G.: Experimental comparison of three methods for constructing ensembles of decision trees: bagging, boosting, and randomization. Mach. Learn. **40**(2), 139–157 (2000)
51. Ferreira, A.J., Figueiredo, M.A.T.: Boosting algorithms: a review of methods, theory, and applications. Ensemble Mach. Learn. 35–85 (2012)

ANN HW-Accelerators

A Scalable Binary Neural Associative Memory on FPGA

Marius Kortekamp[ID], Sarah Pilz[✉][ID], Jens Hagemeyer[ID], and Ulrich Rückert

CITEC - Bielefeld University, Inspiration 1, 33619 Bielefeld, Germany
{mkortekamp,spilz,jhagemey,rueckert}@techfak.uni-bielefeld.de

Abstract. The human brain and its ability to associate is one of the most fascinating things in nature. The long-known concept of binary neural associative memory offers the possibility to build a very simple hardware architecture, that allows direct association. In a BINAM the presented input is associated with the stored content of the memory, without the need for addressing. Hereby the BINAM is a fault-tolerant concept, which allows that erroneous input vectors will usually result in a correct output. In this work, we propose a modern hardware architecture of a BINAM on the VCU1525 FPGA board. We implemented the architecture in VHDL as a scalable, modular, generic, and easy to use design. For the evaluation designs in the range of 8,000 to 740,000 neurons, with equal input and output vector size, have been generated and tested on the FPGA board. Currently, a maximum clock frequency of ∼200 MHz with a resource utilization of only ∼33% CLBs, ∼22% LUTs, and ∼10% registers can be achieved. Maximum times for the storage and association process for all designs are stated in this paper.

Keywords: BINAM · FPGA · Scalable

1 Introduction

One fascinating object is the human brain. It is the central organ of our nervous system and its exact functioning is one of the greatest mysteries of modern times, which scientists around the world have been trying to solve for years. On average, the human brain weighs 1508 g and contains about 86.1 (+/− 8.1) billion neurons and 84.6 (+/− 9.8) billion "non-neuronal" cells (e.g. glial cells) [1,4]. The cerebral cortex weighs about 1233 g and thus makes up 82% of the total brain mass. At the same time, only 19% of the neurons are located in the cerebral cortex. All neurons in the brain are interconnected. On average, every neuron has about 7,000 synapses, but the actual number can vary greatly. In the cerebral cortex, about 0.15 quadrillion synapses have been found [2]. The human brain is also the most efficient associative memory occurring in nature.

M. Kortekamp and S. Pilz—These authors contributed equally to this work.

© The Author(s), under exclusive license to Springer Nature Switzerland AG 2023
I. Rojas et al. (Eds.): IWANN 2023, LNCS 14134, pp. 381–393, 2023.
https://doi.org/10.1007/978-3-031-43085-5_30

The ability to associate is not only extremely valuable in daily life, but also a key part of many learning processes and can therefore be seen as an important prerequisite for intelligence [5, 8].

Today, artificial intelligence is becoming increasingly present and is now being applied in all sorts of applications. The most central element for being able to generate and use modern AI algorithms is a large amount of data. Ever-increasing amounts of data are being generated and stored. Data about machines, our surroundings, and about us. Every bit of data is thereby helping to generate digital representations, e.g., so-called "digital twins", which are already very successfully applied in industry for machines and which are increasingly also generated and used for humans [15]. These steadily increasing amounts of data have to be stored and processed. This requires not only the knowledge of how to do the processing but also a lot of physical resources and energy. Sustainability is a topic that concerns everyone; ideas about saving energy and green coding are becoming more and more popular. Considering all this, wouldn't it be a logical idea to use associative processes already in the storing process of data? So to say, use the memory in a way that stores everything depending on the content and associates it with already stored data in memory, like our brain. Sure, this approach differs from that of today's common storage systems. Usually, every new memory input is written to the next empty address and later accessed again by looking up what is stored at the specific location. But the idea of memory systems that are accessible by content instead of addresses, so-called content addressable memory (CAM) systems, is not new. Pioneers in computer science, like Konrad Zuse and Teuvo Kohonen, already addressed this topic. From Konrad Zuse, there is a hand-drawn sketch from 1943 in which he depicts an associative memory in relay technology [21]. About 40 years later, Teuvo Kohonen referred to over 1,200 publications about CAMs [6]. It was found that in neural associative memories, binary weight vectors offer great advantages. Although only one bit needs to be stored for each weight, very good results can be achieved with only minimal loss of precision. This very efficient form of storage is used in a special type of CAM called "binary neural associative memory"(BINAM), which is particularly suitable for usage in hardware implementations. The concept of a binary neural associative memory in combination with sparse coding was thoroughly researched by Günther Palm [9,10,14].

The BINAM model has very interesting properties. Due to the redundant way in which data is stored, it offers fault tolerance to erroneous input data, as described in [7,12,19]. This also allows the BINAM to be used for pattern completion, pattern recognition, and pattern extraction. Therefore, the BINAM does not simply map objects statically but learns to generalize so that similar inputs have similar outputs. All existing hardware implementations of BINAMs we were able to find, are already 20 years old, but already offer very interesting implementation approaches. In the year 1991 a BINAM was realized on a digital special-purpose hardware by Rückert et al. [17]. A modular design consisting of slice chips were created which could be combined to create BINAMs with size of $8,192 \times 8,192$. The hardware could be run with a clock frequency of 12 MHz.

Another implementation was done in the year 1991 on a RISC processor by Rückert et al. [18]. For this the "overflow algorithm" was created. Furthermore, an design on FPGAs in year 1993 was implemented by Rückert et al. [16]. The implementation achieved a maximum BINAM size of 64,000 × 16,000 and a clock frequency of 10 MHz. A student work from our research group, which was done in 2003 used the same approach but already achieved significantly better results on a more modern hardware platform. A module of this implementation achieved a size of 32,768 × 32,768, whereby a total of six modules were used simultaneously. A higher clock frequency of 25 MHz was also achieved.

As the topic is so promising to solve the current challenges of big data and resource-efficient processing and storage, it is interesting to evaluate the possibilities of today's technology. In this work, we propose an implementation of the BINAM on a modern FPGA. The design was written in VHDL in a generic manner to support different BINAM and memory sizes. We used the FPGA to achieve a highly parallel design, which supports an arbitrarily high number of neurons and is connected to a large memory. Our design was developed and tested on the Xilinx Virtex UltraScale+ FPGA VCU1525 Acceleration Development Kit, utilizing a VU9P Virtex UltraScale+ FPGA and 4 × 16 GB DDR4 memory [20], using the Xilinx Vivado Design Suite Version 2019.1.

1.1   Neural Associative Memory

An associative memory can be considered as a content-addressable device that maps a set of input patterns to a set of output patterns [13]. The input patterns are used instead of addresses that are applied to the memory input port to retrieve the output patterns. Therefore, the input is henceforth denoted as address patterns S^A and the outputs are denoted as content patterns S^C. The mapping from address patterns to output patterns is defined by the set of pairs of address pattern x^k and content pattern y^k: $\{(x^1, y^1), ..., (x^M, y^M) : x^k \in S^A, y^k \in S^C\}$.

A straightforward example of an associative memory is a phone book that maps 'names' to 'phone numbers'. The 'names' act as the address patterns and the 'phone numbers' as the content patterns. The associations are then given by the mapping of the 'names' to their corresponding 'phone numbers'.

The most powerful associative memory occurring in nature is the brain. Accordingly, it makes sense to derive a model of a neural associative memory from it. The structure of our neural associative memory is shown in Fig. 1. It consists of multiple processing elements called neurons. Similar to the dendrites of a biological neuron the model neuron receives synaptic inputs via input signals x_1 to x_n. The weights w_1 to w_n are used so that each input x_i can have a different influence on the respective neuron. In this way, the strength of the connection between the ith row and the jth column and thus the influence on the value of y_j can be specified. The threshold function f is similar as in the biological neuron. The threshold prevents every input from resulting in an output. Instead, the input must be large enough to exceed the threshold so that only large inputs (strong stimuli) result in an output.

Now several model neurons can be arranged in a grid, where the input signals are the same for all of them but the weights are different. The result is a neural associative memory, where content is stored in the weights. The weights can then be trained by the learning rule (1). The number of weights per neuron determines the height of the associative memory and therefore the size of the address patterns. Also, the number of model neurons specifies the width of the associative memory and the size of the output patterns. The result for every output neuron is determined with the threshold function f in the association process (see association rule (2)).

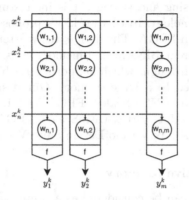

Fig. 1. Basic structure of a neural associative memory (based on [18])

Learning Rule. To train the association between an address pattern and a content pattern, a learning rule is used which adjusts the weights of the association matrix W. Given a set of pairs of address and content pattern S^P, the association matrix must be adjusted so that the address patterns are mapped to their corresponding content patterns. This is achieved by applying the learning rule (1) to all pairs $(\mathbf{x}^k, \mathbf{y}^k) \in S^P$. In simple terms, a pair is learned by writing the content pattern in each active row, where a row i is active if $\mathbf{x}_i^k = 1$. The learning rule uses the Hebb rule [3] meaning that the weights increase in strength only for coinciding firing of the pre-synaptic and post-synaptic neuron.

$$W = (W_{ij}) := \sum_{k=1}^{M} w_{i,j}^k \vee \mathbf{x}_i^k \mathbf{y}_j^k \qquad (1)$$

Storage Process/Storage Capacity. For reasons of efficiency, the memory of the weights should be as small as possible. Palm has analyzed this thoroughly and has come to the conclusion that even for the extreme case when only one bit is available per weight, information can be stored very effectively using sparsely coded input and output vectors [9–11]. Sparsely coded means that the number of active elements l in the input vector and the number of active elements k in the output vector is significantly less than the number of inactive elements. With these assumptions we obtain an asymptotic storage efficiency of 0.69 bits/synapse, given that $l = log_2(n)$ and $k = log_2(m)$. Here, n denotes the height of the BINAM, i.e., the number of rows and m denotes the width of the BINAM, i.e., the number of columns. Also, it is important that similarity-preserving encoding schemes are used to convert data into binary sparse vectors. This means that the encodings map patterns similar in meaning to similar zero-one sequences.

Association Rule. The retrieval of content from the BINAM can be achieved by using the association rule (2). For this, an address pattern is applied to the input port. For each Neuron N_j, the activation strength d_j is calculated by summing over all incoming activities. Finally, a threshold function is applied to the intermediate result d_j yielding the final result $\tilde{y}_j$, the queried content pattern. The choice of the threshold Θ plays an important role and therefore must be chosen wisely. A logical and simple choice for the threshold is the number of active elements of the address pattern. Alternatively, the maximum number of active ones from the summation of the column results can be selected for Θ. The implementation effort is higher since all results must be known to calculate the maximum. However, the error tolerance of the BINAM can increase because there is no direct dependence on the number of active ones in the address vector.

$$d_j = \sum_i W_{ij} x_i$$

$$\tilde{y}_j = f(d_j - \Theta) \tag{2}$$

$$f(x) = \begin{cases} 1, & \text{if } x > 0 \\ 0, & else \end{cases}$$

Application Options. Though the BINAM system concept is very simple, it has desirable characteristics regarding ultra large scale integration (ULSI):

- The modular and regular circuit design allows to integrate a $10^5 * 10^5$ BINAM (10 GBit) on a single chip (in-memory computing)
- The asymptotic storage capacity is 0.69 Bit per synapse
- The number of sparsely coded patterns that can be stored is about $z \sim ln_2(m * n)/(k * l)$ and much larger than the number of columns (artificial neurons). In the case of the $10^5 x 10^5$ BINAM, about 10^7 associations can be stored with low error probability ($l = k = 16$)
- The BINAM concept shows robustness to noisy inputs or defect synapses
- The required operations per association is only $O(log(n) * m)$ and almost constant for all associations
- Energy requirements per association are low compared to alternative associative memory concepts

Because of these attractive characteristics, the BINAM concept is attractive for embedded applications demanding low energy requirements, fast responses, and fault tolerance. Examples are embedded information retrieval in hand-held devices or sense-actor mappings in mobile robotics. In general, a distinction is made between two types of associations: hetero- and auto-association. In hetero-association applications, the input and output vectors are different and can have different lengths. In auto-association applications, the input and output vectors are equal. Auto-association matrices are particularly interesting for determining whether the existing input vector is correct. If this is not the case, the association process will return the appropriate stored value of the matrix.

2   Implementation

The architecture was designed according to the following principles:

1. Scalable to allow large BINAMs that even cross Super Logic Regions or multiple FPGA modules
2. Modular for separation of functions to ensure maintainability and enable options of design space exploration
3. Generic so that different configurations of BINAMs can be easily created that differ in width, height, and the sizes of the address and content patterns
4. Easy to handle for enhanced usability

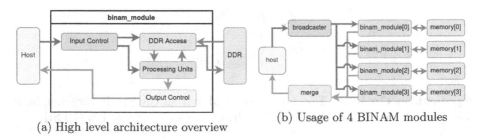

(a) High level architecture overview   (b) Usage of 4 BINAM modules

Fig. 2. BINAM modules

A high-level overview of the architecture and periphery is shown in Fig. 2a, which is described in more detail below. Three components are shown, namely a host, a BINAM module, and a DDR. The host runs on a CPU and is used to control the BINAM module, supplies it with data, and retrieves its results. The DDR is a large memory used for the synaptic weights. The BINAM module resides in the reconfigurable fabric of the FPGA and consists of multiple components. First, there is the *input control*, which manages the transmitted commands and address data from the host by controlling the current state (learn state, associate state, ...) and making the address data accessible to the *DDR access* and *processing units*. Second, the *DDR access* provides an easy and efficient connection to the DDR and thus to the weights. Third, the calculations for the learning, association, and reading operations are done by the *processing units*. For that, it accesses the address data from the *input control* and the weights from the *DDR access* and processes them according to the current state of the *input control*. Finally, the results from the association and read operations are forwarded to the *output control*. The main task of the *output control* is the encoding of the results and their transmission to the host.

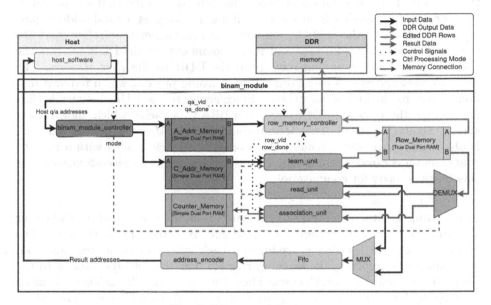

Fig. 3. Binam architecture

A more detailed overview of the BINAM architecture is shown in Fig. 3, which further elaborates the high-level overview. In fact, the components shown are an extension of the four components of the high-level overview and are color-coded accordingly. Also, this overview is consistent with the VHDL implementation. The following describes the components, as well as the strategies employed, that are most important to achieving the enumerated principles.

Input Control. Commands as well as address and content patterns are received and processed by the *binam_module_controller*. Since the patterns are sparse, only the addresses of the active elements are transmitted. This significantly reduces the amount of data to be transmitted and the amount of required memory. Forwarded by the *binam_module_controller*, the address and content patterns are separately stored in the *A_Addr_Memory* and *C_Addr_Memory*. Both memories are split into two blocks to enable double buffering. This technique allows that one block can be filled with addresses while the other block is used for learning/associating/reading. As soon as one block is filled with addresses the *binam_module_controller* signals the *row_memory_controller* that this block is ready for computation. Then the other block can be filled with the next addresses.

DDR Access. In the DDR the weights are stored in the following order: $w_{1,1}$ to $w_{1,m}$, $w_{2,1}$ to $w_{2,m}$, ..., $w_{n,1}$ to $w_{n,m}$. This means that the ith weights of all neurons are stored as a batch, which from now on will be denoted as the ith row of the association matrix. This type of storage enables the parallel processing of weights from many neurons. As soon as a valid address pattern is available in the *A_Addr_Memory*, the *row_memory_controller* starts to sequentially read the addresses and issues commands to the DDR to start the transfer of the corresponding rows from the DDR to the *Row_Memory*. The *Row_Memory* serves as a cache during the learning process, which requires random access to the individual weights. Since access to the DDR memory is time consuming, the rows are first moved to a BRAM, a cache, which allows faster access. Similar to the *A_Addr_Memory* and *C_Addr_Memory*, double buffering is also used for the *Row_Memory*. As soon as one block is filled with a row the *row_memory_controller* signals the *learn_unit*, *read_unit*, or *association_unit* that this block is ready for computation.

Processing Units. According to the current mode of the *input_control* either the *learn_unit*, *association_unit*, or *read_unit* is active. In learn mode, the *learn_unit* starts a learning procedure for each valid pair of address and content pattern by applying the learning rule (1). For that, it needs access to the *Row_Memory* and *C_Addr_Memory*. Then, the row in the *Row_Memory* is modified bit by bit according to the addresses in the *C_Addr_Memory*. Thereafter, the updated row is written back to the DDR. In association mode, the *association_unit* starts an association procedure for each valid address pattern by applying the association rule (2). For that, it only needs access to the *Row_Memory*. Since several rows are summed up when associating, the *Counter_Memory* is needed to store the intermediate results. Finally, the threshold function is applied and the final result is passed to the *address_encoder*. During read mode, the *read_unit* starts a reading procedure for each valid address pattern. Similar to the association procedure, access to the *Row_memory* is required. However, instead of applying a specific rule, the rows are simply passed to the *address_encoder*.

Ideally, the architecture supports as many processing engines in the *learn_unit* and *association_unit* as there are neurons. However, since the rows of the BINAM can become quite large, it is on the one hand rather unlikely that the FPGA will provide enough resources and memory with sufficient width to implement all neurons. On the other hand, even if enough resources are available, the maximum DDR data rate will limit the amount of neurons that can be provided with data in every clock cycle. Here only 512 bit can be retrieved from the DDR, therefore only 512 neurons per BINAM module have been implemented. This subset of processing engines will be referred to as physical neurons. Accordingly, the neurons of the BINAM must share them and since they are not implemented they are henceforth referred to as virtual neurons.

Output Control. The calculated results from the *association_unit* and *read_unit* are forwarded to the *address_encoder*. Since the association results are also sparsely coded, they are encoded by converting the active elements to their addresses. Furthermore, the association results are not encoded as a whole but in chunks and therefore it is worth discarding all chunks with only zeros. Finally, the encoded results are sent back to the host.

Multiple BINAM Modules. The architecture allows easy expansion of the total BINAM size by using multiple BINAM modules, each connected to its own memory. This is shown in Fig. 2b for four BINAM modules, which also matches our implementation on the VCU1525 board. Since the neurons are independent of each other, they are distributed among the different BINAM modules so that they are also independent of each other. This means that each module can run in parallel. Furthermore, it is possible to create even larger BINAMs using multiple FPGAs.

3 Analysis and Tests

To test the implementation, designs with different BINAM sizes were synthesized and tested on the Xilinx VCU1525 FPGA-board. To cover as many variants as possible, designs with 8 k, 16 k, 32 k, 64 k, 128 k, and 740 k neurons were created for evaluation. All designs consist of 4 BINAM modules, each connected to its own DDR4 memory onto which the neurons have been evenly distributed. For each neuron, the number of weights to be stored was chosen to be identical to the existing number of neurons. This means that the evaluated BINAMs always have a square size, depending on their number of neurons. For example with 8 k neurons a total of $8,192 * 8,192 \approx 67$ million weights are stored. The different configurations examined can be seen in Table 1. With 740,000 neurons as the largest configuration, the 64 GB of DDR4 memory is nearly fully utilized.

First, the maximum possible clock frequency was determined for the different design sizes. This was done using a script that automatically created designs in the specified sizes with different clock frequencies until the maximum possible clock frequency was found iteratively, without timing errors according to Vivado. The maximum clock frequencies determined in this way can also be seen in Table 1. It should be noted that slight deviations in the maximum possible clock frequency are possible due to the non-deterministic synthesis process. Furthermore, it can be seen that the frequencies of the different BINAM variants remain relatively constant at 200 MHz. Only for the 740 k variant, there is a significant decrease in clock frequency.

Table 2 shows the resource utilization of the BINAM variants. The resource utilization remains approximately the same for Configurable Logic Blocks (CLBs), LookUp Tables (LUTs), and registers. This means that the design is relatively static for different sizes concerning these resources. It is therefore reasonable that the clock frequencies are similar. Only in the block RAM utilization are larger steps because the Row_Memory and Counter_Memory increase with

Number of Neurons	Neurons per DDR4	Total BINAM Size	Max. Clock Frequency
8,192	2,048	67,108,864	202 MHz
16,384	4,096	268,435,456	202 MHz
32,768	8,192	1,073,741,824	204 MHz
65,536	16,384	4,294,967,296	208 MHz
131,072	32,768	17,179,869,184	209 MHz
740,000	185,000	547,600,000,000	182 MHz

Table 1. Maximum Achieved Clock Frequency for different BINAM Sizes

the number of neurons. But there is also a difference in the 740 k variant, namely that ULTRA RAM is used. Overall, resource utilization is quite low and does not pose any challenges. However, very wide BINAMs (significantly more than 740 k neurons) will eventually exhaust the block RAM/Ultra RAM.

Neurons	CLB	LUTs	Registers	Block RAM	Ultra RAM
8,192	50,241 (34.0%)	267,686 (22.6%)	256,228 (10.8%)	300.5 (13.9%)	0
16,384	49,260 (33.3%)	267,994 (22.7%)	256,283 (10.8%)	300.5 (13.9%)	0
32,768	50,022 (33.9%)	268,012 (22.7%)	256,327 (10.8%)	300.5 (13.9%)	0
65,536	48,780 (33.0%)	258,419 (21.9%)	239,863 (10.1%)	528.5 (24.5%)	0
131,072	49,019 (33.2%)	258,959 (21.9%)	239,958 (10.1%)	528.5 (24.5%)	0
740,000	48,019 (32.5%)	260,358 (22.0%)	240,092 (10.1%)	300.5 (13.9%)	228 (23.8%)

Table 2. Resource utilization for different BINAM Sizes

In addition, a timing analysis was performed for each design with a total of 1 million test patterns for auto-association (address pattern equals content pattern). The test patterns were generated sparsely coded according to the findings of Palm [11]. Accordingly, the number of 1's per vector was determined with $log_2(vector\_length)$ (see also Table 3 and Sect. 1.1). To measure the learning and association times, an IP core was created that supplies the BINAM within the FPGA with data at maximum speed. In this way, inaccuracies that could occur during the transfer from the host are circumvented and the measured results indicate the actual computational speed of the implemented design. We have already optimized the design by using all four available DDR4 in parallel, which increased the access times for learning by a factor of 4. The maximum learning and association rates achieved (measured in operations per second) for this test set-up can be found in Table 3. As expected, the learning rate is significantly higher for small BINAMs and decreases almost linearly anti-proportionally with increasing BINAM size (see Table 3). This clearly shows that the learning rate is limited by the access time to the DDR4 memory. For each row learned, a read and a write access to the DDR4 must be performed. With more neurons, the rows become larger, so more time is needed for the learning process. The association rate is also significantly higher for small BINAMs and decreases almost linearly anti-proportionally with increasing BINAM size (see Table 3), which is due to the fact that larger rows have to be read and processed. The data transfer rate could be optimized by increasing the clock frequency of the DDR4 interface to 300 MHz so that the maximum data transfer rate of 19.2 GBps can be achieved (currently 11.712/12.8 GBps for 182/200 MHz) [20]. In addition, by increasing the clock frequency for the design, the learning itself (updating the rows) and the association rate can be accelerated. Whereby the association rate is mainly limited by the access times to the DDR4 memory. The reason lies in the double buffering, so that the association time is given by: $T_{asso} = l * max(T_{DDR}, T_{add\_row})$, with T_{DDR} = DDR access time, T_{add_row} = time to add a row.

Neurons	Active 1's	Learning (OP/s)	Associate (OP/s)
8,192	13	331,674	561,797
16,384	14	222,222	456,621
32,768	15	166,250	340,135
65,536	16	106,837	193,423
131,072	17	60,734	94,473
740,000	20	11,395	15,383

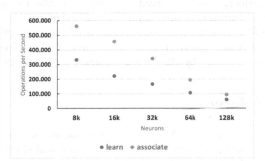

Table 3. Hardware Test of BINAM with different Sizes - association and learning times; All Tests were performed with 1 Million Test Vectors. The performance was measured in operations per second (OP/s). Exact results are presented in the table and visualized in a graph

4   Summary and Future Work

In this work, a modern architecture of a binary neural associative memory was designed and tested on the VCU1525 FPGA board. The design was implemented in VHDL and is scalable, modular, generic, and easy to use. Several variants of BINAMs with different sizes (number of neurons, number of weights) were generated for evaluation. Sizes from 8 k to 740 k were obtained, with the 740 k variant utilizing the complete 64 GB DDR memory. It was found that the resource consumption with ∼33% CLB, ∼22% LUTs, and ∼10% register utilization is relatively low and static for all designs, while the required BRAM/URAM grows with the number of neurons. Accordingly, the measured clock frequencies are also similar at ∼200 MHz, with the exception of the 740 k variant with 182 MHz. Furthermore, the time required for learning and association was measured and analyzed. The main bottleneck for this design is the access time to the DDR4 memory. Therefore, all four available DDR4 memories of the used FPGA board are used in parallel. For further optimization, the critical path is currently analyzed and optimized, to achieve a clock frequency of 300 MHz for DDR access. This will allow a maximum access rate to the DDR4 with 19.2 GBps. Initial tests show, that this is possible, but our optimized design is not running stable at the moment. We expect the optimized design to run stable soon and will then publish the updated results. The architecture has been designed to easily allow the distribution of the BINAM on multiple FPGAs. In the next step, we are going to test this in order to analyze the optimal distribution of memory and calculation units. We are also planning to test the BINAM with real data, which possibly will not completely fulfill Palm's sparsity requirements, and test how well the BINAM can be used for different applications.

Acknowledgments. This publication incorporates results from the VEDLIoT project, which received funding from the European Union's Horizon 2020 research and innovation programme under grant agreement number 957197. Also, Sarah Pilz and Ulrich Rückert were members of the research programme 'Design of Flexible Work Environments-Human-Centric Use of Cyber-Physical Systems in Industry 4.0', which is supported by the North-Rhine-Westphalian funding scheme 'Forschungskolleg'.

References

1. Azevedo, F.A., et al.: Equal numbers of neuronal and nonneuronal cells make the human brain an isometrically scaled-up primate brain. J. Comparat. Neurol. **513**(5), 532–541 (2009)
2. Drachman, D.A.: Do we have brain to spare? (2005)
3. Hebb, D.O.: The organization of behavior: a neuropsychological theory. Psychology Press (2005)
4. Herculano-Houzel, S.: The human brain in numbers: a linearly scaled-up primate brain. Front. Hum. Neurosci. **3**, 31 (2009)
5. Kaufman, S.B., DeYoung, C.G., Gray, J.R., Brown, J., Mackintosh, N.: Associative learning predicts intelligence above and beyond working memory and processing speed. Intelligence **37**(4), 374–382 (2009)
6. Kohonen, T.: Content-addressable memories. 1st Edn. Springer Berlin, Heidelberg (1980). https://doi.org/10.1007/978-3-642-96552-4
7. Kohonen, T.: Self-organization and associative memory, vol. 8. 3rd Edn. Springer Berlin, Heidelberg (2012). https://doi.org/10.1007/978-3-642-88163-3
8. Neisser, U., et al.: Intelligence: knowns and unknowns. Am. Psychol. **51**(2), 77 (1996)
9. Palm, G.: On associative memory. Biol. Cybern. **36**, 19–31 (1980). https://doi.org/10.1002/andp.19053221004
10. Palm, G.: Assoziatives gedächtnis und gehirntheorie. Spektrum der Wissenschaft **6**, 54–64 (1988)
11. Palm, G.: On the asymptotic information storage capacity of neural networks. In: Eckmiller, R., van der Malsburg, C. (eds.) Neural Computers. Springer Study Edition, vol. 41, pp. 271–280. Springer, Heidelberg (1989). https://doi.org/10.1007/978-3-642-83740-1_29
12. Palm, G.: Neural assemblies. 1st Edn. Springer, Heidelberg (1982). https://doi.org/10.1007/978-3-642-81792-2

13. Palm, G., Sommer, F.T.: Associative data storage and retrieval in neural networks. Models of Neural Networks III: Association, Generalization, and Representation, pp. 79–118 (1996)

14. Palm, G.: Neural associative memories and sparse coding. Neural Netw. **37**, 165–171 (2013). https://doi.org/10.1016/j.neunet.2012.08.013. https://www.sciencedirect.com/science/article/pii/S0893608012002298. Twenty-fifth Anniversay Commemorative Issue

15. Pilz, S., Hellweg, T., Harteis, C., Rückert, U., Schneider, M.: who will own our global digital twin: the power of genetic and biographic information to shape our lives. In: Gräßler, I., Maier, G.W., Steffen, E., Roesmann, D. (eds.) The Digital Twin of Humans, pp. 11–35. Springer, Cham(2023). https://doi.org/10.1007/978-3-031-26104-6_2

16. Rückert, U., Funke, A., Pintaske, C.: Acceleratorboard for neural associative memories. Neurocomputing **5**(1), 39–49 (1993)

17. Rückert, U., Kleerbaum, C., Goser, K.: Digital VLSI implementations of an associative memory based on neural networks. In: Delgado-Frias, J.G., Moore, W.R. (eds.) VLSI for Artificial Intelligence and Neural Networks, pp. 275–284 . Springer, Boston, MA (1991). https://doi.org/10.1007/978-1-4615-3752-6_27

18. Rückert, U., Rüping, S., Naroska, E.: Parallel implementation of neural associative memories on RISC processors. In: Delgado-Frias, J.G., Moore, W.R. (eds.) VLSI for Neural Networks and Artificial Intelligence, pp. 167–176. Springer, Boston, MA (1994). https://doi.org/10.1007/978-1-4899-1331-9_17

19. Rückert, U., Surmann, H.: Tolerance of a binary associative memory towards stuck-at-faults. In: Artificial Neural Networks, pp. 1195–1198. Elsevier (1991)

20. XILINX: VCU1525 reconfigurable acceleration platform - UG1268 (v1.5) (2019). https://docs.xilinx.com/v/u/en-US/ug1268-vcu1525-reconfig-accel-platform. Accessed 14 May 2022

21. Zuse, K.: Der plankalkül (1972)

STANN – Synthesis Templates for Artificial Neural Network Inference and Training

Marc Rothmann[✉] and Mario Porrmann

Osnabrück University, Osnabrück, Germany
{mrothmann,mporrmann}@uni-osnabrueck.de

Abstract. While Deep Learning accelerators have been a research area of high interest, the focus was usually on monolithic accelerators for the inference of large CNNs. Only recently have accelerators for neural network training started to gain more attention. STANN is a template library that enables quick and efficient FPGA-based implementations of neural networks via high-level synthesis. It supports both inference and training to be applicable to domains such as deep reinforcement learning. Its templates are highly configurable and can be composed in different ways to create different hardware architectures.

The evaluation compares different accelerator architectures implemented with STANN to showcase STANN's flexibility. A Xilinx Alveo U50 and a Xilinx Versal ACAP development board are used as the hardware platforms for the evaluation. The results show that the new Versal architecture is very promising for neural network training due to its improved support for floating-point calculations.

Keywords: Deep Learning · FPGA · Hardware Accelerators

1 Introduction

Deep Learning accelerators have been an active area of research for a long time. The usual approach is designing a monolithic general-purpose architecture to accelerate the inference of neural networks, especially large convolutional neural networks. FPGA-based accelerators for neural network training have received comparatively little attention, but more research in this domain is starting to emerge.

Accelerating Deep Reinforcement Learning (DRL) algorithms on FPGAs is a promising research domain. Existing implementations of state-of-the-art DRL algorithms have shown that FPGA-based accelerators can even outperform GPU-based DRL training. A good overview of FPGA-based DRL accelerators can be found in this survey [11]. The most compute-intensive part of DRL algorithms is the training of one or more neural networks. Thus, supporting the realization of efficient accelerators for neural network training is highly relevant.

I. Rojas et al. (Eds.): IWANN 2023, LNCS 14134, pp. 394–405, 2023.
https://doi.org/10.1007/978-3-031-43085-5_31

In this publication, we present STANN[1] (Synthesis Templates for Artificial Neural Networks), a C++ template library for high-level synthesis that implements the inference and training of artificial neural networks. It provides flexible C++ templates for the implementation of neural networks, where general network parameters, as well as parameters that determine the architecture generated by the High-Level Synthesis (HLS) are configurable. Additionally, STANN supports integer and floating-point computations, and the templates can be used to generate different general architectures for the neural networks, focusing on either performance or hardware reuse.

The structure of the paper is as follows: Sect. 2 summarizes related work in the domain of neural network acceleration, followed by a discussion of the theoretical background related to neural network training in Sect. 3. After that, the implementation of STANN is described in Sect. 4 and evaluated in detail in Sect. 5. Finally, Sect. 6 concludes the paper.

2   Related Work

Existing work in the field of neural network acceleration can be classified into two main groups: monolithic accelerators distributed as IP cores with a fixed (but still configurable) architecture and accelerator libraries that supply essential functions that can be used to design accelerators in a more flexible way.

One example of an accelerator IP core is the Xilinx DPU [1]. The DPU, like many existing accelerators, targets neural network inference and does not support training. While one-size-fits-all accelerator IPs like the DPU are easy to use, accelerator libraries are more adaptable for specific applications, giving the accelerator designer many more options while designing the optimal accelerator. With accelerator libraries, the accelerator can be tailored to the neural network model to achieve the desired results.

2.1   Neural Network Accelerator Libraries

FINN [13] is a framework for accelerator design developed by Xilinx. Based on python descriptions of the neural network, FINN generates HLS code which can be synthesized to implement neural network accelerators. In contrast to the DPU, which uses a single compute engine that is used sequentially for all layers of the network, FINN generates one hardware module for each layer. The advantage of this approach is that the layers can be pipelined, leading to higher throughput. The disadvantage is the increased resource consumption for the different compute modules and the additional buffers needed to connect them. One of the main use cases of FINN is binary neural network acceleration. Here, the increased resource requirements are more manageable because of the small memory footprint of the network parameters and simple processing elements. The HLS code generated by FINN is based on the FINN-hls library. In addition to CNNs, FINN has also been extended to RNNs based on LSTMs [3].

---

[1] Source code available at: https://github.com/ce-uos/STANN.

Hls4ml [5] is a hardware-software co-design workflow that is in many ways similar to FINN. Based on python descriptions of neural network models, using, for example, Keras or Tensorflow, hls4ml generates HLS code to execute the network on FPGAs. Accelerators built with hls4ml achieve very low latency on the scale of $1\,\mu s$. In addition to the FPGA workflow, hls4ml also supports an ASIC workflow [4].

Both FINN and hls4ml only support inference. Recently, more research has been published regarding training on FPGAs, but it is usually based on specifically designed accelerators and not flexible accelerator libraries. The following section gives a brief overview of the proposed accelerators.

2.2   Neural Network Training on FPGAs

Venkataramanaiah et al. [14] implemented a training accelerator for CNNs with high performance on small batch sizes. Exploiting the advantages of high-band-width memory, their accelerator outperforms a CPU implementation and achieves better energy efficiency than a GPU implementation. Tang et al. [12] designed an accelerator architecture based on data reshaping to mitigate the problem of different data access patterns between forward path, backward path, and weight update in neural network training. While these two architectures use 16-bit and 32-bit floating-point parameters, respectively, Lu et al. [9] explored the implementation of low-precision training on FPGAs, achieving high performance for multiple state-of-the-art CNNs. Another architecture implementing 8-bit training on FPGAs is called DarkFPGA [10]. It exploits batch-level parallelism to achieve an $11\times$ speed up compared to a CPU implementation and achieves a $3\times$ reduction of energy consumption compared to a GPU implementation.

The existing implementations of neural network training for FPGAs are monolithic accelerators, which are less flexible than tool flows based on accelerator libraries like FINN or hls4ml. However, the existing libraries do not support neural network training. STANN combines the modularity of accelerator libraries with training capabilities, enabling the flexible and application-specific design of neural network training accelerators.

3   Theoretical Background

Before the next section describes the implementation of STANN, this section serves as an introduction to the theoretical background of neural network inference and training.

3.1   Fully-Connected Layers and Training

Neural networks that consist of a sequence of fully connected layers are called Multi-Layer Perceptrons (MLPs). Each fully connected layer multiplies its input vector x with its weight matrix W, adds a bias vector b, and applies an activation function f, as shown in the following equation:

$$out^1 = f^1(W^1 x + b) \tag{1}$$

These layers can be composed into a multi-layer neural network by using the output of one layer as the input of the subsequent layer.

Training the neural network means optimizing a cost function C by adapting the parmeters of the network. To optimize the network, it is necessary to compute the gradient of the cost function with respect to the network parameters:

$$\nabla_{W^l} C = \delta^l (a^{l-1})^T \tag{2}$$

Here, δ^l is the vector of partial errors in layer l. With the backpropagation algorithm, the errors at the output layer can be propagated backward into earlier layers:

$$\delta^{l-1} := (f^{l-1})' \circ (W^l)^T \cdot \delta^l \tag{3}$$

The partial errors can then be used by optimizers like Stochastic Gradient Descent (SGD), RMSProp, or Adam to adapt the weights and improve the outputs of the neural network. In the SGD update, the partial errors are multiplied by the inputs of the layer to compute the partial gradients for the weight update:

$$\Delta W = ((\delta^l)^T \cdot out^l) \cdot \alpha \tag{4}$$

To summarize, the training process consists of three main steps: First, the neural network inference needs to be computed. Second, the errors need to be computed with the cost function and backpropagated to compute the partial errors with Eq. 3. Finally, the partial errors can be used to compute the values for the weight update with Eq. 4 [6].

3.2   Convolutional Layers

Convolutional layers can be implemented in a variety of ways, ranging from direct convolution to the Winograd algorithm [7], or im2col-based approaches [2]. The latter is especially interesting because lowering convolutions to matrix multiplication makes the operation easily scalable and allows mapping it to the same compute resources as fully connected layers. Two algorithms from this class are especially notable: im2row and kn2row.

The im2row algorithm divides the input matrix into patches of the size of the convolution kernel and arranges these patches as rows of a new input matrix. This pre-processing step transforms the input matrix of dimension $(H \times W) \times C$ into a new matrix of dimension $(K^2 \times C) \times (H \times W)$. multiplying this matrix with the convolution kernel $M \times (K^2 \times C)$ gives the same output as direct convolution. The advantage of executing the convolution as a matrix multiplication comes with the drawback of additional memory requirements to store the im2row matrix [2].

The kn2row algorithm can mitigate this drawback to some extent by rearranging the kernel instead, which can be done ahead of time and needs no additional memory resources. It is based on the idea that convolution can be seen as the sum of K^2 (1×1)-convolutions. Multiplying the rearranged kernel with dimensions $(K^2 \times M) \times C$ with the input of dimensions $C \times (H \times W)$ results in a temporary output matrix $(K^2 \times M) \times (H \times W)$ which needs to be processed by a shift-add function to get to the actual output matrix with dimensions $M \times (H \times W)$. This memory overhead of the output matrix can be reduced if the kernel is seen as K^2 separate kernels of size $M \times C$ and executed as separate matrix multiplications. In this case, the results of the smaller matrix multiplications can be accumulated, reducing the memory requirements [2].

To summarize, both fully connected and convolutional layers are based on matrix multiplications. However, convolutional layers need an additional pre- or post-processing step, depending on which method is chosen to lower convolutions to matrix multiplication.

4   Implementation

STANN is a library of C++ templates for high-level synthesis. Each supported neural network layer type consists of a namespace with sub-namespaces for each supported data type. These namespaces contain two functions: a forward function for inference and a backpropagation function to compute the gradients needed for training. Currently, supported layers are fully connected, convolutional, pooling, and activation layers.

The layer functions expose template parameters to configure the basic structure of the layer, e.g., the number of inputs and outputs, as well as template parameters that influence the hardware architecture, e.g., the number of processing elements. In addition to the layer templates, STANN includes various other template functions for neural network training and inference – for example, different activations and losses, their respective derivatives, and optimizers like stochastic gradient descent.

Since floating-point calculations may be desirable for neural network training, but quantized neural networks are the state-of-the-art of inference, STANN supports both. Where possible, the data type can be changed simply with a template parameter. In cases where the implementation of a function is also dependent on the data type used for calculations, the different implementations of these functions are separated into different namespaces. This makes switching between data types simple and allows mixing data types if desired.

4.1   Three Architectures

A neural network accelerator can have one of three possible architectures, illustrated in Fig. 1. The main architecture supported by STANN is the dataflow architecture. The dataflow architecture allows for the highest throughput and fastest latency with mini-batch sizes larger than one since the modules for the layers can be pipelined. STANN also supports compute unit and hybrid architectures. The systolic array based matrix multiplication can be used as a generic compute unit. The fully connected layers are just a matrix multiplication, and convolutional layers with the im2row or kn2row algorithms are based on matrix multiplication as well, with an additional pre- or post-processing step. STANN can also be used to build hybrid architectures, e.g., by starting with a dataflow architecture for the first few layers and then leaving the remaining layers to a generic compute unit.

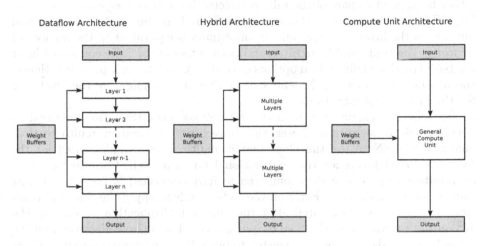

Fig. 1. Three possible accelerator architectures: the dataflow architecture, the hybrid architecture, and the compute unit architecture.

4.2   Block Matrix Multiplication

Both fully connected and convolutional layers can be implemented in terms of matrix multiplication, which makes matrix multiplication the most important building block for neural network accelerators. Implementing this operation as block matrix multiplication, which slices the matrices into small tiles, which are then multiplied, allows fine-grained control of the amount of parallelism.

Given two matrices $A \in \mathbb{R}^{M \times N}$ and $B \in \mathbb{R}^{N \times K}$, the multiplication $A \cdot B = C$ results in a matrix $C \in \mathbb{R}^{M \times K}$. A can be sliced into tiles $a_{ij} \in \mathbb{R}^{m \times n}$ and B into tiles $b_{jk} \in \mathbb{R}^{n \times k}$. The tiles $c_{ik} \in \mathbb{R}^{m \times k}$ can be computed as follows:

$$c_{ik} = \sum_j a_{ij} \cdot b_{jk} \tag{5}$$

STANN supports implementing the small matrix multiplications of the tiles as either a systolic array or fully unrolled basic matrix multiplication. Thus, increasing the size of the tiles increases the parallelism of the block matrix multiplication.

4.3   Implementation of Neural Network Inference

A dataflow architecture for a neural network can be built by calling the forward function of the layers in sequence, as shown in Fig. 2. The layers are connected via HLS streams, and dataflow optimization can be applied to the network function to pipeline the network between the layers. The inference of the fully connected layer is implemented as matrix multiplication, followed by adding the bias vector and applying the activation function. These three steps are pipelined and can happen partially in parallel. For convenience, the activation function can either be applied as part of the fully connected layer or as a separate activation layer. The hyperparameters that can be configured are the number of inputs and outputs of the layer, and the number of outputs is equivalent to the number of neurons. The configurable architectural parameters of the fully connected layer are based on the configuration options of the block matrix multiplication. Hence, the number of processing elements can be chosen by configuring the block size for the matrix multiplication.

STANN implements the im2row and kn2row algorithms, as described in Sect. 3.2. Both algorithms lower the convolution to matrix multiplication, enabling STANN to use the block matrix multiplication implemented for the fully connected layers for the convolutional layers as well. Convolutions need an additional processing step compared to fully connected layers. The kn2row implementation, shown in Fig. 2 on the right, needs to apply the shift-add function as a post-processing step after the matrix multiplication to sum up the K^2 output matrices into the final output matrix. The im2row algorithm, on the other hand, needs a pre-processing step to bring the input matrix into its im2row patch matrix form. The kernel size and the number of kernels of a convolutional layer can be configured as template parameters. As in the fully connected layer,

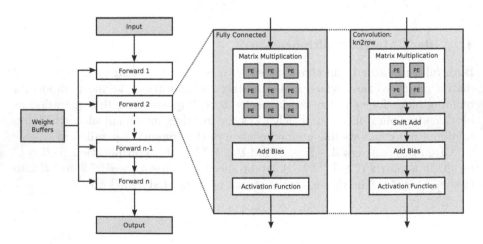

Fig. 2. Inference of a neural network with a dataflow architecture. The white hardware modules are the forward functions of the respective layers, e.g., a fully connected layer or a convolutional layer.

the architectural parameters of the convolutional layer are based on the block matrix multiplication, determining the number of processing elements based on the tile size.

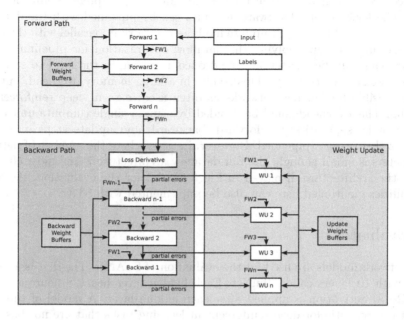

Fig. 3. The training process based on a dataflow architecture. The forward and backward hardware modules are pipelined, and the weight update units can work in parallel with the backword path computations. Pipclining this dataflow architecture introduces a memory overhead due to the necessity of storing the weights three times.

4.4 Implementation of Neural Network Training

For fully connected layers, the backpropagation can be implemented, like the inference, in terms of matrix multiplication, multiplying the partial errors with the network weights. And similar to the forward path, using block matrix multiplication for this computation ensures scalability. The main difference to the inference is the usage of the activation function: the result of the matrix multiplication of the backward path is multiplied by the derivative of the activation function applied to the (inference-)output of the layer.

In addition to the backpropagation, the weight update is the last significant piece for neural network training. The weight update uses optimizers like stochastic gradient descent to move the weights along the computed gradients. Here, another matrix multiplication is needed, multiplying the partial errors with the input of the layer to compute the gradient. These gradients are then multiplied by the learning rate, divided by the mini-batch size, and finally subtracted from the weights as a gradient descent step.

The overall architecture for neural network training becomes much more complex compared to the simple inference of a neural network. The dataflow implementation of neural network training with STANN is shown in Fig. 3. The architecture consists of a forward, backward, and weight-update module for each layer. The forward and backward modules can be pipelined with the dataflow optimization, and the weight update modules can work in parallel with the backward modules. When applying the dataflow optimization for pipelining, each memory can only be written and read once, resulting in the need to store the network weights three times. This results in a large memory overhead, but this is still feasible for smaller networks as often used, e.g., in deep reinforcement learning. This overhead could be avoided by using a single compute unit architecture or by separating the forward, backward, and update steps into three modules that are not pipelined. However, in cases where the number of network parameters is small enough, the full dataflow architecture is the most efficient. While the architecture displayed in Fig. 3 is used for the evaluation, the other possibilities mentioned here can also be implemented with STANN.

5   Evaluation

Two network models are used for the evaluation of STANN. The first is a simple MLP with 10 layers called TenNet. Each hidden layer has 128 neurons, and a leakyReLU activation is applied after each hidden layer. A model of this type and size is realistic for deep reinforcement learning tasks that are not based on image input. The second model is LeNet-5, as described in [8], but with ReLU activations to increase performance. All design in this evaluation are synthesized with a 400MHz clock frequency.

Table 1. TenNet inference with different Architectures

Architecture	PEs	BRAM	LUTs	DSPs	Latency
Dataflow	1	158(5%)	23820(2%)	142(2%)	1.230 ms
Dataflow	16	158(5%)	77926(8%)	188(3%)	14.193 μs
Compute Unit	1	158(5%)	11461(1%)	13(<1%)	12.350 ms
Compute Unit	16	158(5%)	65355(5%)	20(<1%)	0.199 ms

5.1   Training and Inference with STANN on Xilinx Alveo

The first hardware platform for which the STANN-based architectures are evaluated is a Xilinx Alveo U50 accelerator card. Table 1 shows the resource usage and latency of TenNet inference with different architectures and PEs (Processing Elements) per layer. As expected, the dataflow architecture is more than 10× as fast as the compute unit architecture but also requires almost 10× the number of DSPs. The large size of the accelerators, in general, is due to the fact that all TenNet implementations in this evaluation are based on floating-point data. The reasoning behind using 32-bit floats is that it is most suitable for neural network training, and the TenNet model is intended as a realistic model used for deep reinforcement learning training.

Table 2 gives an overview of the resources required for training and the latency that can be achieved. For this evaluation, the training batch size was set to 32. An additional memory overhead is introduced in the training because the network parameters need to be stored three times, as described in the previous section. The resource usage only increase moderately when increasing the PEs because there are other constant factors of the design that require a lot of resources. A dramatic decrease in latency can be seen when increasing the PEs from 1 to 4 because of more efficient usage of the systolic array.

Table 2. TenNet training with different numbers of processing elements.

PEs	BRAM	LUTs	DSPs	Latency
1	456(16%)	156220(17%)	418(7%)	40.090 ms
4	456(16%)	172624(19%)	418(7%)	5.848 ms
8	456(16%)	199012(22%)	418(7%)	2.407 ms
16	456(16%)	270134(30%)	525(8%)	1.718 ms

After this floating-point-based evaluation of TenNet, Table 3 contains data about LeNet-5 with different implementations of the convolutions and different data types. Three implementations are evaluated for convolutions: im2row, kn2row, and an accumulating variant of kn2row, as described in the theoretical background. While the accumulating variant of kn2row has the least memory overhead, im2row achieves the lowest latency. All three implementations are included in the STANN library to give the accelerator designer the choice of the algorithm depending on the needs of the application.

Table 3. LeNet inference with different kinds of convolution and different data types.

Convolution	data type	BRAM	LUTs	DSPs	Latency
im2row	float	196(7%)	22263(3%)	112(1%)	4.542 ms
im2row	uint8	72(2%)	18361(2%)	65(1%)	0.342 ms
kn2row	float	160(6%)	22330(3%)	108(1%)	2.888 ms
kn2row (acc)	float	144(5%)	19888(2%)	98(1%)	2.913 ms

5.2   Evaluation of Xilinx Versal ACAP Floating-Point DSPs

The Xilinx Versal ACAP architecture consists of an ARM core as the Processing System, FPGA fabric, and a grid of vector processors called AI engines. This new architecture is extremely interesting for deep learning accelerators. This evaluation is restricted to the FPGA fabric of the Versal, but in the future, an extension of STANN will also support AI engines. Even without the AI engines, the Versal architecture is promising because the FPGA fabric contains floating-point DSPs and is better suited for floating-point-based accelerators than other FPGA architectures. To evaluate the potential of these floating-point DSPs, Table 4 compares TenNet training with floats and LeNet-5 (im2row) inference with different data types on the Alveo U50 and Versal (vck190). The table shows that for floating-point implementations, the architectures on the Versal need less resources while also achieving much lower latency. As expected, the resources usage and latency for the integer implementation stays approximately the same.

Table 4. Comparison of the Alveo U50 and Versal ACAP hardware platforms.

HW Platform	Network	data type	BRAM	LUTs	DSPs	Latency
Alveo U50	TenNet (Training)	float	456(16%)	270134(30%)	525(8%)	1.718 ms
Alveo U50	LeNet (Inference)	float	196(7%)	22263(3%)	112(1%)	4.542 ms
Alveo U50	LeNet (Inference)	uint8	72(2%)	18361(2%)	65(1%)	0.342 ms
Versal	TenNet (Training)	float	154(7%)	245001(27%)	425(21%)	0.973 ms
Versal	LeNet (Inference)	float	78(4%)	13637(2%)	80(4%)	1.605 ms
Versal	LeNet (Inference)	uint8	42(2%)	24021(3%)	63(3%)	0.342 ms

6   Conclusion

The STANN library is a valuable basis for further research regarding neural network training on FPGAs, as well as machine learning accelerators that rely on the training of the network being accelerated. For example, FPGA-based accelerators for deep reinforcement learning algorithms often rely on neural network training on the FPGA. STANN-based accelerators could be used as part of such a deep reinforcement learning accelerator.

In future work, STANN will be extended in a variety of ways. Support for the training of convolutional layers will be added, and state-of-the-art CNNs, like YOLO and MobileNet, will be supported. Additionally, templates for multi-head self-attention will be added to support transformer architectures. The supported data types will be extended to smaller floating-point formats, like 8-bit floats.

One of the main use cases intended for STANN is deep reinforcement learning, where we will use it to implement state-of-the-art algorithms like DQN, DDPG, or SAC. Especially in this domain, STANN's support for neural network training will be important and enable efficient implementations and effective design space exploration.

Acknowledgment. This publication incorporates results from the VEDLIoT project, which received funding from the European Union's Horizon 2020 research and innovation programme under grant agreement No 957197.

References

1. Agiakatsikas, D., et al.: Evaluation of xilinx deep learning processing unit under neutron irradiation (2022). https://doi.org/10.48550/ARXIV.2206.01981. https://arxiv.org/abs/2206.01981
2. Anderson, A., Vasudevan, A., Keane, C., Gregg, D.: Low-memory GEMM-based convolution algorithms for deep neural networks. ArXiv abs/1709.03395 (2017)
3. Blott, M., et al.: FINN-R: an end-to-end deep-learning framework for fast exploration of quantized neural networks. ACM Trans. Reconfig. Technol. Syst. (TRETS) **11**(3), 1–23 (2018)
4. Duarte, J., et al.: Fast inference of deep neural networks for real-time particle physics applications. In: Proceedings of the 2019 ACM/SIGDA International Symposium on Field-Programmable Gate Arrays, p. 305. FPGA 2019, Association for Computing Machinery, New York, NY, USA (2019). https://doi.org/10.1145/3289602.3293986
5. Fahim, F., et al.: hls4ml: an open-source codesign workflow to empower scientific low-power machine learning devices. CoRR abs/2103.05579 (2021). https://arxiv.org/abs/2103.05579
6. Goodfellow, I., Bengio, Y., Courville, A.: Deep Learning. MIT Press (2016). http://www.deeplearningbook.org
7. Lavin, A., Gray, S.: Fast algorithms for convolutional neural networks. In: 2016 IEEE Conference on Computer Vision and Pattern Recognition (CVPR), pp. 4013–4021 (2016). https://doi.org/10.1109/CVPR.2016.435
8. Lecun, Y., Bottou, L., Bengio, Y., Haffner, P.: Gradient-based learning applied to document recognition. Proc. IEEE **86**(11), 2278–2324 (1998). https://doi.org/10.1109/5.726791
9. Lu, J., Ni, C., Wang, Z.: ETA: an efficient training accelerator for dnns based on hardware-algorithm co-optimization. IEEE Trans. Neural Netw. Learn. Syst. **PP**, 1–15 (2022). https://doi.org/10.1109/TNNLS.2022.3145850
10. Luo, C., Sit, M.K., Fan, H., Liu, S., Luk, W., Guo, C.: Towards efficient deep neural network training by FPGA-based batch-level parallelism. In: 2019 IEEE 27th Annual International Symposium on Field-Programmable Custom Computing Machines (FCCM), pp. 45–52 (2019). https://doi.org/10.1109/FCCM.2019.00016
11. Rothmann, M., Porrmann, M.: A survey of domain-specific architectures for reinforcement learning. IEEE Access **10**, 13753–13767 (2022). https://doi.org/10.1109/ACCESS.2022.3146518
12. Tang, Y., Zhang, X., Zhou, P., Hu, J.: EF-Train: enable efficient on-device CNN training on FPGA through data reshaping for online adaptation or personalization. ACM Trans. Design Autom. Electron. Syst. **27**, 3505633 (2022). https://doi.org/10.1145/3505633
13. Umuroglu, Y., et al.: FINN: a framework for fast, scalable binarized neural network inference. In: Proceedings of the 2017 ACM/SIGDA International Symposium on Field-Programmable Gate Arrays, pp. 65–74. FPGA 2017. ACM (2017)
14. Venkataramanaiah, S.K., et al.: FPGA-based low-batch training accelerator for modern CNNs featuring high bandwidth memory. In: 2020 IEEE/ACM International Conference On Computer Aided Design (ICCAD), pp. 1–8 (2020)

Digit Recognition Using Spiking Neural Networks on FPGA

Shamini Koravuna[1]($\boxtimes$)(iD), Sanaullah[2](iD), Thorsten Jungeblut[2](iD),
and Ulrich Rückert[1]($\boxtimes$)

[1] Bielefeld University, Bielefeld, Germany
{skoravuna,rueckert}@techfak.uni-bielefeld.de
[2] Hochschule Bielefeld – University of Applied Sciences and Arts, Bielefeld, Germany
{sanaullah,thorsten.jungeblut}@hsbi.de

Abstract. This paper presents the results of our first assessment on the emulation of spiking neural networks (SNNs) on Field Programmable Gated Arrays (FPGAs). Three fundamental modules are designed to obtain a fully functional neural network. First, a module that provides the simulation of a single neuron's properties, characteristics, and behavior using an accurate but, at the same time, a computationally efficient mathematical model is designed. Second, a communication or transmission system called Address Event Representation (AER) is developed to manage the neural network's information flow between the neurons. Third, a training algorithm - Spike-Timing Dependent Plasticity (STDP), is designed to provide functionality to the neural network. This modular approach provides the necessary flexibility and scalability for simulating various SNNs and the different numbers of neurons. The modules have been implemented using multiple combinational blocks and flip-flops; The network description is performed using Very High-Speed Integrated Circuit Hardware Description Language (VHDL) in the Xilinx Vivado simulator. The SNN model for digit recognition is implemented and emulated on the Basys3 FPGA development board to demonstrate the accuracy of the model's operation.

Keywords: Spiking Neural Networks · Neuron Models · STDP · AER · Digit Recognition · FPGA · Neuromorphic Hardware

1 Introduction

Spiking Neural Networks (SNNs) are bio-inspired neural networks that model the behavior of biological neurons and are designed to operate using spikes, or discrete time-varying signals that represent the activity of individual neurons [1,2]. Unlike traditional artificial neural networks that use continuous activation functions, SNNs use the timing and frequency of spikes to encode information

Supported by organization "Dataninja" (Trustworthy AI for Seamless Problem Solving: Next-Generation Intelligence Joins Robust Data Analysis).

and perform computations. One of the key advantages of SNNs is their ability to operate on event-based or asynchronous data, making them well-suited for processing real-time data streams such as those generated by sensors or other time-varying signals [3]. This is because SNNs are able to process information in a highly parallel and distributed manner, allowing them to efficiently and accurately process large amounts of data and offer a runtime simulation environment [4].

SNNs have been applied to a wide range of applications, including image [5] and speech recognition [6], robotic control [7], and more recently, in the field of neuromorphic computing [8–10], which aims to develop computer systems that emulate the functionality of biological nervous systems. One of the main challenges in working with SNNs is the lack of a widely accepted training algorithm that can be used to optimize the weights and parameters of the network. However, there have been several developments in this area, including the use of unsupervised learning methods such as spike-timing-dependent plasticity (STDP) [11] and backpropagation-through-time (BPTT) [12] algorithms, which have shown promising results for training SNNs.

Implementing SNNs on Field Programmable Gate Arrays (FPGAs) has become an active area of research in recent years, due to the potential for high performance, low power consumption, and real-time operation [7]. FPGAs are reconfigurable hardware devices that can be programmed to implement custom logic circuits, making them well-suited for the implementation of SNNs. One common approach to implementing SNNs on FPGAs is to use the Address-Event Representation (AER) model [13], which is an SNN architecture that uses a distributed event-based communication scheme. AER networks consist of a large number of neurons, each of which communicates with other neurons through address events, or packets of information that contain the source and destination addresses of the neuron and the time at which the spike was generated. To implement AER networks on FPGAs, several design considerations must be taken into account, including the choice of hardware architecture, the choice of programming language and tools, and the implementation of efficient algorithms for spike processing, weight updates, and other network operations [14].

One of the main challenges in implementing SNN-based digit recognition on basic FPGA devices is the complex nature of SNNs. They require specialized algorithms for spike processing, weight updates, and other network operations. These algorithms must be designed to minimize resource usage while maintaining high accuracy, which can be challenging given the limited resources of basic FPGAs [15]. Another challenge is the limited resources of basic FPGAs, including limited memory, logic cells, and routing channels. This can limit the size and complexity of the SNN that can be implemented, as well as the size of the input data that can be processed. To overcome this challenge, it is important to carefully design the SNN architecture and algorithms to minimize resource usage while maintaining high accuracy.

This paper outlines a modular approach to designing key modules for developing SNNs models, with the goal of increasing their scalability and flexibility in future research. The modules were created using VHDL, and their operation was verified through functional simulations. We then integrated these modules to implement a digit recognition application and finally emulated the network on the Basys3 FPGA board to demonstrate the model's accuracy.

2   Related Work

Designing a digit recognition system using SNN simulation is a promising area of research that has the potential to achieve high accuracy while being energy-efficient. These related works demonstrate that it is possible to design efficient SNN-based architectures for digit recognition using specialized hardware platforms such as FPGAs and neuromorphic hardware. Therefore some popular systems are,

- SpiNNaker [16]: SpiNNaker is a digital system that has been designed for simulating large-scale SNN on a specialized hardware platform. It features a highly parallel architecture with 18 cores, each with its own memory and communication system, and can simulate up to 1 billion neurons and 1 trillion synapses. It has been used for various applications, including digit recognition.
- Loihi [17]: Loihi is a neuromorphic hardware platform designed for SNN simulation. It features a low-power design that is optimized for running SNNs efficiently and has been used for various applications, including digit recognition. Loihi has been shown to achieve high accuracy on the MNIST dataset while consuming significantly less power than traditional computing systems.
- BrainScaleS [18]: BrainScaleS is a neuromorphic hardware platform designed for simulating SNNs using analog circuits. It features a scalable design that can be configured for different applications, including digit recognition. BrainScaleS has been used to demonstrate the feasibility of using analog circuits for efficient and accurate SNN simulation.
- TrueNorth [19]: TrueNorth is a digital chip designed for simulating large-scale SNNs. It features a highly parallel architecture with 1 million neurons and 256 million synapses and has been used for various applications, including digit recognition. TrueNorth has been shown to achieve high accuracy on the MNIST dataset while consuming significantly less power than traditional computing systems.

  Therefore, there is a growing body of research focused on designing digit recognition systems using SNN simulation, and many promising approaches have been proposed. These systems offer the potential for high accuracy while being energy-efficient, which is particularly important for applications such as mobile and embedded devices. Additionally, specialized hardware platforms such as FPGAs and neuromorphic systems are being developed specifically for SNN simulation, which is expected further to improve the efficiency and accuracy of these systems.

3   Functional Blocks

We developed three fundamental modules for designing a fully functional SNN. The first module provides a simulation of the properties, characteristics, and behavior of a single neuron using an accurate but, at the same time, computationally efficient mathematical model, which can be adapted for its further implementation into the FPGAs. There are several different neuronal models of SNNs, including the integrate-and-fire model, the Hodgkin-Huxley model, and the Izhikevich model, among others. We choose the Izikivich neuron model for implementing the digit recognition application as it is a simple neuron model, and it aims to be both biologically plausible, similar to the Hodgkin-Huxley model, and computationally efficient, like the Integrate-and-Fire model. The second module is the communication or transmission system that can manage all the information flow of the neural network between an undetermined number of neurons. A training algorithm using STDP rules is designed to provide functionality to the neural network. Finally, the interconnection between them with the correct configuration will allow the simulation of a neural network with the given functionality.

3.1   Izikivich Neuron

In 2003, Eugene M. Izhikevich [20] proposed the Izhikevich neuron as a mathematical model of a spiking neuron. This simple two-dimensional model can precisely replicate a diverse range of spiking behavior observed in actual neurons. The Izhikevich neuron model defines the membrane potential of a neuron over time using two variables: the membrane potential (v) and a recovery variable (u) that represents the activity of an ionic current restoring the membrane potential to its resting state. The dynamics of the Izhikevich neuron are governed by a set of ordinary differential equations. Once the membrane potential surpasses a certain threshold, the neuron discharges a spike, and the membrane potential resets to a lower value. The neuron model is represented with the equations:

$$v' = 0.04v^2 + 5v + 140 - u + I \tag{1}$$

$$u' = a(bv - u) \tag{2}$$

$$\text{if } v \geq 30\,mV, \text{then } (v \leftarrow c \ \& \ u \leftarrow u + d) \tag{3}$$

where v and u are dimensionless variables, a, b, c, and d are dimensionless parameters, and '$=d/dt$' where t is the time. In relation to the behavior of neurons, the presence of positive synaptic currents (denoted as I) from other neurons serves to increase the membrane potential. If these currents alone fail to trigger an impulse or spike, the membrane voltage is reset to its initial value. Conversely, if the neuron generates a spike ($+30$ mV) as a result of the cumulative input current, both the membrane voltage (v) and the recovery variable (u) are reset according to Eq. 3. In this model, the resting membrane voltage typically ranges between −70 and −60 mV, with the specific value depending on the parameter b. Moreover, similar to real neurons, this model lacks a fixed threshold. Consequently, the potential threshold for spike generation can vary, reaching as low as −55 mV or as high as −40 mV based on the history of the membrane potential preceding the spike's occurrence. To simplify the implementation process, we utilized fixed point arithmetic to implement the IZH model. Furthermore, utilizing powers of two arithmetic for multiplication and division can offer substantial implementation benefits. Therefore, we adjusted Eq. 1 by multiplying the coefficients by 10 to achieve an approximate power of two representations for two coefficients in the equation [21]. Thus, we have derived a set of equations for the adapted model as follows:

$$v' = \frac{1}{250}v^2 + 5v + 1400 - u + I \tag{4}$$

$$u' = \frac{1}{50}\left(\frac{1}{5}v - u\right) \tag{5}$$

$$\text{if } v \geq 300, \text{then}(v \leftarrow 650 \,\&\, u + 80) \tag{6}$$

When utilizing the parameters in this model, it is essential to consider the following factors:

- The parameter a determines the time scale for the recovery of the variable u. Smaller values result in a slower recovery process. A typical value for a is 0.02.
- The parameter b governs the sensitivity of the recovery variable u to subthreshold fluctuations in the membrane potential v. A typical value for b is 0.2.
- The parameter c specifies the reset value of the membrane potential v after a spike occurs. A typical value for c is −65 mV.
- The parameter d denotes the reset value of the recovery variable u following a spike. A typical value for d is 8.

These considerations guide the appropriate selection and usage of the model's parameters. Additionally, the neuron contains a small Random Access Memory (RAM) to store the various weights of its synaptic connections with other neurons.

3.2   Address Event Representation (AER)

The neuron's design enables it to read information from the AER (Address Event Representation) bus and utilize its own RAM to apply itself to the corresponding synaptic weight of the firing neuron's link. The AER system is designed with an encoder that can interpret the spikes of all SNN neurons, translating them into their corresponding neural address. This address is then written onto the AER communication bus to indicate which neuron fired. Additionally, the design addresses the issue of processing multiple events simultaneously, which is hindered by the inability to transmit more than one address through the AER communication bus. To resolve this, the encoder is programmed with a priority condition.

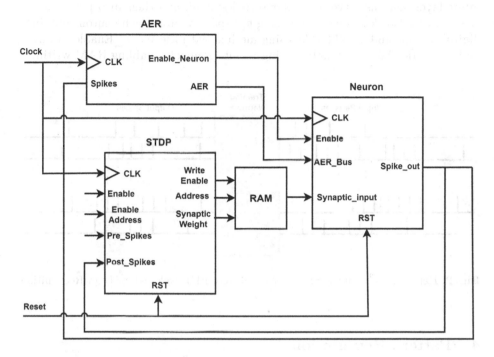

Fig. 1. Interconnection of Different Blocks for the Emulation of SNNs on FPGAs.

3.3 Spike Timing Dependent Plasticity (STDP)

The Spike-Timing-Dependent Plasticity (STDP) learning rule alters the synaptic weights between neurons based on the synchronicity of their firing. In other words, synapses that contribute to the production of output spikes should be strengthened, whereas those that do not contribute to output spikes must be weakened. The responsibility of modifying the weights of all neural network connections lies with the training system. Given the complexity of developing a module to manage all synaptic weights, an alternative approach has been suggested: to create a training module dedicated to the connections of each individual neuron. Consequently, each neuron in the neural network would have its own learning module, responsible for handling all incoming links and maintaining and updating its corresponding RAM.

3.4 SNN Structure

The block diagram in Fig. 1 consists of one neuron, but the architecture permits the incorporation of numerous additional neurons. The AER system is responsible for enabling communication among all the neurons, reading spikes, translating them into the proper address, and transmitting them via the AER bus. All neurons are linked to the Enable_Neuron signal, which can halt neuron activity since the AER bus can transmit only one address per clock cycle. The input layer neurons serve only as external stimuli, and their incorporation into the neural network involves processing the spikes vector. Each neuron contains a digital module and an STDP learning module, with the Write_Enable, Address, and Synaptic_Weight signals serving as interconnects enabling RAM writing.

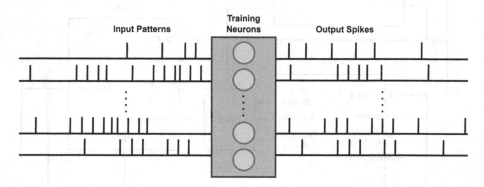

Fig. 2. Representation of the Spiking Neural Network Developed for Digit Recognition.

4   Pattern Recognition

Figure 2 illustrates the SNN model for pattern recognition in 5 × 7 pixel images. The input and output layers form the functional neural network, with 35 input neurons representing each pixel and 6 output neurons corresponding to distinct patterns. Additionally, there is a training layer with 6 neurons for training each pattern. The neural network can recognize up to six different patterns from the digits presented in Fig. 4 based on nerve impulses emitted from the output neurons in response to stimuli from the input layer's 35 pixels. Additionally, as each pixel is either 0 or 1 in binary representation, this represents the presence or absence of nerve stimulation for the input neurons. Furthermore, to verify the neural network's functionality, some noise is added to the digits shown in Fig. 4a. Despite this, the neural network should still be capable of recognizing the corresponding numbers when stimulated with the digits from Fig. 4b after completing its training.

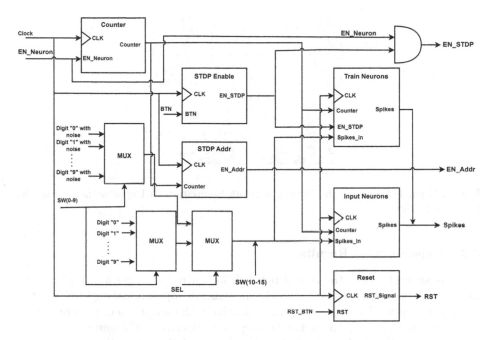

Fig. 3. Implementation of the design on Basys3 FPGA Development Board.

4.1   Implementation on FPGA

A counter governs the entire structure, providing a 100-clock cycle period for introducing stimuli to the neural network and allowing all output neurons to return to their resting state. To select input stimuli, two multiplexers generate a 35-bit vector at their output. Together with a third multiplexer controlled by the SEL button, the user can choose between normal digits and digits with noise. Additionally, the 6 bits from the training neurons are concatenated to this vector to indicate which neurons should fire with the selected image. The Input Neurons block transmits the 35-bit vector corresponding to the image pixels to the AER system when the counter's value is zero. The Train Neurons block transmits the 6-bit vector of the training neurons at the appropriate time intervals to strengthen or weaken the network's synapses. The STDP_Enable block translates a BTN button keystroke from the FPGA board to deliver a training pulse for approximately 150 input stimuli. Finally, the STDP_Addr block is responsible for changing the address of the STDP modules for each output neuron.

Hardware Platform (Basys 3 FPGA board). For our study on designing a digit recognition system using SNN simulation, we chose to use the Basys 3 FPGA board [22] as our hardware platform. This board is based on the Xilinx Artix-7 FPGA, which provides a high level of performance and flexibility for implementing digital systems. Additionally, the board offers a range of input and output options, including VGA output, which is useful for visualizing the output.

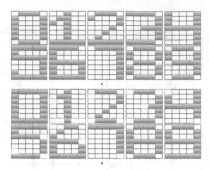

Fig. 4. Input Patterns of the Neural Network Simulation a) Digits without Noise b) Digits with Noise.

4.2   Experimental Results

Various simulations of the neural network for pattern recognition implementation have been created using Vivado to showcase the operational capability of the designed network for digit recognition. Initially, the output neurons are taught patterns corresponding to digits 0 through 5, respectively. The generated output is depicted in Fig. 5. As indicated by the EN_STDP signal pulses, six distinct learning phases are executed, with each output neuron of the SNN undergoing its own learning process. During the initial training phase, where digit 0 is paired with the training neuron assigned to output neuron one, it is observed that the output neuron does not initially generate any spikes. However, as the training progresses over time, the synapses that contribute to the output neuron's firing are adjusted, eventually enabling output neuron 1 to learn and generate spikes for the designated pattern. Moving on to the third phase of training, where digit two is paired with the training neuron assigned to output neuron 3, it can be observed that at the beginning of this phase, output neuron three is firing. However, as the training progresses, output neuron 3 ceases to generate spikes, and it is only output neuron three that produces spikes for the designated pattern. This process continues in the subsequent training phases until completion, which occurs in approximately 3 ms. Once the training is complete, the neural network's ability to recognize patterns correctly is tested. During this testing phase, digits 0 through 5 are presented for one millisecond each, and the corresponding output neurons only generate spikes for their respective trained digit, demonstrating successful pattern recognition by the neural network. Similarly, the network trained for digits 2-7 is shown in Fig. 6. In Fig. 7, it can be observed that the neuron assigned to recognize digit 8 generates spikes not only for digit 8 but also for digit 9 with some noise. This outcome is expected since the shapes of these two digits are quite similar, causing the SNN to recognize them as essentially the same pattern. The purpose of this study was to understand the working methodology using a custom-based dataset, and that's the reason the comparison with the available models is not considered in this point. The implemented SNN model consists of 46 neurons and 210 synapses and serves to demonstrate the concept of a one-layer unsupervised network capable of pattern classification. From Table 1, it can be observed that the resource utilization for emulation of this model on FPGAs is less than 10%. The modular implementation and the achieved results indicate that the network can be readily scaled for applications requiring larger and more complex networks. Along with its hardware efficiency, the proposed network exhibits a high level of unsupervised classification accuracy. Additionally, the network demonstrates rapid learning capabilities, as it can fully learn the patterns in as low as 3 ms.

Fig. 5. Spikes generated from the network trained for recognizing digits 0-5.

Fig. 6. Spikes generated from the network trained for recognizing digits 2-7.

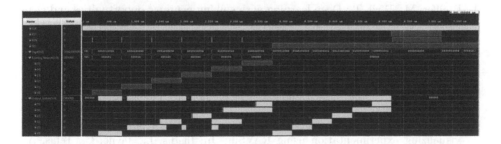

Fig. 7. Spikes generated from the network trained for recognizing digits 4-9.

Table 1. Resource Utilization.

Resource	Utilization	Available	Utilization Percentage
LUT	1885	20800	9.06
LUTRAM	228	9600	2.38
FF	1558	41600	3.75
DSP	6	90	6.67
IO	26	106	24.53
BUFG	1	32	3.13

5   Conclusion

This study demonstrates a real-time FPGA implementation of a proposed Spiking Neural Network (SNN) for digit recognition. The network utilizes AER and event-driven STDP rules to learn various features from input patterns and classify them in an unsupervised manner. The learning process is stable, and the results indicate that the trained network can accurately recognize characters even with some random noise introduced to the patterns. The hardware implementation of the proposed network is highly efficient and cost-effective, and it can be easily reconfigured while maintaining generality and accuracy. Additionally, the network can be scaled up conveniently using similar building blocks. Importantly, the implementation reduces hardware resource costs through a multiplier-less approximation.

Acknowledgement. This research was supported by the research training group "Dataninja" (Trustworthy AI for Seamless Problem Solving: Next-Generation Intelligence Joins Robust Data Analysis), funded by the German federal state of North Rhine-Westphalia.

References

1. Pfeiffer, M., Pfeil, T.: Deep learning with spiking neurons: opportunities and challenges. Front. Neurosci. **12**, 774 (2018)
2. Furukawa, S., Middlebrooks, J.C.: Cortical representation of auditory space: information-bearing features of spike patterns. J. Neurophysiol. **87**(4), 1749–1762 (2002)
3. Chakraborty, I., Jaiswal, A., Saha, A., Gupta, S., Roy, K.: Pathways to efficient neuromorphic computing with non-volatile memory technologies. Appl. Phys. Rev. **7**(2), 021308 (2020)
4. Sanaullah, Koravuna, S., Rückert, U., Jungeblut, T.: SNNs model analyzing and visualizing experimentation using RAVSim. In: Iliadis, L., Jayne, C., Tefas, A., Pimenidis, E. (eds.) Engineering Applications of Neural Networks. EANN 2022. Communications in Computer and Information Science, vol. 1600, pp. 40–51. Springer, Cham (2022). https://doi.org/10.1007/978-3-031-08223-8_4

5. Ahmadi, M., Sharifi, A., Hassantabar, S., Enayati, S.: QAIS-DSNN: tumor area segmentation of MRI image with optimized quantum matched-filter technique and deep spiking neural network. BioMed. Res. Int. **2021**, 6653879 (2021)
6. Loiselle, S., Rouat, J., Pressnitzer, D., Thorpe, S.: Exploration of rank order coding with spiking neural networks for speech recognition. In: Proceedings. 2005 IEEE International Joint Conference on Neural Networks, 2005, vol. 4, pp. 2076–2080. IEEE (2005)
7. Pearson, M.J., et al.: Implementing spiking neural networks for real-time signal-processing and control applications: a model-validated FPGA approach. IEEE Trans. Neural Netw. **18**(5), 1472–1487 (2007)
8. Hagras, H., Pounds-Cornish, A., Colley, M., Callaghan, V., Clarke, G.: Evolving spiking neural network controllers for autonomous robots. Proceed. IEEE Int. Conf. Robot. Autom. **5**, 4620–4626 (2004)
9. Yang, S., et al.: Real-time neuromorphic system for large-scale conductance-based spiking neural networks. IEEE Trans. Cybern. **49**(7), 2490–2503 (2018)
10. Kasabov, N.: To spike or not to spike: a probabilistic spiking neuron model. Neural Netw. **23**(1), 16–19 (2010)
11. Caporale, N., Dan, Y.: Spike timing-dependent plasticity: a Hebbian learning rule. Annu. Rev. Neurosci. **31**, 25–46 (2008)
12. Lillicrap, T.P., Santoro, A.: Backpropagation through time and the brain. Curr. Opin. Neurobiol. **55**, 82–89 (2019)
13. Zhao, B., Ding, R., Chen, S., Linares-Barranco, B., Tang, H.: Feedforward categorization on AER motion events using cortex-like features in a spiking neural network. IEEE Trans. Neural Netw. Learn. Syst. **26**(9), 1963–1978 (2014)
14. Schuman, C.D., et al.: A survey of neuromorphic computing and neural networks in hardware. arXiv preprint arXiv:1705.06963 (2017)
15. Yang, J.Q., et al.: Neuromorphic engineering: from biological to spike-based hardware nervous systems. Adv. Mater. **32**(52), 2003610 (2020)
16. Furber, S.B., Galluppi, F., Temple, S., Plana, L.A.: The spinnaker project. Proc. IEEE **102**(5), 652–665 (2014)
17. Davies, M., et al.: Loihi: a neuromorphic manycore processor with on-chip learning. IEEE Micro **38**(1), 82–99 (2018)
18. Schmitt, S., et al.: Neuromorphic hardware in the loop: training a deep spiking network on the brainscales Wafer-scale system. In: 2017 International Joint Conference on Neural Networks (IJCNN), pp. 2227–2234. IEEE (2017)
19. Akopyan, F., et al.: TrueNorth: design and tool flow of a 65 mW 1 million neuron programmable neurosynaptic chip. IEEE Trans. Comput. Aided Des. Integr. Circuits Syst. **34**(10), 1537–1557 (2015)
20. Izhikevich, E.M.: Simple model of spiking neurons. IEEE Trans. Neural Netw. **14**(6), 1569–1572 (2003)
21. Cassidy, A., Andreou, A.G.: Dynamical digital silicon neurons. In: IEEE Biomedical Circuits and Systems (2009)
22. Basys 3 Artix-7 FPGA Board Academic. https://shop.trenz-electronic.de/en/26083-Basys-3-Artix-7-FPGA-Board-Academic. Accessed 16 Mar 2023

Applications of Machine Learning in Biomedicine and Healthcare

Brain Tumor Segmentation Using Ensemble Deep Neural Networks with MRI Images

Miri Weiss Cohen[✉]

Braude College of Engineering, Karmiel, Israel
miri@braude.ac.il

Abstract. The work proposes an automated segmentation method for brain tumors using MRI scans and a convolutional neural network (CNN) ensemble. The method accurately identifies the tumor's size and location, crucial for treatment planning and monitoring disease progression. The method uses the YOLOv5 object detection algorithm to identify the tumor region in the MRI scan and then applies the U-Net architecture to segment the tumor into the whole tumor, tumor core, and enhanced tumor core. The study employs a multi-class loss function to handle class imbalance in the dataset and evaluates different MRI modalities to optimize the U-Net training process. The results indicate high accuracy in identifying no tumor, necrotic tumor core (NCR), peritumoral edematous/invaded tissue, and enhancing tumor (ET). This proposed method can potentially reduce human errors and assist radiologists in accurately detecting brain tumors.

Keywords: Brain Tumor Segmentation · YOLOv5 · U-Net · Deep Neural Network

1 Introduction

Brain cancer, also known as primary brain tumors, is a complex and devastating disease that affects a significant number of individuals worldwide. It arises from the abnormal growth and division of cells within the brain or surrounding tissues, leading to the formation of a mass or tumor that interferes with normal brain function. Brain cancer can affect people of all ages, including children, and its prognosis and treatment options depend on various factors, such as the type of tumor, location, size, and stage of the disease.

According to the American Brain Tumor Association (ABTA) [10], brain cancer is the leading cause of cancer-related deaths in children and young adults under the age of 39. In the United States alone, an estimated 87,000 new cases of primary brain tumors are expected to be diagnosed in 2022, and about 18,000 people are expected to die from brain tumors. The incidence of brain tumors has been increasing over the years, and this may be due in part to improved diagnostic tools and increased awareness.

I. Rojas et al. (Eds.): IWANN 2023, LNCS 14134, pp. 421–432, 2023.
https://doi.org/10.1007/978-3-031-43085-5_33

The most common types of primary brain tumors are Gliomas [14], which arise from the Glial cells that support and nourish the neurons in the brain. Other types of brain tumors include Meningiomas, which develop from the meninges that cover the brain and spinal cord, and pituitary tumors, which arise from the pituitary gland at the base of the brain. The prognosis for brain cancer depends on various factors, such as the type and grade of the tumor, its location, and the patient's age and overall health. Treatment options may include surgery, radiation therapy, chemotherapy, and targeted therapies [7].

AI - Computer-aided diagnosis (CAD) is an emerging field that utilizes machine learning and image analysis techniques to assist healthcare professionals in the diagnosis of various diseases, including brain cancer [5]. In the context of brain cancer, CAD systems typically consist of several components that work together to provide an accurate and efficient diagnosis. In this study, Magnetic Resonance Imaging (MRI) is used since it is a non-invasive imaging technique using X-rays and advanced computer algorithms to produce detailed images of the brain. Moreover, it is an essential tools in the diagnosis, management, and treatment of brain cancer. In the segmentation process, specific regions of interest within an image are identified and separated. Specifically, brain tumor segmentation involves identifying and separating the tumor from surrounding healthy brain tissue, Fig. 1 depicts four examples of images of brain tumers. An accurate segmentation can assist in several ways, including improving the accuracy of tumor size and location measurements, assisting in treatment planning, and aiding in monitoring disease progression. Currently, manual segmentation by trained experts is time-consuming and subject to variations between observers. There are a number of automated segmentation algorithms available in the literature, among which U-Net is a popular and effective algorithm. Mlynarski et al. [8] proposed a deep learning approach for tumor segmentation that uses both fully annotated and weakly annotated medical images during training. Their model produces both voxelwise and image-level outputs and includes a classification branch to exploit the information contained in weakly annotated images. [4] proposes a dual-path network for brain tumor segmentation in multi-modal medical images. It combines large-scale perceptual domain and non-linear mapping features, reduces overlapping frequency and vanishing gradients, and establishes a dual-path model for fusion of low-level, middle-level, and high-level features. The model achieves high segmentation precision. A different approach suggested by [12] analyzes the effectiveness of Otsu's threshold method with different classes or bins for brain tumor segmentation in MRI images. It uses median filtering as pre-processing and morphological operations to achieve accurate tumor regions. Transformers and is researched by [21] proposes a brain tumor segmentation involving edge information in multimodal MRI. It consists of a semantic segmentation module, an edge detection module, and a feature fusion module. The method uses Swin Transformer for semantic features, CNNs with edge spatial attention for edge detection, and a multi-feature inference block based on graph convolution for feature fusion. Detailed reviews are found in [3,17].

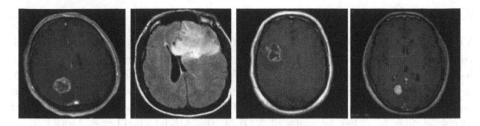

Fig. 1. MRI images showing brain tumors [9]

2 Background

2.1 Magnetic Resonance Imaging

Magnetic Resonance Imaging (MRI) is a non-invasive imaging technology that produces three-dimensional detailed anatomical images [1]. An MRI scan produces detailed images of the brain using powerful magnets and radio waves. Different types of tissues can be distinguished from each other using these images, including white matter, gray matter, and Cerebrospinal fluid. By analyzing these images, doctors can identify the location and extent of a brain tumor, as well as any surrounding tissue that may be affected. The four major MRI modalities are T1-weighted MRI (T1), T2-weighted MRI (T2), T1-weighted MRI with Gadolinium contrast enhancement (T1-ce), and Fluid Attenuated Inversion Recovery (FLAIR) [6]. A sample of the different MRI modalities is shown in Fig. 2. Healthy tissues are generally distinguished from cancer-prone areas using T1, whereas tumors with bright signal intensity are distinguished using T2. By accumulating contrast agents (Gadolinium ions) in the region of active cells on T1-CE images, the tumor border can be clearly identified. Similarly to the T2 images, FLAIR images are primarily used to differentiate the tumor region from the Cerebrospinal Fluid (CSF) by suppressing signals from the water molecules in the brain.

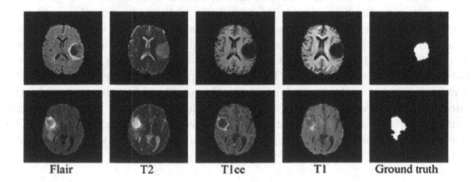

| Flair | T2 | T1ce | T1 | Ground truth |

Fig. 2. MRI modalities [1]

2.2   YOLOv5

YOLOv5 (You Only Look Once) is an object detection algorithm used for real-time detection and recognition of objects in images and videos [20], and is an improvement over previous versions of YOLO. The main goal of YOLOv5 is to provide a faster, more accurate, and more efficient object detection model that can be easily deployed on a wide range of devices. YOLOv5 has three important parts, similar to other versions of YOLO: the Model Backbone, Model Neck, and Model Head. The Model Backbone is responsible for extracting important features from the input image. In YOLOv5, Cross Stage Partial Networks (CSP) are used as a backbone to extract rich and informative features from the input image. The Model Neck generates feature pyramids, which help the model to generalize well on object scaling. Feature pyramids enable the model to identify the same object with different sizes and scales, making them very useful for performing well on unseen data [11].

2.3   U-Net

U-Net is a deep learning architecture introduced in 2015 by Ronneberger et al. [15] for segmenting neuronal structures in electron microscopy images. U-Net is comprised of an encoder-decoder network, where the encoder extracts the high-level features of the input image, and the decoder generates the segmentation mask. In contrast to other segmentation methods that rely on handcrafted features, U-Net acquires its features directly from the data through a combination of convolutional and pooling layers.

The network consists of encoders and decoders linked by skip connections. U-Net's encoder network is similar to that of a conventional CNN. A series of convolutional and pooling layers reduces the spatial resolution of the input image while increasing the number of feature maps. Typically, each convolutional layer is followed by a rectified linear unit (ReLU) activation function, which introduces nonlinearity into the network. During the learning process, the encoder network learns the high-level features of the input image. The U-Net decoder network consists of a series of convolutional and upsampling layers that progressively increase the spatial resolution of feature maps while decreasing the number of feature maps. The decoder network is designed to generate the segmentation mask by fusing the high-level features learned by the encoder with the low-level features of the input image. Additionally, U-Net includes skip connections between encoders and decoders. These connections allow the network to fuse the low-level features of the input image with the high-level features learned by the encoder. It is accomplished by concatenating the feature maps from the encoder network with those from the corresponding layer in the decoder network. The network is therefore capable of segmenting small and complex objects accurately [16,18].

3   Proposed Approach

In medical imaging, brain tumor segmentation is an important task that facilitates diagnosis, treatment planning, and monitoring of patients with brain tumors. The low contrast and high levels of noise present in brain tumor images make it challenging to accurately and efficiently segment them. In this work, we propose a Two-stage approach for segmenting brain tumors in MRI scans as depicted in Fig. 3. The first step involves identifying the tumor region in the MRI scan by using the YOLOv5 object detection algorithm. YOLOv5 is trained and tested on a large dataset of scans, therefore it can accurately detect tumor regions with high precision and recall.

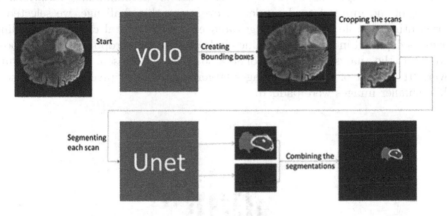

Fig. 3. Proposed approach for segmenting brain tumors from MRI scans

In the second stage, the tumor region is segmented into three distinct areas. the whole tumor, the tumor core, and the enhanced tumor core, utilizing the U-Net architecture. Taking into account the characteristics of MRI scans, such as the presence of noise and artifacts, we modify and research the optimal hyperparameter of the U-Net architecture.

For both stages of the development process, two datasets were used, BraTS 2020 and BraTS 2021 datasets [9]. The dataset contains 4 versions of MRI scans segmented by 3D scans of 1250 patients. The BraTS MRI scans were acquired using various clinical protocols and scanners. These scans are available as NIfTI files (.nii.gz), thus allowing for the generation of a wide range of image examples. All of the imaging datasets were manually annotated by one to four raters using the same annotation protocol, and their annotations were approved by neuroradiologists with extensive experience. In addition, we implemented various data augmentation techniques, including rotation, scaling, and flipping, to increase the model's robustness to variations in the input data in order to avoid overfitting the training process.

To evaluate the performance of our method, various evaluation metrics were used, including accuracy coefficients, sensitivity, and specificity.

Bounding Box Object Detection

There are three reasons for using Bounding Box with YOLOv5:

- Reduces the number of parameters transferred to the U-Net by focusing exclusively on the tumor area
- Computation time for relevant areas within the image is significantly lower than the computation time for the entire image
- Using only U-Net, some small areas of the image were overlooked

To achieve more accurate segmentation results, we suggest capturing all areas using YOLOv5. Since the bounding boxes differ in size (height × width), this output is problematic as an input to the second stage. It was necessary to provide a certain equal input size to U-Net, since it cannot accommodate variations in the size of the input data. In order to overcome this challenge, we selected a representing sample of 40 3D images and created 243 total cropped bounding boxes as depicted in Fig. 4. Each blue circle represents a cropped bounding box, and due to the analysis, a crop size of 80 × 80 pixels was chosen. This size measure covers 91 percent of the cases. Images larger than 80 × 80 pixels were not used, while smaller images were padded.

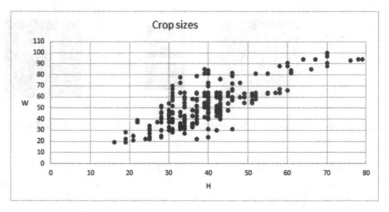

Fig. 4. Bounding box crop sizes results from YOLOv5

MRI Modality

In order to determine the most suitable MRI modality for training the U-net, the different modalities available, namely T1, T1CE, T2, and FLAIR, were evaluated by the authors. This evaluation was conducted using a dataset of 20 3D brain tumors, where each modality was used to obtain MRI images of the same tumors. The same segmentation algorithm was then applied to all the images, and the resulting segmentation were compared against the ground truth, which is typically delineated manually. A comparative analysis was employed for the segmentation performances of each modality to determine the best option for training the U-net. The results showed that the T1CE (contrast enhancement) modality yielded the most accurate segmentation. This evaluation process is crucial for training the U-net accurately, as choosing an inappropriate MRI modality for segmentation can negatively impact the model's performance. By conducting this comparative analysis, we were able to identify the optimal modality for training the U-net and improve the accuracy of the complete brain tumor segmentation model.

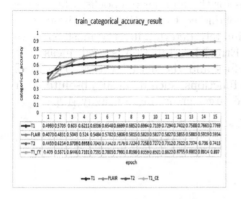

Fig. 5. Training Accuracy

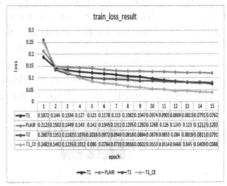

Fig. 6. Training Loss

Training and validation results are evaluated in terms of accuracy and loss, as shown in Figs. 5, 6, 7 and 8. The figures are presented in percentages corresponding to epochs and each modality is indicated by a different color. Figs. 5 and 6 provide an overview of the comparative research conducted in the training phase, while Figs. 7 and 8 present the validation phase results. Having achieved the best results in terms of both accuracy and loss, T1Ce was selected for this work.

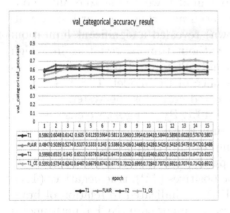

Fig. 7. Validation Accuracy

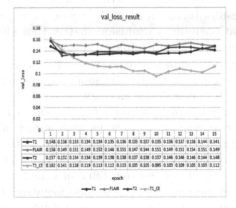

Fig. 8. Validation Loss

4 Hyper-parameter Optimization

During the training phase, we conducted an investigation on the impact of hyper-parameters on the accuracy of the model by searching for the minimum loss value

[2]. Specifically, we focused on the evaluation of the learning rate and epoch size through simulations and training. The batch size was fixed at 32, as this resulted in comparable performance to a batch size of 64, but with lower computational requirements. We tested different values of the learning rate, including 5e–4, 5e–5, and 5e–6, as well as different values for the number of epochs, which were 50 and 100. Multiple experiments were conducted to identify the optimal combination of these hyper-parameters, ultimately determining the most effective combination. The MSE (Mean Squared Error) loss function is used, measuring the difference between the predicted values and the actual values. The MSE loss function is preferred over other loss functions, such as mean absolute error (MAE), when the objective is to penalize larger errors more heavily [19].

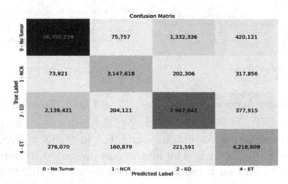

Fig. 9. Confusion Matrix: no tumor, necrotic tumor core (NCR), peritumoral edematous/invaded tissue, and enhancing tumor (ET)

In order to assess the performance of our model, we sought to evaluate our results based on multiple metrics, not just accuracy. However, our initial evaluation using a confusion matrix (not shown) revealed a significant limitation: the model predominantly identified areas without tumors, resulting in a high proportion of true negatives. This made it difficult to use basic metrics such as precision or recall to evaluate the performance of the model. To address this, we implemented a strategy in which we created crops for the system in which the percentage of tumor regions was at least 40%. By doing so, we ensured that the system received crops with sufficient tumor regions to learn to accurately identify areas with tumors, which led to improved performance.

A confusion matrix after the U-net cropping performed is depicted in Fig. 9. The results of the four different values: true positives (TP), true negatives (TN), false positives (FP), and false negatives (FN) are indicated to relation of brain tumor segmentation groups: no tumor, necrotic tumor core (NCR), peritumoral edematous/invaded tissue, and enhancing tumor (ET). By analyzing these values, we can calculate different metrics that help evaluate the model's performance, such as accuracy, precision, recall, and F1-score [13].

Figures 10 and 11 illustrate the process of training and validation for various values of learning rates (5e–4, 5e–5, and 5e–6) and number of epochs (50 and 100). To illustrate how the learning process has progressed through the epochs, each case is represented by a different color, and for clarity, only the first 10 epochs are shown. Table 1 presents the best results for all cases. According to our analysis, the combination of a learning rate of 0.0001, 100 epochs, and a batch size of 32 produced the highest accuracy (see case 6). Moreover, we note that the combined hyperparameters in case 5 achieved a similar level of accuracy, with the only difference being a reduction in the number of epochs. We determined that the difference in accuracy (0.0004) was not significant enough to justify the additional computation cost since this change resulted in a significant reduction in runtime.

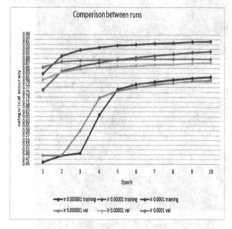

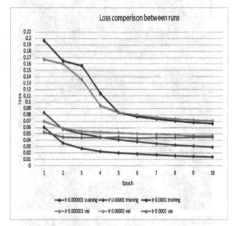

Fig. 10. Validation Accuracy   Fig. 11. Validation Loss

Table 1. Test results for various hyperparameter cases.

No.	Learning rate	Epochs	Accuracy	Loss
1	0.000001	50	0.854	0.053
2	0.000001	100	0.8661	0.04957
3	0.00001	50	0.8876	0.04472
4	0.00001	100	0.8876	0.04659
5	0.0001	50	0.8912	0.0445
6	0.0001	100	0.8916	0.04602

Therefore, the faster run with the combination of hyperparameters in line 5 was chosen. As a summary of the best results for developing the optimized system, we use MRIs of type T1CE learning rate = 0.0001, 50 epochs, and bach size = 32.

5   Results

Four examples of segmentation are shown in 12. The raw MRI data is shown on the left, and the segmentation results are shown on the right. Clearly, the developed system recognizes tumors and segments them according to the following colors: black - no tumor, red - necrotic tumor core (NCR), green - Peritumoral Edematous/invaded tissue, and blue - GD-enhancing tumor (ET).

Input scan   **Output Result**

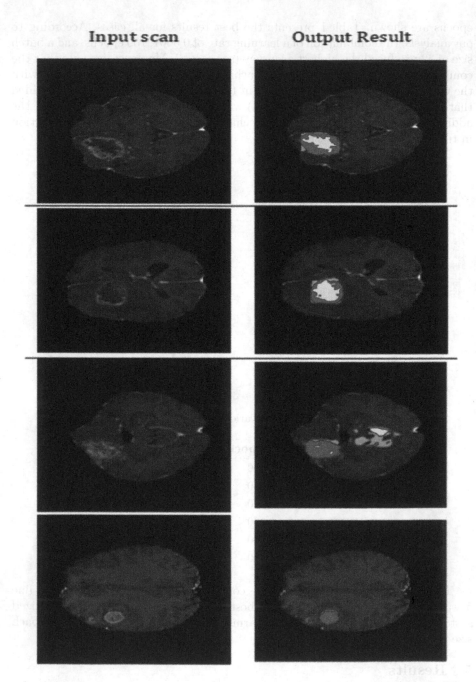

Fig. 12. Black - no tumor, red - necrotic tumor core (NCR), green - Peritumoral Edematous/invaded tissue, blue - GD- enhancing tumor (ET). (Color figure online)

6   Conclusions

n this study, a deep learning approach was developed for detecting cancerous brain tumors. The method utilized two-stage deep networks to identify the location of the tumor and then perform segmentation. To optimize the training process of the U-Net model, several parameters were tested and evaluated, including Learning Rate, Epochs sizes, Batch Size, and Loss function. Through experimentation, the optimal combination was found to be a learning rate of 0.0001, 50 epochs, and a batch size of 32, resulting in an accuracy of 0.8916. The proposed method demonstrated a high level of accuracy in segmenting various tumor areas, including those with no tumor, necrotic tumor core (NCR), Peritumoral Edematous/invaded tissue, and enhancing tumor (ET). This approach has the potential to improve the accuracy and efficiency of brain tumor segmentation, which can aid in treatment planning and disease progression monitoring. Therefore, the results of this study show that the proposed method is a promising technique for automated brain tumor segmentation

Acknowledgement. This work was partially financed by the Foundation for Research, Braude College of Engineering, Grant no. 5001.1.18.12

References

1. Aghalari, M., Aghagolzadeh, A., Ezoji, M.: Brain tumor image segmentation via asymmetric/symmetric UNet based on two-pathway-residual blocks. Biomed. Signal Process. Control **69**, 102841 (2021)
2. Andonie, R.: Hyperparameter optimization in learning systems. J. Membr. Comput. **1**, 279–291 (2019). https://doi.org/10.1007/s41965-019-00023-0
3. Das, S., Nayak, G.K., Saba, L., Kalra, M., Suri, J.S., Saxena, S.: An artificial intelligence framework and its bias for brain tumor segmentation: a narrative review. Comput. Biol. Med. **143**, 105273 (2022)
4. Fang, L., Wang, X.: Brain tumor segmentation based on the dual-path network of multi-modal MRI images. Pattern Recogn. **124**, 108434 (2022)
5. Fujita, H.: AI-based computer-aided diagnosis (AI-CAD): the latest review to read first. Radiol. Phys. Technol. **13**(1), 6–19 (2020)
6. Işın, A., Direkoğlu, C., Şah, M.: Review of MRI-based brain tumor image segmentation using deep learning methods. Procedia Comput. Sci. **102**, 317–324 (2016)
7. McNeill, K.A.: Epidemiology of brain tumors. Neurol. Clin. **34**(4), 981–998 (2016)
8. Mlynarski, P., Delingette, H., Criminisi, A., Ayache, N.: Deep learning with mixed supervision for brain tumor segmentation. J. Med. Imag. **6**(3), 034002 (2019)
9. Schettler, D.: BRaTS 2021 Task 1 Dataset (2021). https://www.kaggle.com/datasets/dschettler8845/brats-2021-task1
10. American brain tumor association (2023). https://www.abta.org/
11. Nepal, U., Eslamiat, H.: Comparing yolov3, yolov4 and yolov5 for autonomous landing spot detection in faulty UAVs. Sensors **22**(2), 464 (2022)
12. Nyo, M.T., Mebarek-Oudina, F., Hlaing, S.S., Khan, N.A.: Otsu's thresholding technique for MRI image brain tumor segmentation. Multimedia Tools and Applications, pp. 1–13 (2022)

13. Ohsaki, M., Wang, P., Matsuda, K., Katagiri, S., Watanabe, H., Ralescu, A.: Confusion-matrix-based kernel logistic regression for imbalanced data classification. IEEE Trans. Knowl. Data Eng. **29**(9), 1806–1819 (2017)
14. Omuro, A., DeAngelis, L.M.: Glioblastoma and other malignant gliomas: a clinical review. JAMA **310**(17), 1842–1850 (2013)
15. Ronneberger, O., Fischer, P., Brox, T.: U-Net: convolutional networks for biomedical image segmentation. In: Navab, N., Hornegger, J., Wells, W.M., Frangi, A.F. (eds.) MICCAI 2015. LNCS, vol. 9351, pp. 234–241. Springer, Cham (2015). https://doi.org/10.1007/978-3-319-24574-4_28
16. Siddique, N., Paheding, S., Elkin, C.P., Devabhaktuni, V.: U-Net and its variants for medical image segmentation: a review of theory and applications. IEEE Access **9**, 82031–82057 (2021)
17. Soomro, T.A., et al.: Image segmentation for MR brain tumor detection using machine learning: a review. IEEE Reviews in Biomedical Engineering (2022)
18. Wang, R., Lei, T., Cui, R., Zhang, B., Meng, H., Nandi, A.K.: Medical image segmentation using deep learning: a survey. IET Image Proc. **16**(5), 1243–1267 (2022)
19. Wang, Z., Bovik, A.C.: Mean squared error: love it or leave it? a new look at signal fidelity measures. IEEE Signal Process. Mag. **26**(1), 98–117 (2009)
20. Zhu, X., Lyu, S., Wang, X., Zhao, Q.: TPH-YOLOv5: improved YOLOv5 based on transformer prediction head for object detection on drone-captured scenarios. In: Proceedings of the IEEE/CVF International Conference on Computer Vision, pp. 2778–2788 (2021)
21. Zhu, Z., He, X., Qi, G., Li, Y., Cong, B., Liu, Y.: Brain tumor segmentation based on the fusion of deep semantics and edge information in multimodal MRI. Inf. Fusion **91**, 376–387 (2023)

Denoising Low-Dose CT Images Using Noise2Noise and Evaluation of Hyperparameters

Or Man and Miri Weiss Cohen$^{(\boxtimes)}$ ⓘD

Braude College of Engineering, Karmiel, Israel
ormn1996@gmail.com, miri@braude.ac.il

Abstract. In computed tomography (CT), the quality of the image is directly related to the exposure of the patient during the scan. A reduction in exposure reduces the health risks for patients, however, an increase in noise compromises the image quality. This work examines the Noise2Noise framework, which requires only noisy image pairs for network training in order to minimize the noise in CT images. This study examines the effects of varying learning rates, batch sizes, epochs, and encoder-decoder network depths on a variety of loss functions and their parameters.

Keywords: CT scans · Noise2Noise · U-Net · Hyper-parameter Optimization

1 Introduction

Radiation exposure during a Computed Tomography (CT) examination affects the image quality in CT images [15,18]. By reducing exposure, patients' health risks are reduced, but image quality is compromised. Furthermore, an increase in noise leads to low quality images. In real-life medical practices, doctors require accurate CT images in order to diagnose and treat their patients appropriately. Research is needed to achieve low-dose image acquisition without sacrificing information, and a method of artificially implementing denoising that preserves image features is one of the most promising approaches.

There are several problems caused by noisy CT images, including the following: First, the edges of the organs may appear blurry, making it difficult to determine their sizes. Furthermore, excessive noise can result in misdiagnosis of fat, muscle, water, gray matter, and white matter as a result of incorrect identification of some parts in the image. Lastly, small features may appear as noise if the noise levels are too high.

A CT scan uses its high contrast sensitivity to distinguish between soft tissues within the body. Despite the fact that low contrast structures can be visualized well, they are hampered by noise, which adversely affects their visibility. An understanding of the sources of the noise, the types of noise, as well as their general properties and characteristics is essential before considering methods of noise reduction. An in-depth description can be found in Diwaka et al. [3,4] with the following summary:

- Random Noise: Random matter can cause some of the X-rays to behave differently, resulting in different densities in various areas
- Statistical Noise: The energy levels of X-ray rays are not constant and vary statistically. In order to reduce the effects of statistical noise, it is necessary to increase the number of detected X-rays. In most cases, this can be accomplished by increasing the X-ray dose to increase the number of X-rays transmitted.
- Electronic noise: Noise generated by the electric circuits that receive the CT's analog signals. CT scanners of the latest generation have been specifically designed to reduce electronic noise.
- Round-off errors: Noise created when analog signals cannot be transformed into digital signals. In order to achieve this transformation, rounding of the data is required, which may lead to rounding errors.

In terms of noise distribution, CT images can be accurately described by the Poisson distribution. On the other hand, multidetector CT scanners are better described by a Gaussian distribution known as Additive Gaussian White Noise (AWGN). Figure 1 depicts three images: noise AWGN (left), ground truth clean image (center), and noisy image derived from both (right).

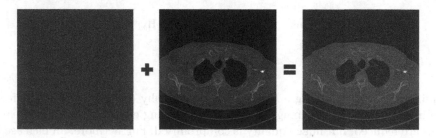

Fig. 1. Noise (left), clean image (middle), AWGN-derived noisy image (right)

In this study, the purpose is to investigate a learning strategy that is able to study real noise as well as synthetic datasets in order to evaluate how effective it is. Various network architectures and loss functions were used to examine the effect of hyperparameters on the accuracy of each model.

2   Denoising

Denoising CT images can be more effectively accomplished with prior knowledge of the CT images and the noise associated with them. A large clean dataset is required for most deep learning methods, which is difficult to obtain. Furthermore, most deep-learning methods require access to the desired output in order to train and validate the learning process, and clean images are not always possible [26]. Despite the best efforts of CT image denoising, it is difficult to significantly improve its accuracy in the absence of prior knowledge [20]. In the process of reducing noise in digital images, conventional filters such as smoothing and sharpening are the most popular. Considering the smoothness factor, smoothing filters do not provide effective noise reduction, especially when noise levels are high. In addition, smoothing filters may also damage details such as edges [5]. An important challenge in medical image analysis is the absence of structures, which can lead to inaccurate results.

Noise reduction in CT images faces the following challenges:: Areas that are flat should remain flat; Image boundaries should be preserved (no blurring); texture details should not be lost; global contrast should be preserved; and new artifacts should not be created.

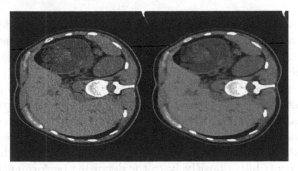

Fig. 2. Noisy CT (left) and blurry denoised output (right).

Figure 2 provides an example of a noisy CT and blurry denoised output. The denoising criteria that examine the measure of image quality include the structural similarity index (SSIM), the entropy difference metric (ED), and the Peak Signal-to-Noise Ratio (PSNR) [9]. The PSNR is the most widely used and is also used in research. Denoised images with higher PSNR values are more similar to their original counterparts. A PSNR is expressed in terms of the difference between a clean image (X) and a denoised image (R) as follows:

$$PSNR = 10 \cdot \log_{10}(\frac{255 \cdot 255}{MSE})$$ (1)

$$MSE = \frac{1}{m \cdot n} \sum_{i=0}^{m-1} \sum_{j=0}^{n-1} [X(i,j) - R(i,j)]^2$$ (2)

Deep learning techniques for image denoising were initially focused on learning how to denoise a picture using a fully supervised model. This image was generated by combining a clean image with a noisy input (noise2clean: N2C). According to [6] it was possible to obtain the denoising model and process by training a CNN model with a large number of pairs of noisy inputs $(x, \hat{x})$ and clean images. In such a scenario, the noisy images are used as training inputs, and the clean images are used as training targets. Based on a loss function, the network is trained to produce an image that is as close to the clean input image as possible. The method is known as Noise2Clean (N2C). A disadvantage of this approach is the difficulty in finding a sufficient number of clean targets for training [10]

A different denoising model can also be obtained by training a CNN model with a large number of training pairs $(\hat{x}, \hat{y})$ of noisy inputs $\hat{x}$ and another noisy realization of the training instances as targets $\hat{y}$ [13]. In such a case, the network is forced to learn to reproduce the second noisy realization, which is a challenging task because of the random behavior of the noise. In spite of this, through a large number of training samples, the network is able to reproduce an average representation rather than a discrete representation, which results in the desired denoising effect. This method is known as Noise2Noise (N2N) [13]. In their work, the authors showed that, for noise removal in Magnetic Resonance Imaging, the results produced with Noise2Noise are comparable with the ones produced with Noise2Clean.

As part of Noise2noise training, Wu et. al. [24] determine independent measurements by calculating a zero-mean noise by splitting the measurements into independent sets. They propose to consider the same loss function for all parts and use Noise2clean framework to create a cleaner image.

In order to denoise PET images, Yie et. al. [25] compare two self-supervised methods (noise2noise: N2N and noiser2noise: Nr2N) and their performance with the conventional N2C model. The Noise2Clean strategy quickly achieves higher overall metric values. A reduction in training data limits Noise2Noise's effectiveness in practical situations, but according to Calvarons [2] using the Noise Surrogate technique contributes positively to the denoising performance. It results in a dissociation of learning trends from other Noise2Noise techniques, eventually reaching Noise2Clean levels. Multiagent environments are used in [8], which combines competition and collaboration. Competition between CT image denoising tasks can be analogous to the adversary principle between the generator and discriminator in various generative adversarial networks (GANs) and is proposed by Isola et. al. [11].

The following research question that we are interested in pursuing is what can be achieved to improve the N2N framework and the PSNR values as a result of our research? In order to address this question, we conducted a study on various hyperparameters influencing the model and developed an improved system based on our findings. The answer to this question can therefore be very useful in enabling low-dose images to be acquired without losing significant information. By doing so, we are able to maintain the characteristics of the image while simultaneously reducing denoising.

3   Proposed Approach

In order to provide good medical care without unnecessary risks, it is necessary to develop a CNN for denoising low-quality images without removing any medical information. We propose to use an Noise2Noise [13] Encoder-Decoder neural network as well as data augmentation to increase the size of existing datasets. The number of possible training pairs is $I \cdot (N \cdot (N - 1))$ if N is the number of noisy realizations per image and I is the total number of available images.

This research aims to identify the contribution of each hyperparameter and value to the resultant best network. In order to achieve this goal, it is necessary to determine the impact of the hyperparameters on the network as well as to understand the results in light of these principles.

Network Architecture Encoder-Decoder

CNNs are typically used for classification tasks, where the output of each image is a vector indicating the likelihood that the image belongs to each class. Nevertheless, in some tasks, such as segmentation it is necessary to classify not only an image, but also each pixel. Segmenting images in this manner is known as semantic segmentation [14,19].

The Encoder-Decoder CNN replaces the fully connected part of the CNN (used for classification) with convolutional layers, which are referred to as the decoder part of the network, while the existing convolutional layers are referred to as the Encoder part. The encoder part reduces the image size while retaining more information per pixel. Decoding is used to increase the size of the image back to its original size while retaining only the necessary information. In the Decoder part, pooling operators are replaced by upsampling operators to create the opposite network. Due to these layers, the output has a higher resolution. High-resolution features from the encoder path are combined with upsampled output to achieve localization. The information can then be used by a subsequent convolution layer to produce a more precise output [1,12]. Both semantic segmentation (which uses the class as the necessary data) and image denoising (which uses the clean image as the necessary data) can be accomplished using the network architecture.

Hyper-Parameters

This research tested three network depths Network depth $\{3, 4, 5\}$. The output image size is the same as the input image, which is accomplished through a padding process. In the bottom layer, a dropout of 0.5 is used. Additionally, in order to maximize network potential, we used Learning rate decay to gradually reduce the learning rate to achieve the best results [22]. The learning rate decay used in this work is detailed in Table 1.

To achieve the best results and to understand the influence of each hyperparameter, the network is trained by comparing a number of hyperparameters. These include the following: Learning rate - in the range of $\{0.000001, \ldots, 0.0001\}$, Number of Epochs varying between $\{40,400$ (small epochs)$\}$, a Batch Size of $\{8\}$. The Loss functions $\{$MAE, MSE, Huber Loss$\}$ were tested and detailed in the next section, where the Huber Loss parameter are δ, with values $\{1.5, 3\}$.

Table 1. Learning rate scheduler

epoch number out of the total	Learning rate based on the initial Lr
$epoch\_id < \frac{1}{4} \cdot max\_epoch$	$initial\_Lr$
$\frac{1}{4} \cdot max\_epoch \leq epoch\_id < \frac{1}{2} \cdot max\_epoch$	$\frac{1}{2} \cdot initial\_Lr$
$\frac{1}{2} \cdot max\_epoch \leq epoch\_id < \frac{3}{4} \cdot max\_epoch$	$\frac{1}{4} \cdot initial\_Lr$
$\frac{3}{4} \cdot max\_epoch \leq epoch\_id$	$\frac{1}{8} \cdot initial\_Lr$

Loss Functions:
By analyzing the loss function, it is possible to determine how well the encoder-decoder models the denoising process. A model's loss measures its accuracy, and the goal is to train and optimize it so that the loss is as low as possible. We evaluated several different loss functions in our study since there is no clear definition or resolute opinion regarding which Loss function measurement is the best. The following section discusses loss functions measurements and their advantages and disadvantages.

- Mean Squared Error (MSE) MSE is defined as:

$$MSE = \frac{1}{m \cdot n} \sum_{i=0}^{m-1} \sum_{j=0}^{n-1} [X(i,j) - R(i,j)]^2 \qquad (2)$$

With the advantage of MSE [21], as with the encoder-decoder model, we can ensure that our trained model does not contain outlier predictions with large errors. This is due to the fact that the advantage of MSE squares the errors. In the event that our model makes a single incorrect prediction, the square part of the function magnifies the error. In many practical cases, these outliers are not taken into account and have no significant impact.

- Mean Absolute Error (MAE) MAE is commonly used in model evaluation studies, and it is defined as follows [23]:

$$MAE = \frac{1}{m \cdot n} \sum_{i=0}^{m-1} \sum_{j=0}^{n-1} |X(i,j) - R(i,j)| \qquad (3)$$

MAE has several advantages over MSE in terms of training the Encoder Decoder Architecture. Due to the fact that the absolute value of the error is used, all errors are weighted equally. Thus, unlike the MSE, the loss function measures the performance of the model in a generic and even manner. A disadvantage of MAE is that it is ineffective if the outlier predictions are poor. A large error caused by an outlier is weighted the same as a smaller error.

- Huber loss: The Huber loss is one of the most suitable criteria as it balances the MSE as well as the MAE [7]. In order to get a well-rounded model, we use the MAE for larger loss values. This mitigates the weight that we place on outliers. In addition, we use the MSE for the smaller loss values in order to maintain a quadratic function near the center. Among the disadvantages of the Huber loss is the need to select the parameter δ, which adds another hyper-parameter [16]. Huber Loss is defined as follows:

$$Huber_{(\delta)} = \begin{cases} 0.5 \cdot MSE & \text{if } MAE \leq \delta \\ \delta \cdot MAE - 0.5 \cdot \delta^2 & \text{if } MAE > \delta \end{cases} \qquad (4)$$

- Root Mean Squared Error (RMSE): An alternative formalization of errors. This method tries to get the same benefits as *MSE* of outlier, but with loss values of MAE:

$$RMSE = \sqrt{MSE} = \sqrt{\frac{1}{m \cdot n} \sum_{i=0}^{m-1} \sum_{j=0}^{n-1} [X(i,j) - R(i,j)]^2} \qquad (5)$$

Data-sets:
In this work we used the DeepLesion dataset [17] which contains 32,735 clean CT images, from 4,400 patients. We add synthetic noise, and use the clean version to assert the network quality. Images are in the format UInt16 (0–65536), and using Additive White Gaussian Noise (AWGN), each image is normalized (to a range of [0,4096]). The baseline Peak Signal-to-Noise Ratio (PSNR) is 36.53 decibels. The normalization is based on the Typical Hounsfield unit (HU) values, of a CT image being in the range of [1024,3071]. Each image is normalized and then an Additive White Gaussian Noise (AWGN) is applied, resulting in noisy images. Figure 3 illustrates a normalization process. Left: the image before normalization; right: the image after normalization.

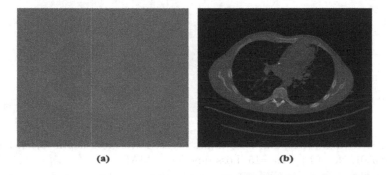

(a)   (b)

Fig. 3. Example of normalization, (a) image before normalization, (b) image after normalization.

4   Hyper-parameter Optimization

In order to achieve the purpose of the research process, which is to denoise a CT image, the work evaluates different hyperparameters. We conducted numerous simulations and research tests, but only a small portion of them is presented here.

4.1   Network Depth

In Fig. 4 fifteen epochs' results were compared with different depths of the network. In this test, the following configurations were tested: Learning rate = 0.0001, No. of Epochs = 15, Loss function = MAE. The figure depicts the accuracy of the training process on the left and the development of the loss function through the training epochs on the right. It can be seen how differences in the architectures of encoders-decoders have a direct impact on the results. The results show that after 15 epochs, the three network depths merge into a single value (in both accuracy and loss). For example in the accuracy rate, network Depth 3 is PSNR-45.73, Depth 4 is PSNR-45.75, and Depth 5 is PSNR-45.72. The results of the analysis indicate that network depth does not have a significant impact on the accuracy rate. Figure 5 depicts a comparison between the depth and learning rate of encoder-decoder combinations. On the left is the accuracy of the training process and the right is the loss function through the epochs. In the figure, depending on the learning rate (0.000001 for example), different network depths can be distinguished, and higher network depths lead to better results (in the learning rate range, but not overall). It appears that at higher learning rates, the different network depths converge and do not seem to affect the result.

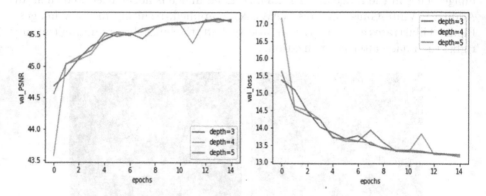

Fig. 4. PSNR values (left) and Loss values (right) in Network depth=3,4,5, Learning Rate =0.0001, No of Epochs =15, Loss function = MAE

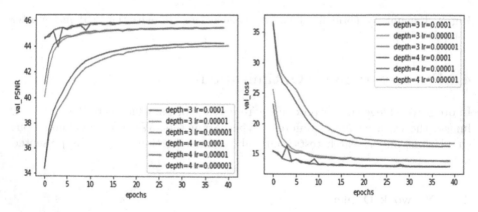

Fig. 5. Learning rate influence with 3 and 4 depth encoder-decoder, No of Epochs =40, and Loss Function=MAE, PSNR values (left) and Loss values (right)

4.2   Loss Functions

In Fig. 6, four runs are shown comparing only their loss functions. The effect of the *MAE, MSE and huber loss* with δ values of 1.5 and 3.0 was tested. Loss functions were tested at a learning rate of 0.0001. As can be depicted in the figure, Loss Function= MAE resulted Best PSNR = 45.74, Loss Function= MSE resulted Best PSNR = 45.6, Loss Function= Huber loss $\{\delta = 1.5\}$ resulted Best PSNR = 45.75, Loss Function= Huber loss $\{\delta = 3.0\}$ resulted Best PSNR = 45.78.

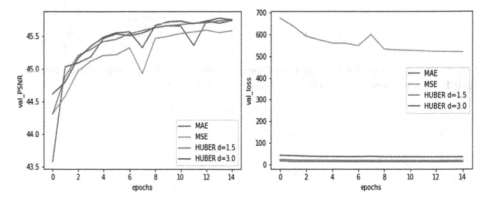

Fig. 6. PSNR values (left) and Loss values (right) simulations of loss function influence MAE, MSE, and Huber loss with learning rate = 0.0001

The Loss value depicted in Fig. 6 (right) shows the different values and effectiveness of the loss function. The *MAE* and *Huber losses* value results in very close values, while for the *MSE* the loss function is not an effective criterion. After the initial simulations, additional simulations were conducted, increasing the epochs to 40 as well as adding the RMSE loss function to the model. As a result it was found that all five loss functions oscillate around the same values when additional epochs are added to the training process, indicating that the learning process had plateaued. Figure 8 illustrates simulations of the five loss functions at three different learning rates. This is done to determine whether there is a correlation between a specific loss function and learning rate or whether they are combined. In the figure on the left, the PSNR values are displayed in relation to the epochs, while in the figure on the right, the loss value is presented in relation to the epochs. Based on the PSNR values, the results can be grouped into three groups as shown in the figure on the left. These groups all share the same learning rate as a common characteristic.

In PSNR results, the loss function appears to be less effective than the learning rate, and there does not appear to be any correlation between any specific loss function and any learning rate that is more effective. If we examine the lowest learning rate, we can gain some insights into the fact that MSE may result in slightly lower PSNR, but at higher learning rates it may not be as significant. Figure 8 on the right shows that most runs are clustered at the bottom (around a validation loss of 0), except for the group representing MSE loss function. MSE

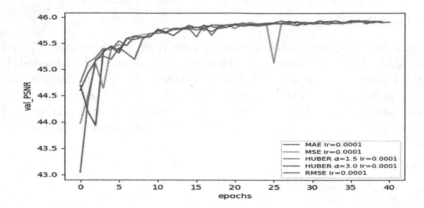

Fig. 7. PSNR values for various Loss function influence: MAE, MSE, Huber loss and RMSE with learning rate=0.0001, expending to 40 epochs

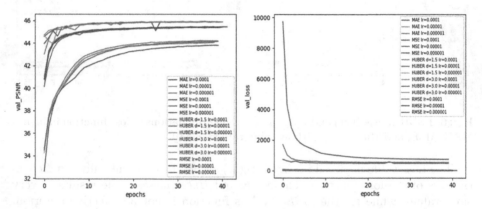

Fig. 8. PSNR values (left) and Loss values (right) with loss function influence MAE, MSE, Huber loss and RMSE

has a numerical value corresponding to the second power of the error, whereas the others fall primarily within the range of its absolute value.

4.3 Learning Rate

As depicted in Fig. 9, we found that learning rate had a significant impact on the results. In order to further investigate this, we disabled the Learning Rate Scheduler and evaluated its effects on the results at different learning rates.

Figure 9 (left) compares three runs with different learning rates without and with a scheduler. At the lowest learning rate (0.000001), the curve without the scheduler shows an incline at the position where the scheduler converged. There is a possibility that longer runs conducted at low learning rates without learning rate schedulers may show better results than those conducted with learning rate schedulers, and even better than those conducted at higher learning rates.

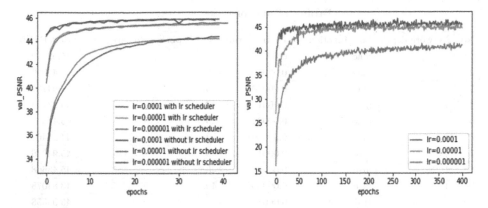

Fig. 9. Comparison of PSNR values between different learning rates with and without a learning scheduler (left) and Different learning rates with small batches and 400 epochs (right)

Most results so far show that the network stabilized in the end. This raises the question of whether the stabilization may happen in the middle of our long epoch. To verify this hypothesis we divided each training epoch so it will contain only 100 batches(instead of 2811) and each validation epoch into 30 batches, we did not reduce the number of training or validation data, only splitting the epoch. Because of the shrinking of the batches, we increase the number of epochs to 400. Figure 9 (right) the results reveal an inability to create the noise that is necessary for better denoising. Just like the hypothesis, the changes in these small batches disappeared in the large epochs. From a comparison of the three learning rates, it is evident that the 0.0001 learning rate is the most appropriate. However, contrary to the hypothesis, the lower learning rate (0.000001) did not receive a higher PSNR than the higher learning rate. This may be due to the use of N2N, so the training to create random noise to clean the data may cause the learning process to differ from what is expected.

A summary of some of the major findings of this study can be found in Table 2, as well as an overview of some of their implications outlined in the table. As can be seen from the table, several parameters contribute to the results, including the Network Depth, Learning Rate, Number of Epochs, and Loss function. This results in the highest PSNR being highlighted for each case achieved.

4.4 Results and Examples

Figure 10 shows an image with average noise and a denoised version. The clean version revealed some missing artifacts from the noisy source. The artifacts are actually a result of noise formation, which was eliminated by the N2N method. Further, the texture of the bone was not affected and was present in the original image.

Table 2. Table of various results

No.	Network depth	Loss function	Learning rate	No of epochs	Remarks	Best PSNR
1	3	mae	0.0001	40	–	45.92699
2	3	mae	0.00001	40	–	45.43896
3	3	mae	0.000001	40	–	44.04781
4	4	mae	0.0001	40	–	45.93173
5	4	mae	0.00001	40	–	45.47377
6	4	mae	0.000001	40	–	44.25464
7	4	mse	0.0001	40	–	45.94246
8	4	mse	0.00001	40	–	45.43668
9	4	mse	0.000001	40	–	43.84078
10	4	Huber $\{\delta = 1.5\}$	0.0001	40	–	45.93558
11	4	Huber $\{\delta = 1.5\}$	0.00001	40	–	45.46615
12	4	Huber $\{\delta = 1.5\}$	0.000001	40	–	44.06553
13	4	huber $\{\delta = 3.0\}$	0.0001	40	–	45.93676
14	4	huber $\{\delta = 3.0\}$	0.00001	40	–	45.44586
15	4	huber $\{\delta = 3.0\}$	0.000001	40	–	44.09719
16	4	rmse	0.0001	40	–	45.93617
17	4	rmse	0.00001	40	–	45.4799
18	4	rmse	0.000001	40	–	44.18776
19	4	mae	0.0001	40	no lr scheduler	45.9272
20	4	mae	0.00001	40	no lr scheduler	45.60669
21	4	mae	0.000001	40	no lr scheduler	44.42599
22	4	mae	0.0001	400	small epochs	46.80384
23	4	mae	0.00001	400	small epochs	45.80757
24	4	mae	0.000001	400	small epochs	41.67177
25	4	huber $\{\delta = 3.0\}$	0.0001	400	small epochs	**46.93232**
26	4	huber $\{\delta = 1.5\}$	0.0001	400	small epochs	46.66901
27	4	mse	0.0001	400	small epochs	46.47902

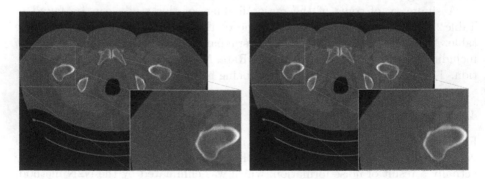

Fig. 10. Input-noisy image (left), Output-clean image(right)

Table 3. Result with high noise inputs.

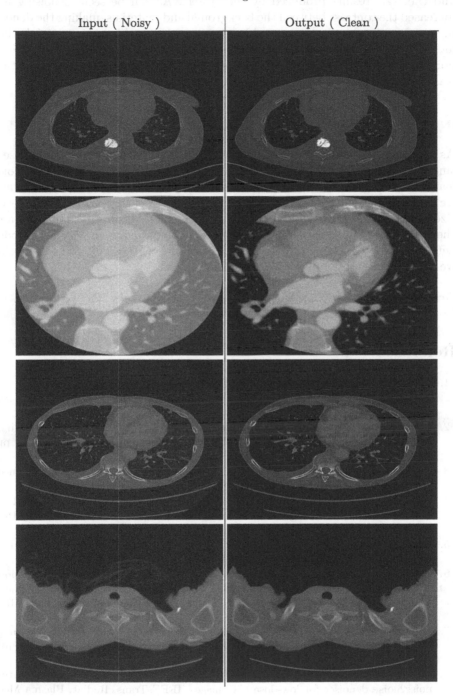

| Input (Noisy) | Output (Clean) |

Table 3 provides results using high noise images (low quality/low dose CT) and the N2N results produced by the system. As can be seen, denoising has increased the contrast between the background and materials, making the denoising more noticeable. In the second image, denoising enabled two materials that looked similar in the source image to be distinguished. Despite the high levels of noise, the delicate texture of the lungs was retained even when the bone texture was barely visible.

5   Conclusions

As demonstrated in our study, deep learning is effective when used to denoise a single image by learning pairs of noisy images. An analysis of the Noise2Noise learning strategy is conducted using real and synthetic datasets. In this study, we examined the effects of encoder-decoder network depth, learning rates, batch sizes, and epochs on the ability to denoise CT images. Various loss functions and their parameters were investigated. As a result of applying an encoder-decoder with four levels, Learning rate of 0.0001, Huber loss of $\{\delta = 3.0\}$, and 400 epochs, we obtained the best PSNR value of 46.93232.

Acknowledgements. This work was partially financed by the Foundation for Research, Braude College of Engineering, Grant no. 561417/2021-2.

References

1. Badrinarayanan, V., Kendall, A., Cipolla, R.: Segnet: a deep convolutional encoder-decoder architecture for image segmentation. IEEE Trans. Pattern Anal. Mach. Intell. **39**(12), 2481–2495 (2017)
2. Calvarons, A.F.: Improved noise2noise denoising with limited data. In: Proceedings of the IEEE/CVF Conference on Computer Vision and Pattern Recognition, pp. 796–805 (2021)
3. Diwakar, M., Kumar, M.: A review on CT image noise and its denoising. Biomed. Sig. Process. Control **42**, 73–88 (2018)
4. Diwakar, M., Singh, P.: Ct image denoising using multivariate model and its method noise thresholding in non-subsampled shearlet domain. Biomed. Sig. Process. Control **57**, 101754 (2020)
5. Fan, L., Zhang, F., Fan, H., Zhang, C.: Brief review of image denoising techniques. Vis. Comput. Ind. Biomed. Art **2**(1), 1–12 (2019)
6. Gnudi, P., Schweizer, B., Kachelrieß, M., Berker, Y.: Denoising of x-ray projections and computed tomography images using convolutional neural networks without clean data. In: The 6th International Conference on Image Formation in X-Ray Computed Tomography, pp. 590–593 (2020)
7. Gupta, D., Hazarika, B.B., Berlin, M.: Robust regularized extreme learning machine with asymmetric huber loss function. Neural Comput. Appl. **32**(16), 12971–12998 (2020)
8. Hasan, A.M., Mohebbian, M.R., Wahid, K.A., Babyn, P.: Hybrid-collaborative noise2noise denoiser for low-dose CT images. IEEE Trans. Radiat. Plasma Med. Sci. **5**(2), 235–244 (2020)

9. Hore, A., Ziou, D.: Image quality metrics: PSNR vs. SSIM. In: 2010 20th International Conference on Pattern Recognition, pp. 2366–2369. IEEE (2010)
10. Ilesanmi, A.E., Ilesanmi, T.O.: Methods for image denoising using convolutional neural network: a review. Complex Intell. Syst. **7**(5), 2179–2198 (2021)
11. Isola, P., Zhu, J.Y., Zhou, T., Efros, A.A.: Image-to-image translation with conditional adversarial networks. In: Proceedings of the IEEE Conference on Computer Vision and Pattern Recognition, pp. 1125–1134 (2017)
12. Ji, Y., Zhang, H., Zhang, Z., Liu, M.: CNN-based encoder-decoder networks for salient object detection: a comprehensive review and recent advances. Inf. Sci. **546**, 835–857 (2021)
13. Lehtinen, J., et al.: Noise2noise: learning image restoration without clean data. arXiv preprint arXiv:1803.04189 (2018)
14. Li, S., Yang, C., Sun, H., Zhang, H.: Seismic fault detection using an encoder-decoder convolutional neural network with a small training set. J. Geophys. Eng. **16**(1), 175–189 (2019)
15. Mculepas, J.M., et al.: Radiation exposure from pediatric CT scans and subsequent cancer risk in the Netherlands. JNCI: J. Natl. Can. Inst. **111**(3), 256–263 (2019)
16. Meyer, G.P.: An alternative probabilistic interpretation of the huber loss. In: Proceedings of the IEEE/CVF Conference on Computer Vision and Pattern Recognition, pp. 5261–5269 (2021)
17. NIH: clinical dataset (2018). https://www.nih.gov/news-events/news-releases/nih-clinical-center-releases
18. Rawashdeh, M.A., Saade, C.: Radiation dose reduction considerations and imaging patterns of ground glass opacities in coronavirus: risk of over exposure in computed tomography. La radiologia medica **126**(3), 380–387 (2021)
19. Ronneberger, O., Fischer, P., Brox, T.: U-net: convolutional networks for biomedical image segmentation. In: Navab, N., Hornegger, J., Wells, W.M., Frangi, A.F. (eds.) MICCAI 2015. LNCS, vol. 9351, pp. 234–241. Springer, Cham (2015). https://doi.org/10.1007/978-3-319-24574-4_28
20. Song, Y., Zhu, Y., Du, X.: Grouped multi-scale network for real-world image denoising. IEEE Signal Process. Lett. **27**, 2124–2128 (2020)
21. Wang, Z., Bovik, A.C.: Mean squared error: Love it or leave it? a new look at signal fidelity measures. IEEE Signal Process. Mag. **26**(1), 98–117 (2009)
22. Wen, L., Gao, L., Li, X., Zeng, B.: Convolutional neural network with automatic learning rate scheduler for fault classification. IEEE Trans. Instrum. Meas. **70**, 1–12 (2021)
23. Willmott, C.J., Matsuura, K.: Advantages of the mean absolute error (MAE) over the root mean square error (RMSE) in assessing average model performance. Clim. Res. **30**(1), 79–82 (2005)
24. Wu, D., Gong, K., Kim, K., Li, X., Li, Q.: Consensus neural network for medical imaging denoising with only noisy training samples. In: Shen, D., et al. (eds.) MICCAI 2019. LNCS, vol. 11767, pp. 741–749. Springer, Cham (2019). https://doi.org/10.1007/978-3-030-32251-9_81
25. Yie, S.Y., Kang, S.K., Hwang, D., Lee, J.S.: Self-supervised pet denoising. Nucl. Med. Mol. Imaging **54**(6), 299–304 (2020)
26. Zhou, S., Nie, D., Adeli, E., Yin, J., Lian, J., Shen, D.: High-resolution encoder-decoder networks for low-contrast medical image segmentation. IEEE Trans. Image Process. **29**, 461–475 (2019)

SIA-SMOTE: A SMOTE-Based Oversampling Method with Better Interpolation on High-Dimensional Data by Using a Siamese Network

Rahmat Izwan Heroza$^{(\boxtimes)}$, John Q. Gan, and Haider Raza

School of Computer Science and Electronic Engineering, University of Essex,
Colchester CO4 3SQ, UK
{rh22708,jqgan,h.raza}@essex.ac.uk

Abstract. SMOTE is an effective method for balancing imbalanced datasets by interpolating between existing samples in the minority class. However, if the synthetic samples generated through interpolation are based on noisy data points, then they may also be noisy and can lead to overfitting and reduced performance on unseen data. This paper proposes a new method SIA-SMOTE, which uses SMOTE for oversampling the minority class and a siamese network for synthetic image selection. SIA-SMOTE also explores the decision boundary to better capture data distribution of the minority class. The proposed method has been compared to random oversampling, SMOTE, and ASN-SMOTE on MNIST, FMNIST, and three medical image datasets. The results show that SIA-SMOTE achieved the best overall performance in terms of three evaluation metrics.

Keywords: SMOTE · siamese network · oversampling · image data augmentation

1 Introduction

Deep learning is a technique that is now widely used for problems in image analysis and pattern recognition. Deep learning even has a strong reputation in image classification tasks, especially when dealing with unconstrained large-scale databases [1]. Many balanced image datasets are publicly available for use in training deep neural networks, such as the MNIST handwritten digit dataset which has 70,000 color images in 10 classes and CIFAR-10 which has 60,000 color images in 10 classes. However, there are other cases with an imbalanced amount of data between classes, such as in agriculture [2], medical [3,4], hyperspectral data [5], remote sensing [6], construction [7], etc. In these cases, the accuracy of the deep learning technique for classifying images tends to degrade [8]. This occurs because the model is more likely to learn features that belong to the major class than the minor class [9].

© The Author(s), under exclusive license to Springer Nature Switzerland AG 2023
I. Rojas et al. (Eds.): IWANN 2023, LNCS 14134, pp. 448–460, 2023.
https://doi.org/10.1007/978-3-031-43085-5_35

There are two approaches for handling the imbalanced dataset problem: 1) the data approach and 2) the algorithm approach. The data approach resamples the dataset such that the amount of data in each class will be the same. Examples of resampling techniques include random over/under-sampling and Synthetic Minority Oversampling Technique (SMOTE) [10]. In recent years, generative adversarial network (GAN) has been successfully used for data augmentation [11]. In the algorithm approach, there is a cost-sensitive method that penalizes more on the minority class data prediction, leading to better performance compared to the conventional algorithms [12]. Examples of the algorithm approach for handling imbalanced datasets include Class Rectification Loss (CRL) [13] and Large Margin Local Embedding (LMLE) [14]. One of the most well-known techniques also in this approach is the ensemble method that uses a combination of base classifiers [15].

This research focuses on the oversampling technique and studies how the SMOTE [16] method oversamples high-dimensional datasets such as images. A new method named SIA-SMOTE is proposed in this paper to overcome the limitations in some existing oversampling methods for handling the issue of imbalanced image classification. The method for image synthesis developed in this paper relies on the concept of interpolation near the decision boundary, which involves generating new images by traversing the border between existing images from the minority class and the nearest neighbor from the majority class. By doing this, the space near the decision boundary will be explored effectively for synthesizing new images. To avoid creating a bad synthesized image, a siamese network [17], a type of neural network that compares and matches images based on their features, is used for image selection. By filtering new images through this network, it can be ensured that the selected synthesized images would be consistent with the features of the original images in the minority class and maintain their visual quality. On the other hand, SMOTE is also used to synthesize new images by interpolating original images in the minority class, which are added to the filtered synthesized images nearby the decision boundary and the original minority samples to form an augmented training dataset. Overall, the proposed approach to image synthesis through interpolation and filtering with a siamese network offers a powerful tool for generating new, diversified, high-quality images closely related to existing ones.

Several comparative studies were carried out to assess the performance of SIA-SMOTE, in comparison with three other oversampling techniques, namely (1) Random oversampling, (2) SMOTE [16], (3) ASN-SMOTE [18], in terms of three performance metrics: G-Mean, F1 Score, and AUC. The proposed approach was tested on the MNIST, FMNIST, and three medical image datasets: pneumonia dataset (ChestXray), histopathological breast cancer dataset (Breakhis), and skin cancer dataset (ISIC2018).

The remaining sections of the paper are structured as follows. In the "Related Work" section, previous studies and other current oversampling techniques are reviewed. The proposed oversampling method is described in detail in the section "The Proposed Method". The experimental design, results, and discussion are presented in the "Experiment and Results" section, followed by the "Conclusion" section.

2   Related Work

Random oversampling is the simplest oversampling method to address the problem of imbalanced datasets [10], in which samples from the minority class are randomly duplicated so that the class distribution becomes more balanced. The process of random oversampling involves randomly selecting samples from the minority class and duplicating them until the desired level of oversampling is achieved. One drawback of this method is that it can lead to overfitting, where the model becomes too specialized to the training data and performs poorly on unseen data.

SMOTE [16] is a popular technique used to address the problem of imbalanced datasets. SMOTE works by generating synthetic samples of the minority class by interpolating between existing samples in the minority class. This approach helps avoid overfitting that can occur when simply duplicating existing samples. The main advantage of SMOTE over random oversampling is that it generates synthetic samples rather than just duplicating existing ones. This means that the generated new samples are not exact copies of the original samples, but are located somewhere along the line connecting two existing samples. This can help improve the generalization of the model by reducing the risk of overfitting. Another advantage of SMOTE is that it is effective at preserving the overall distribution of the minority class. In contrast, random oversampling can lead to the creation of clusters of identical samples, which can introduce bias into the model.

While this technique can address the class imbalance issue, it can also introduce noise into the data if not done carefully. The reason for this is that interpolation involves creating new data points by taking a linear combination of existing data points. This process assumes that the existing data points are representative of the true underlying distribution of the minority class. However, if there is noise in the data, such as outliers or mislabeled samples, these points may not be truly representative of the minority class. If the synthetic samples generated through interpolation are based on noisy data points, then they may also be noisy and not truly representative of the minority class. This can lead to overfitting and reduced performance on unseen data. Therefore, it is important to be aware of noise in the data when using oversampling techniques such as interpolation and to carefully select the data points to use for interpolation to minimize the introduction of noise into the synthetic samples. Trying to resolve the above-mentioned problem, Yi et al. proposed ASN-SMOTE [18].

ASN-SMOTE works by filtering out noises in the minority class, which are from samples whose nearest neighbors are in the majority class. It creates synthetic samples by interpolating between remaining samples in the minority classes. When selecting neighbors from the minority class, it only selects ones that are closer than the nearest majority class. ASN-SMOTE achieved the best performance among nine state-of-the-art oversampling methods namely, random oversampling, SMOTE, ADASYN, Borderline1-SMOTE, Borderline2-SMOTE, k-means SMOTE, SVM-SMOTE, SWIM-RBF, and LoRAS using KNN, SVM, and Random Forests respectively on 24 public datasets from the KEEL repository and the UCI Machine Learning Repository.

3 The Proposed Method

The proposed method SIA-SMOTE consists of four sequential steps: selecting borderline images as the base samples, creating new images using interpolation between the borderline samples and nearest majority samples, selecting new images as samples for training data augmentation, and SMOTE oversampling to make the training data balanced.

3.1 Base Sample Selection

In contrast to several other SMOTE-based oversampling techniques, which consider the samples around the decision boundary as noise, the proposed method makes borderline samples as the base samples instead. Borderline samples should not be considered as noise because they contain important information about the distribution of the data and the decision boundary between the two classes. These samples can be crucial in training machine learning models to accurately classify data points or images, as they represent cases where the classification decision is most uncertain. Ignoring borderline samples could lead to a biased model that is overfitting to the data, as it fails to capture the nuances of the decision boundary. Therefore, it is important to carefully consider borderline samples and use appropriate methods to handle them. The method proposed in this paper will adjust the decision boundary to better accommodate the borderline samples by oversampling. The base samples are selected from the minority class where the nearest neighbor is a sample from another class. This is similar to how ASN-SMOTE defines noise [18]

3.2 Image Synthesis

The image synthesis stage in the proposed method is carried out by interpolating images from different classes as depicted in Fig. 1. The base sample is the minority sample nearby the decision boundary and the neighbor samples are the closest k samples from another class, with $k = 1$ being the default value. Because of this, the determination of the interpolation points is not completely random as in SMOTE. A hyper-parameter d is introduced in this method, which is the maximum distance of the interpolation point from the base sample to the neighbor sample. For example, $d = 0.5$ means that the interpolation is done randomly between the base sample and the midpoint between the base sample and the neighbor sample, making the resulting new images more similar to the base sample.

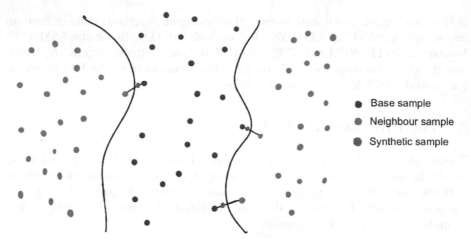

Fig. 1. Interpolation between images from different classes in decision boundary

3.3   Image Selection

Noise contained in the training data can affect the performance of the model. Because the interpolation in the proposed method is done between a base sample and its neighbor samples from other classes, the selection of new synthesized images is very important. A siamese network [17] is used in the proposed method for this purpose, which is a type of neural network architecture that is specifically designed to identify similarities between pairs of images. The network consists of two identical subnetworks, each taking one of the input images as input. The two subnetworks share the same weights and architecture and are trained jointly. To identify similar images, the siamese network compares the feature vectors generated from each of the input images using a similarity metric. The feature vectors are extracted by passing the input images through the subnetworks and retrieving the output from one of the final layers of the subnetworks.

During training, the siamese network is trained to minimize the distance between feature vectors of similar images in the same class and maximize the distance between feature vectors of dissimilar images in different classes. This is achieved through a contrastive loss function that penalizes the network when it incorrectly identifies similar images as dissimilar or dissimilar images as similar. During inference, when the siamese network is presented with a pair of base samples and the newly generated sample resulting from the interpolation process, it computes the distance between the feature vectors of the images and returns a similarity score indicating how similar the two images are. At this selection stage, there is also a hyper-parameter t, which is the threshold value of the confidence level of the siamese network when it performs interference. Newly generated samples that pass the siamese selection will be used as an addition to the existing training dataset.

3.4 Oversampling

Because the number of newly generated samples after the selection process can be still relatively small, SMOTE oversampling is then applied to produce a balanced augmented training dataset. The number of newly generated samples required from SMOTE oversampling is the difference between the number of samples in the minority class and the number of samples in the majority class after taking into account the number of selected samples from the siamese network. The synthesized images created by SMOTE will be filtered using the same siamese network used in the previous step to ensure bad images from interpolation are not included in the new training dataset.

4 Experiment and Results

4.1 Datasets

This study uses several datasets to validate the proposed method, including MNIST and FMNIST, as well as three medical imaging datasets, namely skin cancer images from ISIC 2018 dataset, pneumonia images from Chest-Xray dataset, and breast cancer images from BreakHis dataset. For comparison with other methods, only binary classification has been considered in the experiments. For the datasets with more than two classes, only subsets with data from two classes are used.

MNIST (Modified National Institute of Standards and Technology) is a dataset of handwritten digits, consisting of greyscale images with a resolution of 28 X 28 pixels, with a total of 60,000 training images and 10,000 test images of 10 classes. For this study, the class labeled 3 is used as the majority class and the class labeled 8 is used as the minority class. Random selection of images is carried out in the minority class such that the imbalance ratio in this dataset is 1:25.

FMNIST is a greyscale image dataset on Zalando articles with a resolution of 28 X 28 pixels, which originally consists of 10 classes with a total of 60,000 training images and 10,000 test images. For this study, the class labeled 0 (T-shirt) is used as the majority class and the class labeled 1 (Trousers) is used as the minority class. Random selection of images is carried out in the minority class such that the imbalance ratio in this dataset is 1:40.

The skin cancer dataset used in the experiment is from ISIC Challenge Dataset 2018, which consists of 10,015 color images for training and 193 color images for testing. These images are classified into 7 classes, namely Melanocytic nevus, Melanoma, Benign keratosis (solar lentigo/seborrheic keratosis/lichen planus-like keratosis), Basal cell carcinoma, Actinic keratosis/Bowen's disease (intraepithelial carcinoma), Vascular lesion, Dermatofibroma. For this study, the melanocytic nevus class is used as the majority class and the melanoma class is used as the minority class with an imbalance ratio of 1:6.

The ChestXray dataset [19] consists of 5,863 X-Ray images of 2 categories (Pneumonia/Normal). 80% and 20% of the dataset are used as training and testing sets respectively. On this dataset, pneumonia images are the majority class while normal images are the minority class with an imbalance ratio of 1:1.28.

The last dataset used in this experiment consists of breast cancer color images from Breast Cancer Histopathological Image Classification (BreakHis) [20], which includes 9,109 microscopic pictures of 82 patients' breast tumor tissues magnified by different factors(40X, 100X, 200X, and 400X), with 2,480 benign and 5,429 malignant samples. All the medical image datasets are resized to 128 x 128 pixels.

4.2   Performance Metrics

Specific metrics are used to evaluate the model performance in image classification with imbalanced datasets, namely geometric mean (G-mean), F-measure (F1 score), and AUC. G-mean is defined as the squared root of the product of the sensitivity and specificity for binary classification:

$$G - mean = \sqrt{Sensitivity * Specificity} \tag{1}$$

F1 score, also known as the Dice similarity coefficient (DSC), is defined as the harmonic mean of precision and recall:

$$F1 - score = \frac{2TP}{2TP + FP + FN} \tag{2}$$

The Area Under the Receiver Operating Characteristic Curve (AUC-ROC) is a commonly used performance metric for evaluating the performance of a binary classifier, while the ROC curve is a plot of TPR (sensitivity) against FPR (1-specificity) for different threshold values.

4.3   Experimental Configuration

In our experiment, SIA-SMOTE is compared with three different oversampling techniques, i.e., random oversampling, SMOTE, and ASN-SMOTE [18] on five datasets. Imblearn's module for Python is used for SMOTE implementation and random oversampling. While ASN-SMOTE implementation is taken from the author's GitHub[1]. The number of nearest neighbors is set to 3, 5, 7, or 9 respectively for SMOTE and ASN-SMOTE.

Resnet-50 is used as the base network for the three-channel images in the siamese network in SIA-SMOTE, while the VGG network is used for images with a single channel. A VGG block is a convolutional neural network consisting

---

[1] https://github.com/yixinkai123/ASN-SMOTE/.

of a sequence of convolution layers with small filters and max pooling layers used for image classification tasks [21] Three VGG blocks with 32, 64, or 128 channels respectively are used for images with a single channel. The euclidean distance is used to measure how similar two images are to each other. The number of epochs for training the network is set to 100 and the batch number is 64, with contrastive loss as the cost function and RMSProp as the optimizer during model training. The following threshold values are used during the siamese network inference to choose new images created by the oversampling method: 0.02, 0.05, 0.1, 0.3, and 0.5.

The same classifiers as used in the siamese network are used for the classification task. Before oversampling, all images are flattened into a one-dimensional array. Using the scikit-learn package, five-fold stratified cross-validation is carried out to train and assess models with preserved class distribution for all datasets. Each fold is repeated three times to get a more reliable estimate of the overall performance. The classifiers and all oversampling methods are optimized using five-fold stratified cross-validation in all image datasets in terms of performance metrics. The SIA-SMOTE implementation is available at https://github.com/heroza/sia-smote.

4.4 Results

The average performances and the standard deviation of the four oversampling methods in terms of the three performance metrics for image classification on the validation dataset and testing dataset are shown in Table 1. From the results, it can be seen that SIA-SMOTE achieved the best overall performance among all methods and the best performance ranking on both validation and testing datasets as shown in Fig. 2, where the ranking is from 1 to 4, with 1 being the best.

To demonstrate how SIA-SMOTE can avoid selecting poor quality synthetic images for training data augmentation, Fig. 3 shows some examples of filtered-in and filtered-out synthetic images using the siamese network in SIA-SMOTE. It can be seen that some poor-quality synthetic images are filtered out by the siamese network so they will not become noise in the augmented training dataset.

4.5 Discussion

To make a more convincing conclusion about the performance of our proposed method, a statistical significance test is done on the performance of the validation set. The performance rankings are tested using Friedman Test as a nonparametric statistical test to analyze the differences among multiple methods' performance. If the test statistic is found to be significant, there is evidence to suggest that at least one of the groups or treatments differs significantly from the others, thus a post hoc test will be performed.

According to Friedman's test results in Table 2, p-values on all of the metrics are less than 0.05. Thus there is evidence to suggest that there are significant

Table 1. Comparison of performance of oversampling methods

Dataset	Method	G-mean	F-measure	AUC
MNIST				
Validation set	Random oversampling	99.134 ± 0.652	94.704 ± 2.577	99.137 ± 0.649
	SMOTE	99.184 ± 0.619	92.479 ± 4.584	99.187 ± 0.617
	ASN-SMOTE	99.001 ± 0.904	95.385 ± 2.918	99.007 ± 0.896
	SIA-SMOTE	**99.786 ± 0.320**	**96.818 ± 2.185**	**99.787 ± 0.318**
Test set	Random oversampling	99.752 ± 0.411	**97.899 ± 1.580**	99.753 ± 0.409
	SMOTE	99.614 ± 0.673	96.655 ± 3.434	99.617 ± 0.665
	ASN-SMOTE	99.249 ± 0.988	97.478 ± 1.740	99.257 ± 0.971
	SIA-SMOTE	**99.826 ± 0.329**	97.763 ± 2.169	**99.827 ± 0.328**
FMNIST				
Validation set	Random oversampling	**99.592 ± 0.658**	**96.875 ± 1.344**	**99.594 ± 0.652**
	SMOTE	99.579 ± 0.545	92.736 ± 4.821	99.581 ± 0.541
	ASN-SMOTE	98.073 ± 0.961	93.335 ± 3.164	98.092 ± 0.945
	SIA-SMOTE	99.247 ± 0.806	96.282 ± 2.617	99.253 ± 0.800
Test set	Random oversampling	96.325 ± 1.400	91.157 ± 2.387	96.393 ± 1.347
	SMOTE	97.564 ± 0.689	92.517 ± 4.515	97.590 ± 0.668
	ASN-SMOTE	99.059 ± 0.956	94.477 ± 5.266	99.067 ± 0.947
	SIA-SMOTE	**99.933 ± 0.065**	**97.464 ± 2.441**	**99.933 ± 0.065**
ISIC2018				
Validation set	Random oversampling	81.243 ± 2.101	73.734 ± 3.107	82.589 ± 1.783
	SMOTE	82.346 ± 1.454	74.406 ± 1.252	83.471 ± 1.179
	ASN-SMOTE	81.446 ± 3.131	67.523 ± 1.935	82.338 ± 2.473
	SIA-SMOTE	**83.034 ± 1.468**	**74.934 ± 2.374**	**84.036 ± 1.264**
Test set	Random oversampling	86.267 ± 5.103	80.969 ± 6.930	87.093 ± 4.501
	SMOTE	88.857 ± 3.436	83.170 ± 4.763	89.311 ± 3.118
	ASN-SMOTE	85.846 ± 4.761	74.000 ± 5.077	86.438 ± 4.226
	SIA-SMOTE	**89.045 ± 4.479**	**83.750 ± 4.015**	**89.578 ± 3.969**
ChestXray				
Validation set	Random oversampling	98.200 ± 0.516	97.499 ± 0.684	98.207 ± 0.513
	SMOTE	98.463 ± 0.519	97.633 ± 0.540	98.468 ± 0.515
	ASN-SMOTE	97.579 ± 0.623	95.324 ± 1.758	97.590 ± 0.616
	SIA-SMOTE	**98.477 ± 0.509**	**97.713 ± 0.644**	**98.480 ± 0.505**
Test set	Random oversampling	66.748 ± 2.845	61.498 ± 3.655	72.265 ± 1.916
	SMOTE	65.516 ± 2.194	59.994 ± 2.878	71.458 ± 1.487
	ASN-SMOTE	**70.626 ± 7.193**	**64.885 ± 8.209**	**74.722 ± 4.916**
	SIA-SMOTE	66.927 ± 3.550	61.779 ± 4.531	72.433 ± 2.380
Breakhis				
Validation set	Random oversampling	95.895 ± 0.525	94.602 ± 0.633	95.916 ± 0.515
	SMOTE	96.010 ± 0.308	94.581 ± 0.641	96.027 ± 0.306
	ASN-SMOTE	94.313 ± 1.394	91.599 ± 1.585	94.371 ± 1.336
	SIA-SMOTE	**96.073 ± 0.532**	**94.887 ± 0.815**	**96.095 ± 0.529**
Test set	Random oversampling	96.627 ± 0.394	95.224 ± 0.522	96.633 ± 0.391
	SMOTE	96.401 ± 0.527	94.758 ± 0.774	96.406 ± 0.526
	ASN-SMOTE	94.750 ± 1.348	91.880 ± 2.176	94.813 ± 1.295
	SIA-SMOTE	**96.648 ± 0.464**	**95.300 ± 0.663**	**96.655 ± 0.461**

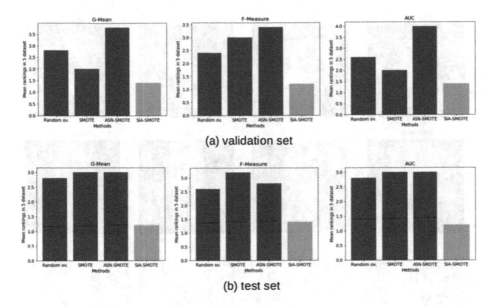

Fig. 2. Ranking of oversampling methods in terms of three performance metrics

Table 2. Friedman's test result

Metric	p-value
G-Mean	0.02110
F-Measure	0.04057
AUC	0.010891

Table 3. Conover test result

Method to be compared with SIA-SMOTE (sig. level after correction: 0.0083)	G-Mean		F-Measure		AUC	
	t-stat	p-value	t-stat	p-value	t-stat	p-value
Random oversampling	2.5849	0.0238	1.9639	0.0731	2.5980	0.0233
SMOTE	1.1078	0.2896	2.9459	0.0122	1.2990	0.2183
ASN-SMOTE	4.4312	**0.0008**	3.6005	**0.0036**	5.6291	**0.0001**

differences among the methods' performance. To further determine pairwise differences between our proposed method and the other methods, the Conover test is performed as a post hoc test with the significance level after Bonferroni correction 0.0083. According to Table 3, our proposed method is only significantly better than ASN-SMOTE with a p-value of 0.0008. While the other two tests produce p-values bigger than the significance level. This latter result might be due to our method's bad performance on the FMNIST dataset which needs to be analyzed further. Despite this, our method is still the best in terms of ranking for overall performance.

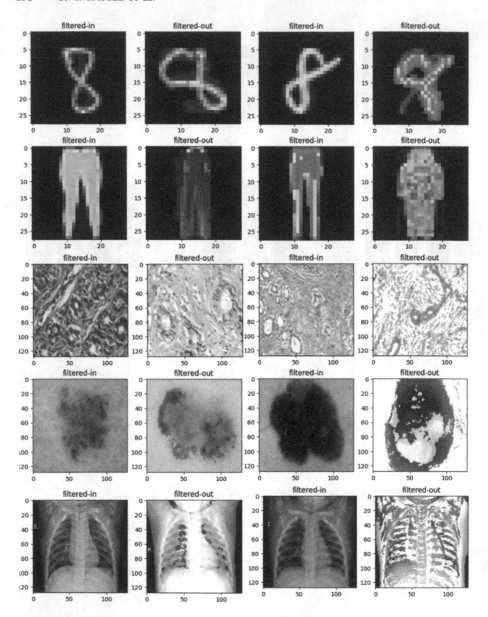

Fig. 3. Synthetic images filtered by the siamese network. The images in the first two columns are synthesized by interpolating in decision boundaries while images in the last two columns are synthesized by SMOTE

SIA-SMOTE successfully achieved better performance due to its ability to filter in synthetic samples of good quality and filter-out bad ones using the siamese network. This ability is reflected in two different occasions. The first occasion is in filtering synthetic images created by SMOTE oversampling. SMOTE is popular due to its effectiveness in oversampling imbalanced datasets. However, in a high-dimensional dataset, the interpolation step in SMOTE has a chance to generate noisy synthetic samples. Examples of bad images synthesized by interpolation in SMOTE can be seen in Fig. 3 in the fourth column. By using the siamese network, only good images, such as those in the third column, are selected for augmenting the training dataset. The second occasion is that the siamese network explores a new area that is not covered by SMOTE, such as in the decision boundary. By interpolating between the minority class and its nearest neighbor from the majority class, new images can be synthesized to capture data distribution of the minority class more effectively, as shown in Fig. 3, with filtered-in good synthetic images in the first column and filtered-out bad synthetic images in the second column.

5   Conclusion

This paper proposes SIA-SMOTE, an oversampling technique based on SMOTE that uses a siamese network to filter synthetic images. This method not only takes advantage of SMOTE oversampling, which is effective at maintaining data distribution of the minority class but also explores new areas in decision boundaries by using the siamese network in order to guarantee that only high-quality images are chosen as synthetic images for training data augmentation. Using the MNIST, FMNIST, and three medical image datasets, SIA-SMOTE is compared with random oversampling, SMOTE, and ASN-SMOTE. The experimental results demonstrate that SIA-SMOTE ranks top in the overall performance evaluation in terms of the G-mean, F-measure, and AUC metrics.

References

1. Dai, W., Li, D., Tang, D., Wang, H., Peng, Y.: Deep learning approach for defective spot welds classification using small and class-imbalanced datasets. Neurocomputing **477**, 46–60 (2022)
2. Sambasivam, G., Opiyo, G.D.: A predictive machine learning application in agriculture: cassava disease detection and classification with imbalanced dataset using convolutional neural networks. Egyptian Inf. J. **22**, 27–34 (2021)
3. Gao, L., Zhang, L., Liu, C., Wu, S.: Handling imbalanced medical image data: a deep-learning-based one-class classification approach. Artif. Intell. Med. **108**, 101935 (2020)
4. Zhao, C., Shuai, R., Ma, L., Liu, W., Wu, M.: Improving cervical cancer classification with imbalanced datasets combining taming transformers with T2T-ViT. Multimedia Tools Appl. 1–36 (2022). https://doi.org/10.1007/s11042-022-12670-0
5. Feng, W., et al.: Dynamic synthetic minority over-sampling technique-based rotation forest for the classification of imbalanced hyperspectral data. IEEE J. Sel. Top. Appl. Earth Obs. Remote Sens. **12**, 2159–2169 (2019)
6. Wang, Y., Cui, L., Zhang, C., Chen, W., Xu, Y., Zhang, Q.: A two-stage seismic damage assessment method for small, dense, and imbalanced buildings in remote sensing images. Remote Sens. **14**(4), 1012 (2022)

7. Chen, J., Huang, H., Cohn, A.G., Zhou, M., Zhang, D., Man, J.: A hierarchical DCNN-based approach for classifying imbalanced water inflow in rock tunnel faces. Tunnelling Undergr. Space Technol. **122**, 104399 (2022)

8. Buda, M., Maki, A., Mazurowski, M.A.: A systematic study of the class imbalance problem in convolutional neural networks. Neural Netw. **106**, 249–259 (2018)

9. Ozdemir, A., Polat, K., Alhudhaif, A.: Classification of imbalanced hyperspectral images using smote-based deep learning methods. Exp. Syst. Appl. **178**, 114986 (2021)

10. Branco, P., Torgo, L., Ribeiro, R.P.: A survey of predictive modelling under imbalanced distributions. Assoc. Comput. Mach. **49**, 5 (2016)

11. Sandfort, V., Yan, K., Pickhardt, P.J., Summers, R.M.: Data augmentation using generative adversarial networks (cyclegan) to improve generalizability in CT segmentation tasks. Sci. Rep. **2019** 9:1–9:1–9 (2019)

12. Wang, S., Liu, W., Wu, J., Cao, L., Meng, Q., Kennedy, P.J.: Training deep neural networks on imbalanced data sets. In: Proceedings of the International Joint Conference on Neural Networks, pp. 4368–4374, October 2016

13. Dong, Q., Gong, S., Zhu, X.: Class rectification hard mining for imbalanced deep learning. In: IEEE International Conference on Computer Vision, pp. 1869–1878 (2017)

14. Huang, C., Li, Y., Loy, C.C., Tang, X.: Learning deep representation for imbalanced classification. In: Proceedings of the IEEE Conference on Computer Vision and Pattern Recognition (2016)

15. Tanha, J., Abdi, Y., Samadi, N., Razzaghi, N., Asadpour, M.: Boosting methods for multi-class imbalanced data classification: an experimental review. J. Big Data **7**(1), 1–47 (2020). https://doi.org/10.1186/s40537-020-00349-y

16. Chawla, N.V., Bowyer, K.W., Hall, L.O., Kegelmeyer, W.P.: Smote: synthetic minority over-sampling technique. J. Artif. Intell. Res. **16**, 321–357 (2002)

17. Bromley, J., Guyon, I., LeCun, Y., Säckinger, E., Shah, R., At&t Bell, and Laboratories Holmdel.: Signature verification using a "siamese" time delay neural network. In: Advances in Neural Information Processing Systems, vol. 6 (1993)

18. Yi, X., Xu, Y., Hu, Q., Krishnamoorthy, S., Li, W., Tang, Z.: ASN-SMOTE: a synthetic minority oversampling method with adaptive qualified synthesizer selection. Complex Intell. Syst. 1–26 (2022). https://doi.org/10.1007/s40747-021-00638-w

19. Kermany, D.S., et al.: Identifying medical diagnoses and treatable diseases by image-based deep learning. Cell **172**, 1122–1131.e9 (2018)

20. Spanhol, F.A., Oliveira, L.S., Petitjean, C., Heutte, L.: A dataset for breast cancer histopathological image classification. IEEE Trans. Biomed. Eng. **63**, 1455–1462 (2016)

21. Simonyan, K., Zisserman, A.: Very deep convolutional networks for large-scale image recognition. In: 3rd International Conference on Learning Representations, ICLR 2015 - Conference Track Proceedings, September 2014

On the Use of First and Second Derivative Approximations for Biometric Online Signature Recognition

Marcos Faundez-Zanuy[1]([✉]) [iD] and Moises Diaz[2] [iD]

[1] Tecnocampus, Universitat Pompeu Fabra, 08302 Mataró, Spain
faundez@tecnocampus.cat
[2] Universidad de Las Palmas de Gran Canaria, Las Palmas de Gran Canaria, Spain

Abstract. This paper investigates the impact of different approximation methods in feature extraction for pattern recognition applications, specifically focused on delta and delta-delta parameters. Using MCYT330 online signature database, our experiments show that 11-point approximation outperforms 1-point approximation, resulting in a 1.4% improvement in identification rate, 36.8% reduction in random forgeries and 2.4% reduction in skilled forgeries.

Keywords: online handwriting · e-security · dynamic time warping · derivatives

1 Online Signature Biometric Recognition

Signatures are a widely utilized biometric modality in e-security systems based on the premise of "something you can do" [1, 14]. Probably one of its main advantages is that the user can decide to change his signature when it is compromised. Unfortunately, this is not possible with most biometric traits such as face, speech, iris, etc. In addition, it has a long tradition of centuries as an authentication method in legal contracts, paintings, etc., and it can play an important role in health assessment too [2, 14]. Several different handwritten tasks can be used [3].

Online biometric recognition can operate in two different ways:

a) Identification (1:N): The goal is to compare a given signature with the N models stored in a database of N users. Usually, the model that best fits the input signature indicates the identified user.

b) Verification (1:1): A user provides his signature. Then his claimed identity and the system tries to guess if he is a genuine or forger user [14]. Signature databases may contain two distinct types of forgeries, namely, random and skilled. In the skilled forgery type, the forger deliberately attempts to imitate the genuine signature. Conversely, in the random forgery type, the forger utilizes their own signature as a replacement for the genuine signature..

In this paper, we evaluate the relevance of delta and delta-delta parameter approximation in the feature extraction block.

I. Rojas et al. (Eds.): IWANN 2023, LNCS 14134, pp. 461–472, 2023.
https://doi.org/10.1007/978-3-031-43085-5_36

The structure of the paper is as follows: Sect. 2 provides a comprehensive review of the relevant literature. Section 3 details the essential components of the recognition system, with a specific focus on feature extraction, including the employed databases, normalization techniques, and distance computation algorithm. Section 4 presents the empirical findings, while Sect. 5 summarizes the key outcomes and conclusions derived from this study.

2   Related Works

The use of derivatives in parametrization, i.e., delta and delta-delta parameter, is a common technique in online signature verification. It can provide additional information about the signature and help to improve recognition accuracies.

There are several different methods that have been proposed for the use of derivatives in online signature verification, including the use of velocity, acceleration, and jerk features [14].

In [15], the authors used delta coordinate differences between two consecutive points in x-y coordinates, achieving the lowest error rates with the delta parameter. This represented the velocity in x-y coordinates independently. In [16], the benefits of using the kth order derivative were described, where the authors only worked out the first and second derivative sequences of vectors. In the last signature verification competition [17], some participants used delta, such as DLVC-Lab or SIG team, who used the first and second-order derivatives. Similar features were selected by the authors of [18], who chose 12 features including the first- and second-order derivatives (delta and delta-delta, respectively), and applied a DTW algorithm for time sequence matching.

Typically, the derivatives were computed by subtracting two consecutive sampling points. However, [8] found good results using a second-order regression, which had the advantage of output features having the same length as the non-derivative ones. This regression formula was also followed in [19], where it was applied to robotic features estimated from x-y.

To the best of our knowledge, related works mostly use the first and sometimes the second derivative, but only few works have gone beyond a simple derivative operation with delta and delta-delta parameters.

3   Experimental Setup

The general pattern recognition system depicted in Fig. 1 is indeed suitable for signature recognition applications. It consists of four blocks, described next.

Block 1 (registered signature): digitizing tablet is used for online signature acquisition. In our case, we have used a well-known pre-existing database, which is MCYT [4], and is summarized in Sect. 3.1. The database is split into two parts: the training set, used for user model computation, and the testing set, used to provide experimental recognition rates.

Block 2 (Feature extraction): The digitizing tablet provides x, y, p, al, az information as well as a time stamp code. Knowing that the sample rate is 200 samples per second, the feature set can be extended as described in Sect. 3.2.

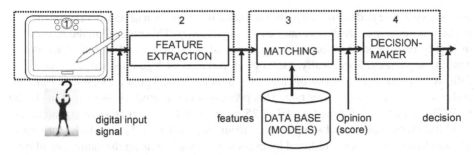

Fig. 1. General pattern recognition system

Block 3 (Matching): In this paper, we use the dynamic time warping (DTW) distance computation [5], which has been widely used as a successful recognition technique and is summarized in Sect. 3.3.

Block 4 (Decision maker): We used the DETWare V2.1 NIST toolbox [6, 7] for identification and verification assessments.

3.1 MCYT Database

The MCYT signature database was acquired using a WACOM graphic tablet, with a sampling frequency of 100 Hz. Each sampled instance of the signature contains the following information:

1) Position along the x-axis, x: [0–12 700], equivalent to 0–127 mm;
2) Position along the y-axis, y: [0–9700], equivalent to 0–97 mm;
3) Pressure, p, applied by the pen: [0–1024];
4) Azimuth angle, az, of the pen with respect to the tablet (as shown in Fig. 2): [0–3600], equivalent to 0–360°;
5) Altitude angle, al, of the pen with respect to the tablet (as shown in Fig. 2): [300–900], equivalent to 30–90°.

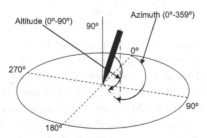

Fig. 2. Azimuth and altitude angles of the pen with respect to the plane of the WACOM tablet.

The MCYT signature database comprises a total of 330 users, each of whom contributed 25 genuine signatures and 25 skilled forgeries. The skilled forgeries were produced by the subsequent five target users, who were provided with static images of the genuine signature and instructed to imitate them (for at least ten times) in a natural manner without introducing any discontinuities or irregularities. This resulted in the acquisition of highly realistic skilled forgeries that captured the shape-based natural dynamics of genuine signatures. Specifically, user n produced a set of 5 genuine signature samples, followed by 5 skilled forgeries of client n–1, another set of 5 genuine signature samples, and 5 skilled forgeries of user n–2. This process was repeated for users n–3, n–4, and n–5. Thus, user n contributed a total of 25 samples of their own genuine signature and 25 skilled forgeries (5 final samples each from users n–1 to n–5). Similarly, for user n, 25 skilled forgeries were produced by users n + 1 to n + 5 using the same procedure.

3.2 Feature Extraction and Normalization

Each signature followed the next process:

1. The spatial coordinates (x, y) are normalized by the centroid or geometric center $([\bar{x}, \bar{y}])$ (1) of each signature. Thus, the center of each signature is displaced to $(0, 0)$ coordinate (2) as follows.

$$[\bar{x}, \bar{y}] = \frac{1}{L} \sum_{l=1}^{L} [x_l, y_l] \qquad (1)$$

$$x_l = x_l - \bar{x}, y_l = y_l - \bar{y} \qquad (2)$$

2. Feature set (f) is enlarged by working out the delta (first derivate) and delta-delta (second derivate) parameters. Thus, from the five feature set provided by the Intuos Wacom digitizing tablet $f = [x, y, p, az, al]$, we obtain an eight dimension feature set $f = [x, y, p, \dot{x}, \dot{y}, \dot{p}, \ddot{x}, \ddot{y}]$. This optimal feature set was obtained in [8], which discards the angles information: az, al.

Delta parameters $(\dot{f_i})$ and delta-delta $(\ddot{f_i})$ features are the first and second derivative, respectively, for $i \in [1, 5]$. The delta parameters are obtained in the following way [9]:

$$delta = \frac{\sum_{k=-M}^{M} k \cdot x[k]}{\sum_{k=-M}^{M} k^2} \qquad (3)$$

The delta value of a feature x is an estimation of the local slope of a region that is centered on sample $x[k]$ and spans M samples before and after the current sample. This approximation is obtained through the method of least squares. The size of the region is defined by the delta window length, which extends from $-M$ to M. The length of the delta window is determined by an odd integer that is equal to or greater than three. On the other hand, we can calculate a simple derivative based on one sample difference $\dot{f_i}[l] = f_i[l] - f_i[l - 1]$, which can be obtained with de MATLAB *diff* function and is equivalent to the basic definition of derivative Eq. (4):

$$\dot{f_i} = \lim_{h \to 0} \frac{f_i(l + h) - f_i(l)}{h} \qquad (4)$$

for h equal to one sample (5):

$$\dot{f}_i = \frac{f_i(l+1) - f_i(l)}{1} = f_i(l+1) - f_i(l) \tag{5}$$

In this case, we add one zero at the beginning of the sequence in order to obtain the same length L for the feature set and its derivative.

Delta-delta is obtained by applying two consecutive times the delta equation.

We have used the audioDelta function from MATLAB in the audio toolbox. It is worthy to mention that delta and delta-delta parameters have a long tradition in speaker and speech recognition [9]. For this reason, they were probably included in the audio toolbox. However, their potential is probably underexploited in handwritten analysis.

3. Features are normalized through a z-score using the following equation, where each feature f_i is subtracted by its mean and divided by its standard deviation.

$$\hat{f}_i = \frac{f_i - \overline{f}_i}{\text{std}(f_i)} \tag{6}$$

Figure 3 shows a sample signature coordinates (x, y) before and after normalization. We can observe that the shape of the signature is preserved. In addition, this z-score normalization makes unnecessary the centroid normalization described before.

Figure 4 shows the dynamic information of the signature of one user. From top to bottom: x-axis values (x), y-axis (y), pressure (p), azimuth (az) and altitude (al). In this specific case, the signature length is 786 samples. We observe that the angles az and al are defined with less bits and they show small variations. Thus, lesser discriminative information can be expected from them.

In order to emphasize the importance of bilateral implementation of derivative rather than relying on a single difference (Eq. 5), we have generated a synthetic signal and applied the single difference approach and bilateral with 15 points over the signal x generated in MATLAB with the next Eq. (7):

$$x = [1 : 200, 200 * \text{ones}(1, 200)] + \text{rand}(1, 400); \tag{7}$$

Equation 7 consists of a ramp of 200 samples followed by a flat region of 200 samples. Then, a Gaussian noise is added to each sample.

Figure 5 shows from top to bottom: the original signal x (Eq. 7), the normalized signal after applying Eq. 6 (xn), the normalization of the first derivative obtained with Eq. (4), and the first derivative with the audioDelta function and a window of 15 points. From this figure, we can clearly see the high impact of noise on the simple derivative approximation.

Worth to mention that fractional order derivatives are an active research field that provides good experimental results, especially in e-health based on handwritten tasks [10–12]. However, a deeper analysis must be done to discover if the improvement is due to the fractional order or the enlarged window analysis for derivative computation.

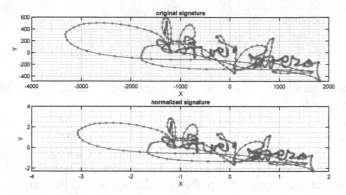

Fig. 3. Shape of a signature with raw coordinates and after the z-score normalization

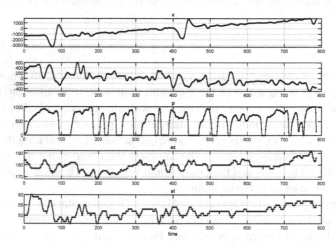

Fig. 4. Dynamic information of the signature of one user (x, y, p, az, al).

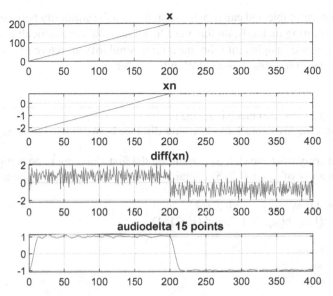

Fig. 5. Example of a synthetic signal x, normalized signal, first derivative approximated by two consecutive samples subtraction and first derivative with a 15-point window.

3.3 Dynamic Time Warping

In this study, feature matching with models is conducted by employing dynamic time warping (DTW), a well-known template matching algorithm that is highly effective in handling the random variations that arise due to intra-user variability [13]. DTW employs a dynamic programming approach to generate an elastic distance metric between two samples, regardless of any differences in their lengths. The algorithm is summarized in Fig. 6.

In addition, the DTW is normalized by the length of the first signature. Otherwise, those users with shorter signatures would tend to obtain smaller distance values. For this paper, we have used the standard DTW algorithm found in the signal processing toolbox in MATLAB.

Dynamic Time Warping (DTW) has several advantages over a machine learning approach for online signature recognition:

1. DTW is a non-parametric method, which means that it does not assume a specific functional form for the signature. This makes it more flexible than machine learning methods, which typically rely on a fixed model structure or distribution assumption.
2. DTW is robust to variations in speed and timing between signatures. In contrast, machine learning methods may require careful preprocessing or feature engineering to account for variations in timing or speed.
3. DTW is computationally efficient and does not require large amounts of training data. It can be applied to new signatures in real-time, making it suitable for online signature recognition applications. Machine learning methods, on the other hand, may require large amounts of training data and can be computationally expensive during both training and prediction phases.

4. DTW is interpretable and can provide insights into the similarity between signatures. Machine learning methods, on the other hand, may be more difficult to interpret and may not provide insights into the underlying similarity structure of the data.This is one of the main reasons to use DTW in this paper.

```
function dist = dtw_distance(sig1, sig2)
% Computes the Dynamic Time Warping distance between two online signatures.

% sig1: an array of shape (L1, 5) containing the first signature to be compared
% sig2: an array of shape (L2, 5) containing the second signature to be compared

% Initialize DTW matrix with zeros
dtw = zeros(length(sig1)+1, length(sig2)+1);

% Initializations
dtw(:, 1) = ∞;
dtw(1, :) = ∞;
dtw(1, 1) = 0;

% Compute DTW matrix
for i = 2:length(sig1)+1
   for j = 2:length(sig2)+1
      dist = norm(sig1(i-1,:) - sig2(j-1,:));
      dtw(i,j) = dist + min([dtw(i-1,j), dtw(i,j-1), dtw(i-1,j-1)]);
   end
end

% DTW distance (dist) is the bottom-right element of the DTW matrix
dist = dtw(end, end);
```

Fig. 6. DTW algorithm.

4 Experimental Results

We have obtained a user model based on the first five training signatures. Then we performed three sets of experiments:

1. Identification: we have used five different testing signatures not used during the training process. Each testing signature was matched against all the models, which consisted of five different signatures. The minimum DTW distance from each testing signature to the five training signatures is selected (see Eq. 8). Then, the user that provides minimum distance is selected as the identified user (see Eq. 9), where $user \in [1, 330]$.

$$Distance_{user} = min(DTW(train1_{user}, test), \cdots, DTW(train5_{user}, test)) \quad (8)$$

$$user = min(Distance_1, \cdots, Distance_{user}, \cdots, Distance_{330}) \qquad (9)$$

2. Verification with random forgeries: genuine user scores are obtained with DTW distances from a given user to its own model, while impostor scores are obtained with the DTW distances from a given user to the other's model. As the largest the DTW distance, the more different the signatures, we changed the sign of the DTW distance to convert from distance to score, as follows:

$$score_{user} = -Distance_{user} \qquad (10)$$

The score set is the same as in the identification mode, but the addressed question is different. In verification, no identity is provided. Therefore, we modified the acceptance/rejection response, based on a decision threshold, which is adjusted by trial and error. This procedure implies a total amount of 330×5 genuine signatures, which produce the same amount of genuine distances, and $330 \times 329 \times 5$ random forgeries (distances).

3. Verification with skilled forgeries. The process is analogous to the random forgeries. The genuine distances are the same of previous section. A new set of 330×25 distances are computed using the skilled forgeries.

We have repeated the whole set of experiments for a variable number of points used in the computation of the derivatives ranging from 1 to 15 (only odd numbers).

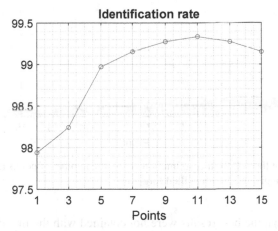

Fig. 7. Identification rate as function of the number of points used to work out the delta and delta-delta parameters.

Figure 7 shows the identification rates as a function of the number of points. In this case, we measured the accuracy of the identification task. Therefore, the higher the rate, the more precise the identification. As we can see, the highest value was achieved by using a window with 11 points.

On the other hand, Fig. 8 shows the verification errors (minimum of the detection cost function or min(DCF) [6, 7]) for random forgeries, and Fig. 9 the min(DCF) for skilled forgeries.

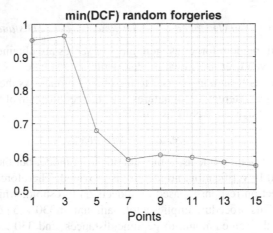

Fig. 8. Minimum of the detection cost function versus the number of points used to work out the delta and delta-delta parameters for random forgeries.

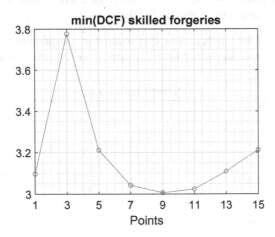

Fig. 9. Minimum of the detection cost function versus the number of points used to work out the delta and delta-delta parameters for skilled forgeries.

We observe that the best results were not obtained with the first delta in both verification cases. In the case of random forgery, an elbow effect was achieved in the plot. Although seven points change the rhythm of the plot, we can see a slight improvement when the number of points in the windows increases.

Observing the skilled forgeries results, we notice an "U" plot, which describes a minimum performance when nine points were used. In addition, a strange result was obtained when the delta and delta-delta were configured with a single point. It can confirm the extended use of one point in the derivates to extract features in different signature verification proposals in the literature.

Nevertheless, increasing the number of points in the delta and delta-delta negatively affects the execution time. For this reason, this is a trade-off problem between performance and execution optimization.

5 Conclusions

This article has shown the importance of configuring the delta (first derivative) and delta-delta (second derivative) to extract features in biometric signature recognition. Furthermore, we have analyzed the performance in several cases of delta and delta-delta for identification and verification problems, including random and skilled forgeries.

Looking at Figs. 7, 6 and 7, we observe that the experimental implementation of the derivative significantly impacts the final performance. The simplest one based on a single point is not good enough. On the other hand, there is a limit on the number of points. After a certain value, the inclusion of more points in the computation window produces a drop in the results.

In view of previous results, we recommend using nine or eleven points for the computation of derivatives. Using eleven points we obtain a relative improvement of 1.4% in identification rate, 36.8% in random forgeries and 2.4% in skilled forgeries over the one point approximation.

Acknowledgments. This work has been supported by MINECO Spanish grant number PID2020-113242RB-I00, and PID2019-109099RB-C41.

References

1. Faundez-Zanuy, M.: Biometric security technology. IEEE Aerosp. Electron. Syst. Mag. **21**(6), 15–26 (2006). https://doi.org/10.1109/MAES.2006.1662038
2. Faundez-Zanuy, M., Fierrez, J., Ferrer, M.A., et al.: Handwriting biometrics: applications and future trends in e-security and e-health. Cogn. Comput. **12**, 940–953 (2020). https://doi.org/10.1007/s12559-020-09755-z
3. Faundez-Zanuy, M., Mekyska, J., Impedovo, D.: Online handwriting, signature and touch dynamics: tasks and potential applications in the field of security and health. Cogn. Comput. **13**, 1406–1421 (2021). https://doi.org/10.1007/s12559-021-09938-2
4. Ortega-Garcia J., et al.: MCYT baseline corpus: a bimodal biometric database. IEE Proc. Vis. Image Sign. Process. **150**(6), 395–401 (2003). https://doi.org/10.1049/ip-vis:20031078
5. Faundez-Zanuy, M.: On-line signature recognition based on VQ-DTW. Pattern Recogn. **40**(3), 981–992 (2007). ISSN 0031-3203, https://doi.org/10.1016/j.patcog.2006.06.007
6. Martin, A., Doddington, G., Kamm, T., Ordowski, M., Przybocki, M.: The det curve in assessment of detection performance. In: Proceedings of the European Conference on Speech Communication and Technology, pp. 1895–1898 (1997)
7. https://www.nist.gov/itl/iad/mig/tools
8. Fischer, A., Diaz, M., Plamondon, R., Ferrer, M.A.: Robust score normalization for DTW-based online signature verification. In: 2015 13th International Conference on Document Analysis and Recognition (ICDAR), Tunis, Tunisia, pp. 241–245 (2015). https://doi.org/10.1109/ICDAR.2015.7333760

9. Rabiner, L.R., Schafer, R.W.: Theory and Applications of Digital Speech Processing. Upper Saddle River, NJ: Pearson (2010)
10. Mucha, J., et al.: Identification and monitoring of Parkinson's disease dysgraphia based on fractional-order derivatives of online handwriting. Appl. Sci. **8**, 2566 (2018). https://doi.org/10.3390/app8122566
11. Mucha, J., et al.: Fractional derivatives of online handwriting: a new approach of parkinsonic dysgraphia analysis. In: 2018 41st International Conference on Telecommunications and Signal Processing (TSP), Athens, Greece, pp. 1–4 (2018). https://doi.org/10.1109/TSP.2018.8441293
12. Mucha, J., et al.: Analysis of Parkinson's disease dysgraphia based on optimized fractional order derivative features. In: 2019 27th European Signal Processing Conference (EUSIPCO), A Coruna, Spain, pp. 1-5 (2019).https://doi.org/10.23919/EUSIPCO.2019.8903088
13. Deller, J.R., Proakis, J.G., Hansen, J.H.L.: Dynamic time warping. In: Discrete-time processing of speech signals, New York: Macmillan Publishing Co. (1993)
14. Diaz, M., Ferrer, M.A., Impedovo, D., Malik, M.I., Pirlo, G., Plamondon, R.: A perspective analysis of handwritten signature technology. Acm Comput. Surv. (Csur) **51**(6), 1–39 (2019)
15. Kholmatov, A., Yanikoglu, B.: Identity authentication using improved online signature verification method. Pattern Recogn. Lett. **26**(15), 2400–2408 (2005)
16. Sae-Bae, N., Memon, N.: Online signature verification on mobile devices. IEEE Trans. Inf. Forensics Secur. **9**(6), 933–947 (2014)
17. Tolosana, R., et al.: ICDAR 2021 competition on on-line signature verification. In: Lladós, J., Lopresti, D., Uchida, S. (eds.) ICDAR 2021. LNCS, vol. 12824, pp. 723–737. Springer, Cham (2021). https://doi.org/10.1007/978-3-030-86337-1_48
18. Jiang, J., Lai, S., Jin, L., Zhu, Y.: DsDTW: local representation learning with deep soft-DTW for dynamic signature verification. IEEE Trans. Inf. Forensics Secur. **17**, 2198–2212 (2022)
19. Diaz, M., Ferrer, M.A., Quintana, J.J.: Robotic arm motion for verifying signatures. In: 2018 16th International Conference on Frontiers in Handwriting Recognition (ICFHR), pp. 157–162. IEEE (2018)

A Deep Transfer Learning Approach to Support Opportunistic Wearable Activity Recognition

Oresti Banos[1]([✉])([iD]), David Gil[2]([iD]), Javier Medina[1]([iD]), Adrian Sanchez[1], and Claudia Villalonga[1]([iD])

[1] Research Center for Information and Communication Technologies, University of Granada, Granada, Spain
{oresti,javiermq,cvillalonga}@ugr.es, adrisanchez@correo.ugr.es
[2] Department of Computer Technology and Computation, University of Alicante, Alicante, Spain
dgil@dtic.ua.es

Abstract. Most wearable activity recognition systems are defined to be used for a specific sensor setup. However, changes in the body sensor network are sometimes experienced due to sensor failures or upgrades. In such cases, the default activity recognition models are no longer applicable and a complete retraining of the system is normally needed, which is both time and resource-consuming. In this work, we present a deep transfer learning approach to automatically instruct new unseen wearable sensors by using the recognition capabilities of the existing activity recognition models used for the default sensor setup. The proposed approach is validated in a popular wearable activity recognition dataset, yielding quite promising results.

Keywords: Deep learning · Transfer learning · Wearable sensors · Activity recognition

1 Introduction

Wearable activity recognition refers to the process of using sensor data from wearable devices to detect and classify physical activities performed by individuals. With the increasing popularity of wearable devices such as smartwatches, fitness trackers and smartphones, the ability to use these devices to monitor and track physical activity has become more accessible than ever before. The recognition of human activities through wearable devices has numerous applications, such as health monitoring [16], sports performance analysis [13], and workload tracking [12]. By analysing sensor data, such as accelerometer and gyroscope measurements, activity recognition from wearable devices can provide valuable information about a person's daily physical activity patterns and help them make more informed decisions about their health and overall well-being.

The wearable activity recognition sensor setup is subject to changes. For example, sensors can suffer from failures or faults that may be irrecoverable.

I. Rojas et al. (Eds.): IWANN 2023, LNCS 14134, pp. 473–482, 2023.
https://doi.org/10.1007/978-3-031-43085-5_37

In such case, a replacement of the affected sensor by a new one of similar or close characteristics may be required to restore the system to full operating capacity. Wearable sensors could also be newly incorporated into the default sensing infrastructure. The addition of new sensors could be part of a specific system upgrade, e.g. to enhance the recognition accuracy or provide network redundancy. More commonly, users may acquire new gadgets or devices to benefit from other services not supported by the currently employed systems. This poses a new sensor setup configuration that is hardly foreseeable during the design phase. Consequently, default activity recognition systems may not directly leverage the data obtained through the new sensor devices. Moreover, different sensor configurations are envisioned during a user's normal day. Depending on the particular context, users may wear specific garments (e.g. at work) or specific accessories (e.g. at the gym). During all these situations, sensors may be removed, substituted or newly added.

Most wearable activity recognition systems are trained on sensor data streams from datasets collected at design time with predefined and optimal sensor configurations. Thus, accounting for these variations would require collecting as many datasets as possible sensor configurations, which happens to be unfeasible. Hence, activity recognition systems should implement mechanisms to autonomously adapt and self-configure their models to the actual sensing configuration. Activity recognition systems should be also defined to intelligently leverage those sensors that happen to be available to the user. Moreover, depending on the particular application, the use of part of the sensing infrastructure could be preferred. This may help to minimize energy and resource consumption, as well as reduce systems simplicity during execution time.

Some approaches have been proposed to deal with the challenging scenarios posed by the varying wearable sensor setups. For example, auto-calibration techniques have been demonstrated fairly applicable for some cases involving sensor displacement [7]. Combinations of multiple sensor modalities can also help tolerate sensor displacement [11] or to substitute sensor modalities [10]. Ensemble approaches have been proposed to exploit the recognition capabilities of the remaining unaffected sensor network to cope with variations in the infrastructure due to sensor failures [2] and sensor displacement [3,6]. An important limiting factor of the above approaches is that they are generally constrained to foreseen run-time variations.

In our prior work, we proposed an approach to transfer recognition capabilities even for unseen variations [1]. The approach relies on the learning of a mapping between existing and new sensor signals through the use of system identification techniques. Once the mapping is learned, the existing activity recognition models are translated to be used on the signals estimated from the new sensor data. While this approach showed good results, it is somewhat limited by the generalisation capacity of the considered system identification models to estimate the signals of the newcomer sensor. In this work, we aim to increase such generalisation capacity by using deep neural networks (DNN) as universal approximators. In doing so, we intend to augment the capabilities of the signal mapping model as to provide a more precise estimation of the signals of the existing sensor from the new sensor data, which in turn potentially leads to a more accurate use of the existing recognition models.

The remainder of this paper is structured as follows. In Sect. 2, we present the methods used in our study for both learning the sensor signal mappings and performing the activity classification. Section 3 describes the evaluation of the proposed model, including the experimental setup and report on the results. In Sect. 4, we provide a discussion of our results, highlighting their implications and limitations. Finally, in Sect. 5, we present our conclusions, summarising the main contributions of this study and their significance for the field, as well as suggestions for future research.

2   Methods

The proposed transfer method works in two steps. First, a model is learned to find a function that maps the signals of the new sensor to the signals of an existing or default sensor. In other words, the signal mapping learning is defined as the process of obtaining the transfer function that transforms the signals of the new to the existing sensor. Based on this mapping, the activity recognition system is then transferred. This process basically consists in conveying or copying the activity model trained for the existing sensor to the new one. To operate on this activity recognition model the new system needs to map its signals to look alike the existing sensor signals for which the activity model was originally defined. In machine learning terms, we could say that the first step corresponds to a regression problem while the second step refers to a classification problem.

As has been said before, the signal mapping is approached as a regression problem in which the signals of the existing sensor are estimated from the signals measured by the newly added sensor. To that end, DNN are used given its potential as universal approximators. In particular, Long Short-Term Memory (LSTM) are considered, which is a type of recurrent neural network architecture. Unlike traditional feedforward neural networks that process data sequentially, LSTM networks have a memory mechanism that allows them to selectively store, retrieve, and update information over time. This makes them particularly suitable for modelling sequential data such as time series, thus of interest for our signal mapping problem. Four popular DNN architectures implementing LSTM are considered: (a) LSTM; (b) CNN+LSTM; (c) LSTM+Dense; and (d) DeepConv+LSTM. LSTM is the realisation of a plain and simple LSTM neural network. CNN+LSTM adds convolutional neural network (CNN) layers at the input to split the signal sequences into smaller chunks from which features are extracted and then fed into the LSTM. LSTM+Dense adds fully connected layers that map the output from the previous layer to a set of output neurons, with the Dense layers used for the final prediction. DeepConv+LSTM is a custom-made DNN which comprises convolutional, recurrent and softmax layers, which has been shown to offer good activity recognition results from wearable sensor data [14]. The above architectures are shown in Fig. 1.

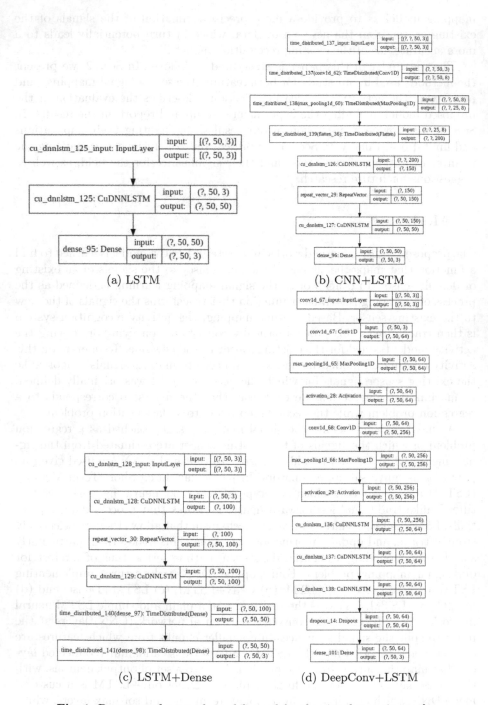

Fig. 1. Deep neural network models used for the signal mapping task.

The classification task is performed via a classic machine learning model. Namely, a k-nearest neighbor (KNN) classifier is considered. The main reason for using this model is due to the high accuracy already proven in previous activity recognition problems.

3 Evaluation

3.1 Experimental Setup

To evaluate the proposed approach, we used REALDISP, one of the most widely used publicly available wearable activity recognition datasets [4]. This dataset contains motion data, including acceleration, rate of turn, and magnetic field orientation, recorded from 17 volunteers performing 33 physical activities. The motion was captured using nine inertial sensors (Sk) attached to different body parts ($S1$: right lower arm; $S2$: right upper arm; $S3$: back; $S4$: left upper arm; $S5$: left lower arm; $S6$: right calf; $S7$: right thigh; $S8$: left thigh; $S9$: left calf). For the sake of simplicity, we considered in our study a subset of 10 activities similar to the one used in [6]: "Walking"; "Jump up"; "Jump rope"; "Trunk twist"; "Waist rotation"; "Upper trunk and lower body opposite twist"; "Frontal crossing of arms"; "Arms inner rotation"; "Knees bending (crouching)"; and "Cycling". We also strictly focused on the acceleration data as it has been the most commonly used sensor modality in previous activity recognition works. One important strength of this dataset lies in the variety of activities, body parts, intensities, and dynamicity captured during data collection in the wild, with no constraints imposed on how the activities should be executed, therefore representing a perfect test bench for our approach.

The sensor signal regression and activity classification models devised for evaluation are described next. No preprocessing of the data is applied to avoid the removal of any relevant information. This is normal practice when the activities are of a diverse nature. A sliding window approach is considered for the partitioning of the sensor data streams. A window size of 1 s is particularly used as recommended in [5]. These windows are used both for learning the mappings and also as instances for the activity classification task. The regressors used for the sensor signal mappings are the ones already presented in Fig. 1. A total of 20 epochs are considered for learning the DNN models to avoid overfitting. A model is learned for each of the following mappings ($Sk \rightarrow Sp$ designates that the signals of the sensor Sp are estimated from the signals of the sensor Sk): $S1 \rightarrow S2$; $S9 \rightarrow S7$; $S4 \rightarrow S8$; and $S3 \rightarrow S2$. These combinations are selected to evaluate the capacity of the models to learn a mapping in a favourable situation (i.e. sensors located close-by) and also in a more challenging scenario (i.e. sensors located in non-adjacent positions). For the activity classifiers, a model similar to the one proposed in [9] is used here. The k-value for the KNN model is set to three given the good results shown in prior related works [2,3,6].

The evaluation of the transfer learning approach is based on a leave-one-subject-out cross-validation both for the signal mapping and activity classification. To avoid data imbalancement artifacts, the $F_1\text{-}score$ metric [15] is used to evaluate the performance of the activity classification systems while the root mean square (RMSE) is used for assessing the quality of the signal mappings.

The models are implemented using Python standard data processing and machine learning libraries, namely Pandas, Numpy and Keras. In particular, the Nvidia cuDNN library [8] is used to speed up the learning of the DNN models.

3.2   Results

Firstly, we evaluate the performance of the signal mapping models. However, before getting there, we would like to provide the reader with a grasp on how similar the estimated signals may look like as compared to the actual or original signals one would measure for the targeted body sensor. Figure 2 shows an example illustrating how fit the acceleration signal estimated by the LSTM regression model is to the actual sensor signal for all three axes. The figure shows in particular the case in which the signals of $S2$ (right upper arm) are estimated based on the signals of $S1$ (right lower arm), with the latter being used as input to the learned regression model. As it can be observed, the acceleration signals estimated by the regressor for the right upper arm align quite closely to the acceleration signals actually measured in that position.

From the above qualitative example one could say that the estimation attained via the signal mapping is quite promising. However, a quantitative analysis is required to really comprehend the capabilities of the signal mappings learned in our study. Table 1 shows the RMSE values resulting from the leave-one-subject-out cross-validation. To fully appreciate what the reported RMSE values represent to this problem, it should be noted that the acceleration signals used are given in SI units (i.e. m/s^2) and the values measured are approximately in the range [−70, 70]. Thus, an RMSE of 0.1 represents a deviation between the estimated and actual measurements of approximately 0.07%. Now, coming back to the results, and taking into account the above, it could be affirmed that all the learned regression models yield quite good results. These findings pretty much hold regardless of the attempted sensor mapping. No relevant differences are observed among DNN models.

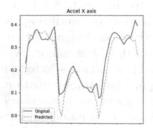 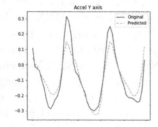 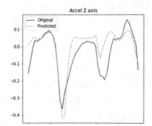

Fig. 2. Example of sensor signal estimations based on the learned mappings. The signals of the sensor S2 (right upper arm) are estimated from the signals measured by the sensor S1 (right lower arm). Both original (blue) and estimated (orange) acceleration signals are shown for each of the three axes. (Color figure online)

Table 1. Regression results for various sensor signal mappings and deep neural network models. The values correspond to the root mean square error (RMSE) calculated between the estimated and actual signals.

Sensor mapping \ DNN model	LSTM	CNN+LSTM	LSTM+Dense	DeepConv+LSTM
$S1 \rightarrow S2$	0.127	**0.123**	0.137	**0.123**
$S9 \rightarrow S7$	0.104	0.105	**0.100**	0.102
$S4 \rightarrow S8$	0.108	0.103	**0.102**	0.114
$S3 \rightarrow S2$	0.116	0.112	0.128	**0.109**

Table 2. Classification results for various sensor signal mappings and deep neural network models. The classification model is learned on the actual signals measured for the targeted sensor and tested on the estimated ones. The reported values correspond to the F_1-*score*. The last column includes the scores of the baseline model trained and tested on actual signals measured for the targeted sensor.

Sensor mapping \ Model used for estimation	LSTM	CNN+ LSTM	LSTM+ Dense	DeepConv+ LSTM	Baseline
$S1 \rightarrow S2$	**0.708**	0.635	0.598	0.671	0.628
$S9 \rightarrow S7$	0.537	0.529	0.545	**0.654**	0.659
$S4 \rightarrow S8$	0.314	0.249	0.267	**0.457**	0.728
$S3 \rightarrow S2$	0.416	0.633	0.549	**0.457**	0.628

Next, we report on the activity classification results after applying the leave-one-subject-out cross-validation. Based on the approach followed in this work, the estimated data for the targeted sensor is used to test the activity classification model originally trained on data of such sensor. For example, if the newly added sensor is $S1$, and we intend to use the existing classification model used by $S2$, a signal mapping $S1 \rightarrow S2$ must be applied to convert the $S1$ signals to look like what is measured by $S2$. Thus, the activity classification model used both as a baseline and for testing is the one learned on $S2$. Table 2 presents the F_1-*score* values obtained for the different sensor mappings and models described above. As it can be observed, the performance results for the $S1 \rightarrow S2$ mapping are quite in line with the ones obtained for the actual sensor data, irrespective of the model. Worse results are however seen for the remainder of cases. In general, the DeepConv+LSTM yields the best results for most scenarios.

4   Discussion

The proposed approach heavily depends on the potential of DNN to learn a good signal mapping. Should such a good signal estimation be possible, the capacity to recognise activities would simply rely on the reliability of the pretrained classification models. In this work, we have opted for the use of LSTM neural networks, which have been widely used in regression problems because they can capture the temporal relationships in the signals, which is essential for accurate predictions. In contrast to traditional regression models that rely on hand-crafted features and assumptions about the data, LSTM networks can learn representations directly from the raw input data, making them more flexible and adaptable to different types of data, or in our case, different types of sensor body positions. Moreover, by incorporating multiple layers of LSTM cells, these networks can capture increasingly complex patterns in the data, leading to improved performance on a variety of regression tasks. These niceties are clearly reflected in the results obtained for the various sensor signal mappings tested. The large volume of data and varying dynamicity of the considered activities possibly explain the good performance of the models in general. This is surprisingly observed even for pairs of sensors that are placed in non-adjacent body locations, such as $S9$ (left calf) and $S7$ (right thigh) or even sensors positioned on different limbs such as $S4$ (left upper arm) and $S8$ (left thigh). A possible explanation is that the neural network fairly captures the structure of the movements carried out for the considered activities, which are essentially of a symmetrical nature. Nonetheless, we cannot conclude from these results how well the mapping can be learned should the sensors be located in other body positions.

The good regression results do not directly translate to the classification level for most cases. In fact, the performance attained is quite comparable to the one achieved by the original sensor when the mapping is learned in between adjacent sensors (see $S1$ and $S2$) or even from one leg to the other (see $S9$ and $S7$). However, the accuracy drops significantly when the mappings are based on sensors placed on an arm and leg respectively. Certainly, this is a more challenging scenario, and even when the signal mapping may be reasonably accurate, the estimated and actual signals may differ enough to result into significant differences at the classification level. This may presumably become more of an issue for low intensity activities, where small estimation errors potentially represent important variations in the signal with respect to reality.

In this work, we have used deep neural networks for learning the mapping between signals of different sensors in a sort of regression problem. One could argue why not using the capacity of the DNN to directly classify the physical activity based on the new sensor data. This is certainly possible since the neural network can simply map the signals to recognised activity classes or labels. However, this poses some constraints on how the default classification model is defined. For example, if the existing sensor is used for a non-neural network model, only by estimating its signals the new sensor could be of use. Furthermore, if the signals of various sensors are processed to extract some features and then combined to be input to an activity classification model, once again estimating directly the activity label would not be of much use. In simple words, defining an all-in-one solution where both signal mapping and activity classification are carried out at once forces the system to completely operate using only neural networks, which may not be preferred.

5   Conclusions

Wearable activity recognition systems are normally trained on sensor data streams from datasets collected at design time with predefined and optimal sensor configurations. To deal with unseen changes in the sensor setup, transfer learning principles allowing a trained system to transfer activity recognition capabilities to another system have been proposed. These approaches normally operate on long time scales as they require all the relevant activities to be observed several times (e.g. timescale of days or more). These models present some other important limitations such as the need of predefining allowed runtime variations or not being defined for adaptation across sensor modalities. The approach proposed here goes beyond previous contributions by fulfilling these requirements of real-world activity recognition systems. The developed transfer model is capable of learning despite sensor variations such as rotations or translations, inherently present in the considered sensor deployment but also in most wearable activity recognition scenarios. Future work should explore the use of combination of sensors (i.e. two or more) as well as cross-modality sensors (e.g. acceleration to gyroscope) as to assess the generalisation capabilities of the proposed approach.

Acknowledgments. This research was funded by the Andalusian Ministry of Economic Transformation, Industry, Knowledge and Universities under grant P20_00163. This research was also supported by the Spanish Government under grant PID2021-127275OB-I00, FEDER "Una manera de hacer Europa" and the Spanish Institute of Health ISCIII under grant DTS21-00047.

References

1. Banos, O., Calatroni, A., Damas, M., Pomares, H., Roggen, D., Rojas, I., Villalonga, C.: Opportunistic activity recognition in IoT sensor ecosystems via multimodal transfer learning. Neural Process. Lett. **53**(5), 3169–3197 (2021)
2. Banos, O., et al.: Multi-sensor fusion based on asymmetric decision weighting for robust activity recognition. Neural Process. Lett. **42**(1), 5–26 (2015)
3. Banos, O., Damas, M., Pomares, H., Rojas, I.: On the use of sensor fusion to reduce the impact of rotational and additive noise in human activity recognition. Sensors **12**(6), 8039–8054 (2012)
4. Banos, O., Damas, M., Pomares, H., Rojas, I., Toth, M.A., Amft, O.: A benchmark dataset to evaluate sensor displacement in activity recognition. In: Proceedings of the 2012 ACM Conference on Ubiquitous Computing, UbiComp 2012, pp. 1026–1035. ACM, New York, NY, USA (2012)
5. Banos, O., Galvez, J.M., Damas, M., Pomares, H., Rojas, I.: Window size impact in human activity recognition. Sensors **14**(4), 6474–6499 (2014)
6. Banos, O., Toth, M.A., Damas, M., Pomares, H., Rojas, I.: Dealing with the effects of sensor displacement in wearable activity recognition. Sensors **14**(6), 9995–10023 (2014)
7. Chavarriaga, R., Bayati, H., Millán, J.D.: Unsupervised adaptation for acceleration-based activity recognition: robustness to sensor displacement and rotation. Pers. Ubiquit. Comput. **17**(3), 479–490 (2013)
8. Chetlur, S., et al.: cudnn: efficient primitives for deep learning. arXiv preprint arXiv:1410.0759 (2014)
9. Cover, T., Hart, P.: Nearest neighbor pattern classification. IEEE Trans. Inf. Theor. **13**(1), 21–27 (1967)

10. Kunze, K., Bahle, G., Lukowicz, P., Partridge, K.: Can magnetic field sensors replace gyroscopes in wearable sensing applications? In: International Symposium on Wearable Computers (2010)
11. Kunze, K., Lukowicz, P.: Dealing with sensor displacement in motion-based onbody activity recognition systems. In: International Conference on Ubiquitous Computing, pp. 20–29 (2008)
12. Manjarres, J., Narvaez, P., Gasser, K., Percybrooks, W., Pardo, M.: Physical workload tracking using human activity recognition with wearable devices. Sensors **20**(1), 39 (2019)
13. Nithya, N., Nallavan, G.: Role of wearables in sports based on activity recognition and biometric parameters: a survey. In: 2021 International Conference on Artificial Intelligence and Smart Systems (ICAIS), pp. 1700–1705. IEEE (2021)
14. Ordonez, F.J., Roggen, D.: Deep convolutional and LSTM recurrent neural networks for multimodal wearable activity recognition. Sensors **16**(1), 115 (2016)
15. Sokolova, M., Lapalme, G.: A systematic analysis of performance measures for classification tasks. Inf. Process. Manage. **45**(4), 427–437 (2009)
16. Wang, Y., Cang, S., Yu, H.: A survey on wearable sensor modality centred human activity recognition in health care. Expert Syst. Appl. **137**, 167–190 (2019)

Random Ensemble of Extended CNN Structures for Medical Image Recognition

Bartosz Swiderski[1], Stanislaw Osowski[2(✉)], Jaroslaw Kurek[1], and Cezary Chudobinski[3]

[1] University of Life Sciences, Warsaw, Poland
bartosz_swiderski@sggw.edu.pl
[2] Warsaw University of Technology and Military University of Technology, Warsaw, Poland
stanislaw.osowski@pw.edu.pl
[3] Copernicus Regional Multi-Speciality Oncology and Trauma Centre, Lodz, Poland

Abstract. The paper presents a new approach for creating an ensemble of deep neural CNN networks for medical image recognition. The idea is to add the extra layer of a limited number of neurons to two different CNN structures. The added layers in both networks are fully connected to the previous flattened layer of the CNN. The outputs of these two layers are randomly mixed to form the set that supplies the final classifier. Averaging the results of many repetitions of such a procedure with random weights provides the final classification decision. Such a form of signal processing is equivalent to implicit regularization. The proposed system has been applied to the recognition of images representing two classes of USG images of lymph nodes. The extensive experiments have shown an advantage over the results obtained by using the same standard CNN structures. The proposed method can be easily adapted to other deep network structures and different tasks.

Keywords: generalization · regularization method · deep neural networks · random ensemble of classifiers · recognition of USG images

1 Introduction

The good generalization of artificial neural networks is the key practical problem in artificial intelligence. It means transferring the knowledge acquired in the adaptation stage on the learning data to the new test data, not used in the learning process. The generalization ability of a network requires recognition of the mechanism of creating the data in the analyzed process and not memorizing the teaching examples [1–3]. Since the testing data might be understood as the learning data buried in the noise the process of reducing such noise is an important direction in improving the generalization ability [4, 5].

The problem is especially difficult for complex problems of medical image recognition, where large neural structures are needed with a relatively small population of available measured data following from limitation of resources. This problem is typical in deep networks applying millions of adapted parameters at very scarce resources of

© The Author(s), under exclusive license to Springer Nature Switzerland AG 2023
I. Rojas et al. (Eds.): IWANN 2023, LNCS 14134, pp. 483–493, 2023.
https://doi.org/10.1007/978-3-031-43085-5_38

available data. The generalization ability of such networks depends on the population of learning data and the complexity of the network structure [1, 6, 7].

A lot of different methods have been developed in the past to improve the generalization ability of neural networks [8–10]. The typical form is to apply the augmentation of data by applying such forms as flipping, translations, rotations, scaling, cropping, adding the noise to the learning samples, non-negative matrix factorization, application of generative adversarial network (GAN) or variational autoencoder (VAR), etc. [8, 9, 11, 12]. Such approaches to generalization improvement are based on artificial increasing the population of learning data. However, their application is of limited success, since the artificially created samples represent either some sort of duplication of the originals (translation, rotation, scaling) or represent the noisy versions of the true images (GAN, VAR), which introduce the noise to the learning process, not necessarily resulting in generalization improvement.

The other techniques are based on the regularization of the system implemented either by the modification of the architecture of the neural networks (weight decay, dropout) or using different methods of learning [6]. There are also implicit forms of regularization, built into the process of creating and adapting the neural architectures [2]. To such form belongs the stochastic gradient descent algorithm, which tends to the solution of the small norm of errors. Another form of implicit regularization in the learning procedure is the early stopping, as well as batch normalization, applied in each cycle of adaptation [1, 2, 10, 11].

A good direction to increase the generalization ability is to apply the ensemble of networks, acting independently [12]. Each member of an ensemble develops its own decision and all such decisions are integrated by majority voting into the final verdict. Different approaches have been proposed to achieve the independent operation of team members. They include such methods as a random choice of learning data for the units of an ensemble, application of randomly selected mini-batches in the learning process, diversification of network structures, or application of different feature selection method for each member of an ensemble [13]. The interesting approach is to create an ensemble composed of randomly selected members [14]. All such approaches are usually combined to develop a better generalizing system. However, the application of many members of the team increases the learning time of the whole system and is computationally expensive.

In this work, a new approach to the implicit regularization of deep CNN neural structure is proposed. The main idea is to apply only two different parallel CNN architectures. The additional layer of a limited number of neurons is added to both CNN structures (directly following the flattening layer). The additional layers of the same size allow the joining of the networks of different architectures and the size of the flattened layer. The layers in both networks are fully connected to the previous layer. Two parallel structures are created and learned simultaneously. The added layers are compatible and generate the output information in the same format. The signals of these two layers are mixed randomly and create the set, which supplies the final classifier. Averaging many repetitions of such mixing procedures with random weights generates many results estimating class probability. The final classification result of the ensemble depends on their average value.

The numerical experiments performed on the medical data representing two classes of USG images of lymph nodes (cancer versus normal) have shown significant improvement in system performance with very limited resources of data. Such a form of the proposed signal processing is equivalent to the implicit regularization of the system.

The numerical experiments aimed at the recognition of two classes of USG images of lymph nodes have confirmed the superiority of the proposed structure over standard deep architectures performing the same task.

The main innovation of the paper is the idea of randomly merging several deep networks (not necessarily two). The additional layer with the same number of nodes introduced in the middle of the nets allows completely different CNN architectures to cooperate. The introduced layer is subjected to adaptation in the learning process of the network, which is aimed at minimizing the cross-entropy function. The signals of the added layer must lead to the minimization of the loss function in different networks but are interpreted differently from the point of view of the individual network structures that form the ensemble. This is what we call implicit regularisation, which leads to an increased generalization capability of the system.

2 Materials

The numerical experiments have been performed using the database of USG images of lymph nodes collected in Copernicus Regional Multi-Speciality Oncology and Trauma Centre in Lodz. They represent two types of images. The set of 172 images represents the normal class and the other set of 226 samples represents cancer. The total number of images is only 398. The database of images was created by medical experts from Oncology and Trauma Centre. Their opinions have been supported by the pathomorphological inspection histological analysis.

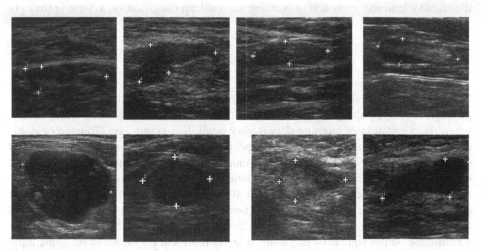

Fig. 1. The exemplary USG images of lymph nodes represent normal cases (upper row) and cancerous cases (bottom row).

The images are represented in greyscale and their original size was 256×256 pixels. Some exemplary images representing both classes are presented in Fig. 1. Cancer lymph nodes on grey scale ultrasound tend to be round in shape, rather well-defined, and appear hypoechoic without echogenic hilus features. On the other side, the normal lymph node is usually ovoid, hypoechoic to the adjacent muscle, and frequently contains an echogenic fatty hilum. In both classes, a large variety of shapes is observed. Moreover, some samples of the cancerous class are like normal images. Table 1 presents the values of the structural similarity index within the images representing each class and the similarity of the normal class to the cancerous one [15].

Table 1. The mean values of the structural similarity indices of images within the normal class, cancerous class, and between both classes

	Mean value (mean)	Standard deviation (std)	Ratio std/mean
Similarity within normal class	0.117	0.036	0.308
Similarity within cancer class	0.124	0.041	0.331
Similarity between normal and cancer classes	0.116	0.038	0.328

The similarity of images within the same class is very limited (mean value below 0.13). Moreover, the similarities between samples representing two opposite classes are of the same level as within the same classes. In all cases, we observe large values of standard deviation. The ratio of mean/std is high and above the value of 0.3). These facts confirm the difficulties in building an efficient system of automatic class recognition. The second problem is the small population of data representing both classes. Therefore, the transfer learning approach in building such a deep classification system is strongly suggested in the first phase of learning.

3   Methods

Nowadays the most efficient image recognition systems apply the deep learning strategy [6, 16]. However, the single CNN network is of limited generalization ability, especially, when the population of learning data is very small, like in our case. To overcome this problem we propose the application of a special structure of an ensemble. It is composed of two classical CNN architectures, each enhanced with the additional layer composed of a small number of neurons directly following the flattened layer of classical CNN. The preliminary experiments have shown that 10 neurons was enough to provide efficient cooperation of both channels. The added neurons in both networks are fully connected to the previous layer and take part in the learning procedure of the whole system. As a result, the output signals of these added layers represent responses to the same input images delivered to both networks. Due to the different architecture of both CNNs, their responses are also different and are independent to a high degree.

In the next step, they are fused randomly, by using weighted summation. If the output signals of these additional layers are denoted by X_1 and X_2, respectively, their fusing is done by applying the formula

$$Y = a \cdot X_1 + (1 - a) \cdot X_2 \tag{1}$$

where a is a random number drawn in the range [0,1]. The signals defined by vector Y represent the input attributes to the final classification stage.

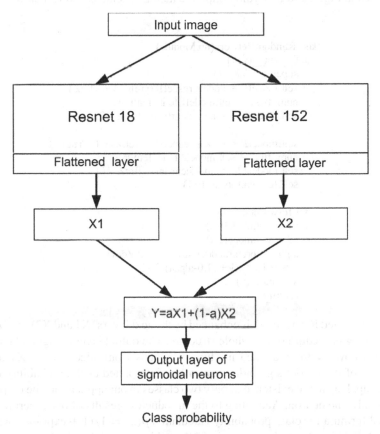

Fig. 2. The general structure of the proposed system. Two pre-trained Resnet architectures (Resnet 18 and Resnet 152) have been enhanced by an additional layer containing 10 neurons fully connected to the flattened layer of the original pre-trained Resnet structures. The output signals of these layers X1 and X2 are mixed randomly using the relation (1) and delivered as the input attributes to the final sigmoidal classifier.

The two CNN systems applied in the solution might be of any architecture. In this work, two different implementations of the Resnet architecture have been applied [17]. They were chosen because such networks apply an efficient solution to the problem of vanishing gradient by introducing the so-called residual blocks. The architecture of

Resnet stacks multiple identity mappings (the layers that do nothing at first), skips those layers, and reuses the activations of the previous layer. Resnet models skip two or three layers at a time with nonlinearity and batch normalization in between. Skipping speeds up initial training by compressing the network into fewer layers.

Moreover, the procedure allows training much deeper networks without the deterioration of the learning procedure. To achieve significant independence of both CNN models we have applied the pre-trained architectures of two significantly different networks: Resnet 18 and Resnet 152 [15, 16]. The general structure of the proposed system is presented in Fig. 2 and its Python implementation scheme is shown below.

```
class RandomNetwork(nn.Module):
    def __init__(self):
        super().__init__()
        self.model1 = models.resnet18(pretrained=True)
        num_ftrs1 = self.model1.fc.in_features
        self.model1.fc = nn.Linear(num_ftrs1, 10)

        self.model2 = models.resnet152(pretrained=True)
        num_ftrs2 = self.model2.fc.in_features
        self.model2.fc = nn.Linear(num_ftrs2, 10)
        self.fc = nn.Linear(10,1)

    def forward(self, x):
        x1 = self.model1(x)
        x2 = self.model2(x)
        alpha = torch.rand(x1.size(0), 1).to(device)
        x = alpha * x1 + (1.0-alpha) * x2
        x = self.fc(x)
        return x
```

The pre-trained Resnet models enhanced by the next layers (X1 and X2) are subjected to the learning procedure of the whole structure. The output layer of sigmoidal neurons is supplied now by the weighted mixture of signals from added layers X1 and X2. The signals of the output sigmoidal neurons are interpreted as the probability of class membership. Due to the existence of only two classes in our application, the output layer contains only one neuron. According to the introductory results of our experiments, the sigmoidal formula for class probability calculation $p(u) = 1/(1 + \exp(-u))$ was more efficient than softmax, used typically in CNN architectures.

In the learning phase of the whole extended structure, all network layers are subjected to adaptation with the random choice of parameter a in each minibatch. As a result of it both parallel CNN architectures experience many modifications in the learning conditions and are adapted in some way to such changes. The loss function E in the learning process is defined as cross-entropy [6, 16]

$$E = -\sum_i \left[d_i \log(y_i) + (1 - d_i) \log(1 - d_i) \right]$$ (2)

where d_i represents the destination class (1 for cancer and 0 for normal) and y_i is the actual class probability value generated by the system for all learning pairs $(\mathbf{X}_i, d_i)$. The stochastic gradient descent with momentum (SGDM) learning algorithm with an initial learning rate of 0.1 and variable step was applied [16, 18] in experiments. The step length was reduced 10 times every 10 epochs. Total number of epochs used in learning phase was equal to 30.

In the testing mode, the same input image is classified many times by changing only the value of the random parameter a in the mixing mode. The classification results following these repetitions (in the form of class probability) are averaged. The highest value indicates the class winner and is treated as the final classification decision of the system.

The random choice of the mixture parameter is the key factor in the implicit regularization and provides the increased generalization ability of the ensemble. The proposed system operates in a continuous range of values of a, from 0 to 1, affecting the final probability of class membership. This is consistent with the performance of the system on noisy data, as averaging such responses is a known method to reduce the influence of noise

4 Results of Numerical Experiments

The numerical experiments have aimed at the recognition of the medical images representing two classes of USG of lymph nodes: cancer versus normal. The numerical simulations have been organized in the form of 5-fold cross-validation and repeated 10 times. This form used in assessing the quality of the classification systems is regarded as the most objective since it uses all data in the testing stage. The final testing results presented in the paper correspond to the mean of all 10 runs in a 5-fold cross-validation procedure. They are given in the form of the typical quality values: average accuracy, sensitivity in recognizing cancer cases, specificity, precision in recognition of cancer and normal classes, F1 measure, the Matthews correlation coefficient, and the area under the ROC curve [19].

The performance of the ensemble system was checked in many repetitions of drawing the random value a in the mixing procedure described by (1). The introductory experiments have shown that 50 repetitions of the value a in the testing procedure deliver the stabilized results on the highest level.

To assess properly the quality of the proposed procedure the experiments have been conducted also for the individual classifiers: classical Resnet18 and Resnet152, as well as their extended individual versions, however, without fusion (notations Resnet18_ext and Resnet152_ext). Table 2 presents the statistical results of all these experiments. It depicts the average accuracy (ACC), sensitivity, specificity, and precision in recognition of cancerous (Prec_cancer), and normal class (Prec_normal), F1 measure for both classes (F1_cancer and F1_normal), Matthews correlation coefficient (MCC), and area under ROC curve (AUC).

The results show a significant role in the cooperation of both Resnet architectures organized in the ensemble form, proposed in the paper. All quality factors (accuracy, sensitivity, specificity, precision, MCC, and AUC) have been improved. For example

Table 2. The statistical results of different configurations of the deep networks. They represent the average of 10 repetitions of experiments for individual classical structures of Resnet18 and Resnet152, as well as the classical structures extended by an additional layer of 10 neurons (Resnet18_ext and Resnet152_ext) and for the proposed system.

	Resnet18	Resnet152	Resnet18_ext	Resnet152_ext	Proposed system
ACC	0.8106	0.8189	0.8352	0.8330	**0.8435**
Sensitivity	0.7839	0.8134	0.8270	0.8235	**0.8415**
Specificity	0.8459	0.8262	0.8463	0.8455	**0.8466**
Prec_cancer	0.8730	0.8623	0.8784	0.8776	**0.8803**
Prec_normal	0.7518	0.7754	0.7918	0.7873	**0.8059**
F1_cancer	0.8240	0.8354	0.8502	0.8483	**0.8587**
F1_normal	0.7937	0.7976	0.8159	0.8135	**0.8236**
MCC	0.6273	0.6387	0.6718	0.6670	**0.6870**
AUC	0.8948	0.8978	0.9109	0.9051	**0.9138**

accuracy of individually operating Resnet18 and Resnet152 equal to 81.06% and 81.89%, respectively, have been increased to 84.35%. The highest ratio of improvement has been observed for the MCC measure, the quality factor very sensitive to the population of data (62.73% of single Resnet18 and 68.70% in the proposed system).

The additional advantage of the proposed system is its resistance to overlearning at a very small population of learning data (396 samples of ROI images used in experiments). This is well seen in the graphs presenting the change in the quality measures for testing data with the increased number of learning cycles.

Figure 3 presents such graphs for ACC, AUC, MCC, sensitivity, specificity, and average precision in recognition of both classes at the application of all arrangements of the system (individual CNN networks and the proposed ensemble). The results of the proposed ensemble system (notation ResNET-18-152_mean) have been obtained from 50 repetitions in the testing phase (ensemble composed of 50 team members) with the random choice of the coefficient a. As is seen in all cases the proposed ensemble system has delivered the best results for all quality measures, irrespective of the applied number of learning cycles. The advantage of the ensemble over the individual CNN models in most cases is in the range of a few percentage points.

5   Discussion

A novel approach to increase the generalization capability of deep networks by introducing a special form of implicit regularisation has been proposed. For this purpose, an ensemble consisting of an extended structure of different CNN networks is proposed. The extension takes the form of a layer with a limited number of nodes. The input signals for this additional layer are taken from the flattened layer of the standard CNN

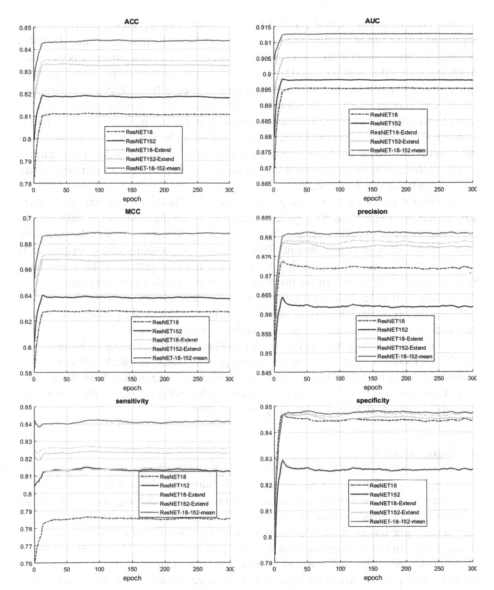

Fig. 3. The graphs present the chosen quality values (ACC, AUC, MCC, average precision of both classes, sensitivity, and specificity) of the individual Resnet architectures (classical and extended) and the proposed system at the prolonged time of the learning phase. The horizontal axis represents the learning cycles.

circuits used in the ensemble. The CNN models used can be of any architecture. The introduction of the additional layer with the same number of nodes in each CNN model forces the ensemble members to produce different but compatible signals that lead to the minimization of the loss function of each member.

The number of nodes in the added layer is small. The representation of the large population of signals of the flattened layer (typically 1000) by their very limited number in the added layer (equal to 10 in our solution) makes the whole structure a weak classifier. Applying the random value of mixing coefficient a in many repetitions of testing mode, generates different responses of both structures. It means their separate performance like in the random forest of decision trees. Averaging the probability results of these runs allows for generating more accurate responses to the testing data not taking part in learning. The randomness introduced in the model might be interpreted as the way how the system reacts to the noisy data (testing samples are often interpreted as noisy learning samples). Averaging process is well-known method to improve the performance of the system in a noisy environment.

The important role in the system performance is fulfilled by the random choice of the mixing factor a in each learning cycle. Its function is like the randomness built into the genetic algorithms of optimization. Thanks to it, the system tries to adapt the parameter values to the changing environment. This increases the system's ability to efficiently process test data that differ significantly from the learning patterns (like noisy data). Such a form of a proposed signal processing is equivalent to implicit regularization of the system, which leads to an improvement in its generalization ability. This is very important in problems where only very limited data resources are available. However, the proposed procedure will be also of great help for larger data sets.

The proposed system was built based on two Resnet CNN architectures, however, any other CNN models can be used. The interesting point in further research is to increase the number of different architectures of CNN models cooperating in the ensemble in such a way. Thanks to this it will be possible to increase the independence of their operation, the condition very important in creating the ensemble.

6   Conclusions

The paper has presented a new approach to building an efficient ensemble of deep classifiers. The idea is to introduce some randomness in the learning and testing phases of two parallel CNNs of different structures. The interesting point is the application of the standard pre-trained models in the introductory phase of learning, which makes the problem easy in practical application, especially in the case of a very small population of available data. The real learning process of the system starts from the relatively well-initiated parameter values, resulting in the acceleration of the whole adaptation process.

The proposed solution has been checked in recognition of two classes of images representing the USG of lymph nodes: cancer versus normal. The obtained results have shown its superiority over the non-integrated individual members of the ensemble. The accuracy, sensitivity, specificity, precision, Matthew's correlation coefficient, and area under the ROC curve have been increased. The proposed solution was found the best in all these aspects.

The implicit regularization implemented in our system is very effective in image analysis, especially in the case of a very small population of learning samples. The system has shown high resistance to the overlearning problem. It might find application in many different tasks of signal and image processing, not limited only to medical problems.

References

1. Poggio, T., Liao, Q., Theory, I: Deep networks, the curse of dimensionality. Bull. Polish Acad. Sci. Tech. Sci. **66**, 761–773 (2018). https://doi.org/10.24425/bpas.2018.125924
2. Zheng, Q., Yang, M., Yang, J., Zhang, Q., Zhang, X.: Improvement of generalization ability of deep CNN via implicit regularization in the two-stage training process. IEEE Access **6**, 15844–15869 (2018). https://doi.org/10.1109/ACCESS.2018.2810849
3. Zhou, P., Feng, J.: Understanding generalization, optimization performance of deep CNNs. In: Proceedings of the 35th International Conference on Machine Learning, Stockholm, Sweden, PMLR 80, pp. 1–19 (2018). PMLR 80.5960.5969
4. Yan, C., Gong, B., Wei, Y., Gao, Y.: Deep multi-view enhancement hashing for image retrieval. IEEE Trans. Pattern Anal. Mach. Intell. **43**(4), 1445–1451 (2020). https://doi.org/10.1109/TPAMI.2020.2975798
5. Yan, C., Li, Z., Zhang, Y., Liu, Y., Ji, X., Zhang, Y.: Depth image denoising using nuclear norm and learning graph model. ACM Trans. Multimedia Comput. Commun. Appl. 1–17 (2020). https://doi.org/10.1145/3404374
6. Goodfellow, I., Bengio, Y., Courville, A.: Deep Learning. MIT Press, Massachusetts (2016)
7. Goodfellow, I., et al.: Generative adversarial nets. Adv. Neural Inf. Process. Syst. 1–9 (2014). arXiv:1406.2661. https://doi.org/10.48550/arXiv.1406.2661
8. Kingma, P., Welling, M.: An introduction to variational autoencoders. Found. Trends Mach. Learn. **12**, 307–392 (2019). https://doi.org/10.48550/arXiv.1412.6980
9. Ren, H., Su, J., Lu, H.: Evaluating generalization ability of CNN and capsule networks for image classification via top-2 classification. arXiv:1901.10112v2 [cs.CV], 1–18 (2019). https://doi.org/10.48550/arXiv.1901.10112
10. Neyshabur, B., Bhojanapalli, S., McAllester, D., Srebro, N.: Exploring generalization in deep learning. Adv. Neural Inf. Process. Syst. 1–19. arXiv:1706.08947 (2017)
11. Zhang, C., Bengio, C., Hardt, M., Recht, B., Vinyals, O.: Understanding deep learning requires rethinking generalization. In: International Conference Learning Representations (ICLR), pp. 1–15 (2017). arXiv:1611.03530v2 [cs.LG], https://doi.org/10.48550/arXiv.1611.03530
12. Kuncheva, L.: Combining Pattern Classifiers: Methods and Algorithms. Wiley, New York (2014)
13. Gil, F., Osowski, S.: Fusion of feature selection methods in gene recognition. Bull. Polish Acad. Sci. Tech. Sci. **69**(3), 1–8 (2021). https://doi.org/10.24425/bpasts.2021.136748
14. Swiderski, B., Osowski, S., Gwardys, G., Kurek, J., Slowinska, M., Lugowska, I.: Random CNN structure – tool to increase generalization ability in deep learning. Eurasip J. Image Video Process. 1–18 (2022). https://doi.org/10.21203/rs.3.rs-277475/v1,
15. Matlab user manual, MathWorks, Inc. Natick, USA (2017)
16. Brownlee, J.: Deep Learning for Natural Language Processing. Develop Deep Learning Models for your Natural Language Problems, Ebook (2018)
17. He, J., Zhang, X., Ren, S., Sun, J.: Deep residual learning for image recognition. In: 2016 IEEE Conference on Computer Vision and Pattern Recognition (CVPR). Las Vegas, NV, USA, pp. 770–778 (2016). arXiv:1512.03385, https://doi.org/10.1109/CVPR.2016.90
18. scikit-learn.org/stable/modules/generated/sklearn.model_selection.RepeatedStratifiedKFold.html
19. Tan, P.N., Steinbach, M., Kumar, V.: Introduction to data mining, Pearson Education Inc., Boston (2013)

Bioinspired Reinforcement Learning Control for a Biomimetic Artificial Muscle Pair

Michele Foggetti[(✉)] [iD] and Silvia Tolu [iD]

Department of Electrical and Photonics Engineering,
Technical University of Denmark, 2800 Lyngby, Denmark
{mifo,stolu}@dtu.dk

Abstract. Artificial muscles are recently developed actuators extremely promising for compliant robotic systems. Their accurate closed-loop control is challenging due to their highly nonlinear behavior.

In this work, we model an artificial muscle pair adopting a non-pulley configuration mimicking more realistically the behavior of biological muscles. Inspired by how the brain regulates dopamine-based learning from interaction with the environment, it is possible to design efficient reinforcement learning control algorithms. Therefore, we propose a reinforcement learning-based controller bioinspired by the parallels between the behavior of temporal difference errors and the activity of dopaminergic neurons. Simulated experiments conducted in a virtual scenario show that the control action can accurately tackle the nonlinear control problem.

The proposed solution could be extended to the dynamic control of more realistic and complex anthropomorphic limb systems due to its inherent adaptability and control effectiveness regardless of the complexity of the environment.

Keywords: Bioinspired Reinforcement Learning · Nonlinear Control System · Artificial Muscles

1 Introduction

Manipulators actuated by artificial muscles are gaining popularity in jobs that require contact with humans and do not endanger users, due to their flexible and lightweight bodies. McKibben actuators are one of the most efficient and widely used fluidic artificial muscles and they have become popular in the last decades for the actuation of soft robots [1]. Yet, as time goes on, robotic and mechatronic systems become more and more complex. Their complexity in turn requires innovative technologies and solutions for modeling and effective control algorithms. Mechanical principles are exploited to describe the behavior of artificial muscles [2–5]. These are leveraged to describe the generation of pulling forces. A torque is obtained by assembling two actuators acting one against the other, thus creating an artificial muscle pair [1,3,6].

© The Author(s), under exclusive license to Springer Nature Switzerland AG 2023
I. Rojas et al. (Eds.): IWANN 2023, LNCS 14134, pp. 494–504, 2023.
https://doi.org/10.1007/978-3-031-43085-5_39

In this work, a non-pulley-chain configuration is considered, inspired by [7]. The non-pulley description of the artificial muscle pair is able to mimic more accurately the behavior of the human elbow musculature. The characteristic of these actuators is highly non-linear due to the inherent soft material structure, therefore their accurate closed-loop positioning is a challenge that has been tackled by different control techniques, such as adaptive control [8], sliding mode control [9], fuzzy logic [10] or neural networks [11], over the last two decades.

Nonetheless, this biomimetic artificial muscle control problem has never been addressed by task-oriented control algorithms that recall neural mechanisms within the brain. Indeed, it is possible to design efficient control algorithms, inspired by the way the dopamine system regulates learning from interaction with the environment. In recent years new machine learning paradigms, such as *reinforcement learning* (RL), have been developed with the aim of addressing goal-oriented problems with growing complexity. RL is about learning how to select correct actions by mapping internal states and maximizing a reward function in a trial-and-error process [12]. The main contribution of our approach is the identification of agents that can learn directly from interactions with the environment to achieve long-term goals, regardless of the complexity and the kind of the robotic system.

Previous works [13,14] have implemented RL algorithms for dynamic position control of rigid and soft robots, but they do not faithfully mimic the brain control architecture. In our study, we propose a novel position biomimetic control system of an artificial muscle pair by exploiting the reward prediction error hypothesis of dopamine neuron activity and the optimization process among Actor-Critic elements that emulate the communication between the dorsal and the ventral striatum.

Our study is divided as follows, Sect. 2 presents the motivation leading to the definition of the *Deep Deterministic Policy Gradient* (DDPG) RL algorithm. Thereafter, Sect. 3 describes the modeling phase of the biomimetic artificial muscle pair and the differential equations governing the system's dynamics. The same Section presents the learning mechanisms of the RL agent that enable the correct positioning of the biomimetic actuator. The results of virtual experiments are shown and analyzed in Sect. 4. Finally, in Sect. 5, the main findings are discussed and compared with those obtained in [7].

2   Background

2.1   Mathematical Framework

A *Markov decision process* (MDP) is a mathematical framework for the implementation of an RL algorithm. It describes the interaction between a learning agent and the environment. In such a framework the agent selects the actions to be sent to the environment, as a function of the state and the reward at time t. The environment returns the state vector and the reward to the agent at time $t+1$. The action defines the activation value of the control variable which, in this case, can be thought of as the cumulative alpha motoneuron signal descending from the brain through the spine into the muscle. The state vector s returned to the agent contains information about the relevant sensed muscle variables. The reward R is a customized function of the state vector and it can be thought of as an abstract summary of the overall effect of different neural signals assessing sensations and states about rewarding and punishing qualities. A reinforcement signal δ manages changes in the learning algorithm of an agent's policy π, which is the function that maps states into actions. At the time t, the reinforcement signal for a TD method is the TD error

$$\delta_{t-1} = r_t + \gamma v(s_t) - v(s_{t-1}),\qquad(1)$$

that is the reward signal adjusted by the value estimate $v(s)$. The latter are predictions of the total reward that an agent can expect to be accumulated over the future. The agent takes decisions by choosing actions leading to states with the largest estimated state values. *Reward prediction errors* (RPEs) is a specific measure of the differences between the expected and the received reward signal. It is positive when the reward signal is greater than the expected signal and negative otherwise. TD errors are special classes of RPEs [12].

2.2   Bioinspired Analogy

The biological analogy of RL is formalized in the Actor-Critic architecture shown in Fig. 1 into which the DDPG algorithm is performed. Experimental evidence suggests that the dopamine neurotransmitter signals RPEs and not rewards [15]. Moreover, the phasic activity of dopamine-producing neurons conveys TD error.

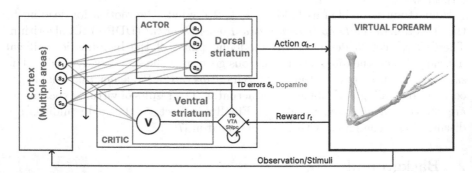

Fig. 1. The Actor-Critic architecture showing the proposed neural topological analogy. Biological elements are highlighted in blue. The actor (dorsal striatum) adjusts its policy based on the TD error δ_t (dopamine signal) it receives from the critic (ventral striatum). The critic adjusts state-value parameters using the same δ_t (the TD error is transmitted by dopamine neurons located in the VTA and SNpc to modulate changes in synaptic efficacy of input from cortical areas to the ventral and dorsal striatum). The critic (VTA, SNpc) produces the TD error from the r_t reward signal and the current estimation change of the $v(s_t)$ state value that is processed by multiple areas of the cortex

This leads to the "reward prediction error hypothesis of dopamine neuron activity". This activity has been observed in two main subdivisions of the striatum, which is a cluster of neurons in the subcortical basal ganglia of the forebrain. The dorsal striatum is primarily implicated in influencing action selection, while the ventral striatum is thought to be critical for different aspects of reward processing, including the assignment of affective value to sensation. For this reason, it is thought that in the striatum an Actor-Critic algorithm is implemented in order to allow the action selection. In an Actor-Critic configuration, both policies and value functions are learned. The policies are learned by the actor while the critic is the component that learns how to criticize the actor's actions. The state-value function for the actor's current policy is learned by the critic through the TD algorithm. TD errors δ are sent to the actor in order to critique its action choices through the critic value function. The actor constantly updates its policy based on these critiques. The same reinforcement signal is used in the learning rules for the critic and the actor and its effect on learning is different for these two components.

Other works have already tried to use an RL framework to describe a possible bioinspired control technique. In [13], the way the areas of the human brain are associated with the elements of an MDP is different from the one made in this study. In the aforementioned research, the building blocks of the actor and the critic are linked to parts of the brain that have a certain function rather than a neurobiological analogy. In such a work, the block that filters observed signals is called 'Proprioceptive Preprocessing' without better identifying which area of the brain it is associated with. Also, the 'Sensory Representation' and 'Motor Command Representation' blocks are introduced as part of the agent. Within our implemented architecture, the blocks constituting the actor and the critic are referred to a specific area of the brain, the dorsal and ventral striatum, respectively. The actor block defined in [13] is replaced by the block that emulates the behavior of the dorsal striatum which receives input from the cortex. The same input is broadcasted to the critic block. The 'Error Evaluation' and 'Global Reward/Punisher Signaller' blocks are respectively replaced by the ventral striatum and ventral tegmental area (VTA), substantia nigra pars compact (SNpc). Furthermore, in that study, the reward signal is processed within the critic while it is a signal sent from the environment and not internally generated. Therefore, the architecture implemented in this work is more coherent from a biological point of view with the description of the elements that make up a (bioinspired) RL algorithm.

3 Methods

3.1 Biomimetic Artificial Muscle Pair Model

Biomimetic actuators are artificial muscles able to mimic the natural behavior of the biological muscles [7, 16]. In order to obtain highly biomimetic McKibben-type artificial muscle a *non-pulley* musculoskeletal geometry must be considered [16, 17]. The adopted configuration, depicted in Fig. 2, considers the insertion point of the muscle attached to the mobile link as is the case of biological muscles, and not to the joint as it happens for classic pulley-chain configuration.

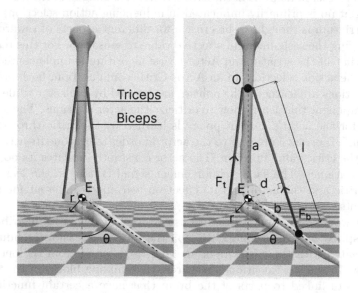

Fig. 2. Left: pulley-chain configuration of an artificial muscle pair. Right: non-pulley configuration of a biomimetic artificial muscle pair. The elbow rotation is constrained in the sagittal plane. Pulling forces are exerted by the biceps and triceps muscle groups. All relevant geometrical quantities (a, b, d, r, l) are shown in the figure.

The static moment T_s generated by the artificial muscles pair is a function of both the forces exerted by the individual muscles F_b and F_t and the relative distance between the joint center and the line containing the direction of the biceps force, d in Fig. 2. The implemented model is composed of two representative muscle groups, the "biceps" and the "triceps" and two links. This is a simplification of the musculoskeletal model of the upper limb. Indeed, the entire group of flexor and extensor muscles acting about the elbow are modeled by two opposing muscles acting at fixed moment arms about the joint axis [18]. The humerus (rigid link) is not actuated by any muscle. The ulna and radius (mobile link) represent the whole forearm to which the biceps muscle is directly attached. The degree of freedom (DOF) of the model is the relative rotation $\theta \in [0, \theta_{max}]$ of the forearm with respect to the arm, positive when the biceps muscle contracts. Both muscles have an origin point in O. The triceps insertion point is on the pulley surface at distance r from the elbow, whose origin point is E. The biceps length is indicated as l and its insertion point I is located on the forearm mobile link. Lastly, a and b represent respectively the distances $\overline{OE}$ and $\overline{EI}$. Geometrical distances l and d can be computed as $l(\theta) = \sqrt{a^2 + b^2 + 2ab\cos\theta}$ and $d(\theta) = ab\sin(\theta)/l(\theta)$ [7]. The static torque generated by the actuation of the artificial muscles can be then computed, as

$$T_s = d(\theta)F_b - rF_t = \frac{ab\sin(\theta)}{\sqrt{a^2 + b^2 + 2ab\cos(\theta)}}F_b - rF_t. \tag{2}$$

There are several models in the literature that describe the behavior of Hill-type muscles and they are always characterized by first an excitation-to-activation function and then an activation-to-force function [19]. The excitation value is related to the use of the variable $\alpha \in (0, 1)$, defined as the action function in the RL framework, which is the control signal. The static force model is a function of the control variable α and the contraction ratio of the muscle ϵ defined as the ratio between the muscle length variation and its initial length, such that

$$F(\alpha, \epsilon) = \alpha F_{max}\left(1 - \frac{c}{\epsilon_{max}}\right). \tag{3}$$

The main limitation of this model is that it does not contain information about the biological muscle passive tension even though it is accurate enough to describe its variable stiffness spring nature.

Supposing that the control variables have symmetrical variations and the same initial length l_0 for both muscles, the resulting forces will be described as $F_b = \alpha F_{max}(1 - \epsilon_b/\epsilon_{max})$ for the biceps and $F_t = (1 - \alpha)F_{max}(1 - \epsilon_t/\epsilon_{max})$ for the triceps with $\epsilon_b = (a + b - l(\theta))/l_0$ and $\epsilon_t = \epsilon_{max} - r\theta/l_0$. Finally, Newton's principles are exploited to obtain the dynamic behavior of the actuator. A linear viscous damping model with damping coefficient β is imposed to describe the kinetic friction behavior inside the actuator. The resulting differential equation that governs the dynamics of the system, moving a given inertia J, is

$$T_s - \beta\dot{\theta} = J\ddot{\theta}. \tag{4}$$

To describe the behavior of a biological arm more realistically, such as shoulder/wrist behaviors, other DOF's might be taken into account. Also, it is supposed that both the biceps and the triceps muscles stand for a larger number of muscles.

In a more realistic setting, each muscle would bring a different control action. This would make the system to be controlled a Multiple-input-Multiple-Output (MIMO) system for which advanced model-free control techniques are required. The computational approach to learn from complex environment interactions is called RL. This approach is ideal for obtaining agents capable of controlling increasingly complex and descriptive systems. In conclusion, this work shows an initial proof of concept for the control of increasingly complex anthropomorphic and biomimetic robotic arms.

3.2   Simulation

The geometrical and physical parameters of the actuator must be defined to allow simulations in a virtual environment and to solve the ordinary differential equations (ODEs) governing the artificial muscle dynamics. The solver is set to *ode4* (Runge-Kutta) with sample time 10^{-3} s. It is considered a max-imum force of $1000\,\mathrm{N}$, a maximum contraction ratio of 0.25, an active initial length of the muscle fibers of $0.5\,\mathrm{m}$, and an elbow joint radius of $0.025\,\mathrm{m}$ accord-ing to values expressed in [7]. This, in turn, allows the computation of the static torque and eventually the implementation of Eq. 4 assuming an inertia $J = 20\,\mathrm{Kg\,m^2}$ and a viscous coefficient $\beta = 9.94\,\mathrm{N\,m^{-1}}$. Computation are per-formed in a MATLAB®/Simulink® framework. The same environment is used to implement the DDPG Actor-Critic RL algorithm. Since the biomimetic actu-ator model, which represents part of the environment that sends clues to the agent, is implemented and tested in the same Simulink® framework, there is no need for additional software interfaces.

3.3   Learning and Control

The DDPG algorithm is a model-free, online, and off-policy reinforcement learn-ing method. The Actor-Critic agent searches for an optimal policy that max-imizes the expected cumulative long-term reward. Moreover, the space of the actuator is continuous both in the observation space and in the action space. At each training step during learning, the DDPG agent first updates the actor and critic properties then stores past experiences using a circular experience buffer, and finally disturbs the action chosen by the policy using a stochastic noise model. The agent, during the learning phase, modifies the parameter values in the critic parameter vector. Such parameters, after the learning phase, remain at their modified value. Critic's parameters are updated by minimizing the loss L across the sampled experiences,

$$L = \frac{1}{m} \sum_{i}^{m} [y_i - q(\boldsymbol{s}_i, \alpha_i, \boldsymbol{w})]^2. \tag{5}$$

The actor's parameters are updated by maximizing the expected discounted reward using the sampled policy gradient,

$$\nabla_\phi J \approx \frac{1}{m} \sum_{i}^{m} \nabla_{\alpha_i} q(\boldsymbol{s}_i, \alpha_i, \boldsymbol{w}) \nabla_\phi \pi(\boldsymbol{s}_i, \boldsymbol{\phi}), \tag{6}$$

where ϕ is the actor parameter vector, $\boldsymbol{w}$ is the critic parameter vector, α_i is the current action, $\boldsymbol{s}_i$ is the current observation vector, $\pi(\boldsymbol{s}_i, \boldsymbol{\phi})$ is the actor policy, $q(\boldsymbol{s}_i, \alpha_i, \boldsymbol{w})$ returns the corresponding expectation of the long-term reward, y_i is the sum of the experience reward r_i and the discounted future reward and m is the number of the stored experiences $(\boldsymbol{s}_i, \alpha_i, r_i, \boldsymbol{s}_i')$, with $\boldsymbol{s}_i'$ the terminal state. A scalar reward signal is chosen after a trial and error procedure,

$$r_i = -|\Delta\theta_i|^2 - 10|\Delta\theta_i| - 100(exceeded\ boundaries). \tag{7}$$

The objective of the agent is to bring the reward value close to zero in order to maximize it. In particular, the first term represents the negative square of the absolute value of the error $\Delta\theta$. It is penalized in order to punish large distances from the target. In fact, if at the beginning of a training episode, the initial position of the actuator is far from the reaching angle, this penalty encourages the agent to move quickly towards the target. But as close as we get to the target, the penalty decreases. In order to further motivate the actuator to reach a closer angle, a penalty depending on the absolute value of the error is added. The last term penalizes exceeded physical boundaries.

4   Results

Every training episode consists of an actuator's step response starting from a random initial rotation angle in the available working space, $\theta_0 \in [0, \pi]$. This condition is specified as the integrators' initial values, such that $[\theta, \dot{\theta}] = [\theta_0, 0]$. Then it learns the action value needed to stay as close as possible to the reference angle to track according to the reward function. The reference angle is any random rotation angle in the available working space, $\theta_{ref} \in [0, \pi]$. Several training sessions are completed and agents' performances are compared in simulation. This is a trial-and-error procedure in which most effort has been spent shaping the reward function and setting hyperparameters. Once an average reward in the last 20 training episodes of $-35.86°$ has been reached for a total number of 2000 episodes, the training phase is considered over.

Experiments in a simulated environment are carried out in order to understand and visualize how the trained agent makes the actuator perform goal-oriented tasks. The following experiment scenarios are defined:

1. The first attempt to evaluate the agent performance is to make it track 100 times a step response with a random target angle $\theta_t \in [0, \pi]$ starting from a random initial value $\theta_0 \in [0, \pi]$. This allows the analysis of the control action's steady state error distribution.
2. Then, a sequence of four rotation angles has to be tracked, in the same simulation. Standard parameters of a dynamical response can be evaluated (overshoot, rise time, settling time, etc.) for each of the four steps in the sequence.
3. Lastly, sine waves are tracked. The first sine wave oscillates in $0° \leq y_1(t) \leq 140°$ with period $T = 5$ s. Since the agent action could not span the full dynamic of the oscillation, sine waves with a lower amplitude are considered ranging in $20° \leq y_2(t) \leq 100°$ and $40° \leq y_3(t) \leq 80°$. The control action's performance is evaluated by computing root mean square error (RMSE) values as

$$RMSE_{y_j(t)} = \left(\sqrt{\frac{1}{n} \sum_{k=1}^{n} (\theta_k - \theta_{ref,k})^2} \right)_j, \quad j = 1, 2, 3 \qquad (8)$$

where j refers to the sine waves $y_1(t)$, $y_2(t)$, and $y_3(t)$. The number of collected samples is $n = 6990$.

In the first experiment scenario (see Fig. 3.a), agent steady-state errors span in the range $[-0.02, 0.02]°$. For 69 times over 100 simulated responses, the error happens to be $\approx -0.001°$ although the stochastic noise affects the system performance. This evaluation quantifies the accomplishment of the reaching task and the effectiveness of the control action at steady-state.

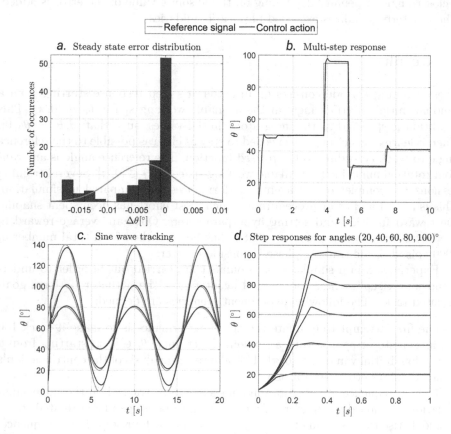

Fig. 3. a. Steady-state error distribution in the first experiment scenario. b. Multi-step dynamic response. c. Sine waves tracking at different amplitudes. d. Step response at different desired angles.

In the second experiment scenario (see Fig. 3.b), it is possible to understand if the actuator is capable of reaching different angles from different positions, that is, correctly flexing/extending both the biceps and the triceps without violating physical bounds. The overshoot σ_i and the 5% settling time $t_{5\%,i}$ are computed for all steps $i = 1, 2, 3, 4$. The pair $(\sigma_i, t_{5\%,i})$ characterizes the dynamic response of the system at each step. The system starts with null initial conditions and reaches the first step at 50° with $(\sigma_1, t_{5\%,1}) = (0.63\%, 0.273\,s)$, the second step at 95° with $(\sigma_2, t_{5\%,2}) = (3.2\%, 0.095\,s)$, the third step at 30° with $(\sigma_3, t_{5\%,3}) = (25.97\%, 0.211\,s)$ and the last step at 40° with $(\sigma_4, t_{5\%,4}) = (5.59\%, 0.130\,s)$.

In the last scenario (see Fig. 3.c), the computed RMSE values are 6.1322°, 3.6826°, 2.9614° for $y_1(t), y_2(t), y_3(t)$, respectively. The computed RMSE value decreases the better the agent's tracking action is. In the first case, the control action is not sufficient to make the system's dynamics follow the sine correctly due to its excessive amplitude. Very large oscillations, as in the case of $y_1(t)$, are impossible to track correctly. As the amplitude to be followed decreases, the performance of the control action improves and therefore the value of the RMSE decreases.

The obtained results are compared with the ones in [7], which implemented an integral control action for the same geometric model of the actuator. In analogy with their experiment, we performed five times step responses at desired angles 20°, 40°, 60°, 80°, and 100° in a 1 s simulation window. Our outcome showed an improvement of ±0.28° in the steady-state error that lies within $[-0.02, 0.02]°$ range, as shown in Fig. 3.d. Besides, the sine wave tracking behaviors appear to be smoother compared to the ones shown in [7]. The highest tracking errors occur at peaks in both works.

5   Conclusion

In this work, we have proposed a novel and effective control action that leverages a bioinspired RL algorithm resembling a brain-based control scheme occurring in the striatum.

The first part of this study regarded the modeling of an artificial muscle pair using a unique mechanical configuration different from the conventional pulley-chain approach found in the literature. By directly connecting the insertion point of the biceps' action to the mobile link representing the forearm, it is achieved a more accurate description and behavior of the artificial muscle pair, similar to that of a biological muscle. After defining the system to be controlled, our study employed a DDPG RL algorithm inspired by the way dopamine manages action selection in the brain to effectively address the nonlinear control problem.

In conclusion, the chosen framework can inherently define a possible procedure for the control of more complex anthropomorphic and biomimetic robotic arms, in terms of the augmented number of DOFs, joints, and artificial muscles.

References

1. Tondu, B.: Modelling of the McKibben artificial muscle: a review. J. Intell. Mater. Syst. Struct. **23** 225–253 (2012). https://doi.org/10.1177/1045389X11435435
2. Caldwell, D., Medrano-Cerda, G., Goodwin, M.: Control of pneumatic muscle actuators. Control Syst. **15**, 40–48 (1995). https://doi.org/10.1109/37.341863
3. Chou, C., Hannaford, B.: Measurement and modeling of McKibben pneumatic artificial muscles. Robot. Autom. IEEE Trans. **12**, 90–102 (1996). https://doi.org/10.1109/70.481753
4. Garbuliński, J., Balasankula, S., Wereley, N.: Characterization and analysis of extensile fluidic artificial muscles. Actuators **10**, 26 (2021). https://doi.org/10.3390/act10020026

5. Schulte, H.: The characteristics of the McKibben artificial muscle (1961)
6. Obiajulu, S., Roche, E., Pigula, F. Walsh, C.: Soft pneumatic artificial muscles with low threshold pressures for a cardiac compression device. In: Proceedings Of The ASME Design Engineering Technical Conference, vol. 6 (2013). https://doi.org/10.1115/DETC2013-13004
7. Tondu, B.: Single linear integral action control for closed-loop positioning of a biomimetic actuator with artificial muscles. In: 2015 European Control Conference (ECC) (2015). https://doi.org/10.1109/ECC.2015.7331088
8. Centurelli, A., Rizzo, A., Tolu, S., Falotico, E.: Open-loop model-free dynamic control of a soft manipulator for tracking tasks. In: 2021 20th International Conference On Advanced Robotics (ICAR), pp. 128–133 (2021). https://doi.org/10.1109/ECC.2015.7331088
9. Braikia, K., Chettouh, M., Tondu, B., Acco, P., Hamerlain, M.: Improved control strategy of 2-sliding controls applied to a flexible robot arm. Adv. Robot. **25**, 1515–1538 (2011). https://doi.org/10.1163/016918611X579510
10. Chiang, C., Chen, Y.: Neural network fuzzy sliding mode control of pneumatic muscle actuators. Eng. Appl. Artif. Intell. **65**, 68–86 (2017). https://doi.org/10.1016/j.engappai.2017.06.021
11. Tian, S., Ding, G., Yan, D., Lin, L., Shi, M.: Nonlinear controlling of artificial muscle system with neural networks. In: 2004 IEEE International Conference On Robotics and Biomimetics, pp. 56–59 (2004). https://doi.org/10.1109/ROBIO.2004.1521751
12. Sutton, R., Barto, A.: Reinforcement Learning: An Introduction. (A Bradford Book, 2018)
13. Chadderdon, G.L., Neymotin, S.A., Kerr, C.C., Lytton, W.W.: Reinforcement learning of targeted movement in a spiking neuronal model of motor cortex. PLoS ONE **7**(10), e47251 (2012). https://doi.org/10.1371/journal.pone.0047251
14. Centurelli, A., Arleo, L., Rizzo, A., Tolu, S., Laschi, C., Falotico, E.: Closed-loop dynamic control of a soft manipulator using deep reinforcement learning. IEEE Robot. Autom. Lett. **7**, 4741–4748 (2022). https://doi.org/10.1109/LRA.2022.3146903
15. Schultz, W., Dayan, P., Montague, P.: A neural substrate of prediction and reward. Science. **275**, 1593–1599 (1997). https://doi.org/10.1126/science.275.5306.1593
16. Hannaford, B., Winters, J., Chou, C., Marbot, P.: The anthroform biorobotic arm: a system for the study of spinal circuits. Ann. Biomed. Eng. **23**, 399–408 (1995). https://doi.org/10.1007/BF02584440
17. Matsutani, Y., Tahara, K., Kino, H., Ochi, H.: Complementary compound set-point control by combining muscular internal force feedforward control and sensory feedback control including a time delay. Adv. Robot. **32**, 411–425 (2018). https://doi.org/10.1080/01691864.2018.1453375
18. Hogan, N.: Adaptive control of mechanical impedance by coactivation of antagonist muscles. IEEE Trans. Autom. Control **29**, 681–690 (1984). https://doi.org/10.1109/TAC.1984.1103644
19. Romero, F., Alonso, F.: A comparison among different hill-type contraction dynamics formulations for muscle force estimation. Mech. Sci. **7**, 19–29 (2016). https://doi.org/10.5194/ms-7-19-2016

Energy-Aware KNN for EEG Classification: A Case Study in Heterogeneous Platforms

Juan José Escobar[1]([envelope]) [ID], Francisco Rodríguez[2], Rukiye Savran Kızıltepe[3] [ID],
Beatriz Prieto[2] [ID], Dragi Kimovski[4] [ID], Andrés Ortiz[5] [ID], and Miguel Damas[2] [ID]

[1] Department of Software Engineering, CITIC, University of Granada,
Granada, Spain
`jjescobar@ugr.es`
[2] Department of Computer Engineering, Automation, and Robotics, CITIC,
University of Granada, Granada, Spain
`cazz@correo.ugr.es` `{beap,mdamas}@ugr.es`
[3] Sofware Engineering, Karadeniz Technical University, Trabzon, Turkey
`rukiye.savrankiziltepe@ktu.edu.tr`
[4] Institute of Information Technology, University of Klagenfurt, Klagenfurt, Austria
`Dragi.Kimovski@aau.at`
[5] Department of Communications Engineering, University of Málaga, Málaga, Spain
`aortiz@ic.uma.es`

Abstract. The growing energy consumption caused by IT is forcing
application developers to consider energy efficiency as one of the fundamental design parameters. This parameter acquires great relevance
in HPC systems when running artificial neural networks and Machine
Learning applications. Thus, this article shows an example of how to
estimate and consider energy consumption in a real case of EEG classification. An efficient and distributed implementation of the KNN algorithm that uses mRMR as a feature selection technique to reduce the
dimensionality of the dataset is proposed. The performance of three different workload distributions is analyzed to identify which one is more
suitable according to the experimental conditions. The proposed approach outperforms the classification results obtained by previous works.
It achieves an accuracy rate of 88.8% and a speedup of 74.53 when running on a multi-node heterogeneous cluster, consuming only 13.38% of
the energy of the sequential version.

Keywords: Parallel and distributed programming · Heterogeneous
clusters · Energy-aware computing · EEG classification · KNN · mRMR

1 Introduction

Greenhouse gas emissions caused by energy consumption associated with the
proliferation of equipment, applications, and computer programs is growing wor-
ryingly. There are estimates that determine that ICT could contribute up to

© The Author(s), under exclusive license to Springer Nature Switzerland AG 2023
I. Rojas et al. (Eds.): IWANN 2023, LNCS 14134, pp. 505–516, 2023.
https://doi.org/10.1007/978-3-031-43085-5_40

23% of global greenhouse gas emissions in 2030 [8]. To reduce consumption, it is necessary to address the problem simultaneously from different approaches. One of them is to analyze the energy consumption of programs and try to run them in the most energy-efficient configuration. The other one, to exploit the qualities of distributed and heterogeneous parallel platforms. Currently, when designing applications, it is not enough to consider only the precision of the results and execution time, but energy efficiency must also be considered as a fundamental parameter. Among the most complex applications in terms of execution time and energy consumption are those related to Machine Learning. These High-Performance Computing (HPC) applications often have to process large amounts of data (Big Data) and are characterized by high algorithmic complexity. Therefore, this work shows the results of an investigation on the energy efficiency of a bioengineering application, based on the KNN (K-Nearest Neighbors) algorithm, capable of exploiting the qualities of a distributed and heterogeneous parallel platform, which also takes into account performance in terms of accuracy of results and execution time.

After this introduction, the rest of the article is structured as follows: Section 2 refers to different works in the literature related to the topic addressed. Section 3 details the proposed KNN implementation for EEG classification. Then, Sect. 4 analyzes the experimental results and discusses the importance of energy awareness in HPC systems. Finally, Sect. 5 provides the conclusions.

2   Background

The use of artificial neural networks and Machine Learning techniques in bioinformatics has experienced exponential growth in recent years. Among other causes is the considerable growth in the size of biological datasets, as in the case of Electroencephalography. Bioinformatics, among other topics, deals with EEG signals, which represent the electrical activity of different parts of the brain. EEG signals are used to aid in the diagnosis of disorders such as schizophrenia [1], epilepsy [4], dyslexia [19], depression [17], autism [11], or sleep problems [18]. They are also used to classify motor functions of the nervous system, such as movement of limbs or eyes [16], and for the classification of human emotions [14]. However, the main problem of working with EEG signals is their high dimensionality, which makes their correct classification difficult since most of them do not contain relevant information. Therefore, it is important to apply feature selection techniques to obtain the most relevant ones. One of the fundamental objectives of the artificial neural networks and Machine Learning is to recognize patterns in these signals for their subsequent classification. Indeed, the EEG classification problem is usually addressed through various Machine Learning methods, such as regression or clustering algorithms.

In this work, a KNN algorithm for instance-based supervised classification has been considered due to its good performance in this type of applications. The KNN algorithm generally tries to classify the instances (patterns) by assigning them to the predominant class among their K nearest neighbors. The steps required to classify each new instance are:

1. Calculate the distance between the instance to classify and the rest of training samples. In this work, the Euclidean distance has been used.
2. Sort the distances in increasing order.
3. Identify the predominant class among the nearest K distances (neighbors).
4. Assign the new instance to the predominant class.

3   The Proposed KNN for EEG Classification

EEG classification has been approached using the KNN algorithm together with the minimum Redundancy Maximum Relevance (mRMR) technique [12], which sorts the N_F features of the dataset from least to most relevant. This approach helps to deal with the dimensionality problem [15] and to reduce computation time by avoiding the evaluation of all 2^{N_F} possible subsets of features. Instead,

---

Algorithm 1: Pseudocode of the approach used by the workers to evaluate all possible values of K for a feature subset.

---

1 **Function** evaluateFeatureSubset(Tr, Te, Idx)

    Input   : Training dataset, Tr

    Input   : Test dataset, Te

    Input   : Index of the last column of the feature subset to evaluate, Idx

    Output: Accuracy of the best value of K, acc

2     $N_I \leftarrow$ getNumberInstances(Te)

    `// Vector with the correct predictions for each value of K`

3     $C_P \leftarrow \{0\}$

    `// Set as many OpenMP threads as logical CPU cores`

4   **#pragma omp parallel for**

5   **for** $i \leftarrow 1$ **to** N_I test rows **do**

        `// Distance between instance te[i] and all training instances`

6         $D \leftarrow$ calculateDistances($Te[i], Tr, Idx$)

7   **for** $k \leftarrow 1$ **to** N_I **do**

8             $P_C \leftarrow$ predominantClass(k, D)

9   **if** prediction P_C is correct **then**

10                $C_P[k]$++

11   **end**

12   **end**

13   **end**

14     $C_P \leftarrow$ sort(C_P, "Descending")

15     $acc \leftarrow \frac{C_P[1]}{N_I}$

16   **return** acc

17 **End**

the proposed algorithm only evaluates N_F of them, where in each one, the next feature from the list provided by mRMR is added. However, for each subset, all possible values of the K parameter are also evaluated to get the best classification accuracy. The mRMR implementation is based on the one proposed in [5] and takes the F-test Correlation Quotient (FCQ) criteria to select the next feature.

KNN algorithm has been parallelized by distributing feature subsets among worker nodes with MPI library and distributing the test instances to classify among CPU threads with OpenMP. The latter can be seen in the `#pragma omp parallel for` directive of Algorithm 1 (Line 4). The evaluation of all values of K for a test instance has been optimized by calculating its distance from all training instances (Line 6). In this way, the array D is reused in the loop of Line 7 to obtain the predominant class according to the value of K (Line 8). If the prediction is correct, the value of the k-th position of the prediction vector, C_P, is incremented by one. Once all instances have been classified, the prediction vector is sorted and the first position is used to calculate the classification accuracy of the best K (Lines 14 and 15).

3.1   A Distributed Master-Worker Scheme for Node-Level Parallelism

The proposed implementation, whose pseudocode can be found in Algorithm 2, follows a master-worker scheme where the master tells each worker which feature subset must use to evaluate a KNN. The execution ends when there is no more work to process and the function returns the best accuracy found (Line 36). The operation is as follows: the master waits in Line 8 for some worker to request its first job or to return the result of one of them, which is also implicitly associated with the assignment of a new job. The message type is identified by the MPI tags described in Table 1. When the master receives a result, it checks if the accuracy of that job (feature subset) is better than the current one. If so, update its value (Lines 9 to 11) and send a new chunk with the `JOB_DATA` tag (Line 14). Before sending work to a worker, the master checks for unprocessed chunks. If there is no availability, the worker will receive the `STOP` tag and stop its execution since there is no more work to do (Line 16).

Regarding the workers (Lines 20 to 35), they apply the mRMR algorithm on the training dataset to obtain the ranked list of features. A worker requests jobs by sending a message with the `FIRST_JOB` tag (Line 23). For each chunk of

Table 1. MPI tags used during communications between master and workers.

MPI tag	Description	Sender	Receiver
`FIRST_JOB`	Request for the first job	Worker	Master
`JOB_DATA`	There is work to do	Master	Worker
`RESULT`	Return the result of the job	Worker	Master
`STOP`	No more work to be done	Master	Worker

Algorithm 2: Pseudocode of the master-worker approach.

---

1 **Function** Main(Tr, Te, C_S, N_{Wk})

    Input : Training and test datasets, Tr and Te

    Input : Maximum number of features to send to workers (chunk size), C_S

    Input : Number of workers nodes, N_{Wk}

    Output: Accuracy of the best feature subset, $bestAcc$

2    $bestAcc \leftarrow -1$

3   **if** *Master* **then**

4        $N_F \leftarrow$ getNumberFeatures(Tr)

5        $indexList \leftarrow \{1, \ldots, N_F\}$

6        $stops \leftarrow 0$

7   **while** $stops \neq N_{Wk}$ **do**

8   MPI::Recv(acc, tag)

9   **if** *tag* is RESULT **and** $acc > bestAcc$ **then**

10 | $bestAcc \leftarrow acc$

11   **end**

            // Next chunk to send, e.g. indices 10,11,12 when $C_S = 3$

12            $chunk \leftarrow$ getNextFeatureChunk($indexList, C_S$)

13   **if** size($chunk$) > 0 **then**

14 | MPI::Send($chunk$, JOB_DATA)

15   **else**

16 | MPI::Send(NULL, STOP)

17 | $stops \leftarrow stops + 1$

18   **end**

19   **end**

20   **else**

        // mRMR. Reordering of datasets for coalesced memory access

21        $rankedFeatures \leftarrow$ mRMR(Tr)

22        $Tr, Te \leftarrow$ sortDatasets($Tr, Te, rankedFeatures$)

        // Start asking the master for workloads

23   MPI::Send(NULL, FIRST_JOB)

24   MPI::Recv($chunk, tag$)

25   **while** *tag* is JOB_DATA **do**

26   **for** $i \leftarrow 1$ **to** size($chunk$) **do**

27               $acc \leftarrow$ evaluateFeatureSubset($Tr, Te, chunk[i]$)

28   **if** $acc > bestAcc$ **then**

29 | $bestAcc \leftarrow acc$

30   **end**

31   **end**

32   MPI::Send($bestAcc$, RESULT)

33   MPI::Recv($chunk, tag$)

34   **end**

35   **end**

36   **return** $bestAcc$

37 **End**

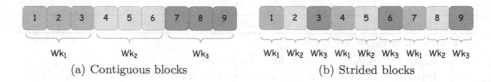

(a) Contiguous blocks   (b) Strided blocks

Fig. 1. The two different static workload distributions used by the master node.

features received (Lines 26 to 31), the worker obtains the accuracy of the corresponding feature subset by calling in Line 27 to the `evaluateFeatureSubset` function of Algorithm 1. If the accuracy of the processed feature subset is greater than the existing one, it will be update. Once all the possible subsets of the received chunk have been processed, the worker returns the best accuracy obtained to the master by sending a message with the `RESULT` tag, and waits for the assignment of a new job (Lines 32 and 33). This process is repeated until the `STOP` tag is received, indicating that the worker can end its execution.

3.2   Ways of Distributing the Workload

As previously seen in Sect. 3.1, feature chunks are sent to workers via the message-passing interface provided by the MPI library. This allows the application to distribute the workload among the different nodes of the cluster. However, in the algorithm proposed here, the workload of each feature subset is asymmetric since the number of features in each one is variable. For example, suppose a dataset with ten features, two worker nodes, and a chunk size of 2. In this scenario, the first chunk that the master will send contains the indices 1 and 2, corresponding to the subsets $\{1\}$ and $\{1, 2\}$. The second worker will receive indices 3 and 4 to compute the subsets $\{1, 2, 3\}$ and $\{1, 2, 3, 4\}$. In other words, a higher index implies computing more features within the KNN and, consequently, a longer execution time. To deal with workload imbalance, by default, the procedure distributes chunks dynamically according to the specified chunk size. Although this has the disadvantage of increasing communications, it is essential in heterogeneous systems to avoid performance drops. If the user wants, the master can also give each worker a chunk of features at the start of the algorithm by dividing the number of total chunks by the number of workers. This can be done in two ways: contiguous or striding features (see Fig. 1). The strided assignment could reduce the workload imbalance [6] present in the contiguous blocks alternative since each node would compute similar subsets.

4   Experimental Results and Discussion

All experiments are repeated ten times to obtain more reliable measurements of the application's behavior. The application has been executed in an HPC cluster composed of eight heterogeneous NUMA nodes whose CPU devices are detailed in Table 2. The cluster runs the *Rocky Linux* distribution (v8.5) and

Table 2. Characteristics of the cluster used in the experiments.

Node	CPU			RAM		
	Model	Total cores/threads	TDP (W)	Frequency (MHz)	Frequency (MHz)	Size (GB)
Master	2x Intel Xeon E5-2620 v2	12/24	160		1,600	
1	1x Intel Xeon E5-2620 v4	8/16	85	2,100	2,133	32
2	2x Intel Xeon E5-2620 v4	16/32	170			
3 to 7	2x Intel Xeon Silver 4214	24/48	170	2,200	2,933	64

schedules the jobs using the SLURM task manager (v20.11.7) [10]. The *C++* source codes have been compiled with the GNU compiler (GCC v8.5.0), the OpenMPI library (v4.0.5), and optimization flags -O2 -funroll-loops. The energy measurements of each node have been obtained from a custom wattmeter, called *Vampire*, capable of capturing in real-time information of instantaneous power (W) and accumulated energy consumed (W · h) for each computing node.

The EEG dataset belongs to the BCI laboratory of the University of Essex and corresponds to a human subject coded as #104 [3]. The dataset includes 178 signals for training and another 178 for testing, each with 3,600 features. As the signals can belong to three different motor imagery movements (left hand, right hand, and feet), the KNN algorithm deals with a 3-class classification problem.

4.1 Classification Analysis

The proposed algorithm achieves a Kappa index of 0.83 using the first 62 features of mRMR and $K = 18$. This solution widely outperforms a run without mRMR (0.34), and other approaches in the literature that use the same dataset: [2] (0.63), [9] (0.70), and [13] (0.76). The result has been validated by replicating its value when executing the KNN with the Python and Matlab languages and the same input parameters. Figure 2(a) shows the corresponding confusion matrix, which reveals an overall accuracy rate of 88.8%. The evolution of the accuracy rate and the Kappa index depending on the number of selected features can be observed in Fig. 2(b). The general trend is that both metrics increase as new features are added until reaching the peak (62), and then progressively fall. It seems that the algorithm's convergence is penalized with the selection of many features, which are also irrelevant. It is also observed that the values of accuracy and Kappa distance themselves for extreme values of the graph.

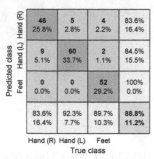
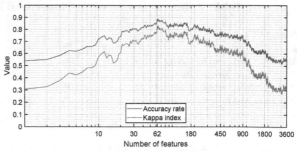

(a) Confusion matrix of the best case (62 features)

(b) Accuracy rate and Kappa index when adding features

Fig. 2. Classification results of the proposed approach when using mRMR and $K = 18$.

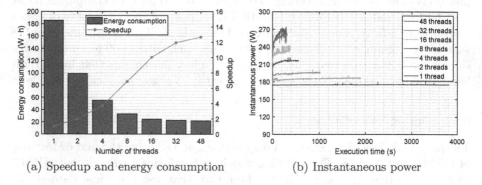

(a) Speedup and energy consumption

(b) Instantaneous power

Fig. 3. Performance of the proposed approach in a single-node configuration when varying the number of OpenMP threads.

4.2 Energy-Time Performance

Figure 3 shows the application's performance after running on Node 3. The goal is to depict the speedup scalability of the first parallelism level, which occurs within each computing node when the number of logical CPU cores is increased. From Fig. 3(a), it can be seen that the maximum speedup of 12.67 is obtained using the 48 threads available in the node. Its behavior is approximately linear, up to four threads, and logarithmic for higher values. The main reason is that the motherboard supports quad-channel memory. Increasing the number of threads above four causes competition for memory accesses since not all of them can do so simultaneously. It is also due, although to a lesser extent, because the workload for each thread decreases and the cost of managing threads becomes important. This means that the speed gain could increase with larger datasets that allow threads to compute for longer periods. It has also been found that distributing the instances to be classified among the threads statically provides the best performance (Line 4 of Algorithm 1). This is expected, as indicated

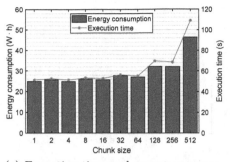 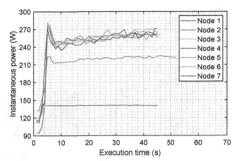

(a) Execution time and energy consumption when the chunk size is increased

(b) Instantaneous power of the best chunk size found (4)

Fig. 4. Performance of the dynamic workload distribution when using all nodes.

in [7], because the computational workload is the same for each thread, so a dynamic distribution has been discarded. Regarding energy consumption, also for the case of using 48 threads, it provides the lowest total energy consumption. This may seem contradictory since the use of more resources is associated with a higher instantaneous power (see Fig. 3(b)). However, energy consumption also depends linearly on execution time, and since speedup increases at a greater rate than energy, total energy consumption is less.

The performance of the hybrid MPI-OpenMP approach that corresponds to the second level of parallelism is shown in Figs. 4 and 5. On the one hand, Fig. 4 exposes the behavior of the application when all nodes are used, and the workload distribution is dynamic. Figure 4(a) reveals that a very large chunk size leads to worse execution time and energy consumption mainly due to workload imbalance. The instantaneous power of each node for a chunk size of 4 is plotted in Fig. 4(b). Despite the fact that the optimal size ranges from 1 to 64, the value 4 has been set as definitive since it works well with few nodes and should do so with more than 7. What is observed in the figure is that most nodes end up simultaneously, which is expected in dynamic workload distributions.

On the other hand, Fig. 5 compares the different workload distributions. The number of computation nodes indicated in Fig. 5(a) does not correspond to the order shown in Table 2. Instead, the nodes in the graph correspond to homogeneous and heterogeneous ones in that order. That is, first Nodes 3 to 7, and later Nodes 1 and 2. In this way, the scalability of the program can be analyzed according to the type of node added. As expected, for all distributions, the observed speedup grows linearly as more nodes are used, but up to 5 and in different magnitudes. From this point on, only dynamic distribution continues to scale its performance, although to a lesser extent since heterogeneous nodes begin to be used. In fact, it can be seen that the increase in speedup is in line with the added heterogeneous node: adding Node 2 boosts speed up more than adding Node 1 (the slowest one), until reaching a maximum speedup of 5.88. With respect to a sequential execution (1 thread), the application achieves a

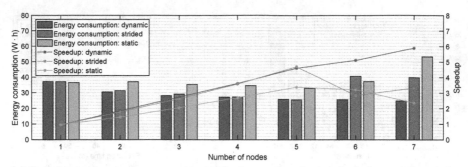

(a) Speedup and energy consumption when increasing the number of computing nodes

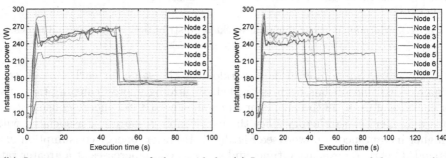

(b) Instantaneous power of the strided distribution when using all nodes

(c) Instantaneous power of the static distribution when using all nodes

Fig. 5. Comparison of performance between the different workload distributions.

speedup of 74.53, consuming only 13.38% of energy. The use of heterogeneous nodes also negatively affects static and dynamic distributions but in different ways. In the case of strided, speedup plummets for 6 nodes and improves slightly after adding the last one. Not so for the static distribution, which worsens its performance for each node added. The instantaneous power in Fig. 5(c) reveals that workload imbalance is responsible. Here, the nodes finish computing in a staggered manner, with a long interval between the first ($t = 30$) and the last ($t = 125$). In the strided case (Fig. 5(b)), only the homogeneous nodes finish at the same time, but before the heterogeneous nodes, causing a bottleneck. Based on the results, it can be affirmed that the dynamic distribution provides the best results in speedup, energy consumption, and scalability since the speed gain is very close to the number of nodes used to compute.

5   Conclusions

This work has proposed to investigate the energy efficiency of a bioengineering application capable of exploiting the qualities of distributed and heterogeneous parallel platforms. The use of mRMR for the selection of features has allowed for the improvement of the performance of existing approaches in the literature that use the same dataset. Another contribution of this article has been to consider energy efficiency as a fundamental parameter, unlike other works that focus only on the accuracy of the results and on the execution time. In addition, different workload distributions for the proposed procedure have been analyzed. The results have verified that a dynamic distribution is the most appropriate option to distribute asymmetric jobs in heterogeneous systems. They have also shown the importance of knowing the architecture when writing parallel code to take advantage of all available resources, reaching speedups of up to 74.53 consuming only 13.38% energy of sequential execution. Even so, the next step is to improve this result using accelerators such as GPUs and increasing data parallelism through vectorization techniques. Another way to reduce energy consumption could be through the use of an energy policy that stops or resumes the execution of the program according to the cost per Megawatt. Consequently, this policy would allow data centers to save energy or money, depending on the user's preferences, but in the latter it would be at the cost of lengthening the execution time and energy consumed.

Acknowledgements. Work funded by the Spanish Ministry of Science, Innovation, and Universities under grants PGC2018-098813-B-C31, PID2022-137461NB-C32 and ERDF funds. Also, the authors would like to thank Dr. Alberto Prieto, from the Department of Computer Engineering, Automation, and Robotics of the University of Granada, Spain, for his valuable collaboration in this work.

References

1. Akbari, H., Ghofrani, S., Zakalvand, P., Tariq Sadiq, M.: Schizophrenia recognition based on the phase space dynamic of EEG signals and graphical features. Biomed. Signal Process. Control **69**, 102917 (2021). https://doi.org/10.1016/j.bspc.2021.102917
2. Aquino-Brítez, D., et al.: Optimization of deep architectures for EEG signal classification: an autoML approach using evolutionary algorithms. Sensors **21**(6), 2096 (2021). https://doi.org/10.3390/s21062096
3. Asensio-Cubero, J., Gan, J.Q., Palaniappan, R.: Multiresolution analysis over simple graphs for brain computer interfaces. J. Neural Eng. **10**(4), 21–26 (2013). https://doi.org/10.1088/1741-2560/10/4/046014
4. Choubey, H., Pandey, A.: A combination of statistical parameters for the detection of epilepsy and EEG classification using ANN and KNN classifier. SIViP **15**(3), 475–483 (2020). https://doi.org/10.1007/s11760-020-01767-4
5. Ding, C., Peng, H.: Minimum redundancy feature selection from microarray gene expression data. In: Computer Society Bioinformatics Conference, CSB 2003, pp. 523–528. IEEE, Stanford, CA, USA, August 2003. https://doi.org/10.1109/CSB.2003.1227396
6. Ding, F., Wienke, S., Zhang, R.: Dynamic MPI parallel task scheduling based on a master-worker pattern in cloud computing. Int. J. Auton. Adapt. Commun. Syst. **8**(4), 424–438 (2015). https://doi.org/10.1504/IJAACS.2015.073191

7. Dong, Y., Chen, J., Yang, X., Deng, L., Zhang, X.: Energy-oriented openMP parallel loop scheduling. In: 6th International Symposium on Parallel and Distributed Processing with Applications, ISPA 2008, pp. 162–169. IEEE, Sydney, NSW, Australia, December 2008. https://doi.org/10.1109/ISPA.2008.68

8. Freitag, C., Berners-Lee, M., Widdicks, K., Knowles, B., Blair, G., Friday, A.: The climate impact of ICT: a review of estimates, trends and regulations. arXiv (2021). https://doi.org/10.48550/ARXIV.2102.02622

9. González, J., Ortega, J., Escobar, J.J., Damas, M.: A lexicographic cooperative co-evolutionary approach for feature selection. Neurocomputing **463**, 59–76 (2021). https://doi.org/10.1016/j.neucom.2021.08.003

10. Gvozdetska, N., Globa, L., Prokopets, V.: Energy-efficient backfill-based scheduling approach for SLURM resource manager. In: 15th International Conference on the Experience of Designing and Application of CAD Systems, CADSM 2019, pp. 1–5. IEEE, Polyana, Ukraine, February 2019. https://doi.org/10.1109/CADSM.2019.8779312

11. Ibrahim, S., Djemal, R., Alsuwailem, A.: Electroencephalography (EEG) signal processing for epilepsy and autism spectrum disorder diagnosis. Biocybernetics Biomed. Eng. **38**(1), 16–26 (2018). https://doi.org/10.1016/j.bbe.2017.08.006

12. Jo, I., Lee, S., Oh, S.: Improved measures of redundancy and relevance for mRMR feature selection. Computers **8**(2), 42 (2019). https://doi.org/10.3390/computers8020042

13. León, J., et al.: Deep learning for EEG-based motor imagery classification: accuracy-cost trade-off. PLoS ONE **15**(6), e0234178 (2020). https://doi.org/10.1371/journal.pone.0234178

14. Li, M., Xu, H., Liu, X., Lu, S.: Emotion recognition from multichannel EEG signals using K-nearest neighbor classification. Technol. Health Care **26**(S1), 509–519 (2018). https://doi.org/10.3233/THC-174836

15. Raudys, S.J., Jain, A.K.: Small sample size effects in statistical pattern recognition: recommendations for practitioners. IEEE Trans. Pattern Anal. Mach. Intell. **13**(3), 252–264 (1991). https://doi.org/10.1109/34.75512

16. Sabancı, K., Koklu, M.: The classification of eye state by using kNN and MLP classification models according to the EEG signals. Int. J. Intell. Syst. Appl. Eng. **3**(4), 127–130 (2015). https://doi.org/10.18201/ijisae.75836

17. Saeedi, M., Saeedi, A., Maghsoudi, A.: Major depressive disorder assessment via enhanced K-nearest neighbor method and EEG signals. Phys. Eng. Sci. Med. **43**(3), 1007–1018 (2020). https://doi.org/10.1007/s13246-020-00897-w

18. Sharma, H., Sharma, K.: An algorithm for sleep apnea detection from single-lead ECG using Hermite basis functions. Comput. Biol. Med. **77**, 116–124 (2016). https://doi.org/10.1016/j.compbiomed.2016.08.012

19. Zainuddin, A.Z.A., Mansor, W., Khuan, L.Y., Mahmoodin, Z.: Classification of EEG signal from capable dyslexic and normal children using KNN. Adv. Sci. Lett. **24**(2), 1402–1405 (2018). https://doi.org/10.1166/asl.2018.10758

A Deep Neural Network
for G-Quadruplexes Binding Proteins
Classification

Francesco Di Luzio[1], Alessandro Paiardini[2], Federica Colonnese[1],
Antonello Rosato[1], and Massimo Panella[1(✉)]

[1] Department of Information Engineering, Electronics and Telecommunications,
University of Rome "La Sapienza", Rome, Italy
{francesco.diluzio,alessandro.paiardini,massimo.panella}@uniroma1.it
[2] Department of Biochemical Sciences "A. Rossi Fanelli", University of Rome
"La Sapienza", Rome, Italy

Abstract. In the last years, a plethora of studies unveiled the fundamental roles of RNA G-quadruplexes (RG4s), unique structural RNA
strands featuring guanine-rich nucleic acid sequences, in basic cellular
processes, as well as in the pathogenesis of important diseases, such as
cancer and neurodegeneration. As the knowledge of the pathological roles
played by RG4s has grown wider, the involvement of RG4-binding proteins in designing new diagnostic and therapeutic strategies has become
increasingly recognized, but the classification of these proteins is still
challenging. In this paper, we describe the architecture and the training
procedure of a deep neural network based on Long Short-Term Memory layers to classify RG4-binding proteins. The impressive classification
accuracy achieved on the test set provides a strong foundation for future
investigations on more data samples and several experimental purposes.

Keywords: G-quadruplexes Binding Proteins · RG4 Proteins
Classification · Deep Learning · Long Short-Term Memory

1 Introduction

Machine Learning (ML) and more in particular Deep Learning (DL) techniques
have already been applied to a large number of Bioinformatics applications and
studies. This is also due to the rapid growth of technology and advanced research
techniques, with the consequent exponentially increased amount of biological
data available [11]. More in general, this fact has created a growing need for
ML advanced tools and methods to extract meaningful insights and information
from a large amount of data in real-world classification tasks [6,12,21]. More
in particular, in Bioinformatics, ML and DL are often used to analyze large
biological datasets, such as genomic data, protein sequences, and clinical data,
to identify patterns and make predictions [26]. In fact, ML and DL have already
been applied to predicting protein structures [28], identifying disease-causing
mutations [20] and analyzing gene expression data [9], with interesting results.

© The Author(s), under exclusive license to Springer Nature Switzerland AG 2023
I. Rojas et al. (Eds.): IWANN 2023, LNCS 14134, pp. 517–528, 2023.
https://doi.org/10.1007/978-3-031-43085-5_41

In this domain, the focus of this paper will be on the construction of a DL model for the classification of G-quadruplexes (G4s) binding proteins. G-quadruplexes are non-canonical DNA and RNA structures (respectively, DG4 and RG4), formed by Guanine-rich sequences, which can occur in genomic DNA stretches such as in both coding and ncRNAs. The G4s forming Guanines are organized in multiple planar G-quartets, each one held together by eight hydrogen bonds and stacked with each other by π-π interactions [13]. In the last twenty years, a plethora of studies described the vast number of roles played in cells by G4s, from telomere stabilization to mRNA translation regulation to riboprotein complexes scaffolding [22]. As our knowledge of the roles played by G4s in physiological mechanisms has grown wider during these years, we also became aware of the fact that abnormal G4s presence drives the development of severe pathologies, leading us to estimate how pivotal G4s are in cellular systems [30]. In recent years, the mechanism of action of G4s, their peculiar roles, and their spread among different genomes is just starting to be unraveled, as those aspects that depend on G4s structural parameters, on their sequences, on their propensity to fold in other structural motifs, on their localization within cell's organelles and on their interacting molecules [4,8]. This has centered the interest in using G-quadruplexes as input for ML and DL algorithms, particularly for the prediction of functional genomic elements.

In this sense, DL and in particular Convolutional Neural Networks (CNNs) is applied in [10] to G4s to extract information on the formation and stability of G4 structures by learning the characteristics of G4 sequences and accurately classify them. Also 'G4Boost' presented in [5] reaches optimal accuracy. G4Boost is a decision tree-based prediction tool created to identify G4 motifs and predict their secondary structure folding probability and thermodynamic stability based on their sequence, nucleotide composition, and estimated structural topology. Prediction of G4s from DNA sequences is made also by the authors of [2] where on top of the sequence information, they improved their prediction accuracy by adding RNA secondary structure information. Also in [7] it is presented an outstanding ANN based on abstract sequence similarity. Their Neural Network (NN) was trained using as a dataset the sequences of the G4RNA database. The purpose of creating their network called 'G4NN' was to enable G4 detection without relying on a specific motif definition and to reduce the influence of expert assumptions. G4NN relies on learning from existing examples and treats both regular and irregular G4s equally.

As for the cellular localization and the chance to interact with different macromolecules, it is clear how G4-forming RNAs could play a wider range of roles inside cells than their DNA counterparts, due to their possibility to exit the nucleus and migrate in different cellular compartments [16]. Despite this, wider attention has been dedicated to DNA G4s rather than RNA G4s, leaving a large part of RG4s roles potentially undiscovered. RG4s play an essential role in many physiological cell functions, and their importance makes it easy to realize how they can also be critical in various pathologies. In fact, studies have shown that the deregulation or abnormal formation of G4s can contribute to the development of different diseases, or be a key player in their pathogenic mechanisms [17,30]. RG4s have been reported to play a significant role in the development of various forms of cancer, such as breast cancer [18], Mixed Lineage Leukemia (MLL) [27], and oral squamous cell carcinoma [14]. RG4s have been associated with neurological disorders caused by triplet expansion of G4-forming sequences, such as Fragile X-chromosome Syndrome (FXS), Amyotrophic Lateral Sclerosis (ALS), and Frontotemporal Dementia (FTD) [29]. As for the role of RG4s in viral pathogenesis, they have been proposed to be involved in cellular and viral replication, recombination, and gene expression control, in various virus genomes (e.g., Human immunodeficiency virus (HIV-1), Epstein-Barr virus (EBV), papillomavirus (HPV)) [15]. Although DG4s have important roles in genome maintenance, stability, and transcription, we have chosen to focus our analysis on RG4s for several reasons. First, RG4s have a higher likelihood of forming alternative, non-canonical structures. Moreover, they can migrate to different cell compartments and function as scaffolds for assembling protein-nucleic acid complexes. Finally, RG4s are specifically implicated in many pathologies where their overexpression can impede RNA G-quadruplexes binding proteins (R4BPs), modify gene expression, or alter mRNA alternative splicing.

These considerations have led us to prioritize RNA G4s in our efforts to better define R4BPs and their binding motifs, and to investigate their involvement in a wider range of physiological and pathological mechanisms. The same situation can be noticed in studies on G4s Binding Proteins (G4BPs), where DNA G-quadruplexes binding proteins (D4BPs) like telomere-binding proteins, promoter region binding proteins, and G4s resolving helicases, received deeper attention compared to R4BPs [4]. As our knowledge of RG4s has grown, we have realized that a significant portion of their roles within the cell involves mediating or regulating the activity of certain proteins. As a result, there has been a gradual increase in the number of studies focused on RG4s and the proteins that can recognize and bind to them [4,16,17,23]. This has enabled us to gain a much better understanding of RG4 formation, its roles, and its potential use as therapeutic targets. To date, at least five different protein sequence motifs or domains have been found to be capable of recognizing and binding to RG4s [1,19,24,25]. However, given the current lack of information regarding RG4-binding proteins, it is possible that there are many other RNA G-quadruplexes Binding Motifs (R4BMs) that have yet to be discovered, and which may be able to specifically recognize some of the many different RG4 structures that have been observed.

While the interest in the field of identifying proteins that can bind to RG4s has been increasing over the years, there has been a noticeable lack of research that utilizes ML and DL architectures for this purpose. In fact, despite the potential benefits of these advanced technologies in accurately detecting RG4-binding proteins, there has been a surprisingly limited exploration in this direction. Further research could help to identify novel RG4-binding proteins, potentially finding also new therapeutic targets for a range of diseases, including cancer, neurodegenerative disorders, and other genetic conditions that have been demonstrated to be associated with RG4s. The focus of our paper on the crucial importance of RG4s and the proteins that recognize and bind to them, combined with the innovative use of DL techniques, sets it apart as a pioneering work in the field of RNA research. Accordingly, the purpose of our study is to shed new light on the dynamic interactions between RG4s and their binding proteins creating a DL-based approach for the binary classification of RG4-binding proteins.

The problem that this paper aims to explore is the one of classifying RG4-binding proteins, an area of research that has been largely unexplored and is considered a significant challenge to the scientific community. In fact, the main approach that has been used until nowadays to identify RG4-binding proteins involves experimental methods such as high-throughput screening or structural biology techniques. However, these methods can be time-consuming and expensive, and may not always be feasible. Because of that, this new open area of research can be studied and analyzed with the implementation of AI technologies that can be trained to detect hidden patterns and relationships in the complex sequences of amino acids. By putting as input to the algorithm relevant data on RG4-binding proteins, along with their classifications, a Supervised Learning NN model can be trained to accurately classify new RG4-binding proteins based on their molecular and structural properties.

Also, Recurrent Neural Networks (RNNs), and in particular Long Short-Term Memory (LSTM), which is a type of NN that is particularly well-suited for analyzing sequential data as the sequences of nucleotides that make up RNA or DNA, have proven to be a powerful tool in solving various computational problems, and also in this area, they could show great potential in predicting and analyzing new RG4-binding proteins. These types of RNNs are designed to process sequential data, making them well-suited to analyze the complex and multi-layered patterns characterizing these types of biological sequences. This is possible thanks to their inherent ability to process sequential data and learn about patterns. Overall, using such models, researchers can potentially uncover novel amino acid sequences that can bind to RG4 structures, unraveling important biological pathways which can potentially accelerate the discovery of new biological insights related to these structures.

2   Materials and Methods

The goal of our research is to advance the current understanding by concentrating on the classification of proteins that bind to RG4. In our opinion, DL can position itself as a significant player in addressing issues within this area. Hence, what we propose is a DNN for the binary classification of RG4-binding amino acidic sequences. In the following, there is an explanation of the preprocessing of the data employed to train this method, as well as a description of the proposed neural architecture.

The dataset employed in this paper is composed of the G-quadruplex bindings proteins of Homo Sapiens listed in [3] merged with other proteins of the human proteome. In particular, each RG4-binding protein contains several amino acidic sequences which could bind RG4. All the sequences in the dataset are composed of 26 residues. This dataset is particularly suitable for this kind of application because it contains a consistent number of data samples (i.e., sequences) that are linked with the relative gene name and with some statistical measurements on their capability to be RG4-binding sequences. When working with DL models, the availability of a large number of samples can represent a really fundamental concept for the success of the experimental activity. Moreover, the dataset is open access and hence can be used to replicate the experiments proposed herein.

At first, each amino acidic sequence x^i in the dataset X was associated with a probability P^i of being RG4-binding. Hence, the first preprocessing step consisted of the selection of a threshold value T of this feature for the sequences to be considered binding. All the sequences x^i such that $P^i < T$ have been deleted from the dataset and not used for the experimental activity while the remaining amino acidic structures have been considered as RG4-binding. After this selection procedure, the dataset was composed of RG4-binding alphabetic sequences in which each letter was associated with an amino acid. To select the control group for the experiments, and for obtaining a balanced dataset, an equal number of non-binding sequences have been selected from the human proteome. At this point, a vectorization method was needed to associate each amino acid with a numerical value. This preprocessing step on the adopted dataset has been conducted with the results shown in Table 1, in which the name of each amino acid associated with an alphabetic letter has been linked with an integer number (from 0 to 19) for computational purposes. After the vectorization procedure, the dataset X is composed of samples that are represented by numerical sequences $x^i = [x_1^i, x_2^i, ..., x_{26}^i]$, which are ready to be fed to the neural network structure described in the following.

One of the main contributions of the approach presented in this work relies upon the employment of a relatively simple DNN architecture for the classification of RG4-binding proteins. This is obtained thanks to the preprocessing procedure described in the previous subsection that leads to a new simple kind of data associated whit each amino acidic structure. The detailed use of numerical sequences as input allows for the potential utilization of RNNs to extract features and analyze not only the entire data sample (a single sequence), but also the relationship between the various amino acids within the sequence. Specifically, the output of the preprocessing step leads to the use of a network based on LSTM to examine the series of adjacent amino acids in the data sets and their relationship.

Let x^i be the i-th numerical sequence feeding the neural model. 'LSTM layer 1' is the first layer of the architecture. At step h, it receives the input integer x_h^i and updates the layer's state basing the computation on the previous inputs of the amino acidic sequence x_{h-1}^i, x_{h-2}^i, etc. The output of this network's layer is obtained with the pair of recurrent state equations described in (1), which are associated with the layer itself:

$$\mathbf{c}_h^i = f\left(x_h^i, \mathbf{q}_{h-1}^i, \mathbf{c}_{h-1}^i; \boldsymbol{\theta}_c\right), \tag{1a}$$

$$\mathbf{q}_h^i = g\left(x_h^i, \mathbf{q}_{h-1}^i, \mathbf{c}_{h-1}^i; \boldsymbol{\theta}_q\right). \tag{1b}$$

Table 1. Amino Acids vectorization.

Amino Acid	1-letter	Numerical Value
Alanine	A	0
Arginine	R	5
Asparagine	N	11
Aspartic acid	D	14
Cysteine	C	17
Glutamic acid	E	8
Glutamine	Q	6
Glycine	G	9
Histidine	H	13
Isoleucine	I	18
Leucine	L	3
Lysine	K	2
Methionine	M	7
Phenylalanine	F	16
Proline	P	10
Serine	S	4
Threonine	T	1
Tryptophan	W	15
Tyrosine	Y	19
Valine	V	12

In this system of equations c_h^i and q_h^i are the 'cell' and 'hidden' state vectors of the LSTM layer, respectively. They are computed for the current h-th step; $f(\cdot)$ and $g(\cdot)$ are general functions obtained by the combination of the gate equations of the LSTM model, depending on the chosen gates and activation functions; θ_c and θ_q are the layer's weights which are selected by the training procedure. The output of this layer is the vector q^i which is the three-dimensional vector of all the hidden states computed for all the different steps. This vector is computed at the end of the i-th amino acidic sequence. The same kind of recurrent system of equations can be used to represent the LSTM layer 2. The difference between these two layers relies on the fact that the second recurrent layer takes in input the hidden state vector output of the previously described LSTM layer 1 and outputs only the last hidden state vector. Afterward, the output of the LSTM layer 2 is fed as input to the fully connected layer 1. The output of this layer is computed as described in (2):

$$\mathbf{d}^i = \rho \left(\mathbf{W}_1 \mathbf{k}^i + \mathbf{b}_1 \right) , \tag{2}$$

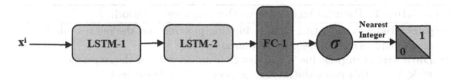

Fig. 1. Graphical representation of the proposed G4-LSTM neural network for classification.

where $\mathbf{k}^i$ is the hidden state vector output of LSTM layer 2, the weight matrix $\mathbf{W}_1$ and the bias vector $\mathbf{b}_1$ are estimated through the learning procedure, and $\rho(\alpha)$ is the activation function of the neurons in this layer. Finally the fully connected layer 2 is the last layer characterized by a single neuron whose output can be computed as shown in (3),

$$\mathbf{L}^i = \sigma\left(\mathbf{W}_2\mathbf{d}^i + \mathbf{b}_2\right), \qquad (3)$$

where the weight $\mathbf{W}_2$ and the bias vector $\mathbf{b}_2$ are estimated by the training algorithm, while $\sigma(\cdot)$ is the activation function of the sole neuron in the layer. As shown in (4), the sigmoid is employed as activation function, returning the final output of the network $\hat{y}^i$ as an integer number between 0 and 1 approximated with the Nearest Integer Function:

$$\hat{y}^i = \left\lfloor \frac{1}{1 + e^{-\mathbf{L}^i}} \right\rceil. \qquad (4)$$

The proposed neural network architecture, which will be denoted in the following as 'G4-LSTM' is illustrated in Fig. 1. The layers involved are described as follows:

- *LSTM layer 1.* It is the first recurrent layer of the architecture. The main hyperparameter of this layer is the number Q_1 of LSTM-cells involved. The layer's output is the vector of hidden states computed at the end of the processing of the entire sequence x^i. The activation function used for this layer is the hyperbolic tangent.
- *LSTM layer 2.* It is the second recurrent layer. The main hyperparameter of this layer is the number Q_2 of cells involved. Also in this case the activation function employed is the hyperbolic tangent.
- *Fully Connected layer 1.* It is the typical feed-forward layer of the network with N neurons connecting the output of the LSTM layer to the succeeding final layer. For this layer, the employed activation is the sigmoid function.
- *Fully Connected layer 2.* It is the last layer, consisting of one only neuron with classification purposes, with a sigmoid activation function for the prediction of the probability of belonging to the RG4-binding class.
- *Nearest Integer function.* The nearest integer function is employed in order to define the predicted class from the probabilities of the previous layer.

The pseudo-algorithm of the entire methodology is summarized in the Algorithm 1 table.

Algorithm 1. Pseudo-Algorithm of the proposed method.

Require: Dataset consisting of RG4-binding proteins and the relative binding amino acidic sequences composed of 26 amino acids.
1: **Data screening:** Define T and delete from the dataset the sequences x^i such that $P^i < T$. At this point the dataset is composed of D elements.
2: **Control group selection:** select D non-binding sequences from the human proteome composed of 26 amino acids and merge the two sets in a sole dataset X with $2D$ samples.
3: **Vectorization:** substitute each letter associated with an amino acid with the corresponding integer according to Table 1.
4: **Hyperparameter definition:** define the set of hyperparameters Q_1, Q_2, N and the number of epochs E for the training procedure.
5: **Training and Test set definition** divide the dataset into training set and test set
6: **Training procedure:** train the network using the data samples in the training set.
7: **Test procedure:** test the network evaluating the performance of the final classification using the data samples in the test set.

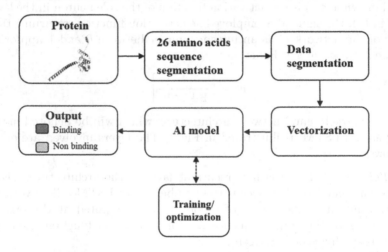

Fig. 2. Schematic representation of the proposed methodology.

3   Experimental Results

The model presented in this paper has been trained on a dataset of 1572 samples and tested on a set of 524 samples. 20% of the training data is employed for validation purposes during the training procedure. The ADAM algorithm has been used to train the model, with the learning rate set to 0.001 and a mini-batch size of 32, selected as default values. The default value used for the parameter T introduced in Sect. 2 is 0.999. The best combination of hyperparameters and additional training options have been identified using a grid search procedure to prevent overfitting and improve the overall ability of the proposed approach.

The optimal configuration of the parameters found with this procedure, referring to Fig. 1, is provided in the following for reproducibility purposes. The number of cells for the first LSTM layer is $Q_1 = 100$, the number of cells of the second LSTM layer is $Q_2 = 50$ and the number of neurons for the Fully Connected layer 1 is $N = 50$. Moreover, also the optimal number of epochs for the training procedure has been set through grid search: $E = 500$.

Python and TensorFlow® were utilized for all experiments on a computer system comprising an AMD Ryzen™ 7 5800X 8-core CPU with a clock speed of 3.80 GHz and 64 GB of RAM. Training, testing, and hyperparameter optimization were performed using an NVIDIA® GeForce™ RTX 3080 Ti GPU with 1.365 GHz clock speed and 12288 MB of GDDR6X RAM. Given that the network parameters are initially chosen randomly, the approach is additionally tested by conducting 10 runs with 10 different random seeds (from 0 to 9) for the initialization, and the average and standard deviation of the error values across these runs are reported to demonstrate the method's resilience to randomness. Given that also the data samples pertaining to the training and validation set are selected randomly after the selection of the random seed, the sets are not perfectly balanced.

The model's performance is evaluated using accuracy, which is the proportion of correctly classified samples to the total number of samples in the training/test set. The optimization procedures employ the binary cross-entropy (BC) loss function, as defined in (5):

$$BC = -\frac{1}{S} \sum_{i=1}^{S} \left[y_i \log(\hat{y}_i) + (1 - y_i) \log(1 - \hat{y}_i) \right], \tag{5}$$

where S is the number of samples in the training/test set, the term y_i represents the true binary label (either 1 or 0) assigned to the i-th sample, while $\hat{y}_i$ denotes the probability estimated by the model being used that the i-th sample represents a protein binding G4. In this particular setting, the binary label 1 has been assigned to the sequences related to RG4-binding proteins, and the binary label 0 to the non-binding amino acidic sequences. With these premises, the trained model can reach an accuracy on the training set equal to 0.9 and an accuracy on the test set equal to 0.9. From a graphical perspective, the numerical results are displayed in Fig. 3, where the confusion matrix achieved by the model on the test set is proposed. The average time needed to train the model with this configuration is 186.97 s and the average time needed for inference on the entire test set is 40.12 s.

Starting with the discussion on the numerical results, it can be seen that the method herein proposed can reach a consistent accuracy on the test set, being able to classify RG4-binding proteins with notable precision. Also looking at the confusion matrix the goodness of the proposed method is notable. This figure is also used to display the light unbalancing of the data samples of the two different classes in both the training and test sets. The lack of benchmarking with other solutions can be a limitation of the proposed approach; however, the consistent level of accuracy reached on the test set with the employment of

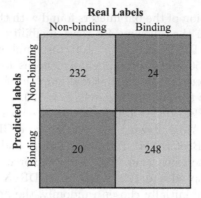

Fig. 3. Confusion matrix obtained by applying the G4-LSTM network on the test set.

several different random seeds is a strong signal of the goodness of the approach and of its resilience to randomness.

Moreover, we decided to keep the model as simple as possible in order to make it easier to interpret and understand the underlying mechanisms that govern its performance. From the applicative point of view, this method is representative of a unique procedure for the recognition and classification of RG4-binding amino acidic sequences and its easy implementation highly increases its usability. The limited training and inference times are proof of the usability of the proposed methodology which can possibly be retrained every few minutes when new data samples are available. In the given context, we are aware that the requirement for real-time working algorithms is not a priority. Instead, the main focus was on creating a model that can perform the given task accurately and efficiently. However, given the variability in training time, it was determined that the best approach would be to keep the model design as simple as possible ensuring that the training process could be completed within a reasonable time frame.

4   Conclusions

Considering the importance of exploring and discovering novel and practical applications of deep learning techniques in the biochemical domain, there is a significant motivation to develop customized solutions for such applications. In this study, we concentrated on the intricate and potentially critical task of categorizing RG4-binding proteins with the employment of an LSTM-based model utilizing accessible and open-source chemical data.

The key element of the research project relies upon the use of a DNN for the classification of RG4-binding amino acidic sequences. In this sense, the study, planning, and implementation of recurrent DNN for the research of new RG4-binding sequences and proteins. After conducting experimental analysis, we were able to confirm the beneficial aspects of the suggested structure, which led to the reporting of high accuracy with a high degree of confidence. As a result, this approach can be considered reliable and practical for use. Moreover, the interesting results obtained prove the goodness of the methodology which can be expanded for even more complex applications such as the computation of the probability of each amino acidic sequence to be RG4-binding.

In our opinion, this work shows the potential to lay the foundation for creating more intricate neural structures that can classify RG4-binding proteins more effectively. Additionally, further experimentation with new data samples could enhance the proposed methodology.

References

1. Antcliff, A., McCullough, L.D., Tsvetkov, A.S.: G-quadruplexes and the DNA/RNA helicase DHX36 in health, disease, and aging. Aging (Albany NY) **13**(23), 25578 (2021)
2. Barshai, M., Aubert, A., Orenstein, Y.: G4detector: convolutional neural network to predict DNA G-quadruplexes. IEEE/ACM Trans. Comput. Biol. Bioinf. **19**(4), 1946–1955 (2021)
3. Brázda, V., Červeň, J., Bartas, M., Mikysková, N., Coufal, J., Pečinka, P.: The amino acid composition of quadruplex binding proteins reveals a shared motif and predicts new potential quadruplex interactors. Molecules **23**(9), 2341 (2018)
4. Brázda, V., Hároníková, L., Liao, J.C., Fojta, M.: DNA and RNA quadruplex-binding proteins. Int. J. Mol. Sci. **15**(10), 17493–17517 (2014)
5. Cagirici, H.B., Budak, H., Sen, T.Z.: G4Boost: a machine learning-based tool for quadruplex identification and stability prediction. BMC Bioinform. **23**(1), 1–18 (2022)
6. Canal, F.Z., et al.: A survey on facial emotion recognition techniques: a state-of-the-art literature review. Inf. Sci. **582**, 593–617 (2022)
7. Garant, J.M., Perreault, J.P., Scott, M.S.: Motif independent identification of potential RNA G-quadruplexes by G4RNA screener. Bioinformatics **33**(22), 3532–3537 (2017)
8. Huppert, J.L.: Structure, location and interactions of G-quadruplexes. FEBS J. **277**(17), 3452–3458 (2010)
9. Hwang, K.B., Cho, D.Y., Park, S.W., Kim, S.D., Zhang, B.T.: Applying machine learning techniques to analysis of gene expression data: cancer diagnosis. In: Methods of Microarray Data Analysis: Papers from CAMDA 2002, pp. 167–182 (2002)
10. Klimentova, E., Polacek, J., Simecek, P., Alexiou, P.: PENGUINN: precise exploration of nuclear G-quadruplexes using interpretable neural networks. Front. Genet. **11**, 568546 (2020)
11. Larranaga, P., et al.: Machine learning in bioinformatics. Brief. Bioinform. **7**(1), 86–112 (2006)
12. Liparulo, L., Zhang, Z., Panella, M., Gu, X., Fang, Q.: A novel fuzzy approach for automatic Brunnstrom stage classification using surface electromyography. Med. Biolog. Eng. Comput. **55**(8), 1367–1378 (2017)
13. Lipps, H.J., Rhodes, D.: G-quadruplex structures: in vivo evidence and function. Trends Cell Biol. **19**(8), 414–422 (2009)
14. Majumder, M., et al.: RNA-binding protein FXR1 regulates p21 and TERC RNA to bypass p53-mediated cellular senescence in OSCC. PLoS Genet. **12**(9), e1006306 (2016)

15. Métifiot, M., Amrane, S., Litvak, S., Andreola, M.L.: G-quadruplexes in viruses: function and potential therapeutic applications. Nucleic Acids Res. **42**(20), 12352–12366 (2014)
16. Millevoi, S., Moine, H., Vagner, S.: G-quadruplexes in RNA biology. Wiley Interdisc. Rev. RNA **3**(4), 495–507 (2012)
17. Mitrasinovic, P.M.: G-quadruplexes: emerging targets for the structure-based design of potential anti-cancer and antiviral therapies. Acta Chim. Slov. **67**(3), 683–700 (2020)
18. Modelska, A., et al.: The malignant phenotype in breast cancer is driven by eIF4A1-mediated changes in the translational landscape. Cell Death Dis. **6**(1), e1603–e1603 (2015)
19. Oyoshi, T., Masuzawa, T.: Modulation of histone modifications and G-quadruplex structures by G-quadruplex-binding proteins. Biochem. Biophys. Res. Commun. **531**(1), 39–44 (2020)
20. Pandey, M., Anoosha, P., Yesudhas, D., Gromiha, M.M.: Identification of potential driver mutations in glioblastoma using machine learning. Briefings Bioinf. **23**(6) (2022)
21. Proietti, A., Liparulo, L., Leccese, F., Panella, M.: Shapes classification of dust deposition using fuzzy kernel-based approaches. Measurement **77**, 344–350 (2016)
22. Rhodes, D., Lipps, H.J.: G-quadruplexes and their regulatory roles in biology. Nucleic Acids Res. **43**(18), 8627–8637 (2015)
23. Ruggiero, E., Richter, S.N.: G-quadruplexes and G-quadruplex ligands: targets and tools in antiviral therapy. Nucleic Acids Res. **46**(7), 3270–3283 (2018)
24. Saito, T., Yoshida, W., Yokoyama, T., Abe, K., Ikebukuro, K.: Identification of RNA oligonucleotides binding to several proteins from potential G-quadruplex forming regions in transcribed pre-mRNA. Molecules **20**(11), 20832–20840 (2015)
25. Selig, E.E., et al.: Biochemical and biophysical characterization of the nucleic acid binding properties of the RNA/DNA binding protein EWS. Biopolymers, e23536 (2023)
26. Shastry, K.A., Sanjay, H.A.: Machine learning for bioinformatics. In: Srinivasa, K.G., Siddesh, G.M., Manisekhar, S.R. (eds.) Statistical Modelling and Machine Learning Principles for Bioinformatics Techniques, Tools, and Applications. AIS, pp. 25–39. Springer, Singapore (2020). https://doi.org/10.1007/978-981-15-2445-5_3
27. Thandapani, P., et al.: Aven recognition of RNA G-quadruplexes regulates translation of the mixed lineage leukemia protooncogenes. eLife **4**, e06234 (2015)
28. Torrisi, M., Pollastri, G., Le, Q.: Deep learning methods in protein structure prediction. Comput. Struct. Biotechnol. J. **18**, 1301–1310 (2020)
29. Wang, E., Thombre, R., Shah, Y., Latanich, R., Wang, J.: G-quadruplexes as pathogenic drivers in neurodegenerative disorders. Nucleic Acids Res. **49**(9), 4816–4830 (2021)
30. Wu, Y., Brosh Jr., R.M.: G-quadruplex nucleic acids and human disease. FEBS J. **277**(17), 3470–3488 (2010)

Identification of Benign Tumor Masses Using Deep Learning Techniques Based on Semantic Segmentation

Mohamed El-Khatib[1], Oana Mihaela Teodor[1], Dan Popescu[1]([✉]), and Loretta Ichim[1,2]

[1] Faculty of Control and Computers, University Politehnica Bucharest, Bucharest, Romania
mohamed.el@stud.acs.upb.ro, oana_mihaela.teodor@stud.fim.upb.ro,
{dan.popescu,loretta.ichim}@upb.ro
[2] "Stefan S. Nicolau" Institute of Virology, Bucharest, Romania

Abstract. Ovarian tumors affect women of all ages and the main challenge for optimal therapeutic management is to determine whether there is a benign or malignant tumor. The main imagistic tool for the evaluation of ovarian tumors is pelvic ultrasonography. To support the diagnosis of clinicians several artificial intelligence applications and ultrasound computer-aided diagnosis systems are emerging in recent years. This paper covers a comparative study between different convolutional neural networks based on semantic segmentation, implemented, and proposed for the identification of four benign ovarian tumor masses (chocolate cyst, mucinous cystadenoma, teratoma, and simple cyst). The semantic segmentation networks used in our comparative study are based on DeepLab-V3+ networks with 5 different encoders and a fully convolutional network. The scope of this study is to present the performances of each network for each of the covered benign classes and to illustrate the ones with the best performances.

Keywords: Benign Tumors · Convolutional Neural Networks · Ovary · Semantic Segmentation · Ultrasonic Imaging

1 Introduction

Ovarian cancer ranks fifth in cancer deaths among women [1]. Its high mortality is partly due to its nontypical symptoms and high invasiveness based on the molecular structure. Early-stage ovarian cancer has a more favorable prognosis and survival rates, compared to advanced stages [2]. For this reason, extensive research focused on identifying serum tumor biomarkers, radiological imaging techniques, and risk stratification algorithms for early diagnosis of ovarian cancer.

Transvaginal pelvic ultrasonography, as the first-line modality for evaluating ovarian tumors, can decrease the number of surgical procedures for benign ovarian pathology and significantly influences the choices of gynecological oncologists in the management of patients with suspected ovarian cancer. There are internationally established ultrasound classification systems with promising results in assessing the rate of malignancy for

© The Author(s), under exclusive license to Springer Nature Switzerland AG 2023
I. Rojas et al. (Eds.): IWANN 2023, LNCS 14134, pp. 529–540, 2023.
https://doi.org/10.1007/978-3-031-43085-5_42

adnexal masses, such as O-RADS, GI-RADS, and IOTA rules [3]. Yet, subjective evaluation of grayscale and Doppler ultrasound findings can make an almost conclusive diagnosis of a few types of ovarian tumors [4].

Recently, image classification through deep learning approaches has proved good results for various medical diagnoses. Well-trained neural networks can classify ovarian tumors on 2D-ultrasound images by extracting comprehensive features automatically [5]. The CNNs (Convolutional Neural Networks) tested were diverse. For example, in [6] VGG 16, GoogLeNet, ResNet 34, MobileNet, and DenseNet were used to classify ovarian tumors into two classes: benign and borderline/ malignant. Similarly, in [7] AlexNet, GoogLeNet, and ResNet were tested for tumors' classification into benign and malignant.

Ultrasound image analysis using deep neural networks can predict ovarian malignancy with a diagnostic accuracy comparable to that of human expert examiners [8] or better [9]. But given the wide spectrum of ovarian tumors' appearance, it is still necessary to extend the studies in the field and to evaluate the performances of different networks and parameter settings for improved results [10].

There are ovarian tumors with pathognomonic imaging features. Evaluation of such ovarian tumors using machine learning solutions could facilitate accurate diagnosis while decreasing the cost of unnecessary additional investigations and patient referral to gynecological oncology clinics. In this regard, we address some benign ovarian tumors that have specific ultrasound appearances (e.g., chocolate cysts, mucinous cystadenomas, teratomas, and simple cysts) and two additional classes (ovary and background).

Functional (simple) cysts like follicular cysts and corpus luteum cysts are the most common benign cystic structures found in healthy ovaries and they usually disappear without treatment within 2 to 3 menstrual cycles [11]. The follicular cysts are less prevalent than follicular cysts but may tend to rupture with intraperitoneal bleeding. In rare situations (e.g., torsion, rupture, and hemorrhage), operative intervention may be needed to treat these cystic masses [12].

Ovarian teratomas, also known as ovarian dermoid cysts, are the most common group of ovarian germ cell tumors. Most mature cystic teratomas are benign, but in about 1–2% of cases, they can become cancerous. They make up 70% of benign masses before menopause and 20% post-menopause [13]. Ovarian teratomas can contain skin, hair, teeth, fat and muscle, and even thyroid and brain tissue.

Ovarian cystadenomas are common benign epithelial neoplasms, the most frequent types including serous and mucinous cystadenomas. From this category, we studied mucinous cystadenoma, which can require a differential diagnosis with a cystic mature teratoma. At the benign end of the spectrum of mucin-containing epithelial ovarian tumors, mucinous cystadenoma can grow much larger than other adnexal masses, measuring up to 30 cm. Recognized as a precursor of ovarian cancer, it may slowly transform into borderline tumors and invasive ovarian cancer [14].

Endometriomas, also known as chocolate cysts, are a localized form of endometriosis usually on the ovary and occur in up to 10% of women of reproductive age [15]. Endometriomas contain dark degenerated blood products following repeated cyclical hemorrhage and are readily diagnosed on ultrasound, with most demonstrating classical radiographic features. The classical example is an avascular unilocular cyst containing low-level, homogeneous "ground-glass" like internal echoes, because of the hemorrhagic debris. This appearance occurs in 50% of cases [16].

Our paper covers a comparative study between different convolutional neural networks based on semantic segmentation, implemented, and proposed for the identification of these benign ovarian tumor masses.

2 Material and Methods

In this study the following semantic segmentation networks were used:

- DeepLabV3+ with ResNet18 as a decoder.
- DeepLabV3+ with ResNet50 as a decoder.
- DeepLabV3+ with MobileNet-V2 as a decoder.
- DeepLabV3+ with Xception as a decoder.
- DeepLabV3+ with InceptionResNet-V2 as a decoder.
- Fully convolutional network with an upsampling factor of 16.

All above-proposed networks are based on encoder-decoder architecture, where features are extracted by encoders and final segmented images are provided via the decoder by upsampling. All experiments were done using Matlab.

2.1 Dataset Used

The dataset on which this study relies could be downloaded from [17] and was introduced and presented in more detail in [18]. The dataset contains 1469 2D ultrasound images and 170 contrast-enhanced ultrasound (CEUS) images.

Our experiments will only cover a subset of the 1469 2D images. The dataset itself already provides manual segmentation done by medical experts. The images from the subset were manually segmented by us using Matlab image labeler, based on the manual segmentation performed by medical experts so that they are represented in the format required by Matlab. Thus, we used a total number of 163 images, out of which 100 were used for training purposes and 63 were used for testing purposes. As we already mentioned, we are going to cover 6 classes, for which each color is represented in Fig. 1. Figure 1 also represents pixel-label distribution for each class over the 63 images used for testing purposes. In Fig. 2, Fig. 3, and Fig. 4 examples of manually segmented different classes using Matlab image labeler are presented.

2.2 Neural Networks Used

- As we mentioned before, two types of CNNs were tested: a) a fully convolutional network (FCN) with an upsampling factor of 16 and b) DeepLabV3+ with different decoders (ResNet18, ResNet50, MobileNet-V2, Xception, and InceptionResNet-V2), 6 networks in total.

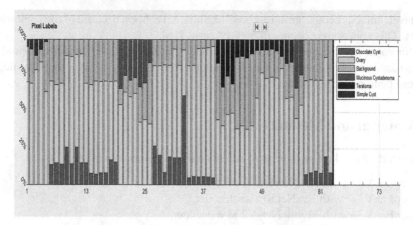

Fig. 1. Labels overview for the 63 images used for testing purposes.

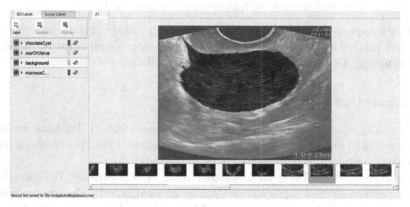

Fig. 2. Example of manual segmentation of background, ovary, and mucinous cystadenoma using Matlab image labeler.

Fully Convolutional Networks

By their names, fully convolutional networks are basically convolutional networks without fully connected layers. They rely on down-sampling and up-sampling operations, and they were first introduced in [19]. In that specific paper, state-of-the-art networks AlexNet, VGG net, and GoogLeNet were adapted into fully convolutional networks and prepared for the segmentation task. In this paper, a fully convolutional network was presented, FCN-16. The final feature map is up sampled with a factor of 16. For size reduction and down-sampling operations, 1×1 convolutions are used.

DeepLab-V3+

DeepLab-V3+ is one of the most successfully used convolutional neural networks in the semantic segmentation field. It is based on an encoder-decoder architecture, as seen in Fig. 5, where the information is encoded by the encoder by applying atrous convolutions, applied at multiple scales, and the final segmented image is output by other convolution

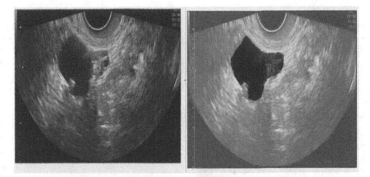

Fig. 3. Example of an original ultrasound image and manually segmented images using Matlab image labeler.

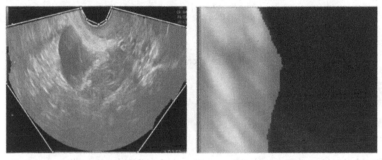

Fig. 4. Example of a fine-tuning techniques using brush/ smart polygon of a manually segmented image using Matlab image labeler.

operations and upsampling by the decoder. This specific model was first introduced in [20]. The reason for using this network with various encoders also on this work is because of previously successful usage in [21], where notable results were obtained. Our study covers DeepLab-V3+ models with ResNet-18, ResNet-50, MobileNet-V2, InceptionResNet-V2, and Xception networks used as encoders.

ResNet-18
ResNet-18 is a convolutional neural network composed of 18 layers, based on residual blocks, of course being able to solve the vanishing gradient problem. In general, based on the latest research, it was demonstrated that residual networks perform better and better, as they get deeper (contain more layers). The same applies to our experimental results, where DeepLab-V3+ together with ResNet-50 used as an encoder, obtained better results compared to the same model, where ResNet-18 was used as an encoder. ResNet-18 has an input size of 244 × 244, is trained on more than one million images from ImageNet and it is able to classify images into 1000 categories.

ResNet-50
ResNet-50 is a 50 layers-deep convolutional neural network, which, like ResNet-18, is

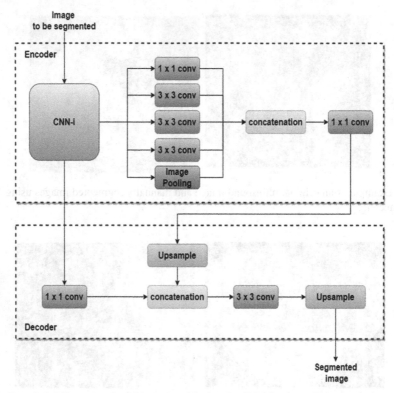

Fig. 5. DeepLab-V3+ network architecture with encoder CNN-i, where "i" in our study represents ResNet-50, ResNet-18, MobileNet-V2, InceptionResNet-V2, and Xception [21].

based on residual/skip connections, being able to solve the vanishing gradient problem. The network is also pre-trained on ImageNet, being able to classify images into 1000 categories and its input is 244×244.

MobileNet-V2
MobileNet-V2, originally introduced in [22], is a 53 layers-deep convolutional neural network designed for mobile applications. Since the main purpose of this network was to be used for embedded vision applications and memory consumption was the main concern, the authors introduced depthwise separable convolutions to reduce the model size and of course the complexity. MobileNets also make use of 1×1 pointwise convolutions and residual blocks to solve the vanishing gradient problem. The input size of MobileNet-V2 is 244×244, the same as for ResNet-18 and ResNet-50.

InceptionResNet-V2
InceptionResNet-V2 is a convolutional neural network composed of 164 layers, pre-trained on the ImageNet dataset, and able to classify images into 1000 categories.

InceptionResNet-V2, for which the input is 299 × 299 makes use of residual connections and inception blocks.

Xception
Xception is a 53 layers-deep convolutional neural network making use of residual connections and depthwise separable convolutions. Like InceptionResNet-V2, the network input is 299 × 299 and like the others, the network has been pre-trained on ImageNet and is able to classify images into 1000 categories. Depthwise separable convolutions are basically 1 × 1 convolutions applied on each RGB channel.

3 Experimental Results and Discussions

In this section, we illustrate the experimental results for the ovary ultrasound images' semantic segmentation. Hyperparameters used for training all networks are presented in Table 1. Table 2 shows all network performances, both for each of the six classes and globally.

The training results (accuracy and loss) obtained for all the proposed networks are presented in Figs. 6, 7, 8, 9, 10 and 11.

Table 3 represents global accuracies presented for all proposed semantic segmentation networks. The global accuracies represent the number of correctly classified pixels as compared with the total number of pixels, regardless of the class.

Table 1. Hyperparameters used for training semantic segmentation networks.

Hyperparameter	Value
Learn rate schedule	Piecewise
Learn rate drop period	Each 3 epochs
Learn rate drop factor	0.3
Mini batch size	8

This accuracy is not associated with each class and there are networks that perform well in some classes and don't perform well in other classes. The global accuracies presented below cover all classes, which means that there is a possibility that a network with lower accuracy might not perform well at all for some classes but perform well in other classes, representing most pixels of the image to be classified. That is why we represented in Table 4 the performances of each proposed semantic segmentation network for each of the 6 classes covered by this study.

The names of classes in Table 4 are the following: C1 - Chocolate cyst, C2 - Mucinous Cystadenoma, C3 - Teratoma, C4 - Simple cyst, C5 - Ovary, and C6 – Background. The names of the networks are the same as in Table 3.

Regarding the global accuracies, we could clearly see that the first places are occupied by DeepLab-V3+ networks having as encoders state-of-the-art networks such as ResNet-18, ResNet-50, InceptionResNet-V2, MobileNet-V2, and Xception, with a global accuracy of >84%. Such accuracies could of course only mean that these networks perform

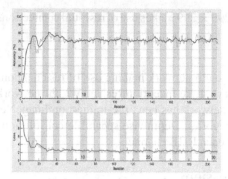

Fig. 6. Training results obtained for FCN-16.

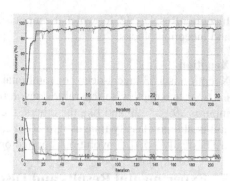

Fig. 7. Training results obtained for DeepLab-V3+ with ResNet-18 used as the encoder.

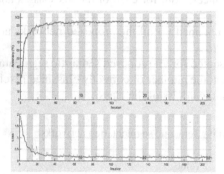

Fig. 8. Training results obtained for DeepLab-V3+ with ResNet-50 used as the encoder.

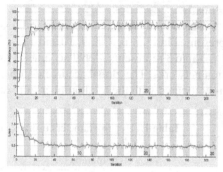

Fig. 9. Training results obtained for DeepLab-V3+ with MobileNet-V2 used as the encoder.

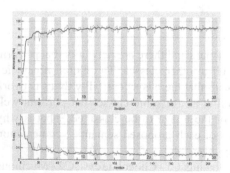

Fig. 10. Training results obtained for DeepLab-V3+ with InceptionResNet-V2 used as the encoder.

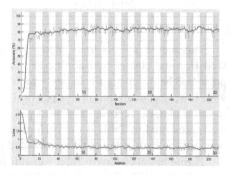

Fig. 11. Training results obtained for DeepLab-V3+ with Xception used as the encoder.

Table 2. Metrics used for evaluating semantic segmentation networks.

Metric	Description
Global Accuracy	Ratio of correctly classified pixels to the total number of pixels, regardless of the class
Mean Accuracy	Ratio of correctly classified pixels in each class to total pixels. The mean is then computed, and the final value is provided
Mean IoU	Average IoU of all classes. The mean is computed, and the final value is provided

Table 3. Global accuracies for all proposed semantic segmentation networks.

Network	Name	Accuracy	IoU
DeepLab-V3+ (ResNet-50)	CNN1	95.55%	83.99%
DeepLab-V3+ (ResNet-18)	CNN2	95.30%	82%
DeepLab-V3+ (InceptionResNet-V2)	CNN3	92.65%	71.11%
DeepLab-V3+ (MobileNet-V2)	CNN4	84.88%	41.44%
DeepLab-V3+ (Xception)	CNN5	84.27%	39.24%
FCN-16	CNN6	71.64%	28.65%

well in almost all classes. We can also see that the fully convolutional network FCN-16 performs less in terms of global accuracy. Another important aspect, clearly seen in the experimental results is also represented by the fact that DeepLab-V3+ based on residual connections performs better, and thus as deep the network with residual connections it is, better performances are obtained. Another important aspect is also represented by the fact that background and ovary classes are the ones for each network to perform well. The reason for this is that these 2 classes cover most of the pixels of the ultrasound image (-75–85%). The other classes, representing the tumoral mass, might be present or not, depending on if the patient is fully healthy or not. Thus, we are going to discuss only performances related to tumoral mass-related classes (chocolate cyst, mucinous cystadenoma, simple cyst, and teratoma).

The network occupying the first position with a global accuracy of 95.55% is DeepLab-V3+ based on ResNet-50 as the encoder (CNN1). In terms of performances for each tumoral mass related classes, the class in which this network performs well is the chocolate cyst, with an accuracy of 92.52%, while the class in which the network performs the worst is the simple cyst, with an accuracy of 73.65%.

The second place is occupied by DeepLab-V3+ with ResNet-18 used as the encoder (CNN2). Thus, again another network based on residual connections performs well. It performs the best on the mucinous cystadenoma with an accuracy of 93.54% and performs the worst on the simple cyst, with an accuracy of 59.46%. Third place is occupied by DeepLab-V3+ with InceptionResNet-V2 used as an encoder, which performs the

Table 4. Accuracies obtained for individual networks per individual class.

CNN	C1	C2	C3	C4	C5	C6
CNN1	92.52%	92.25%	80.09%	73.65%	97.46%	95.09%
CNN2	91.64%	93.54%	85.99%	59.46%	97.11%	94.82%
CNN3	78.87%	72.91%	71.39%	49.55%	97.04%	93.66%
CNN4	47.8%	32.22%	22.06%	0.9%	96.51%	87.37%
CNN5	48%	22.74%	15.07%	0%	96.47%	87.4%
CNN6	22.77%	19.99%	85.37%	2.87%	79.99%	85.37%

best on the chocolate cyst, with an accuracy of 78.87% and performs the worst on the simple cyst with an accuracy of 49.55%. The next place is occupied by DeepLab-V3+ with MobileNet-V2 used as an encoder, which performs the best on the chocolate cyst, with an accuracy of 47.8%, and performs the worst on the simple cyst. The next place is occupied by DeepLab-V3+ with Xception used as an encoder, which performs the best on the chocolate cyst, with an accuracy of 48%, and performs the worst on the simple cyst. The network FCN-16 performs well only on the ovary and background classes and would not be worth using them for semantic segmentation on the tumoral mass-related classes.

It used two initial learning rates, 0.001 for DeepLab-V3+ networks and 0.00001 for the FCN one. The reason for using a smaller initial learning rate for the fully convolutional network was that, during the experiments, the training was reaching a suboptimal result in a very small time, and thus training was failing.

Comparing the best networks obtained in our study with the networks from reference [18], on the same data set (Table 5) the statistical results are better (CNN1) or similar (CNN2).

Table 5. Comparison with results obtained in other paper using the same dataset.

Network	Reference	IoU
CNN1	**ours**	**83.99%**
SegFormer	[18]	82.46%
DANet	[18]	82.2%
PSPNet	[18]	82.01%
CNN2	**ours**	**82%**
TransUNet	[18]	81.31%

4 Conclusions

The paper tested some state-of-the-art neural networks for semantic segmentation of four nonmalignant ovarian tumors. Regarding global accuracies, the most performant neural network is DeepLab-V3+ networks having an encoder ResNet-50. Regarding individual accuracies, they can differ from class to class or from network to network. However, by looking at the actual results for the FCN, it gave good results only for semantic segmentation of ovary or background, and thus couldn't be used for semantic segmentation of the tumoral mass-related classes, subject to which we are interested.

As feature work, we intend to create a multi-neural network system based on the decision fusion of individual networks considering the weights associated with the network performance and the class predicted. Regarding the learning rate we had to increase the initial learning rate and of course increase the number of epochs.

References

1. Siegel, R.L., Miller, K.D., Fuchs, H.E., Jemal, A.: Cancer statistics. CA Cancer J. Clin. **72**(1), 7–33 (2022)
2. Zeppernick, F., Meinhold-Heerlein, I., Meinhold-Heerlein, Á.I.: The new FIGO staging system for ovarian, fallopian tube, and primary peritoneal cancer. Arch. Gynecol Obs. **290**(5), 839–842 (2014)
3. Basha, M.A.A., Metwally, M.I., Gamil, S.A., et al.: Comparison of O-RADS, GI-RADS, and IOTA simple rules regarding malignancy rate, validity, and reliability for diagnosis of adnexal masses. Eur. Radiol. **31**(2), 674–684 (2021)
4. Sokalska, A., et al.: Diagnostic accuracy of transvaginal ultrasound examination for assigning a specific diagnosis to adnexal masses. Ultrasound Obstet Gynecol. **34**(4), 462–470 (2009)
5. Wu, C., Wang, Y., Wang, F.: Deep learning for ovarian tumor classification with ultrasound images. In: Hong, R., Cheng, W.-H., Yamasaki, T., Wang, M., Ngo, C.-W. (eds.) PCM 2018. LNCS, vol. 11166, pp. 395–406. Springer, Cham (2018). https://doi.org/10.1007/978-3-030-00764-5_36
6. Wang, H., et al.: Application of deep convolutional neural networks for discriminating benign, borderline, and malignant serous ovarian tumors from ultrasound images. Front Oncol. **11**, 770683 (2021)
7. Hsu, S.T., Su, Y.J., Hung, C.H., Chen, M.J., Lu, C.H., Kuo, C.E.: Automatic ovarian tumors recognition system based on ensemble convolutional neural network with ultrasound imaging. BMC Med. Inform. Decis. Mak. **22**(1), 298 (2022)
8. Christiansen, F., Epstein, E.L., Smedberg, E., Åkerlund, M., Smith, K., Epstein, E.: Ultrasound image analysis using deep neural networks for discriminating between benign and malignant ovarian tumors: comparison with expert subjective assessment. Ultrasound Obstet Gynecol. **57**(1), 155–163 (2021)
9. Saida, T., et al.: Diagnosing ovarian cancer on MRI: a preliminary study comparing deep learning and radiologist assessments. Cancers (Basel). **14**(4), 987 (2022)
10. Jung, Y., et al.: Ovarian tumor diagnosis using deep convolutional neural networks and a denoising convolutional autoencoder. Sci. Rep. **12**, 17024 (2022)
11. Pavlik, E.J., et al.: Frequency and disposition of ovarian abnormalities followed with serial transvaginal ultrasonography. Obstet Gynecol. **122**(2 Pt 1), 210–217 (2013)
12. Stany, M.P., Hamilton, C.A.: Benign disorders of the ovary. Obstet. Gynecol. Clin. North Am. **35**(2), 271–284 (2008)

13. Louis, M.S., Mangal, R., Stead, T.S., Sosa, M., Ganti, L.: Ovarian dermoid tumor. Cureus, **14**(7), e27233 (2022)
14. Moyon, M.A., et al.: Giant ovarian mucinous cystadenoma, a challenging situation in resource-limited countries. J. Surg. Case Rep. (12), rjz366 (2019)
15. Vercellini, P., Viganò, P., Somigliana, E., Fedele, L.: Endometriosis: pathogenesis and treatment. Nat. Rev. Endocrinol. **10**(5), 261–275 (2014)
16. Van Holsbeke, C., Van Calster, B., Guerriero, S., et al.: Endometriomas: their ultrasound characteristics. Ultrasound Obstet. Gynecol. **35**(6), 730–740 (2010)
17. https://github.com/cv516buaa/mmotu_ds2net. Accessed 21 Jan 2023
18. Zhao, Q., et al.: A multi-modality ovarian tumor ultrasound image dataset for unsupervised cross-domain semantic segmentation. arXiv: 2207.06799 (2022)
19. Long, J., Shelhamer, E., Darrell, T.: Fully convolutional networks for semantic segmentation. In: IEEE Conference on Computer Vision and Pattern Recognition (CVPR), pp. 3431–3440. June 7–12, Boston, MA, USA (2015)
20. Chen, L.-C., Zhu, Y., Papandreou, G., Schroff, F., Adam, H.: Encoder-decoder with atrous separable convolution for semantic image segmentation. In: Ferrari V., Hebert M., Sminchisescu C., Weiss Y. (eds.) Computer Vision – ECCV 2018. ECCV 2018. LNCS, vol. 11211 (2018)
21. El-Khatib, M., Teodor, O., Popescu, D., Ichim, L.: Using combined CNNs for ROI segmentation in early investigation of pregnancy. In: 8th International Conference on Control, Decision and Information Technologies (CoDIT), pp. 897–902. May 17–20, Istanbul, Turkey (2022)
22. Howard, A.G., et al.: MobileNets: Efficient convolutional neural networks for mobile vision applications. arXiv: 1704.04861 (2017)

A Generalized Deep Learning Model for Multi-disease Chest X-Ray Diagnostics

Nabit Bajwa[1,3]([📧]) [ID], Kedar Bajwa[1], Muhammad Faique Shakeel[1] [ID],
Atif Rana[1] [ID], Kashif Haqqi[1], and Suleiman Khan[2] [ID]

[1] Shifa International Hospital, Islamabad, Pakistan
[2] Institute for Molecular Medicine Finland FIMM,
University of Helsinki, Helsinki 00014, Finland
[3] College of Engineering and Computing,
George Mason University, Fairfax, VA, USA
nbajwa4@gmu.edu

Abstract. We investigate the generalizability of deep convolutional neural network (CNN) on the task of disease classification from chest x-rays collected over multiple sites. We systematically train the model using datasets from three independent sites with different patient populations: National Institute of Health (NIH), Stanford University Medical Centre (CheXpert), and Shifa International Hospital (SIH). We formulate a sequential training approach and demonstrate that the model produces generalized prediction performance using held out test sets from the three sites. Our model generalizes better when trained on multiple datasets, with the CheXpert-Shifa-NET model performing significantly better (p-values < 0.05) than the models trained on individual datasets for 3 out of the 4 distinct disease classes.

Keywords: Deep Learning · Health Informatics · Computer Vision · Radiology

1 Introduction

Deep learning image classification models are gaining increasing attention and use in healthcare, with applications ranging from screening for blinding retinal diseases [11] to classifying skin cancer [5] and detecting pneumonia from chest x-rays [14]. Deep learning has consistently outperformed traditional computer vision techniques in accuracy benchmarks for image classification tasks [4,12].

Chest x-rays are a widely used diagnostic tool, accounting for almost 27% of all diagnostic medical x-ray examinations globally. They are often the first test prescribed to diagnose or monitor lung and chest diseases such as pneumonia, lung cancer, heart failure, pneumothorax, tuberculosis, and interstitial lung diseases [16]. Deep learning is a promising technology for analyzing chest x-rays

© The Author(s), under exclusive license to Springer Nature Switzerland AG 2023
I. Rojas et al. (Eds.): IWANN 2023, LNCS 14134, pp. 541–552, 2023.
https://doi.org/10.1007/978-3-031-43085-5_43

and developing computer-aided diagnostic tools, due to the enormous number of existing chest x-ray images in hospital databases worldwide.

However, deep learning techniques require a large number of labeled images to achieve high accuracy, which can be a challenge in medical imaging due to patient confidentiality compliance requirements and the absence of single one-word labels. Medical images are often coupled with a radiologist report that provides details about the patient's background, observations, and findings, but these reports can vary widely in structure and technique, which introduces noise into the labeling process.

To overcome these challenges, researchers typically rely on disease label mining from radiology reports [1], which may introduce noise in the labeling process. Despite this noise, existing work suggests that deep learning models can learn important features for diagnostic purposes much better than classical computer vision techniques [13]. In this work, we focus on studying the generalizability of deep convolutional networks for chest x-ray classification, analyzing their performance across multiple disease classes and datasets from different patient populations.

Our goal is to develop a cost-effective, accurate, and generalizable methodology for training chest x-ray classifiers that can be extended to additional sites sequentially. We aim to gain a better understanding of how weakly labeled medical image data can be effectively used to train deep learning models and how datasets from external sites affect the accuracy and generalizability of these models. By providing a robust evaluation of generalization performance, we hope to contribute to the development of more effective deep learning models for medical image analysis.

2   Related Works

Disease classification from chest x-rays has gained significant attention from researchers, who have shown that deep learning techniques can be used to detect various diseases from chest x-rays [18]. The NIH Chest X-Ray8 dataset [18] is a seminal work in this domain, consisting of over a hundred thousand chest x-rays with labels provided for 14 different disease classes. This dataset has inspired numerous works, ranging from developing different architectures for single and multiple disease detection [6] to localizing and even partially writing the radiology reports [9] using deep learning models.

CheXNet [14] and CheXNeXt [13] claim to achieve radiologist-level disease detection on multiple disease classes from chest x-rays. Both models were trained on the NIH dataset and compared to board-certified radiologists on a held-out test set. The release of these two models led to the development of the CheXpert dataset [8], which, coupled with the MIMIC-CXR19 dataset [10], is the largest dataset of chest radiographs to date. Our work builds upon this massive release of chest radiographs and their corresponding labels, investigating the claim that convolutional neural networks (CNNs) generalize to real-world data if given enough images of that class [19].

In a similar vein, Zech et al. [20] investigated the generalizability of CNNs for pneumonia detection using chest x-ray images from multiple sites. Their work claims that CNNs do not necessarily generalize to unseen test data from external test sites for pneumonia detection. However, our work differs from theirs in two key aspects. First, we investigate generalization across multiple disease classes. Second, we propose a methodology to train and validate models in multiple stages using single datasets to evaluate their generalization ability.

Other studies have also explored various CNN models to quantify their generalization ability and limitations across multiple open source datasets and disease classes [3,15]. Our work confirms the generalizability of deep learning models for chest x-ray datasets, including a large chest x-ray dataset from an entirely different patient population coming from an Asian background (SIH). We also validate our findings with medical experts (radiologists).

3   Methods and Datasets

We conducted experiments to investigate the effect of using X-Ray datasets from multiple sites on the generalizability of deep learning models for various disease classes. Our baseline model was trained on the NIH Chest X-Ray8 dataset [18]. We then trained and tested three models: CheXpert-NET trained on the CheXpert dataset by Stanford, USA [8], Shifa-NET trained on the internal Chest X-Ray dataset by Shifa International Hospital (SIH), Pakistan, and CheXpert-Shifa-NET trained on both the CheXpert dataset and the SIH internal dataset sequentially. The models were evaluated using the Area Under the Receiver Operating Curve (AUROC) and independent test sets from each of the sites, NIH, CheXpert and Shifa, were used to assess their generalizability [8,18].

3.1   Datasets

This section describes the three datasets used in the study, which were collected from independent medical sites in the United States and Pakistan. The data was collected for four disease classes: Atelectasis, Cardiomegaly, Pleural Effusion, and Pneumonia. Additionally, a "No Finding" class was used to indicate a normal chest X-ray without any disease indications. All three training datasets were labeled with an automated labeler, while the test sets of NIH and Shifa datasets were automatically labeled. In comparison, the CheXpert dataset's testing split was radiologist annotated, providing us with a strong ground truth.

Shifa International Hospital (SIH) Chest X-Ray Dataset. The SIH dataset includes 80,035 frontal chest X-ray images from 43,839 patients in Shifa International Hospitals. Images were collected between January 1, 2014, and December 31, 2018, approved by the institutional review board. An automatic labeler was used to classify the dataset based on disease keywords obtained from domain expert radiologists and a manual review of reports. The dataset is weakly

Table 1. Keywords for Shifa International Hospital reports classification

Classes	Keywords
No Finding	'normal study', 'unremarkable study', 'no active lung lesion', 'no active lung disease', 'no acute cardiopulmonary lesion'
Atelectasis	'atelectasis', 'atelectatic bands'
Cardiomegaly	'cardiomegaly', 'enlarged cardiac size'
Pleural Effusion	'bilateral pleural effusion', 'right sided pleural effusion', 'left sided pleural effusion'
Pneumonia	'air space opacification', 'air space consolidation', 'opacification', 'zone infiltrates'

labeled and contains a label for each of the four classes, positive or negative. Table 1 lists the keywords used, and a sample of the reports and corresponding labels is provided in Table 2.

We use 85/10/5 splits for our training, validation, and testing sets. The data splits drawn are completely random. The total number of images and per class prevalence of images in each of the data splits are given in Table 3, and are fairly balanced.

National Institute of Health (NIH) Chest X-Ray8 Dataset. The NIH Chest X-Ray8 [18] dataset consists of 112,120 X-ray images with disease labels from 30,805 unique patients. It provides labeled data for 14 different diseases. The images have been labeled using an automated labeling technique. We use a model pre-trained on NIH dataset as a baseline model. This baseline model is an implementation [2] of the DenseNet-121 [7] model trained on the NIH dataset for 14 classes. Given the wider generalizability of DenseNet-121 model, we preferred this architecture over NIH's original ResNet architecture. Hence, we report the data splitting strategy used by the corresponding implementation of the DenseNet-121 model in our work. Table 4 denotes the per-class prevalence in the data splits.

CheXpert Stanford Dataset. CheXpert [8] is a publicly available dataset from Stanford University consisting of 224,316 chest radiographs and labels for 14 disease classes, collected from both inpatient and outpatient centers between October 2002 and July 2017. In our work, we use the images labeled positive, negative, and uncertain for 4 radiological observations (Atelectasis, Cardiomegaly, Pleural Effusion, and Pneumonia), and employ the U-Ones labeling strategy to map uncertain labels to positive ones. As no public test set is available, we use the radiologist-labeled validation set as our testing set, and manually split the training set 95/5 to create our training and validation sets. Table 5 provides the per-class prevalence in the new training and validation split after applying the U-Ones labeling strategy, along with the test set class prevalence.

3.2   Methods

We propose a method to train a generalized DenseNet-121 model for multi-label disease prediction using three datasets X_1, X_2, and X_3 of X-ray images and corresponding disease labels Y_1, Y_2 and Y_3. DenseNets concatenate feature maps from previous layers into the inputs of future layers, leading to $N(N+1)/2$ connections instead of the N connections in traditional convolutional neural networks. This

Table 2. Sample SIH radiology reports and their corresponding labels

Report Text	Label
In place NG tube and right-sided CVP line are again noted. Minimal left **pleural effusion** with underlying lung consolidation/**atelectasis** is again noted. Infiltrates are again noted in bilateral upper and right mid zones. Minimal atelectatic changes in the right lung base are seen. Rest of the findings are unchanged.	Atelectasis, Pleural Effusion
There are persistent inhomogenous areas of **airspace opacification** in right lung. Mild interval increase in the inhomogenous areas of **airspace opacification** in left lung is noted. There is interval development of mild inhomogenous haziness in bilateral lower zone obscuring both CP angles and diaphragmatic outlines suggesting **pleural effusions** with underlying **atelectasis** on left	Atelectasis, Pleural Effusion, Pneumonia
There is re-demonstration of prominent interstitial markings. No collapse or consolidation is noted. There is no pneumothorax. There is persistent blunting of left CP angle likely due to pleural thickening or **pleural effusion**. Right CP angle is normal. **Cardiomegaly**. cardiac pacemaker with its intact leads are again noted. Rest of the findings are unchanged.	Cardiomegaly, Pleural Effusion
No definite evidence of any consolidation, collapse or pneumothorax is seen. Both hila and cardiomediastinal contours appear normal. Both CP angles are acute. Both hemidiaphragm appear normal. No evidence of any bony lesion seen. CONCLUSION: **Normal study**	No Finding

Table 3. Per class prevalence for disease positive cases in training, testing and validation splits in SIH dataset

	Train	Validation	Test	Total
Atelectasis	14,830	1,730	874	17,434
Cardiomegaly	10,015	1,208	592	11,815
Pleural Effusion	22,350	2,627	1,339	26,316
Pneumonia	14,653	1,758	885	17,296
No Finding	7,948	876	456	9,280

Table 4. Per class prevalence for disease positive cases in training, validation and testing splits in NIH dataset

	Train	Validation	Test	Total
Atelectasis	10,768	652	139	11,559
Cardiomegaly	2,564	155	57	2,776
Pleural Effusion	12,381	737	199	13,317
Pneumonia	1,328	79	24	1,431
No Finding	56,203	3,380	778	60,361

Table 5. Per class prevalence for disease positive cases in training, validation and testing splits in CheXpert dataset

	Train	Validation	Test	Total
Atelectasis	63,810	3,305	80	67,195
Cardiomegaly	33,283	1,804	68	35,155
Pleural Effusion	92,903	4,912	67	97,882
Pneumonia	23,523	1,286	8	24,817
No Finding	21,235	1,146	38	22,419

structure eliminates the need to relearn redundant feature maps and has better feature propagation, without suffering from the vanishing gradient problem [7]. For each dataset, we start by defining train, test and validation splits. For each training instance, we use on the fly data augmentation techniques to generate training instances, and optimize the binary cross entropy loss given as:

$$L(y, \hat{y}) = -\frac{1}{5} \sum_{i=0}^{5} \left(y \log \hat{y}_i + (1 - y) \log (1 - \hat{y}_i) \right) \tag{1}$$

Our model is trained end to end using hyperparameters as defined in the experimental setup, with train, test, and validation splits defined for each dataset. We use on-the-fly data augmentation techniques to generate training instances and optimize the binary cross entropy loss. We train multiple models in two stages, starting with a pretrained DenseNet-121 model as a baseline. In the first stage, we train two models on CheXpert and SIH datasets individually (CheXpert-NET and Shifa-NET, respectively). In the second stage, CheXpert-NET is used as initialization and trained on the SIH dataset to create a final generalized model (CheXpert-Shifa-NET). We evaluate the model using per class AUROC metric on the validation set, and assess its performance on the test set using AUC scores over the four disease classes.

3.3   Experimental Setup

In this section, we provide details on our experimental setup for deep learning model training, evaluation, and testing, summarized in Fig. 1. The datasets are split into train, test, and validation sets. The models are trained end-to-end for 30 epochs, using the Adam optimizer with a binary cross-entropy loss function, and a learning rate of 1e-3 that decays by a factor of 10 if the validation loss does not improve for 5 consecutive epochs. We apply on-the-fly data augmentation during training using the Keras Image Data Generator class, including resizing to 320 by 320 pixels, random horizontal flipping, random rotation up to 20°, and random pixel shifting of up to 20%. We compute validation AUC scores at the end of each epoch, and select the best epoch based on the lowest validation loss and highest validation mean AUC scores for the four disease categories. The fully trained models are evaluated on all three test sets. All training and testing are performed on a local machine with an NVIDIA 1070 Ti GPU, 16 GB of RAM, and an Intel Xeon processor.

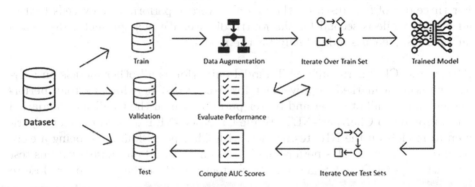

Fig. 1. Experimental methodology

4   Results

We first evaluate our models performance on generalizability task. Specifically, we trained 3 models, CheXpert-NET, Shifa-NET and CheXpert-Shifa-NET, and tested the performance of each of the models on test sets obtained from the three different sources. We first discuss the overall performance, calculated as the average AUC over all four diseases. This is followed by a detailed performance evaluation and discussion of the models on each disease class. Table 6 demonstrates the average AUC scores of the models on the three datasets. We note that the CheXpert-Shifa-NET model shows a significant improvement over the baseline model than the Shifa test set and the CheXpert test. This improvement is greater in magnitude then CheXpert-NET and Shifa-NET on both test sets, validating that the model learns generalizable features useful for disease prediction. On the NIH test set, the improvement is second to CheXpert-NET model. On average,

Table 6. Average testing set results of the 4 models on the 4 disease classes

	Shifa Test Set	CheXpert Test Set	NIH Test Set	Average
NIH Baseline	0.777	0.815	0.792	0.795
CheXpert-NET	0.790	0.824	0.821	0.812
Shifa-NET	0.840	0.825	0.805	0.823
CheXpert-Shifa-NET	0.842	0.834	0.806	0.828

the CheXpert-Shifa-NET model outperforms all the other three models, attesting that the generalized model is most suitable for disease prediction on arbitrary test images. Table 7 lists detailed prediction performance (AUC scores) of all four models across the four disease classes and the three testing data sets. For the average performance on each disease class, the standard deviation is also computed. The results demonstrate that CheXpert-Shifa-NET is the best individual model and outperforms the other models for most disease-dataset combinations. For three out of four diseases, the model's average performance exceeds that of all others, while is second for the fourth disease. We next present a discussion on each disease class individually for the model.

Atelectasis. CheXpert-Shifa-NET model outperforms all other models on average, showing a marked improvement for Atelectasis. It exhibits strong performance on the Shifa test set and better performance on the CheXpert validation set compared to CheXpert-NET. While CheXpert-NET outperforms all experimental models on the NIH testing set, with CheXpert-Shifa-NET being a close second, CheXpert-NET's performance shows high variability across various test sets, whereas CheXpert-Shifa-NET consistently ranks either as best or close to best.

Cardiomegaly. Overall, the CheXpert-Shifa-NET model shows the best performance for Cardiomegaly among all models. It performs particularly well on the Shifa test set and may benefit from being trained on the Shifa training dataset. On the other hand, training on the CheXpert dataset appears to be less valuable for Cardiomegaly predictions, as indicated by the poorer performance of the CheXpert-NET model.

Pleural Effusion. For Pleural Effusion, all models perform similarly well on average, outperforming the NIH baseline. CheXpert-Shifa-NET shows strong performance on the Shifa test set and CheXpert validation set, while CheXpert-NET performs best on the CheXpert test set. The cross-performance of CheXpert-Shifa-NET is a close second to CheXpert-NET, with both models showing significant improvement over the baseline.

Pneumonia. For Pneumonia, Shifa-NET model performs well on all three testing sets, with CheXpert-Shifa-NET model showing similar performance on average.

Table 7. Consolidated results in terms of AUC score. Comparison of the three models trained using our proposed methodology over 4 different disease classes

		NIH Baseline	CheXpert-NET	Shifa-NET	CheXpert-Shifa-NET
Atelectasis	NIH Test	0.767	**0.808**	0.781	0.791
	Shifa Test	0.746	0.741	0.779	**0.781**
	CheXpert Val	0.800	0.826	0.790	**0.833**
	Average	0.771 ± 0.027	0.792 ± 0.045	0.783 ± 0.006	**0.801 ± 0.026**
Cardiomegaly	NIH Test	0.847	0.851	**0.867**	0.860
	Shifa Test	0.826	0.852	0.903	**0.903**
	CheXpert Val	0.813	0.797	0.806	**0.815**
	Average	0.827 ± 0.017	0.833 ± 0.031	0.859 ± 0.049	**0.859 ± 0.044**
Pleural Effusion	NIH Test	0.848	**0.861**	0.848	0.850
	Shifa Test	0.844	0.852	**0.882**	0.88[SAK1] 0
	CheXpert Val	0.857	**0.926**	0.906	0.919
	Average	0.850 ± 0.007	0.880 ± 0.040	0.879 ± 0.029	**0.880 ± 0.049**
Pneumonia	NIH Test	0.707	**0.762**	0.723	0.724
	Shifa Test	0.696	0.716	0.795	**0.797**
	CheXpert Val	0.791	0.748	**0.796**	0.768
	Average	0.731 ± 0.051	0.742 ± 0.023	**0.771 ± 0.042**	0.763 ± 0.037

CheXpert-NET model shows strong performance only on the NIH testing set and exhibits high variation across test sets. It is worth noting that the CheXpert validation set has only nine pneumonia positive cases, making it difficult to draw definitive conclusions.

Our results show that all three models CheXpert-Shifa-NET model, Shifa-NET model and CheXpert-NET perform significantly better than the NIH baseline p-value 0.00441, 0.00785 and 0.0414, Wilcoxon test p-value 0.00379, 0.00639 and 0.04182), confirming that training DenseNet-121 model on additional datasets improves the prediction performance.

4.1 Visualization and Model Interpretability

We qualitatively evaluate CheXpert-Shifa-NET's predictions using Grad-CAM activation maps [17] for disease localization. Figure 2 displays random examples, showing relevant regions identified by the model. Our model accurately predicts pneumonia in Fig. 2A, localizes cardiomegaly in Fig. 2B, and classifies both pneumonia and pleural effusion in Fig. 2C, although it struggles to localize the latter. In Fig. 2D, the model identifies and localizes both pneumonia and cardiomegaly. The radiologist's expert evaluation confirms the model's accuracy. The precision of disease region identification and generalizability of our model indicate the potential clinical applicability of deep learning in radiological diagnosis using chest X-rays.

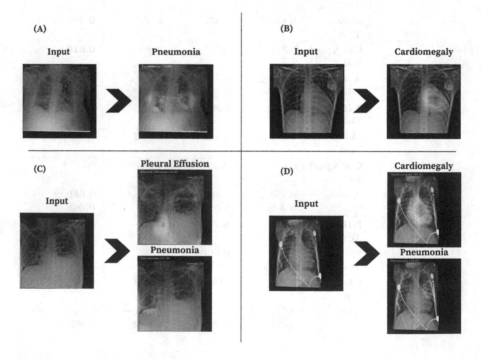

Fig. 2. Grad-CAM results visualizing model performance over different disease classes

5 Discussion

In this work, our primary goal was to investigate the generalization performance of deep learning models trained on datasets from multiple institutions and populations. Our experimental results demonstrate that our proposed methodology of training deep learning models across data from multiple health institutions can lead to models that can generalize to multiple disease classes. Moreover, the sequential nature of our training approach makes it suitable for continual improvements without the need to retrain from scratch. We found that, on average, the final generalized model, CheXpert-Shifa-NET, trained on all three datasets, showed the highest performance across the testing sets from all three healthcare institutions for 3 out of the 4 disease classes. These results are important as they signify the generalization ability of the models, as well as their potential for automatic radiological screening in clinical settings.

It is worth noting that one of the three datasets used in our work was from an entirely different region and population, and accounting for confounding variables in this scenario can be challenging. However, we observed that training on this dataset did not degrade the performance of the model on any of the other datasets, and in fact, it led to increased performance in a majority of cases, attesting to the generalizability of the model. We also observed an improvement in the performance of our model for the Atelectasis and Cardiomegaly disease classes when trained according to our proposed methodology. Furthermore, the performance of our generalized model on the CheXpert test set, which consists of a radiologist-annotated ground truth, did not degrade after training on the SIH dataset, indicating the generalization potential of the CheXpert-Shifa-NET model.

Our findings are in line with recent research within similar population types [3, 15], but our approach extends the scope to multiple health institutions and populations, demonstrating the potential of our proposed methodology for improving the generalization performance of deep learning models for radiological screening.

References

1. Chen, M.C., et al.: Deep learning to classify radiology free-text reports. Radiology **286**(3), 845–852 (2018). https://doi.org/10.1148/radiol.2017171115
2. Chou, B.: CheXNet-Keras, March 2023. https://github.com/brucechou1983/CheXNet-Keras
3. Cohen, J.P., Hashir, M., Brooks, R., Bertrand, H.: On the limits of cross-domain generalization in automated X-ray prediction (2020)
4. Dosovitskiy, A., et al.: An image is worth 16×16 words: transformers for image recognition at scale (2021)
5. Esteva, A., et al.: Dermatologist-level classification of skin cancer with deep neural networks. Nature **542**(7639), 115–118 (2017)
6. Guan, Q., Huang, Y., Zhong, Z., Zheng, Z., Zheng, L., Yang, Y.: Diagnose like a radiologist: attention guided convolutional neural network for thorax disease classification (2018)
7. Huang, G., Liu, Z., Weinberger, K., van der Maaten, L.: Densely connected convolutional networks. arxiv 2017. arXiv preprint arXiv:1608.06993
8. Irvin, J., et al.: CheXpert: a large chest radiograph dataset with uncertainty labels and expert comparison. In: Proceedings of the AAAI Conference on Artificial Intelligence, vol. 33, pp. 590–597 (2019)

9. Jing, B., Xie, P., Xing, E.: On the automatic generation of medical imaging reports. In: Proceedings of the 56th Annual Meeting of the Association for Computational Linguistics (Volume 1: Long Papers). Association for Computational Linguistics (2018). https://doi.org/10.18653/v1/p18-1240

10. Johnson, A.E., et al.: MIMIC-CXR: a large publicly available database of labeled chest radiographs. arXiv preprint arXiv:1901.07042 1(2) (2019)

11. Kermany, D.S., et al.: Identifying medical diagnoses and treatable diseases by image-based deep learning. Cell **172**(5), 1122–1131 (2018)

12. Krizhevsky, A., Sutskever, I., Hinton, G.E.: ImageNet classification with deep convolutional neural networks. Commun. ACM **60**(6), 84–90 (2017). https://doi.org/10.1145/3065386

13. Rajpurkar, P., et al.: Deep learning for chest radiograph diagnosis: a retrospective comparison of the cheXNeXt algorithm to practicing radiologists. PLoS Med. **15**(11), e1002686 (2018)

14. Rajpurkar, P., et al.: CheXNet: radiologist-level pneumonia detection on chest x-rays with deep learning. arXiv preprint arXiv:1711.05225 (2017)

15. Rajpurkar, P., et al.: CheXpedition: investigating generalization challenges for translation of chest X-ray algorithms to the clinical setting (2020)

16. Satia, I., Bashagha, S., Bibi, A., Ahmed, R., Mellor, S., Zaman, F.: Assessing the accuracy and certainty in interpreting chest X-rays in the medical division. Clin. Med. **13**(4), 349 (2013)

17. Selvaraju, R.R., Cogswell, M., Das, A., Vedantam, R., Parikh, D., Batra, D.: Grad-CAM: visual explanations from deep networks via gradient-based localization. In: Proceedings of the IEEE International Conference on Computer Vision, pp. 618–626 (2017)

18. Wang, X., Peng, Y., Lu, L., Lu, Z., Bagheri, M., Summers, R.M.: ChestX-Ray8: hospital-scale chest X-ray database and benchmarks on weakly-supervised classification and localization of common thorax diseases. In: Proceedings of the IEEE Conference on Computer Vision and Pattern Recognition, pp. 2097–2106 (2017)

19. Yosinski, J., Clune, J., Bengio, Y., Lipson, H.: How transferable are features in deep neural networks? (2014)

20. Zech, J.R., Badgeley, M.A., Liu, M., Costa, A.B., Titano, J.J., Oermann, E.K.: Variable generalization performance of a deep learning model to detect pneumonia in chest radiographs: a cross-sectional study. PLoS Med. **15**(11) (2018)

Towards the Identification of Multiclass Lung Cancer-Related Genes: An Evolutionary and Intelligent Procedure

Juan Carlos Gómez-López[1,2]([✉]), Daniel Castillo-Secilla[3], Jesús González[1,2],
Luis Javier Herrera[1,2], and Ignacio Rojas[1,2]

[1] Department of Computer Engineering, Automation, and Robotics (ICAR),
University of Granada, Granada, Spain
[2] Research Centre for Information and Communication Technologies (CITIC-UGR),
University of Granada, Granada, Spain
{goloj,jesusgonzalez,jherrera,irojas}@ugr.es
[3] Fujitsu Technology Solutions S.A., CoE Data Intelligence,
Camino del Cerro de los Gamos, 1, Pozuelo de Alarcón, Madrid, Spain
daniel.castillosecilla@fujitsu.com

Abstract. The amount of available transcriptomic data from which relevant knowledge can be extracted has rapidly increased. Besides, with the advances in areas such as machine learning and high-performance computing, the time and computing efforts for analyzing those data are being reduced, leading to the design of Clinical-Decision Support Systems (CDSS) for the precision medicine paradigm. As a result of this increase, the use of intelligent and evolutionary methods to study and classify cancer diseases has been proposed in the literature showing promising results. This study is aimed at identifying a set of genes able to distinguish between the following types of lung cancer: Adenocarcinoma (ACC), Squamous Cell Carcinoma (SCC), and healthy lung. The Differentially Expressed Genes (DEGs) analysis was carried out through RNA-seq data coming from The Cancer Genome Atlas (TCGA). An optimized evolutionary procedure has been developed with the purpose of finding the optimal combinations among the hundreds of candidate DEGs. Our custom method can maximize multiclass lung cancer recognition while minimizing the number of selected DEGs. The results show an outstanding classification rate with a significantly reduced number of DEGs biologically related to lung cancer.

Keywords: RNA-Seq · Evolutionary Algorithm · Feature Selection · Gene Expression · Transcriptomic Technologies

1 Introduction

Over the last few years, exponential growth in the amount of cancer data resources available has been observed, such as gene-gene or protein-protein interaction data, gene expression experiments, and genome-wide association studies.

I. Rojas et al. (Eds.): IWANN 2023, LNCS 14134, pp. 553–562, 2023.
https://doi.org/10.1007/978-3-031-43085-5_44

These data are complex, heterogeneous, and may be accessible in an unstructured raw status. At this point, effective treatment and analysis of these data entail a technical challenge. Precision medicine usually takes advantage of Feature Selection (FS) techniques, mainly as a result of the intractable high-dimensionality of the available data. In this context, the reduction in the number of selected biomarkers helps clinicians to understand and be able to use the designed decision-making support models, also allowing cheaper medical tests to be manufactured. Furthermore, FS can be considered a powerful tool in biomarkers discovery, becoming an important research topic aiming at identifying possible gene signatures able to differentiate specific biological states [1].

Cancer is one of the utmost mortal diseases nowadays, followed by cardiovascular disease. Every different type of cancer may have one or more gene signatures. A gene signature can be conformed by one or more genes in a concrete cell with a specific gene expression value, which usually happens as a result of an erroneous biological process or pathogenic health condition [12]. Discovering these gene signatures can lead to an early diagnosis and understanding of the underlying cause for developing a multifactorial disease such as cancer. Henceforth, by their usage, discrimination between a patient suffering from cancer and a healthy one can be performed. This discrimination can not be only conducted within healthy or cancer patients but also between different states or sub-types of the same cancer disease.

On the one hand, for the male population, lung cancer is considered to be both the most common and the most deadly type of cancer. On the other hand, it ranks second in terms of mortality and third in terms of incidence for the female population [4]. Lung cancer is divided into two types:

- Small Cell Lung Cancer (SCLC): this type of tumor has an endocrine or paraneoplastic syndrome association because the cells include dense neurosecretory granules, which are vesicles containing neuroendocrine hormones [17]. Mainly, SLCL manifests in the primary and secondary bronchi, i.e., the larger airways.
- Non-Small Cell Lung Carcinoma (NSCLC): this type is also sub-divided into three differentiated sub-types, namely:
    • Adenocarcinoma (ACC): ACC has similar symptoms to other forms of lung cancer. Furthermore, it is the most frequent type of lung cancer in Asian populations and younger women. Adenocarcinoma pathophysiology typically follows a histologic progression from cells residing in healthy lungs to clearly irregular or dysmorphic cells [11].
    • Squamous Cell Cancer (SCC): it originates in the bronchi and is also the second most prevalent type of lung cancer, just after ACC. Its tumor cells are characterized by a squamous appearance. This type of cancer is strongly related to smoker people, more than any other form of NSCLC [11].
    • Large Cell Lung Carcinoma (LCLC): LCLC is a group of undifferentiated malignant neoplasms that difficult the cytologic and architectural features of small cell carcinoma and glandular or squamous differentiation [7].

However, the predominant lung cancer samples in public databases such as TCGA or National Center for Biotechnology Information (NCBI)/Gene Expression Omnibus (GEO) are ACC and SCC [8,18]. To maximize the number of samples while avoiding imbalance, only samples from ACC and SCC were taken into account for this study.

The identification of relevant DEGs related to multiclass lung cancer will allow an early prognostic and the right treatment. This could be the difference between the recovery of the patient and his decease. Thus, it is essential to determine which genes could be causing disorders in one or more biological processes.

In this study, data coming from Large-scale RNA-Sequencing (RNA-Seq) were taken into account. This technology shows the quantity of RNA in a sequence sample at a given moment. RNA-Seq is the predominant transcriptomic technology and is currently used for the creation and development of CDSS to study multiple genetic diseases [19].

RNA-seq datasets usually suffer from the *curse of dimensionality problem*. In this sense, the use of wrapper approaches based on Evolutionary Algorithms (EAs), as an FS technique, to evaluate and reduce the dimensionality of the possible solutions allows to deal with that problem. Thanks to this, these approaches can produce quality solutions that are important in fields like bioinformatics, when data analysis could have a relevant socio-economic impact. In addition, this selection not only allows for reducing the problem's complexity but also keeps the recognition efficiency while eliminating irrelevant or noisy genes.

Specifically, the implemented EA-based wrapper for this study was designed to maximize the multiclass lung cancer recognition while minimizing the number of selected DEGs, as the high cost associated with data sequencing in a real laboratory requires the identification of reduced but robust gene signatures.

Several approaches have been implemented in the literature applied to gene selection and assessment, such as [3,14]. Nevertheless, to the best of our knowledge, there is no other EA-based wrapper procedure that deals with multiclass lung cancer datasets coming from TCGA, reaching a total of 94.35% of mean F1-score on the test dataset with such a reduced set of DEGs biologically related to lung cancer.

After this introduction, Sect. 2 introduces the wrapper developed to select DEGs. Then, Sect. 3 describes the methodology used, while Sect. 4 focuses on the experimental part. Finally, the article is closed with the conclusions in Sect. 5.

2   Description of the EA-Based Wrapper

Wrapper approaches are characterized by searching along with an induction algorithm (classification in this case) that evaluates each of the possible solutions to the problem. Over the last few years, Python has gained special relevance for the development of this type of procedure, mainly due to the great support of the community and its ease of code development. For this reason, the wrapper developed in [9] has been chosen to be developed in Python since it has proven to find quality solutions in high-dimensional problems.

Table 1. Number of samples per class and dataset.

	ACC	SCC	Control	Total
Training	392	398	82	872
Test	97	100	21	218
Total	489	498	103	1090

Basically, the NSGA-II algorithm is in charge of guiding the search, while the *k*-NN classification method has been chosen to evaluate the different solutions to the problem, which take the name of individuals in EAs. Both NSGA-II and *k*-NN have been shown in countless research studies to perform adequately in FS problems for high-dimensional spaces [21]. On the one hand, the *Distributed Evolutionary Algorithm in Python* (DEAP) programming library has been used to develop NSGA-II since it is one of the most widely used libraries for the development of EAs in Python due to the large number of functions it provides. On the other hand, the library par excellence in Python for machine learning procedures, *Scikit-learn*, has been chosen to implement *k*-NN.

3   Proposed Methodology

Transcriptomic RAW data contain an enormous quantity of variables, most of them irrelevant for discerning the addressed states. Launching an EA against all that information will exponentially increase the computational cost to find an optimal solution. To avoid that, a multiclass DEG extraction before the application of the EA has been implemented. Taking that into consideration, the procedure is defined as a multiclass DEGs-driven EA-based wrapper.

Public samples from both the LUng ADenocarcinoma (LUAD) and Lung Squamous Cell Carcinoma (LUSC) TCGA projects have been gathered to develop this study. Both projects were selected because they assembled almost the totality of Lung Cancer samples on the TCGA portal. They are framed within the NSCLC sub-types previously described, and they also are the two most prevalent types of Lung Cancer. To avoid information leaks along the process, training/test datasets were made by defining an 80%-20% sample split, respectively. Table 1 shows the sample distribution between both the training and test datasets.

After data gathering, a multiclass extraction process was only applied to the training dataset to avoid information leaks. For this process, three different parameters were used to ensure the robustness of the extracted DEGs. For each of the expressed genes, three thresholds were imposed: Firstly, the *p*-value must be less than or equal to 0.001, which stands as the threshold for considering a gene as a DEG taking into account the False Discovery Rate, FDR (expected numbers of type I errors). These errors happen when the null hypothesis is incorrectly rejected (False positive). Secondly, the Log-Fold Change (LFC) was set to be greater than or equal to 1.5. LFC is a measure that describes how much a value changes between two different observations by using a logarithmic scale. Equation (1) represents the $Log_2FoldChange$ formulation given two observations A and B.

$$LFC = log_2(B - A)/A \tag{1}$$

Finally, our custom multiclass extraction parameter, Coverage (COV) [5], was tuned to 2. It is of utmost importance to emphasize the difficulty of extracting true multiclass DEGs in studies that address more than two pathological states. COV allows us to expose differentially expressed DEGs between two or more classes by taking into account the number of bi-class comparisons where this differential expression is significant. Taking advantage of this technique, the multiclass DEG detection could be solved efficiently, improving the posterior multiclass machine learning assessment. The COV parameter takes values between 1 and COV_{max}, where COV_{max} is defined as follows:

$$COV_{max} = \frac{N^2 - N}{2} \tag{2}$$

where N is the number of classes. Nevertheless, a value of COV close to COV_{max} is usually too restrictive, while a less restrictive value may introduce DEGs with a poor multiclass potential.

Thanks to that configuration, DEGs that discern not only two classes but also different combinations of pairs of classes were achieved. In addition, this process also filters all these DEGs without any statistical differences between the addressed classes. DEG extraction was made through the use of the KnowSeq R/Bioc package [6].

Once the DEGs are available, the wrapper described above was applied to obtain the most relevant ones, eliminating those considered noise or irrelevant. Finally, an exhaustive search through the literature was carried out to check the relation of the final selected DEGs with lung cancer.

4   Experimentation

After the DEG extraction process from the training dataset, a total amount of 410 DEGs were obtained and used as input for the wrapper. Taking into account the pre-filtered subset of DEGs, the values of the different hyperparameters can be found in Table 2. It is important to highlight that the three selected values for the individuals were chosen to cover all the features of the problem since some studies have shown that this choice is appropriate [9]. Furthermore, three different numbers of generations were set to detect when the algorithm reaches convergence. On the contrary, the P_m hyperparameter refers to the probability that any of the DEGs selected by each individual of the offspring may suffer a mutation, which has been adjusted taking into account the study carried out in [15], where it was stated that a suitable value for this hyperparameter could be $1/N_f$, with N_f being the number of features in the dataset. Finally, the k hyperparameter of k-NN takes the value of the square root of the number of samples in the dataset, as proposed in [9].

Table 2. Hyperparameter values used for the experiments.

Hyperparameter	Value
Number of executions (N_e)	30
Number of generations (N_g)	50, 100, and 150
Number of individuals (N_i)	200, 400, and 600
Crossover probability (P_c)	0.8
Mutation probability (P_m)	0.0024

Table 3. Test Kappa values and number of selected DEGs (avg ± std) when varying N_i and N_g.

N_i	N_g	Test Kappa	Number of selected DEGs
200	50	94.949 ± 0.322	74.908 ± 9.291
	100	94.087 ± 0.789	17.469 ± 4.819
	150	92.505 ± 1.661	11.632 ± 6.099
400	50	94.926 ± 0.238	64.022 ± 7.105
	100	**91.012 ± 0.899**	**5.012 ± 1.797**
	150	91.377 ± 0.759	7.388 ± 7.564
600	50	94.956 ± 0.219	55.397 ± 8.095
	100	91.065 ± 0.895	5.05 ± 2.065
	150	91.015 ± 0.751	4.251 ± 1.901

On the one hand, Table 3 shows the values of the test Kappa coefficient and the number of selected DEGs. Afterward, the Akaike criterion was taken into account, which is defined as a statistical approach to select between different classifiers attending to the complexity of each classifier and their number of parameters [2]. In this sense, a classifier with n genes and similar behavior is

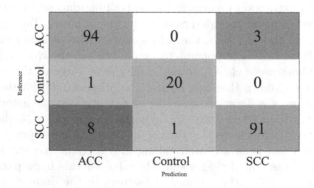

Fig. 1. Test confusion matrix of the best solution achieved by the wrapper.

better than another model with more genes, especially because a short genetic signature is easier and cheaper to validate in a wet lab than a more complex one. Taking into account this criterion, the selected combination was the one with $N_i = 400$ and $N_g = 100$, where the mean test Kappa coefficient value is 91.012 ± 0.899, and the mean number of selected DEGs is 5.012 ± 1.797. As these values represent an average of the N_e executions performed, it was decided to choose the one providing the highest Kappa value as the final solution, which finally selected 7 DEGs.

On the other hand, Fig. 1 contains the test confusion matrix of the best solution achieved by the wrapper. As can be seen, almost all the control samples are tagged correctly, as well as all the ACC and SCC samples. As would have been expected, the majority of mistakes occur between ACC and SCC. Even so, attending to the confusion matrix, the reached F1-score is equal to 94.35%, the specificity equal to 96.5%, and the sensitivity equal to 94.38%. This result reveals the potential of the proposed method when dealing with multiclass and complex gene expression studies.

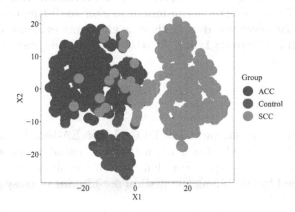

Fig. 2. T-SNE representation of the final 7 DEGs.

In addition, in order to graphically visualize the potential of the seven selected DEGs, a T-Distributed Stochastic Neighbor Embedding (t-SNE) [13] representation was performed. This algorithm makes a two-dimension data transformation of the variables to create a way of separating the different addressed classes into clusters. Figure 2 shows how with our reduced set of 7 DEGs, the three classes have been clearly separated into three clusters. However, some mistaken points go in line with the classification errors observed in the confusion matrix.

Besides the t-SNE representation, it is also very interesting to see the capability of each gene for discerning the addressed classes. In this sense, Fig. 3 plots the gene expression boxplots for each DEG and each class as well. It can be seen how almost all genes can perfectly discern between the three classes.

Finally, to provide a biological profile of the selected genes, an in-depth search was done to determine the relation of each one of the genes with lung cancer or

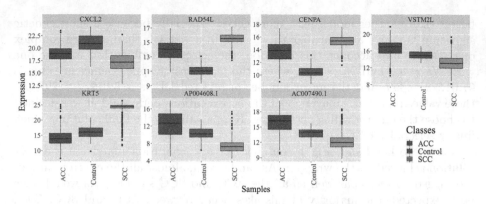

Fig. 3. Gene expression boxplots for each DEG and each one of the addressed states.

any of the addressed sub-types. In this sense, CXCL2, RAD54L, and CENPA are related to worse lung cancer and ACC 5-year overall survival [10,20,22]. Furthermore, KRT5 was detected as an inmunomarker to distinguish between SCC and ACC [16]. However, until now, there is no strong evidence or relationship among VSTM2L, AP004608.1, and AC007490.1 with lung cancer or any of the addressed sub-types.

5 Conclusions

Along this study, a custom multiclass DEGs-driven EA-based wrapper has been proposed to find a reduced set of genes with the potential of discerning the most relevant lung cancer sub-types. For that, a robust evolutionary procedure was carefully designed by combining a statistical pre-filtering of noisy genes with the potential of the EA.

Thanks to that process, outstanding results were achieved using only a reduced subset of 7 DEGs. Concretely, a mean F1-score equal to 94.35%, which indicates that every class was correctly recognized, has been obtained.

In addition, posterior biological research of the selected DEGs was done, concluding that 4 of the 7 DEGs have strong evidence that correlates them with processes related to lung cancer or any of the addressed sub-types.

To sum up, our custom evolutionary procedure can deal with transcriptomic data, achieving excellent results and DEGs biologically related to the addressed problem as well. Moreover, the EA-based wrapper has the capability of reducing the number of biomarkers to the point of making them evaluable in a laboratory.

Acknowledgments. This research has been funded by the Spanish Ministry of Science, Innovation, and Universities (grants PGC2018-098813-B-C31 and PID2022-137461NB-C31) and ERDF fund, in collaboration with the project PID2021-128317OB-I00 from the same funder and the project from Junta de Andalucia P20-00163.

References

1. Abeel, T., Helleputte, T., Van de Peer, Y., Dupont, P., Saeys, Y.: Robust biomarker identification for cancer diagnosis with ensemble feature selection methods. Bioinformatics **26**(3), 392–398 (2010). https://doi.org/10.1093/bioinformatics/btp630
2. Akaike, H.: A new look at the statistical model identification. IEEE Trans. Autom. Control **19**(6), 716–723 (1974). https://doi.org/10.1109/TAC.1974.1100705
3. Arowolo, M.O., Adebiyi, M., Adebiyi, A.A., OKesola, J.: Predicting RNA-SEQ data using genetic algorithm and ensemble classification algorithms. Indones. J. Electr. Eng. Comput. Sci. **21**(2), 1073–1081 (2021). https://dx.doi.org/10.11591/ijeecs.v21.i2
4. Bray, F., Ferlay, J., Soerjomataram, I., Siegel, R.L., Torre, L.A., Jemal, A.: Global cancer statistics 2018: Globocan estimates of incidence and mortality worldwide for 36 cancers in 185 countries. CA J. Clin. **68**(6), 394–424 (2018). https://doi.org/10.3322/caac.21492
5. Castillo, D., et al.: Leukemia multiclass assessment and classification from microarray and RNA-SEQ technologies integration at gene expression level. PLoS ONE **14**(2), e0212127 (2019). https://doi.org/10.1371/journal.pone.0212127
6. Castillo-Secilla, D., et al.: KnowSeq R-Bioc package: the automatic smart gene expression tool for retrieving relevant biological knowledge. Comput. Biol. Med. **133**, 104387 (2021). https://doi.org/10.1016/j.compbiomed.2021.104387
7. Cooper, W.A., et al.: The textbook on lung cancer: time for personalized medicine. Ann. Palliat. Med. **4**(2), 81 (2015). https://doi.org/10.3978/j.issn.2305-5839.2015.04.05
8. Edgar, R., Domrachev, M., Lash, A.E.: Gene expression omnibus: NCBI gene expression and hybridization array data repository. Nucleic Acids Res. **30**(1), 207–210 (2002). https://doi.org/10.1093/nar/30.1.207
9. González, J., Ortega, J., Damas, M., Martín-Smith, P., Gan, J.Q.: A new multi-objective wrapper method for feature selection - accuracy and stability analysis for BCI. Neurocomputing 407–418 (2019). https://doi.org/10.1016/j.neucom.2019.01.017
10. Gu, L., Yao, Y., Chen, Z.: An inter-correlation among chemokine (CXC motif) ligand (CXCL) 1, CXCL2 and CXCL8, and their diversified potential as biomarkers for tumor features and survival profiles in non-small cell lung cancer patients. Transl. Cancer Res. **10**(2), 748 (2021). https://doi.org/10.21037/tcr-20-2539
11. Herbst, R.S., Morgensztern, D., Boshoff, C.: The biology and management of non-small cell lung cancer. Nature **553**(7689), 446–454 (2018). https://doi.org/10.1038/nature25183
12. Liu, J., et al.: Identification of a gene signature in cell cycle pathway for breast cancer prognosis using gene expression profiling data. BMC Med. Genomics **1**(1), 1–12 (2008). https://doi.org/10.1186/1755-8794-1-39
13. Van der Maaten, L., Hinton, G.: Visualizing data using t-SNE. J. Mach. Learn. Res. **9**(11) (2008)
14. Mallik, S., Zhao, Z.: Identification of gene signatures from RNA-SEQ data using pareto-optimal cluster algorithm. BMC Syst. Biol. **12**(8), 21–29 (2018). https://doi.org/10.1186/s12918-018-0650-2
15. Muhlenbein, H., Schlierkamp-Voosen, D.: Optimal interaction of mutation and crossover in the breeder genetic algorithm. In: Proceedings of the Fifth International Conference on Genetic Algorithms, San Francisco, CA, USA, p. 648. Citeseer, Morgan Kaufmann Publishers Inc. (1993)

16. Pan, B., et al.: The value of AGR2 and KRT5 as an immunomarker combination in distinguishing lung squamous cell carcinoma from adenocarcinoma. Am. J. Transl. Res. **13**(5), 4464 (2021). https://doi.org/10.1016/j.lungcan.2012.04.007

17. Rosti, G., Bevilacqua, G., Bidoli, P., Portalone, L., Santo, A., Genestreti, G.: Small cell lung cancer. Ann. Oncol. **17**, ii5–ii10 (2006). https://doi.org/10.1093/annonc/mdj910

18. Tomczak, K., Czerwińska, P., Wiznerowicz, M.: Review the cancer genome atlas (TCGA): an immeasurable source of knowledge. Contemporary Oncology/Współczesna Onkologia **2015**(1), 68–77 (2015). https://doi.org/10.5114/wo.2014.47136

19. Wang, C., et al.: RNA-SEQ profiling of circular RNA in human lung adenocarcinoma and squamous cell carcinoma. Mol. Cancer **18**(1), 1–6 (2019). https://doi.org/10.1186/s12943-019-1061-8

20. Wu, Q., et al.: Expression and prognostic significance of centromere protein a in human lung adenocarcinoma. Lung Cancer **77**(2), 407–414 (2012). https://doi.org/10.1016/j.lungcan.2012.04.007

21. Xue, B., Zhang, M., Browne, W.N., Yao, X.: A survey on evolutionary computation approaches to feature selection. IEEE Trans. Evol. Comput. **20**, 606–626 (2016). https://doi.org/10.1109/TEVC.2015.2504420

22. Zheng, S., et al.: Homologous recombination repair rathway and RAD54L in early-stage lung adenocarcinoma. PeerJ **9**, e10680 (2021). https://doi.org/10.7717/peerj.10680

Prediction of Blood Glucose Levels in Patients with Type 1 Diabetes via LSTM Neural Networks

Ciro Rodriguez Leon[1], Oresti Banos[1], Oscar Fernandez Mora[2],
Alex Martinez Bedmar[2], Fernando Rufo Jimenez[2],
and Claudia Villalonga[1]

[1] Research Centre for Information and Communication Technologies (CITIC-UGR),
University of Granada, Granada, Spain
{crleon,oresti,cvillalonga}@ugr.es
[2] School of Engineering and Technology, Universidad Internacional de La Rioja
(UNIR), Logroño, Spain
{oscar.fernandez007,alex.martinez559,fernando.rufo448}@comunidadunir.net

Abstract. Diabetes is one of the most prevalent diseases of the 21st
century, with more than 500 million people affected. Having tools to
estimate blood glucose levels is critical for these patients in their management of the disease. In this work, we present a comparison of three
neural network architectures based on long short-term memory (LSTM).
Their predictive ability has been evaluated against a longitudinal dataset
with continuous glucose level measurements of patients with type 1 diabetes. All models, trained for different prediction horizons of 30, 60,
90 and 180 min, have generally yielded good prediction results. These
results are further validated using clinical standards resulting in more
than 95% of accurate blood glucose level predictions, mostly leading to
correct treatments.

Keywords: Type 1 Diabetes Mellitus · Deep Learning · Blood glucose
level prediction

1 Introduction

Diabetes, a metabolic disorder that affects the way the body processes and uses
blood sugar, is one of the most common chronic diseases worldwide. In fact, 537
million adults are living with diabetes and according to predictions this number
will rise to 643 million by 2030 and 783 million by 2045. The economic impact
of diabetes in 2021 was at least $966 billion in health expenditures, which was a
316% increase over the last 15 years. In Europe, one in eleven adults has some
form of diabetes, accounting for 61 million in total, even if it is estimated that
one in three adults has undiagnosed diabetes. Furthermore, the number of adults
with diabetes in Europe is expected to grow up to 67 million by 2030 and 69

I. Rojas et al. (Eds.): IWANN 2023, LNCS 14134, pp. 563–573, 2023.
https://doi.org/10.1007/978-3-031-43085-5_45

million by 2045 [7]. This situation will place a large burden on healthcare systems and have a significant economic and social impact.

Although there is no cure for diabetes, it can be controlled through medication, insulin and healthy lifestyles, such as having a healthy diet and exercising regularly. In fact, it is important for patients with diabetes to maintain adequate control of their disease, as this can help prevent long-term complications such as heart disease, stroke, kidney damage and vision loss. The ability to predict the evolution of blood glucose levels in the near future can also be useful for patients with diabetes, as it allows them to anticipate how their blood glucose level will evolve in the coming hours or days. This is specially important for patients with type 1 diabetes mellitus, one of the main types of diabetes which occurs when the body's immune system destroys insulin-producing cells in the pancreas, preventing the body from producing enough insulin to regulate blood sugar levels [9]. Patients with type 1 diabetes mellitus are the ones who find more difficult to maintain their glucose levels in range. Therefore, knowing the future evolution of blood glucose levels will allow them to adapt their lifestyle so that they can maintain these levels close to those of a healthy person and prevent possible complications related to this chronic disease.

One of the applications of IoT is real-time continuous glucose monitoring in diabetic patients. In the first systems marketed in 1999 the devices stored glucose level information and afterwards transmitted and analyzed it. Today systems measure not only glucose, but also blood pressure, temperature, physical activity and dietary data via mobile apps that directly transmit the data to a server [5]. Continuous blood glucose level monitoring has been a major advance in the management of diabetes, as it provides an accurate record of the evolution over time of blood glucose levels in patients with diabetes. This has allowed the development of models that try to predict how blood glucose levels will vary over a short period of time from previous blood glucose level measurements, insulin administered and other collected data. These predictive models can be very useful for patients with diabetes, as they allow them to anticipate how their blood glucose level may be affected by each decision they make in their daily lives, such as the amount of carbohydrates they consume, the amount of exercise they do, or the dose of insulin they take.

This work compares several neural networks used in blood glucose level prediction for patients with type 1 diabetes mellitus. The neural networks are trained with data obtained from continuous glucose monitoring systems. These algorithms use deep learning techniques to process large amounts of data and try to predict how blood glucose levels will vary in the near future. The results of this work may help to improve the accuracy of the proposed neural networks and thus improving the effectiveness of short-term blood glucose level prediction for patients with type 1 diabetes mellitus.

2   State of the Art

Maybe the first approach to glucose prediction in patients with type 1 diabetes using deep learning is presented in 1999 by Tresp, Briegel and Moody [19]

Several Recurrent Neural Network (RNN) models are evaluated and compared with other linear and nonlinear models. Insulin levels, meals, exercise level, and current and previous estimates of blood glucose are used to train the models. An RNN is also trained in [2], but in this case signals from a continuous monitoring devices are used as input. Prediction horizons of 15, 30, 45 and 60 min are compared with the results of a standard feed-forward network and it is found that long-term estimates are more accurate than ones obtained with the RNN.

Long short-term memory (LSTM) networks have been widely used to predict blood glucose levels in patients with type 1 diabetes mellitus. For example, [17] describes a sequential model in which an LSTM and a bidirectional LSTM (BiLSTM) of four units each are combined with three fully-connected layers. 26 datasets from 20 different patients, real and in silico, are used in the evaluation, which shows that the LSTM model improves the predictions of the classical models. [6] presents an LSTM network to characterize the temporal dimensions of the data and two dense layers to extract features. This work tests different combinations of hyperparameters for 10-patient data, obtaining the best results with 50 units in the LSTM and 30 for each dense layer. [12] proposes an LSTM architecture based on a physiological model from which the dependencies between the parameters are extracted. The three-layer architecture, with an LSTM layer and a dense layer for the results, is trained using glucose, insulin, sleep and exercise levels data from real patients in a total of 1600 days. [13], a posterior version of the previous work, proposes an LSTM coupled with a neural attention model. [1] proposes two LSTM networks working in parallel and then connected in a fully connected layer. The first network works with observed data and the second one with estimated data. To improve the model, the weights of the LSTM are adjusted for each patient, obtaining good results both in real patients and in silico at different prediction horizons. [14] proposes four models consisting on an LSTM layer followed by a dense layer, one for each of the inputs: glucose, carbohydrates and fast and slow insulin units. Once the inputs are processed separately, the networks for insulin and carbohydrate concatenate, returning a prediction and then, the glucose information is concatenated to evaluate the final glucose values. [11] also presents an architecture with one LSTM layer that alternates with two fully connected layers, but treats glucose predictions as a classification problem, rather than a classical time series problem. Hypo- and hyperglycemia ranges are normalized and divided into 100 bins, which will be the different classes returned by the model.

LSTM networks are also predominant in the models presented to the second Blood Glucose Level Prediction (BGLP) Challenge, which took place in 2020. In this challange, the OhioT1DM dataset [10] was used by several researchers to train their own models and to compare the efficacy of their different prediction approaches. The results of BGLP are presented in [3], where eight systems that conformed to the challenge rules are ranked based on their errors for 30 and 60 min prediction horizons. The best prediction model is [16], a neural network architecture based on Neural Basis Expansion for Interpretable Time-Series Forecasting (N-BEATS) but replacing the fully connected block structure of

N-BEATS with LSTMs. This winning work presents an architecture which learns to forecast gradually in stages or blocks. Each residual block contains a BiLSTM with a single output layer that produces the forecast and back projection, and additional variables are added as input channels to each block. In fact, this and other ensemble models have been recently used to estimate blood glucose levels in patients with type 1 diabetes mellitus. These approaches train multiple models and combine their independent outcomes into a unified prediction. For example, [8] proposes a system that combines six models called base-learners: two LSTM networks, two Multilayer perceptrons (MLP) and two Partial Least Square Regression (PLSR) models. These base-learners converge into a PLSR layer, the meta-learner, which provides the output prediction of the blood glucose level. Two ensembles based on Bayesian voting to predict the blood glucose level are presented in [18]. These ensembles use three and four LSTM models, respectively, which are selected as the best from a set of ten different neural network architectures. The two proposed ensembles are compared with many of the previously described models. The OhioT1DM dataset is also used to evaluate them under the same conditions at prediction horizons of 30, 60 and 120 min and using the variables glucose levels, basal insulin, insulin dose and carbohydrate intake. The work concludes that there is little difference in predictive capacity since the values of the performance metrics are very close, and the confidence intervals overlap. In fact, although differences have been found statistically between the worst and the best models, from a medical perspective they are irrelevant.

3   Methodology

The objective of this work is to evaluate the performance of three popular recurrent neural network architectures in the field of glucose prediction: long short-term memory (LSTM), bidirectional LSTM (BiLSTM) and convolutional LSTM (ConvLSTM). The evaluation will be performed for different prediction horizons when training the models with a longitudinal dataset of continuous glucose measurements from patients with type 1 diabetes mellitus.

3.1   *T1DiabetesGranada* Dataset

T1DiabetesGranada: a longitudinal multi-modal dataset of type 1 diabetes mellitus [15] is a public dataset which comprises continuous blood glucose levels, demographic and clinical information of 736 patients with type 1 diabetes mellitus. The dataset contains over four years of data collected from patients at the Clinical Unit of Endocrinology and Nutrition of the San Cecilio University Hospital of Granada, Spain. Blood glucose levels are measured every 15 min using *FreeStyle Libre 2*, a flash glucose meter manufactured by *Abbott Diabetes Care, Inc.* The dataset provides more than 22.6 million records that constitute the time series of continuous blood glucose level measurements of the patients during the duration of the study.

3.2 Data Analysis and Preparation

An exploratory analysis of the *T1DiabetesGranada* dataset has been performed and it has been decided to only use the continuous blood glucose level measurements of the patients to train the prediction models. This data has been processed by eliminating possible outliers. The blood glucose level measurements outside the range from 40 to 400 mg/dl have been removed as done in previous works like [18]. Due to the functioning of the flash glucose meter and the interaction of the patient, the time series of blood glucose levels can contain measurements in intervals of less than 15 min. This happens because each time a patient scans the device, the current blood glucose level is measured and it is added as an extra measurement point to the time series. Therefore, the data is processed by eliminating the smaller of the two intervals in the time series, thus obtaining an interval that is closer to 15 min. Furthermore, the time series might also contain data gaps without blood glucose level measurements. This situation occurs in two situations. First, if the patient does not scan the device in less than 8 h, which is the maximum storage time, and the flash glucose meter overwrites the previous measurements. Second, if the patient, does not activate the replacement device early enough after its 14-days life span. To solve this problem, the data is interpolated using the cubic spline method which provides smooth and continuous data, characteristics of blood glucose levels, and generates values adjusted to different data forms. For each patient, the longest sequence of continuous blood glucose level measurements is selected. In order to do so, a tolerance window of 90 min is defined, which is the maximum time allowed in the sequence without data and represents a gap of up to six missing measurements. Cubic spline interpolation is used to obtain the complete time series over the window. The five patients with the longest data sequences are used in this work and the information about their data is presented in Table 1.

3.3 Training, Validation and Test

After the exploratory analysis and data preparation, the data has been separated into the training, validation and test sets, with a split of 70%, 20% and 10%. It is not possible to perform random distributions of the time series, since their temporal correlation must be maintained. Therefore, data windows are implemented to provide the neural network with a set of historical data that can be used to predict future blood glucose levels. The prediction horizon is the time frame within the model is expected to make accurate predictions when trained on a data history of a given size. Since blood glucose levels can vary significantly in a short time due to diet, physical activity and other factors, state-of-the-art prediction horizons of 30 and 60 min are commonly used when the models are trained on a history of 120 min. Considering that the *T1DiabetesGranada* dataset used in this work provides blood glucose level measurements every 15 min, whereas the OhioT1DM dataset provides them every 5 min, it might be necessary to increase the prediction horizon to obtain more accurate predictions. Therefore,

Table 1. Information about the blood glucose level measurements of the patients used to train the prediction models.

	LIB193385	LIB193327	LIB193367	LIB193313	LIB193316
Start date	22/08/2021 23:41:00	13/01/2021 2:28:00	05/04/2021 22:49:00	27/06/2021 22:02:00	27/01/2022 9:47:00
End date	08/11/2021 22:22:00	27/03/2021 1:52:00	17/05/2021 22:31:00	08/08/2021 2:39:00	10/03/2022 9:47:00
Records	746	6989	4025	4008	5059
Interpolated records	34	45	16	18	514
Interpolation percentage	0.46%	0.64%	0.40%	0.45%	10.16%
Mean	128.81	179.35	148.67	132.27	104.58
Deviation	29.90	60.35	46.31	44.91	36.00
Min	52	53	53	52	51
25%	108	135	113	101	76
50%	126	174	141	124	99
75%	146	217	180	156	124
Max	259	400	314	337	268

the prediction models have been trained in four different scenarios: (1) prediction horizon of 30 min with a history of 120 min; (2) prediction horizon of 60 min with a history of 120 min; (3) prediction horizon of 90 min with a history of 360 min; and (4) prediction horizon of 180 min with a history of 360 min. For each scenario, the Mean Absolute Error (MAE) of the trained models has been calculated. In the prediction of blood glucose levels, the MAE is preferred to the Mean Square Error (MSE) and the Root Mean Squared Error (RMSE) because it is considered more robust as it gives less weight to outliers. Although the MAE measures the error between the predictions and the actual values, it does not take into account the clinical context in which the model is used. Therefore, the Clarke Error Grid analysis [4] has been used to represent the expected and estimated values of blood glucose levels and quantify their clinical accuracy. The grid is divided into five zones: zone A represents values clinically accurate thus leading to correct treatments, zone B those leading to a benign or no treatment, zone C to unnecessary treatment, zone D to a failure to detect and treat, and zone E to an erroneous treatment.

3.4   Neural Network Architectures

Three recurrent network architectures based on models presented in the literature have been implemented using Tensorflow. The first model is a 128-unit LSTM recurrent neural network (see Fig. 1a). After the LSTM layer, there are four dense layers with 150, 100, 50 and 20 units and connected to the previous and next layers. Before the second and fourth dense layers, there is a dropout layer with a rate of 0.20 and 0.15, respectively, used to reduce overfitting. ReLu activation function is used in all the layers and the output layer has a single neuron, which returns the predicted value of the blood glucose level. The second model, the BiLSTM network (see Fig. 1b) is a variant of the LSTM network, replacing the recurrent network layer with a 128-unit BiLSTM but leaving the rest of the network unchanged. The third model, the ConvLSTM network (see Fig. 1c) consists of a convolutional layer with 32 filters of kernel size 1. The result of this layer is connected to the original 128-unit LSTM network architecture, with a slight variation in the last dense layer, which has only 16 neurons instead of 20. All three models have been implemented using the same settings. Adam has been used as optimizer and the loss function has been calculated in MSE. The models have been trained for 100 epochs with a batch size of 32, and early stopping is included in some runs to avoid overtraining the model.

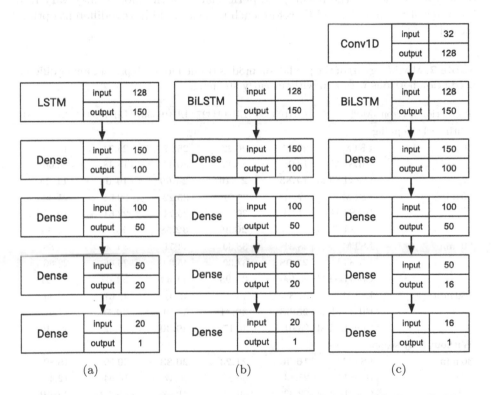

Fig. 1. Neural network architectures: (a) LSTM. (b) BiLSTM. (c) ConvLSTM.

4   Results and Discussion

The three neural network models trained for each patient under the prediction horizons of 30, 60, 90 and 180 min, with and without early stopping, have been evaluated. The MAE of the prediction models are shown in Table 2. The models yielding best results for each scenario have been highlighted. The performance of the models deteriorates as the prediction horizon increases. This is expected since the further out the predicted value is in time, the more complicated is to predict it and the less accurate the prediction will be. The prediction performance of the models varies depending on whether or not early stopping is used during the training phase. With early stopping, the training has been completed in a few epochs, in most cases after six complete training cycles, and in the case of BiLSTM in as few as three training cycles. Without early stopping, the models are trained up to 100 epochs, which can lead to overfitting. For prediction horizons of 30 and 60 min, the ConvLSTM model provides the best results when using early stopping. Without early stopping, the LSTM obtains the best results for three of the patients. For the other two patients, the ConvLSTM performs best for the prediction horizon of 30 min and the BiLSTM for the prediction horizon of 60 min. In view of the results, the performance of the models may vary from one patient to another, and therefore each patient could have a different optimal model.

Table 2. MAE (mg/dl) of the prediction models trained for each patient under different prediction horizons with and without early stopping.

Prediction horizon	Model	LIB193385	LIB193327	LIB193367	LIB193313	LIB193316
With early stopping						
30 min	LSTM	20.67	26.22	25.27	26.74	19.80
	BiLSTM	18.08	27.43	30.44	18.48	15.60
	ConvLSTM	**14.83**	**22.10**	**24.69**	**14.04**	**11.24**
60 min	LSTM	27.31	31.21	35.22	30.01	25.29
	BiLSTM	21.04	30.64	35.74	22.61	19.43
	ConvLSTM	**20.67**	**26.36**	**32.58**	**20.72**	**17.11**
90 min	LSTM	28.31	38.30	44.11	35.66	24.88
	BiLSTM	**22.98**	33.30	45.39	26.66	22.86
	ConvLSTM	23.04	**31.98**	**41.34**	**24.67**	**19.81**
180 min	LSTM	31.89	43.62	49.47	40.07	30.84
	BiLSTM	**25.58**	**39.41**	50.79	**30.78**	31.57
	ConvLSTM	26.01	40.17	**49.30**	31.32	**29.25**
Without early stopping						
30 min	LSTM	**16.46**	**21.23**	**20.33**	20.09	16.80
	BiLSTM	25.14	30.25	26.47	16.94	12.41
	ConvLSTM	22.32	30.40	21.35	**13.79**	**11.40**
60 min	LSTM	**15.62**	**26.06**	**27.64**	24.61	23.45
	BiLSTM	24.48	34.33	28.50	**22.41**	**17.01**
	ConvLSTM	22.56	32.64	29.10	22.65	17.44
90 min	LSTM	**17.92**	**31.09**	**31.58**	33.25	26.50
	BiLSTM	19.50	37.79	35.62	**27.90**	**20.70**
	ConvLSTM	18.29	39.78	33.61	31.62	23.52
180 min	LSTM	21.24	**41.71**	40.73	41.43	31.95
	BiLSTM	**20.26**	43.61	**37.51**	**34.83**	**27.59**
	ConvLSTM	20.85	48.13	38.17	38.94	29.56

To evaluate the performance of the prediction models from a clinical perspective a Clarke Error Grid analysis is performed. Table 3 reports on the percentage of predictions falling in the zones A and B, which lead to clinically correct treatments and those leading to a benign treatment respectively. All percentages are above 94% irrespective of the model, prediction horizon and patient. The best result (99.99%) is obtained for the patient LIB193327 when training the LSTM network under a prediction horizon of 60 min. The worst result (94.02%) is achieved for the patient LIB193313 when training the BiLSTM under a prediction horizon of 180 min. The Clarke Error Grid analysis for these two cases are shown in Fig. 2. Clearly, most of the predictions fall in zones A and B, thus confirming the clinical validity of the developed models even for the worst ones.

Table 3. Percentage of predictions falling in Clarke Error Grid zones A and B.

Prediction horizon	Model	LIB193385	LIB193327	LIB193367	LIB193313	LIB193316
30 min	LSTM	99,95%	99,89%	99,78%	98,96%	95,35%
	BiLSTM	99,89%	99,98%	99,89%	98,01%	97,70%
	ConvLSTM	99,90%	99,95%	99,90%	98,80%	97,79%
60 min	LSTM	99,48%	**99,99%**	99,48%	97,96%	95,36%
	BiLSTM	99,95%	99,90%	99,95%	94,36%	96,65%
	ConvLSTM	99,90%	99,96%	99,90%	96,50%	94,65%
90 min	LSTM	99,95%	99,93%	97,95%	96,35%	94,36%
	BiLSTM	99,95%	99,95%	98,01%	94,65%	95,58%
	ConvLSTM	99,93%	99,89%	98,50%	96,50%	96,66%
180 min	LSTM	99,93%	99,60%	96,54%	95,65%	95,65%
	BiLSTM	99,95%	99,71%	98,88%	**94,02%**	94,36%
	ConvLSTM	99,98%	99,68%	97,01%	96,58%	95,02%

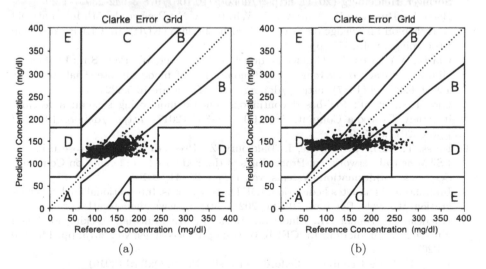

(a)   (b)

Fig. 2. Clarke Error Grid analysis for the best and worst prediction models: (a) LIB193327 - LSTM - 60' (99.99%). (b) LIB193313 - BiLSTM - 180' (94.02%).

5 Conclusions

This work has compared the ability of three neural network models (LSTM, BiLSTM, and ConvLSTM) for predicting blood glucose level measurements in type 1 diabetes patients. The models have been evaluated on four different scenarios with varying prediction horizons (30, 60, 90, and 180 min) and history (120 and 360 min). Few differences are found with respect to the performance of the models, yielding similar prediction errors. Regarding the neural network training strategy, the ConvLSTM stands out as the best model when using early stopping while the LSTM network is found to prevail without early stopping for a majority of patients. According to the experiments, there is no one-fits-all model but rather some models work best for some patients. From a medical point of view, practically all the predictions made by the learned models are in zone A and zone B of Clarke error grid. These results are considered clinically accurate and therefore demonstrate that these models could be used in practice.

Acknowledgements. This research has been funded by the Andalusian Ministry of Economic Transformation, Industry, Knowledge and Universities under grant P20_00163.

References

1. Aiello, E.M., Lisanti, G., Magni, L., Musci, M., Toffanin, C.: Therapy-driven deep glucose forecasting. Eng. Appl. Artif. Intell. **87**, 103255 (2020). https://doi.org/10.1016/j.engappai.2019.103255
2. Allam, F., Nossai, Z., Gomma, H., Ibrahim, I., Abdelsalam, M.: A recurrent neural network approach for predicting glucose concentration in type-1 diabetic patients. In: Iliadis, L., Jayne, C. (eds.) AIAI/EANN -2011. IAICT, vol. 363, pp. 254–259. Springer, Heidelberg (2011). https://doi.org/10.1007/978-3-642-23957-1_29
3. Bach, K., Bunescu, R., Marling, C., Wiratunga, N.: Preface the 5th international workshop on knowledge discovery in healthcare data (KDH). In: CEUR Workshop Proceedings, vol. 2675, pp. 1–4 (2020)
4. Clarke, W.L., Cox, D., Gonder-Frederick, L.A., Carter, W., Pohl, S.L.: Evaluating clinical accuracy of systems for self-monitoring of blood glucose. Diabetes Care **10**(5), 622–628 (1987). https://doi.org/10.2337/diacare.10.5.622
5. Gia, T.N., et al.: IoT-based continuous glucose monitoring system: a feasibility study. Procedia Comput. Sci. **109**, 327–334 (2017). https://doi.org/10.1016/j.procs.2017.05.359
6. Idriss, T., Idri, A., Abnane, I., Bakkoury, Z.: Predicting blood glucose using an LSTM neural network. In: Proceedings of the Federated Conference on Computer Science and Information Systems, vol. 18, pp. 35–41 (2019)
7. International Diabetes Federation: IDF Diabetes Atlas. International Diabetes Federation, Brussels, Belgium, 10 edn (2021). https://www.diabetesatlas.org
8. Khadem, H., Nemat, H., Elliott, J., Benaissa, M.: Multi-lag stacking for blood glucose level prediction. In: CEUR Workshop Proceedings, vol. 2675, pp. 146–150 (2020)
9. Levy, D.: Type 1 Diabetes. Oxford University Press, Oxford (2016)

10. Marling, C., Bunescu, R.: The OhioT1DM dataset for blood glucose level prediction: update 2020. In: CEUR Workshop Proceedings, vol. 2675, pp. 71–74 (2020)
11. Mayo, M., Koutny, T.: Neural multi-class classification approach to blood glucose level forecasting with prediction uncertainty visualisation. In: Proceedings of 5th International Workshop on Knowledge Discovery in Healthcare Data (KDH 2020), vol. 2675, pp. 80–84. Santiago de Compostela, Spain (2020)
12. Mirshekarian, S., Bunescu, R., Marling, C., Schwartz, F.: Using LSTMs to learn physiological models of blood glucose behavior. In: 2017 39th Annual International Conference of the IEEE Engineering in Medicine and Biology Society (EMBC), pp. 2887–2891 (2017). https://doi.org/10.1109/EMBC.2017.8037460
13. Mirshekarian, S., Shen, H., Bunescu, R., Marling, C.: LSTMs and neural attention models for blood glucose prediction: comparative experiments on real and synthetic data. In: 2019 41st Annual International Conference of the IEEE Engineering in Medicine and Biology Society (EMBC), pp. 706–712 (2019). https://doi.org/10.1109/EMBC.2019.8856940
14. Munoz-Organero, M.: Deep physiological model for blood glucose prediction in T1DM patients. Sensors **20**(14), 3896 (2020). https://doi.org/10.3390/s20143896
15. Rodriguez-León, C., et al.: T1DiabetesGranada: a longitudinal multi-modal dataset of type 1 diabetes mellitus (2023). https://osf.io/vd45b/. Accessed 31 Mar 2023
16. Rubin-Falcone, H., Fox, I., Wiens, J.: Deep residual time-series forecasting: application to blood glucose prediction. In: CEUR Workshop Proceedings, vol. 2675, pp. 105–109 (2020)
17. Sun, Q., Jankovic, M., Bally, L., Mougiakakou, S.: Predicting blood glucose with an LSTM and Bi-LSTM based deep neural network. In: 2018 14th Symposium on Neural Networks and Applications (NEUREL), pp. 1–5 (2018)
18. Tena, F., Garnica, O., Lanchares, J., Hidalgo, J.I.: Ensemble models of cutting-edge deep neural networks for blood glucose prediction in patients with diabetes. Sensors **21**(21), 7090 (2021). https://doi.org/10.3390/s21217090
19. Tresp, V., Briegel, T., Moody, J.: Neural-network models for the blood glucose metabolism of a diabetic. IEEE Trans. Neural Networks **10**(5), 1204–1213 (1999). https://doi.org/10.1109/72.788659

Applications of Machine Learning in Time Series Analysis

Photovoltaic Energy Prediction Using Machine Learning Techniques

Gonzalo Surribas Sayago ⓘ, Jose David Fernández-Rodríguez$^{(\boxtimes)}$ ⓘ,
and Enrique Dominguez ⓘ

Department of Computer Science, University of Malaga, 29071 Malaga, Spain
{surribasg,josedavid,enriqued}@uma.es

Abstract. Solar energy is becoming one of the most promising power
sources in residential, commercial, and industrial applications. Solar
photovoltaic (PV) facilities use PV cells that convert solar irradiation
into electric power. PV cells can be used in either standalone or grid-
connected systems to supply power for home appliances, lighting, and
commercial and industrial equipment. Managing uncertainty and fluctuations in energy production is a key challenge in integrating PV systems
into power grids and using them as steady, standalone power sources. For
this reason, it is very important to forecast solar energy power output.
In this paper, we analyze and compare various methods to predict the
production of photovoltaic energy for individual installations and network areas around the world, using statistical methods for time series
and different machine learning techniques.

Keywords: forecasting · photovoltaic energy · machine learning

1 Introduction

In the last years, dramatic drops in the total cost of ownership for many types of
renewable energy power generation have translated into significantly increased
rates of installed power generation, both standalone and connected to the power
grid. In this regard, solar energy has grown enormously, and it is considered
to still have a considerable growth potential, as more and more solar power is
installed to help meet energy demands at a worldwide scale [1]. However, solar
energy comes with serious challenges: its maximum power output is very sus-
ceptible to the amount of solar radiation reaching the solar panels' availability.
As both power grids and standalone facilities require electric power flows to
be as steady as possible, accurate forecasts of available solar radiation become
very important for managing solar facilities. In the case of commercial operators directly selling their output into the electricity market, accurate predictions
are even more relevant, as their profit margins can be significantly affected by
inaccuracies in the predictions [4]. Numerous approaches have been proposed in
the literature to predict the availability of solar radiation [5]. Most are based
on simple, empirical mathematical models that are easy to compute. These

© The Author(s), under exclusive license to Springer Nature Switzerland AG 2023
I. Rojas et al. (Eds.): IWANN 2023, LNCS 14134, pp. 577–587, 2023.
https://doi.org/10.1007/978-3-031-43085-5_46

Year_Month	Power
2016-08-01	2.556462e+06
2016-09-01	1.926004e+06
2016-10-01	1.658863e+06
2016-11-01	1.755318e+06
2016-12-01	1.592755e+06

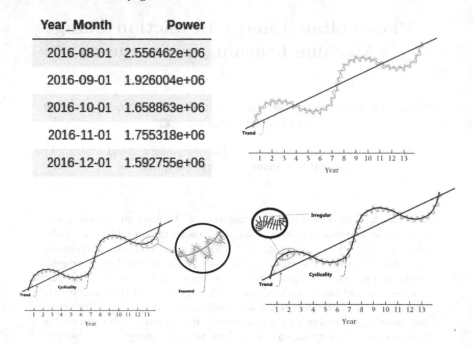

Fig. 1. Elements of a time series (Series, Trend, Seasonal, Irregular)

are widely regarded in the industry as valuable heuristics to predict average daily global solar radiation. Nevertheless, these simple models cannot accurately predict short-term solar radiation availability, as localized and rapid changes in weather conditions (such as cloud cover, intermittent rain, etc.) can significantly impact this availability. Furthermore, these models have been shown to be unable to reflect the complex and nonlinear relationships among dependent and independent variables in humid regions where solar radiation is strongly affected by heavy clouds throughout rainy days [2]. In this work, we propose several approaches based on machine learning to predict short-term solar radiation availability. The proposed techniques are analyzed, and their performance is compared using a common dataset.

2   Dataset and Time Series

The dataset used in this work has been provided by the SunLab platform [7], a collection of on-field PV laboratories installed throughout Portugal with the goal of characterizing the relative performance of various PV technologies. SunLab was set up by Energias De Portugal (EDP), a Portuguese power generation company, in order to support its business units in the acquisition of knowledge in the solar market field. The datasets provided by SunLab are organized by year: from 2014 to 2017. There are two datasets for each year: one with data from weather stations and the other with production and temperature data from

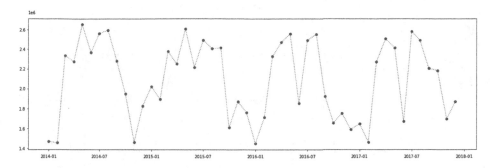

Fig. 2. Per-month SunLab power generation from 2014 to 2017.

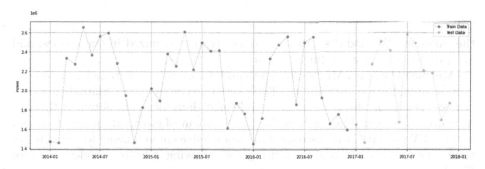

Fig. 3. The data from Fig. 2 has been split into train and test datasets. In this case, the last year is set aside as testing data, and the rest as training data.

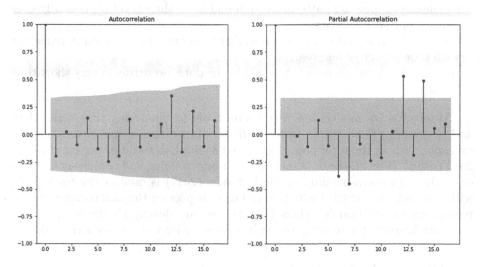

Fig. 4. Autocorrelation and partial autocorrelation functions for the training data in Fig. 3. The highest non-zero value is at 12 in both functions.

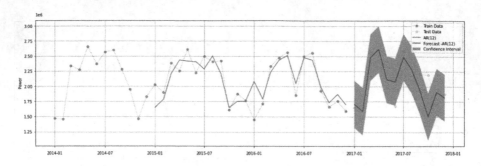

Fig. 5. Autoregressive Model with forecast based on largest autocorrelations (see Fig. 4).

the different PV modules. All these datasets are organized in time series, and all time series have a resolution of 1 min.

Machine learning (ML) techniques can be applied to the time series to forecast solar radiation availability. Time series extracted from SunLab datasets are collections of observations of well-defined data elements obtained through measurements over time, such as the measurement of electricity production in a specific PV module. It is customary in data science environments to characterize time series from real-world data using the following concepts or elements (see Fig. 1):

- Trend: long-term general direction of the time series.
- Cyclicality: repeating patterns of high and low values (cycles), typically over periods of multiple years.
- Seasonality: like cyclicality, but referring to shorter cycles, usually repeating with a frequency of one year.
- Irregularity: rapid changes ("bleeps") in the data, occurring in very short time frames.

Autoregressive models are very useful tools for analyzing time series data and performing forecasts. In order to fit an autoregressive model to a specific example, such as the total power generated at SunLab facilities per month from 2014 to 2017 (Fig. 2), the data must be split into training and testing sets. In this case, the data corresponding to the last year (2017) is used as the testing set, with the rest used as the training set. Then, inspecting the autocorrelation and partial autocorrelation functions (Fig. 4), we can identify the most significant autocorrelation value in order to apply a forecast for the test data (Fig. 5).

3   Proposed Models

In this section, we present four machine-learning models and benchmark them by modeling per-day, fine-grained solar radiation availability, and computing a solar radiation forecast for a specific day used as testing data (May 26, 2015), while

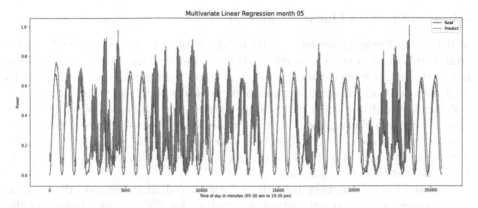

Fig. 6. Multilinear regression forecast for may 2015. The X-axis is in minutes; the Y-axis is the insolation coefficient). The X-axis covers the whole month, but only 14 daylight hours for each day (night periods are omitted from the series).

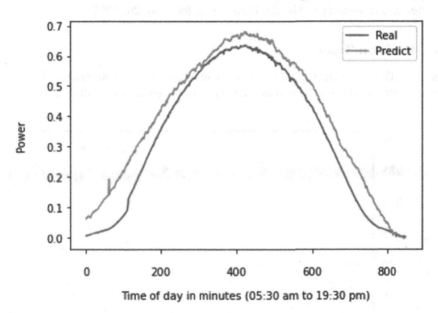

Fig. 7. Multilinear regression forecast for May 26, 2015. The X-axis is in minutes; the Y-axis is the insolation coefficient). The X-axis covers the 14 daylight hours (840 min) for that day.

using the rest of the dataset as training data. The four models are multivariate linear regression (included as a baseline), decision trees, and two ensemble models based on decision trees: random forest regression and XGBoost.

3.1   Multivariate Linear Regression

Multivariate linear regression (MLR) is the simplest regression method considered in this work, used as a baseline. MLR can be regarded as a tool for building linear statistical models that characterize relationships among multiple dependent variables and multiple independent variables, and can be regarded as a collection of multiple linear regressions, each one for a different dependent variable, all of them sharing the same independent variables. Multi-linear regression can be written as $\hat{y} = \beta_0 + \beta_1 * X_1 + \beta_2 * X_2 + ... + \beta_n * X_n$, where $\hat{y}$ is the dependent variable (predicted value), β_0 is the estimated intercept, and β_n is the n-estimated slope coefficient.

Related algorithms have been used for solar radiation forecast. Wang et al. [8] proposed a daily power output forecasting for PV facilities based on the Partial Functional Linear Regression Model (PFLRM) method. The PFLRM was integrated by using both functional and multiple linear regression models.

After fitting an MLR model to the training data, we can use it to forecast solar radiation availability in the testing data. In Fig. 7, we can see the relatively large discrepancies between actual data (blue line) and the forecast (orange line) for the testing data (i.e., the daylight hours for May 26, 2015).

3.2   Decision Trees

Decision Trees are an important type of machine learning algorithm for predictive modeling, where a hierarchy of very simple regression models is built from

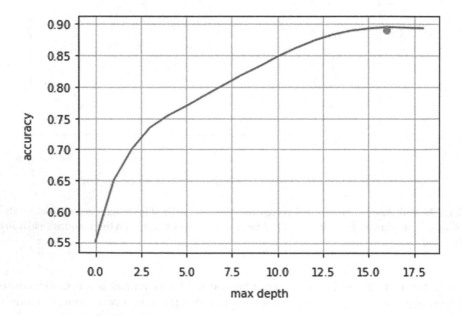

Fig. 8. Accuracy by decision tree depth for our SunLab training data.

Decision Tree

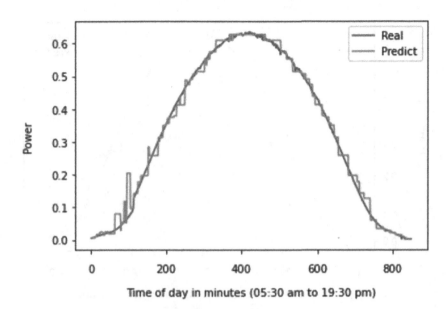

Fig. 9. Decision tree forecast for May 26, 2015. The X-axis is in minutes; the Y-axis is the insolation coefficient). The X-axis covers the 14 daylight hours (840 min) for that day.

the training data, so that samples are broken down into progressively smaller subsets, according to a rule to minimize the prediction error, gradually building up an associated decision tree. In general, as the tree gets deeper, the accuracy increases.

This type of algorithm has already been used by other researchers to predict energy production and its relationship with climatic factors. In [3], authors aimed to predict the output production of solar power plants in kWh and how climatic factors influence that production. The maximum achieved accuracy was around 81% with a maximum tree depth value of 8. However, when using a decision tree to model the SunLab dataset, we found an increased accuracy of 94%, with an optimal depth of 16 (see Fig. 8). With this decision tree model, the predictions for the testing set (May 26, 2015) are significantly better than with the baseline method (see Fig. 9).

3.3 Random Forest Regression

Random Forest Regression can be regarded as a generalization of decision trees: The algorithm works by building multiple decision trees at training time (in parallel, with no interaction between the trees) and using the average of the outputs of the trees as its prediction. When applying this method to our SunLab

Random Forest Regression

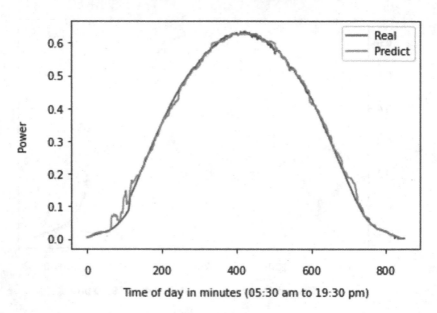

Fig. 10. Random Forest Regression forecast for May 26, 2015. The X-axis is in minutes; the Y-axis is the insolation coefficient). The X-axis covers the 14 daylight hours (840 min) for that day.

dataset, a maximum accuracy of 0.95 was found, averaging results from 300 trees at a tree depth of 16. The forecast of this aggregate model for May 26, 2015 is shown in Fig. 10.

3.4   XGboost (Extreme Gradient Boosting)

Gradient boosting is a family of machine learning algorithms to build an ensemble of models that significantly outperform any models in the ensemble. Typically, the base models are decision trees, so gradient boosting can be regarded as a generalization of these. Broadly speaking, gradient boosting algorithms build decision trees one after another. Crucially, each decision tree is not independent of the rest but is built and fit to correct the prediction errors from previous trees, such that each new tree refines the predictions from previous trees. Models are fit using any arbitrary differentiable loss function and gradient descent optimization algorithm. This gives the technique its name, "gradient boosting," as the loss gradient is minimized as the model is fit, much like a neural network.

XGBoost is a specific implementation of gradient boosting, deploying a wide array of optimizations for speed and performance. XGBoost is regarded as a very competitive implementation of gradient boosting, being used by the winners of many machine learning contests. Obiora et al. [6] obtained very good

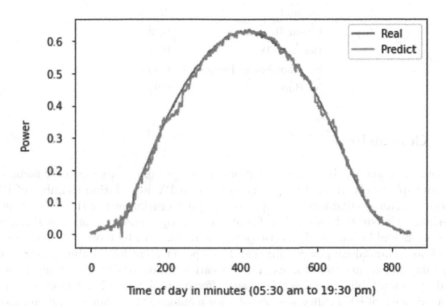

Fig. 11. Extreme gradient boosting forecast for May 26, 2015. The X-axis is in minutes; the Y-axis is the insolation coefficient). The X-axis covers the 14 daylight hours (840 min) for that day.

results predicting solar radiation with this algorithm. When applying XGBoost to our training data, the maximum accuracy is 0.96, using 360 estimators with a maximum tree depth of 8 (see Fig. 11).

4   Experimental Results

Results from the methods discussed in the previous section are gathered together in Table 1. This Table shows the best accuracy achieved with each method. Linear regression models perform significantly worse than the rest; this is to be expected, since solar radiation does not linearly depend on the variables in the SunLab dataset. Accordingly, vanilla decision trees perform significantly better, and the ensemble models provide additional increases in accuracy. While the increases might not seem substantial (Random Forest Regression increases the accuracy over decision trees in 0.01, and XGBoost in 0.02), they are pretty significative, as can be seen by comparing the discrepancies between actual and predicted solar radiation in Figs. 9 (vanilla decision tree), Fig. 10 (Random Forest Regression) and Fig. 11 (XGBoost).

Table 1. Results of the proposed techniques.

Method	R^2 score
Linear Regression	0.73
Decision Tree	0.94
Random Forest Regression	0.95
XGBoost	0.96

5   Conclusions

A lot of research has been directed at optimizing power generation in PV facilities at multiple levels. In particular, the sizing of a PV installation (number of PV modules, storage and inverter capacity, etc.) is a crucial part of the PV system's design, as a correctly sized PV facility with proper energy storage scheduling is a more stable source of electric power, and thus can be more effectively used both as a standalone power source and as a power plant for a public power grid. In this context, accurate forecasting of weather conditions that may affect solar radiation availability can become a very effective tool, not only for more effective optimization of PV facility size but also for managing the balance between power generation and load demand, as balance problems can destabilize the power grid and cause significant economic losses.

This work shows that ensemble algorithms based on decision trees can achieve excellent results in forecasting solar radiation availability. The best accuracy has been achieved using XGBoost with a maximum tree depth of 8 and an ensemble size of 360. While achieving slightly less accuracy, the Random Forest Regression algorithm can achieve results almost as good, with the benefit of being significantly simpler and less computationally intensive to train. In future work, we expect to achieve even better accuracy by using deep learning models built from the ground up to model and effectively generalize time patterns, such as LTSM networks.

References

1. Choudhary, P., Srivastava, R.K.: Sustainability perspectives-a review for solar photovoltaic trends and growth opportunities. J. Clean. Prod. **227**, 589–612 (2019)
2. Fan, J., et al.: New combined models for estimating daily global solar radiation based on sunshine duration in humid regions: a case study in south china. Energy Convers. Manage. **156**, 618–625 (2018)
3. Gupta, A., Bansal, A., Roy, K., et al.: Solar energy prediction using decision tree regressor. In: 2021 5th International Conference on Intelligent Computing and Control Systems (ICICCS), pp. 489–495. IEEE (2021)
4. Gürel, A.E., Ağbulut, Ü., Biçen, Y.: Assessment of machine learning, time series, response surface methodology and empirical models in prediction of global solar radiation. J. Clean. Prod. **277**, 122353 (2020)

5. Liu, Y., Zhou, Y., Chen, Y., Wang, D., Wang, Y., Zhu, Y.: Comparison of support vector machine and copula-based nonlinear quantile regression for estimating the daily diffuse solar radiation: a case study in china. Renewable Energy **146**, 1101–1112 (2020)
6. Obiora, C.N., Ali, A., Hasan, A.N.: Implementing extreme gradient boosting (XGBoost) algorithm in predicting solar irradiance. In: 2021 IEEE PES/IAS PowerAfrica, pp. 1–5. IEEE (2021)
7. Sunlab: EDP open data. https://opendata.edp.com/open-data/en/data.html. Accessed Mar 2023
8. Wang, G., Su, Y., Shu, L.: One-day-ahead daily power forecasting of photovoltaic systems based on partial functional linear regression models. Renewable Energy **96**, 469–478 (2016)

Analysis of the Effect of the Time Interval Between Samples on the Solar Forecasting

Carlos M. Travieso-González[1,2]([✉]) and Alejandro Piñán-Roescher[2]

[1] Signals and Communications Department, University of Las Palmas de Gran Canaria (ULPGC), 35017 Las Palmas de G.C., Spain
ctravieso@dsc.ulpgc.es
[2] Institute for Technological Development and Innovation in Communications (IDETIC), University of Las Palmas de Gran Canaria (ULPGC), 35017 Las Palmas de G.C., Spain

Abstract. This paper analyzes the effect of the choice of the frequency between samples in the field of solar forecasting. To perform the study, the time series of solar radiation is used, in an autoregressive mode, as the only variable, to predict a single time step. Regarding the models used for the tests, the persistent model, which serves as the baseline, and neural networks are used, the most common and increasingly elaborate: Linear, MLP, CNN1D, and LSTM models. To compare the prediction accuracy two error metrics are used: RMSE and MAE. From the results it can be deduced that the analysis of the time interval between samples is a key factor, since a bad choice can result that persistent model being as good as the best predictions of the CNN1D and LSTM models. In addition, it is shown that as the time interval between samples increases, the choice of a model and its input window becomes more important. This paper intends to serve as a first guide that allows selecting parameters to implement predictive models for solar forecasting in an existing infrastructure.

Keywords: Solar forecasting · time series predictions · interval between samples · input steps window to the model · neural networks

1 Introduction

For several decades there has been a need to have a more accurate solar energy forecasting [1], but this has become more important in recent years due to the increase in available solar resources, and its integration into the energy system and/or microgrids. This, together with technological advances, highlighting the reduction of costs in computing capacity, and consequently new artificial intelligence techniques, is giving rise to a dynamic research and development area, which is focused on solar energy forecasting for different time horizons. The state of the art of solar forecasting is quite extensive, and a review can be found in [2–4]. Regarding the prediction methods, the classical statistical and machine learning models are the most used, highlighting the neural networks. In general terms, there are two approaches to realize the solar forecasting: univariate and multivariate. The univariate approach is based only on past PV data, while the multivariate approach uses information from various sources: PV data, meteorological data, weather forecasting, satellite images, and sky cameras [5, 6].

© The Author(s), under exclusive license to Springer Nature Switzerland AG 2023
I. Rojas et al. (Eds.): IWANN 2023, LNCS 14134, pp. 588–600, 2023.
https://doi.org/10.1007/978-3-031-43085-5_47

This paper analyses the effect of the choice of frequency between samples on the solar forecasting. Using only the time series of solar radiation, in autoregressive mode as predictor variables, to predict a single time step. This paper is intended to serve as a first guide to select parameters to implement predictive models for solar forecasting in an existing infrastructure. The process of implementing a predictive model requires setting multiple parameters, plus the added complexity of working with time series, which makes increasingly accurate predictions a real challenge. There are many decisions to be made regarding the design and configuration of a predictive model. Most of these decisions are resolved empirically through trial and error on the working data. As such, it is essential to have a robust way to evaluate the model performance.

The process of implementing a predictive model for time series forecasting can be divided, at a conceptual level, into four large procedures. The first corresponds to collecting a data history, using sensors, of the variables that are intended to be analyzed. A first question that may arise is how much data is needed, and more important at what interval should these data be acquired, since it is later shown that a bad choice can condition the rest of the decisions. The second corresponds to the analysis and preprocessing of the data, in this case the time series. This analysis involves repairing missing data and false values, analyzing trends and seasonality, normalizing values, and finally converting the time series to a supervised learning problem. To do this, an input window to the model must be selected, that is, the number of lags used to make a prediction. This parameter is key factor because it conditions the prediction error. The determination of the value of the input window may require the execution of different experiments to choose the best value for the performance of the model. Due to computational time and cost, it is practically impossible to try all possible sets of values for this parameter. Although an input window value of low lags may provide insufficient information, a large value will increase the complexity of the model and may reduce performance. Having different criteria and proposals to estimate the input window to the model, there is no established solution [7–12]. In most of the articles it is estimated by trial and error, performing multiple simulations, with the consequent effort and time. In addition, some studies analyze the window from 1 sample to N, and most of them stop at the first minimum error. This is better appreciated in seasonal series, because depending on the frequency, for example, an interval of 1 h for one day is 24 samples, up to 24 may have information, but there may also be multiples of this 48, 72, etc., leaving in between worse values, local minima. Other authors use the criterion of mutual information. There are also parallel models, or more elaborate proposals trying to use machine learning directly. However, it also requires testing, albeit less and more focused at best case. In this sense, there is a lot to investigate to obtain the best input window in a simplified way, and not by trial and error. In addition, depending on the prediction problem, a prediction horizon must be decided, that is, the number of steps to predict. The horizon is conditioned by the frequency between samples of the time series and the number of steps that are predicted, that is, a series at an interval of 1 min, a single step is predicting a single minute. A series at 60 min intervals predicts 60 min steps. It is possible that the time series observations are at the wrong frequency, i.e. the time interval may be too small, and simply is the change plus noise, or on the contrary is too large and is missing detail between samples. This is another key factor, how the frequency between samples influences the forecasting result, which is the focus of the analysis in this document. Reviewing the literature there are also publications [13–17], most of them due to the specifications of the problem, they analyze several samples for time series, but none analyze the effect of this on the error in conjunction with the input window to the model, so we think it could be novel in this sense. In addition, as shown below, it can assist the choice of the model by providing a whole set of tests that can finally conclude that persistence model is the best option, with the consequent saving effort and time invested. The third procedure, independently of the prediction algorithm, removing the persistent prediction, is to select a training size, validation, and test. Without forgetting, a cross validation for time series, which is different because the time order itself must be respected, and of course repeating each test n times to see the variance of the models. Once the data and the training strategy are available, the fourth and last procedure is to select a model from the multiple options available. From the most classical ones: AR, ARMA, ARIMA, to the most current neural networks: MLP, CNN, LSTM, or combinations of these forming more elaborate topologies. These models also require parameter adjustment, to finally, after trial and error, be able to select the best option. Four increasingly elaborate models are implemented to test this analysis: persistent, linear, MLP, CNN1D, LSTM. The persistent model, which simply persists the last sample, serves as a baseline, and as will be shown later, for small sample interval it can be as useful as the more elaborate CNN1D or LSTM models. Furthermore, it is shown that as the interval between samples increases, the choice of the model and its corresponding input window becomes more important.

2   Materials, Methods, and Methodology

This section presents the materials, methods, and methodology used in this study. The dataset used, the procedures performed on the data, the models implemented, the validation process, and the metrics used to show the results.

2.1 Materials

Data Set
The present work is based on historical data collected by the ITC (Instituto Tecno-lógico de Canarias), which have been the starting data to start working on the MICROGRID-BLUE project [18]. This historical data represents the solar radiation from 2016-11-01 to 2021-02-26, at 1 min intervals. Figure 1 shows the location of the ITC, as well as the starting data history to carry out the tests. It should be noted that more meteorological variables are collected, but in the present analysis, for simplicity, only solar radiation is used.

Data Preprocessing
The first step has been to analyze the data, recover missing data and outliers, and normalize data between 0 and 1.

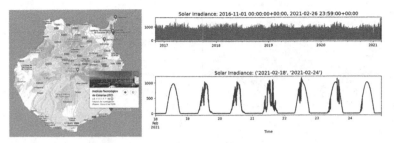

Fig. 1. Location and historical starting data.

2.2 Methods

Time Series Resample

Once the time series is available, without null values, different resamples have been made on the initial time series, at 1 min intervals, to have series of intervals of 1, 5, 15, 30, 60, and 120 min, and perform tests for these intervals and see how the error varies. The resample operation allows to increase, up sampling, and decrease, down sampling, the sample rate, providing the ability to change the sample rate of the original time series. Figure 2 shows the time series for different sample intervals for one day: 1440 samples at 1 min intervals, 288 samples at 5 min, 96 samples in at 15 min, 48 samples at 30 min, and 24 samples 60 min intervals.

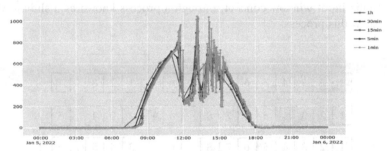

Fig. 2. Solar radiation time series for one day: 1440 samples at 1 m intervals, 288 samples at 5 min, 96 samples at 15 min, 48 samples at 30 min, 24 samples at 60 min.

Autocorrelation (ACF), and Partial Autocorrelation Function PACF

Next, the time series is analyzed using the ACF and the PACF to better estimate the input window of the models to be implemented (Fig. 3).

From the ACF it can be seen the period of the time series, from the first maximum to the second maximum, 96 and 24 samples successively. From the PACF, it can be seen how each delay has less information than the previous one, increasing somewhat

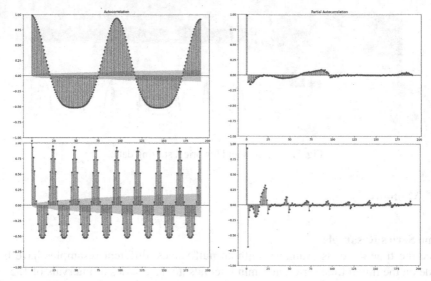

Fig. 3. ACF and PACF for interval between samples of 15 and 60 min.

at periodic intervals of the number of samples per day: for 15 min in 96, 192 samples, and for 60 min in 24,48,72 samples.

Time Series to Supervised Learning, Sliding Window Approach

The next step corresponds to converting the time series to a supervised learning problem, which rise the input window concept, through a process known as a sliding window. Figure 4 show this process.

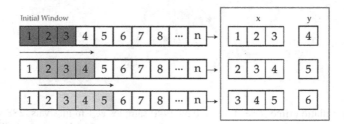

Fig. 4. Sliding Window, 3 input steps window, sliding 1 step, and single time step forecasting.

Implemented Models

Once the data and the steps on them are shown, the next step corresponds to the implementation of the models to carry out the tests. To better see the effect of the frequency between samples, increasingly elaborate models are implemented, where none or few parameters intervene in the first ones, and more in the following ones.

Persistent Model

The persistent model is the simplest model of all, it's not even a neural network, since it simply persists the last delay as the output:

$$Y(t) = X(t-1) \tag{1}$$

The persistent model is used as a baseline to compare the improvement of the following, more elaborate models. It should be noted that the persistent model is independent of the input window, since with a single sample it has the last one, which is its prediction. In this model, the larger the interval between samples, the larger its corresponding error. This technique, on some occasions, especially if the time series has sudden changes in trend, may turn out to be the best option. Small frequencies seconds or a few minutes, as will be demonstrated later.

Linear Model

This model is the simplest to implement with neural networks since it is made up of a single dense layer. This model simply gives each input, lag of the time series, a weight, plus a bias:

$$Y(t) = w1 * X(t-1) + w2 * X(t-2) + \ldots + wn * X(t-n) + bias \qquad (2)$$

A dense layer is a layer that is fully connected to its previous layer. This layer is the most used in artificial neural networks. For this model, the only thing that influences the network topology is the input window of the model. That is, the number of delays used to train the model, and therefore make a prediction.

MLP Model

This model is also made up of dense layers, but it has a greater number of them. This model combines the lags of the time series, but in a non-linear way, and therefore can obtain better results. The models, linear and MLP, do not work on time series (Batch Size, Number of steps, Number of variables) and require a flatten layer to adapt their input to (Batch Size, Input Dimension). This model, in addition to depending on the input window, depends on more topology parameters such as the number of layers, and the number of neurons per layer. For the simulations, two internal layers of 64 and 64 neurons each are used. This model requires more configuration, and more simulations, but in general it gives better results.

CNN1D Model

On top of the MLP model, a first CNN1D layer is added. This model requires the configuration of an MLP, plus the configuration of the CNN1D layer: number of filters, kernel size, and type of padding, in this case 32, 2, and causal, respectively. This layer creates a kernel that convolves with the input over a single spatial, or temporal, dimension to produce an output tensor. This layer by itself is not used to make predictions, but rather to extract features. An important property, in addition to the kernel size, is the type of padding used since it conditions the output size [19]. For practical purposes with time series, it can be thought of as removing noise by filtering the input signal.

LSTM Model

On top of the MLP model, an LSTM layer is added. This model requires the configuration of an MLP, plus the configuration of the LSTM layer, mainly the number of cells, 32 in this case. LSTM networks are a special type of recurrent networks. The main characteristic of recurrent networks is that information can persist by introducing loops in the network diagram, therefore, they can remember previous states and use this information to decide what the next one will be. An advantage of this model is that the parameters do not grow with the length of the input sequence, input window, unlike the MLP. Another advantage is that it allows more complex topologies such as encoder-decoder [20].

2.3 Methodology

Training Strategy

To better observe the variance of the models, and to be able to choose the one with the least error, a cross validation is performed. Figure 5 shows the two options.

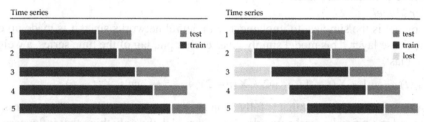

Fig. 5. Validation for time series: Left: Progressively increases the training size, and fixed tests throughout the history. Right: Similar, but in this case the oldest historical is released.

The first option allows to see how the training size affects the prediction. It should be noted that for time series it is essential to respect the order in the validation, unlike non-sequential data. In the other case, once the optimal training size is known, it allows validating how said optimal size behaves over time. To calculate the error, two error metrics are used: RMSE and MAE:

$$RMSE = \sqrt{\frac{1}{N} \sum_{i=1}^{N} (Ypredicted_i - Yreal_i)^2} \tag{3}$$

$$MAE = \frac{1}{N} \sum_{i=1}^{N} |Ypredicted_i - Yreal_i| \tag{4}$$

For each result, 5 repetitions have been made to show the mean and its standard deviation in a result of the form of ERROR ± std.

3 Results

Table 1 shows the simulation results for the persistent model. This persistent model is deterministic and always gives the same results, so its variance is not shown.

Table 1. Persistent model results

Resample	1 min	5 min	15 min	30 min	1 h	2 h
Any Window	RMSE: 34.97 MAE: 7.69	RMSE:43.17 MAE: 13.95	RMSE:57.32 MAE: 24.51	RMSE:74.48 MAE: 37.63	RMSE:113.34 MAE: 64.73	RMSE:189.87 MAE: 117.09

Table 2 shows the simulation results for the linear model, for small intervals, and especially 1 min, the window is indifferent, in addition, the result is like the persistent prediction. This is shown in Tables 2.

Table 2. Linear model results.

RMSE ± std/MAE ± std						
Resample	1 min	5 min	15 min	30 mn	1 h	2 h
Window = 1	34.96 ± 0.002	43.15 ± 0.006	57.19 ± 0.003	74.07 ± 0.013	111.65 ± 0.027	182.71 ± 0.721
	7.69 ± 0.003	13.95 ± 0.002	24.48 ± 0.003	37.46 ± 0.007	63.55 ± 0.016	108.25 ± 0.569
Window = 2	34.81 ± 0.005	43.85 ± 1.053	61.62 ± 0.118	72.96 ± 0.048	84.17 ± 0.061	129.89 ± 1.183
	7.76 ± 0.005	13.71 ± 0.027	21.91 ± 0.004	27.61 ± 0.017	36.59 ± 0.098	71.80 ± 1.144
Window = 5	34.57 ± 0.006	44.89 ± 0.02	58.34 ± 0.058	67.42 ± 0.174	83.25 ± 0.138	121.77 ± 1.687
	7.78 ± 0.008	13.38 ± 0.013	19.51 ± 0.04	25.09 ± 0.131	36.29 ± 0.127	73.19 ± 1.73
Window = 10	34.64 ± 0.004	44.23 ± 0.028	57.05 ± 0.21	67.14 ± 0.254	82.97 ± 0.02	101.64 ± 10.098
	7.75 ± 0.003	12.86 ± 0.013	19.03 ± 0.085	25.06 ± 0.129	36.39 ± 0.131	70.57 ± 8.937
Window = 24	34.66 ± 0.011	43.47 ± 0.064	56.87 ± 0.044	66.51 ± 0.406	62.71 ± 1.374	71.08 ± 0.725
	7.70 ± 0.02	12.55 ± 0.056	18.99 ± 0.031	25.45 ± 0.046	27.62 ± 0.168	29.59 ± 0.095
Window = 48	34.46 ± 0.016	43.45 ± 0.02	56.64 ± 0.054	56.19 ± 0.164	60.72 ± 0.603	66.94 ± 0.557
	7.64 ± 0.048	12.58 ± 0.013	19.12 ± 0.07	22.80 ± 0.13	26.41 ± 0.144	29.29 ± 0.255
Window = 96	34.34 ± 0.01	43.42 ± 0.043	51.65 ± 0.145	55.17 ± 0.244	58.94 ± 0.546	66.89 ± 1.208
	7.68 ± 0.034	12.65 ± 0.054	18.32 ± 0.21	22.49 ± 0.187	25.59 ± 0.419	30.10 ± 0.517
Window = 144	34.31 ± 0.039	43.63 ± 0.312	51.32 ± 0.046	54.43 ± 0.092	58.86 ± 0.596	67.94 ± 1.457
	7.69 ± 0.003	13.95 ± 0.002	24.48 ± 0.003	37.46 ± 0.007	63.55 ± 0.016	108.25 ± 0.569
Window = 192	34.32 ± 0.104	43.69 ± 0.404	51.09 ± 0.111	54.15 ± 0.177	58.96 ± 0.435	69.23 ± 1.921
	7.76 ± 0.005	13.71 ± 0.027	21.91 ± 0.004	27.61 ± 0.017	36.59 ± 0.098	71.80 ± 1.144

Table 3 shows the simulation results for the MLP model. This model at 1 min has similar results to the persistent model. In general terms, it has better results than the linear model, but it requires more configuration in terms of topology parameters (Table 4).

Table 5 shows the simulation results for the LSTM model. This model generally performs well, but not better than the CNN1D model, and sometimes no better than the MLP model, but requires more setup and is more prone to overfitting.

Figures 6 and 7 show the results of plotting the previous tables, but for comparison between models.

Table 3. MLP model results

RMSE ± std/MAE ± std

Resample	1 min	5 min	15 min	30 mn	1 h	2 h
Window = 1	34.97 ± 0.014 7.76 ± 0.019	43.31 ± 0.04 13.96 ± 0.029	57.75 ± 0.093 24.49 ± 0.038	75.73 ± 0.167 37.43 ± 0.034	115.71 ± 0.706 63.09 ± 0.027	192.84 ± 3.718 107.36 ± 0.136
Window = 2	34.63 ± 0.048 7.45 ± 0.053	42.26 ± 0.094 11.86 ± 0.084	53.96 ± 0.191 17.81 ± 0.146	63.35 ± 0.881 22.83 ± 0.323	75.19 ± 1.938 30.52 ± 1.258	107.22 ± 2.112 52.39 ± 2.392
Window = 5	34.12 ± 0.081 7.34 ± 0.114	42.22 ± 0.169 11.99 ± 0.131	53.00 ± 0.081 17.32 ± 0.075	57.98 ± 0.282 21.42 ± 0.374	64.86 ± 1.43 27.65 ± 0.882	95.49 ± 2.528 47.59 ± 3.535
Window = 10	34.04 ± 0.05 7.39 ± 0.023	42.29 ± 0.16 12.09 ± 0.125	51.83 ± 0.378 17.05 ± 0.212	56.16 ± 0.766 21.20 ± 0.473	62.29 ± 1.057 26.48 ± 0.514	66.45 ± 6.564 31.99 ± 7.121
Window = 24	34.08 ± 0.11 7.57 ± 0.068	42.45 ± 0.473 12.18 ± 0.111	50.94 ± 0.203 17.35 ± 0.255	54.61 ± 0.401 20.82 ± 0.532	54.73 ± 0.913 21.81 ± 1.172	60.67 ± 1.177 25.73 ± 0.729
Window = 48	34.37 ± 0.222 7.71 ± 0.089	42.11 ± 0.22 12.30 ± 0.169	51.03 ± 0.157 17.66 ± 0.179	52.71 ± 0.335 19.94 ± 0.414	55.47 ± 0.816 22.06 ± 0.507	61.53 ± 1.801 27.01 ± 1.356
Window = 96	34.30 ± 0.159 7.74 ± 0.17	42.59 ± 0.339 12.76 ± 0.141	49.79 ± 0.371 17.17 ± 0.275	52.37 ± 0.258 20.21 ± 0.296	55.48 ± 0.787 23.19 ± 0.748	61.99 ± 1.95 27.31 ± 0.585
Window = 144	34.17 ± 0.149 7.72 ± 0.026	42.49 ± 0.266 12.74 ± 0.265	49.62 ± 0.199 17.34 ± 0.27	53.06 ± 0.8 20.56 ± 0.591	56.57 ± 0.61 23.77 ± 0.805	63.09 ± 0.854 27.67 ± 0.219
Window = 192	34.19 ± 0.193 7.96 ± 0.1	42.29 ± 0.211 12.52 ± 0.075	49.80 ± 0.161 17.41 ± 0.166	52.98 ± 0.567 20.76 ± 0.426	57.47 ± 0.693 24.57 ± 0.478	65.93 ± 2.988 29.51 ± 1.048

Table 4. CNN1D model results

RMSE ± std/MAE ± std

Resample	1 min	5 min	15 min	30 mn	1 h	2 h
Window = 1	34.95 ± 0.018 7.76 ± 0.059	43.33 ± 0.168 14.04 ± 0.057	58.10 ± 0.135 24.51 ± 0.062	76.17 ± 0.246 37.56 ± 0.183	116.82 ± 0.415 63.09 ± 0.08	194.69 ± 2.324 107.42 ± 0.356
Window = 2	34.69 ± 0.021 7.48 ± 0.155	42.35 ± 0.049 11.96 ± 0.061	54.03 ± 0.193 17.94 ± 0.082	62.99 ± 0.242 22.67 ± 0.217	73.48 ± 1.036 29.32 ± 1.083	104.89 ± 2.004 48.58 ± 3.067
Window = 5	34.07 ± 0.152 7.35 ± 0.104	42.07 ± 0.098 11.97 ± 0.059	52.86 ± 0.342 17.3 ± 0.226	58.94 ± 0.897 21.79 ± 0.454	64.00 ± 1.63 26.77 ± 0.947	96.27 ± 7.581 46.78 ± 7.871
Window = 10	33.92 ± 0.047 7.39 ± 0.105	42.11 ± 0.126 11.91 ± 0.06	51.54 ± 0.189 16.85 ± 0.1	56.12 ± 0.759 20.80 ± 0.733	60.87 ± 0.684 25.68 ± 0.579	62.03 ± 1.528 27.54 ± 0.785
Window = 24	33.81 ± 0.18 7.38 ± 0.047	41.80 ± 0.14 11.90 ± 0.216	50.71 ± 0.239 16.85 ± 0.203	54.82 ± 0.346 20.11 ± 0.274	53.71 ± 0.981 20.67 ± 0.561	59.19 ± 1.239 24.04 ± 0.63
Window = 48	33.76 ± 0.213 7.50 ± 0.068	41.37 ± 0.162 11.98 ± 0.217	50.35 ± 0.212 16.82 ± 0.137	52.21 ± 0.302 18.84 ± 0.319	53.55 ± 0.395 20.38 ± 0.364	61.70 ± 2.049 25.14 ± 1.11
Window = 96	33.95 ± 0.259 7.59 ± 0.027	41.65 ± 0.295 12.1 ± 0.066	49.11 ± 0.265 15.92 ± 0.259	52.02 ± 0.088 18.76 ± 0.136	54.44 ± 1.089 21.06 ± 0.572	62.90 ± 1.584 25.47 ± 0.833
Window = 144	34.22 ± 0.174 7.62 ± 0.058	41.55 ± 0.37 12.09 ± 0.148	49.32 ± 0.557 16.15 ± 0.146	51.94 ± 0.398 18.94 ± 0.165	54.68 ± 0.907 21.16 ± 0.512	62.23 ± 1.325 25.85 ± 0.939
Window = 192	34.34 ± 0.211 7.72 ± 0.062	41.72 ± 0.559 11.99 ± 0.139	48.99 ± 0.3 16.40 ± 0.082	51.72 ± 0.353 19.03 ± 0.306	56.07 ± 1.678 22.10 ± 0.711	64.14 ± 1.085 27.14 ± 0.61

Table 5. LSTM model results

RMSE ± std/MAE ± std						
Resample	1 min	5 min	15 min	30 mn	1 h	2 h
Window = 1	34.99 ± 0.048 7.83 ± 0.122	43.22 ± 0.041 13.99 ± 0.042	57.74 ± 0.128 24.48 ± 0.053	75.78 ± 0.488 37.48 ± 0.079	114.6 ± 1.41 63.20 ± 0.041	186.87 ± 2.801 108.21 ± 0.871
Window = 2	34.58 ± 0.081 7.60 ± 0.15	42.16 ± 0.051 11.99 ± 0.1	54.59 ± 0.691 18.09 ± 0.275	63.39 ± 0.984 23.00 ± 0.495	74.16 ± 2.291 29.97 ± 1.685	103.33 ± 4.068 47.19 ± 3.771
Window = 5	34.09 ± 0.086 7.39 ± 0.137	42.38 ± 0.08 12.23 ± 0.109	53.76 ± 0.151 17.5 ± 0.124	60.47 ± 0.625 22.51 ± 0.479	66.28 ± 0.786 28.27 ± 0.621	96.95 ± 3.868 49.21 ± 4.583
Window = 10	34.01 ± 0.162 7.37 ± 0.111	42.41 ± 0.194 12.13 ± 0.188	52.66 ± 0.294 17.69 ± 0.383	57.09 ± 1.12 21.69 ± 0.913	62.67 ± 0.765 26.73 ± 0.784	67.84 ± 1.051 31.78 ± 0.944
Window = 24	34.02 ± 0.091 7.39 ± 0.066	42.05 ± 0.031 12.06 ± 0.054	51.17 ± 0.195 17.14 ± 0.306	54.98 ± 0.869 20.62 ± 0.547	56.45 ± 1.302 21.94 ± 0.586	63.68 ± 0.87 25.68 ± 0.958
Window = 48	34.12 ± 0.212 7.44 ± 0.091	41.68 ± 0.159 12.04 ± 0.239	50.63 ± 0.084 16.78 ± 0.159	52.56 ± 0.169 19.04 ± 0.092	56.78 ± 0.487 21.84 ± 0.35	67.99 ± 7.726 27.56 ± 3.392
Window = 96	34.25 ± 0.359 7.54 ± 0.051	41.74 ± 0.158 11.99 ± 0.186	49.87 ± 0.152 16.36 ± 0.144	53.09 ± 0.425 19.69 ± 0.473	57.13 ± 1.944 22.65 ± 0.818	69.19 ± 2.902 28.01 ± 1.24
Window = 144	34.16 ± 0.226 7.56 ± 0.089	41.76 ± 0.122 11.97 ± 0.079	49.70 ± 0.2 16.79 ± 0.168	53.58 ± 3.113 20.17 ± 1.892	58.24 ± 1.905 23.07 ± 0.932	66.43 ± 1.708 28.07 ± 0.76
Window = 192	34.14 ± 0.309 7.64 ± 0.18	41.77 ± 0.22 12.09 ± 0.175	50.45 ± 0.894 17.30 ± 0.327	53.68 ± 1.202 20.26 ± 0.828	58.95 ± 1.571 24.26 ± 0.486	74.85 ± 2.45 31.00 ± 1.012

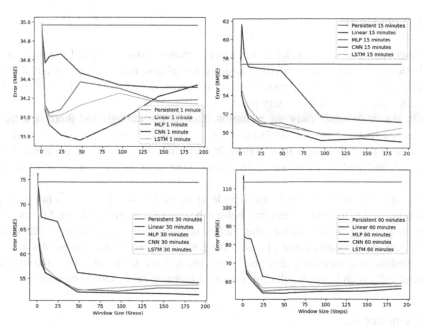

Fig. 6. Error vs. input window size, in function of the sample interval.

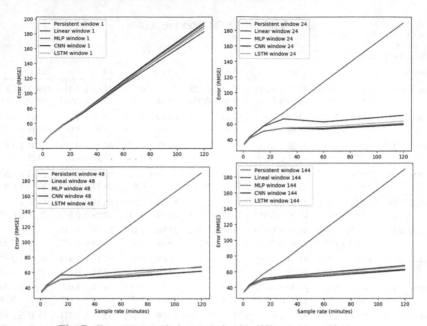

Fig. 7. Error vs. sample interval, for the different input windows.

4   Discussions

Is it worth implementing an elaborate model to predict a single step at 1 min intervals, or is it better to use persistent prediction directly with almost no configuration and therefore effortlessness. In the best of cases the improvement does not reach 1 w/m^2, resulting insignificance for the solar energy field.

On the other hand, only the radiation itself is not enough and there is a lack of information that allows the model to anticipate a fall, since only the trend can be followed well, but with problems in its changes, bigger error. However, it was intended to be analyzed, and adding more variables opens more options and makes analysis more difficult.

To end, one of the strengths of deep learning is the fact that it does not need pre-processing and extraction of features before the process, the network itself takes care of everything. If it is true that, for other fields, such as image classification, this is more than achieved, the same cannot be said for time series, since the best results are those that combine preprocessing and prior analysis, and we have not seen none that do it automatically with better results. In this sense more research should be done.

5   Conclusions

To conclude, the sampling frequency is a key parameter in that it determines the prediction error and the choice of a model. It can be said that for small sample intervals the choice of the model is indifferent, and even persistent prediction can be used with the same success as more elaborate models. However, as the frequency between samples increases, the choice of a model becomes more important, as well as an input window to it.

Regarding the models implemented, the MLP and CNN1D should be highlighted for the speed of training. The CNN1D model obtains the best results. Regarding the LSTM model, it should be noted that it is more complex to implement, since it takes longer to simulate and requires a greater amount of data, it also works better when it is multivariable, and from the results it can be said that it is not the best one.

Acknowledgment. This work is supported under Grant MAC/1.1.b/278 (MICROGRID-BLUE), from the Framework of the 2nd Call of the INTERREG V-A MAC 2014–2022, co-financed with FEDER funds. Too, this work is supported under the Grant CEI2021-06, from direct agreement SD-21/08 by Consejería de Economía, Industria, comercio y conocimiento from Gobierno de Canaria to ULPGC.

References

1. Pelland, S., Remund, J., Kleissl, J., Oozeki, T., De Brabandere, K.: Photovoltaic and solar forecasting: state of the art (2013)
2. Massaoudi, M., et al.: Convergence of photovoltaic power forecasting and deep learning: state-of-art review. IEEE Access **9**, 136593–136615 (2021). https://doi.org/10.1109/ACC ESS.2021.3117004
3. Ahmed, R., et al.: A review and evaluation of the state-of-the-art in PV solar power forecasting: techniques and optimization. Renew. Sustain. Energy Rev. **124** (2020). https://doi.org/10.1016/j.rser.2020.109792
4. Rahimi, et al.: A comprehensive review on ensemble solar power forecasting algorithms. J. Electr. Eng. Technol. (2023). https://doi.org/10.1007/s42835-023-01378-2
5. Moreira, M.O., et al.: Multivariate strategy using artificial neural networks for seasonal photovoltaic generation forecasting. Energies **16**, 369 (2023). https://doi.org/10.3390/en16010369
6. Song, S., et al.: A novel sky image-based solar irradiance nowcasting model with convolutional block attention mechanism. Energy Rep. **8**(Suppl. 5), 125–132 (2022). ISSN 2352-4847. https://doi.org/10.1016/j.egyr.2022.02.166
7. Cheng, M., Fang, F., Kinouchi, T., Navon, I.M., Pain, C.C.: Long lead-time daily and monthly streamflow forecasting using machine learning methods. J. Hydrol. **590**, 125376 (2020). ISSN 0022-1694. https://doi.org/10.1016/j.jhydrol.2020.125376
8. Ballestrín, J., et al.: Soiling forecasting of solar plants: a combined heuristic approach and autoregressive model. Energy **239**, 122442 (2022). https://doi.org/10.1016/j.energy.2021.122442
9. Surakhi, O., et al.: Time-lag selection for time-series forecasting using neural network and heuristic algorithm. Electronics **10**, 2518 (2021). https://doi.org/10.3390/electronics1020 2518
10. Crone, S.F., Kourentzes, N.: Feature selection for time series prediction - a combined filter and wrapper approach for neural networks. Neurocomputing **73**(10–12), 1923–1936 (2010). https://doi.org/10.1016/j.neucom.2010.01.017
11. Polyzos, E., Siriopoulos, C.: Autoregressive random forests: machine learning and lag selection for financial research (2022). https://ssrn.com/abstract=4118546 or https://doi.org/10.2139/ssrn.4118546

12. Karasu, S., Altan, A.: Recognition model for solar radiation time series based on random forest with feature selection approach. In: 2019 11th International Conference on Electrical and Electronics Engineering (ELECO), Bursa, Turkey, pp. 8–11 (2019). https://doi.org/10.23919/ELECO47770.2019.8990664

13. Xiao, X., et al.: Condition monitoring of wind turbine main bearing based on multivariate time series forecasting. Energies **15**, 1951 (2022). https://doi.org/10.3390/en15051951

14. Halyal, S., Mulangi, R., Manjunath, H.: Forecasting public transit passenger demand: with neural networks using APC data. Case Stud. Transp. Policy **10** (2022). https://doi.org/10.1016/j.cstp.2022.03.011

15. Liu, J., Han, D.: On selection of the optimal data time interval for real-time hydrological forecasting. Hydrol. Earth Syst. Sci. Discuss. **9**, 10829–10875 (2012). https://doi.org/10.5194/hessd-9-10829-2012

16. Guo, J., Williams, B., Smith, B.: Data collection time intervals for stochastic short-term traffic flow forecasting. Transp. Res. Rec. J. Transp. Res. Board **2024** (2008). https://doi.org/10.3141/2024-03

17. Remesan, R., Ahmadi, A., Shamim, M., Han, D.: Effect of data time interval on real-time flood forecasting. J. Hydroinform. **12**, 396–407 (2010). https://doi.org/10.2166/hydro.2010.063

18. MICROGRIDBLUE project. Activity 2.1.3. Energy prediction and state of the electrical network. https://www.microgrid-blue.com/es/

19. Lara-Benítez, P., Carranza-García, M., Luna-Romera, J.M., Riquelme, J.C.: Temporal convolutional networks applied to energy-related time series forecasting. Appl. Sci. **10**, 2322 (2020). https://doi.org/10.3390/app10072322

20. Ghimire, S., et al.: LSTM sequence-to-sequence autoencoder with feature selection for daily solar radiation prediction: a review and new modeling results. Energies **15**, 1061 (2022). https://doi.org/10.3390/en15031061

Time Series Classification of Electroencephalography Data

Aiden Rushbrooke$^{(\boxtimes)}$, Jordan Tsigarides, Saber Sami, and Anthony Bagnall

School of Computing Sciences and School of Medicine,
University of East Anglia, Norwich, UK
Aiden.Rushbrooke@uea.ac.uk

Abstract. Electroencephalography (EEG) is a non-invasive technique used to record the electrical activity of the brain using electrodes placed on the scalp. EEG data is commonly used for classification problems. However, many of the current classification techniques are dataset specific and cannot be applied to EEG data problems as a whole. We propose the use of multivariate time series classification (MTSC) algorithms as an alternative. Our experiments show comparable accuracy to results from standard approaches on EEG datasets on the UCR time series classification archive without needing to perform any dataset-specific feature selection. We also demonstrate MTSC on a new problem, classifying those with the medical condition Fibromyalgia Syndrome (FMS) against those without. We utilise a short-time Fast-Fourier transform method to extract each individual EEG frequency band, finding that the theta and alpha bands may contain discriminatory data between those with FMS compared to those without.

Keywords: Time series classification · EEG · Fibromyalgia

1 Introduction

The use of electroencephalography (EEG) for brain activity monitoring has become increasingly popular due to its high temporal resolution, non-invasive nature and low cost. With this has come an increased interest in the use of machine learning algorithms to assist in tasks involving EEG data, such as classification. However much of the focus has been on the processing and feature extraction steps, with standard classifiers being applied to derived features: a recent report found that 40% of studies use support vector machine or nearest-neighbour models [24]. Other popular methods include deep learning or linear models such as ridge classification. Whilst these models often perform well when applied to EEG tasks, they are often used without any adaption and require dataset specific features to be used. EEG datasets are multivariate time series recorded at fixed frequencies and often used in classification tasks. There has recently been a boom in publication of classification algorithms designed to be applied directly to time series from any problem domain [1,22]. Time series classification (TSC) aims to classify datasets consisting of instances of one or more dimensions containing evenly spaced time-points, and can be applied to a wide variety of fields. For example, they have been

© The Author(s), under exclusive license to Springer Nature Switzerland AG 2023
I. Rojas et al. (Eds.): IWANN 2023, LNCS 14134, pp. 601–613, 2023.
https://doi.org/10.1007/978-3-031-43085-5_48

successfully applied to human activity recognition, audio classification [12] and the analysis of spectrographs [17].

TSC internalise and automate the process of feature extraction, and are based on different types of discriminatory patterns such as repeating patterns or common segments. The most accurate approaches combine multiple representations in an ensemble to avoid a weakness of any individual method. Our aim is to investigate whether applying these time series specific algorithms can improve EEG classification over standard approaches. Our contributions are to assess a range of TSC algorithms on some archive EEG problems, identify the most promising approaches then conduct a case study to demonstrate how TSC could help differentiate individuals with a chronic pain medical diagnosis (Fibromyalgia Syndrome) based on their EEG characteristics.

The remainder of the paper is as follows. Section 2 provides background information into EEG analysis and TSC. Section 3 describes nine EEG classification datasets in the time series archive[1] and Sect. 4 evaluates how TSC models perform compared to existing results on these datasets. Section 5 contains a case study into a specific EEG dataset, looking at if TSC methods can find discriminatory data between subject with and without the Fibromyalgia Syndrome (FMS), a medical diagnosis characterised by chronic widespread pain. Finally, Sect. 6 provides a summary of the results found and suggests some future areas of research for further improvement.

1.1 List of Commonly Used Acronyms

EEG: Electroencephalography, a way to measure brain activity by recording electrical signals produced by neurons.
MEG: Magnetoencephalography, similar to EEG but using magnetic fields rather than electrical activity.
(M)TSC: (Multivariate) Time Series Classification, a form of classification where the input data takes the form of a number of evenly spaced data points.
BCI: Brain Computer Interfacing, ways to map brain activity to an external device, commonly using EEG.
FMS: Fibromyalgia Syndrome, a medical condition characterised by a general feeling of generalised chronic pain.

2 Background

2.1 Electroencephalography

EEG is a technique used to measure the brain's electrical activity. It uses electrodes placed on the scalp to measure changes in voltage over time produced by cells of the brain (neurons). EEG data is commonly used in medicine for diagnosis assistance, computer science for human-computer interaction and psychology to further understand disorders such as Narcolepsy [28] or Insomnia [32].

---

[1] https://tsc.com.

Due to the relatively low cost of equipment, ease of use, non-invasive recording method and speed, EEG has become one of the most popular and well used brain imaging methods. Electrodes at different points on the scalp measure different sections of the brain which are responsible for different areas of information processing. EEG data is usually recorded at a high frequency with EEG devices commonly recording at 1000 or more observations per second, allowing for good temporal resolution.

EEG data can be broken down into distinct frequency bands, representing clearly defined frequencies of neural oscillation. Usable information in EEG usually falls between the range of 1 and 50 Hz where 1 Hz represents one oscillation per second, and can be broken up into each band using spectral analysis (including use of Fourier Transforms). Each band related to different levels of brain activity with the import important bands being context-specific for any given study. The frequency ranges for these bands is provided in 1, and a example of splitting an EEG signal into each band in 1.

Table 1. EEG band to frequency range

Band	Delta	Theta	Alpha	Beta	Gamma
Frequency range(Hz)	0.5–4	4–8	8–12	12–30	30+

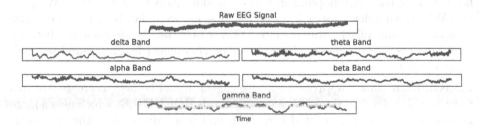

Fig. 1. An example of splitting an EEG signal from one channel into individual frequency bands

Magnetoencephalography (MEG) is another method of measuring brain activity similar to EEG, but measuring changes in magnetic fields rather than the electrical current in the brain directly. MEG has better spatial resolution, meaning it can localise more effectively, as the magnetic fields are less affected by the skull [7]. However, MEG requires a considerable equipment infrastructure, making it less mobile and more expensive. Therefore, EEG tends to be more commonly used.

One of the main uses of EEG is within the clinical setting to support diagnosis of various neurological disorders including epilepsy. Within the context of epilepsy diagnosis [23], EEG data is collected whilst a patient is exposed to specific sensory stimuli (such as flashing lights) or during an active seizure to look for 'epileptiform' features that are indicative of abnormal electrical activity.

Another use for EEG is to understand how the brain works and reacts to different environments. For example, much of our current understanding on how rapid eye movement (REM) sleep works has come from studies using EEG. EEG is also used to further our understanding of human emotions and behaviours [27].

A growing area of use for EEG is within the brain-computer interface (BCI), allowing the brain to communicate with an external device such as a prosthetic limb or computer. EEG provides a simple, non-invasive method for measuring brain activity in real-time, making it perfectly suited to mobile BCIs. One example of an EEG-based BCI [15] demonstrated that EEG data could be used to control a robotic quad copter in a 3D environment.

2.2   Time Series Classification

Time series classification (TSC) is a special case of traditional classification where each instance is a time series target variable pair [1]. For multivariate TSC (MTSC), each observation of the time series is a vector. For EEG data, an observation at a given time point represents a vector of values for each channel. Alternatively, a multivariate time series can be considered as a set of two or more time series aligned in time, one for each channel. A number of different approaches have been proposed for the MTSC task.

Distance based methods utilise distance functions to calculate similarity between two time series, then use a classification method such as nearest neighbour. One of the more popular distance calculations is Dynamic Time Warping (DTW), which allows for a level of warping to occur between the two series to adapt to any offset. Multivariate DTW can either be dependent (distance calculation is pointwise over channels) or independent (distance found for each channel then summed).

Convolution based approaches utilise convolutional kernels to create features to be used for classification. The most popular is ROCKET [8], which uses a large number of randomly generated kernels to find many different feature representations before applying a ridge regression classifier. ROCKET's strength comes from its speed, achieving high accuracy at a significantly reduced computation time. There have also been extensions to ROCKET, such as multi-ROCKET [30], which adds three additional features per kernel, mini-ROCKET [9], which optimises the convolutions used, and the Arsenal [22], an ensemble of ROCKET classifiers.

Deep learning based methods are ever increasing in popularity, and are becoming more viable for TSC. InceptionTime [11] is generally considered the best performing deep learning algorithm. The original InceptionTime uses an ensemble of five Inception networks, an adaption of a residual neural network (ResNet), although this number can be changed.

Feature based classifiers are simple pipeline approaches that extract global summary features then apply a standard classifier. The FreshPRINCE classifier [20], which combines the TSFresh transform [6] with a rotation forest [25], was found to be the most effective combination of transform and classifier.

Dictionary based methods, such as the Temporal Dictionary Ensemble (TDE) [19], adapt bag-of-words approaches used in computer vision. Histograms are formed based on the number of occurrences of discretised sub-sequences, or words.

Shapelet based methods are based on the presence or absence of a pattern, or shapelet. For the Shapelet Transform Classifier [4], large numbers of short subseries are selected from the training data and their discriminatory power is estimated. The best shapelets are retained and used to transform the data from the time domain to vector based distance to shapelet features.

Interval based classifiers are ensemble approaches that combine summary features from random intervals. DrCIF [21] derives Catch22 features [18] over different random intervals for each base classifier.

Hybrid approaches combine classifiers built on different representations. The HIVE-COTEv2 (HC2) [22] classifier combines classifiers from the shapelet, dictionary, interval and convolution domain in a heterogenious meta ensemble. HC2 is currently state of the art for MTSC.

3 Existing EEG Datasets

3.1 MTSC Archive EEG Classification Problems

There are currently nine EEG datasets in the TSML archive of TSC datasets[2]. Five of the nine datasets were used in BCI competitions, while the other four were collected from published research. A breakdown of the data characteristics of the nine datasets is shown in Table 2.

Eyes Open Shut [26] problem is a 2 class dataset on detecting whether a subject has open or shut eyes. One subject was recorded with both open and shut eyes for 117 s at 128 Hz, using 14 channels. In the original paper each time point was treated as a separate instance, equalling 128*117, or 14976 instances with 14 attributes. Their experiments found the best classification accuracy came from instance-based methods such as kstar, with a best accuracy of around 98%. However, their experiments contained biases through their use of k-fold cross validation without considering the temporal ordering of the data, so results may not be directly comparable. In the archive this dataset has been transformed into a time series format by first removing any outliers ($x > 5000$ or $x < 3000$), then segmenting the data into 1 s intervals of 128 observations each. 19 cases were also removed for containing both open and shut eyes. Finally a test train split was created, with test containing the last 21 observations.

Face Detection[3] involved 16 subjects being shown either a face or a scrambled face, with EEG data recorded. Each trial was recorded for 1.5 s from 306 channels, then down-sampled to 250 Hz and high-pass filtered at 1 Hz. Subjects 1 to 10 were used to form the training data, while 11 to 16 formed the test dataset.

---

[2] https://www.timeseriesclassification.com.

[3] https://www.kaggle.com/c/decoding-the-human-brain/data.

Finger Movements [3] dataset was collected by getting one subject to sit in a standard typing position, then press keys in a self chosen order. The goal is to predict if the next key pressed was with the subjects left or right hand. The data was recorded in three 6 min sessions on the same day with breaks. The data was initially recorded at 1000 Hz on 28 channels for 0.5 s, using a band pass filter at 0.05 and 200 Hz to remove outliers. The data was then downsampled to 100 Hz, so each instance contains 50 observations. In a classification competition using this dataset, an error rate of 16% was achieved by extracting features using common spatial subspace decomposition and Fisher discriminant before classifying using a neural network.

Hand Movement Direction[4] dataset was gathered by having 2 subjects move a joystick either up, down, left or right of their choosing. From this a 4 class problem was created. For each trial the subject was given 0.75 s to move the joystick and reach a target, then hold in position for 1 s. The data was recorded with 10 channels at 625 Hz and band pass filtered at 0.5 and 100 Hz, then resampled at 400 Hz. In a competition, the highest accuracy found was 46.9% by first extracting various features, using a genetic algorithm to select relevant features, then classifying using a linear SVM and LDA.

Motor Imagery [16] dataset is from an electrocorticography (ECoG) experiment where a single patient was tasked with imagining moving either their left small finger or tongue. An ECoG is placed directly on the brain rather than externally. Each recording lasted 3 s, starting 0.5 s after a visual cue has ended, at 1000 HzHz, and with 64 dimensions recorded. The training data was recorded on one day, then the test data a week later. The best classification result was 91% accuracy by combining various feature extraction methods such as CSSD and Fisher discriminant analysis, before using a linear SVM classifier.

Self Regulation SCP 1 and 2 [2] datasets are a pair of EEG datasets based on the use of EEG data to provide a method of communication for people paralysed with Amyotrophic Lateral Sclerosis (ALS) The patients were trained to voluntarily produce positive and negative changes in their Slow Cortical Potential (SCP), which was then used to move a cursor up and down on a screen, whilst receiving feedback. SCP 1 was recorded with a healthy patient over 2 d. Each trial was 6 s long and a total of 268 trials were performed. The data was sampled at 256 Hz from 6 channels with 2 classes, either positive or negative. SCP 2 performed the same experiment, but on a subject with ALS. For this experiment, 380 trials, 200 train and 180 test, were performed in total on the same day, each of length 8 s with 4.5 s used. Both of the datasets were used in a BCI classification competition. For SCP 1 the lowest error rate found was 11.3% by first extracting features using spectral analysis, then feeding into a linear classifier. For SCP 2 the lowest error rate was 45.6% using continuous wavelet transform and a linear discriminant analysis classifier. However, this dataset was found to contain very little data relevant to the classification problem.

Blink[5] dataset was formed by getting multiple subjects to blink in two second intervals, as either short or long blinks. The data as recorded at 255 Hz with

[4] http://bbci.de/competition/iv/.

4 channels. In the original dataset the trials were formed into 20 sets of 50 instances, 10 for each class. For the version used in the archive these sets have been joined together, then split into train and test portions.

MindReading [14] involves classifying 5 different visual stimulus shown to a participant using MEG data. The data was originally recorded at 330 Hz with 306 channels before being down-sampled to 200 Hz. Other processing steps include low-pass filtering at 50 Hz, removing noise caused by head movements and removing likely artefacts by applying trend removal. Finally, the data was segmented into 1 s intervals. This dataset was then given to 9 different research groups to partake in a competition to find the best accuracy, which was 68% using a logistic regression with the LASSO regression method.

Table 2. Description of EEG datasets in the tsml archive

Dataset	Classes	Channels	Series Length	Train Size	Test Size	Sample rate (Hz)
Blink	2	4	510	500	450	255
EyesOpenShut	2	14	128	56	42	128
FingerMovements	2	28	400	316	100	100
HandMovementDirection	4	10	400	160	74	400
MindReading	5	204	200	727	653	200
MotorImagery	2	64	3000	278	100	1000
SelfRegulationSCP1	2	6	896	268	293	256
SelfRegulationSCP2	2	7	1152	200	180	256

4 Results

Our experimental goal is to assess how useful TSC algorithms are for EEG classification with no hand crafting of features and no preprocessing beyond that done automatically by band pass filtering. We perform a series of experiments using the TSML archive data using a range of different TSC models. Due to its high number of channels and number of time-points the FaceDetection dataset was excluded leaving 8 datasets. 11 of the most popular and high performing classifiers were used in the experiments. In each experiment the model was trained on a training portion of the dataset, then performance measured against an unseen test set. Experiments were performed using the aeon time series machine learning toolkit[5]. This was repeated on 30 resamples of each experiment to get an average accuracy. These average accuracy scored are shown in Table 3.

Overall, ROCKET based classifiers performed the best, being the best performing classifiers for 5 of the 8 datasets used in the experiment. MiniROCKET has the best average rank (4.25), and the top three ranked classifiers are all ROCKET based. However, for the majority of the datasets there is not a large difference between the best and worst classifiers in terms of accuracy. These

---

[5] https://github.com/aeon-toolkit/aeon.

results can also be compared to results found in the competitions using these datasets, or the papers they originated from. Whilst direct comparison is not valid due to differences in experimental methodology, it does provide a good indication as to how well time series classifiers can perform compared to conventional approaches. These comparisons are shown in Table 4.

Table 3. Accuracy scores for 11 classifiers on 8 EEG/MEG datasets. The best result for each dataset has been underlined.

	Mini-ROCKET	ROCKET	Arsenal	HIVE-COTE 2	FreshPRINCE	Multi-ROCKET	DrCIF	InceptionTime	TDE	1NN-DTW	STC-2Hour
Blink	0.998	_1.000_	1.000	1.000	0.997	0.998	0.999	0.991	1.000	0.946	0.998
EyesOpenShut	0.570	0.514	0.512	0.496	0.540	0.551	0.528	_0.695_	0.489	0.664	0.490
FingerMovements	_0.581_	0.576	0.577	0.550	0.553	0.557	0.548	0.564	0.530	0.546	0.541
HandMovement	0.399	0.450	0.436	0.419	0.383	0.354	_0.467_	0.426	0.351	0.303	0.375
MindReading	_0.737_	0.675	0.678	0.685	0.697	0.726	0.571	0.211	0.332	0.606	0.538
MotorImagery	0.528	0.519	0.518	0.535	0.541	0.522	0.518	0.513	_0.542_	0.518	0.529
SelfRegulationSCP1	0.907	0.867	0.868	0.883	0.898	_0.911_	0.873	0.847	0.838	0.819	0.854
SelfRegulationSCP2	0.514	0.536	_0.546_	0.532	0.517	0.516	0.503	0.521	0.521	0.542	0.512
Average Rank	4.25	4.625	4.75	4.9375	5.25	5.375	6.5	6.875	7.5	7.8125	8.125

Table 4. Comparison between best existing accuracy and best from our experiments

Dataset	Existing results	Our best result	Our worst result
Blink	0.980	1.00	0.991
EyesOpenShut	0.980	0.695	0.489
FingerMovements	0.840	0.581	0.541
HandMovementDirection	0.469	0.467	0.350
MindReading	0.680	0.737	0.212
MotorImagery	0.910	0.542	0.513
SelfRegulationSCP1	0.887	0.911	0.838
SelfRegulationSCP2	0.544	0.546	0.500

For 5 of the 8 datasets used, the results found from our experiments are comparable to the best results found in competitions or papers, with time series classifiers performing better for Blink, MindReading and both SelfRegulation datasets. This is with no bespoke processing: we have simply given the EEG to the classifiers in the format provided. This suggests that at the very least, TSC can provide a useful benchmark for more bespoke, problem specific, classification approaches. They are no panacea though: TSC algorithms performed worse on the datasets, EyesOpenShut, FingerMovements, and MotorImagery.

For the EyesOpenShut dataset this difference could be explained by the biases in experimental set up in generating the original results. Poor performance on the other two is harder to explain. It is probable that the hand crafted approaches for these problems are genuinely discovering discriminatory features the generic approaches cannot automatically discover. In these situations, TSC algorithms offer the opportunity of providing a strong lower bound for performance.

5   VIPA Dataset Case Study

The VIPA study is an EEG dataset from an experiment designed to investigate EEG characteristics in patients with chronic pain. The investigation involved looking at how chronic pain may influence EEG data, and if virtual reality could be utilised in chronic pain treatment. Participants with the Fibromyalgia Syndrome (FMS) were asked to complete various tasks in a virtual reality environment whilst recording EEG data, with their clinical and feasibility outcome variables being recorded before and after each task. A secondary control experiment was completed on subjects without FMS (healthy controls).

For subjects with and without FMS, eyes-closed resting state data was collected at baseline (before any tasks were undertaken). The resulting dataset consisted of 27 individuals with FMS and 14 healthy controls. The data consisted of 64 EEG channels and 3 non-EEG channels (accelerometers), with 58091 time-points recorded at 500 Hz, representing slightly more than 116 s. As each recording lasted slightly different amounts of time, each was truncated to the shortest signal so that all 41 were of equal length.

We have defined a classification problem within this dataset related to the diagnosis of FMS, but this is not the ultimate use case we envisage will be important for classifiers built on FMS EEG data. We are interested in exploring whether we can give insight into the best way to treat FMS using, for example, emerging digital tools such as virtual reality. There is conflicting evidence regarding EEG-based 'biomarkers' in FMS with studies outlining the importance of the theta [10], alpha [31] and beta [13] bands. These studies mainly focus on the frequency domain alone and average data from across the entire electrode array over the total recording time. There is a lack of research investigating changes in EEG microstates and looking more carefully at changes in oscillatory information over time. We hypothesise that achieving improved classification of FMS patients based on alterations in particular frequency bands will support the ability to more closely define these alterations, leading to biomarkers of the future. Ultimately, our follow-on work will explore these potential biomarkers.

We construct classifiers on the full data, and for each individual band. Each channel is transformed independently into a bandwidth using standard methodology. A short-time fast Fourier transform (STFT) [29] method was used to extract individual bands. A one second overlapping sliding window was passed over the raw EEG signal, applying a Fourier transform and extracting an approximation of the absolute power for each band in each window using Simpson's rule. From this, five new multivariate time series can be extracted, each showing how the band differs over time.

Table 5. Accuracy comparing raw data to individual frequency bands

Experiment	Mini-ROCKET	ROCKET	Arsenal
Raw data	0.512	0.584	0.584
Delta	0.512	0.61	0.756
Theta	0.707	0.634	0.634
Alpha	0.707	0.634	0.683
Beta	0.634	0.61	0.659
Gamma	0.61	0.683	0.634
Ensemble	0.610	0.634	0.683

Due to the small number of subjects for each class, and to avoid any bias, a leave one subject out strategy was employed in each experiment. All but one subject were used to train the model, then the remaining subject was used as a test case. This was then repeated for each subject in the study, training a new model in each cross validation. The predicted class is then compared against the true class, and an overall accuracy calculated. This was done for the raw EEG data, each band, and an ensemble of all five bands, where the predicted class is the average prediction for each band. We have done no other pre-processing, such as artefact removal or data validation.

Based on the results presented in Sect. 4 and the relatively large size of the data set, the three top ranked ROCKET classifiers were selected for experiments. The accuracy scores over all subjects are displayed in Table 5. Given the very small sample size and the absence of any preprocessing, we believe these results are promising. Firstly, higher accuracy is generally observed when using any of the 5 frequency bands compared to the raw data. It can also been seen that, aside from the likely outlier for Arsenal with the delta band, accuracy was highest when using the theta and alpha bands. This suggests that important information for EEG classification can be found within the frequency domain, and that time-frequency analysis would likely be the best approach for EEG analysis. It offers some supporting evidence to the importance of alpha [31] and theta [10] bands being discriminatory for FMS. A contingency table for mini-ROCKET on the theta and alpha bands are shown in Tables 6a and 6b. False negatives are more common than false positives, and this could be due to the imbalance in the data towards individuals with FMS.

Table 6. Contingency tables for two EEG bands

(a) mini-ROCKET theta band

	True pos	True negative
Predicted positive	23	8
Predicted negative	4	6

(b) mini-ROCKET alpha band

	True pos	True negative
Predicted positive	22	7
Predicted negative	5	7

We also formed a naive ensemble over all five bands. Whilst it still performed better than classifiers built on the raw data, it was also notably worse than all the individual bands except for mini-ROCKET with delta. This implies that only a minority of bands contain useful information for FMS classification, and so a weighting system would be required to find which bands are most useful.

6   Conclusion

The aim of this study was to show whether time series classification models can usefully be applied to EEG problems. We have shown that not only can these models work well for EEG data, but can do so without needing any dataset-specific feature selection. The experiments on the datasets in the UCR time series archive showed that for 5 of the 8 datasets used, TSC models matched or exceeded results found in competitions involving these datasets. ROCKET based classifiers performed particularly well on these datasets. However, for 2 datasets our results were significantly worse than other studies, indicating that a generalised approach may still need some considerations before becoming viable for all EEG problems. We have also demonstrated the use of TSC models on a new EEG problem, discriminating individuals with a diagnosis of Fibromyalgia. Our findings show that time-frequency analysis increases accuracy over the time domain alone, with the alpha and theta bands the most discriminatory in FMS.

Whilst the results of the experiments performed do show that time series classifiers have potential, more testing on a larger variety of EEG datasets would need to be performed before any full conclusions can be drawn. This approach also has a significant drawback from an increased training time due to the size of raw EEG data. However, we believe that this could be avoided through the use of channel selection algorithms and automated processing techniques.

References

1. Bagnall, A., Lines, J., Bostrom, A., Large, J., Keogh, E.: The great time series classification bake off: a review and experimental evaluation of recent algorithmic advances. Data Min. Knowl. Disc. **31**(3), 606–660 (2017)
2. Birbaumer, N., et al.: A spelling device for the paralysed [4]. Nature **398**, 297–298 (1999)
3. Blankertz, B., Curio, G., Müller, K.R.: Classifying single trial EEG: towards brain computer interfacing. In: Dietterich, T., Becker, S., Ghahramani, Z. (eds.) Advances in Neural Information Processing Systems, vol. 14. MIT Press (2001)
4. Bostrom, A., Bagnall, A.: Binary shapelet transform for multiclass time series classification. Trans. Large-Scale Data Knowl. Centered Syst. **32**, 24–46 (2017)
5. Chicaiza, K.O., Benalcázar, M.E.: A brain-computer interface for controlling IoT devices using EEG signals. In: 2021 IEEE Fifth Ecuador Technical Chapters Meeting (ETCM), pp. 1–6 (2021)
6. Christ, M., Braun, N., Neuffer, J., Kempa-Liehr, A.W.: Time series feature extraction on basis of scalable hypothesis tests (tsfresh-a python package). Neurocomputing **307**, 72–77 (2018)
7. Cohen, D., Cuffin, B.N.: Demonstration of useful differences between magnetoencephalogram and electroencephalogram. Electroencephalogr. Clin. Neurophysiol. **56**(1), 38–51 (1983)

8. Dempster, A., Petitjean, F., Webb, G.: ROCKET: exceptionally fast and accurate time series classification using random convolutional kernels. Data Min. Knowl. Disc. **34**, 1454–1495 (2020)

9. Dempster, A., Schmidt, D.F., Webb, G.I.: Minirocket: a very fast (almost) deterministic transform for time series classification. In: Proceedings of the 27th ACM SIGKDD International Conference on Knowledge Discovery and Data Mining (2021)

10. Fallon, N., Chiu, Y., Nurmikko, T., Stancak, A.: Altered theta oscillations in resting EEG of fibromyalgia syndrome patients. Eur. J. Pain **22**(1), 49–57 (2018)

11. Fawaz, H., et al.: InceptionTime: finding AlexNet for time series classification. Data Min. Knowl. Disc. **34**(6), 1936–1962 (2020)

12. Flynn, M., Bagnall, A.: Classifying flies based on reconstructed audio signals. In: Yin, H., Camacho, D., Tino, P., Tallón-Ballesteros, A.J., Menezes, R., Allmendinger, R. (eds.) IDEAL 2019. LNCS, vol. 11872, pp. 249–258. Springer, Cham (2019). https://doi.org/10.1007/978-3-030-33617-2_26

13. González-Villar, A.J., Triñanes, Y., Gómez-Perretta, C., Carrillo-de-la Peña, M.T.: Patients with fibromyalgia show increased beta connectivity across distant networks and microstates alterations in resting-state electroencephalogram. NeuroImage **223**, 117266 (2020)

14. Klami, A., Ramkumar, P., Virtanen, S., Parkkonen, L., Hari, R., Kaski, S.: Icann/pascal2 challenge: MEG mind reading-overview and results (2011)

15. LaFleur, K., Cassady, K., Doud, A., Shades, K., Rogin, E., He, B.: Quadcopter control in three-dimensional space using a noninvasive motor imagery-based brain-computer interface. J. Neural Eng. **10**(4), 046003 (2013)

16. Lal, T., et al.: Methods towards invasive human brain computer interfaces. In: Saul, L., Weiss, Y., Bottou, L. (eds.) Advances in Neural Information Processing Systems, vol. 17. MIT Press (2004)

17. Large, J., Kemsley, E.K., Wellner, N., Goodall, I., Bagnall, A.: Detecting forged alcohol non-invasively through vibrational spectroscopy and machine learning. In: Phung, D., Tseng, V.S., Webb, G.I., Ho, B., Ganji, M., Rashidi, L. (eds.) PAKDD 2018. LNCS (LNAI), vol. 10937, pp. 298–309. Springer, Cham (2018). https://doi.org/10.1007/978-3-319-93034-3_24

18. Lubba, C., Sethi, S., Knaute, P., Schultz, S., Fulcher, B., Jones, N.: catch22: CAnonical time-series characteristics. Data Min. Knowl. Disc. **33**(6), 1821–1852 (2019)

19. Middlehurst, M., Large, J., Cawley, G., Bagnall, A.: The temporal dictionary ensemble (TDE) classifier for time series classification. In: Hutter, F., Kersting, K., Lijffijt, J., Valera, I. (eds.) ECML PKDD 2020. LNCS (LNAI), vol. 12457, pp. 660–676. Springer, Cham (2021). https://doi.org/10.1007/978-3-030-67658-2_38

20. Middlehurst, M., Bagnall, A.: The FreshPRINCE: a simple transformation based pipeline time series classifier. In: El Yacoubi, M., Granger, E., Yuen, P.C., Pal, U., Vincent, N. (eds.) Pattern Recognition and Artificial Intelligence, ICPRAI 2022. Lecture Notes in Computer Science, vol. 13364, pp. 150–161. Springer, Cham (2022). https://doi.org/10.1007/978-3-031-09282-4_13

21. Middlehurst, M., Large, J., Bagnall, A.: The canonical interval forest (CIF) classifier for time series classification. In: IEEE International Conference on Big Data, pp. 188–195 (2020)

22. Middlehurst, M., Large, J., Flynn, M., Lines, J., Bostrom, A., Bagnall, A.: HIVE-COTE 2.0: a new meta ensemble for time series classification. Mach. Learn. **110**(11), 3211–3243 (2021). https://doi.org/10.1007/s10994-021-06057-9

23. Pillai, J., Sperling, M.R.: Interictal EEG and the diagnosis of epilepsy. Epilepsia **47**(s1), 14–22 (2006)
24. Pahuja, S.K., Veer, K.: Recent approaches on classification and feature extraction of EEG signal: a review. Robotica **40**(1), 77–101 (2022)
25. Rodriguez, J., Kuncheva, L., Alonso, C.: Rotation forest: a new classifier ensemble method. IEEE Trans. Pattern Anal. Mach. Intell. **28**(10), 1619–1630 (2006)
26. Roesler, O., Suendermann, D.: A first step towards eye state prediction using EEG. In: Proceedings of the AIHLS 2013, Istanbul, Turkey (2013)
27. Samavat, A., Khalili, E., Ayati, B., Ayati, M.: Deep learning model with adaptive regularization for EEG-based emotion recognition using temporal and frequency features. IEEE Access **10**, 24520–24527 (2022)
28. Sasai-Sakuma, T., Inoue, Y.: Differences in electroencephalographic findings among categories of narcolepsy-spectrum disorders. Sleep Med. **16**(8), 999–1005 (2015)
29. Sejdić, E., Djurović, I., Jiang, J.: Time-frequency feature representation using energy concentration: an overview of recent advances. Digit. Signal Process. **19**(1), 153–183 (2009)
30. Tan, C.W., Dempster, A., Bergmeir, C., Webb, G.: MultiRocket: multiple pooling operators and transformations for fast and effective time series classification. Data Min. Knowl. Discov. **36**, 1623–1646 (2022)
31. Villafaina, S., Collado-Mateo, D., Fuentes, J., Cano-Plasencia, R., Gusi, N.: Impact of fibromyalgia on alpha-2 EEG power spectrum in the resting condition: a descriptive correlational study. Biomed. Res. Int. **1–6**(04), 2019 (2019)
32. Zhao, W., et al.: EEG spectral analysis in insomnia disorder: a systematic review and meta-analysis. Sleep Med. Rev. **59**, 101457 (2021)

Acid Sulfate Soils Classification and Prediction from Environmental Covariates Using Extreme Learning Machines

Tamirat Atsemegiorgis[1], Leonardo Espinosa-Leal[1], Amaury Lendasse[2],
Stefan Mattbäck[3,4], Kaj-Mikael Björk[1], and Anton Akusok[1(✉)]

[1] Arcada University of Applied Sciences, Jan-Magnus Janssonin aukio 1,
00560 Helsinki, Finland
{atsmegit,espinosl,bjork,akusok}@arcada.fi
[2] University of Houston, Houston, TX 77004, USA
alendass@central.uh.edu
[3] Geology and Mineralogy, Åbo Akademi, Domkyrkotorget 1, Åbo, Finland
stefan.mattback@abo.fi
[4] Geological Survey of Finland, PO Box 97, 67101 Kokkola, Finland

Abstract. This paper explores the performance of the Extreme Learning Machine (ELM) in an acid sulfate soil classification task. ELM is an Artificial Neuron Network with a new learning method. The dataset comes from Finland's west coast region, containing point observations and environmental covariates datasets. The experimental results show similar overall accuracy of ELM and Random Forest models. However, ELM implementation is easy, fast, and requires minimal human intervention compared to conventional ML methods like Random Forest.

Keywords: ELM · Acid Sulfate Soil · Environmental Covariate

1 Introduction

It is a common understanding; the soil of our biosphere houses various natural resources needed for life on this planet. For humans, the soil is a means of agricultural production and a source of raw materials required to build infrastructures. Yet, it is a finite resource that needs proper attention regarding its usage [7,16]. Since 1930, environmental awareness has grown toward soil conservation and prevention methods to maintain a balanced ecosystem [7,13]. Soil composition on the earth's surface is diverse, and knowing its property is essential to implement appropriate soil conservation and management strategies [6,8]. The continued soil degradation of our planet occurs because of man-made activities and naturally occurring phenomena. Acid sulfate soil (ASS) is a naturally occurring phenomenon causing soil acidification. ASS is sulfide-bearing sediment usually found in coastal regions around the globe. Studies show around 17 million hectares of ASS or potential ASS exist in Australia, Africa, Asia, Latin America, and Europe [4,12]. The Baltic basin contains most of Europe's ASS, and the highest concentration is found in Finland, with 130,000 hectares of arable land [8,11,18]. A research article from GTK (Geological Survey of Finland) [5] indicated that AS soil is the major environmental problem in Finland.

The oxidation of acid sulfate soil (ASS) above sea level generates sulfuric acid, leading to the acidification of soils and the release of large amounts of heavy metals into the environment. The volume of metal released through this process is about 10 to 100 times more than that produced by Finish industries [11]. This acidification of soil and the large concentration of metals in the watercourses create a toxic environment for the aquatic plants and animals, which affects the country's socio-economic development [10,15]. Four documented incidences of massive-scale fish mortality exhibited in Finland; occurred in 1834, 1969–1971, 1996, and 2005-2006 due to acidification and leaching of heavy metals into rivers and watercourses. Studies done on crops grown and milk harvested from ASS land have shown one form of heavy metal more than the expected level [4,15]. The study and the fish mortality incidences prove the negative impact of the creation and expansion of acidic soil on the country's overall economy. Therefore, it is essential to implement policy-based soil conservation and soil management strategies to prevent damage caused by ASS.

The first step to minimize the impact of ASS is to get knowledge or information about the distribution of ASS and then create an ASS map. Since the 19501950ss, researchers have been engaging in ASS mapping projects by manually taking sample soils from the target area and then analyzing the PH level in the laboratory to determine whether the sample soil is ASS or not. Mapping with this method is very libraries and time-consuming. Since 2009, GTK (Geographic Survey of Finland), with other collaborators, has been working hand in hand to develop a systematic way of ASS mapping and making it accessible to the public [9,10]. These days, artificial intelligence (AI) and machine learning (ML) methods are used widely to classify soil as ASS or not.

This project study aims to build a Machine learning system for classifying soil as acid sulfate soil (ASS) or not by adopting an Extreme Learning Machine (ELM) model. Secondly, it aims to assess the comparative advantage of using the ELM model over other conventional models, namely Random Forest (RF). A randomized grid search cross-validation hyperparameter tuning was used to get the best result for the RF model. The next section of this paper, research database, is about preparing point observations and environmental covariate layers datasets. The project framework section focuses on the model development process, which includes data preparation, parameter tuning, model selection, model training, and model evaluation. The project experiment runs on Jupyter Notebook and uses current cutting-edge libraries; GeoPandas, PySpark (a python library engine for large-scale parallel data processing) [14] and others. The experimental result section presents the comparative results and evaluations of the proposed models, and the last section presents the conclusions of the project works.

2   Research Database

The database under study is a combination of two spatial datasets; a vector
dataset (sample soil observations), some examples are shown in Table 1, and a
raster dataset of image tiles (environmental covariates layers). A spatial dataset
is a collection of observational attributes of phenomena organized in a tabular
format with its unique ability to represent a geographical location worldwide.
GTK is the provider of the points' dataset. However, the covariates layers map
tiles were generated using the QGIS tool based on a remote sensing dataset.
The points spatial dataset has geographical location information: longitudes,
latitudes, and binomial classes of soil types. The classification of soil types as
acid sulfate soil (ASS) or noon acid sulfate soil (non-ASS) depends on the PH
level of a sample soil. Figure 1 shows a heatmap for acidic soil distribution on
the west coast of Finland.

Table 1. Some sample soil types from the west coast of Finland.

X	Y	class
25.768938	64.777988	ASS
25.776304	64.793496	ASS
25.784691	64.786808	ASS
25.315115	64.988732	ASS

The covariates layers used in the experiment are composed of three distinct
groups of layers:

1. Terrain: includes Slope, Aspect, Hillshade, Topographic Wetness Index
   (TWI), Topographic Position Index(TPI), Normalized Difference Vegetation
   Index (NDVI), and Topographic Ruggedness Index (TRI) layers,
2. Geophysics (magnetism or electric conductivity data): includes electromagnetic real, electromagnetic imaginary and electromagnetic resistivity layers,
3. Quaternary map: 41 soil types in our case "bedrocks" and 49 different classes
   of land cover classes(Corine land cover) were included.

The combined database consists of 5824 rows and 104 columns. Among the
features, the "class" variable is a binomial class of soil types (ASS, non-ASS)
and it is a target feature for the experiment. There are 3490 ASS and 2334 non-
ASS soil types, and their frequency distribution in percentages is 60% and 40%,
respectively. The database contains 104 features; 2 coordinate points (x and y),
13 environmental covariates layers, and in total, 15 features were used. Among
the covariant layers, features "corine-land-cover" and "bedrock" covariates layers
consist of 49 and 41 distinct values, respectively. A one-hot encoding technique
was used to represent those values. The encoding method adds extra 88 features:
corine1 ... corine49 from corine-land-cover and bedrock1 ... bedrock41 from the
bedrock layers.

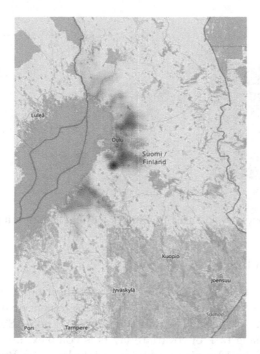

Fig. 1. Heatmap of acidic soils distribution on the west coast of Finland.

3   Project Framework

The project framework illustrates the whole end-to-end machine learning process of model building, from data ingestion to model evaluation in predicting soil types. As Fig. 2 below demonstrates, the model development process that includes many complicated data analysis and image processing tasks. The tasks are dataset collection (point observations and environmental covariates layers), model selection, model training, parameter tuning, and assessing the performance of the choice model.

3.1   Data Preparation

The sample point dataset is prepossessed by creating GeoPandas dataframe with a new variable of point geometry. Since Finland is located near the North Pole, the point dataset's longitude and latitude need to be set to the regional standard coordinate reference system (CRS) "WGS84" to avoid image distortion that occurs near the pole. The next step will be extracting pixel coordinates (x, y) and tile coordinates (z, x, y) using the geographical location coordinates of the sample point (longitude and latitude) and zoom level z. Tile image (z, x, y) is a 256×256 pixel-sizes of multiple neighboring pixel points (x, y) for a given zoom level z.

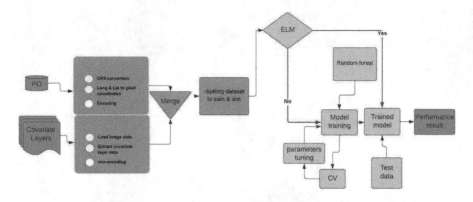

Fig. 2. Model development framework for acid sulfate soil prediction

3.2 Data Encoding

This research project is a supervised binary classification of ASS and non-ASS. The target feature "class" is a categorical feature that needs to be converted to a numerical feature of 0 and 1 for soil type ASS and non-ASS, respectively.

The prepossessing task of covariate layers is complex, and it starts by loading tiles images of the 13 environmental covariate layers. A Python function was developed to load the image tiles and then extract the covariate layer's data pixel for a given sample point location and zoom level. The two covariate layers, corine-land-cover, and bedrock are categorical variables; hence, the one-hot encoding technique was used to represent the categorical values of the variables. The encoding method helps to avoid the ordinal relationship of integral values between values used to distinguish the categorical attributes of the given feature. Other prepossessing works include merging the two datasets (points and covariate layers datasets), splitting the database into train and validation sets, and scaling the training dataset using the Scikit-learn RobustScaler module.

3.3 Feature Selections

Feature selection is one of the important data processing techniques for selecting features that contribute most to model building. In most cases, incorporating irrelevant or less significant features in model training has a negative impact on the generalization role of the model on unseen data. Removing the irrelevant or less informative features reduces over-fitting, improves performance, and decreases training time. Figure 3 shows mean-test-accuracy (features importance) of the Random Forest model for 103 features for the research database.

As we know, more features in the model mean more complexity and more time to train a model. It is vital to reduce the size of the features to a level that would not harm the model's performance. This experiment employed a Scikit-learn tool called RFECV (Recursive Feature Elimination Cross-Validation), with a 10 folds cross-validation features selection method. RFECV works on a subset

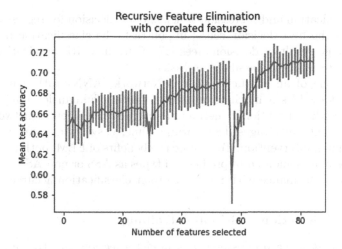

Fig. 3. Accuracy of Random Forest model regarding the number of features

of all possible space of features, recursively training a model and pruning the less significant one on all possible subsets until the optimal number of features is reached. Deploying the RFECV method using a random forest model on the study database reduces the feature number to 22. Figure 4 shows the top 5 features and their importance.

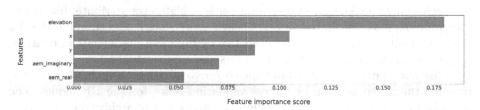

Fig. 4. Random Forest feature importance for the top 5 informative features

It is apparent that elevation, x, y, and aem_imaginary features are the most significant features and account for about 44% of the RF model's predictive power. Features aem_real, aspect, NDVI, TRI, and TWI, in aggregate accounts for about 23%, and features hillshade, slope, TPI, and aem_apparent contribute about 15% of the predictive power of the RF model. Therefore, out of 103 features, 12 features contribute about 82% of the total predictive power of the RF model.

4   Experimental Results

The experiment employs Random Forest (RF) and Extreme Learning Machine (ELM) models. Random Forest is a popular ensemble-supervised machine learning algorithm, which is a combination of many tree-based predictors. RF is widely used in classification and regression tasks, and the decision for regression is made by selecting the best decision score; in contrast, the classification decision will be the average result of decision trees [17]. Currently, RF application is broad and widely used in soil classification tasks.

ELM is one of an Artificial Neuron Network (ANN) with a new learning method that enables a machine system to imitate human learning behaviors. Compared with conventional methods and traditional neural networks, ELM offers significant advantages: less learning time, easy to implement, and takes minimal human intervention. To explore the benefits of ELM, both models (ELM and RF) were implemented to predict soil types as ASS or non-ASS and compare the models' performance with respect to their classification accuracy.

4.1   Model Selection and Model Training

The next phase after prepossessing is model selection and training using the training dataset, along with parameter tuning. There are several machine learning algorithms to choose from. Generally, the selection process relies on the dimension of the dataset, the required accuracy, the interpretability of the output, the time needed to train, and the linearity of the training dataset. This experiment employed two classification models: Extreme Learning Machine (ELM) and Random Forest (RF). The dataset was partitioned into train and test sets to train the RF model, and the train-to-test ratio choice was 3 to 1. The test set is meant to validate the predictive performance of the fitted model.

Technically speaking, a hyperparameter is a high-level attribute: like n_job, n_estimators, max_depth, that a practitioner sets before model training. Besides that, the model learns other characteristics by finding a mathematical relationship between the training dataset (features and target variable).

This research employed a randomized cross-validation search for hyperparameter tuning to get the best cross-validation score for the RF model. The optimization process of parameter tuning is carried out to archive a better accuracy model prediction of soil types, ASS or non-ASS. The parameters tuning phase depends on manually seated parameters space with a randomized cross-validation search of 10 folds. That means, 10 randomly categorized subset groups were created from the training dataset, and each of the subsets of the train groups was used to validate the model performance; the rest 9 subsets were used for training the model. Therefore, because of the deployment of the cross-validation search technique, 10 different models were fitted with the corresponding 10 sets of validation estimators. Through this process, the best score's parameters among the 10 fitted model parameters are selected and used for building RF model prediction of soil types, ASS or non-ASS.

In the case of ELM, there is no need to carry out the time-consuming hyperparameters tuning task. That is one of the benefits of using ELM solutions for machine learning prediction tasks. Scikit-ELM toolbox was used because of its flexibility and usability; the reader is directed to the canonical papers for more detail [1–3].

4.2   Evaluation Metrics

The overall objective of building a predictive machine learning model is to deliver a high accuracy score for unseen data. Measuring how robust the model prediction is, and explaining the performance before deployment is essential. Several evaluation metrics exist, and their selection depends on the model types as well as on the implementation plan. Some of the evaluation matrices are discussed as follows:

- precision or Sensitivity: the proportion of positive cases correctly identified, i.e.:

$$\frac{TP}{(TP + FP)}$$

- Recall or specificity (negative predictive value): is the proportion of correctly identified negative case, i.e.:

$$\frac{TP}{(TP + FN)}$$

- Classification Accuracy: the proportion of correct predictions to the total number of input samples, i.e.:

$$\frac{(TP + TN)}{(TP + TN + FP + FN)}$$

- The AUC_ROC (Area Under Curve – Receiver Operating Characteristics) curve: represents the true positive rate (Sensitivity) as a function of the false positive rate (1-specificity). The AUC value is between 0 and 1; if the value is above threshold 0.5, the model can identify the classes very well.

4.3   Results and Discussion

The research experiment used RF and ELM models for class prediction of soil as acid sulfate soil (ASS) or normal soil (non-ASS). The evaluation statistics are described as follows. In the case of RF, Fig. 5 compares mean_test_score versus tree size (param_n_estimators) for tree depth (param_depth) of 10 and 20. The model offers a higher mean_score for 20 param_depth than 10 for a given number of trees (param_n_estimator). Fig. 6 shows that the higher the tree depth, the better the performance is, but with a more extended period to process.

As discussed in the previous sections, randomized grid searches for hyper-parameter tuning and feature selection were deployed on the RF model to get the best performance. The best results are shown in the first part of Fig. 7, and this figure presents a confusion matrix and related metrics called classification reports. The confusion matrix present; the model correctly classified 322 sample points as Normal soils (True Positive) and 705 as Acid Sulfate soil (True Negative). However, the model mispredicted 171 sample soils as Normal and 258 as Acid Sulfate soil. Both classes' correct predicted soil types are greater than

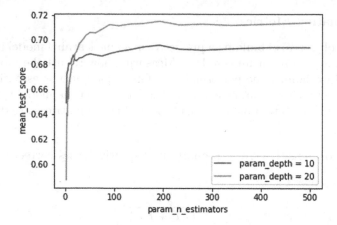

Fig. 5. Random Forest's test score line chart with 10 and 20 tree depths

the incorrectly predicted classes. Therefore, the RF model built is capable of classifying unseen location soil with significant accuracy.

The classification report at the top of the confusion matrix of the first chart of Fig. 7 present, the overall accuracy of the RF model is 71%, i.e., 71% of the prediction is correct. The precision for Acid Sulfate and Normal soil types are 73% and 65%, respectively, i.e., 73% and 65% of the respective soil types classified as such are accurate. However, the recalls are 80% and 56% for Acid Sulfate and Normal soil types, respectively, i.e., 80% and 56% of all soil types of the respective soils are classified correctly. The prediction for each class is presented by F1-score 77% for Acid Sulfate and 60% for the Normal, so the model classifies Acid Sulfate soils with better accuracy than the Normal.

Figure 8 is the ROC curve for the Rf and ELM models. In the figure, the green curve for RF is above the non-discrimination line in blue, which means

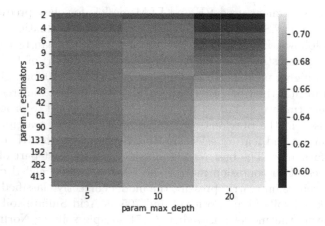

Fig. 6. Random Forest's test scores Heatmap with decision trees versus tree depths.

the classification model works well. The area under the curve ROC is 0.78, which is not a perfect case where ROC is equal to 1; the model prediction rank of 0.78 explains that the RF model prediction is significant.

ELM model Implementation of acid soil classification using Scikit-elm liberty is very easy [2]. As shown in Fig. 2, the ELM model doesn't need to train using a training dataset. However, the model is fitted with features, and the target variable, then the class prediction of soil type ASS or Normal, is delivered. The statistics results are shown in the second chart of Fig. 7; the figure presents the confusion matrix and the classification report of the ELM model. The overall model accuracy is 71%, the same percentage as RF model prediction; the precision values are 72% and 67%, and the recalls are 83% and 51% for Acid Sulfate and normal soils, respectively. F1-scores are 77% and 58% for Acid Sulfate and Normal soil, respectively, so as the RF model, ELM classifies the Acid Sulfate soil type better than the Normal soil type.

The ELM ROC curve shown in the second chart of Fig. 8 is above the non-discrimination line, which means the model works well in classifying the soil types. The area under the curve ROC is 0.76, it is not a perfect result, but the rank explains that the ELM model prediction is significant.

Comparing the two models based on the statistical results presented above, the precision values of both models indicate that both models performed almost equally the same in predicting acid sulfate soils (Acid Sulfate). Still, the ELM model performs better predicting noon acid sulfate soil (Normal) type. About recalls: concerning prediction on own class, ELM performed better in predicting acid sulfate soil (Acid Sulfate), and RF performed better in predicting noon-acid sulfate soil (Normal) type. However, the overall performance of both models is the same. On the other hand, comparing the processing time and complexity in the implementation process, the EML model is very fast and user-friendly.

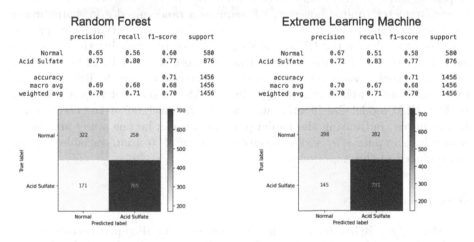

Fig. 7. Classification report and confusion matrix for Random Forest and ELM models

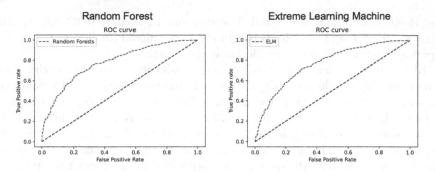

Fig. 8. ROC curve for Random Forest and ELM models

5　Conclusions

This project is designed to classify soil as Acid Sulfate Soil (ASS) or not (non-ASS) based on environmental covariates layers tiles using Extreme Learning Machine (ELM). Besides that, the research focuses on identifying the most relevant features and exploring the performance of the two models: ELM and RF. Sample point observations dataset from GTK and environmental covariates layers of images were used to run the experiment. Now a day, it is a common understanding that the environmental covariates variables can characterize the soil type; hence for this experiment, 13 environmental covariates layers and two location variables (longitude and latitude) were used. During the prepossessing phase through one-hot encoding techniques, additional features were created, making the total number of features 103.

After data processing, RF and ELM classification models were implemented to predict soil type. The Scikit-learn ELM model implementation doesn't require hyperparameter tuning; however, RF requires a randomized grid search hyperparameter tuning to get the best performance. Feature selection techniques are also used to enhance the RF model's performance by removing the less significant and irrelevant features in predicting soil class. As a result, 22 features were selected by their level of influence on the predictive power of the RF model. The two models work well in classifying soils into ASS or not, but they are far from perfect. ELM model classification was found to be much faster and easier to use. In the future, optimizing the model performance can be one study area that a researcher can pursue by adding more environmental covariate variables that are not included in this experiment.

References

1. Akusok, A., Björk, K.M., Miche, Y., Lendasse, A.: High-performance extreme learning machines: a complete toolbox for big data applications. IEEE Access **3**, 1011–1025 (2015)

2. Akusok, A., Leal, L.E., Björk, K.-M., Lendasse, A.: High-performance ELM for memory constrained edge computing devices with metal performance shaders. In: Cao, J., Vong, C.M., Miche, Y., Lendasse, A. (eds.) ELM 2019. PALO, vol. 14, pp. 79–88. Springer, Cham (2021). https://doi.org/10.1007/978-3-030-58989-9_9
3. Akusok, A., Leal, L.E., Björk, K.-M., Lendasse, A.: Scikit-ELM: an extreme learning machine toolbox for dynamic and scalable learning. In: Cao, J., Vong, C.M., Miche, Y., Lendasse, A. (eds.) ELM 2019. PALO, vol. 14, pp. 69–78. Springer, Cham (2021). https://doi.org/10.1007/978-3-030-58989-9_8
4. Andriesse, W., van Mensvoort, M.: Acid sulfate soils, distribution and extent, p. 6. Marcel Dekker (2002)
5. Auri, J., et al.: From a general survey to risk management - acid sulfate soils are Finland's most persistent environmental problem, but research can mitigate the harms they cause (2022). https://www.gtk.fi/en/current/from-a-general-survey-to-risk-management-acid-sulfate-soils-are-finlands-most-persistent-environmental-problem-but-research-can-mitigate-the-harms-they-cause
6. Brevik, E.C., et al.: Soil and human health: current status and future needs. Air, Soil Water Res. 13 (2020)
7. Eash, N.S., Sauer, T.J., O'Dell, D., Odoi, E.: Soil Science Simplified. Wiley, Hoboken (2015)
8. Epie, K., Virtanen, S., Santanen, A., Simojoki, A., Stoddard, F.: The effects of a permanently elevated water table in an acid sulphate soil on reed canary grass for combustion. Plant Soil 375(1–2), 149–158 (2014)
9. Estévez, V., et al.: Machine learning techniques for acid sulfate soil mapping in southeastern Finland. Geoderma 406, 115446 (2022)
10. Estévez Nuño, V.: Machine learning methods for classification of acid sulfate soils in virolahti. Master's thesis (2020)
11. Fältmarsch, R., Åström, M., Vuori, K.M.: Environmental risks of metals mobilised from acid sulphate soils in Finland: a literature review. Boreal Environ. Res. 13, 444–456 (2008)
12. Huang, P., Li, Y., Sumner, M.: Handbook of Soil Sciences: Resource Management and Environmental Impacts, 2nd edn. Taylor & Francis (2011)
13. Lal, R., et al.: Soils and sustainable development goals of the united nations: an international union of soil sciences perspective. Geoderma Reg. 25, 398 (2021)
14. Li, Y., Yang, R., Guo, P.: Spark-based parallel OS-ELM algorithm application for short-term load forecasting for massive user data. Electr. Power Compon. Syst. 48(6–7), 603–614 (2020)
15. Palko, J.: Mineral element content of timothy (phleum pratense l.) in an acid sulphate soil area of tupos village, northern finland. Acta Agriculturae Scand. 36(4), 399–409 (1986)
16. Sarangi, S.K., Mainuddin, M., Maji, B.: Problems, management, and prospects of acid sulphate soils in the Ganges delta. Soil Syst. 6(4), 95 (2022)
17. Schonlau, M., Zou, R.Y.: The random forest algorithm for statistical learning. Stand. Genomic Sci. 20(1), 3–29 (2020)
18. Yli-Halla, M.: Classification of acid sulphate soils of Finland according to Soil Taxonomy and the FAO/Unesco legend. Agric. Food Sci. 6(3), 247–258 (1997)

Deep Learning-Based Approach for Sleep Apnea Detection Using Physiological Signals

A. R. Troncoso-García[1]([✉]), M. Martínez-Ballesteros[2], F. Martínez-Álvarez[1],
and A. Troncoso[1]

[1] Data Science and Big Data Lab, Pablo de Olavide University, 41013 Seville, Spain
{artrogar,fmaralv,atrolor}@upo.es
[2] Department of Computer Science, University of Seville, 41012 Seville, Spain
mariamartinez@us.es

Abstract. This paper explores the use of deep learning techniques for detecting sleep apnea. Sleep apnea is a common sleep disorder characterized by abnormal breathing pauses or infrequent breathing during sleep. The current standard for diagnosing sleep apnea involves overnight polysomnography, which is expensive and requires specialized equipment and personnel. The proposed method utilizes a neural network to analyze physiological signals, such as heart rate and respiratory patterns, that are recorded during sleep to authmatic sleep apnea detection. The neural network is trained on a dataset of polysomnography recordings to identify patterns that are indicative of sleep apnea. The results compare the use of different physiological signals to detect sleep apnea. Nasal airflow seems to have the most accurate results and higher specificity, whereas EEG and ECG have higher levels of sensitivity. The best model concerning accuracy is compared to bias models previously applied to sleep apnea detection in literature, achieving greater results. This approach has the potential to provide automatic sleep apnea detection, being an accessible solution for diagnosing sleep apnea.

Keywords: Sleep apnea · Time series · Deep learning · classification · forecasting

1 Introduction

Sleep apnea is a sleep disorder that causes repeated pauses in breathing or shallow breathing during sleep. These pauses could last from a few seconds to several minutes and could occur many times throughout the night. These interruptions in breathing can cause a significant reduction in sleep quality and can lead to a variety of significant health consequences [12]. Obstructive sleep apnea (OSA) is the most common form of sleep apnea. Patients with OSA usually experience loud snoring, gasping, or choking during sleep, and fatigue during the day. Hypoapnea is a related term used to describe a partial reduction in airflow to the

I. Rojas et al. (Eds.): IWANN 2023, LNCS 14134, pp. 626–637, 2023.
https://doi.org/10.1007/978-3-031-43085-5_50

lungs during sleep. Causes breathing to become shallower or slow for a period of time, typically lasting at least 10 s. Hypoapnea is often associated with OSA. Other symptoms of sleep apnea could include headaches in the morning, difficulty in concentrating, mood changes, and irritability. These symptoms directly affect the patient's daily life: sleep apnea contributes to motor vehicle accidents and reduced productivity, creating problems at work or school. Sleep apnea is closely related to obesity, smoking, alcohol consumption, and family history. Men are more likely to develop sleep apnea than women, which is more common in older adults. Sleep apnea also plays an increasing role in cardiovascular disease, particularly hypertension and congestive heart failure [12]. Treatment for sleep apnea can include lifestyle changes such as weight loss, exercise, and the elimination of alcohol and sedatives. There are also medical interventions, such as continuous positive airway pressure therapy [4]. Effective treatment can improve sleep quality, reduce symptoms, and improve overall health and well-being.

Polysomnography (PSG) is a widely used clinical test to diagnose sleep apnea and other sleep disorders. During a PSG test, a person is monitored while sleeping to measure various physiological parameters, such as heart rate or breathing rhythm. PSG is considered the standard for diagnosing sleep apnea because it provides a comprehensive assessment of the severity and frequency of breathing disturbances during sleep [1]. However, the PSG test has several disadvantages. PSG involves spending a night in a hospital or a sleep laboratory. The patient must be connected to various sensors, which is uncomfortable. In addition, PSG could be expensive. The costs may be high for public health systems and could be inaccessible for some patients. The test are also not accurate in certain situations. For example, the PSG test only provides information about the patient's sleep patterns during the stay in the hospital. It may not capture typical sleep patterns or account for the variability in sleep over time. PSG may also disrupt the patient's natural sleep patterns, as they are sleeping in an unfamiliar environment and connected to various sensors. Finally, the interpretation of PSG results requires specialized training and knowledge of relevant experts [8].

PSG records are considered as time series data because they involve the recording of physiological signals over time. During a PSG recording, various physiological signals such as brain waves (EEG), eye movements (EOG), muscle activity (EMG), heart rate (ECG), and breathing patterns are continuously monitored and recorded. These signals change over time and are typically sampled at a fixed frequency, resulting in a sequence of data points that can be analyzed as a time series. Analyzing PSG data as a time series reveals patterns and trends in physiological signals and provides insights into sleep disorders and other conditions that affect sleep. In this work, time series analysis techniques are used to study and interpret PSG data and extract relevant features for the diagnosis and treatment of sleep apnea. Artificial intelligence and deep learning (DL) techniques in particular are applied to detect sleep apnea. DL models analyze large amounts of data quickly and accurately. They represent a tool for doctors to help make decisions, leading to more reliable diagnoses and personalized treatment plans for patients. In addition, technology is applied to provide a noninvasive and cost-effective way to detect sleep apnea. Portable devices, such as wearable sensors and smartphone apps, could also be used at home, making sleep monitoring easy for patients. This paper proposes a DL approach to detect sleep apnea events. A neural network is applied to PSG data including ECG, EEG, blood pressure (BP), and nasal respiration. Sleep apnea detection methods using the four signals are compared. Then, the best method is compared to bias models using the same input data.

The remainder of the paper is structured as follows. First, in Sect. 2, the latest developments in deep learning regarding the detection of sleep apnea are discussed. Then, in Sect. 3, the experiments conducted are explained in detail, while Sect. 4 presents the results obtained. Lastly, the paper is concluded in Sect. 5.

2   Related Work

As previously introduced, computer-based sleep apnea detection is useful to help physicians diagnose the disease. Several examples of neural networks applied to the diagnosis of sleep apnea are found in the literature. The paper [5] revised existing algorithms that have been applied to the detection of obstructive sleep apnea using various sensors and the combination of different approaches. The paper presented 84 original research articles published between 2003 and 2017. The articles were selected to provide valuable information to researchers who want to implement potential signal-processing algorithms on hardware. The contributions of the article in [6] regarding automatic sleep apnea scoring processes are also discussed. Another review is found in [7]. The goal of the paper is to analyze the research published in the last decade, examining how different deep networks are implemented, what preprocessing or feature extraction is necessary, and the advantages and disadvantages of different types of networks. In the field of classifiers, neural networks are the most used models for the detection of sleep apnea. Namely, these models are deep vanilla neural network (DVNN), convolution neural network (CNN), and recurrent neural network (RNN).

The paper [9] discussed the usefulness of ML and DL models as a diagnosis-decision-support tool for the detection of sleep apnea. The article then focused on obstructive sleep apnea. ML models were applied to the analysis of the respiratory signal waveform to aid in its diagnosis. Local Interpretable Model-Agnostic (LIME) library was used to explain the results obtained from a PSG study for automatic detection of sleep apnea. The results obtained help humans to understand the importance of each feature. The study carried out in [2] proposed a CNN model to detect sleep apnea. The input data are four different types of sleep study that focus on portability and signal reduction. The CNN model used the level of oxygen saturation (SpO_2) as input. The results showed that it is a valid and cost-effective alternative to PSG. The study used 190000 samples from SPO2 sensors from 50 patients, and the overall accuracy of sleep apnea detection was 91.3%, with a loss rate of 2.3 using the cross-entropy cost function using the deep convolutional neural network.

The study in [10] presented a new approach for automatically detecting sleep-disordered breathing events, such as sleep apnea. RNN was used to analyze nocturnal electrocardiogram (ECG) recordings. The proposed RNN model included recurrent layers with LSTM and a gated-recurrent unit (GRU). The model was trained and tested on ECG recordings from 92 patients, resulting in a F1-score of 98.0% for LSTM and 99.0% for GRU. These results showed that the proposed method outperformed conventional methods and could be used as a screening and diagnostic tool for patients with sleep breathing disorders. Furthermore, the study in [11] proposed an algorithm based on DL models to automatically detect sleep apnea events in respiratory signals. The algorithm improved the scoring per patient when assigned to the apnea-hypopnea index. The proposed algorithm was proved to be a useful tool for trained staff to quickly diagnose sleep apnea. Finally, the study in [13] explored an alternative to PSG for detecting sleep apnea and hypopnea syndrome using ECG and SpO_2 signals. The paper proposed a combination of classifiers to improve classification performance by using complementary information from individual classifiers.

3 Methodology

This paper proposes the application of a neural network to time series classification in the problem of detecting sleep apnea. Figure 1 shows the methodology carried out in this article. As introduced, several signals are used to characterize sleep apnea in the clinical scope. Here, the focus is on comparing nasal respiration, BP, EEG, and ECG signals to see which time series is better for the detection task of sleep apnea. A grid search is carried out to optimize the hyperparameters of the neural network model. Then, the signal obtaining best results is selected, and this model is compared to bias models using the same input data.

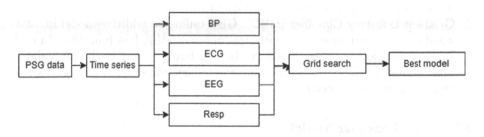

Fig. 1. Purposed methodology.

3.1 Data Preprocessing

The PSG data are in a standardized format that is commonly used in sleep studies. PSG data includes a variety of signals, such as EEG, ECG, respiration, and body movement, among others. The data also include annotations that provide information on sleep stages, arousal, and other events that occur during the sleep study. These annotations are created by experienced physicians who visually inspect signals and label them according to established guidelines. The Waveform Database (WFDB) is an open source software package developed by Physionet that provides tools for reading, writing, and processing physiological signals. WFDB allows users to easily access and manipulate large databases of physiological signals, including PSG data. The WFDB Python package is a Python interface to the WFDB software, providing a convenient way to access and analyze PSG data using Python [3]. In this paper, WFDB has been used to covert waveform and signal data to time series data. The signals have been processed and filtered to detect errors and outliers. PSG data is divided into four different dataset, each of them containing one physiological signal such as EEG or ECG.

3.2   Bias Models

The results of the purposed DL methodology are compared to the benchamark algorithms. Bias models have been used with default configuration and have been applied to the same input data. The bias models are presented in a previous work for the International Conference KES 2022 [9]. The models are detailed as follows:

1. Logistic Regression (LR) is a basic model used for binary classification that predicts targets using a linear approximation.
2. K Nearest Neighbors (KNN) classifier. The model uses a k-nearest neighbor vote for classification. The parameter **k** is set to 5 in this case.
3. Decision Tree (DT) classifier. The model is a nonparametric supervised learning method that creates a model by learning simple decision rules from the data features to predict the target variable.
4. Gradient Boosting Classifier (GBC). GBC builds an additive model in a forward stage-wise manner and optimizes differentiating loss functions. In each stage, regression trees are fitted on the negative gradient of the binomial or multinomial deviance loss function. For binary classification, only a single regression tree is induced.

3.3   Deep Learning Model

The neural network used in this study is a dense neural network with several layers with a certain number of neurons. To avoid overfitting, a dropout layer has been added to the network. Dropout is a regularization technique that randomly removes some of the neurons in a layer during training, helping to prevent the network from relying too heavily on any one neuron or feature.

The neural network architecture has been implemented using grid search. In particular, the following hyperparameters have been tuned: number of layers, number of neurons per layer, and dropout. The range of the hyperparameters is presented in Table 1. Dropout with a value equal to 0 means that there is no dropout layer.

Table 1. Hyperparameters range.

Hyperparameter	Range	Optimal value
Number of layers	2, 3, 4, 5	4
Neurons	100, 200, 300, 400, 500	[100, 200, 300, 400]
Dropout	0, 0.1, 0.2, 0.3, 0.4, 0.5	0.2

The best model obtained includes four dense layers and a dropout layer after each of them. Other hyperparameters such as batch size, learning rate, or training epochs have not been tuned. The batch size determines how many samples are processed at once during each training iteration, and the learning rate controls how much the network weights are updated during each training iteration. The network was trained with a batch size of 64, a learning rate of 0.1 and during 10 epochs. In general, the architecture of the neural network and the training parameters have been carefully selected to optimize the performance of the model for the specific task of detecting apnea events using PSG data.

4   Results

The results of sleep apnea detection using the neural network model are presented in this section. First, the input data is characterized. Then, the performance of the model is evaluated using several quality measures, such as accuracy, sensitivity, specificity, and F1 score. Finally, the best model is compared to bias model previously used in sleep apnea detection.

4.1   Input Data

This study uses data from the MIT-BIH Polysomnographic Database [3], which contains recordings from 16 patients. Different physiological signals are presented, including ECG, invasive BP, EEG, and nasal respiration airflow. Health professionals carefully study the signals and annotate them based on the existence of apnea events and the stage of sleep. The PSG data are processed to create a dataset for classification tasks, where each instance is a 30-second window of the four datasets of pyhsiological signal labeled as either an apnea or hypoapnea event (1), or normal breathing (0). The final dataset has 7500 attributes per instance due to the 250 Hz sampling rate. Therefore, the dataset is treated as a time series where each instance represents a 30-second interval measurement. Figure 2 illustrates an example of a complete PSG record with four signals and their measure unit. Namely, the signals are: ECG (mV), blood pressure (BP mmHg), EEG (mV) and nasal respiration (l). Data must be normalized because of the different scales in the four signals.

The classification task is performed using this dataset. The four different signals recorded during 30 s are every single instance of the input data. The classification model achieves a binary classification task. One of the biggest problems

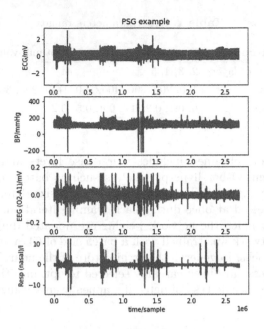

Fig. 2. PSG example record.

when performing sleep apnea detection is that the data is clearly unbalanced. Figure 3 shows the distribution of apnea events: normal (meaning that there is no apnea event, normal breathing) and anomalous. Anomalous events mean hypoapnea (partial breathing interruption), obstructive apnea (total obstruction), and central apnea. In this paper, the target value has been summarized as "apnea" (1) and "no apnea" (0).

4.2   Quality Measures

Classification refers to the task of automatically assigning input data to one of several predefined categories or classes based on a set of characteristics. In this paper we work in the field of binary classification, meaning that there are two classes: "No apnea" (0) and "Apnea" (1). The LSTM algorithm learns to identify patterns in the input data that are characteristic of each class and then uses these patterns to classify new unlabeled data. Classification aims to create a model that can accurately predict the class of new data based on the features provided.

The quality measures used to evaluate the proposed methodology are presented as follows.

– Accuracy. It measures the percentage of correct predictions made by the model out of all predictions.

$$Accuracy = \frac{TP + TN}{TP + TN + FP + FN} \tag{1}$$

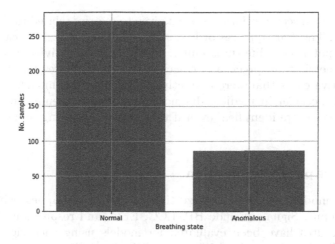

Fig. 3. Types of apnea distribution.

- Precision. It is the fraction of true positive predictions (correctly classified as positive instances) out of all positive predictions (instances classified as positive). It is calculated as the ratio of the number of true positives (TP) to the sum of true positives and false positives (FP).

$$Precision = \frac{TP}{TP + FP} \qquad (2)$$

- Sensitivity. This measure refers to the fraction of true positive predictions out of all actual positive instances. Calculated as the ratio of the number of true positives to the sum of true positives and false negatives. It could also be called recall.

$$Sensitivity = Recall = \frac{TP}{TP + FN} \qquad (3)$$

- Specificity. It is a metric that measures the ability of a classification model to correctly identify negative instances as negative. Specifically, it is the fraction of TN predictions (correctly classified as negative instances) out of all negative predictions (instances classified as negative). It is calculated as the ratio of the number of true negatives to the sum of true negatives and false positives. High specificity indicates that the model is good at avoiding false positives.

$$Specificity = \frac{TN}{FP + TN} \qquad (4)$$

- F1 Score. This metric is known as the harmonic mean of precision and recall. It is a weighted average of precision and recall, where the weights are equal and ranges from 0 to 1, with 1 being the best possible score. The F1 score is calculated as 2 times the product of precision and recall, divided by the sum of precision and recall.

$$F1 = \frac{2 * Precision * Recall}{Precision + Recall} = \frac{2 * TP}{2 * TP + FP + FN} \qquad (5)$$

In the clinical scope, priority is often given to metrics related to the detection of TP cases. The consequences of FN, such as missing a positive case, could be severe and potentially life-threatening. In this way, sensitivity can be a more important metric than precision or accuracy, as it measures the proportion of actual positive cases that were correctly identified by the model. For example, recall can be crucial in medical diagnosis or disease screening to ensure that all positive cases are identified, even if that means sacrificing some precision or accuracy.

4.3   Sleep Apnea Classification

The LSTM model is used to analyze time series of sleep apnea using different signals as input. Signals include BP, ECG, EEG, and respiration (Resp). The quality measures have been evaluated for models using each signal as input data including accuracy, sensitivity, specificity, and F1 score that have been introduced in the previous Sect. 4.2. Table 2 presents the quality measures for the four physiological signals.

Table 2. Quality measures of the LSTM model using different signals as input.

Signal	Accuracy	Sensitivity	Specificity	F1
BP	0.583	0.438	0.641	0.238
ECG	0.593	**0.688**	0.576	0.334
EEG	0.574	0.625	0.565	0.303
Resp	**0.712**	0.125	**0.913**	**0.370**

Table 2 shows that the model achieved the highest accuracy (0.712) when using respiration as input. However, the sensitivity values was lower than the other signals, indicating that the model tended to classify more instances as negative. On the contrary, the model that used EEG as input achieved higher sensitivity values to detect apnea events (positive class). Similar results are obtained with ECG. Regarding specificity, the best results are again obtained with Resp signal, with greater differences with the others. Finally, concerning F1 score the four signals have similar values although Resp has a little better result.

The confusion matrix for each experiment is presented in Fig. 4d. The confusion matrix compares the predicted values with the actual values. It summarizes the number of true positive (TP), true negative (TN), false positive (FP), and false negative (FN) predictions made by the model.

In this case, positive means the presence of an apnea event, whereas negative means that there is not apnea. The rows of the matrix represent the actual class labels, while the columns represent the predicted class labels. The diagonal of the matrix shows the number of correct predictions, while the off-diagonal elements represent the incorrect predictions. On the one hand, concerning the predictions

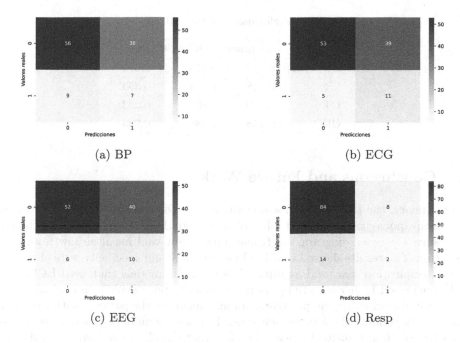

Fig. 4. Confusion matrix of the four experiments.

obtained using nasal respiration, the highest accuracy is reached. However, this results are not ideal, as the model tends to classify every instance as a 'no apnea' (0) event, as shown in Fig. 4d. This can be attributed to the imbalanced classes. On the other hand, the models that use ECG and EEG perform better in detecting apnea events (1) with a sensitivity of 0.688 and 0.625, respectively. However, these models also tend to classify several cases as positive (1) that are not apnea events. This is shown in the lower levels of specificity. The results of the experiments highlight the importance of choosing the appropriate input signal for the model to achieve the best performance. It also shows that while accuracy is important, other measures, such as sensitivity, should also be considered when evaluating the performance of the model, specially in clinical scope.

4.4 Comparation with Bias Model

The better model which is trained with nasal respiration signal which obtains the best results, is compared to the bias algorithms previously detailed in Sect. 3. Quality measures, namely accuracy, ROC-AUC score and F1 score for both bias models and the purposed model are presented in Table 3. The neural network outcomes clearly LR and DT models. Concerning GBC, results are quite similar.

Table 3. Performance of bias ML models.

Model	Accuracy	ROC-AUC	F1
NN - Resp	0.712	0.675	0.370
LR	0.698	0.514	0.363
DT	0.750	0.567	0.292
GBC	0.758	0.686	0.333

5   Conclusions and Future Work

In summary, the LSTM model was evaluated for the classification of time series of sleep apnea using different physiological signals as input, including blood pressure, electrocardiogram, electroencephalogram, and nasal airflow from respiration. The results showed that the best accuracy and specificity were obtained when respiration was used as input. However, the models that used ECG and EEG achieved higher sensitivity values to detect apnea events. In clinical scope is essential identifying the positive instances, meaning the patients with a certain disease. In general, the four models tended to classify more instances as negative (no apnea) than positive (apnea). The fact that the classes are imbalanced may have contributed to this bias. Nasal respiration has been proved as the most useful signal to detect sleep apnea.

Therefore, the results in the paper are useful in choosing the appropriate input signal to evaluate the model performance, particularly in clinical applications. An accurate automatic detection system is capable of improving sleep apnea management by providing a more objective measure of treatment efficacy and an objective feedback on treatment efficacy. The existence of these systems could help both patients and clinicians. Automatic detection systems are used as a diagnostic support tool for doctors. These systems could also monitor treatment progress and adjust therapy as needed, moving towards personalized medicine.

Future works could investigate the performance of ensemble DL models. These models could be applied to improve the accuracy and reliability of classification results. Another approach is to combine multiple physiological signals to improve the robustness of the classification model.

Acknowledgements. The authors would like to thank the Spanish Ministry of Science and Innovation the support within the projects PID2020-117954RB-C21 and TED2021-131311B-C22.

References

1. Brockmann, P.E., Schaefer, C., Poets, A., Poets, C.F., Urschitz, M.S.: Diagnosis of obstructive sleep apnea in children: a systematic review. Sleep Med. Rev. **17**(5), 331–340 (2013)

2. Chaw, H.T., Kamolphiwong, S., Wongsritrang, K.: Sleep apnea detection using deep learning. Tehnički glasnik **13**(4), 261–266 (2019)
3. Goldberger, A.L., et al.: PhysioBank, PhysioToolkit, and PhysioNet: components of a new research resource for complex physiologic signals. Biomedicallation **101**(23), e215–e220 (2000)
4. Kakkar, R.K., Berry, R.B.: Positive airway pressure treatment for obstructive sleep apnea. Chest **132**(3), 1057–1072 (2007)
5. Mendonca, F., Mostafa, S.S., Ravelo-Garcia, A.G., Morgado-Dias, F., Penzel, T.: A review of obstructive sleep apnea detection approaches. IEEE J. Biomed. Health Inform. **23**(2), 825–837 (2018)
6. Mostafa, S.S., Mendonça, F., Ravelo-García, A.G., Morgado-Dias, F.: A systematic review of detecting sleep apnea using deep learning. Sensors **19**(22), 4934 (2019)
7. Sherstinsky, A.: Fundamentals of recurrent neural network (RNN) and long short-term memory (LSTM) network. Physica D **404**, 132306 (2020)
8. Tan, H.L., Kheirandish-Gozal, L., Gozal, D.: Pediatric home sleep apnea testing: slowly getting there! Chest **148**(6), 1382–1395 (2015)
9. Troncoso-García, A., Martínez-Ballesteros, M., Martínez-Álvarez, F., Troncoso, A.: Explainable machine learning for sleep apnea prediction. Procedia Comput. Sci. **207**, 2930–2939 (2022)
10. Urtnasan, E., Park, J.U., Lee, K.J.: Automatic detection of sleep-disordered breathing events using recurrent neural networks from an electrocardiogram signal. Neural Comput. Appl. **32**, 4733–4742 (2020)
11. Van Steenkiste, T., Groenendaal, W., Deschrijver, D., Dhaene, T.: Automated sleep apnea detection in raw respiratory signals using long short-term memory neural networks. IEEE J. Biomed. Health Inform. **23**(6), 2354–2364 (2018)
12. White, D.P.: Sleep apnea. Proc. Am. Thorac. Soc. **3**(1), 124–128 (2006)
13. Xie, B., Minn, H.: Real-time sleep apnea detection by classifier combination. IEEE Trans. Inf. Technol. Biomed. **16**(3), 469–477 (2012)

Deep Learning for the Analysis of Solar Radiation Prediction with Different Time Horizons and Data Acquisition Frequencies

Carlos M. Travieso-González[1,2(✉)] and Alejandro Piñán-Roescher[2]

[1] Signals and Communications Department, University of Las Palmas de Gran Canaria (ULPGC), 35017 Las Palmas de G.C., Spain
ctravieso@dsc.ulpgc.es
[2] Institute for Technological Development and Innovation in Communications (IDETIC), University of Las Palmas de Gran Canaria (ULPGC), 35017 Las Palmas de G.C., Spain

Abstract. This study aims to develop and compare different AI systems for predicting solar radiation and evaluate their performance across different prediction horizons. Predicting solar radiation is of crucial importance for harnessing renewable energy sources. The models were designed to predict radiation over 6-h and 15-min horizons with the lowest possible error. The impact of prediction horizon and data acquisition frequency on prediction accuracy is discussed, emphasizing the need to consider the number of parameters and training time when comparing models. To improve the accuracy of short-term solar radiation predictions, five deep learning models, including classical, convolutional, and recurrent neural networks, were analyzed. The accuracy of the predictions was compared using two error metrics: root mean square error and mean absolute error.

Keywords: Solar irradiance prediction · prediction horizon · forecast · data acquisition frequencies · deep learning · renewable energy forecasting

1 Introduction

1.1 Motivation

Electricity is a fundamental aspect of nearly every part of our daily lives, society, and economy. The electrical system is constantly changing. Each day, new technologies and new ways of sourcing electrical energy are created and used to try to maximize the production of electricity at the lowest cost possible. The need for such an electricity that is safe, environment-friendly, reliable, and low-cost is paramount.

Perhaps, solar energy is one of the most environmentally safe and efficient sources of electrical energies. In fact, the decrease in the cost of photovoltaic (PV) panels and associated technology in recent years [1], as well as the improvement of forecasting solar power with different methods, have contributed to solar energy becoming a popular real alternative to traditional energy sources.

© The Author(s), under exclusive license to Springer Nature Switzerland AG 2023
I. Rojas et al. (Eds.): IWANN 2023, LNCS 14134, pp. 638–653, 2023.
https://doi.org/10.1007/978-3-031-43085-5_51

Unfortunately, most of the electricity that we consume as a society is generated by nonrenewable sources. In fact, nearly 80% of the global demand for energy is met by these fossil fuels [2]. This rate of consumption, along with the rapid industrialization and urbanization of third world countries, has led to more and more emission of pollutants harmful to the environment, and arguably are a leading cause of climate change.

As such, there is an increasing demand for forecasting solar radiation. Currently, accurately forecasting the power of the solar generation is one of the major problems facing this renewable source of energy. Apart from the typical drawbacks of the volatility and unpredictability of sunlight, those who use solar energy (such as businesses and homeowners) are unable to easily or precisely predict the energy they have available to use in the short term. This problem creates costly setbacks for the solar energy consumers because they cannot depend solely on this kind of energy and will need to supplement with other conventional and more expensive sources like fuel and gas. Therefore, to avoid this problem, it is vital to have a method that enables us to forecast more accurately the energy we have at our disposal in the short term.

1.2 Relative Works

Generally, the methods for forecasting solar radiation [3] can be separated into three categories—physical, statistical, and artificial intelligence (AI) technologies. The physical and statistical forecasting methods are limited because of their complex models and possibility of lower prediction accuracy [4]. However, due to the advancement in technology facilitated by the so-called "Fourth Industrial Revolution," artificial intelligence can help provide more accuracy and optimization for the forecasting methods.

Thanks to the advancements in artificial intelligence, many forecasting methods can now utilize neural networks to interpret the data for the most accurate predictions. Some models are based on Artificial Neural Networks (ANNs) [5].

Certain models, such as neural networks, can predict multiple values of a time sequence simultaneously, in one-shot or at one time. These types of models are known as *Multiple Output forecasting* [24]. *Multiple Output forecasting* has the advantage over the previous strategies because only one model is needed to predict multiple values at the same time. In fact, this model requires less training and configuration because the model itself learns to relate the inputs and output based on the nature of the data it is given.

Given the three different strategies to perform predictions, the *Multiple Output forecasting* [24] is the best method to use in forecast analyses. In this analysis, different models of *Multiple Output forecasting* will be looked at to determine which model produces the most optimal outcome for the given circumstances studied.

A popular method that uses ANNs is the Multilayer Perceptron (MLP) [6]. This forecasting model is frequently used in this field to predict solar irradiance.

Another option to forecast solar energy in the short-term (up to 6 h in advance) is with recurrent neural networks (RNNs). An example of this model's capacities is shown in this article [7]. For this study, seven months of data were chosen for training, which reduces the mean square error (RMSE) from 47 to 30.1, the mean absolute error (MAE) from 56 to 39 and, more importantly, the mean absolute percentage error (MAPE) from 18 to 11% overall.

Apart from the two principal networks, ANNs and Recurrent Neural Network (RNNs), another popular method for forecasting solar radiation is the Lorenz's block matching technique and four optical flows (OF). This method tries to perform weather predictions up to four-five hours ahead and forecast the solar radiation [8]. A study using the Lorenz's method compared different satellite-based approaches to obtain hourly irradiance forecasts up to five hours in advance using a 2-year dataset. The strategies used in this study achieved successful predictive estimations of solar power in the short term.

The advantage of applying AI to different methods can also be witnessed in the following study, [9], where AI was used in following methods: linear, feed-forward, Elman recurrent and radial basis neural networks. Together with the adaptive neuro-fuzzy inference scheme and AI, the analysis revealed that these models outperform conventional approaches. More specifically, the multivariate LM model proved to be the most complete in terms of prediction error and training time with a 74% improvement in mean square error over traditional line methods.

In [10], the objective was to compare various Machine Learning techniques for forecasting meteorological variables, such as Multiple Linear Regression, Polynomial Regression, Random Forest, Decision Tree, XGBoost, and Multilayer Perceptron Neural Network (MLP). The evaluation of these techniques was based on different metrics including (RMSE), (MAPE), (MAE), and coefficient of determination (R2). Random Forest and XGBoost were found to be the most efficient techniques for almost all the variables studied. Overall, Random Forest was identified as the best performing technique for forecasting temperature, relative humidity, solar radiation, and wind speed.

It should be noted that AI does not solve all the problems and that solar radiation is not easy to capture. Many times, it becomes difficult to capture in adverse conditions. As such, one model is typically not enough to use to obtain accurate data. The best way to achieve the most accurate results are through a hybrid model. In this article [11] an autoregressive model (AR) is combined with a dynamic system model. Thanks to this combination, more accurate predictions with less error were achieved. For this current research, as will be detailed below, a combination of different methods were tested, since it was shown that these were more successful.

As discussed above, the adoption of AI models for the prediction of solar energy has been proven in literature to have a wider application and higher accuracy in comparison to other models [12]. In plenty of research works, artificial intelligence techniques are integrated with different models, sky imagers and satellite imagers to improve the data handling algorithm, which implicitly results in forecasting accuracy [13]. It is indisputable that various models using AI have increased the accuracy of solar radiation estimation in short term periods.

1.3 Proposal and Innovation

The aim of this work is to study, develop and analyze distinct AI systems used in solar radiation [14] prediction and compare these models over different prediction horizons.

This study introduces a novel approach to data analysis through the development of five neural network models. Specifically, we will investigate two horizon models for short-term solar radiation forecasting. The first model will utilize a frequency of 30 min to predict 12 samples, providing a prediction horizon of 6 h. The second model will use a frequency of 1 min to predict 15 samples, offering a prediction horizon of 15 min. By analyzing the performance of these different models, we aim to identify the most reliable and accurate option for solar radiation forecasting in the short-term.

2 Materials

The present work starts from all the data recollected by the ITC, (Instituto Tecnológico de Canarias). All this data represents the solar radiation from 2016-11-01 to 2021-02-26, at one-minute intervals. Figure 1 shows the location of the ITC (red point). It should be noted that more meteorological variables are collected, although only solar radiation is used in this analysis.

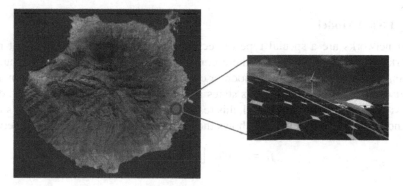

Fig. 1. Location of Instituto Tecnológico de Canarias in Gran Canaria

3 Methods

3.1 Preprocessing

Two different prediction horizons were considered in this study. The first one used a 30-min frequency to predict 12 samples, while the second one used a 1-min frequency to predict 15 samples over a 15-min prediction horizon. To obtain these predictions, the data used for prediction were arranged in four different ways. For the 1-min sampling frequency, a window size of 15 (Fig. 2a) and 55 (Fig. 2b) were used. For the 30-min sampling frequency, a window size of 15 (Fig. 2c) and 55 (Fig. 2d) were used as well.

3.2 Models

The next step involves the application of several neural network models. Specifically, we will discuss five different models, namely: Long Short-Term Memory (LSTM) [15], Bidirectional Long Short-Term Memory (LSTM Bidirectional) [16], Multilayer Perceptron (MLP) [17], Convolutional Neural Network (CNN) [18], and Convolutional Neural Network - Long Short-Term Memory (CNN-LSTM) [19]. By explaining and analyzing the different models, we aim to gain a comprehensive understanding of their respective strengths and limitations for the task at hand.

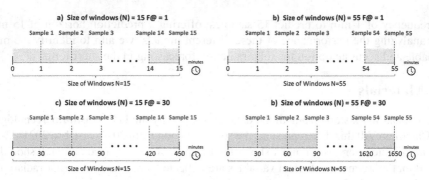

Fig. 2.   a, b, c, d. Different Windows size used in the study

3.2.1 LSTM Model

LSTM networks are a special type of recurrent neural networks. A recurrent neural network is an artificial network where connections can be made between inputs and outputs, creating an informational loop or cycle in the network. This connection allows the networks to remember previous states and use that previous information to decide the next step in the series. In general, this feature makes recurrent networks very suitable for handling time series. Figure 3 shows the operational scheme of an LSTM network.

$$f_t = \sigma(W_f \cdot [h_{t-1}, x_t] + b_f \tag{1}$$

$$i_t = \sigma(W_i \cdot [h_{t-1}, x_t] + b_i \tag{2}$$

$$o_t = \sigma(w_o \cdot [h_{t-1}, x_t] + b_o \tag{3}$$

$$\tilde{C}_t = tanh(w_c \cdot [h_{t-1}, x_t] + b_c) \tag{4}$$

$$C_t = f_t \odot C_{t-1} + i_t \odot \tilde{C}_t \tag{5}$$

$$h_t = o_t \odot \tanh(C_t) \tag{6}$$

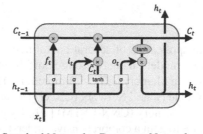

Fig. 3. Standard Networks, Recurrent Network and LSTM

The architecture of this neural network includes an LSTM layer, which is used to model sequential data. This layer has 128 LSTM units and uses a ReLU activation function, with "return_sequences" set to False, meaning only the last output of the sequence is returned. After the LSTM layer, there are three dense layers. The first dense layer is fully connected with 256 units and ReLU activation, while the second dense layer also is fully connected and has 128 units with ReLU activation. Finally, the output layer of the network is another dense layer, this time with 15 units and no specific activation function.

The advantage of this model are the parameters. Unlike other models such as MLP (which will be later discussed), the parameters do not grow with the length of the input sequence or input window. Additionally, LSTMs are designed for sequences and time steps, and therefore do not require an additional *flatten* layer. LSTMs also allow for more complex topologies like those of the encoder-decoder model or the LSTM Bidirectional model.

3.2.2 Bidirectional LSTM

Bidirectional LSTM is one of the complex topologies that can be derived from the basic LSTM network. It is the process of creating bidirectional memory, that is, making the LSTM neural network have the sequence information in both directions. In bidirectional memory, the sequence information should flow backwards from the future to the past and forwards from the past to the future. In other words, the inputs are analyzed in two directions. This model is generally used for forecasting series (Fig. 4).

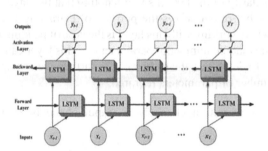

Fig. 4. LSTM Bidirectional Layer

This architecture is based on a Bidirectional LSTM model, which incorporates two layers of LSTMs. The first layer comprises 64 neurons/units and processes the input data in a forward direction. The second layer, with the same number of neurons, processes the input data in a backward direction. The output of both layers is then combined by the Bidirectional layer, resulting in a sequence that is passed to a Dense output layer. The number of neurons in the output layer equals the number of steps to be predicted.

3.2.3 CNN Model

The CNN model is a type of deep learning for processing data that has grid pattern. In this study, this kind of model is used to construct a layer that creates a convolution kernel that convolves with the input of the layer over a single dimension in space or time to outrun an output tensor (Fig. 5).

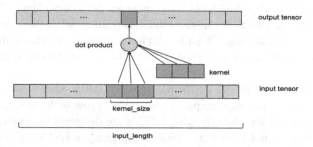

Fig. 5. Convolution process in one dimension

The architecture of the model used in this study is based on a one-dimensional convolutional layer with 128 filters, which employs a kernel size of either 15 or 55 (depending on the model size), with ReLU activation. The subsequent flatten layer reduces the dimensionality of the data from three dimensions to two dimensions, which is necessary for the dense layers that follow. These dense layers consist of a first layer with 128 neurons and ReLU activation, followed by a second layer with 64 neurons and ReLU activation. The output dense layer has either 15 or 12 neurons (depending on the number of steps predicted) and uses a linear activation function.

To enhance the clarity of the text, it's worth noting that this layer is not designed for making predictions, but rather serves the purpose of extracting features. Another key factor to consider when working with this layer is the type of padding used. Additionally, it's important to note that the number of parameters for this layer can be calculated using a specific formula. To delve into the technicalities further, we can apply this formula to obtain the exact number of parameters required.

Parameters numbers = output_channels ∗ (input_channels ∗ window_size + 1)

3.2.4 MLP Model

The MLP model, short for Multi-layer Perceptron, is a type of neural network that consists of multiple layers of nodes, including at least two dense layers. The dense layers are made up of interconnected neurons, each of which applies a linear function followed by a non-linear activation function to the input data. This process helps the network learn complex patterns and relationships in the data. In addition to the dense layers, an MLP model typically includes an output layer, which produces the final output of the network. MLP models are widely used in a variety of machine learning tasks, including classification, regression, and prediction. In the context of the given text, Table 1 displays a two-layer MLP model, along with the output layer.

Table 1. Two-layer MLP model

Model: "model"		
Layer	Output shape	Param #
input_1 (inputlayer)	[(None, 55, 1)]	0
flatten (Flatten)	(None, 55)	0
dense (Dense)	(None, 64)	3584
dense_1 (Dense)	(None, 32)	2080
dense_2 (Dense)	(None, 12)	396
Total params6060		
Trainable params 6060		
Not trainable params 0		

Furthermore, these kinds of models do not work on time series (Batch size, numbers of steps, numbers of variables) (none, 55, 1) and require a flatten layer to adapt their input. (Batch size, input dimension), (none, 55).

This model architecture varies depending on the frequency of the data. For the 1-min frequency data, the MLP model has a Flatten layer followed by three Dense layers with ReLU activation, where the number of units for each layer is 32, 16, and 15, respectively. On the other hand, for the 30-min frequency data, the MLP model has the same Flatten layer followed by three Dense layers with ReLU activation, but with 256, 128, and 12 units, respectively.

3.2.5 CNN-LSTM Model

The CNN-LSTM model is a hybrid model that combines the best features from CNN and LSTM and gives a brilliant perform for the forecasting studies. This combined model is composed by convolutional layers that extract features and LSRM layers that help the model to propagate them recursively.

This model architecture consists of five layers: a 1D convolutional layer, an LSTM layer, and three dense layers. The 1D convolutional layer utilizes 128 convolutional filters with a kernel size of 5 and a ReLU activation function. The LSTM layer has 128 LSTM units and "return_sequences" set to False. The first dense layer is fully connected with 256 units and a ReLU activation function, followed by another dense layer with 128 units and a ReLU activation function. The last layer is a dense layer with 8 units, possibly used for a classification task.

3.3 Model Setup

This section discusses and provides details of the models utilized in this case study. A brief explanation will be given of each model and how the results of each model were obtained. It should be noted that when testing a specific model, each simulation was repeated 5 time to observe the variation and ensure the best possible outcome with the least amount of error.

To obtain the results of each specific model, two simulations were run:

- Simulation 1: Training 2017, Validation 2018, Test 2019 (Repeated 5 times)
- Simulation 2: Training 2018, Validation 2019, Test 2020 (Repeated 5 times)

3.4 Validation Metrics

After running both simulations, a mean of all the results and their standard deviation was calculated and represented with the form ERROR ± std. To calculate the ERROR, which represents the difference between the actual value and the predicted value, two error metrics, namely the MAE and RMSE, were used. An important aspect to highlight in this study is the utilization of RMSE and MAE as the primary metrics, as they are widely used and recognized in the field. Nevertheless, it is worth mentioning that alternative metrics such as the Coefficient of Determination (R-squared) or Mean Bias Error (MBE) could also have been employed in the analysis.

The Mean Absolute Error (MAE) ± std measures the average magnitude of errors in a set of forecast predictions but does not consider their direction. In short, it measures the accuracy for continuous variables [20].

$$MAE = \frac{1}{n} \sum_{i=1}^{n} |y_i - y| \tag{7}$$

The RSME (Root Mean Square Error) ± std is a quadratic scoring equation which measures the average magnitude of the error. This type of metric use is especially useful when large errors are undesirable [21]

$$RMSE = \sqrt{\frac{\sum_{i=1}^{n} \|y(i) - \hat{y})i)\|^2}{N}} \tag{8}$$

4 Results

The following tables and figures show the results of the two methods for the different time horizons. Analyzing these results, various points can be drawn.

The main implication from the above tables is seen in the studies for the 30-min frequency window. Here, it becomes very apparent the method used to obtain the results. Indeed, the method used has a considerable influence for accuracy and prediction. Specifically, the results obtained through RMSE are worse compared to those obtained through MAE. As the data shows, for $N = 15$, the best model is the CNN-LSTM, both in windows 1 and 12. On the other hand, for $N = 55$, the best model is the Bi-LSTM. This model is also in the two windows.

Conversely, when analyzing the data obtained from the 1-min frequency, the results obtained through the MAE method are better than those obtained by RMS. It is important to highlight that the results in this time window are not so easily influenced by the choice of model. In fact, all the models in this time window produce practically the same results. As such, the choice of the model is not so important. All these points can be seen more visually in the following chart (Tables 2, 3, 4, 5, 6, 7, 8, 9 and Figs. 6, 7).

Table 2. Obtained results for two different windows of input data and a sample frequency of 30 min and size of windows (N) = 15, using RMSE.

Size of windows (N) = 15 F@ = 30

n + i	MLP	LSTM	Bi-LSTM	CNN	CNN-LTSM
1	54,40 ± 0,30	54,42 ± 0,37	54,38 ± 0,23	54,46 ± 0,39	54,10 ± 0,21
2	70,22 ± 0,65	69,56 ± 0,51	69,68 ± 0,73	70,66 ± 0,69	69,83 ± 0,95
3	79,83 ± 0,62	79,65 ± 1,06	79,33 ± 0,86	80,63 ± 0,71	79,06 ± 1,05
4	90,28 ± 0,85	90,37 ± 1,34	89,54 ± 1,12	90,94 ± 0,79	89,10 ± 1,18
5	102,78 ± 0,93	102,88 ± 1,37	102,02 ± 1,53	103,18 ± 1,34	100,94 ± 1,01
6	116,23 ± 0,90	115,84 ± 1,74	115,28 ± 1,47	115,63 ± 1,39	113,04 ± 1,16
7	127,68 ± 1,09	127,22 ± 2,03	126,54 ± 1,81	126,85 ± 1,48	123,49 ± 1,36
8	137,10 ± 1,35	136,49 ± 2,17	135,58 ± 1,76	136,12 ± 1,77	131,87 ± 1,65
9	144,54 ± 1,65	143,71 ± 1,84	142,52 ± 1,88	143,76 ± 2,52	138,73 ± 2,00
10	149,13 ± 1,86	147,71 ± 1,51	146,36 ± 1,88	148,61 ± 2,71	142,96 ± 1,81
11	150,85 ± 2,82	149,27 ± 2,22	147,85 ± 2,92	150,78 ± 3,60	144,68 ± 2,53
12	149,32 ± 3,91	147,42 ± 3,34	146,28 ± 3,87	149,77 ± 4,70	143,54 ± 3,23

Table 3. Obtained results for two different windows of input data and a sample frequency of 30 min and size of windows (N) = 55, using RMSE.

Size of windows (N) = 55 F@ = 30

n + i	MLP	LSTM	Bi-LSTM	CNN	CNN-LTSM
1	53,40 + 0,19	53,56 ± 0,52	53,04 ± 0,29	53,06 ± 0,24	53,32 ± 0,31
2	67,04 ± 0,52	66,48 ± 0,41	66,59 ± 0,50	66,51 ± 0,40	66,66 + 0,37
3	74,00 ± 0,41	73,31 ± 0,49	73,30 ± 0,37	73,35 ± 0,35	73,72 ± 0,58
4	79,46 ± 0,48	78,24 ± 0,73	78,65 + 0,40	78,93 ± 0,69	78,74 ± 0,83
5	84,051 ± 0,95	82,49 ± 1,57	82,95 ± 1,13	83,27 ± 1,20	83,41 ± 1,61
6	87,76 ± 1,31	86,09 ± 1,94	86,34 ± 1,52	86,68 ± 1,40	86,69 ± 1,78
7	90,51 ± 1,85	88,89 ± 2,75	89,20 ± 2,35	89,91 ± 1,83	89,82 ± 2,71
8	93,04 ± 2,14	90,93 ± 2,99	91,28 ± 3,09	92,05 ± 2,10	92,34 ± 3,37
9	94,79 ± 1,80	93,02 ± 3,35	93,07 ± 3,12	94,05 ± 2,05	94,39 ± 3,72
10	96,27 ± 2,17	94,39 ± 3,29	94,69 ± 3,44	95,56 ± 2,34	96,38 ± 4,01
11	97,34 ± 2,40	95,52 ± 3,54	95,88 ± 3,42	96,42 ± 2,07	97,87 ± 4,52
12	97,77 ± 2,11	96,34 ± 3,23	96,63 ± 3,36	97,08 ± 2.01	98,93 ± 4,51

Table 4. Obtained results for two different windows of input data and a sample frequency of 30 min and size of windows (N) = 15, using MAE.

Size of windows (N) = 15 F@ = 30				
n + i	MLP	LSTM	Bi-LSTM	CNN	CNN-LTSM
1	19,54 ± 0,98	19,57 ± 0,81	19,35 ± 1,00	19,70 ± 0,94	18,96 ± 0,93
2	27,30 ± 1,16	26,83 ± 1,01	26,97 ± 1,27	27,78 ± 1,17	26,31 ± 1,09
3	33,61 ± 1,24	33,12 ± 1,31	33,08 ± 1,52	33,96 ± 1,31	32,03 ± 1,23
4	40,02 ± 1,21	39,45 ± 1,36	39,26 ± 1,66	40,46 ± 1,38	38,05 ± 1,48
5	47,25 ± 1,37	46,76 ± 1,51	46,42 ± 1,76	47,88 ± 1,40	44,76 ± 1,41
6	55,41 ± 1,49	54,59 ± 1,43	54,26 ± 1,83	55,84 ± 1,48	52,33 ± 1,53
7	63,15 ± 1,25	62,18 ± 1,33	61,72 ± 1,84	63,49 ± 1,59	59,32 ± 1,71
8	69,71 ± 1,46	68,47 ± 1,35	67,85 ± 2,04	69,83 ± 1,77	65,13 ± 1,95
9	74,50 ± 1,81	73,00 ± 1,56	72,28 ± 2,26	74,35 ± 2,52	69,34 ± 2,19
10	77,26 ± 2,12	75,41 ± 1,51	74,69 ± 2,17	77,11 ± 2,71	71,91 ± 1,99
11	77,91 ± 2,54	75,98 ± 1,89	75,37 ± 2,51	77,88 ± 3,13	72,56 ± 2,40
12	77,08 ± 2,95	75,02 ± 2,20	74,57 ± 2,56	77,29 ± 3,73	71,90 ± 2,59

Table 5. Obtained results for two different windows of input data and a sample frequency of 30 min and size of windows (N) = 55, using MAE.

Size of windows (N) = 55 F@ = 30				
n + i	MLP	LSTM	Bi-LSTM	CNN	CNN-LTSM
1	19,79 ± 0,91	19,02 ± 0,80	18,40 ± 0,72	18,95 ± 0,64	18,69 ± 0,80
2	25,30 ± 0,89	24,03 ± 0,73	24,21 ± 1,08	24,19 ± 0,85	23,94 ± 0,81
3	28,28 ± 1,18	27,03 ± 1,18	27,51 ± 1,42	27,33 ± 0,93	26,98 ± 1,14
4	30,59 ± 1,12	29,14 ± 1,23	29,96 ± 1,52	29,66 ± 1,12	29,10 ± 1,39
5	32,31 ± 1,30	30,83 ± 1,42	31,80 ± 1,69	31,53 ± 1,20	30,90 ± 1,65
6	33,82 ± 1,32	32,29 ± 1,58	33,32 ± 1,61	32,85 ± 1,33	32,25 ± 1,85
7	34,93 ± 1,66	33,49 ± 1,82	34,35 ± 1,90	34,14 ± 1,56	33,54 ± 2,19
8	35,89 ± 1,74	34,32 ± 1,99	35,46 ± 1,90	34,96 ± 1,75	34,68 ± 2,34
9	36,49 ± 1,65	35,04 ± 2,01	36,23 ± 1,90	35,59 ± 1,76	35,32 ± 2,61
10	37,24 ± 2,02	35,66 ± 2,31	36,87 ± 2,09	36,21 ± 1,93	36,03 ± 2,98
11	37,66 ± 2,19	36,37 ± 2,22	37,20 ± 2,14	36,60 ± 2,00	36,73 ± 3,24
12	38,20 ± 2,23	36,72 ± 2,32	37,64 ± 2,26	37,25 ± 2,13	37,14 ± 3,31

Table 6. Obtained results for two different windows of input data and a sample frequency of 1 min and size of windows (N) = 15, using RMSE.

Size of windows (N) = 15 F@ = 1				
n + i	MLP	LSTM	Bi-LSTM	CNN	CNN-LTSM
1	29,77 ± 0,88	29,52 ± 0,81	29,35 ± 0,79	29,81 ± 0,84	29,36 ± 0,79
2	43,06 ± 1,21	42,84 ± 1,13	42,84 ± 1,13	43,07 ± 1,19	42,86 ± 1,32
3	49,61 ± 1,31	49,49 ± 1,44	49,52 ± 1,28	49,64 ± 1,36	49,55 ± 1,33
4	53,76 ± 1,37	53,67 ± 1,34	53,69 ± 1,37	53,72 ± 1,40	53,72 ± 1,47
5	56,83 ± 1,39	56,73 ± 1,37	56,73 ± 1,38	56,73 ± 1,41	56,82 ± 1,50
6	59,37 ± 1,41	59,25 ± 1,37	59,22 ± 1,42	59,17 ± 1,41	59,24 ± 1,44
7	61,39 ± 1,36	61,35 ± 1,37	61,33 ± 1,43	61,25 ± 1,44	61,31 ± 1,41
8	63,15 ± 1,40	63,16 ± 1,42	63,19 ± 1,47	63,02 ± 1,40	63,13 ± 1,46
9	64,85 ± 1,38	64,78 ± 1,46	64,82 ± 1,51	64,72 ± 1,38	64,81 ± 1,51
10	66,46 ± 1,34	66,39 ± 1,51	66,47 ± 1,53	66,34 ± 1,36	66,46 ± 1,49
11	68,01 ± 1,36	67,95 ± 1,51	68,01 ± 1,49	67,92 ± 1,30	67,96 ± 1,45
12	69,45 ± 1,31	69,37 ± 1,45	69,45 ± 1,42	69,35 ± 1,21	69,44 ± 1,44

Table 7. Obtained results for two different windows of input data and a sample frequency of 1 min and size of windows (N) = 55, using RMSE.

Size of windows (N) = 55 F@ = 1				
n + i	MLP	LSTM	Bi-LSTM	CNN	CNN-LTSM
1	30,67 ± 0,89	29,66 ± 0,92	29,59 + 0,90	30,65 ± 0,91	29,38 ± 0,86
2	43,44 ± 1,19	41,93 ± 1,31	42,93 ± 1,32	43,34 ± 1,16	42,71 ± 1,19
3	49,79 ± 1,29	49,45 ± 1,41	49,41 ± 1,48	49,63 ± 1,26	49,23 ± 1,38
4	53,83 ± 1,27	53,55 ± 1,63	53,47 ± 1,61	53,70 + 1,29	53,32 ± 1,47
5	56,82 ± 1,33	56,22 ± 1,57	56,45 ± 1,68	56,68 ± 1,25	56,33 ± 1,59
6	59,27 ± 1,32	59,19 ± 1,42	58,86 ± 1,69	59,15 ± 1,31	58,71 ± 1,52
7	61,33 ± 1,37	61,53 ± 1,57	60,92 ± 1,69	61,21 ± 1,32	60,73 ± 1,53
8	63,12 ± 1,42	63,06 ± 1,34	62,68 ± 1,70	62,95 ± 1,27	62,54 ± 1,60
9	64,79 ± 1,36	64,14 ± 1,51	64,32 ± 1,82	64,63 ± 1,38	64,18 ± 1,57
10	66,40 ± 1,37	67,01 ± 1,37	65,85 ± 1,85	66,26 ± 1,35	65,69 ± 1,61
11	67,93 ± 1,37	65,99 ± 1,32	67,34 ± 1,82	67,83 ± 1,31	67,20 ± 1,55
12	69,36 ± 1,31	69,13 ± 1,52	68,69 ± 1,84	69,21 ± 1,31	68,48 ± 1,55

Table 8. Obtained results for two different windows of input data and a sample frequency of 1 min and size of windows (N) = 15, using MAE.

Size of windows (N) = 15 F@ = 1

n + i	MLP	LSTM	Bi-LSTM	CNN	CNN-LTSM
1	6,67 ± 0,41	6,62 ± 0,44	6,46 ± 0,36	6,73 ± 0,29	6,49 ± 0,28
2	10,30 ± 0,50	10,20 ± 0,48	10,20 ± 0,49	10,33 ± 0,56	10,12 ± 0,45
3	12,46 ± 0,64	12,29 ± 0,62	12,28 ± 0,62	12,46 ± 0,63	12,32 ± 0,64
4	13,92 ± 0,66	13,82 ± 0,72	13,82 ± 0,66	13,92 ± 0,67	13,79 ± 0,69
5	15,11 ± 0,30	15,01 ± 0,77	15,00 ± 0,75	15,08 ± 0,74	14,98 ± 0,76
6	16,14 ± 0,77	16,01 ± 0,82	15,99 ± 0,78	16,07 ± 0,77	16,00 ± 0,75
7	17,05 ± 0,82	16,87 ± 0,79	16,90 ± 0,81	16,99 ± 0,85	16,88 ± 0,80
8	17,79 ± 0,89	17,71 ± 0,85	17,74 ± 0,87	17,83 ± 0,86	17,68 ± 0,83
9	18,54 ± 0,90	18,45 ± 0,94	18,48 ± 0,87	18,60 ± 0,95	18,42 ± 0,91
10	19,30 ± 0,92	19,16 ± 0,97	19,25 ± 0,91	19,29 ± 1,00	19,22 ± 0,97
11	20,02 ± 0,89	19,92 ± 1,04	19,90 ± 0,96	20,01 ± 0,92	19,86 ± 0,98
12	20,67 ± 0,96	20,52 ± 1,02	20,69 ± 1,04	20,62 ± 0,98	20,50 ± 1,01

Table 9. Obtained results for two different windows of input data and a sample frequency of 1 min and size of windows (N) = 55, using MAE.

Size of windows (N) = 55 F@ = 1

n + i	MLP	LSTM	Bi-LSTM	CNN	CNN-LTSM
1	7,21 ± 0,38	6,37 ± 0,39	6,66 ± 0,41	7,15 ± 0,47	6,74 ± 0,38
2	10,61 ± 0,50	9,90 ± 0,56	10,31 ± 0,59	10,59 ± 0,54	10,38 ± 0,21
3	12,61 ± 0,66	12,61 ± 0,74	12,40 ± 0,66	12,60 ± 0,68	12,44 ± 0,65
4	14,08 ± 0,78	14,40 ± 0,66	13,93 ± 0,82	14,12 ± 0,80	13,90 ± 0,79
5	15,25 ± 0,83	15,43 ± 0,69	14,97 ± 0,88	15,19 ± 0,76	15,01 ± 0,93
6	16,17 ± 0,86	16,59 ± 0,95	16,00 ± 0,96	16,17 ± 0,81	15,91 ± 0,85
7	16,97 ± 0,83	16,56 ± 0,81	16,78 ± 0,95	16,99 ± 0,84	16,77 ± 0,94
8	17,73 ± 0,90	17,46 ± 0,71	17,62 ± 1,02	17,73 ± 0,94	17,56 ± 1,05
9	18,42 ± 0,98	18,65 ± 0,88	18,29 ± 1,03	18,45 ± 0,90	18,25 ± 1,01
10	19,19 ± 0,94	18,98 ± 0,86	18,92 ± 1,05	19,05 ± 0,99	18,87 ± 1,02
11	19,75 ± 0,95	19,61 ± 1,20	19,54 ± 1,10	19,75 ± 1,01	19,49 ± 1,09
12	20,38 ± 1,03	20,75 ± 1,12	20,12 ± 1,15	20,33 ± 1,00	20,11 ± 1,17

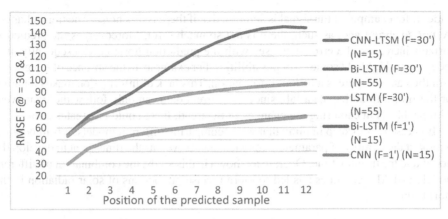

Fig. 6. Graphical results of the most efficient RMSE models

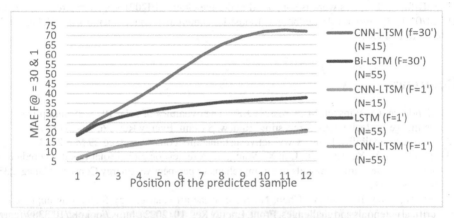

Fig. 7. Graphical results of the most efficient MAE models

5 Conclusions

The objective of this study was to develop different models with the purpose of implementing two final models for: 6h@30m and 15m@1m, one for each specification with the lowest error possible.

It should be noted that as the time horizon increases, the model choice becomes a crucial aspect. For example, for a frequency of @1m, since the difference is minimal, the system tends to do an ingenuous prediction. This prediction is as good as the best one obtained by any of the models. However, as the time horizon increases, the models play a more important role. For example, for 30@m, the models used had huge differences between them. As such, it is important to recognize how time horizons influence the prediction outcomes of the models.

Apart from the time horizons, another key factor to this study is the sample rate. As the sample rate increases, it becomes better, but within a limit. Limiting the rate is important because if data is increased without range, the data will stop being accurate. This is evident, for example, if the signal is smoothed or if there is less noise. The trend changes are not so abrupt and, as such, there is not so much error. Moreover, as the prediction horizon increases, the error increases with the prediction horizon. Consequently, it is necessary to highlight some error possibilities. For example, in the absence of integration with the camera, there is a lack of information. This lack of information makes it harder for the model to anticipate a fall, since the current data can only follow the trend well. With problems in capturing the changes of the trend, there could be greater error.

It is essential to keep in mind that when comparing different models, it is imperative to look at the number of parameters for each one, as well as the time spent to develop the model and the training time. Overall, as shown in this study, the combination of different models and AI techniques has led to more precise predictions of solar radiation in the short term.

Acknowledgement. This work is supported under Grant MAC/1.1.b/278 (MICROGRID-BLUE), from the Framework of the 2nd Call of the INTERREG V-A MAC 2014–202, co-financed with FEDER funds. Too, this work is supported under the Grant CEI2021-06, from direct agreement SD-21/08 by Consejería de Economía, Industria, comercio y conocimiento from Gobierno de Canaria to ULPGC.

References

1. Candelise, C., Winskel, M., Gross, R.J.K.: The dynamics of solar PV costs and prices as a challenge for technology forecasting. Renew. Sustain. Energy Rev. **26**, 96–107 (2013)
2. Activesustanaibility. https://www.activesustainability.com/. Accessed 29 Mar 2023
3. Zhou, Y., Liu, Y., Wang, D., Liu, X., Wang, Y.: A review on global solar radiation prediction with machine learning models in a comprehensive perspective. Energy Convers. Manag. **235**, 113960 (2021)
4. Ye, H., Yang, B., Han, Y., Chen, N.: State-of-the-art solar energy forecasting approaches: critical potentials and challenges. Front. Energy Res. **10** (2022). https://doi.org/10.3389/fenrg. 2022.875790
5. Nguyen, T.-A., et al.: A recent invasion wave of deep learning in solar power forecasting techniques using ANN. In: IEEE International Future Energy Electronics Conference (IFEEC), Taipei, Taiwan, pp. 1–6 (2021). https://doi.org/10.1109/IFEEC53238.2021.9661747
6. Gardner, M.W., Dorling, S.R.: Artificial neural networks (the multilayer perceptron)—a review of applications in the atmospheric sciences. Atmos. Environ. **32**(14–15), 2627–2636 (1998). https://doi.org/10.1016/S1352-2310(97)00447-0
7. Babbar, S.M., Yong, L.C.: Short term solar power forecasting using deep neural networks. In: Arai, K. (ed.) FICC 2023. LNNS, vol. 652, pp. 218–232. Springer, Cham (2023). https://doi.org/10.1007/978-3-031-28073-3_15
8. Aicardi, D., Musé, P., Alonso-Suárez, R.: A comparison of satellite cloud motion vectors techniques to forecast intra-day hourly solar global horizontal irradiation **223**, 42–60 (2022). https://doi.org/10.1016/j.solener.2021.12.066
9. Sfetsos, A., Coonick, A.H.: Univariate and multivariate forecasting of hourly solar radiation with artificial intelligence techniques. Sol. Energy **68**(2) (2000). ISSN 0038-092X. https://doi.org/10.1016/S0038-092X(99)00064-X
10. Segovia, J.A., Toaquiza, J.F., Llanos, J.R., Rivas, D.R.: Meteorological variables forecasting system using machine learning and open-source software. Electronics **12**, 1007 (2023). https://doi.org/10.3390/electronics12041007

11. Huang, J., Korolkiewicz, M., Agrawal, M., Boland, J.: Forecasting solar radiation on an hourly time scale using a coupled autoregressive and dynamical system (CARDS) model. Sol. Energy **87**, 136–149 (2013)
12. Bamisile, O., Cai, D., Oluwasanmi, A., et al.: Comprehensive assessment, review, and comparison of AI models for solar irradiance prediction based on different time/estimation intervals. Sci. Rep. **12**, 9644 (2022). https://doi.org/10.1038/s41598-022-13652-w
13. Krishnan, N., Ravi Kumar, K., Inda, C.S.: How solar radiation forecasting impacts the utilization of solar energy: a critical review. J. Clean. Prod. **388** (2023). ISSN 0959-6526. https://doi.org/10.1016/j.jclepro.2023.135860
14. Obando, E.D., et al.: Solar radiation prediction using machine learning techniques: a review. IEEE Lat. Am. Trans. **17**(04), 684–697 (2019)
15. Van Houdt, G., Mosquera, C., Nápoles, G.: A review on the long short-term memory model. Artif. Intell. Rev. **53**, 5929–5955 (2020). https://doi.org/10.1007/s10462-020-09838-1
16. Graves, A., Fernández, S., Schmidhuber, J.: Bidirectional LSTM networks for improved phoneme classification and recognition. In: Duch, W., Kacprzyk, J., Oja, E., Zadrożny, S. (eds.) ICANN 2005. LNCS, vol. 3697, pp. 799–804. Springer, Heidelberg (2005). https://doi.org/10.1007/11550907_126
17. Gardner, M.W., Dorling, S.R.: Artificial neural networks (the multilayer perceptron)—a review of applications in the atmospheric sciences. Atmos. Environ. **32**(14–15), 2627–2636 (1998). ISSN 1352-2310. https://doi.org/10.1016/S1352-2310(97)00447-0
18. Albawi, S., Mohammed, T.A., Al-Zawi, S.: Understanding of a convolutional neural network. In: 2017 International Conference on Engineering and Technology (ICET), Antalya, Turkey, pp. 1–6 (2017). https://doi.org/10.1109/ICEngTechnol.2017.8308186
19. Huang, W., et al.: Spatio-spectral feature classification combining 3D-convolutional neural networks with long short-term memory for motor movement/imagery. Eng. Appl. Artif. Intell. **120**, 105862 (2023). ISSN 0952-1976. https://doi.org/10.1016/j.engappai.2023.105862
20. Karunasingha, D.S.K.: Root mean square error or mean absolute error? Use their ratio as well. Inf. Sci. **585**, 609–629 (2022). ISNN 0020-0255. https://doi.org/10.1016/j.ins.2021.11.036
21. Chai, T., Draxler, R.R.: Root mean square error (RMSE) or mean absolute error (MAE)? – arguments against avoiding RMSE in the literature. Geosci. Model Dev. **7**, 1247–1250 (2014). https://doi.org/10.5194/gmd-7-1247-2014
22. Mohajerin, N., Waslander, S.L.: Multistep prediction of dynamic systems with recurrent neural networks. IEEE Trans. Neural Netw. Learn. Syst. **30**(11), 3370–3383 (2019). https://doi.org/10.1109/TNNLS.2019.2891257
23. Taieb, S.B., Sorjamaa, A., Bontempi, G.: Multiple-output modeling for multi-step-ahead time series forecasting. Neurocomputing **73**(10–12), 1950–1957 (2010). ISSN 0925-2312. https://doi.org/10.1016/j.neucom.2009.11.030
24. Venkatraman, A., Hebert, M., Bagnell, J.: Improving multi-step prediction of learned time series models. In: Proceedings of the AAAI Conference on Artificial Intelligence, vol. 29, no. 1 (2015). https://doi.org/10.1609/aaai.v29i1.9590

Random Forests Model for HVAC System Fault Detection in Hotel Buildings

Iva Matetić[1][iD], Ivan Štajduhar[1,2][iD], Igor Wolf[1][iD], Darko Palaić[1][iD], and Sandi Ljubic[1,2]([✉])[iD]

[1] Faculty of Engineering, University of Rijeka,
Vukovarska 58, 51000 Rijeka, Croatia
{iva.matetic,ivan.stajduhar,igor.wolf,darko.palaic,sandi.ljubic}@riteh.hr
[2] Center for Artificial Intelligence and Cybersecurity,
University of Rijeka, R. Matejčić 2, 51000 Rijeka, Croatia

Abstract. Heating, ventilation, and air conditioning (HVAC) systems are essential for maintaining a comfortable indoor environment in modern buildings. However, HVAC systems are known to consume a lot of energy, which can account for up to 50% of a building's energy consumption. Therefore, it is important to detect and troubleshoot problems in HVAC systems timely. Fault detection and diagnosis (FDD) techniques can help with HVAC monitoring and optimizing system performance for efficient use of energy. In this paper, we demonstrate how to create efficient fault detectors using physics-based modeling and machine learning. We show how to build a simulation model of a hotel building, which we then use to sample augmented data with typical faults commonly found in HVAC systems. We train predictive models using random forests (RFs). The results suggest that RFs can be used as stand-alone detectors for FDD, albeit their performance depends heavily on the data quality.

Keywords: Fault detection and diagnosis · HVAC systems · Fan coil unit · TRNSYS · Random Forests

1 Introduction

One of the main challenges associated with heating, ventilation, and air conditioning (HVAC) systems is their high energy consumption. This can be caused by a number of factors, including poor system design and sizing, lack of maintenance, outdated equipment, and inefficient controls, to name a few. HVAC systems are an essential part of buildings but can also be a significant source of high CO_2 emissions. Furthermore, it is anticipated that as the urban population grows, there will be a high demand for HVAC systems as new buildings are constructed [9]. By addressing the abovementioned issues, building owners can ensure that their HVAC systems operate at peak efficiency and create a comfortable indoor environment.

I. Rojas et al. (Eds.): IWANN 2023, LNCS 14134, pp. 654–665, 2023.
https://doi.org/10.1007/978-3-031-43085-5_52

The possible solution to typical HVAC problems is applying fault detection and diagnosis (FDD) methods. There are many approaches to these methods, which can be categorized as: physics-based, data-driven, knowledge discovery, and hybrid [7]. The data-driven approach can be easier to implement than other approaches because it does not require much information about the building, expert knowledge in the field of energetics and thermodynamics, and is not as complex to develop.

Data-driven techniques are usually based on machine learning (ML) algorithms. In the field of HVAC systems, ML-based solutions designed for FDD have proven effective in providing valuable insights into HVAC system performance, enabling building operators to optimize system settings and improve overall building performance. These solutions provide robust and reliable FDD models for HVAC systems that can improve energy efficiency, reduce maintenance costs and extend the life of HVAC equipment.

Some notable examples of data-driven solutions for FDD in HVAC systems are based on the following models: the ensemble of k-nearest neighbor (KNN), support vector machine (SVM), and random forests (RF) [5], generative adversarial networks coupled with SVM [12], multi-class neural network [3], domain adversarial neural network [14], isolation forests [13], RF [1,2,6,11], and deep recurrent neural networks [10]. In our work, we place emphasis on RF models that have demonstrated very good performance in various implementations. As demonstrated in [1,2], RFs were used as a main classifier and had the best performance when compared to KNN, SVM, and decision trees (DT). RFs have also been used multiple times in combination with other models – such as SVM – where feature extraction was mainly done by RF because of its low computational expense and capability of working with high dimensional data [6,11].

In this paper, we wanted to utilize the possibilities of RF modeling for FDD over a dataset that predominantly contains information from the real-world HVAC system of a hotel building. We have developed RF-based detectors that are designed to detect four types of problems commonly found in hotel HVAC systems, specifically related to Fan Coil Units (FCU). These common defects include a stuck control valve in fan coil units ($F1$), a clogged air filter or a fan problem ($F2$), a total system failure ($F3$), and a faulty window status sensor ($F4$). It is important to point out that we have assumed a 50% stuck position for $F1$ and $F2$. Moreover, we looked for a way to preprocess large amounts of data to reduce computational effort and to establish temporal relationships between data points, which has not been addressed in previous research regarding RF models.

2   Materials and Methods

The development of the predictive models and the experimental setup were done in Python using Pandas library for data cleaning and preprocessing and Scikit-learn library for machine learning. Finally, TRNSYS was used to artificially augment collected data by inducing occurrences of four distinct faults. In the next subsections, we present the research and development pipeline details.

2.1 Measured Data

In this work we used data from a hotel in Zagreb (Croatia) gathered within a study to optimize the system for efficient energy consumption, while retaining or even improving user comfort. Our FDD method is to be implemented as part of the hotel's smart room concept, which monitors all sensors and controls the room temperature by setting the *Set_temp* accordingly. For more information about the related smart room concept, we refer the reader to our previous work [8]. Data was collected for 166 hotel rooms between years 2013 and 2021 with a sampling rate of five minutes. For this reason, not all possible events are included, but it can still give an insight into how the system works under certain conditions. Although the original dataset contains data for all 166 rooms, in our research, we limited ourselves to the largest group of rooms with similar characteristics, 100 in total. A complete list of all features gathered is shown in Table 1.

Table 1. Dataset features and their respective descriptions and data types.

Feature	Description	Data type
Datetime	Date and time	datetime
Hvac_mode	1 = FCU heating/0 = FCU cooling	binary
Hvac_state	1 = FCU on/0 = FCU off	binary
Room_occupancy	1 = room occupied/0 = room unoccupied	binary
Window	1 = window open/0 = window closed	binary
Hvac_state_manual	1 = HVAC control guest regulated	binary
FS_0	1 = FCU off	binary
FS_1	1 = FCU speed 1	binary
FS_2	1 = FCU speed 2	binary
FS_3	1 = FCU speed 3	binary
Orientation_S	1 = room orientation south	binary
Orientation_W	1 = room orientation west	binary
Orientation_N	1 = room orientation north	binary
Orientation_E	1 = room orientation east	binary
Is_weekday	1 = current day weekday	binary
Set_temp	Set temperature of a room ($^\circ C$).	int
Room_temp	Measured room temperature ($^\circ C$)	int
Room_temp_up	Measured temp. of room upstairs ($^\circ C$)	int
Room_temp_down	Measured temp. of room downstairs ($^\circ C$)	int
Room_temp_cw	Measured temp. of right-neighboring room ($^\circ C$)	int
Room_temp_ccw	Measured temp. of left-neighboring room ($^\circ C$)	int
Outside_temp	Outside weather temperatures ($^\circ C$)	float
Humidity	Relative air humidity (%)	float
Irradiation	Solar energy reaching a surface (W/m^2)	float

During the manual inspection of the collected data, we noticed that some portions recorded during earlier years involved a significant number of missing or noisy data. Moreover, the pandemic period in 2020 was not included, as this period does not reflect the normal operating environment of the hotel. Therefore, we have limited our analysis only to the data recorded from 2015 to 2019 and during 2021.

2.2 Augmenting and Labeling Measured Data

TRNSYS software was employed to construct a physical model of the system in order to generate labeled data with anomalies. This model was developed using the architectural descriptions of the building. This model simulates the thermal behavior of one part of the building consisting of six thermal zones, each representing a guest room. The central thermal zone is monitored in detail, and the other thermal zones are created to achieve boundary conditions. The model can be used to simulate any room within the building since the physical properties and functions of each room are consistent. To ensure the accuracy of the simulation, the model was calibrated and validated using real-world data from the hotel building. Additionally, the simulation environment was designed to replicate the conditions in the rooms, such as the desired temperature, window openings, and occupancy. Furthermore, the FCU was implemented with the same control logic as the building and can provide either heating or cooling as needed. Data inputs to the model include weather data collected from a nearby weather station as well as fault insertion to activate faults at specific times. The outputs of the model are the air temperature of the room ($Room\_temp$) and the speeds of the FCU (FS_0 – FS_3). As illustrated in Fig. 1, the thermal zones defined in Google SketchUp 3D depict the physical model of the system.

Subsequently, the physical model was used to generate and label data with marked faulty periods of the system. The simulations were conducted using the data collected from the building. The first simulations were run under normal conditions, wherein no faults were inserted, resulting in a healthy dataset. In

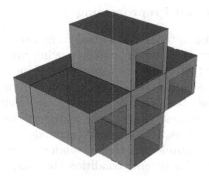

Fig. 1. Geometric shapes of thermal zones in Google SketchUp 3D. The thermal response of the central room is influenced by its neighboring rooms.

subsequent simulations, problems of varying lengths that resulted in faulty system operation were inserted in the system at randomly picked moments. This was done by generating interchanging continuous spans of normal and faulty periods. The respective lengths of normal and faulty periods were uniformly sampled from specific time spans, and each simulation had unique generated periods. In normal periods, there were no conditions under which faults could occur. In faulty periods, problems were inserted into the system so that the system would behave anomalously. For fault $F1$, the length of both the normal and faulty periods was arbitrarily selected to be between 20 and 40 days. For fault $F2$, the normal and faulty periods' lengths were chosen from ranges 10–30 days and 10–40 days, respectively. For $F3$, which typically occurs less frequently, we used a 5–15 days range for the normal period, and a 1–10 days range for the faulty period. Lastly, for fault $F4$, the length of the normal period was picked between 3 days and 7 days, whereas the length of the faulty period was determined at random between 1 h and 24 h. The duration of the periods was set to allow sufficient time for faults to occur. Since faults are not always detected immediately, a comparison was made between the healthy and faulty datasets based on air temperature ($Room\_temp$). This is because certain conditions must be met for faults to be detected. For example, faults cannot be detected if the FCU system is not working or operating at lower power. Therefore, only periods with large deviations in air temperature were marked as fault periods. In this way, the periods where faults were inserted in simulation could be labeled as truly faulty periods.

Four types of faults were included in the model to simulate possible problems with the system. The first fault ($F1$) represents a fan coil valve stuck at 50%, the second fault ($F2$) reflects reduced airflow in the room where the fan is providing 50% less air to the room, the third fault ($F3$) represents a total failure of the system (FCU is not working), and the fourth fault ($F4$) shows that the window sensor is not working. These faults can be identified by temperature deviations, as they affect both the heating and the cooling system, as well as the sensing equipment.

2.3   Data Cleaning and Preprocessing

The dataset, comprising data on 100 rooms, from 2015 to 2019, and 2021, was subjected to a rigorous cleaning process, removing missing data and outliers, such as Not-a-Number (NaN) data or zero values of the set temperature feature (Set_temp) which represent an error in sensory measurements.

Next, to ensure that temporal relationships between data points are preserved, we used a 12-h sliding window with a step size of 6 h as a preprocessing step to extract both positive and negative (anomaly present/absent) examples from the dataset. The example of a sliding window state can be seen in Fig. 2.

In addition, we added a specific condition that skips windows whose index is non-monotonous. With this condition, we excluded transition periods in the output data by minimizing the impact of different seasonal variations or different

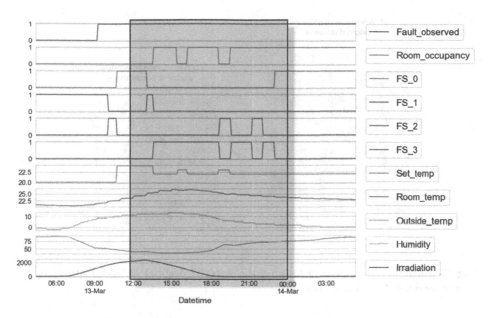

Fig. 2. A 12-h sliding window used in the data preprocessing procedure. Only a few of the used features are visualized for illustration purposes.

room sensory measurements. Furthermore, each window of 144 rows (corresponding to 12 h of data sampled every 5 min) and 23 feature values (*Datetime* value was left out) was compressed into a single vector by combining the values for each feature. This resulted in a reduction in memory usage and computational costs while preserving the important temporal relationships between features.

Figure 3 illustrates the preprocessing step of compressing windowed data features into a new set of summarized features. The illustration shows an example time window that contains a series of continuous values for outdoor temperature (*Outside_temp*), and two series of binary values indicating room occupancy (*Room_occupancy*) and fault detection in the HVAC system (*Fault_observed*), respectively. The size of the window is arbitrary. Descriptive statistics functions are applied to the values from the window of selected original features. Namely, the minimum value, maximum value, sum of values, mean value and standard deviation, and median are calculated for features with continuous values. For binary-typed features, only the minimum, maximum, and number of sequences containing consecutive values of one (*groups*) are determined. It is important to note that we used only the maximum value of the *Fault_observed* feature in the compressed window. This means the corresponding vector would be flagged as faulty even if only a single anomaly was detected between two HVAC system state samples 5 min apart. Following the described procedure, the resulting window was reduced from 144 rows and 23 features to a single vector consisting of 96 features.

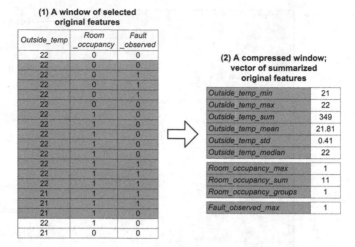

Fig. 3. An illustration of transforming the selected original dataset features to the compressed window of new features.

2.4   Model Development

In the search for the best classification model, our aim was to experiment with traditional models: RF, SVM, DT, logistic regression, and multilayer perceptron. Early on, RF models performed better and were less computationally intensive. This corresponds to the recent findings presented in [4]. For this reason, we focused exclusively on the development of RF models. We aimed to develop four separate classification models – one for each simulated fault. In the following text, we will use terms such as model $M1$ in correspondence to a dataset with generated $F1$, and so on. Furthermore, RF modeling has some advantages over other modeling techniques in that it handles high-dimensional datasets and noisy data well. These advantages are important because sensory measurements are usually subject to a lot of noise and outliers.

To mitigate the class imbalance problem, we set the number of faults in the data during the TRNSYS simulations relatively high. Compared to the real world, faults occurred less frequently than we needed, and so we went overboard with labeling faults to get a representative dataset with distributed faults that the model could learn from. Additionally, we conducted iterative experiments to determine the optimal cost function adjustments through trial and error. Thus we achieved better results in the detection of individual anomalies. We applied sample weights with a ratio of 1 : 10, giving ten times higher priority to faulty classes. Next, the hyperparameters were determined through empirical analysis using a dataset of simulated fault $F2$, which was chosen because it exhibited the most balanced fault distribution. These same hyperparameters were then applied to other models. Hyperparameters set up were the maximum depth of the trees at 30, the number of trees at 100, and the maximum features parameter at 0.35. The maximum depth of the trees was the most impactful parameter in improving model performance. Although we obtained better results using deeper trees, there were still some potential issues concerning computational cost and overfitting. The number of trees parameter was set to a relatively small value to overcome these issues since its impact on model performance was negligible. Next, the potential overfitting of the model was controlled by the maximum feature parameter, which added randomness to the model (bagging with random subsampling). Its value corresponds to a random subset of 35% of the total features for the model to consider when doing region partitioning.

In all our tests, we used leave-one-year-out cross-validation to asses the generalization properties of the developed models. This particular cross-validation is implemented so that there is no overlap (data leakage) between our data from different years. We evaluated model performance using a confusion matrix, receiver operating characteristic (ROC) curve, area-under-the-curve (AUC) score, accuracy, precision, recall, and F1 score, which are standard metrics for binary classification tasks.

3   Results and Discussion

Class distribution of the final datasets was as follows for each respective fault: $F1$ (0: 89.72%, 1: 10.28%), $F2$ (0: 73.34%, 1: 26.66%), $F3$ (0: 71.29%, 1: 28.71%), and $F4$ (0: 83.85%, 1: 16.15%). Experimental results are shown in Table 2 and in Fig. 4.

Table 2 shows the final mean and standard deviation values for all 6 cross-validation tests of accuracy, precision, recall, F1, and AUC score. The ROC curves for training and test data and the confusion matrix for all models are shown in Fig. 4. The results indicate that $M3$ has the best performance with an AUC of 0.91, followed by $M2$ with an AUC of 0.87. $M4$ has an AUC of 0.86, while $M1$ is considerably worse than the previously mentioned models with an AUC of 0.74.

We have a highly accurate model, $M3$, which detects complete system failure ($F3$), and comparably good models, $M2$, which detects airflow blocked at 50% ($F2$), and $M4$, which detects window status sensor failure ($F4$). The recall rates of these models are 0.84, 0.78, and 0.73, respectively, indicating that $M3$

Table 2. Simulation study results. A model for each fault was evaluated using classification accuracy, precision, recall, F1 score, and area under the ROC curve (AUC). The mean cross-validation score and its respective standard deviation are shown for each metric-model pair.

Model	Accuracy	Precision	Recall	F1	AUC
$M1$	0.93 ± 0.02	0.77 ± 0.07	0.51 ± 0.04	0.61 ± 0.01	0.74 ± 0.01
$M2$	0.92 ± 0.03	0.89 ± 0.03	0.78 ± 0.03	0.83 ± 0.01	0.87 ± 0.01
$M3$	0.94 ± 0.01	0.94 ± 0.03	0.84 ± 0.02	0.89 ± 0.01	0.91 ± 0.01
$M4$	0.95 ± 0.01	0.93 ± 0.05	0.73 ± 0.03	0.82 ± 0.02	0.86 ± 0.01

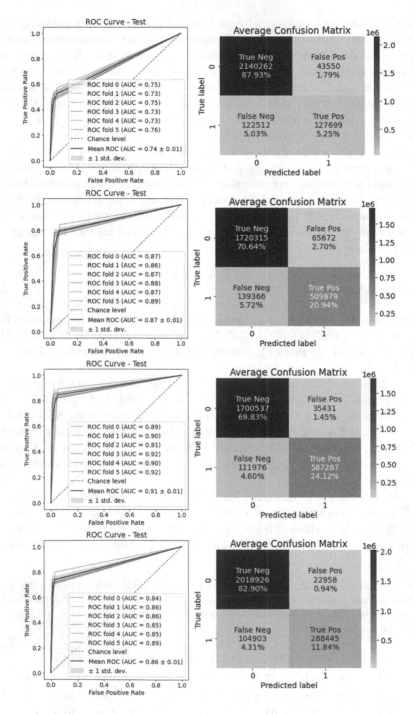

Fig. 4. ROC curves calculated from the test data and the mean confusion matrices for models $M1$ to $M4$ (from top to bottom, respectively).

successfully detects the faulty occurrence 84% of the time, $M2$ detects it 78% of the time, and $M4$ detects it 73% of the time. However, $M1$'s performance in detecting fan coil valve stuck at 50% ($F1$) is the worst, with a recall rate of 0.51, suggesting that the model correctly detects only 51% of the observed defects. A closer look shows that its highly imbalanced dataset greatly impacted the model's classification performance.

The results indicate that all four RF models achieved high accuracy (ranging from 0.92 to 0.95) in detecting faults. The observed differences can be attributed to the datasets' diverse nature of simulated faults. Therefore, it is important to

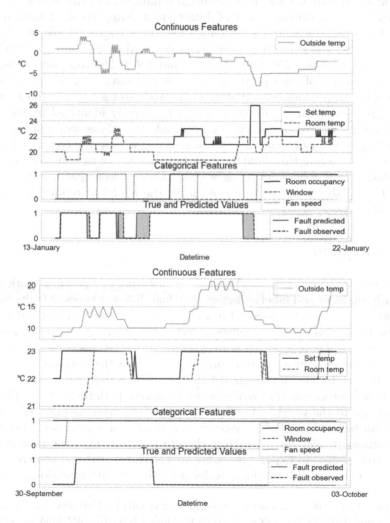

Fig. 5. Two cherry-picked examples of $M3$ performance (fault predicted vs. fault observed), one illustrating commonly occurring detection errors (top) and the other illustrating flawless detection (bottom). Only selected features are displayed, and their values correspond to the values obtained after feature extraction.

664   I. Matetić et al.

employ comprehensive fault simulation techniques, encompassing a wide range of fault scenarios, to improve the models' robustness and generalization capabilities.

One crucial factor to consider is the difficulty of detecting the $F1$ fault, which can be characterized as subtle. Namely, subtle faults pose challenges for detection algorithms since they may not exhibit substantial variations in output temperatures compared to normal instances. The lack of significant differences in output temperatures between fault and non-fault conditions further complicates the detection task.

In Fig. 5, we show how $M3$ performs on sensory measurements of one room. The features shown are not from the original dataset, but the ones we obtained after feature extraction as a part of data preprocessing. We used median values for continuous features and maximum values for categorical features for visualization. In addition, the fan speed was calculated by combining the values of the following features: FS_0, FS_1, FS_2, and FS_3. The top part of the figure shows an example of false positives and false negatives. Our models often misclassified the transitions immediately before and after a fault. The bottom part of the figure shows an example of the fault correctly identified by the model. If we consider both situations, we can conclude that these shortcomings can be further improved by determining the target label with a new controlled variable that would serve as a threshold for classifying a window as faulty or non-faulty. This should be a good solution as we have already seen that a lack of control of the target label can lead to problems for the model to report anomalies on non-faulty data.

4   Conclusion

To conclude, we have used the TRNSYS model from our previous work to synthetically augment and label a dataset with four different types of faults induced. The dataset was preprocessed with a sliding window, and we derived new features based on descriptive statistics, which we used to summarize each dataset window into a characteristic feature vector. This step was critical because we wanted RF to recognize the temporal trends within the datasets and run efficiently by not taking too much RAM. Four RF models were then developed for each fault type, and the results showed different performances, ranging from the best $M3$ to the worst $M1$. We have successfully demonstrated that the RF model can be used in HVAC systems as a stand-alone fault detector. Still, it needs consistent, high-quality labeled data to achieve the best performance. Our future work plan is to expand research with convolutional and recurrent neural networks, as we are still reaching performance limits with traditional models.

Acknowledgements. This work was supported in part by European Regional Development Fund (ERDF) under grant agreement number KK.01.2.1.02.0303, project *Adria Smart Room*.

References

1. Abdollah, M.A.F., Scoccia, R., Aprille, M.: Data driven fault detection and diagnostics for hydronic and monitoring systems in a residential building. J. Phys: Conf. Ser. **2385**(1), 012012 (2022). https://doi.org/10.1088/1742-6596/2385/1/012012
2. Aldrich, C., Auret, L.: Fault detection and diagnosis with random forest feature extraction and variable importance methods. IFAC Proc. Volumes **43**(9), 79–86 (2010). https://doi.org/10.3182/20100802-3-ZA-2014.00020
3. Dey, M., Rana, S.P., Dudley, S.: Smart building creation in large scale HVAC environments through automated fault detection and diagnosis. Futur. Gener. Comput. Syst. **108**, 950–966 (2020). https://doi.org/10.1016/j.future.2018.02.019
4. Grinsztajn, L., Oyallon, E., Varoquaux, G.: Why do tree-based models still outperform deep learning on typical tabular data? In: 36th Conference on Neural Information Processing Systems (NeurIPS 2022) Track on Datasets and Benchmarks (2022)
5. Han, H., Zhang, Z., Cui, X., Meng, Q.: Ensemble learning with member optimization for fault diagnosis of a building energy system. Energy Build. **226**, 110351 (2020). https://doi.org/10.1016/j.enbuild.2020.110351
6. Masdoua, Y., Boukhnifer, M., Adjallah, K.H.: Fault detection and diagnosis in AHU system with data driven approaches. In: 8th International Conference on Control, Decision and Information Technologies (CoDIT), pp. 1375–1380. IEEE (2022). https://doi.org/10.1109/CoDIT55151.2022.9803907
7. Matetić, I., Štajduhar, I., Wolf, I., Ljubic, S.: A review of data-driven approaches and techniques for fault detection and diagnosis in HVAC systems. Sensors **23**(1), 1 (2022). https://doi.org/10.3390/s23010001
8. Matetić, I., Štajduhar, I., Wolf, I., Palaić, D., Ljubic, S.: Data visualization tool for smart buildings HVAC systems. In: Daimi, K., Al Sadoon, A. (eds.) ICR 2022. Advances in Intelligent Systems and Computing, vol. 1431. Springer, Cham (2022). https://doi.org/10.1007/978-3-031-14054-9_41
9. Pérez-Lombard, L., Ortiz, J., Pout, C.: A review on buildings energy consumption information. Energy Build. **40**(3), 394–398 (2008). https://doi.org/10.1016/j.enbuild.2007.03.007
10. Taheri, S., Ahmadi, A., Mohammadi-Ivatloo, B., Asadi, S.: Fault detection diagnostic for HVAC systems via deep learning algorithms. Energy Build. **250**, 111275 (2021). https://doi.org/10.1016/j.enbuild.2021.111275
11. Tun, W., Wong, J.K.W., Ling, S.H.: Hybrid random forest and support vector machine modeling for HVAC fault detection and diagnosis. Sensors **21**(24), 8163 (2021). https://doi.org/10.3390/s21248163
12. Yan, K., Chong, A., Mo, Y.: Generative adversarial network for fault detection diagnosis of chillers. Build. Environ. **172**, 106698 (2020). https://doi.org/10.1016/j.buildenv.2020.106698
13. Zeng, Y., Chen, H., Xu, C., Cheng, Y., Gong, Q.: A hybrid deep forest approach for outlier detection and fault diagnosis of variable refrigerant flow system. Int. J. Refrig **120**, 104–118 (2020). https://doi.org/10.1016/j.ijrefrig.2020.08.014
14. Zhu, X., Chen, K., Anduv, B., Jin, X., Du, Z.: Transfer learning based methodology for migration and application of fault detection and diagnosis between building chillers for improving energy efficiency. Build. Environ. **200**, 107957 (2021). https://doi.org/10.1016/j.buildenv.2021.107957

Time Series Forecasting with Quantum Neural Networks

M. P. Cuéllar[1]($\boxtimes$) (ID), M. C. Pegalajar[1] (ID), L. G. B. Ruiz[2] (ID), and C. Cano[1] (ID)

[1] Department of Computer Science and Artificial Intelligence,
University of Granada, Granada, Spain
{manupc,mcarmen,bacaruiz}@decsai.ugr.es
[2] Department of Software Engineering, University of Granada, Granada, Spain
http://www.ugr.es

Abstract. In this work we explore the use of Quantum Computing for Time Series forecasting. More specifically, we design Variational Quantum Circuits as the quantum analogy of feedforward Artificial Neural Networks, and use a quantum neural network pipeline to perform time series forecasting tasks. According to our experiments, our study suggests that Quantum Neural Networks are able to improve results in error prediction while maintaining a lower number of parameters than its classical machine learning counterpart.

Keywords: Quantum Neural Networks · Quantum Machine Learning · Time Series Forecasting

1 Introduction

Quantum Computing (QC) [15] was born in 1982 after Richard Feynman pointed out the complexity of simulating a quantum system with a classic computer. Since then, QC has been growing as a research area until nowdays, where contemporary applications of QC are varied and include cryptography, finance, game theory, chemical modelling, or machine learning [5,10,12,17], to mention just a few. The recent reality of QC hardware and the existence of quantum computer simulators able to run in classical computers have contributed significantly to improve the state-of-the-art in QC, although quantum supremacy (understood as a significant speedup from exponential time to polynomial time) have not been achieved but for a handful of applications such as search in unordered sets in $\mathcal{O}(\sqrt{n})$ with Grover's search, finding if a function is balanced or not with Deutsch-Jozsa method, or integer factorization with Shor's algorithm [15], to mention the most sounded examples.

This article was supported by the project QUANERGY (Ref. TED2021-129360B-I00), Ecological and Digital Transition R&D projects call 2022, Government of Spain, and Grant PID2021-128970OA-I00 funded by MCIN/AEI/10.13039/501100011033/FEDER.

I. Rojas et al. (Eds.): IWANN 2023, LNCS 14134, pp. 666–677, 2023.
https://doi.org/10.1007/978-3-031-43085-5_53

In we focus on Quantum Machine Learning (QML) [4], all types of supervised [11], unsupervised [13], and reinforcement learning [1,6] tasks have been explored with this new computing paradigm. Most of the approaches show the benefits of QML to solve these tasks, and often they provide benefits in either time complexity and/or performance. In this work, we adopt the supervised learning paradigm under QC, and propose a Quantum Neural Network [8] to solve tasks of time series forecasting, with the goal of performing a preliminar evaluation of the scope of QC to tackle this type of problem.

A Time Series is a sequence of measurements or observations of a given phenomenon, sampled periodically and indexed in time. Time series forecasting is an ubiquitous problem to almost all areas in science, and attempts to predict future values of the data series $x(t+1)$ with historical time series data $x(t), x(t-1), x(t-2), ...$, and a hypothesis model f often parameterized with parameters θ, i.e. $x(t+t) = f(x(t), x(t-1), x(t-2), x(t-3), ..., \theta)$. Although the number of models used for time series forecasting is wide and the proposals come from different areas (statistics, electronics, computer science, economics, etc.), in this work we focus in the special case of neural networks for forecasting [16]. Both feedforward and recurrent neural network models have been extensively tested in a wide variety of problems, as it is described in the survey [3].

On the other hand, if we focus on the problem of time series forecasting with Quantum Computing, the reference literature is scarce and very few research articles have addressed the problem. In particular, the work [2] proposed a new neural network model whose computation units are inspired in quantum amplitude and phase operations, and applied the proposal for stock market forecasting. Lately, in [14] it is proposed an adaptation of the Quantum-inspired Optimization Algorithm (QOA) for fuzzy sets, and the approach was tested over the TAIFEX stock market time series and temperature time series, among others. We remark that these two approaches are not purely from QC, although they are inspired by elements of QC to build classical models. Pure QC approaches are the work [9], that proposes a hybrid classical/quantum neural network containing layers from both computing paradigms and trained the model to predict the Sun Spot time series; and the article [7] which develops a framework for quantum machine learning temporal tasks using reservoir computing and quantum neural networks (QNNs), applied to S&P 500 stock market time series prediction problems. In these cases, all papers in the literature report an increase in performance/accuracy of the QC models with respect to classic computing ones.

In this manuscript, we design a pure QC neural network using Variational Quantum Circuits, and test the approach in classic benchmark time series prediction datasets. Since loops are not allowed in a quantum algorithm, the designed quantum neural network has a feedforward structure, and contribute to the existing literature by means of the proposal of how to encode time series data into quantum states, the QNN design and measurement of results, and a proof-of-concept experimentation to assess limitations and future possible ways to address the problem of scalability.

This work is structured as follows: Sect. 2 makes an overview of QC and QML to make this article self-contained. After that, Sect. 3 review the concept of Quantum Neural Network and describes our approach. Then, Sect. 4 describes the experimentation performed, and Sect. 5 concludes.

2   Quantum Computing and Machine Learning

This section contains a brief introduction to Quantum Computing and Quantum Machine Learning for article self-completeness. The main references used to write this section, and for further information, are [10,15] for QC and [5] for QML.

There is no need to argue that Quantum Computing is a completely different computer programming paradigm to the traditional (classic) computer programming. However, we may find some common elements in both: Classic computing is based on computation over binary digits (bits), and the output of a classic computing algorithm is a set of n bits with values in $\{0,1\}^n$ whose underlying mathematical model is $\mathbb{Z}_2^n$, i.e. the cartesian product of $\mathbb{Z}_2$ n times or $\times^n \mathbb{Z}_2$. Similarly, QC is based on operations over quantum binary digits (qubits), and the output of a QC system is also a set of n binary values in $\{|0\rangle, |1\rangle\}$. However, the underlying mathematical model of QC is a vector subspace of $\mathbb{C}^{2^n}$ obtained by means of the tensor product of $\mathbb{C}^2$ n times, i.e. $\bigotimes^n \mathbb{C}^2$. The values $|0\rangle = (1,0)^t$ and $|1\rangle = (0,1)^t$ are the orthonormal basis column vectors of the vector subspace of a system with one qubit, also refered to as the *computational basis*. For systems with a larger number of qubits, the computational basis is calculated by means of the tensor product of the qubit basis vectors, as for instance $\{|0\rangle \otimes |0\rangle = |00\rangle = (1,0,0,0)^t, |0\rangle \otimes |1\rangle = |01\rangle = (0,1,0,0)^t, |1\rangle \otimes |0\rangle = |10\rangle = (0,0,1,0)^t, |1\rangle \otimes |1\rangle = |11\rangle = (0,0,0,1)^t\}$ for the case of 2 qubits. It is worth noting that, when a new bit is included into a classical system, the dimension of the computational space increases by one as a consequence of the cartesian product operation; however, when a new qubit is included into a quantum system the dimension of the computational space doubles its size.

As a member of a vector subspace in $\mathbb{C}^2$, the value of an arbitrary qubit $|\psi\rangle$ can be modelled as a linear complex combination of the basis states $\{|0\rangle, |1\rangle\}$ as $|\psi\rangle = \alpha_0|0\rangle + \alpha_1|1\rangle, \alpha_i \in \mathbb{C}$ with the additional constraint that $\sum_i |\alpha_i|^2 = 1$, so that a qubit can potentially hold an infinite number of values. The coefficients α_i are called *amplitudes* and, when the user retrieves the output of a quantum algorithm through the *measurement* operator, the qubit collapses to value $|0\rangle$ with probability $|\alpha_0|^2$ or to value $|1\rangle$ with probability $|\alpha_1|^2$. This also holds for systems with a larger number of qubits, as for instance $|\psi\rangle = \alpha_0|00\rangle + \alpha_1|01\rangle + \alpha_2|10\rangle + \alpha_3|11\rangle$ for a 2-qubit system.

Operations in a classic algorithm are implemented at a fundamental level using a sequence of logic gates such as AND, NOT, OR, etc., whose mathematical model relies on the addition and product over the field with two elements $\mathbb{Z}_2$. Similarly, a quantum algorithm can also be implemented using quantum gates; however, these gates are a reversible linear transformation over the complex space $\mathbb{C}^{2^n}$ and are modelled as unitary matrices that multiply the quantum state. For example, the σ_x gate/operation in Eq. 1 is the quantum analog to the NOT classical gate, and its application over qubits with values $|0\rangle$ or $|1\rangle$ is written as $\sigma_x|0\rangle = |1\rangle$ or $\sigma_x|1\rangle = |0\rangle$, respectively. Another example is the Hadamard (H) gate in Eq. 2, where $H|0\rangle$ moves the qubit into the standard superposition of the basis states $|\psi\rangle = \frac{\sqrt{2}}{2}|0\rangle + \frac{\sqrt{2}}{2}|1\rangle$. A final example of a gate with 2 inputs is shown in Eq. 3 and it corresponds to the Controlled-NOT (CNOT) gate which switches the second qubit from $|0\rangle$ to $|1\rangle$ or vice versa if the first qubit is $|1\rangle$, of leaves it unchanged otherwise. In the general case, for an arbitrary 2-qubit state $|\psi\rangle = \alpha_0|00\rangle + \alpha_1|01\rangle + \alpha_2|10\rangle + \alpha_3|11\rangle$, the CNOT exchanges the amplitudes of the last basis vectors so that $CNOT|\psi\rangle = \alpha_0|00\rangle + \alpha_1|01\rangle + \alpha_3|10\rangle + \alpha_2|11\rangle$.

$$\sigma_x = \begin{pmatrix} 0 & 1 \\ 1 & 0 \end{pmatrix} \tag{1}$$

$$H = \frac{1}{\sqrt{2}} \begin{pmatrix} 1 & 1 \\ 1 & -1 \end{pmatrix} \tag{2}$$

$$CNOT = \begin{pmatrix} 1 & 0 & 0 & 0 \\ 0 & 1 & 0 & 0 \\ 0 & 0 & 0 & 1 \\ 0 & 0 & 1 & 0 \end{pmatrix} \tag{3}$$

Of special interest to our work are the set of parameterized gates, i.e. gates whose behaviour depends of input parameters θ, such as the rotation gates $R_x(\theta), R_y(\theta), R_z(\theta)$, whose unitary matrices for one qubit are described in Eqs. 4–6, respectively. These gates allow to change the output of a quantum algorithm depending on the parameter θ, and are the fundamental block to build the quantum neural network used in this work.

$$R_x(\theta) = \begin{pmatrix} cos(\theta/2) & -isin(\theta/2) \\ -isin(\theta/2) & cos(\theta/2) \end{pmatrix} \tag{4}$$

$$R_y(\theta) = \begin{pmatrix} cos(\theta/2) & -sin(\theta/2) \\ sin(\theta/2) & cos(\theta/2) \end{pmatrix} \tag{5}$$

$$R_z(\theta) = \begin{pmatrix} e^{-i\theta/2} & 0 \\ 0 & e^{i\theta/2} \end{pmatrix} \tag{6}$$

Quantum algorithms are described as a sequence of operations over a quantum state, as for instance the operations $CNOT((H|q_0\rangle) \otimes |q_1\rangle)$, and are implemented into quantum circuits. Figure 1 shows the implementation of the previous algorithm, which starts from state $|q_0 q_1\rangle = |00\rangle$ and measures state $|00\rangle$ or $|11\rangle$ with probability 0.5 into classical bits (line c). In a circuit, each horizontal line is assigned to a single qubit, and gates are organized sequentially until measurement. Thus, loops are not allowed in a quantum program. A special case of a quantum circuit is the Variational Quantum Circuit (VQC), whose main feature is that it contains parameterized gates such as the aforementioned $R_x(\theta), R_y(\theta), R_z(\theta)$.

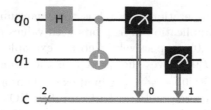

Fig. 1. Example of circuit implementing the algorithm $CNOT((H|q_0\rangle) \otimes |q_1\rangle)$.

Quantum Machine Learning has been a research area of growing interest for the last two decades; however, it has obtained a special focus in the last few years thanks to the advances in Quantum Computer Hardware and Quantum Computing simulators. The main goal of QML is to design and implement methods able to run in a quantum computer to solve the traditional supervised, unsupervised and reinforcement learning tasks of classic Machine Learning, taking advantages of quantum operations that are not present in a classic computer such as superposition, tunneling, entanglement, or quantum parallelism, coming from Quantum Computing (QC). In this work, we focus on the case of Quantum Neural Network (QNN) design. A QNN is the quantum analog of a classic neural network. Each layer of a QNN is a VQC containing parameterized gates, where the parameters are the quantum analog of the classic network weights. It also contains a mechanism to transfer information among the existing qubits as an analogy to a classic connection between neurons of different layers. Usually, this information transfer is implemented as entanglements using operators such as the CNOT gate.

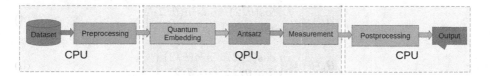

Fig. 2. Usual pipeline of Quantum Machine Learning, where CPU stands for operations performed on a classical computer and QPU operations over a quantum hardware.

The process to create a QML model usually involves the following steps (see Fig. 2): First, the dataset is loaded and preprocessed into a classical CPU. After that, the classic data are encoded into quantum states on a quantum hardware of QPU using a quantum embedding technique. Once the classic data has been represented into quantum states, the core model implemented in the antsatz is executed and its results measured into classical bits. Finally, these results are post-processed if necessary in CPU to provide the expected model output. In this work, we follow this general pipeline to study how a Quantum Neural Network can be used for time series forecasting.

3 Variational Quantum Circuits for Forecasting

A Quantum Neural Network [8] can be typically organized as a sequence of layers:

- The **input layer**, in charge of transforming the classic input data into a quantum state.
- The **antsatz**, containing a Variational Quantum Circuit whose structure is concatenated L times to create the quantum analog of L network layers.
- The **output layer**, which performs measurement operations over qubits to return the expected outcome.

The **input layer** is usually implemented as a parameterized variational circuit with rotation and controlled-rotation gates that help to set the desired quantum state for a given input classic data. This process is called the *quantum embedding* procedure, and it encompasses a set of techniques such as basis encoding, amplitude encoding, hamiltonian encoding, or tensor product encoding, to mention just a few. In this work, we use the tensor product encoding consisting of a single X-rotation gate for each qubit, where the gate parameter is the classic data scaled to $[-\pi, \pi]$. This is a simple and fast encoding technique which requires operations in $O(1)$ to perform the creation of a quantum state; however, it has the limitation that the number of qubits must increase with the number of input classic data linearly. In addition, quantum embedding can be influenced by the bias and scale of the input dataset and, for that reason, we have modified the tensor product classic scheme to include further learnable parameters to scale and bias the input data. Figure 3 shows an example of the input layer for a network containing 3 qubits. Values I_i are the classic data features, θ_i are the input scale parameter and b_i the bias parameter.

With respect to the **antsatz**, we may notice that the literature does not offer a set of fixed quantum layer structures as there are in the classic neural network domain (fully connected, recurrent, etc.). The number of possible gates used for quantum information transfer between qubits is wide, and the organization of these gates to make the data transfer has not been extensively studied yet. In this work, we use the Real Amplitudes antsatz which has been used previously in other domains with success such as policy estimation for quantum reinforcement

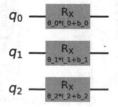

Fig. 3. Tensor product embedding for a classic data containing 3 features in $[-\pi, \pi]$.

learning and classification. The antsatz starts with full rotation X/Y/Z parameterized gates as the quantum analog of connection weights, followed by a set of
CNOT gates organized with a ring structure for the qubit information transfer.
Figure 4 shows the implementation of the described antsatz as the analogy of a
quantum network layer for a 3-qubit network. Thus, a quantum network layer in
our work contains a number of $3 * n$ parameters, where n stands for the number
of qubits.

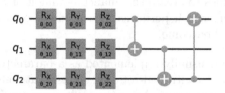

Fig. 4. *Real Amplitudes* antsatz for a 3-qubit network.

$$\sigma_z = \begin{pmatrix} 1 & 0 \\ 0 & -1 \end{pmatrix} \tag{7}$$

With respect to the **output layer**, *measurement* is often performed over
a selected observable. A typical observable is the σ_z operator over the computational basis (see Eq. 7). The network output can be calculated by means of
the expectation of the observable over a quantum state, i.e. $\langle \psi | \sigma_z | \psi \rangle$, where $\langle \psi |$
stands for the conjugate transpose of $| \psi \rangle$, so that the output is in the range
$[-1, 1]$. This must be taken into account for the QML system design, since the
output data must be scaled if the target patterns in our dataset have a different
range. In this work, we use the expectation of the σ_z observable over the first
qubit q_0 of our network as the network output. However, we append a final scale
parameter and bias to be learned, so that the network output is less sensitive to
dataset bias and scale. The whole designed model is depicted in Fig. 5.

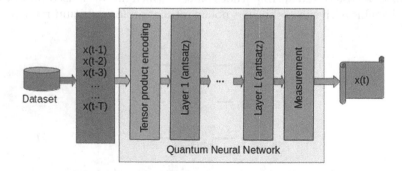

Fig. 5. Proposed Quantum Neural Network model.

As we are using Quantum Computing simulation software, the training of the proposed QNN is performed in CPU using classic algorithms such as the Adam optimizer. Gradient computation is performed in CPU using the classic propagation rules, while the gradient in the QPU is calculated using the parameter-shift rule. Figure 6 shows the training process pipeline, where θ^1 stand for the scale/bias parameters in the input layer, θ^2 are the parameters of the layers containing the antsatz, and θ^3 are the scale/bias parameters for the network outputs.

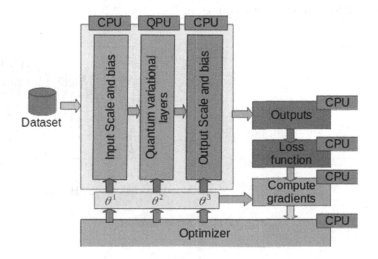

Fig. 6. Proposed Quantum Neural Network model.

The use of the proposed QNN model for time series forecasting is as follows: Since a QNN is a feedforward model, we first set a time horizon T, and the time series must be transformed to tabular data where the target is the time series value at time $t, x(t)$, and the inputs are the values $x(t-1), x(t-2), ..., x(t-T)$, as described in Fig. 5.

4   Experiments

The experimentation in this paper is a proof-of-concept regarding the capabilities of QNNs to perform time series prediction. For that reason, we used two classic and well-tested time series: The **laser** time series from the far-infrared laser dataset A of the Sata Fe Time Series competition, and the synthetic **Henon** map time series. Figure 7 shows the time series used, containing 150 data points each. As a QNN has a feedforward structure, we compare the results with a classic Multilayer Perceptron (MLP) instead of more complex recurrent neural networks, in order to make the most fair comparison possible. The time series were divided into the first 75% of data for training/test, and the remaining 25% for validation. To ease QNN learning, we scaled the time series to the interval $[-1, 1]$ as a preprocessing step.

The MLP model was implemented in tensorflow 2.7, containing 1 layers with 10 neurons, the *tanh* activation function in the hidden layer and the identity in the output layer. On the other hand, the QNN was implemented in Tensorflow Quantum 0.6.1 and, in both datasets, a single layer was included in the network structure. The training algorithm was Adam with a learning rate of 0.01 for both MLP and QNN, and 30 different executions were performed using 4-fold cross-validation, to make a statistical analysis of results over a desktop computer Intel(R) Core(TM) i5-9600 K CPU at 3.70 GHz with 32 GB RAM with a NVidia GeForce RTX 2060 GPU. The source code for this experimentation is available at https://github.com/manupc/qnn_tsp.

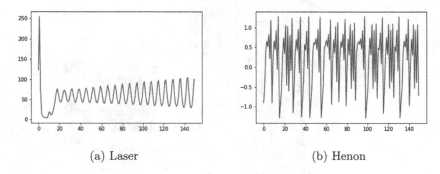

(a) Laser   (b) Henon

Fig. 7. Time Series datasets

Table 1 summarizes the results of the experiments performed. Column 1 prints the metric under study, Columns 2–3 the values of the corresponding metric for the **Laser** dataset obtained by MLP and QNN, respectively, and Columns 4–5 the values for the **Henon** dataset. On the other hand, rows 2–4 show the MSE obtained for the training, test and validation sets; and the rows 5–6 the minimum and maximum MSE obtained in the 30 experiments, respectively. Finally, the last row contains the average computational time in seconds for each run. A Mann-Whitney U test was applied over the validation MSE of all 30 executions, and we remark results of QNN in row 3 with (+) if the test

Table 1. Summary of results.

Metric	Laser		Henon	
	MLP	QNN	MLP	QNN
Avg. Tr. MSE	0.0380	0.0168	0.0761	0.0044
Avg. Ts. MSE	0.0478	0.0199	0.0734	0.0063
Avg. Val. MSE	0.0863	0.0476 (+)	0.0977	0.0145 (+)
Min. Val. MSE	0.05774	0.01461	0.0655	0.0096
Max. Val. MSE	0.1215	0.0683	0.1416	0.02145
Avg. Time	10.36	30.92	10.44	30.89

concluded that there are significant differences between MLP and QNN and the latter outperformed the former, which occurs in both cases. Boxplots in Fig. 8 help to analyse these results.

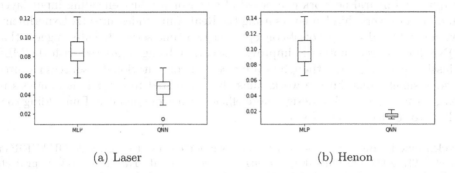

(a) Laser   (b) Henon

Fig. 8. Boxplots of MSE for Laser and Henon datasets in our experiments

According to Table 1, the QNN was able to outperform the classic MLP in the both datasets studied, both in training/test and validation data splits. Also, both the best solution ans worst were better with QNN. However, the average time required to perform the experimentation is x3 times slower in QNN, which makes the model less scalable than the classical counterpart for a large number of qubits. This could be expected, since the experiments were performed over a QC simulation software instead than in a true quantum hardware. Boxplots in Fig. 8 support this conclusion, and also suggest that the robustness of QNN is better than in MLP since the difference betweeen the first and third quartiles are lower in the former case.

If we analyze the results in terms of model complexity (number of parameters), the MLP with 10 hidden neurons contains (T+1)*10 parameters in the hidden layer (weights and biases) and 11 parameters in the output layer. On the other hand, the number of parameters of the QNN is 2*T in the input layer (input scale and bias), 3*T parameters in the hidden layer, and 2 parameters (scale and bias) in the output layer. In the case of T=7, the MLP contains 91 parameters while the QNN has 37. Thus, the network model in the quantum proposal is significantly smaller than the classic counterpart. This fact suggests that QML can contribute not only with an improvement in accuracy of models, but also in model complexity.

Despite of these results, it is important to note that the computational space of a QNN increases by power of two everytime a qubit is included into the network. For that reason, we believe that the proposed model cannot be used under simulation for large time series. This fact opens the doors to future works where a more efficient way to embed time series could be analyzed to reduce the quantum network size.

5 Conclusions

In this work we have studied how Quantum Neural Networks can be used to perform Time Series forecasting tasks in quantum computers. We have designed a quantum neural network composed of a tensor product encoding input layer, and one or several hidden layers using the Real Amplitudes antsatz. Experiments were conducted as a proof-of-concept over two time series. Results suggest that QNN have a great potential improving accuracy being compared with the MLP classical counterpart, although at a big cost of computational resources required under simulations. Future works must be conducted to extend the analysis to larger time series and to design more efficient ways of Quantum Embedding that allow to reduce the QNN size.

Acknowledgements. This article was supported by the project QUANERGY (Ref. TED2021-129360B-I00), Ecological and Digital Transition R&D projects call 2022 funded by MCIN/AEI/10.13039/501100011033 and European Union NextGenerationEU/PRTR, and Grant PID2021-128970OA-I00 by MCIN/AEI/10.13039/501100011033/FEDER.

References

1. Andrés, E., Cuéllar, M.P., Navarro, G.: On the use of quantum reinforcement learning in energy-efficiency scenarios. Energies **15**(16), 6034 (2022). https://doi.org/10.3390/en15166034

2. Azevedo, C.R.B., Ferreira, T.A.E.: Time series forecasting with qubit neural networks. In: Proceedings of The Eleventh IASTED International Conference on Artificial Intelligence and Soft Computing, ASC '07, pp. 13–18. ACTA Press, USA (2007)

3. Bryan, L., Stefan, Z.: Time-series forecasting with deep learning: a survey. Phil. Trans. R. Soc. A **379**, 20200209 (2021). https://doi.org/10.1098/rsta.2020.0209

4. Ciliberto, C., et al.: Quantum machine learning: a classical perspective. Proc. R. Soc. A: Math. Phys. Eng. Sci. **474**, 20170551 (2017). https://doi.org/10.1098/rspa.2017.0551

5. Ganguly, S.: Quantum Machine Learning: An Applied Approach. Springer, Cham (2021)

6. Jerbi, S., Gyurik, C., Marshall, S., Briegel, H., Dunjko, V.: Parametrized quantum policies for reinforcement learning. In: Ranzato, M., Beygelzimer, A., Dauphin, Y., Liang, P., Vaughan, J.W. (eds.) Advances in Neural Information Processing Systems, vol. 34, pp. 28362–28375. Curran Associates, Inc. (2021)

7. Kutvonen, A., Fujii, K., Sagawa, T.: Optimizing a quantum reservoir computer for time series prediction. Nat. Sci. Rep. **10**, 14687 (2020). https://doi.org/10.1038/s41598-020-71673-9

8. Kwak, Y., Yun, W.J., Jung, S., Kim, J.: Quantum neural networks: concepts, applications, and challenges. In: 2021 Twelfth International Conference on Ubiquitous and Future Networks (ICUFN), pp. 413–416 (2021). https://doi.org/10.1109/ICUFN49451.2021.9528698

9. Li, X., Cheng, C.T., Wang, W.C., Yang, F.Y.: A study on sunspot number time series prediction using quantum neural networks. In: 2008 Second International Conference on Genetic and Evolutionary Computing, pp. 480–483 (2008). https://doi.org/10.1109/WGEC.2008.76
10. Norlén, H.: Quantum Computing in Practice with Qiskit and IBM Quantum Experience. Packt, Birmingham (2020)
11. Schuld, M., Petruccione, F.: Supervised Learning with Quantum Computers, 1st edn. Springer Publishing Company, Cham (2018). Incorporated
12. Senekane, M.: Hands-on Quantum Information Processing with Python. Packt, Birmingham (2021)
13. Shrivastava, P., Soni, K.K., Rasool, A.: Classical equivalent quantum unsupervised learning algorithms. Procedia Comput. Sci. **167**, 1849–1860 (2020). https://doi.org/10.1016/j.procs.2020.03.204. International Conference on Computational Intelligence and Data Science
14. Singh, P.: FQTSFM: a fuzzy-quantum time series forecasting model. Inf. Sci. **566**, 57–79 (2021). https://doi.org/10.1016/j.ins.2021.02.024
15. Sutor, R.: Dancing with Qubits. Packt, Birmingham (2019)
16. Torres, J.F., Hadjout, D., Sebaa, A., Martínez-Álvarez, F., Troncoso, A.: Deep learning for time series forecasting: a survey. Big Data **9**(1), 3–21 (2021). https://doi.org/10.1089/big.2020.0159
17. Wittek, P.: Quantum Machine Learning: What Quantum Computing Means to Data Mining. Elsevier, Amsterdam (2014)

Intra- and All-Day PV Power Forecasting Using Expansion PDE Models Composed of the L-Transform Components in Nodes of Step-by-Step Evolved Polynomial Binary-Nets

Ladislav Zjavka[✉] and Václav Snášel

Department of Computer Science, Faculty of Electrical Engineering and Computer Science, VŠB-Technical University of Ostrava, Ostrava, Czech Republic
{ladislav.zjavka,vaclav.snasel}@vsb.cz

Abstract. Photovoltaic (PV) power is one of the most important energy sources available in backcountry regions or developing southern countries with missing infrastructure. Intra- or all-day statistical models, using the latest environmental and power data records, can predict PV power for a plant-specific location and condition on time. Numerical Weather Prediction (NWP) systems are run every 6 h to produce free prognoses of local cloudiness with a considerable delay and usually not in operational quality. Differential binomial neural networks (D-BNN) are a novel neurocomputing technique that can model the characteristics of the weather. D-BNN decomposes the n-variable Partial Differential Equation (PDE), allowing a complex representation of the near-ground atmospheric dynamics, into a set of 2-input node sub-PDEs. These are converted and substituted using the Laplace transform formulations of Operation Calculus. D-BNN produces applicable PDE components, one by one using the selected binary nodes to extend its sum models. Historical spatial data are examined to pre-assess daily training samples for a specific inputs->output time shift used in forecasting the Clear Sky Index. Iterative 1–9 h and 24-h sequence PV power prediction models using machine learning (ML) and statistics are compared and evaluated. Daily modelling allows for sequence predictions of full PV power (PVP) cycles in operational quality. Reliable PV forecasting is required in load management in plant power supply and consumption.

Keywords: Uncertainty modelling · Partial differential equation · Binomial neural network · Operation calculus · PDE conversion · Laplace transformation

1 Introduction

Chaotic irradiance fluctuations and transmission are caused by uncertain interactions of atmospheric factors, which can be represented by differential equations. Efficient planning and use of intermittent PV power supply along with energy integration require reliable forecasting models adaptable to local specifics and anomalies. Weather prediction methods can be broadly classified into 3 main approaches:

© The Author(s), under exclusive license to Springer Nature Switzerland AG 2023
I. Rojas et al. (Eds.): IWANN 2023, LNCS 14134, pp. 678–689, 2023.
https://doi.org/10.1007/978-3-031-43085-5_54

- NWP systems using physical considerations
- statistical regression or soft-computing approach using historical data
- hybrid or ensemble models based on combined methods

NWP deterministic models are unable to recognise detailed physical events at the surface level, as they use a fixed grid scale for the numerical simulations in several atmospheric layers. These systems simplify atmospheric circulation by solving sets of primitive physical equations for ideal gas flow. NWP models can combine their outputs with results from satellite- or ground-based sky image methods that analyse and detect cloud structures in an initial time period. Their solutions using an actual time step and changes (in cell size) only provide approximate estimates, where errors increase exponentially in longer time horizons and extent [5] . Artificial intelligence (AI) can model or analyse local atmospheric phenomena to plan the operation of a photovoltaic plant [3]. Statistical intra- or day-ahead predictions of generated PV Power (PVP) are challenging because of many uncertain processes with chaotic parameter fluctuations in the near-ground layer dynamics, resulting in local irregularities. The input solar radiation and output PVP time series over the estimated daily periods can be used to generate 24-h prediction models. Statistical techniques, using historical records, can predict stochastic PVP supplies a few hours ahead or post-process NWP data on the daily prediction horizon. Their conversion of radiation forecasts is mainly influenced by the precision of NWP data [7]. Their 24-h data processing uses the output of NWP models to compute PVP output series at the related times. The AI iterative training, e.g. using artificial neural networks (ANN), starts from random weights yielding different output values in uncertain situations. This ML approach shows some shortcomings; e.g. model over-fitting, simplification, or adaptation to a local situation, etc. AI can process additional information, for example, cloud motion vectors or spectral parameters, that characterise variances and classes of cloud structures at atmospheric levels, to improve the model performance. The final ML model is evaluated according to test criteria. Additional statistical properties can be extracted from autocorrelated data to capture periods in cycles [6].

The novel design of D-BNN applies self-organising principles in the node-by-node extension of binary structures of binomial neural networks (BNN) to produce applicable sub-PDE components applicable at nodes, included in the sum model [4]. It decomposes the n-variable PDE into a set of 2-variable node converted sub-PDEs using OC. The principles of complex-valued neural networks (CNN), using the real and imaginary representation of data, are applied in the adapted OC to solve 2-variable PDEs. This hybrid maths-computing approach allows modelling of local weather patterns in an adequate operational complexity. D-BNN overcomes the problem of data pre-processing or feature extraction by effective selecting search for relevant 2-input variables in each node. This step-by-step model expansion is based on Goedel's incompleteness theorem, which states: The self-decomposition of input-output data relations into an expanding multilayer BNN is analogous to the convolution principles of deep learning. No data transformation is necessary, as the PDE derivatives are L-transformed in the initial model conversion to be inversely restored in the node searched function originals. The first pre-assessed training samples allow us to elicit models operable in various situations. If the complexity of a model gradually increases, the objective functions of a problem definition are transferred to the minima [1].

2 Iterative Intra- and Sequenced All-Day PV Power Forecasting

The multistage intra-hour AI modelling approach uses increasing time inputs->output delay to compute the Clear Sky Index (CSI) separately for each predicted time [2]. The self-evolved hourly models process the last available data input to compute the CSI output in the training horizon time from 8am to 5pm. The day-ahead strategy estimates the output CSI in a 24-h series processing, using one model in the 24-h input-output time delay of fixed training. It processes the last-day series input in a sequence iterative procedure to compute the related time complete CSI series in all overall PVP cycles from 6am to 7:30pm. The intra-hourly separate modelling approach applies series input from the early morning hours, so that the PVP supply predictions on a day basis. Day-ahead models are applied sequentially to the complete last-day input series in predicting all-day CSI series. This approach is usually not as accurate, using the fixed 24-h input delay, with respect to the iterative strategy based on a step-by-step increasing 1–9 hourly horizon (Fig. 1).

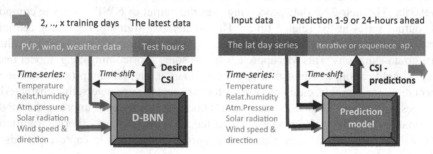

Fig. 1. Intra-hour and day-ahead CSI prediction (red right) using models developed in the estimated day training intervals (blue left) in increasing 1–9 or fixed 24-h input delay.

The initialisation time assessment identifies the training data range series for the last 1-x days. These models are tested to obtain the test error minima that indicate the optimal day intervals [7]. The predetermined training days allow the evolution of models applicable to the last unseen data in the final prediction stage (Fig. 2).

3 PDE Partition Using Backward D-BNN Binary Tree-Structures

D-BNN inserts its nodes one by one into the last layer added. Node blocks (Fig. 3) in the next layers can produce Composite terms (CT) in addition to simple neurones (8). CTs are the products of selected neurones, that is, converted sub-PDEs at nodes, formed by the back-connected blocks in the actual and previous layers (Fig. 3). The number of CT combinations in blocks doubles in each back-joined layer in formation of high-composite models in selected node PDE-components (see the Appendix).

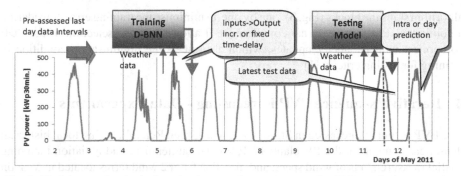

Fig. 2. Training, testing in the iterative intra- and one-sequence day PVP prediction procedure

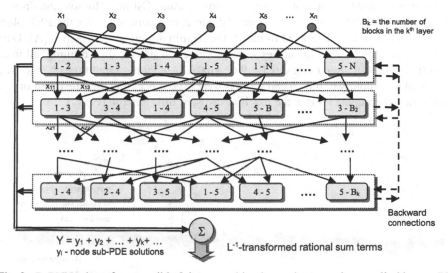

Fig. 3. D-BNN selects from possible 2-input combination nodes to produce applicable sum PDE model components of neurons and CTs in blocks solving node sub-PDEs

The D-BNN output Y is the arithmetic mean of the selected active neurones or CT outputs in node blocks, included in the total output sum (1).

$$Y = \frac{1}{k} \sum_{i=1}^{k} y_i \quad k = \text{ the number of active neurons or CTs (node PDE solutions)} \quad (1)$$

The D-BNN structure is dynamically developed/modified in each training cycle, allowing appropriate changes in the number of layers and the member nodes. All input combination couples are examined by the node blocks in each iteration step to find an applicable neurone or CT, which can be inserted/removed in/from the sum model to better approximate the desired output. Their binomial parameters and weights are gradient pre-optimised using [9]. An appropriate 1st or 2nd order binomial and its members are selected to form and solve a specific 2-variable PDE in the BNN nodes. This 1-block iteration algorithm skips the blocks (Fig. 3), one by one, to minimise the training error considering a continuous test using external complement. It allows only the selection of the PDE component with parameter updates, which also meets the minimisation condition in training for testing criteria [1].

4   Iterative/Sequence PVP Forecasting - Data Experiments

The PVP output is forecasted in Starojcka Lhota, Czech Republic, using local historical data measurements of the PV plant: PVP, ground-out.temper., and radiation including spatial measurements of wind speed and direction from 2 wind farms located in Maletin and Vesel nad Moravou. The additionally applied data: meteorological observations (ground temper, humidity, see pressure, wind speed/direction, and visibility state) of the 3 weather stations located at the airports Brno-Turany, Ostrava-Mornov, and Prerov-Bochor (Fig. 4). PVP plant & wind farm 10-min data record series for the PVP plant and the wind farm were averaged to match the 30-min airport weather series [A]. Data record samples of 26 input quantities were processed in statistical AI training, all from the initially assessed time ranges of 1…9 or a fixed 24-h input-output time change. The developed intra-hourly or day-ahead models were used processing the latest hour or day input series to compute PVP in the machine-assisted time horizon (Figs. 1, 2).

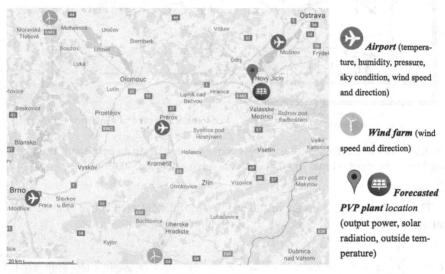

Fig. 4. The observational localisation of the PVP plant, airport station and wind farms

Figures 5, 6 present intra-hourly and day-ahead PVP prediction comparisons of the D-BNN, Matlab Statistics and Machine Learning Toolbox (SMLT) for regression [B], and persistent benchmark models on selected demonstrative days in the 2-week experimental spring interval May 12–25, 2011. The statistics ML models were evolved in relation to the pre-determined training intervals, where the number of applicable days is selected considering testing error minima. SMLT models of Gaussian Processed Regression (GPR),

Support Vector Machine (SVM) and Ensemble Boosted or Bagged Tree (EBT), processing the 26 data inputs to compute the output CSI series, get with the RMSE minima in testing, involving slight error variability only. Simple persistent comparative models, average CSI over the pre-optimised time intervals of the last 1, 2, 3, .., x days to be evaluated their PVP series at the related hours in the maximal approximation. Both the intra-hour and day prediction models are tested with the last 6-h available series data.

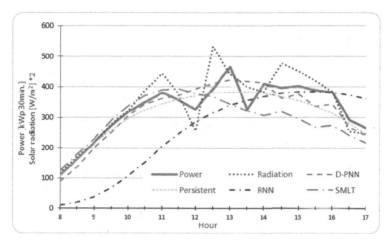

Fig. 5. St.Lhota, May 25, 2011 – intra 1–9-hourly predictions (mostly clear). RMSE: D-BNN = *36.4*, Persistent = *37.6*, RNN = *119.8*, SMLT = *65.0* kWp 30 min.

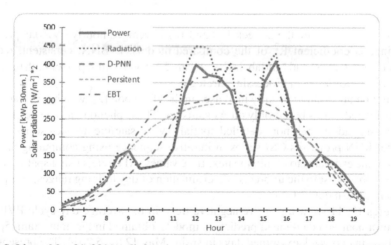

Fig. 6. St.Lhota, May 21, 2011 – all-day ahead predictions (variable cloudiness). RMSE: D-BNN = *66.2*, Persistent = *71.7*, SMLT (GPR) = *88.3* kWp 30 min.

PVP production is related to specific pattern radiation day cycles dependent on variable cloud structure type, progress dynamics, wind gust character, visibility index, etc. (Fig. 7). The most correlated quantity is humidity. These parameters were identified to determine primarily the applicability of the data to predict the following day hours, considering the data intervals [C].

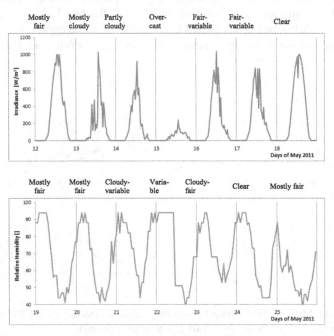

Fig. 7. Radiation and humidity correlation cycle day specific patterns in 2 week predictions.

Figures 8, 9 compare the 2-week 1–9 and 1–14 averaged hourly PVP day prediction determination coefficient R^2 of the compared models: D-BNN, persistent, recurrent neural network (RNN) and Matlab SMLT at 8am–5pm and 6am–7pm for both the hourly-intra- and day-to-day data processing strategies.

Radiation uncertainty usually increases in the afternoon [2], which may cause inaccuracies in predictions based only on AI statistics. These phenomena can be eliminated by processing additional more detailed spatial sky- or satellite-assisted series of radiation data. RNN processes CSI series in iterative sequence using computing data from previous time estimates to obtain the next time values. AI models using spatial selective data are more robust to the unprecise determination of day training ranges, computed in the initial assessment test stage.

Figures 10, 11 show a positive/negative trend in the average day cycle PVP errors of the intra-hour and day-ahead prediction models, obtained in the hours 8am–5pm and 6am–7pm,, in two week examined days in spring May 12 to 25 2011. The presented 14-day PVP aver, error trends are balanced (zero) in most cases using the AI methods, except RNN based on computing CSI sequence series in the nest time data output estimation, which show more or less positive bias as compared to the SMLT models.

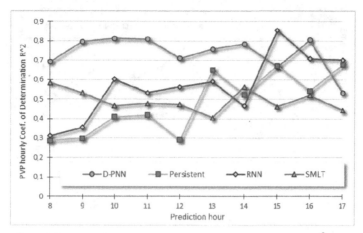

Fig. 8. 2-week intra-day hourly 1–9 average 8am–5pm PVP prediction R^2: D-BNN $= 0.74$, Persist. $= 0.48$, RNN $= 0.57$, SMLT $= 0.49$.

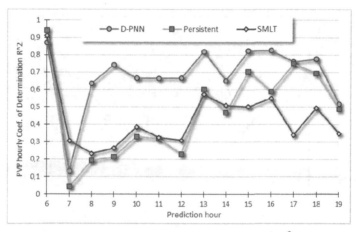

Fig. 9. 2-week intra-day 2 week hourly 1–14 average PVP prediction R^2: D-BNN $= 0.68$, Persist. $= 0.47$, SMLT $= 0.43$.

5 Evaluation of PVP Prediction Data Experiments Based on AI

Patten similarity in day PVP cycle series of a type of settled weather primarily affects the CSI prediction results based on AI or regression statistics. The D-BNN, SMLT machine learning can predict PVP in changeable cloud conditions (Fig. 6), following the previous period of clear or partly sunny (windy) weather with an acceptable accuracy compared to NWP. The designed and developed AI day-ahead predictions are sometimes more accurate than models of the intrahourly processing and forecasting strategy in some specific cases, in spite of the fixed 24-h time output horizon. The results are related to specific conditions in the environment and pattern characteristics in training - testing, prediction, which leads to differences and variability in the applicability of data series in training

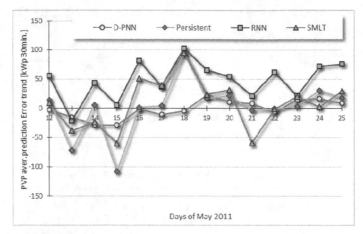

Fig. 10. 2-week 1–9 h average PVP prediction error trends [kWp 30 min]: D-BNN $= -1.03$, Persist. $= 2.10$, RNN $= 48.28$, SMLT $= 8.65$.

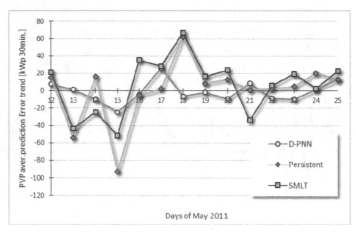

Fig. 11. 2-week 1–14 h average PVP prediction error trends [kWp 30 min]: D-BNN $= -1.19$, Persist. $= 0.45$, SMLT $= 6.33$.

and final prediction stage. These uncertain parameters vary considerably according to the applied input time change and the dynamics. NWP systems [B] using an increasing time initialisation interval can eliminate uncertain forecasting parameters and obtain better precision in the simulation results in specific cases [5]. The estimated optima in ML sequenced periods reduce the variability and uncertainty in the dynamics and large-scale disturbances in local weather states and essentially determine the prediction accuracy and reliability. The adequate approximation of day-cycle PVP series is usually related to a type of more-less settled and similar weather patterns in the subsequent day time, in contrast to PVP ramp events in a changeable cloudiness. The most uncertain situations in prediction statistics are days close to the frontal breakovers (Fig. 6). Training - Testing data with PVP patterns might be totally different as those on prediction computing times.

The error test minima can define a threshold that cannot be exceeded without using the converted NWP cloudiness or radiation forecasts [A] [7].

6 Conclusions

D-BNN has the most accurate average day prediction results compared to statistical regression and advanced AI solutions, whose models mostly only simplify the pattern representation and complexity of weather-dependent systems. The progressive PDE-modelling approach is based on dynamic structure composition and modification using several synchronised optimisation algorithms to match radiation fluctuant changes and chaotic processes. Optimal training-time initialisation improves the evolution and robustness of day-ahead PVP models. Intra-hourly predictions depend on the latest hour data processing in a short time horizon to detail day PVP forecasts in more accurate approximations. The multistage sequence strategy is significantly time-consuming, requires the evolution of models for each time step, and does not enable operational day supply planning of PVP plants. The compared day PVP models predict the full-day cycles in one processing sequence step, obtaining the required accuracy primarily in the morning and afternoon. Applicable training/testing sample records can be identified in larger historical weather archives according to pattern similarity between real and forecast data series at the same time to reduce unacceptable uncertainty resulting from frontal disturbance zones, wind gusts, and atmospheric waves. The proposed modelling approach can process data input in a 24-h delay in attempt of possible improvements in the approximation in day-cycle meteo-variables, for instance, humidity, electrical power load, etc. The C++ application D-BNN software including power and meteo- data is possible to use to reproduce or compare AI predictions [C].

Acknowledgements. This work was supported by SGS, VSB - Technical University of Ostrava, Czech Republic, under the grant No.\SP2023/12 'Parallel processing of Big Data X'.

APPENDIX - Node PDE Conversion Using the L-Transform

D-BNN develops a BNN structure to decompose the n-variable PDE into sub-PDEs in its two input nodes. These are converted using OC into pure rational terms corresponding to the L-transformed node functions. The sum of selected inverse L-transformed rational terms gives the PDE model of an n-variable output function [8].

$$a + bu + \sum_{i=1}^{n} c_i \frac{\partial u}{\partial x_i} + \sum_{i=1}^{n}\sum_{j=1}^{n} d_{ij} \frac{\partial^2 u}{\partial x_i \partial x_j} + \ldots = 0 \qquad u = \sum_{k=1}^{\infty} u_k \qquad (2)$$

$u(x_1, x_2, , \ldots, x_n)$ − *unknown separable function of* n − *input variables*

$a, b, c_i, d_{ij}, \ldots$ − *weights of terms* $\qquad u_i$ − *partial sum functions*

The general linear PDE (2) can describe an unknown separable u function of n inputs, which can be expressed in convergent sum series (2) of partial u_k function solutions of

2-variable simple sub-PDEs defined by the equality of 8 variables.

$$L\left\{f^{(n)}(t)\right\} = p^n F(p) - \sum_{k=1}^{n} p^{n-i} f_{0+}^{(i-1)} \quad L\{f(t)\} = F(p) \tag{3}$$

$f(t), f'(t), \ldots, f^{(n)}(t) - originals\ continuous\ in <0+, \infty> \quad p, t - complex\ and\ real\ variables$

The OC binomial conversion of the $f(t)$ function n^{th} derivatives in an Ordinary differential equation (ODE) is based on the proposition of their Laplace transformation (L-transform) taking into account the initial conditions (3).

$$F(p) = \frac{P(p)}{Q(p)} = \frac{Bp + C}{p^2 + ap + b} = \sum_{k=1}^{n} \frac{A_k}{p - \alpha_k} \tag{4}$$

$B, C, A_k - coefficients\ of\ elementary\ fractions \quad a, b - binomial\ parameters$

The conversion of ODE results in algebraic equations from which the L-transform $F(p)$ can be expressed in the complex form of a pure rational term (4). The pure rational function represents the original function $f(t)$. It can be expressed in the form of elementary sum fractions (4), which are transformed into inverse L images using the OC definitions (5) to obtain the original $f(t)$ of a real variable t in the ODE solution.

$$F(p) = \frac{P(p)}{Q(p)} = \sum_{k=1}^{n} \frac{P(\alpha_k)}{Q_k(\alpha_k)} \frac{1}{p - \alpha_k} \quad f(t) = \sum_{k=1}^{n} \frac{P(\alpha_k)}{Q_k(\alpha_k)} e^{\alpha_k \cdot t} \tag{5}$$

$a_k - simple\ real\ roots\ of\ the\ multinomial\ Q(p) \quad F(p) - L - transform\ image$

D-BNN composes pure rational terms, for example (7), from GMDH binomials (6) [4] in its block nodes (Fig. 9) to convert specific 2-variable sub-PDEs (2) into the L-transforms of unknown summation u_k functions (2) (Fig. 12).

$$y = a_0 + a_1 x_i + a_2 x_j + a_3 x_i x_j + a_4 x_i^2 + a_5 x_j^2 \tag{6}$$

$x_i, x_j - 2\ input\ variables\ of\ neuron\ nodes$

The inverse L-transform is applied to rational terms in the selected nodes (7), according to OC (5). The sum of the two-variable u_k original, produced in the nodes (Fig. 9), gives a model of the separable output u function (2). Each node block forms simple neurones, for example, (7), to convert and solve specific 2-variable PDEs.

$$y_i = w_i \frac{b_0 + b_1 x_1 + b_2 sig(x_1^2) + b_3 x_2 + b_4 sig(x_2^2)}{a_0 + a_1 x_1 + a_2 x_2 + a_3 x_1 x_2 + a_4 sig(x_1^2) + a_5 sig(x_2^2)} \cdot e^\varphi \tag{7}$$

$\varphi = arctg(x_1/x_2) - phase\ representation\ of\ 2\ input\ variables\ x_1, x_2$

$a_i, b_i - binomial\ parameters \quad w_i - weights \quad sig - sigmoidal\ transform$

The Euler notation of complex numbers (8) corresponds to the expression OC $f(t)$ (5). The radius amplitude r represents a rational term, while the angle phase $= arctg(x_2/x_1)$ of 2 variables of real value can give an inverse L transformation of $F(p)$.

$$p = \underbrace{x_1}_{Re} + i \cdot \underbrace{x_2}_{Im} = \sqrt{x_1^2 + x_2^2} \cdot e^{i \cdot \arctan\left(\frac{x_2}{x_1}\right)} = r \cdot e^{i \cdot \varphi} = r \cdot (\cos\varphi + i \cdot \sin\varphi) \tag{8}$$

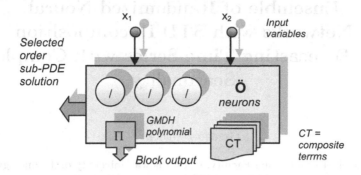

Fig. 12. 2-input node blocks produce simple (/) and composite neurons to solve PDEs

References

1. Anastasakis, L., Mort, N.: The Development of Self-Organization Techniques in Modelling: A Review of the Group Method of Data Handling (GMDH). The University of Sheffield (2001)
2. Coimbra, C., Kleissl, J., Marquez, R.: Overview of Solar Forecasting Methods and A Metric for Accuracy Evaluation, pp. 171–194. Elsevier, Amsterdam (2013)
3. Liu, Z.-F., Luo, S.-F., Tseng, M.-L., Liu, H.-M., Li, L.: Short-term photovoltaic power prediction on modal reconstruction: a novel hybrid model approach. Sustain. Energy Technol. Assess. **45**, 1–12 (2021)
4. Nikolaev, N.Y., Iba, H.: Adaptive Learning of Polynomial Networks. Genetic and Evolutionary Computation. Springer, New York (2006). https://doi.org/10.1007/0-387-31240-4
5. Vannitsem, S.: Dynamical properties of MOS forecasts: analysis of the ECMWF operational forecasting system. Weather Forecast. **23**, 1032–1043 (2008)
6. Wan, C., Zhao, J., Song, Y., Xu, Z., Lin, J., Hu, Z.: Photovoltaic and solar power forecasting for smart grid energy management. CSEE J. Power Eenergy Syst. **1**, 38–46 (2015)
7. Zjavka, L., Krömer, P., Mišák, S., Snášel, V.: Modeling the photovoltaic output power using the differential polynomial network and evolutional fuzzy rules. Math. Model. Anal. **22**, 78–94 (2017)
8. Zjavka, L., Mišák, S.: Direct wind power forecasting using a polynomial decomposition of the general differential equation. IEEE Trans. Sustain. Energy **9**, 1529–1539 (2018)
9. Zjavka, L., Snášel, V.: Constructing ordinary sum differential equations using polynomial networks. Inf. Sci. **281**, 462–477 (2014)
10. Weather underground historical data series. www.wunderground.com/history/airport/LKTB/2016/7/22/DailyHistory.html
11. Matlab - Statistics and Machine Learning Tool-box for Regression. www.mathworks.com/help/stats/choose-regression-model-options.html
12. D-BNN application C++ parametric software & Power data. https://drive.google.com/drive/folders/1ZAw8KcvDEDM-i7ifVe_hDoS35nI64-Fh?usp=sharing

Ensemble of Randomized Neural Networks with STD Decomposition for Forecasting Time Series with Complex Seasonality

Grzegorz Dudek[✉]

Electrical Engineering Faculty, Czestochowa University of Technology,
Al. AK 17, 42-200 Częstochowa, Poland
grzegorz.dudek@pcz.pl
http://www.gdudek.el.pcz.pl

Abstract. This paper proposes a novel ensemble forecasting method that combines randomized neural networks (RandNNs) and seasonal-trend-dispersion decomposition (STD) in four different ways to construct ensembles for time series with multiple seasonal patterns. We evaluate the performance of the proposed ensemble methods on short-term load forecasting problems with triple seasonality, and demonstrate that the hybridization of RandNN and STD decomposition results in highly reliable accurate forecasting.

Keywords: Ensemble forecasting · Multiple seasonal patterns · Randomized neural networks · Seasonal-trend-dispersion decomposition · Short-term load forecasting

1 Introduction

Time series data expressing real-world phenomena and processes such as stock prices, weather conditions, and electricity demand are often complex with time-varying trends, a significant random component, structural breaks and multiple seasonal patterns. This complexity can make it difficult to identify the best single model to accurately forecast the data. To address this issue, a common approach is to use ensemble forecasting. This involves using multiple forecasting models to capture the various drivers of the data generating process and mitigate uncertainties regarding model form and parameter specification [1].

Ensemble forecasting has been shown to be effective in improving the accuracy and reliability of time series forecasts [2]. By combining forecasts, the aim is to take advantage of the strengths of multiple models while reducing the impact

Supported by grant 020/RID/2018/19 from the Polish Minister of Science and Higher Education titled "Regional Initiative of Excellence", 2019-23.

I. Rojas et al. (Eds.): IWANN 2023, LNCS 14134, pp. 690–702, 2023.
https://doi.org/10.1007/978-3-031-43085-5_55

of their individual weaknesses. For example, one model may perform well in capturing short-term fluctuations while another model may be better at capturing long-term trends. The main advantage of combining forecasts is that it often leads to a more accurate prediction than any single model can provide. By averaging or weighting the forecasts from multiple models, we can reduce the impact of outliers and errors in individual forecasts and arrive at a more accurate and robust prediction. Combining forecasts can also help to reduce the uncertainty associated with any single prediction. By considering multiple models, we can get a better sense of the range of possible outcomes and the likelihood of each outcome. Using an ensemble of models can also increase our confidence in the prediction, particularly if the models are diverse and provide different perspectives on the same problem. By comparing and contrasting the predictions of multiple models, we can gain a more nuanced understanding of the problem and build a more robust solution.

There are several approaches for combining forecasts, including more complex weighting schemes, but the arithmetic average of forecasts based on equal weights is a popular and surprisingly robust method that often performs well [4,5]. In this study, we choose to focus on this method to keep the analysis simple and easy to interpret. An essential issue in ensembling is to ensure diversity among the individual models [3]. If the models are too similar, the ensemble may not capture all the relevant features of the time series, and as a result, may not improve the predictive performance. To address this issue, we utilize randomized neural networks (RandNNs) with effective mechanisms for controlling diversity, as described in Sect. 4.

This work builds upon our previous research on ensemble forecasting using RandNNs [6,7]. RandNN is an ideal candidate for an ensemble member due to its ability to control diversity and fast training [8]. Recently, several papers have proposed ensemble methods based on different types of randomized NNs. For instance, in [9], an ensemble deep random vector functional link network was proposed, while [10] presented a bagging ensemble approach based on NN with random weights for online data stream regression. Another approach was presented in [11], where an ensemble based on decorrelated random vector functional link networks was proposed. In [12], a selective ensemble of randomization-based NNs using a successive projections algorithm was proposed, and in [13], a framework for constructing an ensemble model by selecting appropriate representatives from a set of randomization-based NNs was described.

To tackle complex seasonality of time series, we utilize the seasonal-trend-dispersion decomposition method (STD), which simplifies the forecasting problem by extracting three components: trend, seasonal, and dispersion components. The main contribution of this study is the combination of RandNN with STD decomposition in four different ways to construct ensembles for forecasting time series with multiple seasonal patterns. We evaluate the performance of our proposed ensemble methods on challenging short-term load forecasting problems with triple seasonality, and our results show that the hybridization of RandNN and STD decomposition produces high accuracy and reliability in forecasting.

2   Related Works

2.1   Seasonal-Trend-Dispersion Decomposition

STD was proposed in [14] in two versions: with and without a remainder component. In this study, the latter version is used, as it is more convenient for our purpose. STD differs from existing methods such as STL (Seasonal and Trend decomposition using Loess), wavelet-based multi-resolution analysis, and empirical mode decomposition, among others, because it extracts a separate dispersion component that captures the short-term variability of the series and can be useful for analysing and forecasting time series with varying variance.

STD produces interpretable components that express a trend, dispersion, and seasonal patterns. The seasonal component consists of centered and normalized seasonal patterns, making it easy to compare and analyze seasonal cycles. STD is parameter-free for seasonal time series and requires only one parameter for non-seasonal time series. The algorithm is straightforward, fast, and easy to implement. STD can be used to encode input and output variables for forecasting models, simplifying the relationship between variables and facilitating the forecasting of complex heteroscedastic time series with multiple seasonality.

Assuming that the time series $\{z_t\}_{t=1}^N$ has a seasonality of period n and that the length of the series is a multiple of the seasonal period, i.e., $N/n = K$ for some $K \in \mathbb{N}$, the time series can be represented as a series of successive seasonal sequences: $\{\{z_{i,j}\}_{j=1}^n\}_{i=1}^K = \{\{z_{1,j}\}_{j=1}^n, ..., \{z_{K,j}\}_{j=1}^n\}$ where $i = 1, ..., K$ denotes the running number of the seasonal cycle, and $j = 1, ..., n$ denotes the time index inside a given seasonal cycle.

The mean and diversity of the i-th seasonal sequence are:

$$\bar{z}_i = \frac{1}{n} \sum_{j=1}^n z_{i,j}, \qquad \tilde{z}_i = \sqrt{\sum_{j=1}^n (z_{i,j} - \bar{z}_i)^2} \tag{1}$$

Based on the above defined mean and diversity, the trend and dispersion components are defined as follows: $\{T_t\}_{t=1}^N = \{\{\underbrace{\bar{z}_i, ..., \bar{z}_i}_{n \text{ times}}\}\}_{i=1}^K$, $\{D_t\}_{t=1}^N = \{\{\underbrace{\tilde{z}_i, ..., \tilde{z}_i}_{n \text{ times}}\}\}_{i=1}^K$.

A seasonal component is defined as:

$$S_t = \frac{z_t - T_t}{D_t} \tag{2}$$

The proposed STD decomposition can be expressed as:

$$z_t = S_t \times D_t + T_t \tag{3}$$

Note that in contrast to standard time series decomposition methods that are either additive or multiplicative in nature, STD has a mixed character. Figure 1 illustrates an STD decomposition. The trend and dispersion components are step functions with step length equal to the seasonal period n ($n = 12$ in this example). The trend component expresses the level of the time series in successive seasonal periods, while the dispersion component represents the variation of the series in these periods. The seasonal component comprises centered seasonal patterns of length n with zero average and equal variance. Although the seasonal patterns are unified, they differ in shape, reflecting the series' unified variations in successive seasonal periods. Unlike standard decomposition methods, the shapes of the seasonal patterns are not smoothed or averaged.

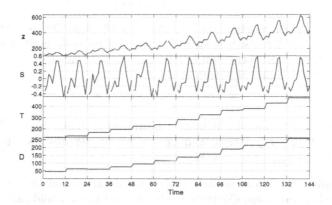

Fig. 1. Example of STD decomposition.

2.2 Randomized NN

In this study, we adopt a randomization-based NN, which is a type of NN that uses a single hidden layer feedforward architecture with randomly initialized hidden nodes. Unlike traditional, fully trained NNs, in RandNNs, the weights of the hidden nodes are not updated during training. The only weights that are learned are the ones connecting the hidden nodes to the output nodes, and this is done using a linear regression technique. This makes RandNN faster to train compared to standard NNs since it does not require an iterative optimization process to update the weights.

Assume RandNN consisting of n inputs, m hidden nodes with the standard sigmoidal nonlinearity, and q linear output nodes. Assume training set $\Phi = \{(\mathbf{x}_i, \mathbf{y}_i)\}_{i=1}^{N}, \mathbf{x}_i \in \mathbb{R}^n, \mathbf{y}_i \in \mathbb{R}^q$.

The improved training algorithm of RandNN that we adopt from [15] consists of four steps:

1. Select randomly hidden node weights: $a_{k,j} \sim U(-u, u)$, where $j = 1, ..., n$ and $k = 1, ..., m$ refer to the input number and the hidden node number, respectively.

2. Select randomly m training patterns from Φ and calculate biases for the hidden nodes as:

$$b_k = -\mathbf{a}_k^T \mathbf{x}_k^* \tag{4}$$

where $\mathbf{x}_k^*$ is the training x-pattern selected for the k-th hidden node.

3. Calculate hidden node responses to input patterns:

$$\mathbf{H} = \begin{bmatrix} \mathbf{h}(\mathbf{x}_1) \\ \vdots \\ \mathbf{h}(\mathbf{x}_N) \end{bmatrix} \tag{5}$$

where $\mathbf{h}(\mathbf{x}) = [h_1(\mathbf{x}), h_2(\mathbf{x}), \ldots, h_m(\mathbf{x})]$ denotes a nonlinear random projection of the input pattern from n-dimensional to m-dimensional space, and $h_k(\mathbf{x})$ represents an activation function of the k-th node.

4. Calculate the weights of output nodes as:

$$\boldsymbol{\beta} = \mathbf{H}^+ \mathbf{Y} \tag{6}$$

where $\boldsymbol{\beta} \in \mathbb{R}^{m \times q}$ denotes a matrix of output weights, $\mathbf{Y} \in \mathbb{R}^{N \times q}$ denotes a matrix of target patterns, and $\mathbf{H}^+ \in \mathbb{R}^{m \times N}$ denotes the Moore-Penrose generalized inverse of matrix $\mathbf{H}$.

Hyperparameter u corresponds to the maximum inclination angle of the sigmoid activation functions. Controlling the slope of the sigmoids affects the flexibility of the model. A larger slope allows the model to fit more complex patterns but can also lead to overfitting. To prevent overfitting, we can regularize the model by reducing interval U from which the weights are selected. For better interpretability, we use maximum angle α_{max} as the hyperparameter in the RandNN model, instead of u. The value of u can be determined from α_{max} using the formula $u = 4 \tan \alpha_{max}$.

The biases determined according to (4) based on the training points and node weights ensure the introduction of the sigmoid's steepest fragments into the input hypercube. This avoids the saturation problem and improves the approximation properties of the model.

3   Forecasting Models

Four scenarios are employed to combine STD with RandNNs. In two of these scenarios, STD serves as an encoder, extracting the key components of the time series. The components are then predicted by RandNN for the next period, and the decoder combines them to generate the forecast for the desired horizon (referred to as RandNN-STD2 variant). Alternatively, the components obtained from STD are used to encode the output data (RandNN-STD1 variant). RandNN predicts the encoded data, and the decoder transforms it back into a time series sequence. The variants RandNN-STD3 and RandNN-STD4 directly predict time series sequences based on STD components, utilizing slightly different RandNN architectures.

3.1 RandNN-STD1

Figure 2 illustrates the architectures of the proposed RandNN-STD models. An encoder decomposes time series $\{z_t\}_{t=1}^N$ using STD, and defines three components for each seasonal sequence, namely, mean value $\bar{z}_i$, diversity, $\tilde{z}_i$ (1), and seasonal pattern:

$$\mathbf{x}_i = \frac{\mathbf{z}_i - \bar{z}_i}{\tilde{z}_i} \tag{7}$$

Seasonal patterns correspond to (2), which in fact is a sequence of the seasonal patterns of the consecutive seasonal sequences. The seasonal patterns are unified versions of the sequences with zero mean and the same diversity. Moreover, vectors $\mathbf{x}_i$ have unity length.

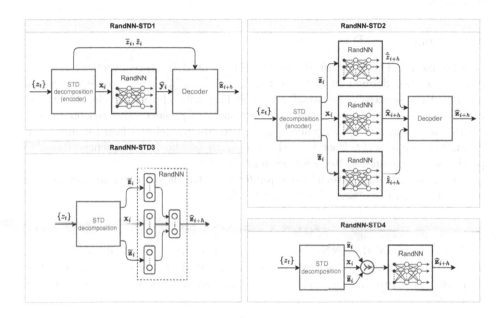

Fig. 2. RandNN-STD architectures.

Seasonal patterns (7) are the inputs for RandNN. RandNN produces forecasts of output patterns defined as follows:

$$\mathbf{y}_i = \frac{\mathbf{z}_{i+h} - \bar{z}_i}{\tilde{z}_i} \tag{8}$$

where h denotes a forecast horizon.

The y-patterns represent the forecasted seasonal sequences $\mathbf{z}_{i+h} = [z_{i+h,1}, ..., z_{i+h,n}]$. Note that y-patterns are defined similarly to the x-patterns, but instead of using the mean and diversity for sequence $i+h$, they use the mean and diversity for sequence i. This is because these values for sequence $i + h$ are not known at time i when this sequence is predicted.

To transform the predicted y-pattern into the actual sequence $\mathbf{z}_{i+h}$, a decoder performs an inverse transformation of (8):

$$\hat{\mathbf{z}}_{i+h} = \hat{\mathbf{y}}_i \tilde{z}_i + \bar{z}_i \tag{9}$$

Based on the STD products, the model implements the following mapping:

$$f(\mathbf{x}_i, \bar{z}_i, \tilde{z}_i) = \text{RandNN}_{n,m,n}(\mathbf{x}_i) \cdot \tilde{z}_i + \bar{z}_i \tag{10}$$

where $\text{RandNN}_{n,m,n}(.) = \text{Linear}_{m,n}(\text{RandProj}_{n,m}(.))$, $\text{RandProj}_{n,m}(.)$ denotes a random projection from n-dimensional space into m-dimensional space, and $\text{Linear}_{m,n}(.)$ denotes a linear projection from m-dimensional space into n-dimensional space.

3.2   RandNN-STD2

In this model, the STD products are represented by three vectors: seasonal pattern $\mathbf{x}_i$ (as defined in (7)), vector $\bar{\mathbf{z}}_i = [\bar{z}_{i-p_m+1}, ..., \bar{z}_i]$ representing the means of p_m successive seasonal cycles ending with the i-th cycle, and vector $\tilde{\mathbf{z}}_i = [\tilde{z}_{i-p_d+1}, ..., \tilde{z}_i]$ representing the diversities of p_d successive seasonal cycles ending with the i-th cycle, where p_m and p_d are predefined lags for mean and diversity, respectively. Based on the x-pattern of the i-th sequence, RandNN predicts the seasonal pattern for sequence $i + h$:

To transform this pattern into sequence $\mathbf{z}_{i+h}$, the decoder requires predictions of the mean and diversity of sequence $i + h$:

$$\hat{\mathbf{z}}_{i+h} = \hat{\mathbf{x}}_{i+h} \hat{\tilde{z}}_{i+h} + \hat{\bar{z}}_{i+h} \tag{11}$$

Two additional RandNNs are employed to produce forecasts of the mean and diversity of sequence $i + h$ based on vectors $\bar{\mathbf{z}}_i$ and $\tilde{\mathbf{z}}_i$, respectively. The function mapping STD products into the $(i + h)$-th seasonal sequence can be expressed as follows:

$$f(\mathbf{x}, \bar{\mathbf{z}}, \tilde{\mathbf{z}}) = \text{RandNN}_{n,m_s,n}(\mathbf{x}) \cdot \text{RandNN}_{p_m,m_m,1}(\bar{\mathbf{z}}) + \text{RandNN}_{p_d,m_d,1}(\tilde{\mathbf{z}}) \tag{12}$$

where m_s, m_m, and m_d represent the number of hidden nodes in the RandNNs predicting the seasonal, mean, and diversity components, respectively.

3.3   RandNN-STD3

As for RandNN-STD2, STD extracts three vectors $\mathbf{x}_i$, $\bar{\mathbf{z}}_i$, and $\tilde{\mathbf{z}}_i$, which are then inputted into RandNN. Each input vector is randomly projected by a separate section of hidden nodes, as shown in Fig. 2. The number of nodes in each section can be adjusted independently, depending on the complexity of the relationship between the output variable and a given group of predictors. The random projections performed by the three sections of nodes are then combined using a linear layer to produce the forecast of sequence $\mathbf{z}_{i+h}$.

The mapping function implemented in RandNN can be expressed as follows:

$$
\begin{aligned}
f(\mathbf{x}, \bar{\mathbf{z}}, \tilde{\mathbf{z}}) &= \text{RandNN}_{n+p_m+p_d, m_s+m_m+m_d, n}(\mathbf{x}, \bar{\mathbf{z}}, \tilde{\mathbf{z}}) \\
&= \text{Linear}_{m_s+m_m+m_d, n}(\text{Concat}(\text{RandProj}_{n, m_s}(\mathbf{x}), \\
&\qquad \text{RandProj}_{p_m, m_m}(\bar{\mathbf{z}}), \text{RandProj}_{p_d, m_d}(\tilde{\mathbf{z}})))
\end{aligned}
\tag{13}
$$

3.4   RandNN-STD4

This model extracts STD components and expresses them as vectors $\mathbf{x}_i$, $\bar{\mathbf{z}}_i$, and $\tilde{\mathbf{z}}_i$ (similarly to RandNN-STD2 and RandNN-STD3). It then concatenates these vectors and uses them as input for RandNN to predict seasonal sequence $\mathbf{z}_{i+h}$. The mapping function of RandNN can be expressed as follows:

$$
f(\mathbf{x}, \bar{\mathbf{z}}, \tilde{\mathbf{z}}) = \text{RandNN}_{n+p_m+p_d, m, n}(\text{Concat}(\mathbf{x}, \bar{\mathbf{z}}, \tilde{\mathbf{z}}))
\tag{14}
$$

4   Ensembling

The forecasting models defined above are combined as base learners to form an ensemble of M members. For each model type, a separate ensemble is composed.

To generate diversity among ensemble members, two methods, Ens1 and Ens4, were selected from a set of strategies evaluated in a previous study [6] and merged into a single method. The diversity of each member is generated using the following steps:

- the number of hidden nodes, m, is randomly selected from interval Λ,
- the weights of each node are randomly selected, with the upper bound for the sigmoid slope angles, α_{max}, chosen randomly from set Γ, and
- for each hidden node, the input pattern is randomly selected from the training set, and the node bias is calculated based on this pattern and the node weights using (4).

In the proposed approach, the diversity level of members is controlled by two intervals, namely Λ and Γ. Wider intervals provide the models with more diverse approximation abilities. A higher number of nodes as well as a higher upper bound for sigmoid slope angles increase the flexibility of RandNN and thus the diversity of members.

As a measure of ensemble diversity, we define the average standard deviation of forecasts produced by individual learners [6]:

$$
\text{Diversity} = \frac{1}{n|\Psi|} \sum_{i \in \Psi} \sum_{j=1}^{n} \sqrt{\frac{1}{M} \sum_{l=1}^{M} (\hat{z}_{i,j}^l - \bar{\hat{z}}_{i,j})^2}
\tag{15}
$$

where Ψ is a test set, $\widehat{z}_{i,j}^l$ is a forecast of the j-th element of the i-th seasonal sequence produced by the l-th learner, and $\overline{\widehat{z}}_{i,j}$ is an average of forecasts produced by M learners.

An ensemble prediction is calculated by combining the predictions of individual learners, denoted as $\widehat{\mathbf{z}}^l$, using a simple averaging method:

$$\widehat{\mathbf{z}}_{\text{ens}} = \frac{1}{M} \sum_{l=1}^{M} \widehat{\mathbf{z}}^l \tag{16}$$

5   Experimental Study

We assess the effectiveness of our proposed ensemble models on four real-world short-term load forecasting problems. The data for the hourly electrical load time series is collected from www.entsoe.eu for Poland (PL), Great Britain (GB), France (FR), and Germany (DE) spanning 2012 to 2015. The time series exhibit yearly, weekly, and daily seasonalities, and we forecast the daily load sequence ($n = 24$ hours) with a horizon of $h = 1$ for each day of 2015, excluding atypical days such as public holidays (which occur about 10–20 days per year depending on the dataset). For each forecasted day, a separate model was trained on the selected training patterns, representing the same day of the week as the forecasted pattern, i.e. when the index of the forecasted daily sequence is k, the indexes of the selected training patterns are $k - 7, k - 14, \ldots$

We use a total of $M = 100$ ensemble members to ensure robustness in our evaluation. We set $p_m = p_d = 7$, $m_m = m_d$, and use the same intervals Λ and Γ for predicting both the mean and diversity of the forecasted sequence. To match the statistics of the x-patterns, we normalize the input and output z-patterns and divide them by 10.

The following performance metrics were used: MAPE (Mean Absolute Percentage Error), MdAPE (Median APE), RMSE (Root Mean Squared Error), MPE (Mean PE), and StdPE (Standard deviation of PE).

5.1   Results

In the preliminary studies, we conducted a grid search to determine the optimal intervals for the hyperparameters, Λ and Γ. The number of nodes, m, was varied in the range of 5 to 100 with a step size of 5, while the maximum angle, α_{max}, was explored within the range of 10° to 90°, with increments of 10° up to 70°, followed by increments of 5° up to 90°. Figure 3 presents the results obtained for the PL dataset, while Table 1 summarizes the optimal intervals that yielded the lowest forecast errors.

Table 2 displays the quality metrics for various ensemble approaches using the optimal intervals for the hyperparameters. For comparison, we include the results for Ens1, Ens4, and a single RandNN from [6]. In this study, we propose a method for generating diversity by combining Ens1 and Ens4, as described in

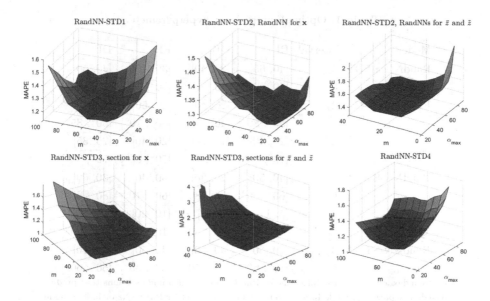

Fig. 3. Hiperparameters impact on MAPE for PL data.

Sect. 4. Since these approaches share an architecture similar to RandNN-STD1, the three methods produce comparable results as shown in Table 2.

As shown in Table 2, RandNN-STD3 and RandNN-STD4 are the most accurate approaches for PL and GB data, whereas RandNN-STD1 (and its predecessor Ens1) and Rand-STD4 are the best performing methods for FR and DE data. The poorest performing method is RandNN-STD2, which involves independently forecasting each component of STD. This method was outperformed by all other ensemble approaches for all datasets, and even by a single RandNN for GB and FR data.

MPE shown in Table 2 allows us to assess the bias of the forecasts. A positive value of MPE indicates underprediction, while a negative value indicates overprediction. For PL and DE all the models underpredicted, while for GB and FR they overpredicted.

Ensemble diversity in Table 2 is expressed in relation to the mean level of the time series to unify the results for different countries. As shown in the table, the diversity values range from 0.89% to 1.95% across the ensemble variants and countries. Among the proposed models, RandNN-STD1 exhibits the lowest diversity for each country.

5.2 Discussion

Instead of utilizing intervals for hyperparameters, Λ and Γ, specific values of the hyperparameters for all ensemble members can be determined based on the results presented in Table 1. For instance, for RandNN-STD1, the number of nodes can be set at 40 for each country, as this value is the common value

Table 1. Optimal intervals for hyperparemeters.

Model	Interval	PL	GB	FR	DE
RandNN-STD1	Λ_x	[40, 65]	[20, 40]	[30, 40]	[40, 50]
	Γ_x	[40°, 70°]	[40°, 70°]	[50°, 70°]	[70°, 80°]
RandNN-STD2	Λ_x	[20, 40]	[20, 40]	[20, 50]	[40, 70]
	Γ_x	[50°, 80°]	[40°, 80°]	[50°, 70°]	[80°, 85°]
	Λ_z	[25, 40]	[10, 20]	[20, 40]	[25, 35]
	Γ_z	[70°, 85°]	[50°, 75°]	[70°, 85°]	[70°, 85°]
RandNN-STD3	Λ_x	[40, 50]	[20, 30]	[30, 40]	[30, 50]
	Γ_x	[60°, 80°]	[70°, 80°]	[60°, 70°]	[80°, 85°]
	Λ_z	[5, 15]	[5, 15]	[10, 20]	[10, 20]
	Γ_z	[10°, 30°]	[10°, 30°]	[10°, 30°]	[10°, 30°]
RandNN-STD4	Λ_x	[50, 60]	[40, 50]	[40, 50]	[50, 70]
	Γ_x	[40°, 60°]	[40°, 50°]	[30°, 35°]	[50°, 60°]

where index x refers to the intervals for RandNNs or hidden node sections that predict based on x-patterns, while index z refers to the intervals for RandNNs or hidden node sections that predict based on z-patterns ($\bar{z}$ and $\tilde{z}$).

within the intervals for these countries. Similarly, α_{max} can be set at a fixed value of 70°. We decided to select hyperparameters from the intervals, randomly for each member, to provide an additional way of controlling diversity.

It is worth noting that RandNN-STD3 and RandNN-STD4 exhibit similar performance results. The only difference in their architectures is that RandNN-STD3 divides the hidden nodes into sections assigned to three types of predictors, while RandNN-STD4 does not. The division into sections does not seem to have a significant impact on the performance of the model.

Typically, random NNs require a large number of nodes, ranging from hundreds to thousands, to model relationships with high accuracy. However, our results from Table 1 show that relatively few hidden nodes are needed in our approach. This may be due to the simplification of the forecasting problem through the use of STD decomposition. By decomposing the time series into its trend, dispersion, and seasonal components, the complex relationships between the variables in the original series (note that in our case these relationships express triple seasonality) are replaced by simpler relationships between the decomposition products and the output pattern.

Among the proposed models, the most complex one is RandNN-STD2, which is composed of three RandNNs. This model requires the calculation of three Moore-Penrose inverse operations, which is the most computationally expensive operation in RandNN. Other models have a similar complexity to each other, although RandNN-STD3 and RandNN-STD4 may require more hidden nodes than RandNN-STD1 due to more inputs, resulting in increased computational cost.

It is worth noting that the training of RandNN is very fast since the training data is presented to the network only once, and there is no need for time-consuming iterative backpropagation that requires multiple calculations of gradients.

Table 2. Forecasting quality metrics.

		RandNN -STD1	RandNN -STD2	RandNN -STD3	RandNN -STD4	Ens1 [6]	Ens4 [6]	RandNN [6]
PL	MAPE	1.14	1.27	**1.07**	1.10	1.14	1.16	1.32
	MdAPE	0.79	0.88	**0.77**	0.80	0.79	0.81	0.93
	RMSE	303	359	**269**	283	304	313	358
	MPE	0.25	**0.06**	0.29	0.34	0.31	0.25	0.40
	StdPE	1.67	1.98	**1.48**	1.54	1.67	1.72	1.94
	Diversity_%	0.89	1.15	1.03	1.06	–	–	–
GB	MAPE	2.49	2.95	**2.44**	**2.44**	2.51	2.49	2.61
	MdAPE	1.74	2.32	**1.73**	**1.73**	1.76	1.77	1.88
	RMSE	1138	1237	**1119**	1121	1151	1131	1187
	MPE	−0.60	−0.42	−0.34	−0.51	−0.53	−0.53	−0.61
	StdPE	3.42	3.84	3.33	**3.31**	3.48	3.42	3.57
	Diversity_%	1.38	1.41	1.93	1.95	–	–	–
FR	MAPE	**1.57**	2.31	1.64	1.61	**1.57**	1.61	1.67
	MdAPE	**1.03**	1.56	1.04	1.08	1.04	1.06	1.15
	RMSE	1390	1912	1480	**1366**	1378	1402	1422
	MPE	−0.29	−0.20	−0.19	−0.19	−0.28	−0.31	−0.42
	StdPE	2.56	3.51	2.66	**2.51**	2.53	2.57	2.60
	Diversity_%	1.06	1.71	1.34	1.63	–	–	–
DE	MAPE	**1.18**	1.38	1.19	1.19	**1.18**	1.20	1.38
	MdAPE	**0.81**	0.97	0.82	0.83	**0.81**	0.82	0.96
	RMSE	1078	1212	1108	**1057**	1077	1097	1281
	MPE	0.11	0.10	0.18	0.21	0.11	**0.09**	0.14
	StdPE	1.91	2.22	1.96	**1.84**	1.89	1.92	2.22
	Diversity_%	1.16	1.21	1.40	1.29	–	–	–

6   Conclusions

This study introduces a novel ensemble forecasting method that combines the strengths of both RandNN and STD. RandNN is well-suited as an ensemble member due to its fast training and random nature, which allows for control of the ensemble diversity. The diversity is managed by adjusting the number of hidden nodes and the width of the interval from which their weights are drawn.

STD simplifies the forecasting problem with multiple seasonal patterns by extracting three components of the series: trend, seasonal, and dispersion components. The experimental results clearly demonstrate that combining RandNN and STD results in highly reliable, accurate forecasting of time series with complex seasonality.

Future research will focus on more sophisticated methods of combining forecasts produced by RandNN-STD through meta-learning. Additionally, a boosted ensemble learning approach based on the opposed response method [7] will be developed.

References

1. Wang, X, Hyndman, R., Li, F., Kang, Y.: Forecast combinations: an over 50-year review. Int. J. Forecast. (2022)
2. Wu, H., Levinson, D.: The ensemble approach to forecasting: a review and synthesis. Transp. Res. Part C: Emerg. Technol. **132**, 103357 (2021)
3. Brown, G., Wyatt, J., Harris, R., Yao, X.: Diversity creation methods: a survey and categorisation. Inf. Fusion **6**(1), 5–20 (2005)
4. Blanc, S., Setzer, T.: When to choose the simple average in forecast combination. J. Bus. Res. **69**(10), 3951–3962 (2016)
5. Genre, V., Kenny, G., Meyler, A., Timmermann, A.: Combining expert forecasts: can anything beat the simple average? Int. J. Forecast. **29**(1), 108–121 (2013)
6. Dudek, G., Pełka, P.: Ensembles of randomized neural networks for pattern-based time series forecasting. In: Mantoro, T., Lee, M., Ayu, M.A., Wong, K.W., Hidayanto, A.N. (eds.) ICONIP 2021. LNCS, vol. 13110, pp. 418–430. Springer, Cham (2021). https://doi.org/10.1007/978-3-030-92238-2_35
7. Dudek, G.: Boosted ensemble learning based on randomized NNs for time series forecasting. In: Groen, D., et al. (eds.) Computational Science, ICCS 2022. LNCS, vol. 13350, pp. 360–374. Springer, Cham (2022). https://doi.org/10.1007/978-3-031-08751-6_26
8. Ren, Y., Zhang, L., Suganthan, P.N.: Ensemble classification and regression - recent developments, applications and future directions. IEEE Comput. Intell. Mag. **1**(1), 41–53 (2016)
9. Shi, Q., Katuwal, R., Suganthan, P.N., Tanveer, M.: Random vector functional link neural network based ensemble deep learning. Pattern Recogn. **117**, 10797 (2021)
10. de Almeida, R., et al.: An ensemble based on neural networks with random weights for online data stream regression. Soft. Comput. **24**, 9835–9855 (2020)
11. Alhamdoosh, M., Wang, D.: Fast decorrelated neural network ensembles with random weights. Inf. Sci. **264**, 104–117 (2014)
12. Mesquita, D.P.P., et al.: Building selective ensembles of randomization based neural networks with the successive projections algorithm. Appl. Soft Comput. **70**, 1135–1145 (2018)
13. Huang, C., Li, M., Wang, D.: Stochastic configuration network ensembles with selective base models. Neural Netw. **264**, 106–118 (2021)
14. Dudek, G.: STD: a seasonal-trend-dispersion decomposition of time series. arXiv:2204.10398 (2022)
15. Dudek, G.: Generating random parameters in feedforward neural networks with random hidden nodes: drawbacks of the standard method and how to improve it. In: Yang, H., Pasupa, K., Leung, A.C.-S., Kwok, J.T., Chan, J.H., King, I. (eds.) ICONIP 2020. CCIS, vol. 1333, pp. 598–606. Springer, Cham (2020). https://doi.org/10.1007/978-3-030-63823-8_68

Author Index

© The Editor(s) (if applicable) and The Author(s), under exclusive license
to Springer Nature Switzerland AG 2023
I. Rojas et al. (Eds.): IWANN 2023, LNCS 14134, pp. 703–707, 2023.
https://doi.org/10.1007/978-3-031-43085-5

Printed in the United States
by Baker & Taylor Publisher Services